PRINCIPLES OF ADSORPTION AND REACTION ON SOLID SURFACES

Wiley Series in Chemical Engineering

PRINCIPLES OF ADSORPTION AND REACTION ON SOLID SURFACES

Richard I. Masel
Department of Chemical Engineering
University of Illinois at Urbana-Champaign
Urbana, Illinois

A WILEY-INTERSCIENCE PUBLICATION

JOHN WILEY & SONS, INC.

New York • Chichester • Brisbane • Toronto • Singapore

Library of Congress Cataloging-in-Publication Data:

Masel, Richard I., 1951–
 Principles of adsorption and reaction on solid surfaces / by
Richard I. Masel.—1st ed.
  p. cm.—(Wiley series in chemical engineering)
 "A Wiley-Interscience publication."
 Includes bibliographical references.
 ISBN 0-471-30392-5 (cloth : alk. paper)
 1. Surface chemistry. 2. Chemical kinetics. 3. Adsorption.
I. Title. II. Series.
QD506.M34 1996
541.3'3—dc20   95-17776
             CIP

Printed in the United States of America

10 9 8 7 6 5 4 3 2 1

CONTENTS

PREFACE

This book arose from a series of notes that were prepared for a course entitled "Principles of Adsorption and Reaction on Solid Surfaces," which is offered by the Chemistry and Chemical Engineering Departments at the University of Illinois. When I began writing this book, there were already several excellent books in the general area of surface physics and chemistry: Somorjai's *Introduction to Surface Chemistry and Catalysis* provides an excellent compendium of results in the general field of surface chemistry; Zangwill's *Surface Physics* provides an excellent qualitative review of surface physics; Gates's *Principles of Catalytic Chemistry* provides a good overview of reactions on catalysts. In the past ten years, however, people have begun to model surface reactions in a quantitative way. The ability to do quantitative calculations has changed the way people think about surface reactions. Older theories about the nature of surface reactions have been discarded and new theories have been developed to replace them. We were interested in presenting this new material to our classes. However, much of the relevant material did not appear in any of the existing texts.

The objective of this book was to provide a quantitative treatment of the application of principles from thermodynamics and reaction-rate theory to adsorption and reaction on solid surfaces. Our objective was to teach our students enough of this material so that they have a solid conceptual basis that they can use to understand adsorption and reaction on solid surfaces. We also wanted to provide enough of the material so that the students could model surface reactions in a quantitative way.

We offer two versions of the course, one for undergraduates and one for graduate students. These courses are taken by a cross section of seniors, first- and second-year graduate students in the chemistry and chemical engineering departments, and graduate students in physics and materials science. Some of these students are actively working in surface chemistry, catalysis, or electronic materials production and need to know this material for their research. The majority of the students, however, are taking the course to broaden their education. Also, the backgrounds of the students vary tremendously. Some of the students are very familiar with statistical mechanics, quantum mechanics, and reaction-rate theory while others need a review of this material.

In our course we try to balance the needs of the experts and the needs of the students who are taking the course to broaden their education. We try to include enough background material in the course so that students get an overview of the field without getting too caught up in the details. We try to include a fair amount of general education material. We want the students to learn about surface chemistry, and we want them to learn enough statistical mechanics and kinetics so they can do things other than surface chemistry when they finish their degrees.

In writing the book, I tried to have a similar presentation. An attempt was made to strike a balance between the needs of the experts and the need to provide a general education. The text emphasizes application of principles from statistical mechanics and reaction-rate theory to reactions on surfaces. I have tried to provide a coherent quantitative picture of the overall field of reactions on surfaces and put in enough background material so that the book would be intelligible to people with a variety of backgrounds. Detailed derivations have been included because I find that without the detailed derivations, many students do not understand many of the key results. A series of solved examples has been included to illustrate how the principles discussed in the book can be applied to real situations. Problems of varying difficulty were included so faculty members could tailor the problem sets to their classes. The book includes more material than we usually include in our course so that faculty at other universities would have the opportunity to choose the topics that they want to treat in more detail. Also, I wanted this book to have lasting value. Therefore, I included a fair amount of reference material.

With any book, one needs to make some compromises to keep the length of the book reasonable. I have chosen to not discuss experimental techniques, except where some detail of the experimental methods was particularly cogent to the arguments. I have also avoided doing a detailed survey of the literature, although I have included many illustrative examples. I decided to concentrate mostly on reactions on metal and elemental semiconductor surfaces and only briefly discuss reactions on oxide and compound semiconductor surfaces. If I had been writing a two-volume set, I would have included much more of this material. I needed to keep the book a reasonable length, however, and so I was forced to leave out some things that I would have included if I had written two volumes. I have written several appendices for the book. I am also distributing software to go with the book. Please write for details.

One of the difficulties in writing this text was that the theory of surface reactions is still an active research area. In many cases, no one yet knows why surface reactions occur in the way they do. In many cases ideas are controversial. Therefore, the approach has been to outline the general principles that can be applied to reactions on surfaces, indicate ways these principles might be applied, and then discuss the limitations of our current understanding. I anticipate that as the field evolves various aspects of this book will have to be revised. However, this book provides a current snapshot of what we know and indicates where the analysis is heading.

In the course of writing this work I have had many useful suggestions from colleagues and students. I am particularly indebted to the helpful comments of Jay Benziger, Andy Gellman, Lanny Schmidt, Ed Seebauer, John Vohs, Bill Millman, Vincent Van Spaendonk, Lee Nigg, Laura Farmer, Linda Cadwell, Amit Sachdev, and Keith Glassford. Part of the work discussed in this book was supported by the Department of Energy, Office of Basic Energy Sciences, under Grant DE-FG02-91ER14178. I am also indebted to my secretary, Nancy Carr, for all of her assistance in preparing the manuscript and to John Sollami for carefully editing the book.

RICHARD I. MASEL

Urbana, Illinois

1 Introduction

1.1 HISTORICAL INTRODUCTION

Surface reactions have an important place in current technology. At present, 90% of all chemicals are produced via a heterogeneously catalyzed process where a reaction occurs on the surface of a catalyst. Integrated circuits are made by using a reaction to deposit films on the surface of a semiconductor. Surface reactions also play a key role in the production of fuels, the disposal of noxious chemicals, the corrosion and lubrication of materials, the processing of metals and ceramics, and the production of photographic films. Surface reactions are even important to bioregulation.

Given the importance of surface reactions, it is not surprising that studies of surface reactions have had a long history. Before recorded history, the ancient Greeks and Romans developed techniques to limit the tarnishing of metals. Medieval alchemists made extensive searches for substances that would transform one substance into another. They termed these substances "philosopher's stones," but now we would call such substances "catalysts." We do not know much about the alchemists' work, because little of it was actually published. However, among their discoveries were the lead chamber process for the production of sulfuric acid, which was used from the sixteenth century to the middle of the twentieth century, and the acid catalyzed conversion of ethanol into diethyl-ether, which is still used today.

The first published study of surface reactions I could find was the work of Priestley [1775, 1790], who discovered that ethanol could be decomposed over a hot copper surface to yield tar and gas. Priestley did not follow up on his discovery. However, van Marum [1796] analyzed the products produced during the decomposition process and found that the alcohol decomposed to yield water, hydrogen, and carbon. This work was the forerunner of catalytic dehydrogenation.

The next major advance came about 20 years later when Davy [1817] discovered that a warm platinum gauze would spontaneously glow when exposed to the mixture of gases present in an 1820s coal mine even though no flame was produced. At the time this work was done, mines often exploded because the candles used to illuminate the mine ignited the gases in the mine. However, a platinum gauze will glow in the absence of a flame. Further, flames are extinguished when exposed to a platinum gauze. As a result Davy found that he could add a platinum gauze flame arrester to a lamp to make the lamp explosion proof.

Davy then designed a safety lamp for use in the mines. The design was similar to a standard oil lamp. However, it had a platinum gauze flame arrestor to prevent explosions. The lamp went into production soon thereafter and saved hundreds of miners' lives within its first few years in use. At the time this work was done, chemistry was still mainly a rich man's pastime. However, Davy's success showed that chemists could contribute to the well-being of society. As a result, people began to fund chemistry and hire chemists.

Henry [1824] did the first quantitative analysis of the reactions in the Davy lamp. He found that main reactions were

$$H_2 + \tfrac{1}{2} O_2 \rightarrow H_2O$$

$$CO + \tfrac{1}{2} O_2 \rightarrow CO_2$$

1

Henry also found that the reaction was more facile on a finely divided platinum powder. Dobereiner [1823] and DuLong and Theonard [1822, 1823] found that the ability of a platinum gauze to facilitate chemical reactions increased with treatments that increase the surface area and porosity of the gauze. Faraday [1834] showed that exposing the bare platinum to small amounts of grease poisoned the reaction under conditions where the grease was held on the surface of the platinum. This work led to the idea that heterogeneously catalyzed reactions are controlled by what happens in the pores of the catalyst. This idea was accepted for about the next 80 years.

During the 1830s there was considerable discussion about how platinum was able to facilitate chemical reactions. Faraday [1834] proposed that reacting gases were held onto the platinum by an electrical force, and that this electrical force helped facilitate the reactions. Berzelius [1836], on the other hand, did extensive measurements of reactions on platinum that led him to suggest that the influence of the platinum was more subtle than a simple electrical force. Berzelius called this more subtle interaction a "catalytic force," which he viewed as "not independent of (the) affinities of matter, but only a new manifestation of the same." Berzelius also adopted the term "catalysis."

A number of catalytic processes were patented in the later part of the nineteenth century. Corenwinder [1852], Debus [1863], and von Wilde [1874] discovered catalytic hydrogenation. The Phillips company [1831] patented the catalytic oxidation of SO_2 to sulfuric acid. On the theoretical side, the most important work was that of Van't Hoff [1871, 1896], Ostwald [1902], and Sabatier [1897, 1918], each of which won Nobel prizes for their work. In the late 1800s most people thought that catalysts allowed reactions to occur. However, Van't Hoff and Ostwald showed that the main effect of the catalyst was to alter the rate of the reaction. Van't Hoff examined rates of both steady-state and non-steady-state reactions on catalytic surfaces and showed that the rate of the reaction increased as the amount of catalyst or surface increased. Ostwald showed that catalysts only affected the rate of the reaction and did not influence the equilibrium constant for the reaction. Together Van't Hoff and Ostwald showed that catalysts change reaction rates and not the final equilibrium. Van't Hoff won the first Nobel prize in chemistry in 1901. Ostwald won the Nobel prize in chemistry in 1909.

The next major advance came from the work of Sabatier. To set this work in perspective, note that Berzelius, Ostwald, Van't Hoff, and their compatriots had assumed that catalysis occurred because the reactants *ab*sorbed into the cavities of the porous catalyst where they were compressed to a high enough density that they would react. Van't Hoff had already shown that reaction rates increase when gases are compressed. However, Sabatier noted that the reaction would occur only on special metals, and if the metal held the reactants too strongly (i.e., compressed it too much) no reaction would occur. Therefore, Sabatier concluded that compression of the reactants was insufficient to explain catalytic action. Sabatier proposed that instead the reactants formed a temporary unstable compound with the catalyst during the catalytic reaction, and it was the production and destruction of these unstable intermediates that led to catalysis. A similar idea had been proposed by Clémemt and Desormes [1806] a hundred years before, but it had been largely ignored. Sabatier's proposal was originally criticized because the idea of an unstable reaction intermediate was unbelievable to most influential chemists at the end of the nineteenth century. However, Sabatier showed that this theory explained a wide range of catalytic data and Sabatier's view won out in the literature. Sabatier also showed that the unstable intermediate was associated with what we now call a chemisorbed state. Sabatier won a Nobel prize for this work in 1912.

Soon thereafter, catalytic processes started to be used in industry. One of the key workers was Haber, who discovered catalysts that were useful for the production of ammonia. Haber won a Nobel prize in chemistry in 1918 for that work.

Up until 1912, there were no general expressions for reaction rates on catalysts. However, in a number of papers between 1912 and 1918, Langmuir developed a general model of reactions on surfaces as part of his studies of the reactions of gases with the filaments of light bulbs. Langmuir showed that gases were adsorbed at specific sites on solid surfaces. This work is discussed in Chapters 3 and 4. Langmuir also introduced the concept of trapping and sticking. This work is discussed in Chapter 5. Langumir's ideas were widely adopted by the surface chemists of his day, and his ideas are still in use today. Langmuir won a Nobel prize for his work in 1932.

There were no Nobel prizes in the general field of surface reactions from 1938 to 1955. However, that period saw an explosive growth of the use of industrial catalytic processes. Industrial researchers systematically examined the interaction of gases with a variety of metal surfaces and tried to correlate their measurements with what was known about the properties of the metals. The key ideas from these studies are discussed in Chapters 10 and 12. These researchers also discovered a large number of new processes, including the basis of most of the processes that are now used to produce industrial chemicals.

The idea that the rate of surface reactions varied with the surface structure also arose from work done between 1925 and 1955. In his original work, Langmuir [1918] treated a reaction as though it were occurring uniformly over the surface of the catalyst. However, a few years after Langmuir's original articles appeared, Pease and Stewart [1925] and Taylor [1925] discovered that when they added poisons to a catalyst, they would reduce the rate of reaction much more than they would reduce the rate of adsorption. This led Taylor to propose that reactions on surfaces occur on special places on the catalyst's surface. Taylor called these special places "active sites." The idea of active sites has been greatly expanded since Taylor's time, and we now talk about reactions in terms of structure-sensitive and structure-insensitive reactions. However, the idea that rates of reactions on surfaces vary with the structure of the surface has continued to this day.

In the 1960s the focus of surface science changed. In the United States, people were beginning to explore outer space. The space program needed to know how gas phase collisions affected the wear in a spacecraft. There also was interest in the momentum transfer that arises during gas/surface collisions in order to estimate the frictional losses a satellite experiences in orbit. People built a number of ultra-high vacuum (UHV) systems so that they could simulate the environment in space close to the ground. They then placed surfaces in these vacuum systems and examined the interaction of the surfaces with beams of gas. This work is discussed in Chapter 5.

The advent of UHV had a secondary benefit as well, which is that it permitted reactions on surfaces to be examined in the absence of gas phase collisions. Up until 1965, most of the models for surface reactions were fairly tentative. The presence of the gas phase interfered with the measurements. Only high surface area materials could be examined. High surface area materials tend to be fairly inhomogeneous, and the reaction often only occurs on a small fraction of the surface. There was no way to make spectroscopic measurements on only the active areas of the surface. As a result there was considerable question whether the things one could measure had anything to do with reactions. However, the use of UHV techniques allowed people to examine reactions under conditions where they could be assured that they were really measuring properties of active areas of their surfaces. As a result people were able to develop a more fundamental understanding of the processes that occur.

The 1960s also saw the start of the integrated circuit industry. Integrated circuits are made by depositing films on semiconductor substrates. Precise control of the properties of the film are essential to the performance of the circuit. A number of surface analysis tools were developed mainly in response to the needs of the integrated circuit manufacturing industry. These tools allowed people to measure surface structure and composition, and correlate these data to measurements of reaction rates. As a result, one could finally examine surface processes directly on a molecular level and really test theories directly.

The current situation is that we are now able to examine reactions on surfaces in greater detail than ever before. This has allowed people to not only determine what happens during a surface reaction, but also begin to ask why it happens and what general principles apply to surface reactions. This is still an evolving field and many of the subjects covered by this book are still under active investigation. However, the object of this book is to present a current snapshot of our understanding of the principles of surface reactions and to indicate how those principles can be applied to real systems.

1.2 OVERVIEW OF THE APPROACH TAKEN IN THIS BOOK

Everything we discuss in this book will build on the principle of Sabatier. Sabatier won the Nobel prize in chemistry in 1912 for his work in catalysis. As noted earlier, Sabatier was the first person to show conclusively that chemisorption of reactants was important to catalysis. Along the way Sabatier developed a general model for catalytic action that has come to be called the Principle of Sabatier. Sabatier's principle is derived by noting that a typical heterogeneously catalyzed reaction involves adsorption of some reactants onto the surface of a catalyst, transformations of the reactants on the catalyst, and then desorption of products. Sabatier noted that if the reactants adsorbed too weakly, there will be few surface complexes. As a result, the reaction rate will be low. Nevertheless if the reactants adsorb too strongly, the surface complex will be too stable to decompose. Hence, the reaction rate will also be low. The implications of Sabatier's principle is that an effective catalyst will bind the reactants strongly, but not so strongly that no products can desorb.

This idea has been widely applied to surface reactions. In catalysis, one chooses a catalyst that forms a bond to the intermediates of intermediate strength. In CVD, one chooses a source gas with ligands that are strongly enough bound to make it down onto the substrate, and yet the metal ligand bond needs to be weak enough that the ligands can desorb. Still, there are exceptions to Sabatier's principle. Based on Sabatier's principle one would expect aluminum and germanium to be good catalysts for the hydrogenation of hydrocarbons. However, no one has observed catalytic hydrogenation on either of these metals. Another exception comes with surface-structure-sensitive reactions. It is difficult to explain structure sensitivity based on Sabatier's principle alone. At present we do not fully understand the exceptions to Sabatier's principle, although some ideas on the subject are discussed in Chapter 10.

In this book we examine adsorption and reaction on solid surfaces. We start with a description of surfaces and their structure, in order to develop some of the language we need to discuss surfaces. We concentrate on metals and semiconductors because that is where our understanding is best. We then examine adsorption of reactants onto the surface and desorption of products and try to describe what is known about the kinetics and thermodynamics of the adsorption/desorption process. We then move onto the surface reactions themselves. Here the principles are less well understood. However, the objective of this book is to lay out the key issues that people are examining in order to understand rates of reactions on solid surfaces.

Along the way we will discuss ways of thinking and many different principles from statistical mechanics, reaction rate theory, and physical-organic chemistry. One of the reasons why so many Nobel prizes went to people who worked in surface reactions is that surface reactions have proved to be a very convenient way to develop principles that could then be applied to other systems. Ostwald's work, which showed that catalysts affected the rate of a reaction but not the equilibrium constant of the reaction, proved that the free energy of a system was a state property. Such a finding had broad implications. Sabatier's finding that reactions proceed via unstable intermediates showed that a species does not have to be observable to exist. This finding also has had broad implications throughout chemistry. Another example where studies of surfaces contributed to our general understanding of science is Davidson and Germer's work, which showed that electrons diffract from surfaces. This work showed, for the first time, that matter could behave like a wave. Davidson won a Nobel prize for this work in 1937. Throughout their history, surfaces and surface reactions have provided interesting examples of general principles that are applicable to many areas of science in addition to reactions on surfaces. In this book we will discuss these principles in some detail. We expect that the reader will learn enough to actually apply the principles to the specific examples of reactions on surfaces that are discussed in this book. However, we hope that the reader will also learn enough that he or she can apply the ideas and principles discussed in this book to disciplines outside of surface chemistry.

This book also provides a historical account of the work. A general theme is that when one is starting to explore a new area, one starts with simple models, sees where the simple models fail, and then tries more complex models. Readers should look for that general theme as they proceed through the book.

1.3 TECHNIQUES OF SURFACE ANALYSIS

The majority of this book is concerned with principles of surface reactions and not experimental techniques. However, we did want to include a brief overview of some of the techniques used in a modern surface science experiment so that when we discuss specific experimental results, the reader will know how they were obtained.

Later in this book, we refer to data taken by low-energy electron diffraction (LEED), high-energy electron diffraction (HEED), ultraviolet photoemission spectroscopy (UPS), X-ray photoemission spectroscopy (XPS or ESCA), infrared spectroscopy (IR), electron energy loss spectroscopy (EELS), scanning tunneling microscopy (STM), atomic force microscopy (AFM), Auger electron spectroscopy (AES), temperature programmed desorption (TPD), and temperature programmed reaction (TPR). Table 1.1 lists each of these techniques, and briefly describes some of the information one can obtain from them.

LEED, HEED, STM, and AFM are all structural probes. In LEED a beam of electrons is directed at a surface, and the diffraction pattern of the electrons that scatter from the surface

TABLE 1.1   Some of the Common Experimental Techniques Used to Examine Surface Reactions

Technique	Acronym	Property Probed
Low-energy electron diffraction	LEED	Surface structure; adsorbate arrangement
High-energy electron diffraction	HEED	Surface structure; adsorbate ordering
Scanning tunneling microscopy	STM	Surface topology; surface electronic structure
Atomic force microscopy	AFM	Surface topology
Temperature programmed desorption	TPD	Binding energy of adsorbates, activation energy of desorption
Temperature programmed reaction, thermal desorption spectroscopy	TPR, TDS	Reaction pathways activation barriers for surface processes
Auger electron spectroscopy	AES	Surface composition
X-ray photoemission spectroscopy	XPS, ESCA	Core electronic levels of adsorbates and of the surface
Ultraviolet photoemission spectroscopy	UPS	Valence electronic structure of the adsorbate and the surface
Secondary ion mass spectroscopy	SIMS	Composition of the surface and the adsorbate
Electron energy loss spectroscopy	EELS, HREELS	Vibration levels of the adsorbate and the surface
Reflection infrared spectroscopy	IR, RAIRS, FT-IRS	Vibrational levels of the adsorbate

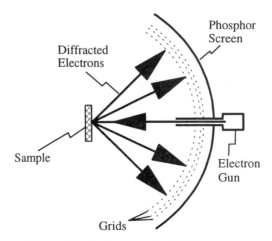

Figure 1.1   A schematic of a LEED apparatus.

is measured using the apparatus shown in Figure 1.1. One then uses an analysis similar to that used for X-ray diffraction to calculate the arrangement of the surface atoms. HEED is much like LEED except that one uses a glancing-angle electron beam and a higher beam energy.

Figure 1.2 shows a LEED pattern of a Si(100) surface. Notice the square symmetry of the picture. From the symmetry of the spots and the number of spots we can obtain information about the symmetry and periodicity of the adsorbed layer. Detailed analysis of the intensity of each of the spots as a function of the voltage of the incident beam (i.e., so-called LEED I-V curves) gives information about the position of all of the atoms in the surface. Hence, detailed structural information can be obtained from LEED.

STM and AFM are also structure probes. In STM and AFM a small tip is scanned over a surface as indicated in Figure 1.3. A circuit is then used to adjust the height of the tip so that it keeps either the current to the surface or the force between the tip and the surface

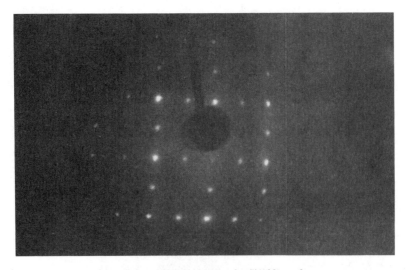

Figure 1.2   A LEED pattern of a Si(100) surface.

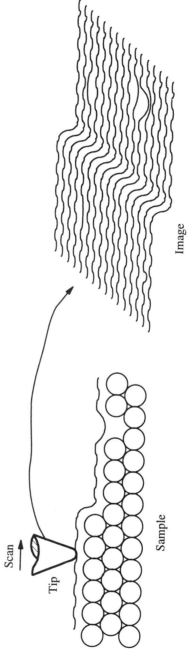

Figure 1.3 A schematic of an STM experiment.

7

constant. The current and force are a function of the distance between the tip and the surface. As a result, a plot of the height of the tip as a function of the position over the surface is a map of the surface topography. Generally, LEED and STM or AFM are complementary techniques. LEED does a good job of measuring the periodicity of adsorbed layers. One can also accurately measure the distance between the adsorbate layer(s) and the surface via LEED I-V curves. However, measurements of the horizontal displacements of atoms within an adsorbate layer are generally less accurate with LEED. Further, LEED is relatively insensitive to defects. STM and AFM, on the other hand, are very good at measuring the arrangement of molecules within a unit cell, and STM is thought to be very sensitive to defects. However, it is difficult to interpret the vertical distances one gets from STM and AFM. As a result, LEED and STM can be used together to get a complete description of the atomic arrangement in the adsorbed layer.

AES, XPS, and SIMS are used to measure the surface composition. In AES, a beam of electrons or X-rays is used to knock out a core electron from an atom either in the surface or in an adsorbed layer. When the core electron leaves the atom, producing what is called a "core hole," all of the other electrons in the atom move in toward the nucleus of the atom to compensate. Later, when the core hole is filled, the electrons move out again. Hence, there is some probability of ejecting an outer shell electron when the core hole is filled. The simultaneous filling of a core hole and ejection of a secondary electron is called an Auger process. Note that from conservation of energy, the energy of the secondary electron must equal the energy gained when the core hole is filled minus the initial energy of the electron before it is ejected, plus some corrections due to incomplete relaxations and so-called final state effects. The energy gained when the core hole is filled varies significantly from one element to the next. As a result, the energy of the secondary electron varies significantly from one element to the next.

In an AES experiment, one uses an energy analyzer (shown in Figure 1.4) to collect the secondary electrons. Typically, one observes a series of peaks corresponding to each of the elements in the surface and adsorbed overlayer. The intensity of the peaks is related to the concentrations of the atoms on the surface. Usually, AES is not completely quantitative because some of the secondary electrons are lost due to impacts with other atoms. Other Auger electrons do not travel in the right direction to reach the detector. Further, the probability that an Auger electron is formed on a given atom varies somewhat with the composition of the surrounding atoms. Hence, quantitative analysis using AES can be imprecise. However, AES has the advantage that it is a relatively simple technique and that it is sensitive to

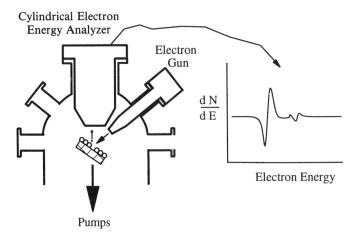

Figure 1.4 A schematic of an AES experiment.

submonolayer levels of impurities. As a result, AES is used to routinely determine the atomic composition of surfaces, and in particular to examine surface cleanliness.

XPS and SIMS are also used to measure surface composition. Figure 1.5 shows a schematic of the XPS experiment. In XPS one directs a beam of X-rays at a solid surface, and then measures the energy distribution of the electrons that leave the surface with an electron energy analyzer. During the XPS process, the X-rays knock electrons out of the core levels in the surface atoms. Each element has its own characteristic core levels. As a result, an XPS spectrum shows a series of peaks corresponding to each individual element in the surface. Generally, XPS is more quantitative than AES for surface analysis, but is usually less sensitive, because X-rays are not as efficient as electrons in creating core holes.

UPS and XPS can also be used to examine the electronic structure and the bonding of molecules on surfaces. In XPS and UPS one is measuring the energy distribution of the electrons that leave the surface. That distribution is directly related to the energy distribution of the electrons in the surface (i.e., the surface electronic structure). Hence, one can get a measure of the electronic structure of the surface from UPS and XPS.

UPS and XPS also provide information about the bonding of molecules in the surface. In Nobel prize winning work, Siegbahn et al. [1967] showed that the core levels of the atoms in a surface shift according to the chemical environment around the surface atoms; an oxygen atom bound to the carbon in adsorbed CO has a slightly different core level than an oxygen atom bound to a metal surface. One can measure the shifts in the core levels with XPS. As a result, one can use the shifts in the XPS peaks to get some information on the bonding of the atoms on the surface. Siegbahn et al. called the use of XPS to determine the bonding of atoms "electron spectroscopy for chemical analysis" (ESCA) to distinguish it from the use of XPS to determine surface composition. However, the acronyms "XPS" and "ESCA" are now used synonymously in the literature.

UPS is much like XPS. One directs an ultraviolet photon at a solid surface and measures the energy distribution of the emitted electrons using an apparatus like that illustrated in Figure 1.5. Ultraviolet photons excite the valence electrons in molecules or the conduction electrons in surfaces. Generally, UPS is used to probe the electronic structure of surfaces and the changes in the electronic levels due to the presence of adsorbates. UPS is probably the best way to measure surface electronic structure. Many investigators have also used UPS to

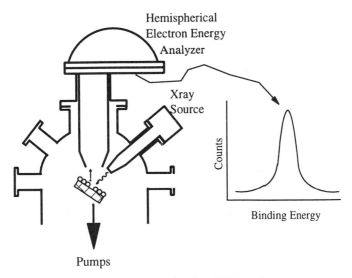

Figure 1.5 A schematic of an XPS experiment.

learn about the bonding in adsorbed molecules. There have been many cases where UPS measurements have yielded some very important information. Unfortunately, UPS measurements of valence levels in molecules are often difficult to interpret. Hence, UPS generally has not proved to be as good of a probe of the bonding of molecules on surfaces as other techniques.

In practice, UPS and XPS are most sensitive to changes in bonding that lead to changes in the oxidation state or electron affinity of the atoms in the surface. There is a big difference between the XPS position of carbon in CO and carbon adsorbed on an iron surface. However, a much smaller difference is seen between carbon bound in a C_2H_5 group and carbon bound in a C_9H_{19} group. Hence, there are limits to how well one can use XPS and UPS to determine the structure of the molecules in an adsorbed layer.

SIMS is also used to measure surface composition. Figure 1.6 shows a schematic of the SIMS experiment. In SIMS, one directs a beam of ions at a solid surface and measures the mass distribution of the ions that are produced when the incident ions collide with the surface. During a SIMS experiment, one chooses a high enough beam energy (~ 1000 eV) that the impact of the incident ions with the surface is sufficient to actually knock off surface atoms and molecules during the collision process. These surface atoms and molecules are collected with a mass spectrometer. As a result, in principle one should be able to measure the composition of the surface with SIMS.

In practice, SIMS is usually very sensitive but not as quantitative as other techniques. The biggest limitation of SIMS is that when the incident ion collides with the surface, a mixture of ions and neutral species is formed. However, during the SIMS experiment only the ions that leave the surface are detected, and not the neutrals. Experimentally, one often finds that as one changes the chemical environment around a species, the fraction of species that leave as ions changes. As a result, SIMS data often require some care to interpret. Still, many investigators have succeded in using SIMS in a semiquantitative way. The advantage of SIMS is that it allows one to obtain a great deal of information in a single experiment. One can identify the presence of one molecule in the presence of other similar molecules. As a result, SIMS is quite useful and it is especially so in systems that show complex behavior.

IR and EELS are much better probes of the structure and bonding of molecules at surfaces.

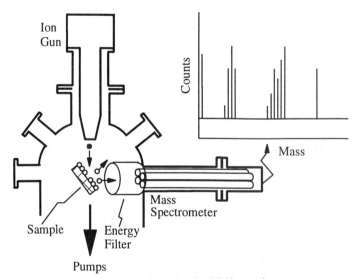

Figure 1.6   A schematic of a SIMS experiment.

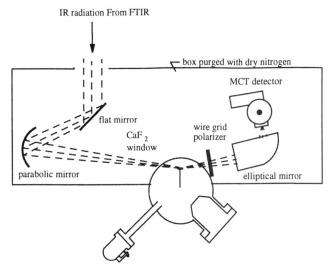

Figure 1.7 An illustration of a typical reflection infrared apparatus.

Figure 1.7 shows a diagram of a typical reflection infrared apparatus. In IR a beam of infrared photons is directed at a solid surface, and measures the fraction of the photons that absorb as a function of frequency. Then the observed IR spectrum is compared to ones from a series of reference compounds to learn about the bonding of molecules to surfaces. In actual practice, reflection IR does have some sensitivity limitations. However, it can give some very useful information about the bonding of adsorbed complexes.

EELS is much like IR. A monoenergetic beam of electrons is directed at a solid surface, and the energy distribution of the electrons that scatter from the surface is measured, as indicated in Figure 1.8. Generally, one can observe distinct peaks for each of the vibrational losses. As a result one can use EELS to obtain the same type of information as one gets from IR. EELS is normally more sensitive than IR and it can measure both the IR-active and IR-inactive transitions. Hence EELS can provide considerable information about surface species.

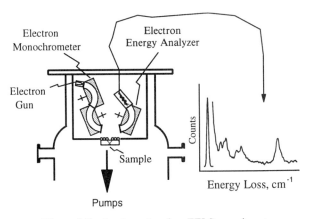

Figure 1.8 A schematic of an EELS experiment.

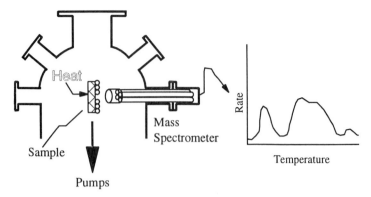

Figure 1.9 A schematic of a TPD/TDS experiment.

However, the resolution of an EELS instrument is usually less than that of an IR machine. As a result, the interpretation of EELS data is less certain than the interpretation of IR results.

We also list TPD, TPR, and TDS in Table 1.1. TPD, TPR, and TDS are all very similar. Gas is adsorbed onto a surface. The surface is then heated in a controlled way and a mass spectrometer is used to measure the rate at which products desorb from the surface (see Figure 1.9). Using TPD, one can tell what reactions occur when the sample is heated. One can also tell how strongly molecules are bound to the surface. Further, one can also learn about reaction pathways. A few investigators use the acronym ''TPD'' to refer to measurements on non-reactive systems while ''TPR'' and ''TDS'' are used for reactive systems. However, generally the acronyms TPD, TPR, and TDS are used interchangeably throughout the literature and so there is no real reason to distinguish between them here.

Later in this book we also present some data taken with a surface molecular beam system. Figure 1.10 is a diagram of a typical apparatus used to do these types of measurements. Generally, one directs a monoenergic beam of molecules at a surface and examines the

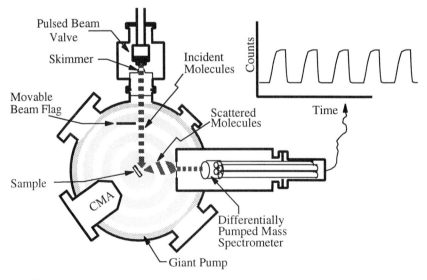

Figure 1.10 A schematic of a typical apparatus used to examine surface reactions with molecular beams.

molecules that bounce from the surface using a differentially pumped mass spectrometer. The beam is formed in a supersonic nozzle embedded in a beam valve. In the apparatus shown, the valve produces a pulsed beam. However, people often instead use continuous beams. Also many investigators use movable mass spectrometers as detectors instead of the fixed detector shown in Figure 1.10.

Beam machines are used for a variety of purposes. One can measure kinetics from the transient waveforms. Activation energies can be examined by varying the energy of the incident beam. Momentum transfer rates between the gas and the surface can be determined by measuring the momentum of the incident and scattered beam.

There are many other techniques that are used less frequently in the surface science literature. One should refer to Woodruff and Delchar [1986] or Ertl and Küpers [1986] for a discussion of these techniques.

REFERENCES

Berzelius, J. J., *Fort. Physic. Wissenshaft, Tubingen*, 243 (1836).

Berzelius, J. J., *Edinburg New Philos. J.* **21,** 223 (1836).

Corenwinder, B., *Annales de Chemie et de phys.* **34,** 77 (1852).

Davy, H., *Philos. Trans.* **107,** 77 (1817).

Debus, H., *Liebig's Annalen der Chemie und Pharmacie* **128,** 200 (1863).

Dobereiner, M., *Annales de Chimie* **24,** 93 (1823).

DuLong, P. L., and Theonard, L. J., *Annales de Chimie* **23,** 440 (1822); **24,** 330 (1823).

Ertl, G., and J. Küpers, *Low Energy Electrons and Surface Chemistry*, Verlag Chemie, Weinheim (1986).

Faraday, M., *Philos. Trans.* **124,** 55 (1834).

Langmuir, I., *J. Am. Chem. Soc.* **35,** 105 (1913); **37,** 1139 (1915); **40,** 1361 (1918).

Ostwald, W., *Revue Scientifique, Paris* **1,** 640 (1902).

Pease, R. N., and R. J. Stewart, *J. Am. Chem. Soc.* **47,** 1235 (1925).

Priestley, J., *Experiments On Different Kinds Of Air*, J. Johnson, London (1775).

Priestley, J., *Experiments On Different Kinds Of Air and Other Branches of Natural Philosophy*, p. 425, Birmingham (1790). (Available from The University of Michigan Library.)

Sabatier, P., *Catalysis in Organic Chemistry* Van Nostrand, NY (1918). (A translation appears in P. A. Emmett, ed., *Catalysis Then and Now*, Franklin Publishing, Englewood, NJ, 1965.)

Sabatier, P., and J. B. Senderens, *Comptes Rendus* **124,** 1359 (1897).

Siegbahn, K., C. Nordling, A. Fahlman, R. Nordberg, K. Hamerin, J. Hedman, G. Johansson, T. Bergmark, S. E. Karlsson, I. Lindgren, and B. Lindberg, *Electron Spectroscopy for Chemical Analysis*, Almqvist and Wiksells, Stockholm (1967).

Taylor, H. S., *Proc. Royal. Soc. London A* **108,** 105 (1925).

van Marum, M., *Scheikundige Biblioteek, Roelofswaar,* **3,** 209 (1796). (See R. J. Forbes, *Martinis Van Marum: Life and Work*, H. D. Tjeenk, Haarlem, 1969.)

von Wilde, M. P., *Ber. Deutsch. Chem. Ges.* **7,** 352 (1874).

Van't Hoff, J. H., *Studies of Chemical Dynamics*, reprinted in English by Williams and Northgate, London (1871, 1896).

Woodruff, D. P., and T. A. Delchar, *Modern Techniques of Surface Science*, Cambridge (1986).

BIBLIOGRAPHY FOR EXPERIMENTAL TECHNIQUES

Benninghoven, A., *Secondary Ion Mass Spectroscopy*, Wiley, NY (1987).

Briggs, D., and N. P. Seah, *Practical Surface Analysis*, Wiley, NY (1983).

Carlson, T. A., *Photoelectron and Auger Spectroscopy*, Plenum, NY (1975).

Chabal, Y. T., *Surface Infrared Spectroscopy, Surface Sci. Rep.* **8,** 221 (1988).

Clark, L. J., *Surface Crystallography*, Wiley, NY (1985).

Czanderna, A. W., and D. M. Hercules, *Ion Spectroscopies for Surface Analysis*, Plenum, NY (1987).

Dror, S., *Scanning Force Microscopy*, Oxford University Press, NY (1991).

Ertl, G., and J. Küpers, *Low Energy Electrons And Surface Chemistry*, Verlag Chemie, Weinheim (1986).

Ferguson, I. F., *Auger Microprobe Analysis*, A. Hilger, NY (1989).

Feuerbacher, B., *Photoemission and the Electron Structure of Surfaces*, Wiley, NY (1978).

Hansson, G. U., and R. I. Uhrberg, *Photoemission Spectroscopy of Surface States, Surface Sci. Rep.* **9,** 197 (1988).

Ibach, H., *Electron Energy Loss Spectroscopy and Surface Vibrations*, Academic Press, NY (1982).
Ibach, H., *Electron Energy Loss Spectrometers*, Springer-Verlag, Berlin (1991).

Luth, H., *Surface and Interfaces of Solids*, Springer-Verlag, Berlin (1993).

Pendry, J. B., *Low Energy Electron Diffraction*, Academic Press, NY (1974).

Plummer, E. W., and W. Eberhardt, *Angular Resolved Photoemission as a Tool for the Study of Surfaces, Adv. Chem. Phys.* **49,** 533 (1985).

Thompson, M., *Auger Electron Microscopy*, Wiley, NY (1985).

Woodruff, D. P., and T. A. Delchar, *Modern Techniques of Surface Science*, Cambridge University Press, Cambridge (1986).

Yates, J. T., and T. E. Madey, *Vibrational Spectroscopy of Molecules on Surfaces*, Plenum, NY (1987).

2 The Structure of Solid Surfaces and Adsorbate Overlayers

PRÉCIS

In this book, we discuss the behavior of gases on surfaces in some detail. The objective of this chapter is to describe the structure of surfaces both when they are clean and when they are covered by an adsorbate overlayer. We will start off by describing the structure of surfaces and bulk lattices in a qualitative way. We will then develop some of the notation used to describe bulk crystals and use that notation to describe the structure of ideal surfaces. We will then move on to nonideal surfaces and describe how the notation needs to be modified. We will then use that information to describe adsorbate layers. Additional references to this material include Kelly and Groves [1970], Nicholas [1965], Wood [1964], Lander [1965], Park and Manden [1968].

2.1 HISTORICAL INTRODUCTION

The study of the structure of solids dates back to the ancients. Burke [1966] reviews the important contributions of Aristotle and Plato in the early description of the solid state. The first quantitative information about the structure of solids came from the work of Hooke and Huygens in the latter part of the seventeenth century. Hooke [1665] and Huygens [1666] examined the properties of crystals using the newly developed technique of optical microscopy. They proposed that crystalline solids were composed of small spherical particles of matter held together in an ordered array. Haüy [1784] refined this idea to suggest that crystals were composed of a periodic array of particles that were not necessarily spherical. Haüy [1801] later identified these particles as the groupings of atoms that we now call molecules.

Surfaces were first studied in a serious way in the mid-nineteenth century (i.e., 1800s). At the time, there was no good way to examine the structure of surfaces on a molecular level. As a result, most investigators initially assumed that surfaces were composed of ordered arrays of atoms as suggested by Haüy.

Figures 2.1 and 2.2 show schematics of the idealized arrays of atoms described by Haüy [1801]. Generally, up until about 1950, people only discussed periodic or quasi-periodic structures, although non-periodic structures were mentioned by Haüy.

Kossel [1927] and Stranski [1928] proposed that the surfaces shown in Haüy [1801] can be divided into three groups: flat surfaces, surfaces with mono-atomic steps, and surfaces with kinks. The simplest surfaces are flat surfaces, where all of the surface atoms are arranged in a plane. These arrangements are characteristic of closed packed surfaces of most transition metals. The term closed packed is defined later in this chapter. Another important kind of surface is called a "stepped surface." Figure 2.2b shows a diagram of a stepped surface. Basically, the atoms start out all in a plane, but every so often there is a place where a new plane of atoms is put above the first. The result is a staircase pattern of atoms. We call the staircase pattern of atoms a **stepped surface.** The long plane of atoms in a stepped surface is called a **terrace,** while the place where the new plane starts is called a **step.**

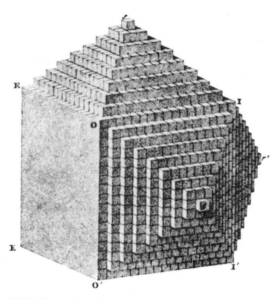

Figure 2.1 Haüy's [1801] illustration of how molecules can be arranged to form a dodecahedron.

A third important structure is called a **kink.** Figure 2.2c illustrates the kink. Basically, a kink is a place in a surface where rather than the step moving along a straight line, there is a jog in the path of step. The result is a more complex structure than a stepped surface.

The first experimental data on the structure of surfaces came in 1927 when Davidson and Germer showed that electrons would diffract from a surface. Davidson and Germer were mainly interested in using these results to confirm the quantum-mechanical prediction that matter sometimes behaves like a wave. However, this work also showed that the atoms in the surface were ordered, although it was difficult to determine the details of the surface structure using the data available from the 1927 experiments.

In 1937 Müller showed that he could produce an image of the atoms on a surface using a technique called field electron microscopy (FEM). Later he improved his pictures using a technique called field ion microscopy (FIM) (Müller [1951]). Figure 2.3 shows a field ion microscope image of a tungsten field emitter tip and a model of the corresponding surface. Notice that one can image individual atoms and observe the arrangements of all of the atoms. Hence one can use a field ion microscope to determine surface structure. If one looks carefully

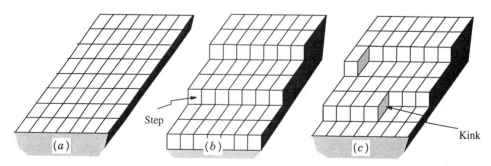

Figure 2.2 A schematic of the structure of some ideal surfaces as described by Haüy [1801].

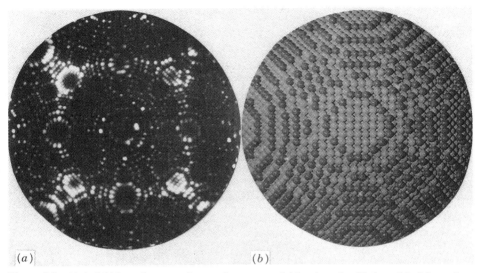

Figure 2.3 (*a*) A field ion microscope image of a tungsten field emitter tip. (*b*) A model of the surface structure of the tip. (From Tsong [1990]. Reprinted with permission of Cambridge University Press. © 1990 Cambridge University Press.)

at Figure 2.3 one can find several steps and kinks. Hence, the field ion microscopy results seem to confirm the model of surface structure proposed by Haüy.

Nevertheless, at about the same time that Müller was publishing his work on field emission microscopy, Farnsworth and coworkers found deviations from Haüy's model using low-energy electron diffraction (LEED). For a review, see Farnsworth [1950]. As noted in Chapter 1, LEED was originally invented by Davidson and Germer [1927] as a way of testing quantum mechanics. However, Farnsworth recognized that LEED could also be used to study surfaces. Farnsworth developed ways to clean extended single-crystal surfaces so that they would give sharp LEED patterns. He then heated surfaces and watched them change. Farnsworth also varied the beam energy in order to get a complete diffraction pattern.

Farnsworth found that surfaces often had the same periodicity as the bulk material. However, occasionally the surface had a different arrangement than the bulk. Farnsworth called these rearrangements of the surface "surface reconstructions." Surface reconstructions are quite a common phenomenon. We discuss them in Sections 2.6 and 2.7.

Since 1960 there has been a rapid increase in the number of papers that have examined surface structure. Work of Farnsworth [1950], Germer [1962], Beebe [1968], Pendry [1974], Duke [1967], Jona [1970], and others converted LEED into a quantitative tool for surface structural analysis. Binning and Rohrer [1986] invented scanning tunneling microscopy (STM) and showed that it could be used to examine the atomic arrangement of surfaces. Cowley [1986], Marks [1986], Smith [1985], and others improved transmission electron microscopy (TEM) so that it could be used to image the rows of atoms in surfaces. The result is that there are now a series of tools that can be used to examine surface structure, and these tools have been used to examine surface structure for many important examples.

2.2 QUALITATIVE DESCRIPTION OF SURFACE STRUCTURE

In general, two types of surface structures have been observed experimentally: periodic and nonperiodic surfaces.

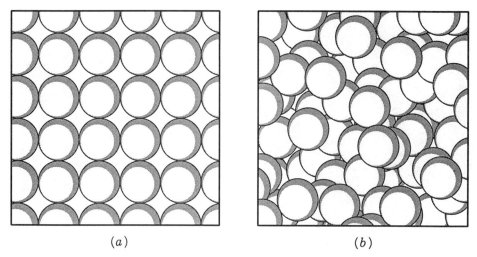

(a) (b)

Figure 2.4   An illustrastion of (a) a periodic surface and (b) a nonperiodic surface.

- We say that the surface structure is **periodic** when the surface structure repeats at regular intervals, as indicated in Figure 2.4a.
- We say that the surface structure is **nonperiodic** when the surface structure does not repeat. Instead one observes an amorphous structure, as indicated in Figure 2.4b.

One way to learn what the surfaces of materials are like is to examine the surfaces with a scanning electron microscope (SEM). Figure 2.5 shows an SEM image of an amorphous iron catalyst. Notice that the iron looks very rough and amorphous in the picture. If we could examine the surface at much higher magnification, we would find that the atomic spacing of

Figure 2.5   A SEM picture of an amorphous iron catalyst. (Courtesy of K. Suslick.)

the iron is fixed, so that the surface looks periodic on an atomic level. However, the surface has no long-range periodicity.

Most real surfaces are like that of iron. The surface has short-range, but no long-range, periodicity or symmetry. Such surfaces are very difficult to describe.

There are special surfaces, called single-crystal surfaces, that are periodic. Single crystal surfaces are made by cutting a bulk single crystal at some particular angle and then polishing appropriately. Anything that forms a crystalline solid can in principle be cut and polished to yield a single-crystal surface. Single crystals of most elements, many semiconductors, and many minerals can be purchased commercially. For example Figure 2.6 shows a picture of a silicon wafer. The wafer is about 6 in. in diameter and 1 mm thick. It is cut from a single crystal rod that may be 2–3 meters long. Metal single crystals are usually 1–2 cm in diameter and up to 5 cm long.

Figure 2.7 shows a STM image of a part of a gold single crystal that is oriented in the (111) direction (see Section 2.4 for a description of this notation). Notice that the atoms in the surface are arranged in a periodic array. We describe the atomic arrangement in great detail later in this chapter. However, for now, the key thing to remember is that while most real surfaces are not periodic, one can produce surfaces that are periodic on an atomic scale.

Periodic surfaces have several advantages in terms of an experiment. Their properties are uniform over the surface of the solid. One of the main thrusts of this book is to understand how the structure and properties of the surface affects the reactivity of the surface. As a result, much of what is discussed in this book focuses on periodic surfaces.

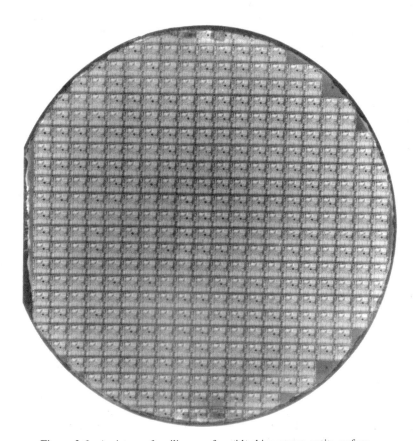

Figure 2.6 A picture of a silicon wafer with chips grown on its surface.

Figure 2.7   An STM image of a Au(111) single-crystal surface. (Courtesy of J. Lyding.)

2.3   QUANTITATIVE DESCRIPTION OF THE STRUCTURE OF SOLIDS AND SURFACES

As noted previously, periodic surfaces are formed by cutting bulk single crystals. The structure of bulk single crystals has been studied in great detail by crystallographers. In the next section we briefly review some of the notation crystallographers have developed to describe structures of bulk single crystals. A more complete review of this material can be found in Kittel [1986] or Kelly and Groves [1970].

2.3.1   Two-Dimensional Bravais Lattices

Crystallographers classify the structure of a crystal in terms of the positions of the atoms in the crystal. In a crystalline material the atoms form a periodic lattice. If one knows the periodicity of the lattice and the structure of the repeat unit, one can calculate the structure of the crystal. It is very convenient to describe a periodic lattice in terms of something called a **unit cell** for the lattice. The unit cell is defined as the basic repeat unit for the lattice. Figure 2.8 shows several possible choices for the unit cell of an oblong two-dimensional lattice. If one starts with one of these unit cells and repeats it to the left and right and up and down, as illustrated in Figure 2.9, one can generate the entire lattice. Figure 2.8 shows that the unit cell may contain one atom or several atoms. However, it has to have the same rotational symmetry as the lattice. One can define a unit cell for any lattice. However, Figure 2.8 shows that the choice of the unit cell is not unique. In the literature, it is common to define the **primitive** unit cell(s), which is the unit cell with the smallest area or volume. In Problem 2.7 we ask the reader to show that unit cells 1, 2, 3, 4, and 5 in Figure 2.8 are all primitive unit cells. Unit cell 6 is a non-primitive unit cell.

Once we have decided on a unit cell, we can define the **lattice vectors** for the lattice, where the lattice vectors are the vectors pointing from one unit cell to an equivalent point in the next adjacent unit cell. In general, the lattice vectors form the sides of the unit cell, as illustrated in Figure 2.8. One needs two lattice vectors for a two-dimensional lattice and three lattice vectors for a three-dimensional lattice. The lattice vectors may or may not be orthogonal. However, the analysis is easier if the lattice vectors are orthogonal.

Crystallographers classify lattices in terms of the symmetry of the arrangement of the atoms in each lattice. For a simple two-dimensional lattice, with only one atom per primitive unit cell, one can specify the periodicity of the atoms in the lattice in terms of the lengths of the unit vectors, a_x and a_y, and the angle between them, γ (see Figure 2.8). One can show

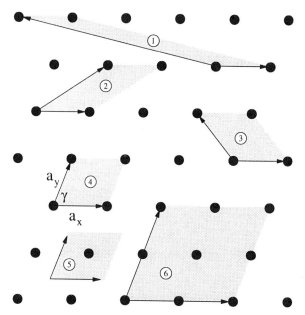

Figure 2.8 A picture of several different possible choices of unit cells (shaded) and lattice vectors (arrows) for a two-dimensional lattice.

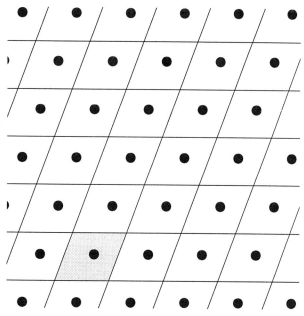

Figure 2.9 An illustration of how a unit cell can be repeated to form a periodic lattice. The shaded area is the original unit cell. The white areas are the repeat units.

that there does not have to be any special relationship between a_x, a_y, and γ except that by convention $0 < \gamma < 180°$. If there is some special relationshp between a_x, a_y, and γ, however, the system will have some symmetry as described on the next page.

Before we proceed, it is useful to review some information about symmetry elements in molecules and surfaces. Generally, the most important symmetry elements are rotation axes, mirror planes, and screw axes. Figure 2.10 illustrates systems with mirror planes, rotation axes, and screw axes. In the figure, open circles were used to designate atoms or molecules. The figure shows that a system that contains an n-fold rotation axis is invariant to a rotation of 360°/n. A system with a mirror plane is invariant to reflection across the plane. Following the standard notation from the International Tables of Crystallography we indicate the reflected molecules by circles with dotted borders since during a reflection, the internal coordinates of the molecule are inverted. Screw axes are more complex. When an atom is rotated around a screw axis, the atom also translates along the axis, so its position is displaced. Two-dimensional systems with screw axes are similar to those with mirror planes, except that in the case of a screw axis the mirror image is displaced along the screw axis by half a unit cell.

Next, we will demonstrate that in two dimensions there are only five possible types of primitive lattices corresponding to five different symmetry groups. To do so, we need to consider what kinds of symmetry can be found in a two-dimensional lattice. From inorganic chemistry we know that a molecule can have many different combinations of rotational axes mirror planes and inversion centers. We limit ourselves to periodic lattices, however, and for the moment we also limit our discussion to lattices with one atom per primitive unit cell. In the supplemental material at the end of the chapter we prove that in a periodic two-dimensional lattice, we can only have twofold, threefold, fourfold, and/or sixfold rotational axes. Hence, if we want the lattice to be symmetric, γ must be such that we get a twofold, threefold, fourfold, or sixfold rotational axis.

We get a twofold axis in all two-dimensional structures with one atom per primitive unit cell. However, we can only get a three- or sixfold axis for a γ of 60° or 120°. We get a fourfold axis with a γ of 90°. As a result, we form a symmetric lattice whenever γ is 120, 90, or 60°.

It happens that two of these values of γ are equivalent. Notice that from Figure 2.8 if we have a γ of 60°, we can define an equivalent unit cell with $\gamma = 120°$. As a result, the cases

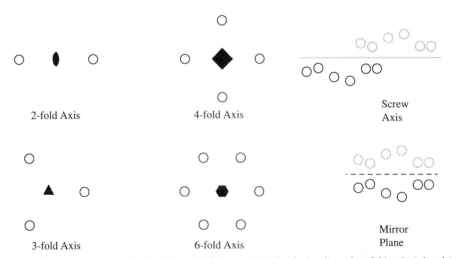

Figure 2.10 An illustration of structures containing a twofold axis (oval), a threefold axis (triangle), a fourfold axis (diamond), and a sixfold axis (hexagon), a mirror plane (dashed lines), and a screw axis (dotted line).

for $\gamma = 60°$ and $120°$ are equivalent. Therefore, special symmetries occur only when γ is $90°$ or $120°$.

We can also get symmetries if there is a special relationship between a_x and a_y. Clearly, a_x and a_y can be different. If $a_x = a_y$ there is some extra symmetry, however.

We can categorize a two-dimensional lattice by the symmetry elements present in the lattice. For example, if $\gamma = 90°$ and $a_x = a_y$, we would have one symmetry, while if $\gamma = 120°$, we would have a different symmetry. A third symmetry would occur if γ were some value other than $90°$ or $120°$. Three more cases occur if $\gamma = 90°$, $120°$, or some other value and $a_x \neq a_y$. That gives a total of six cases, all of which are listed in Table 2.1. Note that two of the six are equivalent. When $\gamma = 120°$ and $a_x = a_y$, the system will have a sixfold axis, and hence some special symmetry. However, if $a_x \neq a_y$ the lattice will not have a sixfold axis. Hence the lattice will not have any special symmetry. The symmetry will be the same as when $a_x \neq a_y$, $\gamma \neq 90°$, or $120°$. Therefore, the case for $a_x \neq a_y$, $\gamma = 120°$ is equivalent to the case for $a_x \neq a_y$, $\gamma \neq 90°$, or $120°$.

We call the five allowed possibilities the five two-dimensional **primitive Bravais lattices.** Table 2.1 summarizes the properties of these lattices. In the literature, it is common to refer to these lattices as **oblique lattices, centered rectangles, primitive rectangles, squares,** and **hexagonal lattices.** The names are defined in Table 2.1. Table 2.1 also lists both the primitive unit cell and the "conventional unit cell" for each of the lattice types. The primitive unit cells are as described previously. The conventional unit cell is the same as the primitive unit cell for the oblique, square, hexagonal, and primitive rectangular lattices. The primitive unit cell for the centered rectangle is the regular parallelogram shown on the left side of Figure 2.11. However, notice that one could also describe the centered rectangle by the rectangular unit cell shown in Figure 2.11. In the next section, we will do matrix operations with the lattice vectors for two-dimensional lattices; the algebra is much easier if one uses a rectangular lattice. As a result, it has become conventional to discuss the centered rectangle lattice via a rectangular unit cell with two atoms per unit cell. This unit cell is also shown in Figure 2.11.

Now, if one allows there to be more than one atom per primitive unit cell one can get some additional symmetries. For example Figure 2.12 shows several different primitive rectangular unit cells. The first case on the left is the simple primitive rectangle with two identical molecules and two reflected molecules per unit cell. This lattice has mirror planes as indicated by the dashed lines in the figures. There are also twofold axes at the intersections of each of the mirror planes. The second case from the left has mirror planes but no twofold axes. The third case has twofold axes but no mirror planes. Notice that each of these cases has a different symmetry even though the Bravais lattice is the same. Hence, if we allow there to be more than one atom per unit cell, we cannot specify the symmetry of the system by just specifying the Bravais lattice.

TABLE 2.1 The Symmetry Properties of the Two-Dimensional Primitive Bravais Lattices

Lattice	Axes of Primitive Cell	Conventional Cell	Axes of Conventional Cell
Oblique	$a_x \neq a_y$, $\gamma \neq 90°$ or $120°$	Parallelogram	$a_x \neq a_y$, $\gamma \neq 90°$ or $120°$
Centered rectangle	$a_x = a_y$, $\gamma \neq 90°$ or $120°$	Rectangle	$a_x' \neq a_y'$, $\gamma' = 90°$
Primitive rectangle	$a_x \neq a_y$, $\gamma = 90°$	Rectangle	$a_x \neq a_y$, $\gamma = 90°$
Square	$a_x = a_y$, $\gamma = 90°$	Square	$a_x = a_y$, $\gamma = 90°$
Hexagonal	$a_x = a_y$, $\gamma = 120°$ with a sixfold axis	Hexagonal	$a_x = a_y$, $\gamma = 120°$
Oblique	$a_x \neq a_y$, $\gamma = 120°$	Parallelogram	$a_x \neq a_y$, $\gamma = 120°$

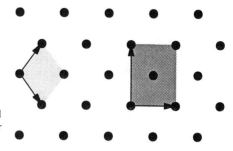

Figure 2.11 A diagram of the primitive unit cell (light gray) and conventional unit cell (dark gray) for a centered rectangle.

The individual arrangements of atoms with different symmetries are called **space groups.** In 1891 Schoenflies showed that there are only 17 unique two-dimensional space groups and 230 unique three-dimensional space groups. Figure 2.13 shows what is called **Schoenflies diagrams** of all of the 17 two-dimensional groups. The diagram on the left of each picture shows the unit cell while the diagram on the right shows the symmetry elements in the picture. In the diagrams two-, three-, four-, or sixfold axes are denoted by ovals, triangles, diamonds, and hexagons, respectively. Mirror planes are denoted by dashed lines, while screw axes are denoted by dotted lines. The positions of molecules around the rotational axes are denoted by open and shaded circles. Reflected molecules are denoted by circles with dotted edges.

Two different notations are used to designate the space groups. One notation, due to Schoenflies, designates each lattice by the symmetry elements in the system. This is also sometimes called the point group of the system. The lattice on the left of Figure 2.12 has a two-fold axis and two perpendicular mirror planes, so it is called point group 2 mm, where the 2 designates the rotation axis in the lattice and the m's denote the first two mirror planes. Similarly, the second lattice in Figure 2.12 with only mirror planes is called point group m, while the third lattice, in Figure 2.12 which only has a twofold axis, has a point group of 2. We also use g's to designate screw axes that translate by half a unit cell. An alternative notation is to list the arrangements in terms of a few key symmetry elements and whether

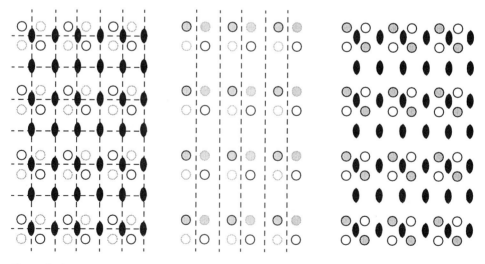

Figure 2.12 A series of primitive rectangular lattices with different space groups. The dashed lines denote mirror planes; the ovals denote twofold axes; the circles are molecular positions.

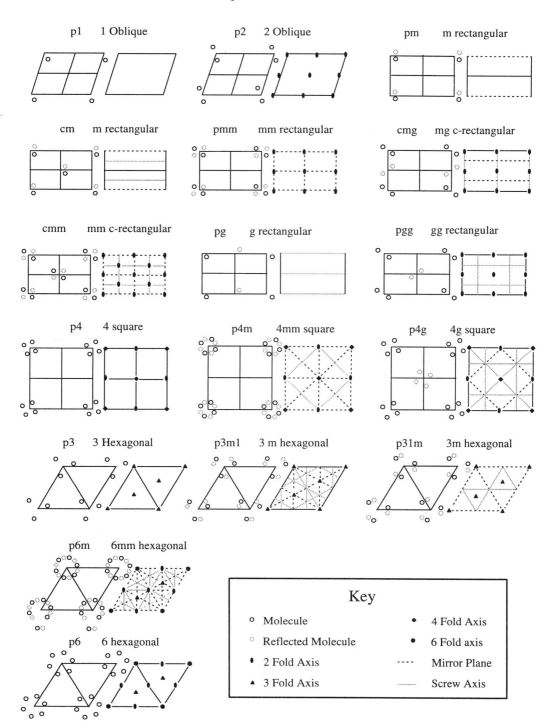

Figure 2.13 Schoenflies diagrams of the 17 two-dimensional space groups. Note people sometimes report glide planes rather than screw axes.

the Bravais lattice is a primitive lattice or a centered rectangle. The primitive lattices are labeled with a p, for example, pmm, while the centered lattices are labeled by a c, for example, cmm. That notation is also listed in Figure 2.13.

In general, one needs to specify both the space group and the Bravais lattice to specify the symmetry of a two-dimensional lattice. Many different combinations are possible. For example, Figure 2.14 shows four different lattices, each with a space group of p2. The point group specifies the symmetry elements, while the Bravais lattice specifies the periodicity. In Problem 2.4 the reader is asked to work out the point group and Bravais lattice for a series of two-dimensional structures.

2.3.2   Extension to Slabs and Surfaces

The analysis in Section 2.3.1 also applies to slabs and surfaces. Slabs and surfaces generally show two-dimensional periodicity. Therefore, they can be described as belonging to one of the two-dimensional primitive Bravais lattice types. The unit cells for surfaces or slabs,

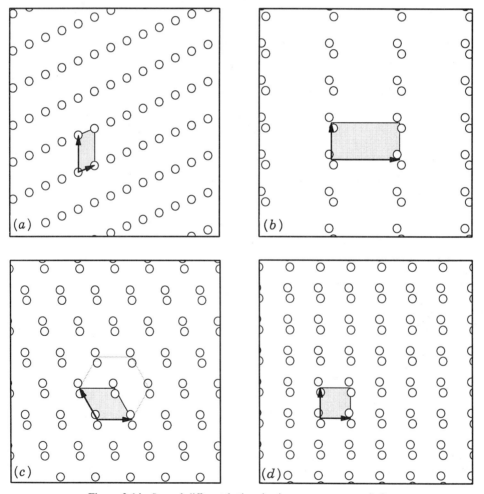

Figure 2.14   Several different lattices having a space group of p2.

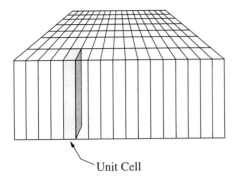

Unit Cell

Figure 2.15 A unit cell in a semi-infinite slab.

however, extend through the material, as illustrated in Figure 2.15. As a result, the unit cells are three-dimensional even though they only have two-dimensional periodicity.

2.3.3 Three-Dimensional Bravais Lattices

One can extend these arguments to three dimensions. With a three-dimensional lattice, there are three lattice vectors, a_x, a_y, and a_z and three important angles, as illustrated in Figure 2.16. Bravais [1850] considered all possible combinations of symmetries to show that one can only have 14 primitive lattices in three dimensions. Schoenflies [1891] showed that there are only 230 three-dimensional space groups. Table 2.2 shows the properties of these primitive lattices. As before, one can describe each of the lattices via a primitive unit cell. However, several of the primitive unit cells have complex shapes. Hence, it is often more convenient to describe the lattices with a conventional unit cell with cubic symmetry. Table 2.2 lists both the conventional unit cell and the primitive unit cell for the 14 lattice types. The conventional unit cells and the primitive unit cells are the same for the primitive triclinic, primitive monoclinic, primitive orthorhombic, primitive tetragonal, primitive hexagonal, primitive rhombohedral, and simple cubic lattices. The primitive and conventional unit cells are different in the other cases. Figure 2.17 shows a diagram of the conventional unit cells for these 14 lattices. Details of these structures can be found in the *International Handbook of Crystallography*. In the problem set the reader is asked to compare the diagrams of the lattices to the numbers in Table 2.2. The most important Bravais lattices for the work in this book are the face-centered cubic (FCC) lattice, and the body-centered cubic (BCC) lattice. We also consider the hexagonal closed packed (HCP) lattice, which is a nonprimitive lattice. These lattices are shown in greater perspective in Figure 2.18. We have also included a more complex structure called the zinc blend/diamond lattice and the NaCl lattice in Figure 2.18 because it will be used extensively in the work later in this book. The reader may want to carefully study Figure 2.18 before proceeding.

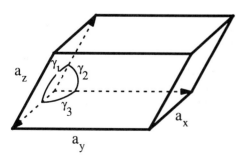

a_z γ_1 γ_2 γ_3 a_x a_y

Figure 2.16 A diagram of a generalized three-dimensional unit cell.

TABLE 2.2 Properties of the Three-Dimensional Bravais Lattices

Lattice	Axes of Primitive Unit Cell	Axes of Conventional Unit Cell
Primitive triclinic	$a_x \neq a_y \neq a_z$ $\gamma_1 \neq 90°, 120°, \gamma_2 \neq 90°, 120°$ $\gamma_3 \neq 90°, 120°\ \gamma_1 \neq \gamma_2 \neq \gamma_3$	$a_x \neq a_y \neq a_z$ $\gamma_1 \neq 90°, 120°, \gamma_2 \neq 90°, 120°$ $\gamma_3 \neq 90°, 120°\ \gamma_1 \neq \gamma_2 \neq \gamma_3$
Primitive monoclinic	$a_x \neq a_y \neq a_z$ $\gamma_1 = \gamma_2 = 90°, \gamma_3 \neq 90°$	$a_x \neq a_y \neq a_z$ $\gamma_1 = \gamma_2 = 90°, \gamma_3 \neq 90°$
Side-center monoclinic	$a_x \neq a_y \neq a_z$ $\gamma_1 \neq 90°, 120°, \gamma_2 \neq 90°, 120°$ $\gamma_3 \neq 90°, 120°$	$a'_x = a'_y \neq a_z$ $\gamma'_1 = \gamma'_2 = 90°\ \gamma_3 \neq 90°, 120°$
Primitive orthorhombic	$a_x \neq a_y \neq a_z$ $\gamma_1 = \gamma_2 = \gamma_3 = 90°$	$a_x \neq a_y \neq a_z$ $\gamma_1 = \gamma_2 = \gamma_3 = 90°$
Side-centered orthorhombic	$a_x = a_y \neq a_z$ $\gamma_1 \neq 90°, 120°, \gamma_2 = \gamma_3 = 90°$	$a'_x \neq a'_y \neq a'_z$ $\gamma'_1 = \gamma'_2 = \gamma'_3 = 90°$
Body-centered orthorhombic	$a_x \neq a_y \neq a_z$ $90° = \gamma_1 \neq \gamma_2 \neq \gamma_3$	$a_x \neq a_y \neq a_z$ $\gamma'_1 = \gamma'_2 = \gamma'_3 = 90°$
Face-centered orthorhombic	$a_x = a_y \neq a_z$ $\gamma_1 \neq 90°, \gamma_2 = \gamma_3 \neq 120°, 90°$	$a'_x \neq a'_y \neq a'_z$ $\gamma'_1 = \gamma'_2 = \gamma'_3 = 90°$
Primitive tetragonal	$a_x = a_y \neq a_z$ $\gamma_1 = \gamma_2 = \gamma_3 = 90°$	$a_x = a_y \neq a_z$ $\gamma_1 = \gamma_2 = \gamma_3 = 90°$
Body-centered tetragonal	$a_x = a_y \neq a_z$ $\gamma_1 = 90°, \gamma_2 = \gamma_3 \neq 54.7°$	$a_x = a_y \neq a'_z$ $\gamma_1 = \gamma_2, \gamma'_3 = 90°$
Primitive hexagonal	$a_x = a_y \neq a_z$ $\gamma_1 = 120°, \gamma_2 = \gamma_3 = 90°$	$a_x = a_y \neq a_z$ $\gamma_1 = 120°, \gamma_2 = \gamma_3 = 90°$
Primitive rhombohedral	$a_x = a_y = a_z$ $\gamma_1 = \gamma_2, \gamma_3 \neq 90°$	$a_x = a_y = a_z$ $\gamma_1 = \gamma_2 = \gamma_3 \neq 90°$
Simple cubic	$a_x = a_y = a_z$ $\gamma_1 = \gamma_2 = \gamma_3 = 90°$	$a_x = a_y = a_z$ $\gamma_1 = \gamma_2 = \gamma_3 = 90°$
Face-centered cubic	$a_x = a_y = a_z$ $\gamma_1 = 90°, \gamma_2 = \gamma_3 = 120°$	$a'_x = a'_y = a'_z$ $\gamma'_1 = \gamma'_2 = \gamma'_3 = 90°$
Body-centered cubic	$a_x = a_y = a_z$ $\gamma_1 = 90°, \gamma_2 = \gamma_3 = 54.7°$	$a'_x = a'_y = a'_z$ $\gamma'_1 = \gamma'_2 = \gamma'_3 = 90°$

An FCC lattice has atoms at the corners of a cube and atoms in the centers of each of the faces of the cube. Metals form **closed packed** FCC lattices, where we say that a lattice is closed packed when all of the atoms touch at their covalent radius. Figure 2.18 shows a diagram of the closed packed FCC lattice. In the view shown the FCC lattice looks like a cube with atoms at each corner of the cube and other atoms in each of the faces of the cube.

For the discussion that follows, it is useful to consider what would happen if we would cut the FCC cube to expose the surface that is shaded in Figure 2.18. In Section 2.5, we show that when we cut the FCC cube to expose the shaded atoms in Figure 2.18, the exposed face will have a hexagonal arrangement of atoms shown in Figure 2.19. We will also show that an identical packing arrangement is seen in the plane of atoms to the right of the shaded plane in Figure 2.18. As a result, we can construct an FCC lattice by stacking hexagonal closed packed planes of atoms.

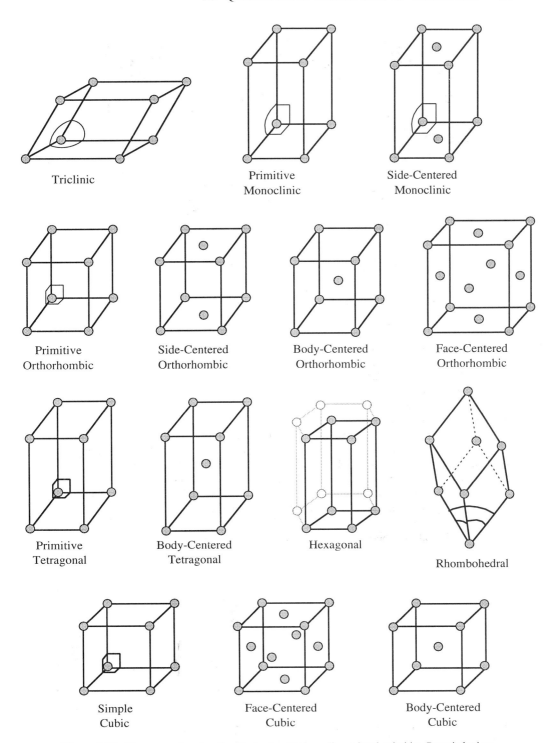

Figure 2.17 The conventional unit cells for the 14 three-dimensional primitive Bravais lattices.

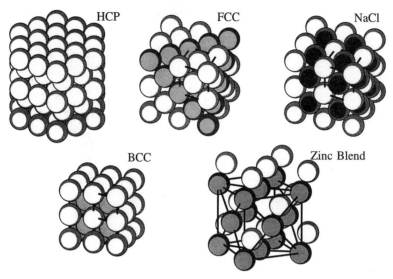

Figure 2.18 A diagram of the crystal structure of the BCC, FCC, HCP, NaCl, and zinc-blend structure.

Now let's consider the HCP lattice. A diagram of the HCP lattice is given in Figures 2.18 and 2.21. Notice that the top plane of the hexagon also has a hexagonal closed packed structure. Hence, we can also construct an HCP lattice by stacking hexagonal closed packed planes of atoms. Thus, the FCC and HCP lattices are quite similar.

The main difference between the FCC and HCP lattice is the stacking pattern. Figure 2.20 shows a diagram of two hexagonal closed packed planes stacked one on top of the other. The planes are arranged so that the atoms in the top layer sit over what is called a **threefold hollow** in the second layer, where the threefold hollow is the opening between three adjacent atoms in the hexagonal closed packed plane. For the discussion that follows, we designate the positions of the atoms in the bottom (dashed) plane in the figure the B sites, and the positions of the atoms in the top plane the A sites, as indicated in Figure 2.20. Now consider adding a third plane of atoms. We can put the atoms in the third hexagonal closed packed plane in one of two threefold hollows: the hollow directly above the center of the dashed circles, site B, and the hollow between the atoms at the centers of the dashed circles. We

Figure 2.19 A schematic of the arrangement of atoms in a hexagonal closed packed plane. This arrangement is characteristic of the (111) face of an FCC metal or the (001) face of an HCP metal.

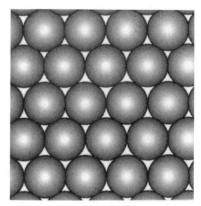

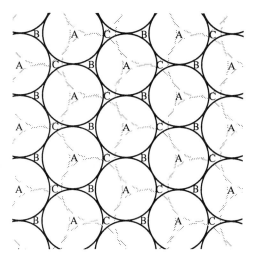

Figure 2.20 A view of two hexagonal closed packed planes stacked one on top of another. The solid circles are the top layer of atoms, while the dotted circles are the second layer of atoms.

will designate the latter site "site C." If we stack the atoms so that we get an ABABAB. . . stacking pattern, we will generate an HCP lattice, while we will generate an FCC lattice if we use an ABCABC. . . stacking pattern. Hence, we can distinguish between FCC and HCP lattices by carefully examining the stacking pattern of the hexagonal closed packed planes.

The BCC lattice is different. There are no closed packed planes in a BCC lattice. The closest packing direction is along the corners of the cube. However, the plane has openings between atoms. The overall packing density of the BCC structure is only 92% of the packing density of the FCC or HCP lattice.

The zinc-blend lattice is more complex than the closed packed FCC, HCP, and BCC lattices. The zinc-blend structure has two atoms per unit cell. The two atoms are arranged in an FCC structure as shown in Figure 2.18. Hence, one can think of the zinc-blend structure as being composed of two interpenetrating FCC lattices. In Problem 2.3, we ask the reader to find the two interpenetrating FCC lattices in Figure 2.18.

There is another important structure called a diamond structure. The diamond structure is almost identical to the zinc-blend structure. The only difference is that the zinc-blend structure

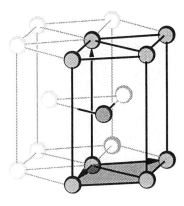

Figure 2.21 The conventional unit cell for an HCP lattice.

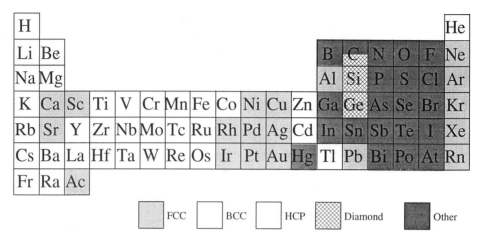

Figure 2.22   The crystal structure of the elements at room temperature and below.

contains two different atoms (e.g., Ga and As), while in the diamond structure all the atoms are the same.

Figure 2.22 shows how the crystal structure of the elements varies over the periodic table. Most metals have a BCC, FCC, or HCP structure. Germanium and silicon have a diamond structure; GaAs and InP have a zinc-blend structure; other compounds have more complex crystal structures. In fact, one can find minerals whose structures correspond to most of the 230 space groups described by Schoenflies [1891].

One can distinguish between these various arrangements of atoms by denoting the positions of all of the atoms in the lattice. The positions of the atoms in the unit cell are called the **basis** of the unit cell. Usually one keeps track of the basis via a series of position vectors $B^j = [B_1^j, B_2^j, B_3^j]$, which indicate the position of each of the atoms in the unit cell in units of the lattice vectors. For example, there are two atoms per unit cell in the HCP lattice illustrated in Figure 2.21. One is at the origin and one is two-thirds of a unit cell in the a_x direction, one-third of a unit cell in the a_y direction, and one-half of a unit cell in the a_z direction. Table 2.3 lists the basis and the lattice vectors for the FCC, BCC, HCP, and zinc-blend lattices. There is one site/unit cell for the simple cubic lattice, two sites per unit cell for the BCC and HCP lattices, four sites per unit cell for the FCC lattice, and eight sites per unit cell for the zinc-blend structure.

It is useful to compare the FCC and zinc-blend structures. The zinc-blend structure has all of the basis positions in the FCC lattice plus a second set of positions that are displaced by $(\frac{1}{4}, \frac{1}{4}, \frac{1}{4})$ from the FCC structure. As a result, it is common to discuss the zinc-blend (GaAs) or diamond (Si) structure as being two interpenetrating FCC lattices, as discussed earlier.

Once the basis and lattice vectors for a given lattice are known, we can calculate the positions of all of the atoms in the structure via

$$\vec{r} = (l_1 + B_1^j)\vec{b}_1 + (l_2 + B_2^j)\vec{b}_2 + (l_3 + B_3^j)\vec{b}_3 \qquad (2.1)$$

where the set of all $\vec{r}$'s is the position of the atoms in the lattice, B_1^j, B_2^j, and B_3^j are the basis for atom j in the lattice; $\vec{b}_1$, $\vec{b}_2$, and $\vec{b}_3$ are the lattice vectors for the lattice, and l_1, l_2, and l_3 are random integers. In Example 2.A we use this equation to calculate the position of the atoms in an FCC lattice.

TABLE 2.3 Some Properties of the FCC, BCC, HCP, and Zinc-blend Lattices

Lattice	Nearest Neighbor Distance in Units of the Lattice Dimension, a	Atom Positions in Units of the Basis Vectors	Lattice Vectors in Units of the Lattice Dimensions, a
Simple cubic	a	[0,0,0]	[1,0,0] [0,1,0] [0,0,1]
FCC	0.707 a	$[0,0,0]$ $[0,\frac{1}{2},\frac{1}{2}]$ $[\frac{1}{2},0,\frac{1}{2}]$, $[\frac{1}{2},\frac{1}{2},0]$	[1,0,0] [0,1,0] [0,0,1]
BCC	0.866 a	$[0,0,0]$, $[\frac{1}{2},\frac{1}{2},\frac{1}{2}]$	[1,0,0] [0,1,0] [0,0,1]
HCP	a	$[0,0,0]$, $[\frac{2}{3},\frac{1}{3},\frac{1}{2}]$	$[0.866, -\frac{1}{2},0]$ [0,1,0] [0,0,1.633]
Zinc blend	0.707/2 a	(A) $[0,0,0]$, $[0,\frac{1}{2},\frac{1}{2}]$ $[\frac{1}{2},0,\frac{1}{2}]$, $[\frac{1}{2},\frac{1}{2},0]$ (B) $[\frac{1}{4},\frac{1}{4},\frac{1}{4}]$, $[\frac{1}{4},\frac{3}{4},\frac{3}{4}]$ $[\frac{3}{4},\frac{1}{4},\frac{3}{4}]$, $[\frac{3}{4},\frac{3}{4},\frac{1}{4}]$	[100] [010] [001]

2.4 MILLER INDICES

As noted earlier, we form a single crystal surface by cutting a bulk crystal. If we take an arbitrary crystal; we can cut it at any angle. Hence, we need some notation to describe the structure of surfaces. In 1839, Miller devised a notation to designate planes and lattice vectors in bulk lattices. Miller's notation, called **Miller indices,** has since been adopted by most investigators.

There are two kinds of Miller indices: the Miller indices for a given direction in a crystal lattice, and the Miller indices for a plane in the lattice. By convention we enclose the Miller indices in parentheses, for example, (111), to designate the Miller indices of a plane while we enclose them in square brackets, for example, [111], to designate a crystallographic direction.

Next, we show how to determine the Miller indices of a plane. Consider the plane shown in Figure 2.23. We can define b_x, b_y and b_z as the intersection of the plane with the x, y, and z axes of the unit cell as shown in Figure 2.23. The Miller indices of the plane, (i j k), are then given by

$$(i \, j \, k) = \left(\frac{cd}{b_x} \frac{cd}{b_y} \frac{cd}{b_z} \right) \qquad (2.2)$$

where cd is a constant. In this book we take cd to be the lowest common denominater of $1/|b_x|$, $1/|b_y|$, and $1/|b_z|$ greater or equal to 1, ignoring infinite values of b_x, b_y, and b_z. For example, a plane that intersects the x, y, and z axes at 2, 3, and 4 (lowest common denom-

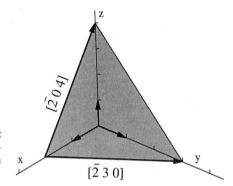

Figure 2.23 A plane that intersects the x, y, and z axes of the crystal at 2, 3, and 4. Following the discussion in the text, the Miller indices of the plane are (643).

inator 12) would be called the (643) plane. A plane that intersects the x, y, and z axes at 2, 1, and ∞ (lowest common denominator 2) would be the (120) plane. A plane that intersects the x, y, and z axes at $\frac{1}{2}$, ∞, ∞ (cd = 1) will be the (200) plane. People use bars to designate negative values. For example, the plane that intersects the x, y and z axes at 2, −1, and ∞ (lowest common denominator 2) would be the ($1\bar{2}0$) plane. Figure 2.24 shows several examples of planes with a variety of Miller indices.

One can easily show that many different planes have the same Miller indices. For example, the planes which intersect the x, y, and z axes at (1,1,1), (2,2,2) and (3,3,3) are all (111) planes. Therefore, people sometimes discuss families of planes with identical indices. Figure 2.25 shows the (111), (100), and (110) family of planes in the cubic lattice. All of the planes in a given family are parallel to each other and differ only in where they intersect the crystallographic axes.

We use three indices in a curly bracket, i.e., {ijk}, to designate an equivalent group of planes. For example, in a cubic lattice, {111} would include the (111), ($\bar{1}$11) ... families of planes.

We use three indices in a square bracket, i.e., [ijk], to designate the Miller indices of a vector. The Miller indices of a given vector are defined as the distance that the vector traverses measured in units of the lattice vector. For example, in order to move in the [123] direction, one needs to move two lattice vectors in the y direction and three lattice vectors in the z

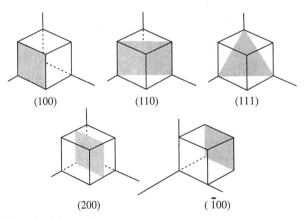

Figure 2.24 A schematic of the (100), (110), (111), (200), and (100) planes in a cubic lattice. (Adapted from Kittel [1986].)

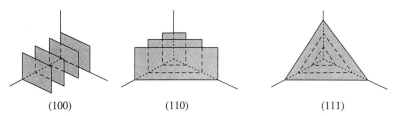

(100)	(110)	(111)

Figure 2.25 The (111), (100), and (110) family of planes in a cubic lattice.

direction for every one lattice vector in the x direction. We use three indices in an angular bracket, i.e., ⟨ijk⟩, to denote an equivalent set of directions. For example, in a cubic lattice, ⟨111⟩ would include the [111], [$\bar{1}$11] . . . , set of directions.

In a cubic crystal the (ijk) plane is perpendicular to the [ijk] direction. However, the (ijk) plane may not be perpendicular to the [ijk] direction in other crystal lattice.

There are some alternative ways to calculate the Miller indices of a plane in a cubic lattice. Consider the plane shown in Figure 2.23. Notice that the [$\bar{2}$04] and [$\bar{2}$30] vectors all lie within the (643) plane. Recall from vector calculus, that if we have two vectors in a plane in a cubic lattice [ijk] and [lmn], the indices of the perpendicular to the plane [opq] can be calculated from

$$[opq] = [ijk] \times [lmn]$$

where the $\times$ denotes a cross product. Therefore, we can calculate the Miller indices of a vector perpendicular to the (643) plane as [$\bar{2}$30] $\times$ [$\bar{2}$04] = [12 8 6] = 2[643]. With a cubic lattice the cross product of any two vectors in the (ijk) plane is always n(ijk) where n is a constant. Therefore, with a cubic lattice we can calculate the Miller indices of a plane as the cross product of two vectors in the plane. This idea is illustrated in Example 2.B.

Unfortunately, these alternate schemes do not work in a noncubic lattice because the (ijk) plane is not perpendicular to the [ijk] direction in a noncubic crystal. Figure 2.26, for ex-

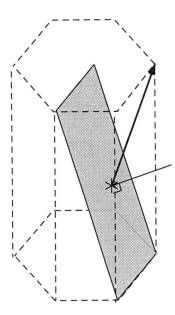

Figure 2.26 The (101) plane and [101] direction in a hexagonal lattice.

ample, shows the (101) plane and [101] direction in a hexagonal lattice. Notice that the [101] direction is not perpendicular to the (101) plane. As a result, in a noncubic crystal one cannot use a cross product to calculate the Miller indices of a given plane.

Another complication is that there are alternatives to the standard Miller indices that are occasionally used for noncubic lattices. For example, one occasionally sees a four-index notation used for an HCP material, e.g., Re(0001), where the (ijlk) refer to the intersection of the plane with the x, y, w, and z axes shown in Figure 2.27. The l index is redundant since, from analytic geometry,

$$i + j = -l \qquad (2.3)$$

The four-index notation, however, has the advantage that the symmetry of the surface is given by the symmetry of the indices. For example, in the four-index notation, the six sides of the hexagon in Figure 2.27c are given by $(1\bar{1}00)$, $(10\bar{1}0)$, $(01\bar{1}0)$, $(\bar{1}100)$, $(\bar{1}010)$, $(0\bar{1}10)$, while in the three-index notation the sides are given by $(1\bar{1}0)$, (100), (010), $(\bar{1}10)$, $(\bar{1}00)$, $(0\bar{1}0)$. It is slightly easier to remember that $(1\bar{1}00)$ is equivalent to $(10\bar{1}0)$ than to remember that $(\bar{1}10)$ is equivalent to (100). Hence, the four-index notation is often used to designate the surface of HCP crystals.

This brings up the issue of equivalent directions. In an FCC, BCC, or zinc-blend material, the x, y, z, −x, −y, and −z directions are all equivalent, which means that we can randomly interchange the order and sign of the indices and arrive at a structure that is equivalent to the first. For example, the Pt(321) and $(2\bar{3}1)$ plane have an identical surface structure. The HCP structure is different in that in the three-index notation, the indices are not equivalent. Re(100) has quite a different surface structure than Re(001). In Problem 2.14 we ask the reader to work out the equivalent planes for an HCP lattice.

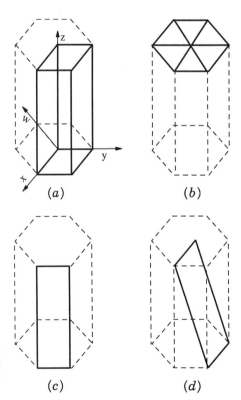

Figure 2.27 (a) The unit cell of the HCP lattice; (b) the (0001) or (001) plane; (c) the (1010) or (100) plane; (d) the (1011) or (101) plane. (Adapted from Gasser [1985].)

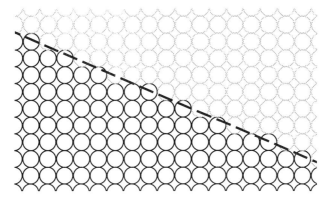

Figure 2.28   A diagram of how one can cut a single crystal to expose a stepped surface. (Adapted from Nicholas [1965].)

2.5   THE STRUCTURE OF SOLID SURFACES

Next we discuss the structure of single-crystal surfaces. A single-crystal surface is produced by orienting a bulk single crystal in an appropriate direction, cutting off excess atoms, and then cleaning, polishing and annealing the resultant sample. The simplest view of this process is that the surface is cleaved uniformly on a molecular level, as indicated in Figure 2.28, and that the surface atoms remain in their original positions. It is by no means obvious that the surface atoms remain in their original positions during a real sample preparation procedure. Typically, one spark-cuts a crystal using a wire that is 50,000,000 Å thick, and then polishes the crystal with diamond paste with an average particle size of 10,000 Å. Hence, it is not clear that one will produce a surface that is flat or uniform on a molecular level. However, the amazing thing is that crystals prepared in this way are often reasonably flat on a molecular level. It is easy to get defect densities below 1%. Properly prepared surfaces are often flat for tens of hundreds of angstroms. Further, the surface structure is often almost just as one would expect if one could simply slice a bulk crystal, as indicated in Figure 2.28. In the remainder of this section we discuss the structure of ideal surfaces, i.e., surfaces that are created by slicing the bulk single crystal in a way that we remove all of the atoms that are above the plane where the crystal is cut. However, the atoms below the plane remain in their original positions. Complications that arise because atoms can move when the surfaces are prepared are discussed in Sections 2.6 and 2.8.

2.5.1   Models of Ideal Single-Crystal Surfaces

The structure of an ideal surface is calculated by starting with the positions of bulk crystals and eliminating atoms to expose the desired plane, as indicated in Figure 2.28. A computer program that allows one to do the calculations is given in Example 2.A. The reader is encouraged to use this program to explore surface structure on his or her own. Generally two kinds of surfaces can be cut: low index, closed packed, or nearly closed packed surfaces and stepped surfaces. Closed packed surfaces are produced by cutting a single-crystal sample along some closed packed direction (i.e., where the atoms are touching), while stepped surfaces are created by cutting along other directions.

Figure 2.29 and 2.30 show diagrams of the (111), (100), and (110) faces of perfect FCC and BCC materials. In an FCC crystal, the atoms in the (111) face are arranged to form a closed packed hexagon, while the atoms in the (100) face are arranged to form a square, and the atoms in the (110) face have an up and down, washboard arrangement. In a BCC crystal,

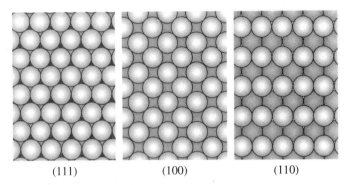

(111)   (100)   (110)

Figure 2.29 The (111), (110), and (100) faces of a perfect FCC crystal.

the atoms in the (110) face have a hexagonal arrangement, but in contrast to what is seen with the (111) face of a FCC metal the hexagon is elongated and there are gaps between the atoms. The atoms in the (100) face of a BCC material show a square atomic arrangement. However, if one models the surface with hard spheres, there are gaps between the atoms. The (111) face of a BCC material shows a more complex atomic arrangement with three layers of exposed atoms, each of which forms a centered rectangle.

Figure 2.31 shows the structure of the (001), (100), and (101) faces of a perfect HCP crystal. Surface structures in HCP materials are more complex than surface structures in FCC or BCC because HCP is not a primitive Bravais lattice. The stacking of the atoms above and below the A sites in Figure 2.19 (i.e., the base sites in Figure 2.21) is not equivalent to the stacking of the atoms above and below the B sites in Figure 2.19 (i.e., the middle sites in Figure 2.21). As a result, if the crystal is cut to expose the A sites, one can sometimes get a different surface structure than when the crystal is cut to expose the B sites. It happens that with the (001) face both arrangements are equivalent. In either case, the atoms are arranged in a hexagonal closed packed structure. However, two different arrangements are seen with the (100) face, which we denote by (100)A and (100)B. Both arrangements show a washboard structure, but in one case, the atoms in the individual facets of the washboard have a rectangular arrangement, while in the other case, atoms are arranged in a triangular pattern.

The differences are even larger with the (101) surface. One arrangement shows a nearly hexagonal arrangement, with every other row sticking up slightly, as indicated in the side view in Figure 2.32. However, the other arrangement is highly stepped. In Problem 2.15 we ask the reader to show that there are two exposed atoms per unit cell in the first case and three exposed atoms per unit cell in the second case. As a result, the properties of these two surface arrangements are quite different, even though they are both HCP (101) surfaces.

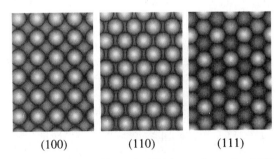

(100)   (110)   (111)

Figure 2.30 The (111), (100), and (110) faces of a perfect BCC crystal.

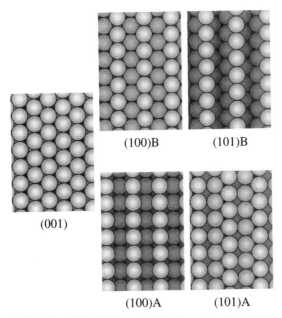

Figure 2.31 The (001), (100), and (101) faces of an ideal HCP crystal.

We will often refer to the surface structures in Figures 2.29, 2.30, and 2.31 later in this book. The reader should carefully study these surface structures before proceeding.

2.5.2 The Stereographic Triangle

Figures 2.29, 2.30, and 2.31 show the most important surface structures in BCC, FCC, and HCP materials. However, BCC, FCC, and HCP crystals can exhibit many other surface structures. In the material that follows, we will discuss some of these other structures. We will organize the discussion by asking how the surface structure varies over something called the **stereographic triangle,** where the term "stereographic triangle" will be defined shortly.

Our discussion is organized as follows: First we define the stereographic triangle and derive an equation for the position of a plane within the stereographic triangle. Then we discuss how surface structure varies throughout the stereographic triangle.

The idea of examining how surface structure varies over the stereographic triangle arose from field ion microscopy (FIM). At the time this work was done, the only way to study surface structure was with FIM. Therefore, it was reasonable to consider how the surface structure varied over the FIM image.

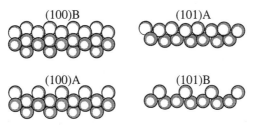

Figure 2.32 A side view of the (100)A, (100)B, (101)A, and (101)B faces of an ideal HCP material.

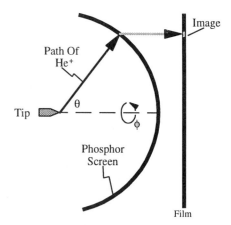

Figure 2.33 Schematic of an FIM experiment.

Let's derive an equation for the position of a given plane in a FIM image. In FIM, one images the planes in a sharp field emission tip by mounting the tip in a UHV system, with about 10^{-4} torr of helium. A high voltage is applied to the tip so that the helium ionizes on the surface of the tip. The ions that are produced in this way fly in a straight line to a phosphor screen, as illustrated in Figure 2.33. One then takes a flat picture of the phosphor screen. Consider a tip that is oriented in the [100] direction, and for the moment assume that the tip is made of a cubic material. The ions produced on the (ijk) plane on the tip will fly off at a direction perpendicular to the plane, which in a cubic material is the [ijk] direction. They will then hit the phosphor screen at some angle (θ_{FIM}, ϕ_{FIM}). In a cubic material θ_{FIM} and ϕ_{FIM} are given by

$$\cos(\theta_{FIM}) = \frac{i}{\sqrt{i^2 + j^2 + k^2}} \tag{2.4}$$

$$\tan(\phi_{FIM}) = \frac{k}{j} \tag{2.5}$$

Equations 2.4 and 2.5 apply to a cubic material. However, in Example 2.A we will provide a subroutine `Rotation_Matrix_Calc` which calculates θ_{FIM} and ϕ_{FIM} for a noncubic material. In an FIM experiment one takes a flat image of the picture. From analytic geometry the radial position, r_{image}, θ_{image} of the (ijk) plane in the FIM image is given by

$$\phi_{image} = \phi_{FIM} \tag{2.6}$$

$$r_{image} = R_{FIM} \sin(\theta_{FIM}) \tag{2.7}$$

where R_{FIM} is the distance from the tip to the phosphor screen in Figure 2.33.

Figure 2.34 shows a typical wide field image of the tip. Note that the image is quite symmetric. There is a fourfold axis at the center of the image, and mirror planes going from the center of the image to the (111) and (110) planes. Notice that if one knows the structure of the planes in the 45° slice of the image between the two mirror planes, then from symmetry one can calculate the structure of all of the other planes in the tip.

We call this slice the **stereographic triangle.** Figure 2.35 shows a diagram of the position of various planes within the stereographic triangle for an FCC material where again the positions of various planes correspond to the positions of the various planes in an FIM image. The (100) plane is at the sharp corner of the triangle, while the (110) and (111) planes lie at the other two corners of the triangle. The positions of other (ijk) planes satisfying i ≥ j ≥

Figure 2.34 An FIM image of a [100]-oriented platinum field emission tip. (From Tsong [1990], Reprinted with permission of Cambridge University Press. © 1990 by Cambridge University Press.)

$k \geq 0$ is given by

$$x_{stereo} = r_{image}\,\sin(\phi_{image}) = \frac{j}{\sqrt{i^2 + j^2 + k^2}}$$

$$y_{stereo} = r_{image}\,\cos(\phi_{image}) = \frac{k}{\sqrt{i^2 + j^2 + k^2}} \tag{2.8}$$

The line extending between the (100) and (111) plane is called the $[01\bar{1}]$ zone axis, since the analysis in Example 2.D shows that one needs to rotate the crystal around the $[01\bar{1}]$ direction to move from the (100) to the (111) plane. These plans have the same last two Miller indices. Similarly, the line between the (100) and (110) plane is called the [001] zone axis. These planes have a zero for their last Miller index. The arc between the (111) and (110) plane is called the $[1\bar{1}0]$ zone axis. These planes have the same first two Miller indices. In the literature, people sometimes plot an alternative version of the stereographic triangle where

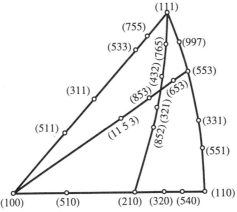

Figure 2.35 A stereographic triangle for an FCC material containing (ijk) planes with $i \geq j > k \geq 0$.

they calculate the radial position r_{stereo}, ϕ_{stereo} of a given plane via

$$\phi_{stereo} = \phi_{image} \tag{2.9}$$

$$r_{stereo} = R_{FIM}\,\theta_{FIM} \tag{2.10}$$

The stereographic triangle provides a way to discuss surface structure in an organized way. In an FCC or BCC (but not HCP) lattice the surface structure does not change as one changes the order or sign of the Miller indices, i.e., the (853) plane has the same structure as the (385) or (5$\bar{3}$8) plane. As a result any plane not satisfing i $\geq$ j $\geq$ k $\geq$ 0 has an equivalent plane in Figure 2.35.

2.5.3   The Structure of Stepped Metal Surfaces

Figure 2.36 shows a diagram of the surface structure of several planes along the [1$\bar{1}$0] zone axis of an FCC material, i.e. the line from (111) to (110). The (111) face has a hexagonal structure. As one moves from the (111) face toward the (110) face, one slowly adds steps to the surface. The step density increases moving toward the (110) face. We have only included a few different planes in Figure 2.36, and they all have a regular pattern of steps. However, Figure 2.28 shows that it is not difficult to cut faces where the step pattern does not repeat at regular intervals.

In the literature, it has become common to denote the structure of the surfaces between (110) and (111) via the arrangement of the terraces and the arrangement of the steps in the overall structure. Note that the terraces of all the planes in Figure 2.36 have a hexagonal arrangement that is similar to the atomic arrangement on a (111) face. It is hard to see, but all of the steps have a similar atomic arrangement. Hence, the surfaces shown in Figure 2.36 are often referred to as having (111) terraces and (111) steps.

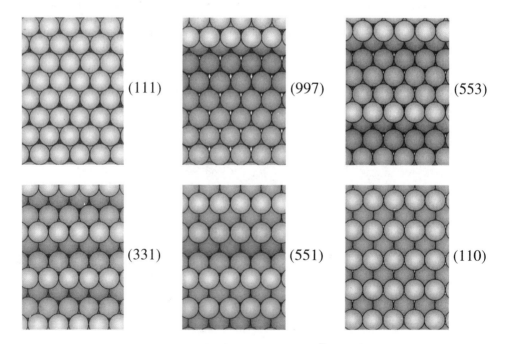

Figure 2.36 The surface structure of several planes along the [1$\bar{1}$0] zone axis of an FCC material.

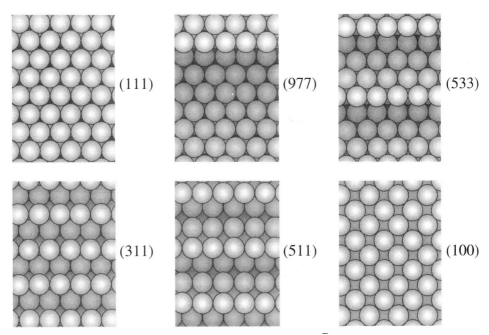

Figure 2.37 The surface structure of several planes along the [01$\bar{1}$] zone axis of an FCC material.

Figure 2.37 shows a diagram of the surface structure of several planes along the [01$\bar{1}$] zone axis of an FCC material. Again the (111) face shows a hexagonal arrangement. As one moves from the (111) toward the (311) and (100) face, one adds (100)-oriented steps to the surface. The surfaces between (111) and (311) can be described as having (111)-oriented terraces and (100)-oriented steps. However, once one gets past the (311) face, the surface structure inverts and one observes (100) terraces and (111) steps.

Figure 2.38 shows a diagram of the surface structure of several planes along the [001] zone axis of an FCC material. The (100) face has a square arrangement of atoms. When one moves from the (100) face toward the (210) and (110) faces, one adds (110)-oriented steps to the surface. The (110)-oriented steps look qualitatively different than (111)- or (100)-oriented steps because the steps are not smooth. Instead, the steps have a washboard pattern. The jobs in the washboard are called **kinks.** The (210) surface is very rough looking, with both (100)- and (110)-like structures. The surfaces between (210) and (110) have (110) terraces and (100) kinks.

Figure 2.39 is a diagram of the surface structure of several faces along the [0$\bar{3}$5] axis of the stereographic triangle, where the [0$\bar{3}$5] axis is a line between the (553) plane and the (100) plane on the stereographic triangle. The (553) plane is a stepped surface with (111) terraces and (111) steps. As one moves along the [0$\bar{3}$5] axis toward the (100) face, one adds kinks to the surface, as shown in Figure 2.39. The kink density increases as one moves along the zone axis; the planes in the middle of the stereographic triangle have very complex atomic arrangements. However, the surfaces near (100) are smooth.

All of the planes within the center of the stereographic triangle have a mixture of terraces, steps, and kinks. The surface structure is often difficult to describe. Hence, one rarely sees work done on these materials. However, it happens that the surface structure is relatively simple along the [1$\bar{2}$1] axis. Surfaces along the [1$\bar{2}$1] axis have (111) terraces and (10$\bar{1}$) steps as shown in Figure 2.40.

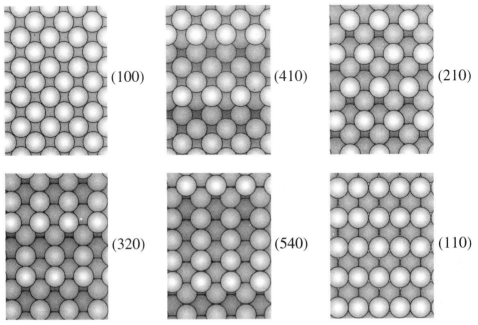

Figure 2.38 The surface structure of several planes along the [001] zone axis of an FCC material.

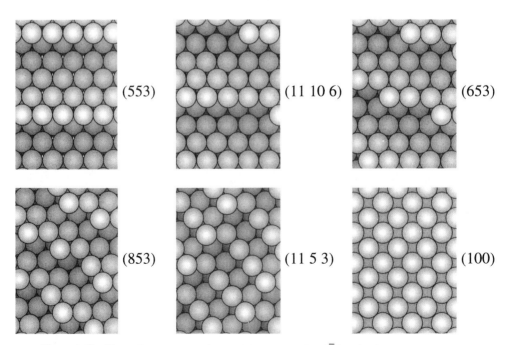

Figure 2.39 The surface structure of several faces along the [0$\bar{3}$5] axis of an FCC material.

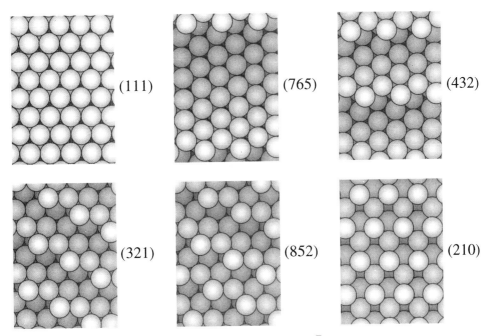

Figure 2.40 The surface structure of several faces along the [1$\overline{2}$1] zone axis of an FCC material.

Figure 2.41 summarizes the variation of the structure of an ideal FCC material over the stereographic triangle. The reader should carefully study this figure before proceeding.

Figure 2.42 shows a selection of the faces of a BCC material. Notice that the pictures look very similar to those for an FCC material, although the Miller indices for similar surfaces are different in BCC and FCC. The (111) face of an FCC crystal has a hexagonal arrangement of atoms, while the (110) face of BCC is composed of elongated hexagons. The (110) face of an FCC material has a washboard structure as does the (211) face of a BCC lattice. The (111) face of the BCC material looks just like the (210) face of an FCC material. The (100) faces of FCC and BCC lattices both have a square arrangement of atoms. In the supplementary material we generalize this idea to show that the surface structure of the (ijk) face of a BCC material with i ≥ j ≥ k is almost the same as the (lmn) face of an FCC material where

$$(lmn) = (i \quad j+k \quad j-k) \tag{2.11}$$

Of course, if one looks in detail, one finds some subtle differences in the atomic arrangements of surfaces in the BCC and FCC lattices. However, the similarities overwhelm the differences. Therefore, one can say that qualitatively, BCC surfaces look quite similar to FCC surfaces, and it is only the Miller indices that are substantially different.

Figure 2.43 shows a modified stereographic triangle for a BCC material. In the figure, the point for each plane has been plotted at the position of the equivalent plane in an FCC lattice where the equivalent points were calculated via

$$x_{stereo} = \frac{m}{\sqrt{l^2 + m^2 + n^2}} \qquad y_{stereo} = \frac{n}{\sqrt{l^2 + m^2 + n^2}} \tag{2.12}$$

The bottom axis in the figure was also stretched so that it goes all the way out to the position of the (120) face in an FCC stereographic triangle. The advantage of this modified stereo-

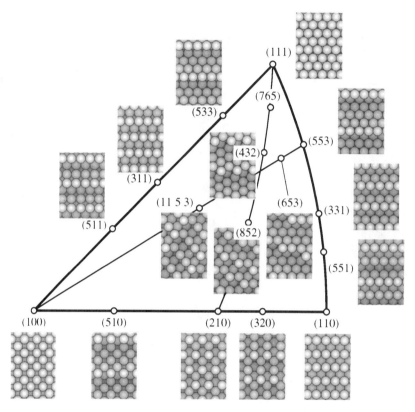

Figure 2.41   The surface structure of an ideal FCC material.

graphic triangle is that the surface structures are almost the same at the same places in Figure 2.41 and Figure 2.43. The disadvantage is that the stereographic triangle in Figure 2.43 no longer corresponds to a slice of an FIM image.

It is not possible to define a similar modified stereographic triangle for an HCP material, since HCP is not a primitive Bravais lattice. The surface structures of the higher Miller index planes of the HCP lattice are much more complex than those in BCC or FCC. All of the faces of HCP materials except the (001) and (110) have two different possible surface structures denoted by (ijk)A and (ijk)B in Figure 2.44. It happens that the surface structures seen when the A sites are the topmost position can often be quite different than when the B sites are the topmost layer. Both A and B atoms are exposed with all of the planes except the (001). These complications make it difficult to discern any trends in the variation of the surface structure of an HCP material with changes in where the plane lies in the stereographic triangle.

2.5.4   The Surface Structure of Semiconductor Crystals

The structure of semiconductors is more complex than that of metals. Gallium arsenide, indium phosphide, gallium phosphide, and so on, show a zinc-blend structure with, for example, Ga on the A sites and arsenic on the B sites. Silicon and germanium show what is called a diamond structure, where a diamond structure is identical to a zinc-blend structure except that the same atom (i.e., silicon) occupies both the A and B sites. Figure 2.45 shows a diagram of several faces of a zinc-blend material. Again there are two different surface

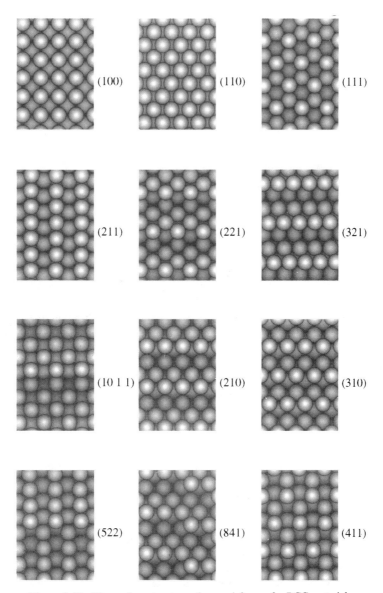

Figure 2.42 The surface structure of several faces of a BCC material.

structures possible, one with the A (i.e., gallium) sites and one with the B (i.e., arsenic) sites exposed. We show diagrams of the faces formed when there is an A site at the middle of the surface. In this configuration, all of the atoms in the outermost plane of the (100) and (111) surfaces are also A type, while a mixture of A and B type atoms are seen with all of the other structures shown. We designate these two structures as (lmn)-A. There also is a series of structures with a B site at the middle of the picture designated (lmn)-B. The B structures have B atoms where A atoms are shown in Figure 2.45 and A atoms where B atoms are shown in Figure 2.45.

In the semiconductor literature one often sees references to a **viscinal** surface, where a

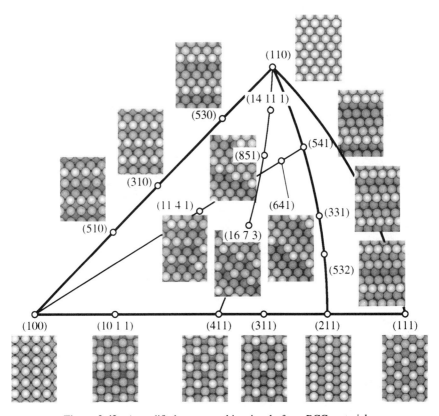

Figure 2.43   A modified stereographic triangle for a BCC material.

viscinal surface is a stepped surface that lies near the (100), the (111), or the (110) plane in the stereographic triangle. For example, the (111) viscinal surfaces are stepped surfaces that lie within a few degrees of the (111) plane. The (111) viscinal planes consist of (111) terraces and some small density of steps and/or kinks.

People often use a special notation to designate viscinal surfaces. The notation is based on how you would cut the crystal. For example, if one starts with a crystal oriented to face the (100) plane, and then rotates it by $3°$ toward the (110) plane, one could designate the plane as a viscinal (100) plane that is cut $3°$ off the (100) plane toward the (110) plane. A second notation is to note that when one rotates toward the (110) plane, one is actually rotating along the [001] zone axis. Therefore, one can designate the plane as a (100) viscinal surface that is cut by rotating $3°$ around the [001] zone axis. Both notations are used in the literature, with the rotation axis being the more common. One also sees the notation (100) $3°$ [001], which is the plane that is formed if one starts with a (100) plane and rotates by $3°$ around the [001] zone axis. In Example 2.E we show how to convert from the microfacet notation to the Miller index notation. In the problem set we ask the reader to show that in this notation the [511] plane would be (100) $15.8°$ [01$\bar{1}$].

2.6   RELAXATIONS AND SURFACE RECONSTRUCTIONS

What makes this all the more complicated is that the ideal surface structures in Figure 2.45 are rarely seen experimentally. For example, Figure 2.46 shows a diagram of an ideal Si(111)

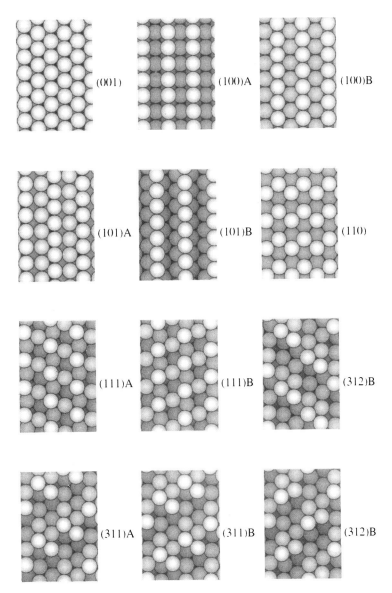

Figure 2.44 A series of surface structures of various faces of an ideal HCP material.

surface. In the bulk, each silicon atom is attached to four other silicon atoms with sp^3 bonds. However, if one could cut an ideal Si(111) surface, each silicon atom in the surface would only be bonded to three other silicon atoms. That leaves an unpaired electron in a dangling bond perpendicular to the surface. Zhener et al. [1981] and Yang and Jona [1983] found that they could produce a nearly ideal Si(111) surface by laser evaporation. However, the ideal Si(111) surface is unstable once the laser is turned off because the dangling bonds would like to pair up. That derives what is called a **surface reconstruction,** where the surface structure changes to reduce the number of dangling bonds. There are several different reconstructions that are possible. The most stable reconstruction is called the Si(111) (7x7). Other reconstruc-

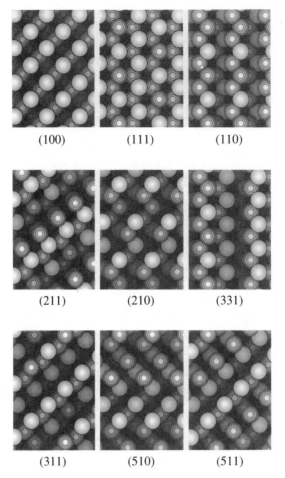

(100)   (111)   (110)

(211)   (210)   (331)

(311)   (510)   (511)

Figure 2.45 The surface structure of several faces of a zinc-blend material. The A sites are indicated as light circles while the B sites are indicated as dark circles.

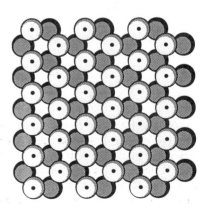

Figure 2.46 A schematic of the ideal Si(111) surface. There are dangling bonds pointing vertically from the surfaces centered at the points indicated by black dots.

tions include what is called a (2x1) and a (3x3) where the notation (2x1) and (3x1) will be defined in Section 2.7. Figure 2.47 shows a diagram and an STM image of the (7x7) reconstruction of Si(111). Notice how different the (7x7) reconstruction is from the ideal Si(111) surface. The details of Figure 2.47 are discussed later in this chapter. However, for now the key thing to remember is that one can get quite complex surface rearrangements when one tries to produce a surface with many dangling bonds.

There are basically two kinds of rearrangements that occur when surfaces are cut, **reconstructions** and **relaxations.** Reconstructions involve wholesale changes in the adsorbed layer. Atoms move from their bulk positions and the periodicity of the surface changes. The Si(111) (7x7) structure in Figure 2.47 is an example of a surface reconstruction. Relaxations, on the other hand, involve motions of whole layers of atoms. Usually the top layer of atoms moves in or out with respect to the bulk, and the second or lower layers move by smaller amounts. The arrangement of atoms within each layer does not change. Experimentally, most surfaces relax. Reconstructions are less common, but they happen in materials that are quite important. For example, the surfaces of all metals relax. However, some of the faces of platinum, iridium, gold, tungsten, and molybdenum also reconstruct. Similarly, all semiconductors relax. However, silicon, gallium, and all III–V (e.g., GaAs) semiconductors also reconstruct.

2.6.1 Rare Gas Crystals

Relaxations were first predicted from calculations of the surface structure of rare gas crystals. Consider the two-dimensional solid in Figure 2.48. Assume that the solid consists of a series of rigid planes of atoms, and assume that each plane of atoms interacts with all of the other planes via a modified Lennard-Jones potential, where V_{ij}, the potential between two planes, is given by

$$V_{ij} = \frac{Well}{2}\left[\left[\frac{Z_o}{Z_{ij}}\right]^9 - 3\left[\frac{Z_o}{Z_{ij}}\right]^3\right] \tag{2.13}$$

where *Well* is the well depth, Z_o is a parameter, and Z_{ij} = the distance between plane i and plane j. Figure 2.49 is a plot of V_{ij} vs. Z_{ij}. Notice that the minimum in the potential occurs for $Z_{ij} = Z_o$. Therefore, if there were only two planes, the equilibrium spacing between the planes would be Z_o. Now consider what would happen if there are many planes in the lattice. Notice that all of the planes that lie at a distance greater than Z_o from a given plane will attract that plane. The attractions produce a compressive force that causes the lattice dimension to be reduced. Numerical calculations similar to those in Example 2.F show that the final lattice dimension is compressed by about 3%, i.e. the interplane spacing is reduced to 97% of the initial value.

Now consider what will happen if one cuts a surface. Notice that when the surface is cut, the surface layer will interact with fewer layers of atoms. That reduces the compressive force. As a result, the spacing between the layers will increase from 97% of Z_o to something close to Z_o. Hence, the surface layer will move away from the atoms below it. One can show that this outward motion is a consequence of the attractive nature of the interaction potential at distances greater than Z_o and is not dependent on the form of the potential. Hence it is a universal phenomenon for surfaces held together by pairwise additive potentials with long-range attractions. In the material that follows we call the outward motion of the surface atoms a surface expansion.

Alder, Vaisnys, and Jura [1959] calculated the expansion of an FCC lattice of atoms that interacts via a generalized Lennard-Jones potential. They found that in general the (100) and (111) surfaces should expand by 2.5% when a surface is cut. Lennard-Jones potentials are

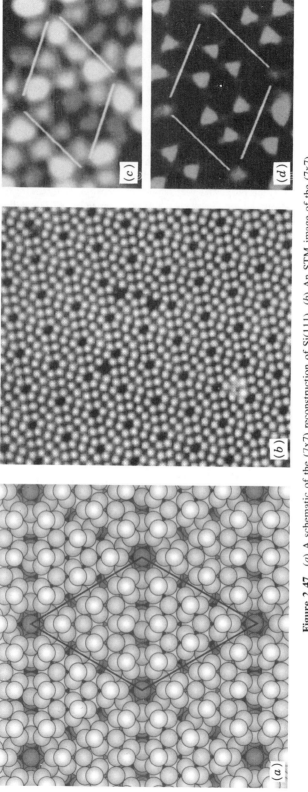

Figure 2.47 (a) A schematic of the (7x7) reconstruction of Si(111). (b) An STM image of the (7x7) surface. (c), (d) Images of the dangling bonds in the (1x1) surface. (From Trump et al. [1988].)

52

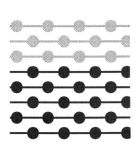

Figure 2.48 Planes in a two-dimensional solid.

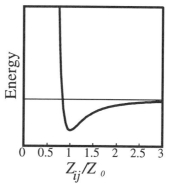

Figure 2.49 A plot of V_{ij} vs. Z_{ij}.

thought to be appropriate for the interactions in the surface of a condensed rare gas (e.g., solid xenon). Therefore, based on calculations, one would expect the surface layer of a rare gas solid to move out from the bulk.

Experimentally, the situation is unclear. Ignatiev, Pendry, and Rhodin [1977] found that the Xe(100) surface does not expand significantly. It is unclear what happens with other rare gas surfaces.

2.6.2 Metals

Metals, however, have been found to behave quite differently than predicted by calculations like those in the last section. Typically, when a metal surface forms, the top layer of atoms relaxes in toward the bulk while one or more of the lower layers of atoms relaxes out slightly. For example, Figure 2.50 shows a diagram of the Pt(210) face. During relaxation, the top layer of atoms moves in by about 21%, the second layer moves in by 12%, while the third layer moves out by 4%. These seem like very large displacements, but note that this is the percentage change in the distance between the light and crosshatched planes of atoms in

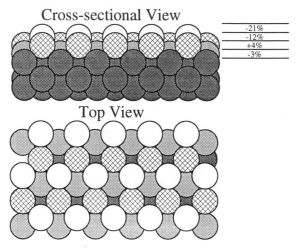

Figure 2.50 A schematic of the surface relaxations when forming a Pt(210) surface as measured by LEED. (Data of Zhang et al. [1991].)

Figure 2.50. The adjacent planes are stacked very closely together in the Pt(210) structure. A 21% contraction represents a displacement of only 0.12 Å.

Inglesfield [1985] notes that in a metal most of the bond occurs via a covalent bond to what is called the conduction band (i.e., the electrons that can move freely through the metal and hence carry electricity). There are no long-range attractions between the cores in the metal. As a result, the analysis in Section 2.6.1 does not apply to metal. In fact, one can show that in general when a surface forms, the atoms in a metal will move in (i.e., toward the bulk) rather than out (i.e., toward the vacuum). For example, imagine two planes of metal atoms, A and B, held together with covalent bonds. Now consider what happens when a third plane of metal atoms C is placed next to B to form an ABC slab. Typically, bonds form between B and C. Electrons flow from the A-B bonds to form the B-C bonds. As a result, the A-B bond is weakened and A is repelled. The net effect is that the A-B distance is increased due to the presence of C.

Now consider starting with an infinite lattice containing A-B-C, and cutting a surface between B and C so that C is no longer on the lattice. In that case, the A-B distance can relax back toward its original equilibrium. The A-B distance decreases. As a result, the B plane of atoms will move in toward the bulk when a surface forms. In the material that follows we call such a motion a lattice compression.

One can understand how metal surfaces reconstruct by examining how the electron density around the metal atoms change as a surface forms. For example, Lee and Masel [1993] calculated the changes in the electron density which occur at the atomic cores when the Pt(210) surface reconstructs. Table 2.4 shows some of their results. In the table, the electron densities were normalized so that an atom in the bulk would have an electron density of 12. In bulk platinum, each atom has 12 nearest neighbors, but in an ideal Pt(210) surface, the topmost atoms only have 6 nearest neighbors. Lee and Masel's calculations indicate that in the unreconstructed surface, the electron density is about six-twelfths of the bulk. Of course, if the platinum atom only has six-twelfths of the electron density of the bulk, there is a large driving force for the atom to increase its overlap with other surface atoms by moving (i.e., contracting) toward the second layer of atoms. If the atoms contract too much, one gets energetically unfavorable core-core repulsions, and the atoms below the first layer will have more electrons than they need. As a result, the extent of relaxation is limited. However, Zhang et al. [1991] found that one gets a 21% contraction in the lattice dimension due to these effects. Lee and Masel [1993] found that with a 21% contraction, the core-core overlap goes up to about nine-twelfths of the bulk value. Hence it appears that the atoms in the Pt(210) surface do relax substantially, due to an enhancement in overlap when they relax.

Experimentally, metal surfaces often show an alternating pattern of contractions and expansions of the various layers of atoms i.e. the first layer moves in while a layer below moves out. Smoluochowski [1941] provided a qualitative explanation of this phenomenon. The idea

TABLE 2.4   The Changes in Electron Density at the Atomic Cores that Occur When the Pt(210) Surface Reconstructs

Layer	Coordination Number of Atoms	Electron Density in the Ideal Lattice	Electron Density in the Relaxed Lattice
1	6	6.1	8.7
2	9	9.0	11.7
3	11	10.9	12.4
4	12	11.9	11.9
5	12	12.0	11.9
6	12	12.0	12.0

Source: Adapted from Lee and Masel [1993].

is that in general atoms would like to have an electron overlap similar to that which they experience in the bulk. When the top layer of atoms moves in, the inward motion increases the electron density at the second layer of atoms. In an open surface like Pt(210) the electron density on the second layer is still below the bulk value so the second layer also moves in. However, in the third layer, the electron density is above that in the bulk, so the third layer moves out. That leaves the fourth layer with a lowered electron density, so the fourth layer moves in. The displacement is largely damped by the fifth layer.

One can extend this idea to closed packed surfaces. The top layer of atoms in a closed packed surface usually moves in, since the top layer is deficient of electrons. The inward motion usually causes the second layer to have an excess of electrons so the second layer expands slightly. That leaves the third layer slightly deficient of electrons, so the third layer moves in slightly. The variations are usually damped past the third layer. There are a few surfaces that do not show this pattern of inward and outward displacements. However, the majority of metal surfaces show a pattern of inward and outward motions of the layers of planes.

It happens that with FCC metals, one can accurately calculate the displacements of the layers using what is called the embedded atom method (EAM). Daw [1986] has shown that E_i, the energy of atom i in a lattice, can be approximated by the sum of two terms: the core-core repulsion, $V_{core}(r)$, due to the interactions with each of the adjacent atom, j, and an electrostatic attraction, $F_{EAM}(\rho_{ei})$, due to the overlap of the atom with the electrons in the metal, i.e.

$$E_i = \sum_j V_{core}(r_{ij}) + F_{EAM}(\rho_{ei}) \tag{2.14}$$

where r_{ij} is the distance from atom i to atom j, and ρ_{ei} is the electron density on i due to overlap with i's neighbors. The functions V_{core} and F_{EAM} are usually fit to bulk data. One then does a simple energy minimization to calculate the surface reconstructions. Generally, the results are somewhat dependent on how one fits the functions. However, one can usually calculate the qualitative features of the surface reconstructions without much difficulty; quantitative predictions are possible when one uses the correct semiempirical functions. One should refer to papers by Jona and Marcus [1988], Foiles, Baskes, and Daw [1986], or Zhang et al. [1991] for details.

Figure 2.51 shows the results of Jona and Marcus's [1988] calculations of the lattice contraction of typical BCC and FCC metals. Relaxations as large as 25% of the lattice dimension have been observed experimentally. Calculations indicate that relaxations as large as 35% should be observed on highly stepped surfaces, but note again that the percentages are deceiving because it is a percentage of a small number. The absolute atomic displacements are usually less than 0.2 Å.

One can occasionally get true reconstructions with metals. Generally, the surface reconstructions occur because of the formation of local bonds due to overlap of d-electrons. Experimentally, the nonclosed packed surfaces of platinum, iridium, and gold reconstruct easily. Reconstructions are also seen with tungsten and molybdenum. However, most other metals show relaxations but no real surface reconstructions.

The most important reconstructions occur with platinum, iridium, and gold. For example, Figure 2.52 shows a diagram of the so-called "(2x1)" reconstruction of Au(110) and a TEM image taken by looking down the Au(110) face of a small gold particle. The (2x1) reconstruction has a missing row structure, with every other row of atoms missing as indicated in Figure 2.52. Pt(110) and Ir(110) also undergo a similar surface reconstruction. In the problem set we ask the reader to find the missing rows in the TEM image in Figure 2.52.

There are two other important reconstructions in transition metal surfaces; the so-called "hex" reconstruction of Au(100), Ir(100), and Pt(100), and the so-called "c(2x2)" reconstruction of W(100). Figure 2.53 shows a diagram of the hex reconstruction of Ir(100). In

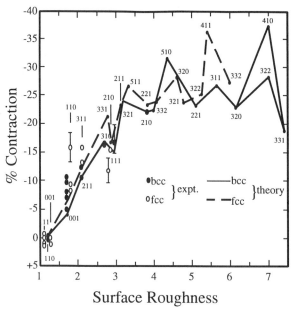

Figure 2.51 The percentage of contraction measured for several faces of FCC and BCC materials. (Adapted from Jona and Marcus [1988].)

the hex reconstruction, the top layer of atoms rearranges to form a hexagonal array. The hex reconstruction is sometimes called a "(5x1)" where the notation "(5x1)" is defined in Section 2.7. Pt(100) undergoes a reconstruction that is very similar to that of Ir(100), but the hexagonal structure is rotated by a few degrees relative to the underlying lattice. This reconstruction is usually called "(5x1)" or "(5x20)" in the literature, although it is sometimes just called the hex reconstruction. Gold also undergoes a similar reconstruction, which is called hex-Au(100).

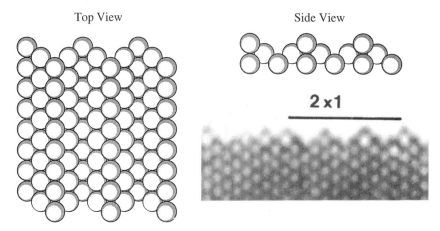

Figure 2.52 A schematic of the (2x1) reconstructon of Au(110) and a TEM image that looks down the surface. (The TEM image is from Marks and Smith [1983]. Reprinted with permission from *Nature* **303,** 316 (1983). © 1983 McMillan Magazines Limited.)

Top View

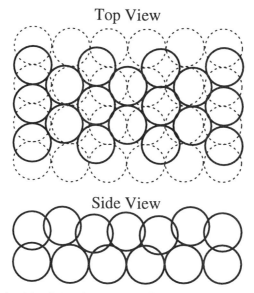

Side View

Figure 2.53 A schematic of the (5x1) or hex reconstruction of Ir(100). Solid circles, top layer of atoms. Dashed circles, second layer of atoms.

W(100) and Mo(100) also show important reconstructions. Figure 2.54 shows a diagram of the unreconstructed and c(2x2) reconstructed surfaces of W(100). The unreconstructed surface has a square structure with gaps between the atoms. However, when the c(2x2) reconstruction forms, the atoms in the top layer move in the direction indicated by the arrows in Figure 2.54 to reduce the number of gaps. That produces a zig-zag chain of atoms in the [011] direction. The reduction of gaps seems to be an important driving force on BCC metals, although it is not seen with FCC metals.

2.6.3 Elemental Semiconductor Surfaces

The pairing up of atoms to reduce gaps also occurs with semiconductor surfaces. As noted previously, when one cuts a semiconductor surface, one exposes many dangling bonds. That drives a surface reconstruction where the dangling bonds try to pair up. For example, consider

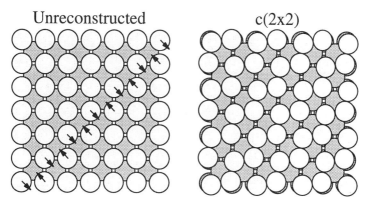

Figure 2.54 The surface structure of the unreconstructed and c(2x2) reconstructed W(100) surface.

Si(111). Figure 2.46 shows that if one would try to form an ideal Si(111) surface, one would produce a surface with many dangling bonds. Experimentally, the ideal Si(111) surface is unstable. Zhener et al. [1981] and Yang and Jona [1983] found that an unreconstructed Si(111) surface can be created by laser evaporation of a silicon crystal at low temperatures. However, when the laser is turned off the dangling bonds quickly pair up to form what is called the (2x1) structure.

Figure 2.55 shows a side view of the changes that occur when the (2x1) reconstruction forms. Notice that in order to pair up the dangling bonds in the lattice, one has to strain some of the other bonds to the surface atoms. The final surface structures are ones that balance the energy gain due to pairing of dangling bonds and the energy loss due to strain.

LaFemina [1992] and Duke [1993] note that the competition between pairing and strain controls the surface structure of most semiconductor surfaces. The competition can lead to some complex reconstructions. For example, when one turns off the laser in the experiment described above the Si(111) sample first undergoes what is called a (2x1) reconstruction. However, upon annealing, the surface further reconstructs into what is called the (7x7) reconstruction. The (7x7) reconstruction of Si(111) is shown in Figure 2.47. It is quite a complex structure. There are vacancies (i.e., missing atoms) in the lattice, some of the atoms are displaced from their normal FCC/diamond lattice positions to HCP sites, and many of the bonds are bent. Yet this is a stable structure.

Experimentally, one does not always get a (7x7) structure. There are other metastable reconstructions that can form during the sample preparation procedure. However, the (7x7) reconstruction is a relatively stable structure once it forms, which illustrates how complex surface reconstructions can be.

Si(100) is a little simpler than Si(111) in that one does not get as much strain when the dangling bonds pair up in the Si(100) lattice as one gets in the Si(111) lattice. Experimentally, the simple (2x1) reconstruction where the dangling bonds on adjacent atoms pair up is the most stable.

At present, there is still some controversy about the structure of the Si(100) (2x1) lattice. Figure 2.56 shows what is called the symmetric dimer model of the surface. Some of the available data can be explained by this structure. However, a buckled dimer model of the surface can also explain the data and appears to have a slightly lower energy (see Garcia and Northrup [1993] or Duke [1993] for details). Weakliem and Carter [1992] and Lin, Low, and Owy [1993] predicted that the Si(100) (2x1) structure fluctuates between two nearly iso-energetic configurations. It remains to be seen if this prediction will be borne out by experiment.

LaFemina [1992] and Duke [1993] review the attempts that people have made to calculate the geometry of the surface reconstructions. One of the things that comes out of this analysis

Ideal Unreconstructed   (2x1)

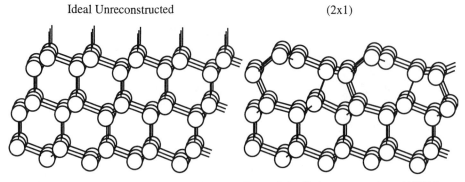

Figure 2.55   A side view of the ideal Si(111) surface and the (2x1) reconstruction of Si(111).

Unreconstructed   (2x1)

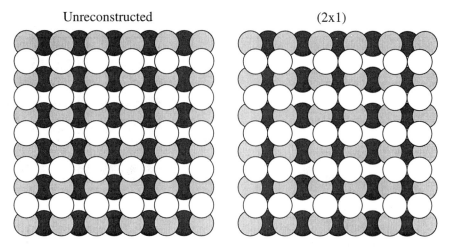

Figure 2.56 The structure of an unreconstructed Si(100) surface and the symmetric dimer model of the Si(100) (2 × 1) reconstruction.

is that there are often a large barrier in moving from one metastable surface reconstruction to another one. As a result, when a semiconductor surface is prepared, the final structure may not be determined by equilibrium. Rather one can get to a metastable surface structure that is quite different than the equilibrium surface structure.

2.6.4 Ionic Crystal

The surfaces of ionic solids (e.g. metal oxides or salt crystals) behave much differently than those of simple metals or semiconductors. An ionic crystal is held together by coulombic forces rather than local bonds, and the balance between those forces and core repulsion determines the final surface configuration. The situation is similar to rare gas crystals in that there is a long-range attraction between the atoms in the crystal. This compresses the lattice. When the surface forms, the compression is released, so the surface of the crystal should expand. However, there are some additional effects.

Tasker [1979] showed that the long-range nature of coulombic forces can lead to interesting surface instabilities. For example, consider the NaCl(111) face depicted in Figure 2.57. The surface can be considered to consist of alternate planes of Na^+ and of Cl^- stacked on top of one another. The top plane of sodium ions is attracted by the first plane of chlorine ions and

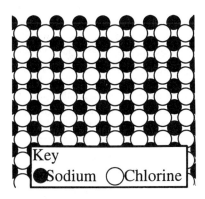

Key
● Sodium ○ Chlorine

Figure 2.57 A cross-sectional view of the NaCl(111) face.

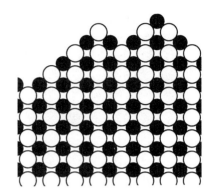

Figure 2.58   A faceted NaCl(111) face.

repelled by the next plane of sodium ions. The net force is attractive since the chlorines are closer than the sodiums. Similarly, the next two planes produce a net attraction. There are thousands of layers in the lattice, and each of them gives a net attractive interaction.

It happens that the net attractive force only decays very slowly with distance. So slowly, in fact, that Bertaut [1958] showed that if one adds up all of the contributions, the net interaction is an infinite force on the surface for an infinitely thick crystal. The surface cannot resist that infinite force. As a result, the NaCl(111) face is unstable. Benson and Yun [1965] showed that if the NaCl(111) face would reconstruct as shown in Figure 2.58 to produce steps and terraces with equal numbers of positive and negative ions, the force would be finite. As a result, the reconstructed surface could be stable.

In reality NaCl(111) **facets** rather than reconstructing into a simple ordered phase. Basically, what happens is that the steps and terraces in Figure 2.58 grow until the surface is covered by finite-size crystals. The result is a faceted surface like that in Figure 2.59.

Tasker [1979] noted that Bertaut's findings are applicable to an arbitrary ionic solid. Tasker divides ionic solids into three classes, as shown in Figure 2.60: Those with alternating positive and negative charges in each layer arranged so that the surface layer does not have net dipole perpendicular to the surface; those with alternating planes of positive and negative charges, but the charges are arranged so that the net dipole of some group of planes is zero; and those planes with alternating layers of positive and negative ions arranged so that there is a net dipole perpendicular to the surface. We refer to these as type I, type II, and type III surfaces, respectively. The structure on the right of Figure 2.60 has alternating positive and negative charges in the top layer, so it is a type I surface. The structure in the middle of Figure 2.60 has a three-layer repeat unit with negative charges in the top layer, positive charges in the second layer, and negative charges in the third layer. There is a net dipole from the top two layers, but it is canceled out by the net dipole from the second and third layers. Therefore, there is no net dipole from the repeating unit. As a result, this is a type II surface. The right structure in Figure 2.60 has a two-layer repeat unit. Each repeat unit has a net dipole perpendicular to the surface. As a result, the right structure is a type III surface. Type I surfaces include the MgO(100) and NaCl(100) faces. Type II surfaces include the VO_2(111) and TiO_2(001) faces. Type III surfaces include the MgO(111), the NaCl(111), and the CoO(111) faces.

Tasker shows that the surface force is infinite with type III surfaces. As a result, type III surfaces should always reconstruct or facet. Type II surfaces can be stable. However, if type II surfaces lose ions by, for example, annealing in vacuum, they become unstable and facet. In contrast, type I surfaces are predicted to be completely stable and not facet or reconstruct.

Experiments have generally supported Tasker's analysis. Type I surfaces are generally stable. Type II surfaces are generally metastable. One can form an ideal type II surface.

Figure 2.59 A photo of a faceted platinum surface

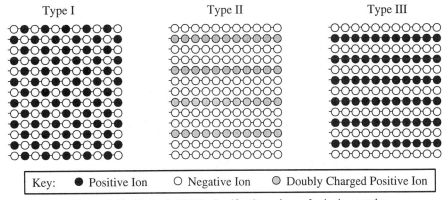

Figure 2.60 Tasker's [1979] classification scheme for ionic crystals.

However, the type II surface facets upon annealing in vacuum due to loss of ions. Type III surfaces are rarely seen. The only example of a type III surface that we know about comes from Ignatiev, Lee, and VanHove's [1977] studies of the oxidation of Co(001). Ignatiev, Lee, and VanHove found that under the right conditions they could grow a thin layer of CoO(111). This assertion was later supported by the detailed LEED analysis of Van Hove and Echenique [1979]. Note however, that CoO(111) has the same structure as NaCl(111). There are alternating rows of Co^+ and O^-. Hence, CoO(111) is a type III oxide. Therefore, CoO(111) should be unstable, at least in an infinite crystal.

Van Hove and Echenique [1979] noted that even though the (111) face of an infinite CoO sample would be unstable, a thin film of CoO(111) could be stable. Alternatively, Prutton and Weldon-Cook [1979] noted that one can also explain the LEED data with a complex reconstruction model. At present, it is unclear which of these explanations is correct. In any case, it is clear that type III surfaces reconstruct of facet for infinitely thick crystals. It is unclear, however, whether thin films of type III surfaces could exist.

Tasker also notes that there is an interesting phenomenon with type I surface called **rumpling.** The idea is that in a mixed surface (e.g., a surface with Na^+ and Cl^-) some of the ions are more easily compressed than others, so the different types of ions can move different amounts or in different directions when a surface is created. For example, Benson, Freeman, and Dempsey's [1963] calculations indicated that when a NaCl(100) surface is created, the sodium atoms in the top layer move *in* by about 9% while the chlorines move *out* by about 4%, as indicated in Figure 2.61.

Experimentally, the situation is unclear. Heinrich [1985] and Coburn [1992] review much of the data. So far there is little evidence for rumpling. However, most of the available data have been on metal oxides where the rumpling is predicted to be small. At present, no one knows whether there will be significant rumpling of alkyl-halide surfaces.

2.6.5 Compound Semiconductors

The analysis in the last section also applies in part to compound semiconductors. For example, Figure 2.62 illustrates the stacking pattern in an ideal GaAs(100), (110), and (111) surface. Notice that an ideal GaAs(110) surface has a type I stacking pattern, while the ideal GaAs(111) and GaAs(100) surfaces have a type III stacking pattern. Thus, based on Tasker's ideas the GaAs(110) surface should be relatively stable, while GaAs(100) and GaAs(111) surfaces should reconstruct or facet.

Experimentally, the GaAs(110) surface is fairly stable. When one cleaves a GaAs crystal, GaAs(110) faces form. GaAs(110) does relax. However, the atomic displacements are small

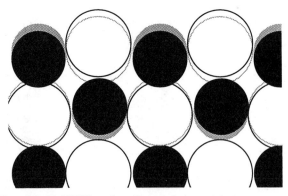

Figure 2.61 Rumpling of the NaCl(100) surface. Dotted circles: original atomic positions; solid circles: reconstructed positions.

GaAs(110)

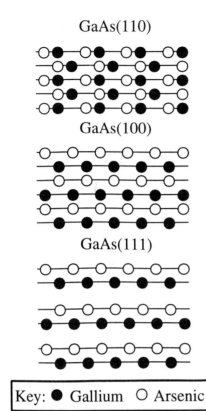

GaAs(100)

GaAs(111)

Key: ● Gallium   ○ Arsenic

Figure 2.62 The stacking sequence in ideal GaAs(100), GaAs(110), and GaAs(111) surfaces. (Adapted from Tasker [1979].)

and there is no change in periodicity. In contrast, the GaAs(100) and GaAs(111) surfaces undergo complex reconstructions.

Duke [1993] notes that the (110) faces of GaAs, GaP, GaSb, AlP, InP, InAs, InSb, CuI, AgI, CuCl, ZnS, ZnSe, ZnTe, and CdTe all show a universal relaxation. Figure 2.63 shows an exaggerated side view of this reconstruction. Basically, the metal atoms rotate out while

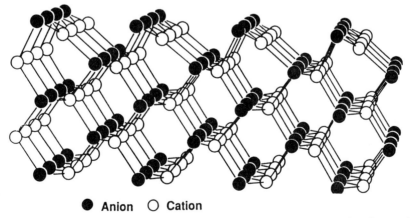

● Anion   ○ Cation

Figure 2.63 An approximate sideview of the relaxed (110) face of a GaAs sample. The atomic displacements are exaggerated for clarity. (After Duke [1993].)

the other atoms rotate in. The bond lengths are approximately conserved during the rotation. Duke shows that the rotation angle is $29 \pm 3°$ for all of the zinc blend materials mentioned earlier.

Such a result is easy to understand in GaAs. Note that in simple inorganic compounds, gallium usually forms an sp_2 hybrid with three electrons and an empty lone pair orbital. Arsenic forms sp_3 hybrids with five electrons. However, in bulk GaAs, the gallium and arsenic are each held in an sp_3 environment with four electrons on average around each gallium and arsenic atom. Thus, in bulk GaAs, the gallium and arsenic atoms are in a different environment than they would be in an inorganic molecule. When the (110) surface forms, the gallium and arsenic would like to move back to their preferred orientation in inorganic molecules, i.e. gallium would like to become sp_2 bonded with three electrons and an empty lone pair orbital; the arsenic would like to become sp_3 bonded with five electrons on a filled lone pair orbital. It happens that the reconstruction in Figure 2.63 does that. Hence, it is extremely stable. A similar argument also holds for all III–V semiconductors. Hence, they show identical reconstructions. From this simple inorganic chemistry analogy, it is not obvious why the (110) face of a II–VI compound such as zinc selenide also goes to the same reconstruction as GaAs(110), since neither zinc nor selenium prefer sp_2 hybrids. However, detailed calculations indicate that the sp_2 configuration is preferred in a ZnSe(110) surface.

The (100) and (111) surfaces of zinc-blend materials show much more complex reconstructions than those described in the last paragraph. In GaAs the galliums and arsenics rehybridize. However, there are additional motions to balance all of the charges. Experimentally the reconstruction varies with how the surface is prepared. GaAs(100), for example, shows seven metastable reconstructions. The structure of many of the reconstructions is still being examined. As a result, it is unclear what can be said about them (see Kahn [1944], Biegelson et al. [1990], LaFemina [1992], or Duke [1993] for details.)

2.7 DESCRIPTIVE NOTATION FOR SURFACE RECONSTRUCTIONS

Next we focus on what we mean by notation such as the (2x1) reconstruction of Pt(110). People started to study surface structure in the late 1950s. The early workers knew that they had reconstructions, but did not have a language to talk about them. As a result, the papers were difficult to follow. In 1964, however, Wood proposed a simple notation to discuss the reconstructed surfaces. Wood's idea is that if one starts with an ideal surface, one can easily find the lattice vectors for the primitive lattice, $\vec{b}_1$ and $\vec{b}_2$, as shown in Figure 2.64, where by convention $\vec{b}_1$ is the shorter of the two lattice vectors. The reconstructed surface has different lattice vectors $\vec{b}'_1$ and $\vec{b}'_2$, where

$$\vec{b}'_1 = \vec{b}_1$$

$$\vec{b}'_2 = 2\,\vec{b}_2 \tag{2.15}$$

Wood proposed that one can call the reconstructed surface Pt(110) (1x2), where the 1 is the ratio of the length of the $\vec{b}_1$ vector to the $\vec{b}'_1$ vector and the 2 is the ratio of the length of the $\vec{b}_2$ vector to the $\vec{b}'_2$ vector. In this notation the unreconstructed Pt(110) surface would be called Pt(110) (1x1). The notation (1x2) is pronounced "one by two."

Wood's notation has been widely adopted in the surface science literature. However, people usually do not follow Wood's suggestions exactly. Generally, when someone first discovers a new reconstruction, they give it a name and subsequent investigators follow the original suggestion. For example, the (2x1) reconstruction of Pt(110) is usually called

Pt(110)(1x1)

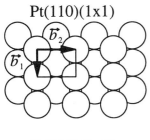

Pt(110)(1x2)

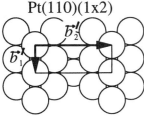

Figure 2.64 The lattice vectors for (1x1) and (2x1) Pt(110).

(2x1) Pt(110) not Pt(110) (1x2) as suggested by Wood. Similarly, the (1x1) reconstruction is called (1x1) Pt(110).

Another difficulty arises when the unit vectors of a primitive lattice for the reconstructed surface do not go in the same direction as the original unit cell. For example, the lattice vectors for the primitive unit cell of the c(2x2) reconstruction of W(100) go at an angle of 45° with respect to the unit cell of the unreconstructed surface as shown in Figure 2.65. One notation for that structure is $(\sqrt{2}x\sqrt{2})$ R45°, where the $\sqrt{2}$ denotes that the original lattice vectors are $\sqrt{2}$ times the lengths of the original lattice vectors, and the R45° denotes that the reconstructed lattice is rotated by 45° relative to the original lattice. Note, however, that if one instead uses the conventional unit cell in Figure 2.65, the unit cell dimensions will each be two times the dimensions of the unit cell for the unreconstructed surface. Hence, the conventional unit cell in Figure 2.65 has a (2x2) structure. People call such a structure a "c(2x2)" overlayer, where the "c" is used to indicate that the (2x2) unit cell is a centered or conventional unit cell rather than a primitive unit cell.

People often list a surface structure via the conventional unit cell rather than the primitive unit cell. Note, however, that there are no accepted guidelines to decide whether a reconstruction should be listed via the primitive unit cell or the conventional unit cell. Occasionally

Primitive   Conventional
Unit Cell   Unit Cell

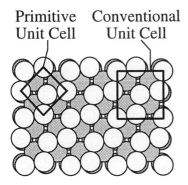

Figure 2.65 The primitive and conventional unit cell for the c(2x2) overlayer on W(100).

different investigators use different labels to discuss the same surface structure. Hence, one needs to be careful when reading the literature, if one wants to know when people are discussing two different structures and when people are using two different notations for the same surface structure.

Some years ago, Lander [1965] and Park and Madden [1968] devised a different notation that avoids these difficulties. Park and Madden's idea was to designate each reconstruction via a matrix, $\overline{m}$, where $\overline{m}$ satisfies

$$\vec{b}_1' = \overline{m}\vec{b}_1$$

$$\vec{b}_2' = \overline{m}\vec{b}_2 \qquad (2.16)$$

Note that the $\overline{m}$ for the p(2x2) and c(2x2) reconstructions of a W(100) surface would be

$$\overline{m}_{p(2x2)} = \begin{bmatrix} 2 & 0 \\ 0 & 2 \end{bmatrix} \qquad \overline{m}_{c(2x2)} = \begin{bmatrix} 1 & 1 \\ -1 & 1 \end{bmatrix}$$

Hence it is easy to specify an overlayer arrangement precisely with the matrix notation. Still, most workers continue to use Wood's notation rather than the matrix notation because Wood's notation is simpler. The major problem with Wood's notation is that there are some structures that one cannot specify by Wood's notation. An example of that is given in Example 2.C. The matrix notation is commonly used for such systems.

Another problem is that according to Equation 2.1 in order to specify the arrangement of a two-dimensional layer, one needs to specify both the lattice vectors and the basis (i.e., atomic positions within the unit cell). When one says that a lattice has a (2x1) reconstruction, one is only specifying the lattice vectors and not the atomic positions. Hence, it is possible for a given value of the Wood's indices to refer to more than one reconstruction of the same surface. In Example 2.C we show that the Wood's notation can be used to label ordered overlayers of adsorbates. There are several examples where two different adsorbate arrangements have the same Wood's indices.

2.8 SITE NOTATION

There is an alternative way that people discuss surface structure, which is to examine the **coordination number** of all of the atoms on the surface. The coordination number of a given atom is defined as the number of nearest neighbors of that atom. For example, consider an atom on one of the A sites in Figure 2.19. The atom has six nearest neighbors in the surface plane and three nearest neighbors below the plane, so the atom has a coordination number of 9. Such atoms are called C_9 atoms. Atoms in hexagonal closed packed layers have 9 nearest neighbors, while atoms in the (100) face of an FCC metal have 8 nearest neighbors. They are called C_9 and C_8 atoms, respectively. This notation has many limitations. There is not a clear definition of what constitutes a nearest neighbor in the literature. For example, in an ideal BCC lattice, atoms in the (110) plane form a distorted hexagon. The distances between adjacent atoms vary by about 15%. At present, there is no consensus in the literature as to which of these atoms to count as nearest neighbors.

The situation is even more difficult when the surface undergoes a complex reconstruction like the (5x1) reconstruction of Ir(100) or the (7x7) reconstruction of Si(111). In both cases there is a wide distribution of atomic distances in the surface layer, so it is by no means clear what constitutes a nearest neighbor.

Lee and Masel [1993] proposed an alternative notation to discuss these situations. The idea is to label each atom by the overlap of the atomic cores of each atom with the electron cloud of its neighbors. The overlaps are normalized so that the overlap is twelve in the bulk. Some of the electron density numbers are given in Table 2.4. In FCC metals the normalized electron density in an unreconstructed surface is almost equal to the number of nearest neighbors. An atom in an unreconstructed surface with seven nearest neighbors has a normalized overlap of about 7, while an atom with 9 nearest neighbors has a normalized overlap of about 9. Of course the overlaps can change significantly when the surface relaxes (see Table 2.4). The advantage of using overlaps to designate sites in surfaces is that one can use the notation in complex situations, and one does not have to make any arbitrary decisions about what constitues a nearest neighbor. Still, it remains to be seen whether this notation will be widely adopted.

2.9 ROUGHENING AND SURFACE MELTING

Another thing to watch out for is that the arrangement of the surface can vary with temperature. It is not unusual for a surface to show one equilibrium reconstruction at low temperature and to show another equilibrium surface structure upon heating.

The two most heavily studied changes in surface structure with temperature are called **surface roughening** and **surface melting.** The idea in surface roughening is that if a perfectly ordered surface is heated above some critical temperature called the roughening temperature, T_R, the surface disorders. Figure 2.66 shows some calculations of the disordering of a stepped simple cubic crystal. At low temperatures, the crystal is nearly flat. However, it disorders at high temperatures. Experimentally, one observes a reversible change in the surface structure. If one heats a surface above the roughening temperature, the surface disorders. However, the process is reversed upon slowly cooling, and one gets back to a smooth surface. At present most of measurements of the roughening process are done by indirect means, for example, X-ray scattering. However, Frenken et al. [1990] have produced some very nice STM images of the roughening transition of Au(511). Note that this transition is unusual in that it occurs just above room temperature (i.e., between 50 and 90°C).

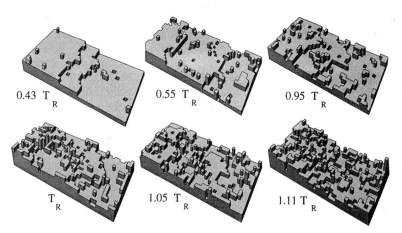

Figure 2.66 A schematic of the roughening of a stepped face of a simple cubic lattice as a function of temperature as calculated by Weeks [1980].

The one thing that one does have to watch out for is that cooling must be done slowly to eliminate defects. For example, Gellman et al. [1992] found that they could deliberately create defects in a sulfur-covered Pd(111) surface by annealing to high temperature, and turning off their heaters and allowing the sample to rapidly cool. This work demonstrates that contrary to common practice, defects are not eliminated from surfaces by simply annealing to high temperatures and quenching.

Surface melting has not been studied as much as roughening. The idea in surface melting is that if one heats a surface to somewhere below the bulk melting temperature, the surface can still form a melted phase. Experimentally, it is sometimes difficult to distinguish roughening from melting. However, there has been considerable interest in the surface melting transition.

Roughening and surface melting are very important in film growth. Growth rates are much higher on rough surfaces than on smooth ones. Hence, when this book was originally being designed the plan was to include a long discussion of roughening and surface melting. This book is already quite long, however, and so there was no room to include such a discussion here. Good descriptions of surface roughening can be found in Conrad [1992], Beijeren and Norden [1987], Saito and Müller-Krumbhaar [1987], or Weeks [1980]. Dietrich [1988] provides a good overview of surface melting and a related phenomenon called wetting.

2.10   SOLVED EXAMPLES

Example 2.A   Calculating the Positions of Atoms in Planes of FCC Materials

In Section 2.5, we provided several pictures of surface structure of some ideal single crystals. These plots were actually calculated using a computer. The objective of this problem is to provide a working computer program so the reader can calculate the atomic positions himself.

Years ago, Nicholas [1965] showed that one could use what he called a stacking method to calculate positions of atoms in surfaces. This method was used for years, but more recently people began using what is called the rotation method. In the rotation method a computer is used to simulate an idealized crystal preparation procedure. The computer first generates a list of the positions of all of the atoms in the crystal. The computer then rotates the positions so that the plane to be plotted faces along the z axis. The computer then eliminates any atoms with a z greater than 0, where z = 0 is the place where the crystal is cut.

A computer program utilizing the rotation method is given on the next few pages. The computer program is organized to simulate an idealized crystal preparation procedure. The program starts by using Equation 2.1 to calculate the positions of all of the atoms in the crystal. The computer then rotates each of the positions so that the desired plane is exposed (i.e., so that it faces the z direction). Next, the computer eliminates atoms that would be removed during an ideal cutting procedure. The computer then prints out all of the atomic positions. The results can then be plotted with a plotting package.

The one difficult part of the program is in the subroutine `rotation_matrix_calc` where we calculate a matrix to rotate the atoms in the crystal so that the appropriate plane is perpendicular to the z axis. In the program we define a vector `norm` which is the x, y, and z coordinates of a vector normal to the plane that needs to be plotted. With a cubic lattice the (ijk) plane is normal to the [ijk] direction so no calculations are necessary. However, in the problem set we ask the reader to modify the program for noncubic lattices. Therefore, the program calculates `norm` as the cross product of two vectors in the plane of the surface. Next, the program calculates a solid angle θ, ϕ between `norm` and the z axis, where θ and ϕ are defined in Figure 2.67. From analytic geometry, one can rotate the [ijk] direction to the z axis by creating a new x, y, and z given by

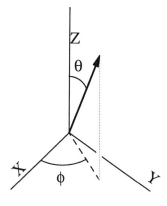

Figure 2.67 A schematic of the solid angle between the [ijk] vector and the z axis.

$$\begin{pmatrix} x_{new} \\ y_{new} \\ z_{new} \end{pmatrix} = T \begin{pmatrix} x \\ y \\ z \end{pmatrix} \tag{2.17}$$

where T is given by

$$T = \begin{pmatrix} \cos\theta & 0 & -\sin\theta \\ 0 & 1 & 0 \\ \sin\theta & 0 & \cos\theta \end{pmatrix} \begin{pmatrix} \cos\phi & \sin\phi & 0 \\ -\sin\phi & \cos\phi & 0 \\ 0 & 0 & 1 \end{pmatrix} \tag{2.18}$$

The matrix on the right rotates the `norm` vector so that it lies in the x-z plane, while the matrix on the left rotates the `norm` vector so it moves to the z axis. The choice of T matrix is not unique; once the `norm` vector points along the z axis, the sample can then be rotated around the z axis and without changing the atoms exposed when the crystal is cut.

From analytic geometry, θ and ϕ can be calculated from

$$\cos\theta = \frac{k}{(i^2 + j^2 + k^2)^{1/2}} \tag{2.19}$$

$$\tan\phi = \frac{i}{j} \tag{2.20}$$

In Problem 2.23 we ask the reader to analyze the program in more detail.

We have included two program listings on the following pages. The first listing is a Microsoft C version of a program to calculate the atomic positions, plot the results to the screen, and provide a list of atomic coordinates. The program was written for clarity rather than efficiency so there are places where the program could be improved. The second program is a FORTRAN program to calculate the atomic positions. It is included in case the reader is unfamiliar with C. If the reader has trouble with the program, a source code and a compiled version of the first program is available from the author. The author's version also creates a CGM file of the plot. CGM files are graphics files that can be read by many plotting routines and word processors on both MACs and PCs, including Microsft Word, WordPerfect, Lotus Freelance, Harvard Presentation Graphics, and Canvas.

```c
#include <stdio.h>
#include <conio.h>
#include <math.h>
#include <signal.h>
#include <process.h>
#include <graph.h>
#include <stdlib.h>

/* program to print atom positions in a FCC lattice
*/

/*1*//*function prototypes for ANSI compatability*/
/*1*/
/*1*/ main(void);
/*1*/ create_filename( int, int, int, char *);
/*1*/ sort_data( int);
/*1*/ rotation_matrix_calc(int,int,int);
/*1*/ screen_plot(int);
/*1*/ cntrl_c_handler(void);
/*1*/ set_interup(void);
/*1*/ rotate_pict(int,double *,double *,double,
        double);
/*1*/ alt_rotate_pict(int,double,double *, double *
        ,double,double);
/*1*/ zmin_choose(int,int,int);
/*1*/
/*1*/ /***************************/

double x[1000],y[1000],z[1000];
    /* x, y, z are the atomic positions */
int order[1000];

float basVEC[3][3]
/*basVEC are the basis vectors for the unit cell*/
    = {{(float)1.,(float)0.,(float)0.},
       {(float)0.,(float)1.,(float)0.},
       {(float)0.,(float)0.,(float)1.} };

float atomRAD = (float)7.07;
    /* atomRAd equals the atomic radius */

float b[4][3]    /* b is the positions of the atons
within the unit cell

                    in units of the basis vectors */

    = {{(float)0.,(float)0.,(float)0.},
       {(float)0.,(float)0.5,(float)0.5},
       {(float)0.5,(float)0.,(float)0.5},
       {(float)0.5,(float)0.5,(float)0.}};

float T[3-00][3]  /* T is a rotation matrix to rota-
te the
unit cell so it lies along the z direction */
    = {{(float)1.,(float)0.,(float)0.},
       {(float)0.,(float)1.,(float)0.},
       {(float)0.,(float)0.,(float)1}};

main()
{
    char fname[13];
    FILE *out1;
    int ix,iy,iz, not_ok, below;
    int natoms, nx, ny,nz,i,i,j,k,l;
    float rlat[3],r[3],rp[3],fact,zmin;

    not_ok=1;
    while(not_ok)
    { printf("\n\nEnter the miller indices of ");
      printf( "the plane to be \nplotted. ");
            printf("Seperate the indices by spaces  ");
      l=scanf("%d %d %d",&ix,&iy,&iz);
      while(l<3)
      {   printf("\n\nERROR IN INPUT -- THE");
          printf(" MACHINE SOMETIMES");
          printf(" HANGS IF PROGRAM");
          printf(" CONTINUES SO ABORT");
          exit(-1);
```

```c
int nbasis = 4; /*nbasis equals the number of
atoms within the unit cell*/
   if(ix>15||iy>15||iz>15||ix<-15||iy<-15||iz<-15||
      (ix==0&&iy==0&&iz==0) ) printf(
"cannot plot the %d %d %d plane",ix,iy,iz);
      else not_ok=0;
      }
   zmin=(float) zmin_choose(ix,iy,iz);
}

/********************************************
* Calculate rotation matrix
*/

rotation_matrix_calc(ix,iy,iz);
printf("\n\nCalculating the atom positions");
printf(" for the (%d %d †) plane"
,ix,iy,iz);
printf("\n\n    PLEASE BE PATIENT \n");
/*******************************************
Search For atoms to be plotted
*******************************************/
nx=15;ny=15;nz=15;natoms=0;
for(l=0;l<nx;l++)
for(k=0;k<ny;k++)
{printf(" ");
for(j=0;j<nz;j++)
for(i=0;i<nbasis;i++)
   { /*calculate atom positions via equ 2.1*/
   rlat[0]=(float)(20.*b[i][0]+20.*((float)(l-nx/2)));
   rlat[1]=(float)(20.*b[i][1]+20.*((float)(k-ny/2)));
   rlat[2]=(float)(20.*b[i][2]+20.*((float)(j-nz/2)));
   for(ii=0;ii<3;ii++)r[ii]=basVEC[0][ii]*rlat[0]
   +basVEC[1][ii]*rlat[1]+basVEC[2][ii]*rlat[2];
   /* rotate */
   for(ii=0;ii<3;ii++)
   rp[ii]=T[ii][0]*r[0]+T[ii][1]*r[1]+T[ii][2]*r[2];
   /* change plot direction */
   rp[0]= -rp[0];rp[1]= -rp[1];
   /*move origin to (50,50,0)*/
   y[natoms]=(double)rp[1];
   z[natoms]=(double)rp[2];
   natoms++;
   }
}
printf("\n\n\nnatoms = %d ",natoms);

/****************************************************
* plot data to the screen
*/
screen_plot(natoms);

/****************************************************
/*write data to file*/
create_filename(ix,iy,iz,fname);
printf("\n\n\n   writing data to %s\n\n"
   ,fname);
out1=fopen(fname,"wb");
for(j=0;j<natoms;j++)
   fprintf(out1,"%f %f %f\n",
      x[order[j]],y[order[j]],z[order[j]]);

return 1;
}
```

```
        rp[0]+=(float)50;rp[1]+=(float)50;
        /*cut*/
        if(zmin<rp[2]&&rp[2]<=(float)1.&&
            rp[0]>=(float)-50.&&rp[0]<(float)150.&&
            rp[1]>=(float)-50.&&rp[1]<(float)150.)
            {
            x[natoms]=(double)rp[0];
/**********************************************
    subroutine to calculate the matrix to rotate the
    crystal to expose the desired plane.
*/
rotation_matrix_calc(ix,iy,iz)
int ix,iy,iz;
{
    float va[3],vb[3],vap[3],vbp[3],norm[3],len,
        costheta,sintheta,cosphi,sinphi;
    int i;
/**********************************************
    * calculate two vectors in the surface plane
    */
va[0]=(float)(iy*iz);va[1]=(float)(-ix*iz);
va[2]=(float)0;
vb[0]=(float)(iy*iz);vb[1]=(float)(-ix*iy)0;
if(ix==0)
{vb[0]=(float)0;vb[1]=(float)-iz;vb[2]=(float)iy;}
if(iy==0)
{vb[0]=(float)-iz;vb[1]=(float)0;vb[2]=(float)ix;}
if(iz==0){va[0]=(float)-iy;va[1]=(float)ix;}
if(ix==0&&iy==0)va[1]=(float)1;
    printf("\nThe surface vectors are [ %5.2f,
%5.2f, %5.2f] and [%5.2f,%5.2f,%5.2f]\n",
va[0],va[1],va[2],vb[0],vb[1],vb[2]);
/**********************************************
    *   Convert to x y z coordinates
    */

        (iy<0&&norm[1]>(float)0)
        ||(iz>0&&norm[2]<(float)0)
        ||(iz<0&&norm[2]>(float)0))
        {norm[0]=-norm[0];norm[1]=-norm[1];
            norm[2]=-norm[2];}
    len=(float)sqrt(norm[0]*norm[0]+
        norm[1]*norm[1]+norm[2]*norm[2]);
    norm[0]=norm[0]/len;norm[1]=norm[1]/len;
    norm[2]=norm[2]/len;
    printf("The normal vector is [%5.3f, %5.3f,
%5.3f] ",norm[0],norm[1],norm[2]);
    costheta=norm[2];
    sintheta=(float)sqrt( ((float)1.)-costheta*costheta);
    if(fabs(norm[0])<1.e-5&&fabs(norm[1])<1.e-5)
    {cosphi=(float)1.;sinphi=(float)0;}
    else
        {
        cosphi=norm[0]/sqrt(norm[0]*norm[0]+
            norm[1]*norm[1]);
        sinphi=norm[1]/sqrt(norm[0]*norm[0]+
            norm[1]*norm[1]);
        }
    T[0][0]=costheta*cosphi;
    T[1][0]=-sinphi;
    T[2][0]=sintheta*cosphi;
    T[0][1]=costheta*sinphi;
    T[1][1]=cosphi;
    T[2][1]=sintheta*sinphi;
    T[0][2]=-sintheta;
    T[1][2]=(float)0;
    T[2][2]=costheta;
```

72

```c
for(i=0;i<3;i++)
{
vap[i]=va[0]*basVEC[0][i]+va[1]*basVEC[1][i]+
    va[2]*basVEC[2][i];
vbp[i]=vb[0]*basVEC[0][i]+vb[1]*basVEC[1][i]+
    vb[2]*basVEC[2][i];
}
norm[0]=vap[1]*vbp[2]-vap[2]*vbp[1];
norm[1]=vap[2]*vbp[0]-vap[0]*vbp[2];
norm[2]=vap[0]*vbp[1]-vap[1]*vbp[0];
if(  (ix>0&&norm[0]<(float)0)||
     (ix<0&&norm[0]>(float)0)
   ||(iy>0&&norm[1]<(float)0)||
```

```c
printf( "\n\n      %5.3f,  %5.3f,  %5.3f "
    ,T[0][0],T[0][1],T[0][2]);
printf( "\n T= %5.3f,  %5.3f,  %5.3f "
    ,T[1][0],T[1][1],T[1][2]);
printf( "\n      %5.3f,  %5.3f,  %5.3f"
    ,T[2][0],T[2][1],T[2][2]);
printf("\n\n");
return 0;
}
```

Fortran Program To Calculate Ideal Surface Structure
© 1992 R. I. Masel

```fortran
block data
real basvec(3,3), atomrad,b(3,4),t(3,3)
integer nbasis
common/data_blk/basvec,atomrad,b,t,nbasis
data basevec /1.,0.,0.,1.,0.,0.,0.,1./
data nbasis/4/
data atomrad/7.07/
data b/0.,0.,0.,0.5,0.5,0.,0.5,0.,0.5,0.,0.5,0./
data t/1.,0.,0.,0.,1.,0.,0.,0.,1/
end
      program fcc_latt
      character fname*14
      integer ix,iy,iz, not_ok, color
      integer natoms, nx, ny,nz, ii,i,j,k,l,nbasis
      real rlat(3),rp(3),zmin,x,y,z
      real basvec(3,3),atomrad,b(3,4),t(3,3)
      common/data_blk/basvec,atomrad,b,t,nbasis
      color=11
```

```fortran
     +Calculating  the atomic positions for the'
     +,i2,i2,   plane)
      write(*,108)
108   format(///,         PLEASE BE PATIENT',///)
      call rotation_matrix_calc(ix,iy,iz)
      nx=15
      ny=15
      nz=15
      do 10 l=1,nx
      do 10 k=1,ny
      do 10 j=1,nz
      do 10 i=1, nbasis
      rlat(1)=b(1,i)+(l-nx/2)
      rlat(2)=b(2,i)+(k-ny/2)
      rlat(3)=b(3,i)+(j-nz/2)
      do 3 ii=1,3
3     r(ii)=basvec(ii,1)*rlat(1)+basvec(ii,2)
     * *rlat(2)+basvec(ii,3)*rlat(3)
      do 5 ii=1,3
```

73

```fortran
      natoms=0
      not_ok=1
      do while( not_ok.eq.1)
        write(*,*) ' Enter the miller indices of the plane'
        write(*,*) ' to be plotted. Seperate the'
        write(*,*) 'indices by spaces'
        read(*,*) ix,iy,iz
        if( (iabs(ix).gt.15).or.(iabs(iy).gt.15).or.
     +      (iabs(iz).gt.15) ) then
          write(*,*)'cannot plot the' ,ix,iy,iz,'plane'
        else
          not_ok=0
        endif
      end do
      zmin=0.05*zmin_choose(ix,iy,iz)
      call create_filename(ix,iy,iz,fname)
      write(*,*) '   writing data to ',fname
      open(unit=4,file=fname)
      write(*,*)
      write(*,107) ix,iy,iz
107   format(

      subroutine rotation_matrix_calc(ix,iy,iz)
      integer ix,iy,iz,nbasis
      real va(3),vb(3),vap(3),vbp(3),norm(3)
      real len,costheta,sintheta
      real basvec(3,3),atomrad,b(3,4),t(3,3)
      common/data_blk/basvec,atomrad,b,t,nbasis
      real cosphi,sinphi
      integer i
      va(1)= iy*iz
      va(2)=-ix*iz
      va(3)= 0.
      vb(1)= iy*iz
      vb(2)=0
      vb(3)=-ix*iy
      if(ix.eq.0) then
        vb(1)=0
```

```fortran
5     rp(ii)=t(ii,1)*r(1)+t(ii,2)*r(2)+t(ii,3)*r(3)
      rp(2)= -rp(2)
      rp(1)= -rp(1)
      if((zmin.lt.rp(3)).and.(rp(3).le.0.1).and.
     +   (abs(rp(1)).le.5.).and.(abs(rp(2)).le.5.) )
     +   then
        x=20.*rp(1)+50.
        y=20.*rp(2)+50.
        z=20.*rp(3)
        natoms=natoms+1
        write(4,100) x,y,z,atomrad,color
100     format(4(f7.4,2x),i3)
      endif
10    continue
      close(4)
      write(*,*)
      write(*,*) ' Found ',natoms,'   atoms'
      stop
      end

      norm(1)=-norm(1)
      norm(2)=-norm(2)
      norm(3)=-norm(3)
      endif
      len=sqrt(norm(1)*norm(1)+norm(2)*
     +   norm(2)+norm(3)*norm(3) )
      norm(1)=norm(1)/len
      norm(2)=norm(2)/len
      norm(3)=norm(3)/len
      costheta=norm(3)
      sintheta=sqrt( 1.-costheta*costheta)
      if( (abs(norm(1)).lt.1.e-5)
     +   .and.(abs(norm(2)).lt.1.e-5))then
        cosphi=1
        sinphi=0
      else
```

```fortran
      vb(2)=-iz
      vb(3)=iy
      endif
      if(iy.eq.0)then
      vb(1)=-iz
      vb(2)=0
      vb(3)=ix
      endif
      if(iz.eq.0)then
      va(1)=-iy
      va(2)=ix
      endif
      if((ix.eq.0).and.(iy.eq.0)) va(2)=1
      do 1 i=1,3
      vap(i)=va(1)*basvec(i,1)+va(2)*
+ basvec(i,2)+va(3)*basvec(i,3)
      vbp(i)=vb(1)*basvec(i,1)+vb(2)*
+ basvec(i,2)+vb(3)*basvec(i,3)
1     continue
      norm(1)=vap(2)*vbp(3)-vap(3)*vbp(2)
      norm(2)=vap(3)*vbp(1)-vap(1)*vbp(3)
      norm(3)=vap(1)*vbp(2)-vap(2)*vbp(1)
      if( ((ix.gt.0).and.(norm(1).lt.0)).or.
+     ((iy.lt.0).and.(norm(2).gt.0)).or.
+     ((iy.gt.0).and.(norm(2).lt.0)).or.
+     ((iz.lt.0).and.(norm(2).gt.0)).or.
+     ((iz.gt.0).and.(norm(3).gt.0)).or.
+     ((iz.gt.0).and.(norm(1).lt.0)) ) then
        cosphi=norm(2)/sqrt(norm(1)*
+ norm(1)+norm(2)*norm(2))
        sinphi=norm(1)/sqrt(norm(1)*
+ norm(1)+norm(2)*norm(2))
      endif
      t(1,1)=costheta*cosphi
      t(2,1)=-sinphi
      t(3,1)=sintheta*cosphi
      t(1,2)=costheta*sinphi
      t(2,2)=cosphi
      t(3,2)=sintheta*sinphi
      t(1,3)=-sintheta
      t(2,3)=0.0
      t(3,3)=costheta
      return
      end

c subroutine to choose the slice to be plotted
c This subroutine is rather empirical and will
c need to be modified for some purposes.
c However,
c it does result in useful plots.
      function zmin_choose(ixi,iyi,izi)
      integer ixi,iyi,izi,tmp
      real zmin, ix, iy, iz, id
      ix=iabs(ixi)
      iy=iabs(iyi)
      iz=iabs(izi)
      do while ( (ix.it.iy).or.(ix.it.iz).or.(iy.it.iz))

C/*********************
C  subroutine to create the filename ix_iy_iz.dat
C*/
      subroutine create_filename(i,j,k,fname)
      integer i, j, k,ia,ja,ka
      character fname*14,is*3,js*3,ks*2
      ia=iabs(i)
      ja=iabs(j)
      ka=iabs(k)
      if(i.lt.0) then
        write(is,10)ia
```

Program to Calculate Atomic Positions (Continued)
© 1992 R. I. Masel

```
if(ix .lt. iy ) then
  tmp=ix
  ix=iy
  iy=tmp
endif
if(ix .lt. iz )then
  tmp=ix
  ix=iz
  iz=tmp
endif
if(iy .gt. iz )then
  tmp=iz
  iz=iy
  iy=tmp
endif
end do
if( ( ((iy+iz)/ix.eq.2).or. ((iy+iz).eq.0) then
  zmin =3
else
  zmin=1.6*atomrad+3
endif
zmin_choose = -zmin
return
end
```

```
10   format(i2.2,b)
     else
       write(is,11)ia
11     format(i2.2,_)
     endif
     if(j.lt.0) then
       write(js,10)ja
     else
       write(js,11)ja
     endif
     write(ks,12)ka
12     format(i2.2)
     fname=is//js//ks//'.dat'
     return
     end
```

Example 2.B   Finding the Miller Indices of a Plane Given a Picture of the Plane

In the preceding example, we showed how to calculate the coordinates of the atoms in a given plane and use those coordinates to calculate a picture of a plane. Occasionally, however, one needs to do this process in reverse, i.e., start with a two-dimensional picture of the surface and calculate the Miller indices of the surface. That is not easy. In general, one does not know the arrangement of the atoms in three dimensions from a two-dimensional picture. Hence, it is difficult to do the calculations except in simple materials. Still, it is possible to go backwards with an FCC or BCC material.

The general scheme is to

Identify the surface unit cell

Calculate the Miller indices of the lattice vectors of the surface unit cell

Calculate the Miller indices of the normal to the surface as the cross product of the lattice vectors. In an FCC material, the Miller indices of the surface are equal to the Miller indices of the normal to the surface multiplied by the common denominator of the indices.

In practice this process is fraught with error. As a result, readers are advised to always check findings by back-calculating the surface structure from the Miller indices.

In this example, we do the calculation for an FCC material. For a BCC material, it is easiest to first calculate indices for an equivalent FCC material and then use Equation 2.11 to convert to BCC.

Figure 2.68 shows a diagram of the surface structure of the (410) faces of an ideal FCC material. We have shown the surface unit cell and the surface lattice vectors in the figure. In order to calculate the Miller indices of the surface one first has to determine the Miller indices of the lattice vectors. The easiest way to do that is to view the surface from the [111], [110] or [111] direction and then define a set of X and Y axes, X_{new} and Y_{new}. It is useful to pick X_{new} and Y_{new} so they correspond to the lattice vectors for the (100), (111) or (110) faces. One then compares the lattice vectors for the surface to the lattice vectors for the (100), (110), and (111) faces shown in Figure 2.69. If one can find a way to express the lattice vectors for the surface as a sum of lattice vectors in the (100), (111), or (110) surface, one can use vector addition to calculate the lattice vectors.

For example, as we have drawn Figure 2.68 the steps in the (410) surface go in the [001] direction. One lattice vector is therefore [001]. The other lattice vector is harder to determine. However, in the right section of Figure 2.68 we have constructed the vector as moving two unit cells in the [010] direction, down half a unit cell in the [100] direction, and then half a

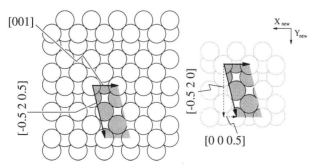

Figure 2.68   A schematic of the surface structure of the (410) face of an ideal FCC material. The unit cell is shaded and the lattice vectors are indicated by arrows. The diagram on the right shows the decomposition of the [−0.5 −2.0 0.5] vector.

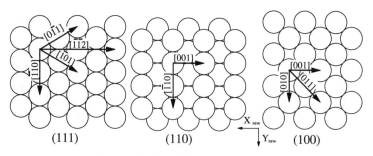

Figure 2.69 Some of the key directions in the (100), (111), and (110) faces of an ideal FCC material.

unit cell in the [100] direction, and then half a unit cell in the [001] direction. Therefore the other lattice vector is [−0.5 2 0.5].

The surface becomes

$$[-0.5\ 2\ 0.5] \times [0\ 0\ 1] = [2\ 0.5\ 0]$$

The common denominator of 0.5 and elements without a denominator is 2. Therefore, the Miller indices of the plane are 2(2 0.5 0) or (4 1 0).

Note that we could have instead rotated the figure so the steps run along the [010] direction. In that case we would have concluded that the surface is (4 0 1).

In more complex cases it is usually helpful to remember that in an FCC metal, vectors moving from one atom to an adjacent atom are in the $\langle \frac{1}{2} \frac{1}{2} 0 \rangle$ family of directions. Hence, a given vector can always be constructed as a series of atom-to-atom displacements. The one trick in that procedure is to be careful in choosing the individual $\langle \frac{1}{2} \frac{1}{2} 0 \rangle$ directions so that the signs are consistent. In a cubic material like FCC, the [lmn] direction is perpendicular to the (lmn) plane. As a result, if a vector [ijk] is in the (lmn) plane, the dot product of [ijk] and (lmn) will be zero. The dot product is non-zero for vectors out of the plane. For example, in a (100) facet, adjacent atoms in the plane are connected by the vectors $[0\ \frac{1}{2}\ \frac{1}{2}]$, $[0\ \frac{1}{2}\ -\frac{1}{2}]$, $[0\ -\frac{1}{2}\ \frac{1}{2}]$, or $[0\ -\frac{1}{2}\ -\frac{1}{2}]$. A vector moving from a step on a (100) plane to an adjacent atom on a (100) terrace will have a first index of $-\frac{1}{2}$ if the vector goes down and $+\frac{1}{2}$ if the vector moves up. In a (111) facet the vectors between adjacent atoms in the plane are $[\frac{1}{2}\ -\frac{1}{2}\ 0]$, $[\frac{1}{2}\ 0\ -\frac{1}{2}]$, $[0\ \frac{1}{2}\ -\frac{1}{2}]$, $[0\ -\frac{1}{2}\ \frac{1}{2}]$, $[-\frac{1}{2}\ \frac{1}{2}\ 0]$, $[-\frac{1}{2}\ 0\ \frac{1}{2}]$. The vectors pointing from a step to an adjacent atom in a (111) terrace are $[\frac{1}{2}\ \frac{1}{2}\ 0]$, $[\frac{1}{2}\ 0\ \frac{1}{2}]$, $[0\ \frac{1}{2}\ \frac{1}{2}]$, $[-\frac{1}{2}\ -\frac{1}{2}\ 0]$, $[-\frac{1}{2}\ 0\ -\frac{1}{2}]$, $[0\ -\frac{1}{2}\ -\frac{1}{2}]$. One can determine which $\langle \frac{1}{2} \frac{1}{2} 0 \rangle$ direction corresponds to a given vector in a picture by using Equation 2.17 to project each of the members of the $\langle \frac{1}{2} \frac{1}{2} 0 \rangle$ directions onto the X_{new} and Y_{new} axes and see which vector goes in the right direction.

Note from Equation 2.17:

$$T_{(111)} = \begin{bmatrix} 0.408 & 0.408 & -0.816 \\ -0.707 & 0.707 & 0.0 \\ 0.577 & 0.577 & 0.577 \end{bmatrix} \quad T_{(100)} = \begin{bmatrix} 0 & 0 & -1 \\ 0 & 1 & 0 \\ 1 & 0 & 0 \end{bmatrix}$$

$$T_{(110)} = \begin{bmatrix} 0 & 0 & -1 \\ -0.707 & -0.707 & 0 \\ 0.707 & 0.707 & 0 \end{bmatrix} \tag{2.21}$$

Figure 2.70 shows several other examples of surface structures. Readers should convince themselves that the other vectors are correct in this figure.

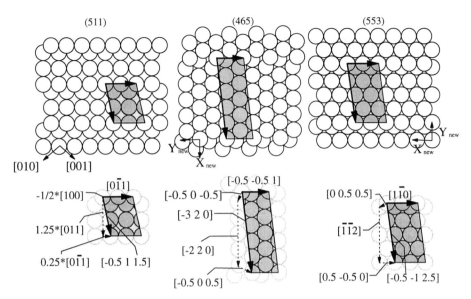

Figure 2.70 A schematic of the surface structure of the (511), (465), and (553) faces of an FCC material with lattice vectors indicated in the figure.

Example 2.C Several Examples of Wood's Notation for Overlayers

Earlier in this chapter, we noted that Wood's notation is commonly used to designate the arrangement of overlayers. Figures 2.71, 2.72, and 2.73 show several examples of overlayers and Wood's notation for each of them. In the figures the open circles are surface atoms and the black circles are adsorbate molecules. The reader should look carefully at each of the diagrams and work out Wood's notation for each of the structures on his/her own. In the problem set we ask the reader to solve several cases on his/her own. The one thing to watch out for in solving these problems is that when the reader draws the unit cell for the system, he or she needs to be careful that it is a unit cell for both the overlayer and the surface. For example, if an atom is put at the center of the p(3x3) unit cell in Figure 2.71n, one might incorrectly conclude that the system has a c(3x3) structure. The primitive unit cell, however, is not a unit cell for the lattice because the site at the center of the unit cell is not equivalent to the site at the corner of the unit cell. Hence the system is not a c(3x3) structure. The system does have (3x3) symmetry, so we say that the system has a p(3x3) structure.

Explanations of specific figures:

Figure 2.71a shows the primitive unit cell of the (100) surface of an ideal FCC material. The primitive unit cell is a square with a lattice dimension equal to the interatomic distance.

Figure 2.71b shows an example of a p(1x1) overlayer on the (100) surface. Again the unit cell is a square with a lattice dimension equal to the interatomic distance. The figure shows one atom per unit cell. However, the layer would still be a p(1x1) structure even if there were more than one atom per unit cell, since the (1x1) unit cell is the smallest (i.e., primitive) unit cell of the lattice.

Figure 2.71c, d, g, and h show several examples of p(2x2) overlayers. In each case the unit cell is a square. Unlike the case in Figure 2.71b, however, the lattice dimension is twice that of the clean surface. Hence, these are all p(2x2) layers. Of course, the positions of the atoms within the unit cell are different in the various cases. This example

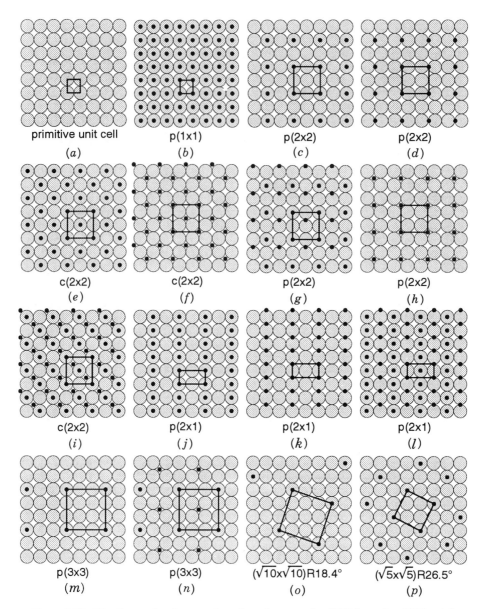

Figure 2.71   Wood's notation for a number of overlayers on the (100) face of an FCC metal.

illustrates that many different overlayer structures are described with the same Wood's indices.

Figure 2.71e and f show c(2x2) overlayers. Figure 2.71e, f, and g all have two atoms per conventional unit cell. However, while the two lattice points are not equivalent in Figure 2.71g, the lattice points are equivalent in Figure 2.71e and f. One can have a smaller unit cell in the latter two cases. As a result, the (2x2) unit cell is not a primitive cell in the pictures. People usually refer to these type of structures as c(2x2) structures. However, an alternative notation for the structure is $(\sqrt{2}x\sqrt{2})R45°$.

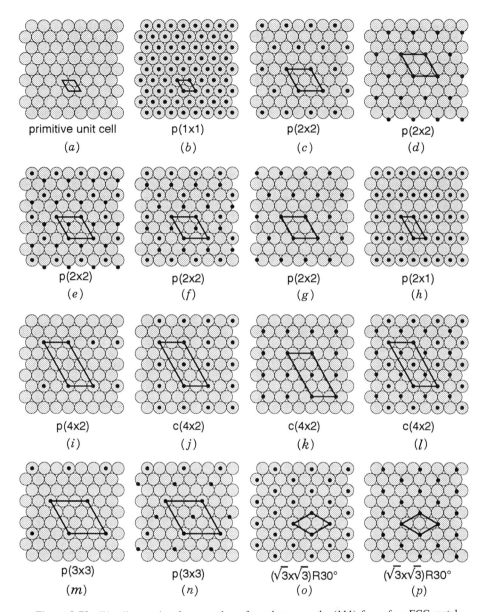

Figure 2.72 Wood's notation for a number of overlayers on the (111) face of an FCC metal.

The remaining cases in Figure 2.71 are fairly obvious. The case in Figure 2.71n is a p(3x3) overlayer not a c(3x3) overlayer for reasons discussed three paragraphs ago. The unit cells in Figure 2.71o and p go two or three atoms to the left and one up. These structures have relatively simple matrix representations

$$\overline{m}_o = \begin{bmatrix} 3 & 1 \\ -1 & 3 \end{bmatrix} \qquad \overline{m}_p = \begin{bmatrix} 2 & 1 \\ -1 & 2 \end{bmatrix}$$

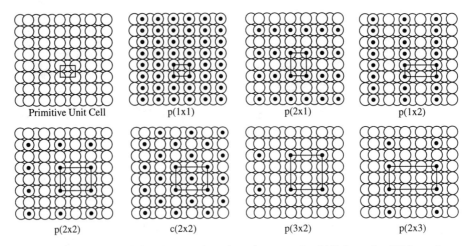

Figure 2.73   Wood's notation for a number of overlayers on the (110) face of an FCC metal.

However, Wood's notation for the two structures is more complex:

$$(\sqrt{2^2 + 1^2} \text{ x } \sqrt{2^2 + 1^2})R \arctan(1/2) = (\sqrt{5} \text{ x } \sqrt{5})R26.5°$$

$$(\sqrt{3^2 + 1^2} \text{ x } \sqrt{3^2 + 1^2})R \arctan(1/3) = (\sqrt{10} \text{ x } \sqrt{10})R18.4°$$

Figures 2.72 and 2.73 show several other examples. The reader may want to refer back to them when he or she is reading Chapter 3.

The one particularly interesting example is the c(4x2) case in Figure 2.72. Note that the primitive unit cell for the c(4x2) structure is rectangle, while the primitive unit cell for the lattice is rhombus. One cannot rotate the primitive unit cell for the lattice so that its axes go in the same direction as the primitive unit cell for the overlayer. As a result, one cannot specify the primitive unit cell for the overlayer via Wood's notation. The matrix notation can be easily defined.

For the purposes of derivation, it is useful to assume that the x axis goes from left to right in Figure 2.72, while the y axis goes from the bottom to the top of the figure. In this notation, the primitive unit cell has two lattice vectors given by

$$\vec{b}_1 = \begin{bmatrix} 1 \\ 0 \end{bmatrix} \qquad \vec{b}_2 = \begin{bmatrix} -0.5 \\ \sqrt{3/2} \end{bmatrix} \tag{2.22}$$

The lattice vectors of the overlayer are given by

$$\vec{b}_1^1 = \begin{bmatrix} 2 \\ 0 \end{bmatrix} \qquad \vec{b}_2^1 = \begin{bmatrix} 0 \\ \sqrt{3} \end{bmatrix} \tag{2.23}$$

Substituting into Equation 2.16 shows that $\overline{m}$, the matrix notation for the overlayer, satisfies

$$\overline{m} \begin{bmatrix} 1 & -1/2 \\ 0 & \sqrt{3/2} \end{bmatrix} = \begin{bmatrix} 2 & 0 \\ 0 & \sqrt{3} \end{bmatrix} \tag{2.24}$$

One can use matrix algebra to show that

$$\begin{bmatrix} 1 & 1/\sqrt{3} \\ 0 & 2/\sqrt{3} \end{bmatrix}$$

is the inverse of

$$\begin{bmatrix} 1 & -1/2 \\ 0 & \sqrt{3}/2 \end{bmatrix}$$

Multiplying Equation 2.24 by

$$\begin{bmatrix} 1 & 1/\sqrt{3} \\ 0 & 2/\sqrt{3} \end{bmatrix}$$

yields

$$\overline{m} \begin{bmatrix} 1 & -1/2 \\ 0 & \sqrt{3}/2 \end{bmatrix} \begin{bmatrix} 1 & 1/\sqrt{3} \\ 0 & 2/\sqrt{3} \end{bmatrix} = \begin{bmatrix} 2 & 0 \\ 0 & \sqrt{3} \end{bmatrix} \begin{bmatrix} 1 & 1/\sqrt{3} \\ 0 & 2/\sqrt{3} \end{bmatrix} \qquad (2.25)$$

Performing the algebra,

$$\overline{m} = \begin{bmatrix} 2 & 0 \\ 0 & \sqrt{3} \end{bmatrix} \begin{bmatrix} 1 & 1/\sqrt{3} \\ 0 & 2/\sqrt{3} \end{bmatrix} = \begin{bmatrix} 2 & 2/\sqrt{3} \\ 0 & 2 \end{bmatrix} \qquad (2.26)$$

Example 2.D   Rotational Axes in Crystals

There were several places in the text where we mentioned rotating around an axis to get from one plane to another. The objective of this example is to show how to determine the Miller indices of the rotational axis carrying a given plane into another.

Figure 2.74 illustrates that if there are two planes in space, then if the system is rotated around the line that is the intersection of the two planes, the rotation will carry one plane into the other. Let's define $[x_1y_1z_1]$ and $[x_2y_2z_2]$ to be the normals to the two planes in x, y, z space, where the x, y, and z axes may or may not correspond to the crystallographic axes.

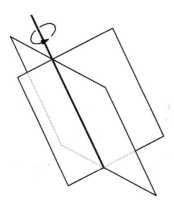

Figure 2.74   Rotation of one plane into another.

Further let's define [ℓmn] to be the x, y, z space indices of the line of intersection of the two planes. Note that [ℓmn] is in plane one. Therefore, it must be perpendicular to [x_1 y_1 z_1]. Similarly, [ℓmn] is in plane 2. Therefore, it must be perpendicular to [x_2 y_2 z_2]. From analytic geometry two vectors are perpendicular when their dot product is zero. Therefore,

$$\ell x_1 + my_1 + nz_1 = 0 \tag{2.27}$$

$$\ell x_2 + my_2 + nz_2 = 0 \tag{2.28}$$

If one sets m = 1, one can solve Equations 2.27 and 2.28 simultaneously for ℓ and n. One then has to scale by the lattice dimension and multiply by the common denominator, cd, to calculate the Miller indices of the planes.

With an FCC, BCC, simple cubic, or zinc-blend lattice there are some tricks that make this easier. In each of these cases the [ijk] direction is perpendicular to the (ijk) plane. Further, one can express all of the directions that are perpendicular to the [ijk] direction as a linear combination of any two of the following three vectors [$j\bar{i}0$], [$k0\bar{i}$], and [$0\bar{k}j$]. For example, if one wants to rotate the (100) plane into the (553) plane, the first index of the rotational axis must be zero, since the rotational axis must be perpendicular to the [100] direction. Any vector that is perpendicular to the [553] direction can be written as a linear combination [$5\bar{5}0$], [$30\bar{5}$], and [$0\bar{3}5$]. Notice that

$$[0\bar{3}5] = [0\bar{3}5] + 0 \cdot [5\bar{5}0] \tag{2.29}$$

It is also perpendicular to the [100] plane. Therefore, if we rotate around the [$0\bar{3}5$] axis, we will carry the (100) plane into the (553) plane.

Similarly, if we want to rotate from the (210) to the (321) plane, we know that any vector perpendicular to the [210] direction can be written as [$1\bar{2}m$], where we still have to solve for m. Substituting [$1\bar{2}m$] into Equation 2.27 yields

$$(3 \cdot 1) + (2)(-2) + (1) \cdot m = 0 \quad \text{or} \quad m = 1$$

Therefore, if we wanted to rotate the (210) plane into the (321) plane, we would rotate around the [$1\bar{2}1$] axis. A similar analysis shows that the [001] vector lies in both the (100) and (110) planes, so that a rotation around the [001] axis carries the (100) plane into the (110) plane. Similarly, a rotation about the [$1\bar{1}0$] axis carries the (111) plane into the (110) plane, while a rotation about the ($01\bar{1}$) axis carries the (111) plane into the (100) plane.

These ideas are used to cut crystals. If we wanted to cut a (321) plane, we could first align the crystal in the [100] direction (the [100] direction has an easily recognized x-ray diffraction pattern), then rotate around the [$01\bar{2}$] direction to the (321) plane. We would have to rotate by an angle θ, where, from the cosine law, θ is given by

$$[100] \cdot [321] = |[100]| \, |[321]| \cos \theta \quad \text{or} \quad \theta = \arccos\left(\frac{3}{\sqrt{14}}\right) = 36.7°$$

Example 2.E   Viscinal Notation

The results in Example 2.D can be used to relate the viscinal notation for a plane to the Miller indices of the plane. For example, the results in the last paragraph show that if one starts with the (100) plane and then rotates by 36.7° around the ($0\bar{1}2$) axis, one gets to the (321) plane. Therefore, the viscinal notation for the (321) plane is (100) 36.7° ($0\bar{1}2$).

Now let's consider another case, which is to assume that we know that a given plane has a viscinal notation of (111) 14.4° ($01\bar{1}$). How would one calculate the Miller indices of the plane in a cubic lattice?

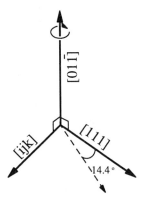

Figure 2.75 Unit vectors for Example 2.E.

Solution: We need to start with the [111] direction and rotate it by 14.4° around the [01$\bar{1}$] axis.

It is useful to define an orthonormal set of basis vectors as shown in Figure 2.75 where one vector is in the [01$\bar{1}$] direction, another is in the [11$\bar{1}$] direction, and the third axis is perpendicular to the first two. The plane 14.4° from the [111] plane will have a normal in the plane of the second and third vector and will lie 14.4° from the [111] vector. Let's calculate the Miller indices of the three orthonormal vectors. One vector is [111]/|[111]| where the notation |[lmn]| denotes the length of a vector [lmn], i.e.,

$$|[\text{lmn}]| = \sqrt{l^2 + m^2 + n^2} \tag{2.30}$$

Similarly, another vector is [01$\bar{1}$]/|[01$\bar{1}$]|. From analytical geometry the third vector is given by

$$[\text{i j k}] = \frac{[01\bar{1}] \times [111]}{|[01\bar{1}] \times [111]|} = \frac{[2\bar{1}\bar{1}]}{\sqrt{6}} \tag{2.31}$$

From analytical geometry the vector 14.4° away from [111] has the Miller indices

$$[\text{i j k}] = \cos(14.4°) \frac{[111]}{\sqrt{3}} + \sin(14.4°) \frac{[2\bar{1}\bar{1}]}{\sqrt{6}}$$

$$[\text{i j k}] = [0.762\ 0.458\ 0.458] = \frac{[533]}{6.55} \tag{2.32}$$

Therefore, the (111) 14.3° [01$\bar{1}$] plane has the Miller indices [533].

Example 2.F A Sample Lattice Expansion Calculation

In Section 2.6.1 we noted that if we have a solid connected with a series of pairwise additive potentials of the form in Figure 2.49, the surface atoms will move out from the bulk whenever a surface is cut. The objective of this example is to quantify the expansion.

Consider an FCC solid and assume that each atom in the solid interacts with every other atom via a Lennard-Jones potential

$$V_{\ell m} = Wl \left[\left(\frac{Z_1}{r_{\ell m}} \right)^{12} - 2 \left(\frac{Z_1}{r_{\ell m}} \right)^6 \right] \tag{2.33}$$

where $V_{\ell m}$ is the interaction between atoms l and m, $r_{\ell m}$ is the distance between the two atoms, and Wl is the well depth.

The total energy of the lattice becomes, E_{total},

$$E_{total} = \sum_{\ell} \sum_{m > \ell} V_{\ell m} \qquad (2.34)$$

where the second sum goes for $m > \ell$ so that the $V_{\ell m}$ interaction is only counted once in the sum. It is useful to rewrite Equation 2.34 as

$$E_{total} = \sum_{\ell} \sum_{m \neq \ell} \left(\frac{V_{\ell m}}{2} \right) \qquad (2.35)$$

where the factor of 1/2 comes into Equation 2.35 to compensate for the fact that $V_{\ell m}$ appears twice in the sum.

It is useful to define a quantity V_{ℓ} by

$$V_{\ell} = \sum_{m \neq \ell} \left(\frac{V_{\ell m}}{2} \right) \qquad (2.36)$$

Assume that the lattice dimension is a_x and that atom ℓ is at the origin. Substituting Equation 2.33 into Equation 2.36 yields

$$V_{\ell} = \frac{Wl}{2} \sum_{m \neq \ell} \left[\left(\frac{Z_1}{r_m} \right)^{12} - 2 \left(\frac{Z_1}{r_m} \right)^6 \right] \qquad (2.37)$$

Next we calculate the lattice dimension a_x by minimizing V_{ℓ} with respect to a_x. According to Equation 2.1, $\vec{r}_m$ the position of an atom m in the lattice is given by

$$\vec{r}_m = (m_1 + B_1^m)\vec{b}_1 + (m_2 + B_2^m)\vec{b}_2 + (m_3 + B_3^m)\vec{b}_3 \qquad (2.38)$$

where m_1, m_2, m_3 are the indices of the unit cell containing atom m and B_1^m, B_2^m, and B_3^m are the positions of atom within the unit cell.

For the material that follows, it is useful to define dimensionless lattice vectors $Б_1$, $Б_2$, $Б_3$ by

$$a_x Б_1 = \vec{b}_1 \qquad a_x Б_2 = \vec{b}_2 \qquad a_x Б_3 = \vec{b}_3 \qquad (2.39)$$

where $Б_1$, $Б_2$, $Б_3$ are the lattice vectors for a lattice with a lattice dimension of unity. Combining Equations 2.38 and 2.39 yields

$$\vec{r}_m = a_x[(m_1 + B_1^m)Б_1 + (m_2 + B_2^m)Б_2 + (m_3 + B_3^m)Б_3] \qquad (2.40)$$

At this point it is useful to look back to Equation 2.37. Note that the sum over m in Equation 2.37 is a sum over all of the atoms in the lattice. We can list the position of the atom by a single index m, or by a series of indices m_1, m_2, m_3, B_1^m, B_2^m, B_3^m. If we sum over all possible values of m_1, m_2, m_3, B_1^m, B_2^m, B_3^m, we get to all lattice positions. Therefore, we can rewrite Equation 2.37 as

$$V_{\ell m} = \frac{Wl}{2} \sum_{m_1, m_2, m_3, j} \left[\left(\frac{Z_1}{r_m} \right)^{12} - 2 \left(\frac{Z_1}{r_m} \right)^6 \right] \qquad (2.41)$$

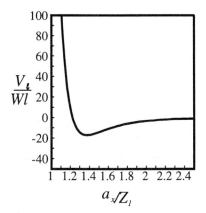

Figure 2.76 A plot of Equation 2.43.

where j is an index of the B_i^m. Rearranging Equation 2.41 yields

$$V_\ell = \frac{Wl}{2}\left[\left(\frac{Z_1}{a_x}\right)^{12}\left(\sum_{m_1,m_2,m_3,j}\left(\frac{a_x}{r_m}\right)^{12}\right) - 2\left(\frac{Z_1}{a_x}\right)^{6}\left(\sum_{m_1,m_2,m_3,j}\left(\frac{a_x}{r_m}\right)^{6}\right)\right] \quad (2.42)$$

Performing the sums in Equation 2.42 using a modification of the program in Example 2.A yields

$$V_\ell = \frac{Wl}{2}\left[776.4\left(\frac{Z_1}{a_x}\right)^{12} - 231.2\left(\frac{Z_1}{a_x}\right)^{6}\right] \quad (2.43)$$

Figure 2.76 shows a plot of V_ℓ as a function of a_x calculated from Equation 2.43. The potential reaches a minimum at $a_x = 1.373\ Z_1$. As a result the equilibrium spacing is 1.373 Z_1. According to Table 2.3 the minimum spacing between the atoms is 0.707 $a_x = 0.97\ Z_1$. Therefore, at equilibrium the atomic spacing in a Lennard-Jones FCC solid is 0.97 Z_1.

Notice that the minimum in the original Lennard-Jones potential (i.e., Equation 2.37) is at $r_{\ell m} = Z_1$. Therefore, if there were only two atoms in the lattice, the lattice spacing would be Z_1 while it is 0.97 of Z_1 in the solid. Hence, the solid is compressed compared to two isolated atoms.

In Section 2.6.1 we noted that with pairwise additive potentials the lattice will be expanded again when a surface forms. Next we will show how to quantify the expansion.

Consider cutting a (100) surface in our FCC lattice. Figure 2.77 shows a side view of the lattice. When the plane is first cut, the atoms start out with a spacing of 0.97 of Z_1. The spacing between the top plane of atoms and the second plane is 0.5 $a_x = 0.687\ Z_1$. Now consider what happens when the surface relaxes. The top plane of atoms moves out from the bulk, while the second plane moves out by a lesser extent. How can one calculate how much the lattice expands?

Solution: Consider a slab of atoms ten layers high. The energy of any atom within the slab can be calculated from Equation 2.41. Now consider the total energy of the slab. It's useful to calculate the total energy of the slab, E_{total} by dividing the slab into slices, as indicated in Figure 2.77, and then summing the slices, i.e.,

$$E_{total} = \sum_{slices}\sum_{\substack{atoms \\ in\ slices}} V_\ell \quad (2.44)$$

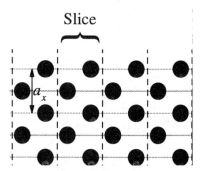

Figure 2.77 A side view of the (100) surface. The vertical dashed lines indicate the boundaries between the slices referred to in the text.

Note that all of the slices are equivalent. Therefore, the total energy of the slab is minimized, when the energy of each individual slice is minimized.

For the moment, it will be assumed that the layers of atoms move together. One can write the energy of a slice as

$$E_{slice} = \sum_{\substack{atoms\ in \\ slice}} V_\ell \qquad (2.45)$$

Therefore, one can compute the energy of the slice as a function of the distance between the top layer of atoms and all of the lower planes of atoms. The program on the next page does that. Optimization of the energy shows that the spacing between the top two layers of atoms expands from $0.5\ a_x - 0.687\ Z_1$ to $0.70\ Z_1$, i.e., the top layer of atoms moves out by 3.5%.

One still has to consider whether the surface rumples, i.e., whether different atoms can move by different amounts. Consider starting with atoms positioned at the equilibrium distance calculated as described in the last paragraph. If one displaces any one of the atoms, that atom will move into a less favorable position since the computer program already optimized the position of the atom. With pairwise additive potentials none of the other atoms will go into a more favorable position when we move the atom. As a result, the surface energy goes up when the surface rumples. Therefore, rumpling will not occur.

These results, of course, only apply to pairwise additive potentials. With a non-pairwise additive potential, when a given atom is displaced, the interactions of all of the adjacent atoms change. It is possible for changes to be such that the energy of the system goes down even though the energy of the given atom goes up. This can cause rumpling of the surface atoms.

Example 2.G   Counting the Atoms in a Unit Cell

Lastly, we include an example showing how to count the number of atoms within the surface unit cell. At first sight this would seem to be a difficult problem because, if one looks at the unit cell on the left of Figure 2.78 one notices that some of the atoms are partly within the unit cell and partly outside of the unit cell. However, there is a trick that makes the counting process much easier.

Consider the unit cell on the right of Figure 2.78 where the edges of the unit cell have been drawn so the edges of the unit cell does not go through the centers of any of the atoms. Now the nuclei of atoms 2, 3, 4, 5, 8, and 9 lie with the unit cell and the nuclei of all of the other atoms lie outside of the unit cell. Therefore, one can immediately conclude that atoms 2, 3, 4, 5, 8, and 9 are within the unit cell and all of the other atoms are outside of the unit cell.

Another way to look at the situation is that atom 2 is only partway within the unit cell.

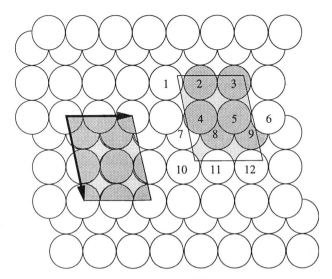

Figure 2.78 The FCC 511 surface.

However, the part of atom 2 that is outside of the unit cell is exactly equal to the part of atom 11 that lies inside of the unit cell. Similarly, the piece of atom 4 outside of the unit cell is exactly equal to the part of atom 6 within the unit cell. The part of atom 9 outside of the unit cell is exactly equal to the part of atom 7 within the unit cell. Atom 3 is a little harder to see, but one can show that the area of atom 3 outside of the unit cell is exactly equal to sum of the areas of atoms 1, 10, and 12 that lie within the unit cell. Therefore, everything exactly cancels and there are exactly six atoms within the unit cell.

Note, however, that this trick works only when none to the edges of the unit cell overlap the nuclei. If one picks a unit cell that overlaps some of the nuclei, we have to be careful about the fact that some of the atoms are only partway within a given unit cell.

2.11 SUPPLEMENTAL MATERIAL

2.11.1 Proof That There Can Only be Two-, Three-, Four-, and Sixfold Axes in an Infinite Periodic Two-Dimensional Lattice

Earlier in this chapter we stated that one can only have two-, three-, four- or sixfold rotational axes in a periodic lattice. The objective of this section is to prove that only two-, three-, four-, and sixfold axes are possible. Bravais and Cauchy provided proofs in the 1840s. The proof below follows one in Bravais [1850].

Consider a periodic lattice such as that in Figure 2.79. We can define a basis vector for the primitive unit cell in the lattice as the lines from point 0 to point 1 and from point 0 to point 2 in the figure.

Let us define R to be the shortest distance between two lattice points, and rotate the figure so that the shortest axis lies parallel to the x axis. Now let's assume that there is an n-fold rotational axis at each lattice point. If there is an n-fold axis, then if we rotate point 1 in Figure 2.79 by $\theta_r = 360°/n$, we will arrive at another lattice point. Now consider the vector $\hat{V}$ in Figure 2.79. From analytic geometry, V, the length of the $\hat{V}$ vector, is given by

$$\frac{V}{2} = R \sin\left(\frac{\theta_r}{2}\right) \qquad (2.46)$$

```
#include <stdio.h>
#include <conio.h>
#include <math.h>
#include <stdlib.h>
/* program to calculate lattice expansions */

/*1*/ /* function prototypes for ANSI compatability */
/*1*/
/*1*/ main(void);
/*1*/ eslice( double *,double, int );
/*1*/ /*********************************/

double x[2000],y[2000],z[2000];
/* x, y, z are the atomic positions */

float ax=1.37365;
float basVEC[3][3]
/*basVEC are the basis vectors for
the unit cell*/
= {{(float)1.,(float)0.,(float)0.},
{(float)0.,(float)1.,(float)0. },
{(float)0.,(float)0.,(float)1.} };
int nbasis = 4;
/*nbasis equals the number of atoms

= {{{(float)0.,(float)0.,(float)0.},
{(float)0.5,(float)0.5,(float)0.},
{(float)0.5,(float)0.,(float)0.5},
{(float)0.,(float)0.5,(float)0.5}};

main()
{ int natoms,nx, ny,nz,i,ii,j,k,l;

        float rlat[3],r[3],energy,zz;
/*******************************/
Set up an array of atomic positions for
the slab positions 0 to 199 are atoms
in the first layer positions 200 to 399
are atoms in the second layer etc
the center slice in the cluster contains
atoms 55, 255, 455, etc.
********************************/
nx=10;ny=10;nz=5;natoms=0;
for(j=0;j<nz;j++) for(i=0;i<nbasis;i++)
{print(" ") ;
  for(l=0;l<nx;l++) for(k=0;k<ny;k++)
  { /*calculate atom positions
                 via equation 2.1 */
rlat[0]=(float)(b[i][0]+((float)(l-nx/2));
rlat[1]=(float)(b[i][1]+((float)(k-ny/2));
rlat[2]=(float)(-b[i][2]-((float)(j)));
```

```
within the unit cell*/

float b[4][3]
/* b is the positions of the atoms
within the unit cell
in units of the basis vector */
        x[natoms]=r[0]*ax;
        y[natoms]=r[1]*ax;
        z[natoms]=r[2]*ax;

        natoms++;
    }} print("\n\n");
for(zz=0.0;zz<0.036;zz+=0.005)
{ eslice(&energy,zz,natoms);
print("%8.5f, %9.5f\n",zz,energy);
}return 1;
}

eslice( energy, zz,natoms)
float *energy, zz;int natoms;
{   int inslice, atom;
    float rx,ry,rz,r2,r6,z1,z2;
    double ener;

    ener =0;
    for(inslice = 55; inslice <
                natoms; inslice += 200)
    {if(inslice==55) z1=zz;
     else z1=z[inslice];
     printf("..");

for(ii=0;ii<3;ii++)
r[ii]=basVEC[0][ii]*rlat[0]
        +basVEC[1][ii]*rlat[1]
        +basVEC[2][ii]*rlat[2];

for(atom=0;atom<natoms;atom++)
{ if(atom<200)z2=zz;
  else z2=z[atom];
  rx= x[inslice]-x[atom];

  ry= y[inslice]-y[atom];
  r2= z1-z2;
  r2=(rx*rx+ry*ry+rz*rz);
  r6=r2*r2*r2;
  if(atom!=inslice)
  ener= ener + 0.5/(r6*r6) - 1./r6;
  rz=z1-z[atom]+5*ax;
  r2=(rx*rx+ry*ry+rz*rz);
  r6=r2*r2*r2;
  ener= ener + 0.5/(r6*r6) - 1./r6;
}}

    *energy = ener; return 1;
}
```

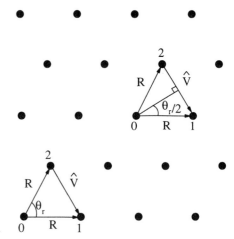

Figure 2.79 An n-fold axis in a periodic lattice.

Note that if

$$\left(\sin\frac{\theta_r}{2}\right) < 0.5 \quad V < R \tag{2.47}$$

In such a case, if we have any lattice vector of length R, we can always find a smaller lattice vector of length $\hat{V}$. We can repeat the analysis on $\hat{V}$ to find an even smaller lattice vector. Therefore, if we repeat the analysis ad infinitum, we find that if $\sin(\theta_r/2) < 0.5$ we need the spacing between adjacent lattice points to approach zero. Therefore, we cannot find a periodic lattice with a finite spacing between lattice points unless

$$\sin\left(\frac{\theta_r}{2}\right) \geq 0.5 \tag{2.48}$$

Substituting $\theta_r = 360°/n$ into Equation 2.48 yields

$$n \leq 6 \tag{2.49}$$

Therefore, based on the analysis so far, we can have a two-, three-, four-, five-, or sixfold axis, but we cannot have a seven-, eight-, or nine- . . . fold axis in a periodic structure.

There is a special constraint that excludes fivefold axes. Consider the part of a lattice shown on the left of Figure 2.80. If we assume that points 0 and 1 are lattice points, then

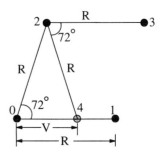

Figure 2.80 A diagram illustrating why a fivefold axis cannot exist in a periodic lattice.

point 2 needs to be a lattice point by rotation, and point 3 needs to be a lattice point by translation. Now consider the effect of a rotational axis around point 2 in the lattice. If we rotate around point 2, we will create a new point, point 4. The distance between point 0 and point 4, V, is given by

$$\frac{V}{2} = R \sin 18° \quad \text{or} \quad V = 0.618 \, R \tag{2.50}$$

Notice that point 4 is closer to point 0 than points 1 or 2. Therefore, if we have a fivefold axis with a lattice dimension, R, we can always find a smaller lattice dimension V. The previous analysis shows that that leads to an impossible situation. Therefore, we cannot have a fivefold axis in an infinite periodic lattice with a finite lattice dimension. We can have a *single* fivefold lattice in a finite crystal, however.

At this point we have proved that one cannot have five-, seven-, eight-, or nine- . . . fold axes in periodic lattices. We can have two-, three-, four-, and sixfold axes, as illustrated in the two-dimensional lattices in Figure 2.13.

2.11.2   The Reciprocal Lattice

In Sections 2.4 and 2.5 and Examples 2.A and 2.B we showed how to calculate the lattice vectors for a given arrangement of atoms. People also sometimes describe planes via something called the reciprocal lattice vectors for the plane. Consider a surface with two lattice vectors $\hat{a}_1$ and $\hat{a}_2$ and a normal $\hat{n}$.

We can define the reciprocal lattice vectors $\hat{q}_1$ and $\hat{q}_2$ by

$$\hat{q}_1 = \frac{\hat{a}_1 \times \hat{n}}{(\hat{a}_1 \times \hat{n}) \cdot \hat{a}_2} \tag{2.51}$$

$$\hat{q}_2 = \frac{\hat{a}_2 \times \hat{n}}{(\hat{a}_2 \times \hat{n}) \cdot \hat{a}_1} \tag{2.52}$$

where the $\times$ denotes a cross product and the $\cdot$ denotes a dot product. There is also something called the reciprocal lattice, where a vector $\hat{q}_{ij}$ in the reciprocal lattice is given by

$$\hat{q}_{ij} = i\hat{q}_1 + j\hat{q}_2 \tag{2.53}$$

where i and j are random integers.

It works out that the set of all reciprocal vectors is simply related to the diffraction pattern from the surface. Consider an electron incident onto the surface from the z direction, with a momentum $\hat{K}_a$ given by

$$\hat{K}_o = \left[0, \, 0, \, \frac{\sqrt{2m_e E_o}}{\hbar} \right] \tag{2.54}$$

where m_e is the mass of the electron, E_o is the electrons incident energy, and $\hbar$ is Planck's constant. One can show that one will get a LEED spot whenever the momentum of the electrons leaving the surface, $\hat{k}_{exit}$, satisfy

$$\hat{k}_{exit} = \hat{K}_o + \hat{q}_{ij} \tag{2.55}$$

Hence, one can calculate a LEED pattern if one knows the reciprocal lattice vectors for the surface. We are not going to be discussing LEED here, and so the reciprocal lattice will not be discussed in detail. One should refer to Pendry [1974] or Kittel [1986] for a detailed discussion of the reciprocal lattice.

2.11.3 Somorjai Step Notation

Somorjai has proposed an alternative notation to designate step surfaces. The idea is to describe a stepped surface by the length and indices of the terrace and the indices of the step. For example, the Pt(443) surface has (111) terraces and $(11\bar{1})$ steps. In Somorjai's notation this surface would be

$$2(111) \times (11\bar{1}).$$

This notation can be confusing because it is unclear how to define the length of the terrace. Is the atom at the top edge of the step, in the step, or in the terrace? Still, the notation is used occasionally, so it needed to be mentioned.

2.11.4 The Correspondence of Planes in BCC and FCC Lattices

In Section 2.5 we noted that there is a close correspondence between the surface structure of planes within the BCC and FCC lattices, and it is only that the Miller indices of the two systems are substantially different. The objective of this example is to quantify the correspondence.

Let's start with a BCC lattice. One usually discusses a BCC lattice using the lattice vectors [100], [010], and [001]. For the purposes of discussion, however, we consider a unit cell bounded by the three vectors [100], $[01\bar{1}]$, and [011]. Next we demonstrate that if we have a point $\vec{r} = [ijk]$ in the original lattice, its Miller index in the [100], [011], and [011] lattice $\vec{r}^{\,1} = [\ell mn]$ is given by

$$\vec{r}^{\,1} = [\ell mn] = [ijk] \begin{bmatrix} 1 & 0 & 0 \\ 0 & 0.5 & 0.5 \\ 0 & 0.5 & -0.5 \end{bmatrix} \tag{2.56}$$

Note

$$\begin{bmatrix} 1 & 0 & 0 \\ 0 & 0.5 & 0.5 \\ 0 & 0.5 & -0.5 \end{bmatrix} = \begin{bmatrix} 1 & 0 & 0 \\ 0 & 1 & 1 \\ 0 & 1 & \bar{1} \end{bmatrix}^{-1} \tag{2.57}$$

The proof starts by considering a point

$$[ijk] = \ell[100] + m[011] + n[01\bar{1}] \tag{2.58}$$

Multiplying each side of Equation 2.58 by the matrix in Equation 2.57 yields

$$[ijk] \begin{bmatrix} 1 & 0 & 0 \\ 0 & 0.5 & 0.5 \\ 0 & 0.5 & -0.5 \end{bmatrix} = [\ell mn] \tag{2.59}$$

QED

There are four lattice points within the unit cell bounded by [100], [011], and [01$\bar{1}$]

$$\vec{r}_1 = [0\ 0\ 0] \quad \vec{r}_2^{\,1} = [\tfrac{1}{2}\ \tfrac{1}{2}\ \tfrac{1}{2}] \quad \vec{r}_3 = [0\ 1\ 0] \quad \vec{r}_4 = [\tfrac{1}{2}\ \tfrac{1}{2}\ -\tfrac{1}{2}]$$

Transforming to the new coordinates using equation 2.56 yields

$$\vec{r}_1^{\,1} = [0\ 0\ 0] \quad \vec{r}_2^{\,1} = [\tfrac{1}{2}\ \tfrac{1}{2}\ 0] \quad \vec{r}_3^{\,1} = [0\ \tfrac{1}{2}\ \tfrac{1}{2}] \quad \vec{r}_4^{\,1} = [\tfrac{1}{2}\ 0\ \tfrac{1}{2}]$$

It is useful to compare these numbers to those for the FCC lattice in Table 2.3. Notice that the locations of all of the BCC atoms within the [100], [01$\bar{1}$], [011] unit cell are identical to the positions of the atoms in a conventional FCC lattice. Hence, if one computed the Miller indices of a given [ijk] plane in the [100], [01$\bar{1}$], [011] lattice, the surface structures would be very similar to those for an FCC lattice. The atomic positions will all be the same. The only difference will be that the unit cell will be stretched by a factor of $\sqrt{2}$ in the y and z directions. For example, the (111) face of an FCC material has regular hexagons, while the "equivalent" (110) faces of a BCC material has "hexagons" where the long axis of the hexagons is $\sqrt{2}$ longer than the short axis. Note, however, that this effect is minimized for (ijk) planes when $i \geq j \geq k \geq 0$. Hence, if one defines (ℓmn) to be the Miller indices of a plane in a BCC material computed using the [100], [01$\bar{1}$], [011] unit cell, one finds that the (ℓmn) plane in BCC has almost the same structure as the (ℓmn) plane in FCC. We leave it to the reader to show that (ℓmn) is related to (ijk) by

$$(\ell mn) = (ijk) \begin{bmatrix} 1 & 0 & 0 \\ 0 & 1 & 1 \\ 0 & 1 & \bar{1} \end{bmatrix} = (i \quad j+k \quad j-k) \tag{2.60}$$

PROBLEMS

2.1 Define the following terms in three sentences or less.

(1) Stepped surface	(20) BCC lattice
(2) Kink	(21) Zinc-blend lattice
(3) Periodic surface	(22) Threefold hollow
(4) Unit cell	(23) Closed packed
(5) Primitive unit cell	(24) Miller indices of a plane
(6) Nonprimitive unit cell	(25) {321}
(7) Lattice vector	(26) (257)
(8) Reciprocal lattice vector	(27) [263]
(9) Basis of unit cell	(28) Ideal surface structure
(10) Mirror plane	(29) Stereographic triangle
(11) Threefold axis	(30) (111) terrace
(12) Screw axis	(31) (111) step
(13) Oblique lattice	(32) (100) step
(14) Centered rectangle	(33) Surface relaxation
(15) Primitive rectangle	(34) Surface reconstruction
(16) Point group	(35) Viscinal surface
(17) Bravais lattice	(36) (111) 6° (100)
(18) HCP lattice	(37) (111) 6° [01$\bar{1}$]
(19) FCC lattice	

2.2 Figure 2.3 shows an FIM image of platinum field emission tip. Make a copy of the image and identify at least ten steps and eight kinks.

2.3 Make a copy of the picture of the zinc-blend lattice in Figure 2.8, and identify the two interpenetrating FCC lattices. What is the Bravais lattice for this system?

2.4 Identify the Bravais lattice for the structures shown in Figure 2.14. Show that each has a p2 point group.

2.5 Table 2.2 lists the lattice of each of the three-dimensional Bravais lattices, while Figure 2.17 shows a diagram of the conventional unit cell for each structure.

   (a) Draw the axes of each conventional unit cell for each of these Bravais lattices onto a photocopy of Figure 2.17 and show that the Table 2.2 is correct.

   (b) Find the primitive unit cell for each of the structures. Note that in some cases the primitive unit cell may include some atoms outside of the conventional unit cell.

2.6 Calculate Miller indices for a plane that intersects the crystallographic axes at (1) 1, 3, and 5. (b) 1, ∞, and 2, (c) 3, ∞, and ∞. Draw a diagram of each of the planes. Your diagram should look like Figure 2.23.

2.7 Use the analysis in Example 2.D to calculate the viscinal notation for the following planes: Pt(511), Pt(551), Pt(997), Si(511), Si(551), Si(997), W(511). Does the procedure in Example 2.D work for Ru(511)? Why not?

2.8 Make copies of Figures 2.36–2.40. Identify the atomic arrangement (e.g., 111) of all of the terraces, steps, and kinks in the figures.

2.9 (a) Draw an FCC stereographic triangle. (b) Find the positions of all of the planes in Figures 2.36–2.40 within this stereographic triangle. (c) Use your findings to describe, in words, how surface structure varies over the stereographic triangle. (d) Repeat for BCC. If you have access to the program SURFSTRU (available from the author) you may want to use it for this problem.

2.10 In LEED one gets beams of electrons moving along various crystallographic directions. Consider a beam of electrons that leaves an FCC crystal along the [431] direction, and assume that the electron continues moving along the [431] direction until it hits a flat plate that sits 2 cm away from the crystal and is oriented in the z direction. (a) Calculate the X and Y coordinates of the spot where the beam hits the plate. (b) Repeat for a beam moving along the [431] direction in an HCP crystal.

2.11 In Figure 2.52, we showed the missing row model of the (2x1)Pt(110) surface. This model of the surface structure is now well established. However, years ago, people were unsure whether the surface had a missing row structure, a sawtooth structure, or a paired row structure. All three structures are shown in Figure P2.11.

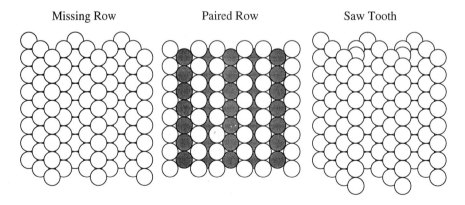

Missing Row   Paired Row   Saw Tooth

Figure P2.11 The missing row, paired row, and sawtooth models of (2x1) Pt(110).

(a) Show that all three structures have a (2x1) atomic arrangement.

(b) Compare all three structures to the TEM photo in Figure 2.47. Which seems to fit best?

(*Note:* In reality the interpretation of TEM data is subtle; one does not know that the surface has a missing row structure just because the image superficially seems to follow the missing row structure. However, ignore those effects for this problem.)

2.12 From analytic geometry you know that three points determine a plane.

(a) Calculate the equation for a plane going through the points (1,0,0) (2,3,1), and (1,0,2). (*Hint:* The equation for a plane is ax + by + cz = d, where a, b, and c are constants and d is 0 or 1.)

(b) Calculate the intersection of the plane with the x, y, and z axes.

(c) Calculate the Miller indices of the plane.

2.13 Figure P2.13 shows a diagram of the Pt(111), Pt(110), Pt(100), Pt(321), and Pt(430) planes.

(a) Identify a primitive surface unit cell in each diagram. (Note there are multiple answers to this question.)

(b) What is the area of the primitive unit cell, assuming that the lattice dimension of platinum is 3.92 Å (nearest neighbor distance 2.77 Å).

(c) How many exposed atoms lie within the primitive unit cell? Be sure to consider that some of the atoms are only partially within the unit cell.

(d) Calculate the atomic density by dividing the number of atoms in the unit cell by the area of the unit cell.

(e) What do your results tell you about the density of the atoms in stepped surfaces compared to that in the (111) and (100) planes?

(f) Calculate the coordination number of all of the atoms within the unit cell.

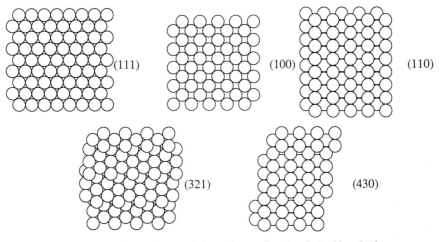

Figure P2.13   A diagram of the surfaces referred to in Problem 2.13

2.14 In Section 2.4 we mentioned that one often defines groups of planes. Find all of the planes (ijk) within the {321} group in an FCC lattice. Only consider planes where i, j, and k are integers, and their absolute value is 6 or less. [*Hint:* Which planes have the same structure as (321)?]

2.15 The (100) faces of BCC and FCC lattices both show a square arrangement of atoms. (a) Draw a diagram of both faces, and indicate the positions of the atoms below the plane. (b) Calculate the distance between the plane and the next lowest layer. (*Hint:* Use Equation 2.1 to calculate the atomic positions.) (c) Compute the ratio of the distances between the planes and between atoms within the plane. What can you say about the differences between FCC and BCC lattices from these results?

2.16 Compare the (100)A and (100)B surfaces of an HCP material. (a) How are the faces the same and how are they different? (b) Repeat for the (101)A and (101)B surfaces. (c) Show that there are two exposed atoms per unit cell in the (101)A and three in the (101)B.

2.17 Compare the (110)A and (110)B surfaces of a zinc-blend material. How are the faces the same, and how are they different?

2.18 In Section 2.3.3 we noted that the relative positions of the atoms above and below the A and B sites in an HCP lattice are different. Start with Figure 2.19, and try to show how the stacking pattern is different.

2.19 Consider the planes along the $[0\bar{7}9]$ axis in FCC [i.e., axis going from (997) to (100)].
   (a) What are the planes like?
   (b) How are the planes different than those along the $[0\bar{3}5]$ axis?

2.20 In Example 2.D we noted that a knowledge of rotational axes is quite useful when one wants to cut single crystals. Assume one wanted to cut a Pt(11 9 7) face.
   (a) Locate Pt(11 9 7) in the stereographic triangle.
   (b) Describe the structure of the Pt(11 9 7) face in words. What are the terraces like? What kinds of steps/kinks are present?
   (c) Assume you cut the crystal by first orienting along the [111] direction, then rotating toward the (11 9 7). What rotational axis should you use?
   (d) How far should you rotate?

2.21 Figures 2.71, 2.72, and 2.73 show the surface structure of several adsorbate overlayers.
   (a) Verify that Wood's notation for each of the layers is correctly listed in the figures.
   (b) Work out the matrix notation for the overlayers.

2.22 Figure P2.22 shows a number of ordered adsorbate layers on an FCC surface. On a photocopy of the figure, identify:
   (a) The surface unit cell in the absence of the adsorbate.
   (b) The unit cell of the adsorbate.
   (c) The lattice vectors for the adsorbate.
   (d) The space group of the system. Note that a Pt(111) surface only has threefold symmetry, because of the orientation of the atoms below the plane. (See Figure 2.19.) Assume that the adsorbate sits 1 Å above the surface.
   (e) Wood's notation and the matrix notation for the adsorbate overlayer.

2.23 Explain how the computer program used to calculate atomic positions in Example 2.A works.
   (a) Outline in words the steps the program executes to calculate the atomic positions in plane. (*Hint:* The first step is to define the lattice vectors and basis; the second is to read in the Miller indices of the plane to be plotted.)
   (b) Identify where the program
      (1) Defines the lattice vectors and basis.

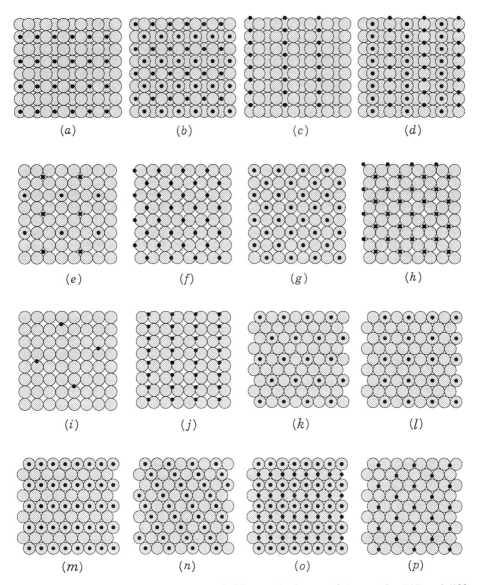

Figure P2.22 The surface structure of several different adsorbate overlayers on the (111) and (100) face of an FCC metal.

      (2) Reads in the Miller indices of the plane.

      (3) Calculates the atomic positions.

      (4) Calculates the T matrix.

      (5) Eliminates excess atoms.

  (c) How does the program decide whether to keep an atom (i.e., print it out).

  (d) Outline, in words, the steps the program goes through to calculate the T matrix. Identify where it

      (1) Calculates the lattice vectors of the plane.

(2) Calculates the normal direction.

(3) Calculates the elements of the T matrix.

(e) Try a number of cases to verify that the program is calculating the normal direction correctly (e.g., try (111), (110), (210), (211), (321), (510), (511)).

(f) Try a number of cases to see that the program is calculating the T matrix correctly. Note that the T matrix for the (ijk) plane rotates the normal to the (ijk) plane so it faces along the Z_{new} direction, while vectors in the (ijk) plane are rotated into the X_{new} Y_{new} direction. The [ijk] vector is the normal to the (ijk) plane, and [j $\bar{i}$ 0] and [k 0 $\bar{i}$] are two vectors in the plane. Therefore,

$$T \begin{bmatrix} i \\ j \\ k \end{bmatrix} = \begin{bmatrix} 0 \\ 0 \\ ? \end{bmatrix} \qquad T \begin{bmatrix} j \\ -i \\ 0 \end{bmatrix} = \begin{bmatrix} ? \\ ? \\ 0 \end{bmatrix} \qquad T \begin{bmatrix} k \\ 0 \\ -i \end{bmatrix} = \begin{bmatrix} ? \\ ? \\ 0 \end{bmatrix}$$

where the ? can have any numerical value.

(g) Verify that our particular choice of T matrix gives

$$T \begin{bmatrix} j \\ -i \\ 0 \end{bmatrix} = \begin{bmatrix} 0 \\ ? \\ 0 \end{bmatrix}$$

for a number of cases so that one of the lattice vectors for the surface always lies in the y direction.

2.24 Use the program in Example 2.A to calculate the atomic positions in the (a) Ni(100); (b) Pt(711); (c) Ag(655) surfaces. In each case identify the Bravais lattice and lattice vectors for the two-dimensional periodic unit cell formed by the exposed surface atoms. Also calculate the coordination number of the exposed atoms and the positions of each of the planes in the stereographic triangle.

2.25 Repeat Problem 2.24 for (1x1)Pt(110) and (2x1)Pt(110). Note that (2x1)Pt(110) is formally Pt(110)(1x2). You will have to modify the way you choose atoms in your program to calculate the arrangement of the atoms in (2x1)Pt(110) surface. It is often easiest to print out the coordinates of the atoms in the (1x1) layer, then change the z coordinate of some of the atoms with a word processor.

2.26 How would the program in Example 2.A have to be modified for BCC and HCP crystals? Use the crystal plotting program SURFSTU to the plot out the atomic positions in (a) W(210); (b) Si(110); (c) Ru(01$\bar{1}$1); (d) Fe(111); (e) Fe(110); (f) Si(111); (g) Os(10$\bar{1}$0). In each case identify the Bravais lattice and basis vectors for the two-dimensional periodic unit cell formed by the exposed surface atoms, and determine the position of each of the exposed surface atoms in the two-dimensional unit cell.

2.27 The BCC lattice is nonprimitive, having two lattice points per unit cell.

(a) Draw a BCC unit cell and indicate the primitive lattice on your drawing.

(b) If the volume of the BCC lattice is 1, what is the volume of the primitive lattice? (*Hint:* Assume that the atomic density is 1 atom per primitive unit cell.) How many atoms are in the conventional unit cell? Calculate the volume of the conventional unit cell by dividing the number of atoms in the conventional unit cell by the atomic density in atoms per primitive unit cell.

2.28 Figure 2.31 is a diagram of the (100), (001), and (101) faces of an HCP material. Now consider the following list of planes:

(a) $(01\underline{0})$ (d) $(1\bar{1}0)$ (g) $(1\bar{\bar{1}}1)$

(b) $(00\bar{1})$ (e) $(01\bar{1})$ (h) $(1\bar{1}1)$

(c) $(1\bar{1}0)$ (f) $(10\bar{1})$ (i) (111)

All but one of these planes has a structure that is identical to that of the (100), (001), or (101) face. Which faces go with which structures? (*Hint:* Work out the four index notation (i, j, l, k) for each of the planes, and use the fact that the w, x, y directions are equivalent, so the i, j and l indices commute. Also note that the crystal is symmetric when (x, y, w) → (−x, −y, −w) or z → −z, so that multiplying each group of indices by −1 has no effect.) Repeat for an FCC lattice using the diagrams in Figure 2.29.

2.29 In Figure 2.31 we showed a diagram of the ideal surface structures of the (100)A, (100)B, (101)A, and (101)B surfaces of an HCP material.

(a) Make a copy of the figure and indicate a primitive unit cell for each of these surfaces.

(b) What is the Bravais lattice and point group for each surface?

(c) Calculate the area of each primitive unit cell, assuming that the nearest neighbor distance is 3 Å. (*Hint:* Use your computer program from Example 2.A or Problem 2.26 to calculate the atomic coordinates.)

(d) Calculate how many exposed atoms lie within each primitive unit cell, accounting for the fact that some atoms are only partially in each unit cell.

(e) Calculate the atomic density of each surface as the number of exposed atoms in a unit cell divided by the area of the cell.

(f) What can you conclude about the similarity of the (100)A and (100)B surfaces? How about the (101)A and (101)B surfaces?

2.30 The results in Problem 2.12 and Example 2.B allow one to calculate the Miller indices of a plane. Figure P2.30 shows a diagram of the (415) and (711) faces of an ideal FCC material. Figure 2.68 and Figure 2.70 show lattice vectors for the (470), (511), (465), and (553) faces of the same material.

(a) Verify that the lattice vectors are correct in Figure 2.68 and Figure 2.70.

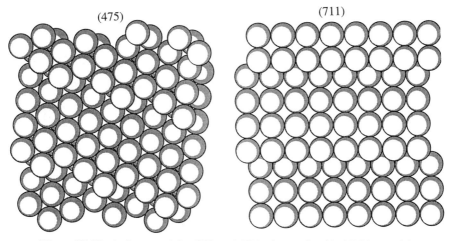

Figure P2.30 A diagram of the (475) and (711) planes of an ideal FCC material.

(b) Draw the unit cell for each of the planes on a photocopy of Figure P2.30.

(c) Identify the lattice vectors for each surface $\vec{v}_1$ and $\vec{v}_2$.

(d) Calculate the Miller indices of the lattice vectors using the methods outlined in Example 2.B. Verify that your lattice vectors are correct by demonstrating that the dot product of the Miller index of each lattice vector and the Miller index of the plane is zero.

(e) Calculate the Miller indices of the plane using the methods outlined in Example 2.B.

(f) Use the results in Problem 2.12 to calculate the Miller indices of the plane going through the points $(1,1,1)$, $(1,1,1) + \vec{v}_1$ $(1,1,1) + \vec{v}_2$. How does this compare to your results in part e?

2.31 If an additional point is added to the FCC lattice at the position $(1/2, 1/2, 1/2)$, what is the Bravais lattice of the new structure that is formed? (*Note:* This problem is a modification of one suggested by Andy Gellman.)

2.32 Figures 2.35 and 2.41 show stereographic triangles for an FCC material. Locate the (a) (111); (b) (100); (c) (110); (d) (410); (e) (210); (f) (320); (g) (511); (h) (211); (i) (322), (j) (331); (k) (332); (l) (321); (m) (975) planes in the diagram and describe qualitatively what each plane is like. Hint: the (511) plane is a stepped surface with (100) terraces and (111) steps.

2.33 Repeat Problem 2.32 for a BCC material. If you have access to the program SURFS-TRU (available from the author) use it to generate the structures.

2.34 Table 2.1 lists the primitive unit cells for each of the two-dimensional Bravais lattices.

(a) Calculate the lattice vectors and reciprocal lattice vectors for each of these lattices, and plot the first Brillouin zone. Assume that the conventional unit cells for the primitive rectangular, centered rectangular, and oblique lattices have $a_x = 1$ $a_y = 4$. Also assume $\alpha_{oblique} = 105°$.

(b) Comment on any similarities or differences you see between the shape of the primitive unit cell and the shape of the first Brillouin zone.

2.35 Find all of the mirror planes and rotational axes in the FCC unit cell. Repeat for the BCC and HCP unit cells.

2.36 In Section 2.5.3 we noted that the surface structure of BCC and FCC metals is very similar, but, that the Miller indices of equivalent planes are quite different. The objective of this problem is for the reader to convince himself or herself that the ideas are right, but that there are limitations to the analogy.

(a) Start with the planes in Figure 2.42. Use Equation 2.11 to compute an "equivalent" plane in FCC. Compare the surface structures in the two cases. If you have access to the program SURFSTRU (available from the author), use it to generate the structures. How are they the same? What are the differences?

(b) Notice that according to Equation 2.11, the (111) face in BCC is equivalent to the (120) face in FCC, while the (411) in BCC is equivalent to the (210) face in FCC. However, the (210) and (120) faces are identical in FCC. Therefore, another implication of Equation 2.11 is that the surface structures 4 (411) and (111) in BCC are similar. How similar are these two surfaces? What are the differences?

(c) What does the fact that two planes with two different Miller indices have quite similar structures tell you about the symmetry of the BCC lattice?

(d) In the diagrams in Sections 2.5.3 and 2.5.4 we provided ball models of the surface structures of several planes in the BCC and FCC lattice. However, atoms are not balls; the electron clouds on adjacent atoms overlap so there are no gaps between the adjacent surface atoms. If you were a molecule, how could you sense the difference between the equivalent faces of a BCC and an FCC material?

2.37 In Section 2.3.3 we noted that although the FCC and HCP lattices can be both constructed from hexagonal closed packed planes. In FCC all of the atoms are the same, while in HCP the A sites are not equivalent to the B sites.

(a) Consider an FCC lattice made by stacking planes, as shown in Figure 2.19. Assume that the FCC lattice has an ABC ABC ABC. . . stacking pattern. Verify that there is a threefold screw axis at the A, B, and C sites so that when one goes from the A plane to the B plane the atoms undergo a counterclockwise rotation of 120°; similarly, there is a 120° counterclockwise rotation in going from B to C and from C to A. Hence, the A, B, and C sites are equivalent to rotation in an FCC lattice.

(b) Now consider the HCP lattice. The HCP lattice has an ABABAB pattern. What is the rotation going from A to B? How about from B back to A? Be sure to consider the direction of the rotation. Are all atoms equivalent in an HCP lattice?

(c) The same result can also be obtained by examining the spaces between the atoms. Start with the FCC case. Compute the vectors going between the atoms in the conventional unit cell. Show that if you have a vector $\overline{V}$ from any atom in the unit cell to any other, then if one adds that vector to any of the atomic positions in the unit cell, one always ends up at another atomic position within the lattice, although not necessarily within the same unit cell.

(d) Repeat part c for an HCP lattice. Define $\overline{V}$ to be a vector that goes between an A site and a B site. Show that if one adds that vector to the B site, there is *no* corresponding point in the lattice. Hence, the A site is not equivalent to the B site.

2.38 We discussed ionic crystals in Section 2.6.4. The objective of this problem is to apply Tasker's analysis to see which faces of a variety of materials will be stable, and which will reconstruct or facet. Consider the following examples:

(1) MgO(100), (2) MgO(111), (3) MgO(110), (4) MgO(544), (5) InP(111), (6) InP(100), (7) InP(110), (8) InP(711), (9) InP(776), (10) InP(540)

(a) Work out the ideal surface structure of each of these materials. Note MgO has the same crystal structure as NaCl (i.e., an FCC lattice of oxygen atoms with magnesiums in the (100) plane between each oxygen), while InP has a zinc-blend lattice.

(b) Which of these ideal surface structures will be stable toward faceting?

(c) Which of these ideal surface structures will reconstruct to balance the surface charge? (Tasker's analysis would say that such faces should reconstruct.)

(d) Which of these ideal structures will reconstruct to allow the surface dangling bonds to pair up?

2.39 Explain in your own words why the surfaces of ionic crystals rumple. Why do they facet?

2.40 Pd(110) normally has a (1x1) reconstruction. However, when Christmann et al. saturated the surface with hydrogen, they found that at massive hydrogen doses, the surface assumed a (2x1) reconstruction. How is it possible for an adsorbate such as hydrogen to induce a surface reconstruction?

2.41 Avouris and Wolkow (*Phys. Rev. B*, **39**:5091, 1984) and S. Y. Tong, H. Huang, C. M. Wei, W. E. Packard, F. K. Men, G. Glander, M. B. Webb (*J. Vac. Sci. Technol. A*, **6**:615, 1988) discuss the different sites available on Si(111) (7x7). Look up these papers. (a) Identify each of the sites available in the diagram in Figure 2.46 or a printout from the program SI1117x7; (b) Use the results in Avouris and Wolkow's paper to describe how the chemistry of each of the sites is different.

More Advanced Problems

2.42   In Section 2.3.1 we said that unit cells 1, 2, 3, 4, and 5 in Figure 2.8 are all primitive unit cells without proof. The objective of this problem is to prove that they are primitive unit cells. By definition, the primitive unit cell is the smallest repeat unit of the lattice. Let's consider a lattice with one atom per primitive unit cell.

   (a) Count the atomic density in unit cells 1, 2, 3, 4, 5, and 6 in Figure 2.8, accounting for the fact that in some cases only part of the atom is in the unit cell. Prove that unit cells 1, 2, 3, 4, and 5 all have an atomic density of one atom per unit cell.

   (b) Calculate the ratio of the area of unit cells 1, 2, 3, 4, and 5 to the area of a primitive unit cell via

$$\frac{A_x}{A_p} = \frac{\rho_x}{\rho_p} \tag{2.61}$$

   where A_x and A_p are the areas of a given unit cell x and of a primitive unit cell p, and ρ_x and ρ_p are the corresponding atomic densities.

   Your results in part b show that the areas of unit cells 1, 2, 3, 4, and 5 are equal to the area of a primitive unit cell. Since, by definition, the primitive unit cell has the minimum possible area, unit cells 1, 2, 3, 4, and 5 all have the minimum possible area and therefore are primitive unit cells.

2.43   We cut single crystal faces by orientating the crystal, rotating along some axis, and cutting. If we want to make a stepped surface, we need to cut at some angle. Consider cutting a platinum crystal by orienting the crystal in the (100) direction and then rotating. Use the data in Nicholas, [1965] to calculate the density of C_6, C_7, C_8, C_9, C_{10}, and C_{11} sites as a function of the angle of rotation.

   If the first letter of your last name is A–L: Rotate around the [001] axis.

   If the first letter of your last name is M–Z: Rotate around the [01$\bar{1}$] axis.

   Do not consider rotations past the (111) or (110) planes. If you have access to the program SURFSTRU (available from the author), use it to generate the structures.

2.44   Equations 2.4 and 2.5 apply to the stereographic triangle for a BCC or FCC material, but not an HCP material. The objective of this problem is to derive an equation for the position of the planes in the stereographic projection of an HCP material.

   Consider a [001]-oriented tip of an HCP material. The (001) face of the HCP material is sixfold symmetry, but the [001] direction only has threefold symmetry because of the orientation of the planes under the (001) face, as indicated in Figure 2.19. We need a 60° slice of the image to get all of the important planes. Hence, the stereographic triangle contains a 60° slice of the FIM image rather than the 45° slice we needed for FCC and BCC. The objective of this problem is to calculate the position of a given (i j k) plane in the stereographic triangle of the HCP material.

   (a) Use an analysis similar to that in the subroutine `rotation_matrix_calc` to calculate ϕ_{image} for the
      (1) (100); (2) (101); (3) (001); (4) (211);
      (5) (221); (6) (342); (7) (111)
      planes of the HCP material.

   (b) Describe the structure of each of these faces. If you have access to the program SURFSTRU (available from the author) use it to generate the structures.

2.45   Repeat Problem 2.44 for a tip oriented perpendicularly to the (101) plane.

2.46 At the end of Section 2.4 we discussed the four-index notation for HCP lattices, and noted that the four-index notation has the advantage that the symmetry of the plane is reflected in the symmetry of the indices. However, the disadvantage of the four-index notation is that we can come up with indices for which no plane exists. Consider the (1) (1010); (2) (2010); (3) (1010); (4) (11$\bar{2}$0) planes.

(a) Which of these planes satisfies Equation 2.3?

(b) Show that the planes which do not satisfy Equation 2.3 are impossible, while the planes which satisfy Equation 2.3 exist. (*Hint:* The equation for a plane is ax + by + cz = d, where a, b, c are constants, d is zero or one, and x, y and z are the perpendicular coordinate system. Start with the (1010) case. The (1010) "plane" would intersect the X, Y, W, and Z axes at (1,0,0,0), (0,0,0,∞), (0,∞,0,0), and (0,0,1,0). Calculate an equation for the plane that passes through the first three points. Does the plane also pass through the fourth point?)

2.47 Crossley and King (*Surface Sci.*, **95**:131, 1980) examined the adsorption of CO on Pt(111) with LEED. They found that the CO mainly adsorbs in a c(4x2) overlayer, but the detailed structure was unclear from the LEED data.

(a) Propose five possible structures for the c(4x2) layer. Assume that the CO can adsorb onto linear, bridgebound, or triply coordinated sites. Linear sites are the sites directly on top of the surface atoms. Bridgebound sites are centered at the intersection of two adjacent surface atoms. Triply coordinated sites lie over three-fold hollows.

(b) Propose a series of experiments to distinguish between these possibilities using the experimental techniques described in Chapter 1. Assume that you can differentiate between linear, bridgebound, and triply coordinated CO with IR.

2.48 In Section 2.5 we only showed surface structures that repeat at regular intervals. However, we noted that it is easy to cut surfaces that show an irregular pattern of steps. The objective of this problem is to develop some criterion for when the surface structure will repeat at regular intervals. Start by considering cutting a two-dimensional crystal by beginning with a two-dimensional array of atoms, putting a line through the crystal, and eliminating atoms above the line as shown in Figure 2.26.

(a) Show that if the slope of the line is a rational number (i.e., the ratio of two integers i and j the surface structure will repeat, and the repeat length will be $\sqrt{i^2 + j^2}$ unit cells. (*Hint:* Show that if you move i unit cells in the y direction and j unit cells in the x direction, your are back to an equivalent point in the surface.)

(b) Show that if the slope is an irrational number, the surface structure does not repeat at a regular interval.

(c) Show that when we cut an (ijk) plane, the slope in the z = 0 plane is i/j, and the slope in the y = 0 plane is i/k.

(d) Use your results to show that if i, j, and k are integers, the surface structure will repeat at regular intervals. However, if you miscut the surface so that i, j, and k are not integers, you create some defects in the stacking pattern, so the surface structure no longer repeats in a regular way.

2.49 In Section 2.6 we noted that surfaces reconstruct because when we cut a surface we leave many dangling bonds. The objective of this problem is to consider the effects of adding an adsorbate to saturate the dangling bonds.

(a) Start with the Si(111) (7x7) surface. Consider adding gallium, arsenic, hydrogen, chlorine, and zenon to the surface. Which of these adsorbates would you expect to lift (i.e., remove) the reconstruction?

(b) Would a metal such as bismuth, which bonds weakly, lift the reconstruction?

(c) Would something that needs to bond to two atoms (e.g., oxygen) lift the reconstruction?

2.50 In Section 2.9 we noted that surface roughening can have an important influence on the rates of surface processes. How would you detect roughening in a kinetic experiment?

REFERENCES

Alder, B. J., J. R. Vaisnys, and G. Jura, *Phys. Chem. Solids* **11,** 182 (1959).

Beebe, J. L., *J. Phys. C,* **1,** 82 (1968).

Benson, G. C., and K. S. Yun, *J. Chem. Phys.* **42,** 3085 (1965).

Benson, G. C., P. I. Freeman, and E. Dempsey, *J. Chem. Phys.* **39,** 302 (1963).

Bertaut, F., *Comp. Rend.* **246,** 3447 (1958).

Biegelsen, D. K., R. D. Brigans, J. E. Northrup, and L. E. Schwartz, *Phys. Rev. B* **41,** 5706 (1990).

Binnig, G., and H. Rohrer, *IBM J. Res. Dev.* **30,** 355 (1986).

Bravais, A., *On the Systems Formed by Points Regularly Distributed on a Plane or in Space*, Paris (1850).

Burke, J. G., *Origins of the Science of Crystals*, University of California Press, Berkeley (1966).

Coburn, E. A., *Surf. Sci. Rep.* **15,** 281 (1992).

Conrad, E. H., *Prog. Surface Sci.* **39,** 65 (1992).

Cowley, J., *Prog. Surface Sci.* **21,** 209 (1986).

Davidson, C., and L. H. Germer, *Physical Rev.* **30,** 705 (1927).

Daw, M. S., *Surface Sci.* **166,** L161 (1986).

Daw M. S., and M. I. Baskes, *Phys. Rev. B.* **29,** 6443 (1989).

Dietrich, S., in C. Domb, ed., *Phase Transitions and Critical Phenomenona*, Academic Press, NY **12,** 1 (1988).

Duke, C. B., *Appl. Surface Sci.* **65,** 543 (1993).

Duke, C., and A. Bennet, *Phys. Rev.* **160,** 541 (1967).

Farnsworth, H. T., *Rev. Sci. Instrum.* **21,** 102 (1950).

Foiles, S. M., M. I. Baskes, and M. S. Daw, *Phys. Rev. B* **33,** 7983 (1986).

Frenken, J. W. M., R. J. Hamers, and J. E. Demuth, *J. Vac. Sci. Tech. A* **8,** 1293 (1990).

Garcia A., and J. E. Northrup, *Phys. Rev. B* **48,** 17350 (1993).

Gasser, R. P. H., *An Introduction to Chemisorption and Catalysis by Metals*, Oxford University Press, UK (1985).

Germer, L. H., *Adv. Catalysis,* **13,** 192 (1962).

Haüy, R., *Essai d'une de la Theorie Sur la Structures des Crystaux*, Paris (1784).

Haüy, R., *Traite de Minérologie*, Paris (1801).

Henrich, V. E., *Rep. Prog. Phys.* **48,** 1481 (1985).

Henrich, V. E., in J. Nowotny and L.-C. Dufour. eds., *Surface and Near Surface Chemistry of Oxide Materials*, Elsevier Material Science Monographs, vol. 47 (1988).

Hooke, R., *Micrographia*, London (1665).

Huygens, C., *Traite de la Lumiere*, London (1666).

Ignatieu, A., B. W. Lee, and M. A. Van Hove, *Proc. 7th Int. Vac. Congr., Vienna* **2,** 1733 (1977).

Inglesfield, J. E., *Prog. Surface Sci.* **20,** 105 (1985).

Jackson, D. P., in P. C. Gehlen, J. R. Beeler, and R. I. Jaffee, eds., *Interatomic Potentials and Simulation of Lattice Defects*, Plenum, NY (1912).

Jona, F., *IBM J. Res.* **14,** 445 (1970).

Jona, F., P. M. Marcus, in J. F. Van der Veen and M. A. Van Hove, eds. *The Structure of Surfaces II*, Springer-Verlag, NY (1988).

Kahn, A., *Surface Sci.* **300**, 469 (1994).

Kelly, A., and G. W. Groves, *Crystallography and Crystal Defects*, Addison Wesley, Menlo Park, CA (1970).

Kittel, C., *Introduction to Solid State Physics*, Wiley, NY (1986).

Kossel, W., *Nachr. Ges. Göttigen, Math. Phys.*, 135 (1927).

LaFemina J. P., *Surface Sci. Rep.* **16**, 133 (1992).

Lander, J. J., *Prog. Solid State Chem.* **2** (1965).

Lee, W. T., R. I. Masel, unpublished.

Lin, H. S., K. C. Low, and C. K. Owy, *Phys. Rev.* **48**, 1595 (1993).

Marks L. D., in W. Schommers, P. Von Blackenhagen, eds., *Structure and Dynamics of Surfaces, I*, Springer-Verlag, NY (1986).

Marks, L. D., and D. J. Smith, *Nature* **303**, 316 (1983).

Miller, W. H., *A. Treatise on Crystallography*, J. J. Deighton, Cambridge (1839).

Müller, E. W., *Z. Physik* **106**, 541 (1937).

Müller, E. W., *Z. Physik* **131**, 136 (1951).

Nicholas, J. F., *An Atlas of Models of Crystal Surfaces*, Gordon and Breach, NY (1965).

Park, R. L., and H. H. Madden, *Surface Sci.* **11**, (1968).

Pendry, J. B., *Low Energy Electron Diffraction*, Academic Press, NY (1974).

Prutton, M., and M. R. Weldon-Cook, *Surface Sci.* **88**, 9 (1979).

Saito, Y., and H. Müller-Krumbharr, in K. Binder ed., *Applications of the Monte Carlo Method in Statistical Physics 2d ed.*, p. 223, Springer-Verlag, NY (1986).

Satterfield, C., *Heterogeneous Catalysis in Industrial Practice*, McGraw-Hill, NY (1991).

Schoenflies, A., *Krystallsysteme und Krystallstructur*, Liepzig (1891).

Smith, D. J., *J. Vac. Sci. Tech.* **3**, 1563 (1985).

Smoluochowski, R., *Phys. Rev.* **60**, 661 (1941).

Stoneham, A. M., and P. W. Tasker, in J. Nowotny and L.-C. Dufour, eds., *Surface and Near Surface Chemistry of Oxide Materials*, Elsevier Material Science Monographs, vol. 47 (1988).

Stranski, I. N., *Physik. Chemie*, **136**, 259 (1928).

Tasker, P. W., *J. Phys. C.* **12**, 4977 (1979).

Trump, R. M., E. J. van Loenen, R. J. Hammers, and J. F. Demuth, in J. F. Van der Veen and M. A. Van Hove, eds., *The Structure of Surfaces II*, p. 282, Springer-Verlag, NY (1988).

van Beijeren, H., I. Norden, in W. Schommers and P. Von Blanchenhagen, eds., *Structure and Dynamics of Surfaces I*, p. 259, Springer Verlag, NY (1986).

Van Hove, M. A., and R. M. Echenique, *Surface Sci.* **82**, L298 (1979).

Van Hove, M. A., and R. M. Echenique, *Surface Sci.* **88**, 11 (1979).

Weakliem, P. C., and E. A. Carter, *J. Chem. Phys.* **96**, 3240 (1992).

Weeks, J. D., in T. Riste, ed., *Ordering in Strongly Fluctuating Condensed Matter Systems*, p. 293, Plenum, NY (1980).

Wood, E., *J. Appl. Phys.* **35**, 1306 (1964).

Yang, W. S., F. Jona, *Phys. Rev. B* **28**, 1178 (1983).

Zhang, X. G., M. A. Van Hove, G. A. Somorjai, P. J. Rous, D. Tobin, A. Gonis, J. M. Maclare, K. Heinz, M. Michl, H. Lindner, K. Muller, M. E. Hasi, and J. H. Block, *Phys. Rev. Lett.* **67**, 1298 (1991).

Zhener D. M., J. R. Noonan, H. L. Davis, C. W. White, *J. Vac. Sci. Tech.* **18**, 852 (1981).

3 Adsorption I: The Binding of Molecules to Surfaces

PRÉCIS

The adsorption of the reactants is usually the first step in a reaction on a surface. Adsorption, therefore, is a key step in the progress of the overall reaction. In Chapters 3, 4, and 5, we discuss adsorption in some detail. In this chapter we describe some of the principles that are used to understand how molecules bind to surfaces. We then describe some of the models that are used to quantify the bonding process. Part of the discussion is incomplete because the binding of molecules to surfaces is still an area being actively investigated. However, an attempt is made to indicate where the analysis is heading.

3.1 HISTORICAL OVERVIEW

Adsorption is a process where molecules from the gas phase or from solution bind in a condensed layer on a solid or liquid surface. The molecules that bind to the surface are called the **adsorbate** while the substance that holds the adsorbate is called the **absorbent.** The process when the molecules bind is called **adsorption.** Removal of the molecules from the surface is called **desorption.**

According to Freundlich [1909] adsorption was independently discovered by Scheele and Fontana in about 1777. Scheele [1773, 1777] found that when he heated charcoal, adsorbed gases desorbed. The gases then readsorbed when the charcoal was cooled. Priestley [1775] made similar measurements. However, he noted that when he adsorbed fresh air, the composition of the air changed upon adsorption/desorption. As a result he suggested that the process was much different than absorption into liquids. Ostwald [1875, 1885] and Morozzo [1783] indicated that Fontana [1776, 1782] also made similar measurements. However, Fontana's papers were published in books that were no longer available.

The nature of adsorption was controversial throughout the nineteenth century. In an early paper, Faraday [1834] discussed the possibility that gases are held onto surfaces by an electrical force and suggested that the gases could react more easily once they were in the adsorbed state. However, Berzelius [1836] noted that the best adsorbents were highly porous materials. Therefore, Berzelius proposed that adsorption was a process where surface tension or some other force caused gas to be condensed into the pores of a porous material. Recall that Laplace [1806] had shown that, due to surface tension, a small drop of fluid has a larger vapor pressure than a bulk fluid. If the drop has a radius R_{drop}, then the vapor pressure of the drop, P_{drop}, is given by

$$kT \ln \left(\frac{P_{drop}}{P_{vapor}} \right) = \frac{2\gamma_{drop} V_{drop}}{R_{drop}} \tag{3.1}$$

where P_{vapor} is the vapor pressure of the bulk fluid, γ_{drop} is the surface tension of the drop, V_{drop} is the molar volume of the drop, k is Boltzmann's constant, and T is temperature. Equation 3.1 is sometimes called the Kelvin equation.

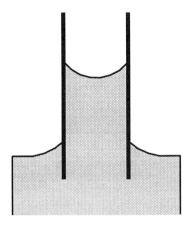

Figure 3.1 A diagram of fluid inside a small capillary.

Young [1855] showed that Equation 3.1 also applies to condensation of fluid in the pores of a capillary. When fluid condenses in a capillary, the radius of curvature of the fluid is usually negative, as indicated in Figure 3.1. As a result, the vapor pressure of a fluid in a small pore is less than the bulk vapor pressure of the fluid. This effect becomes especially important when the dimensions of the pore are less than about 100 Å. According to Equation 3.1 at a pore radius of 20 Å, the vapor pressure of water is a factor of 1.7 less than the vapor pressure of bulk water at 300 K. As a result, water vapor will have some tendency to condense into a small pore.

The idea that most adsorption processes were really just pore condensations was actively debated in the literature in the 1850s to 1920s. Magnus [1825, 1853] and Magnus [1929] showed that pore condensation does occur. However, other investigators found there was some data that were not in accord with the idea that pore condensation alone explained adsorption.

Chappuis [1881] found that if he took a charcoal sample and ground it up into small pieces to slightly increase the surface area of the charcoal, the charcoal would *ad*sorb more CO_2 even though the total amount of solid was constant. The difference was slight in Chappuis's data (Table 3.1) because his original charcoal already had a tremendous surface area so that the grinding process only had a small effect. However, Kayser [1881], expanded on Chappuis's experiment by starting with a glass rod and drawing the rod into a long fiber. Kayser then varied the diameter of the fiber and tried to see how the quantity of ammonia the fiber adsorbed varied with the mass and surface area of his fiber. He found that the surface area of the fiber was crucial. Kayser could not detect any adsorption on his original glass rod; less than 0.02 cc-stp of ammonia adsorbed. However, when Kayser drew the rod into a fiber 0.0261 cm in diameter and 22,325 m long, he found that 8.26 cc-stp of ammonia adsorbed

TABLE 3.1.   The Amount of Carbon Dioxide that Adsorbs on a Charcoal Sample as a Function of the Physical Form of the Charcoal

Adsorbent	Amount Adsorbed (cc/gm)
Solid charcoal	6.71
Pulverized charcoal	6.79

Source: Data from Chappuis [1881].

at a pressure of 538 torr (760 torr = 1 atm). Kayser repeated these measurements on several glass fibers of different lengths and diameters and found that the amount of adsorption per unit surface area was essentially constant on different diameter fibers, but that the amount of adsorption per unit mass varied significantly. The amount of adsorption also varied from one gas to the next. Thus, Kayser suggested that adsorption was a surface phenomenon.

Kayser also proposed that when one discusses sorption of gas by a solid, one should distinguish between **adsorption,** where a gas binds directly to the surface of a solid, and **absorption,** where the gas dissolves directly into the bulk of a fluid. Kayser [1881] noted that *ad*sorption is fundamentally different from *ab*sorption. In *ab*sorption, when the mass of absorbent is doubled, twice as much gas can be *ab*sorbed. In contrast, the total amount of gas that *ad*sorbs onto the surface of a solid is proportional to the total surface area of the solid and not the volume or mass of the solid. Breaking up a solid into small pieces has no effect on the amount of gas that can *ab*sorb into the bulk of the solid but it strongly affects the amount which *ad*sorbs. Hence *ad*sorption of a gas onto a solid is quite different from *ab*sorption of the gas into a solid or liquid in that in adsorption, the quantity of gas that adsorbs scales with the surface area rather than the volume of the adsorbent.

Sabatier [1906] lent support to the hypothesis that adsorption was a surface phenomenon. Sabatier examined the catalytic activity of a number of metal surfaces and showed that he could explain his data by assuming that the surface formed a direct bond with the adsorbate.

Langmuir [1912, 1918] expanded on Sabatier's findings. Langmuir showed that hydrogen and oxygen can adsorb onto clean platinum and tungsten wires. Neither of Langmuir's wires had any detectable pores, and the rate of adsorption was such that there was insufficient time to transfer a significant amount of hydrogen into the bulk of the metal. Thus, it was clear that hydrogen and oxygen can directly bind to clean metal surfaces.

Soon thereafter, Bancroft [1917, 1918] tested Langmuir's models on most of the adsorption data that had been measured prior to that time. Bancroft noted that Henry [1824], Davey [1817], and Döbereiner [1824, 1825, 1834] had found that on platinum, much more hydrogen adsorbed than methane or carbon dioxide. However, on charcoal much more carbon dioxide and methane adsorbs than hydrogen. Bancroft then noted that such a finding could not be explained by the pore condensation model. Note, that according to Equation 3.1, the vapor pressure of a fluid over a small pore is a function of the diameter of the pore and the properties of the fluid, but not the composition of the pore, provided that the contact angle is zero. Thus, if one has two species, A and B, the species with the higher surface tension will have a greater tendency to undergo a pore condensation than the species with the lower surface tension, independent of the type of adsorbent. However, the data on hydrogen, carbon dioxide, and methane adsorption does not show such behavior. Rather, hydrogen has the highest affinity for platinum and the lowest affinity for charcoal. Bancroft examined a number of examples and showed that in general the nature of the adsorbent was a key factor in determining the amount of gas that adsorbs. Bancroft [1918] also noted that Faraday [1834], among others, had shown that trace impurities could inhibit adsorption. As a result, Bancroft [1918] concluded that when gases adsorb on solid surfaces, the gases form a physical or chemical bond to the adsorbent. On the other hand, liquids often show capillary condensations.

Most modern investigators have agreed with Langmuir's and Bancroft's findings. Generally, adsorption involves a mixture of direct adsorbate surface binding and pore condensation. Pore condensation is most important for liquids near or below their boiling point, while direct adsorbate/surface binding dominates at higher temperatures. Generally, gases adsorb on surfaces as illustrated in Figure 3.2 where there is a layer of gas in direct contact with the solid, and then possibly a number of layers of gas condensed onto the first layer. One refers to each layer of a gas as a **monolayer.** The first monolayer (i.e., the atoms that are in direct contact with the solid) usually form a physical or chemical bond to the surface. The next layer still has a weak interaction with the surface, and a stronger interaction with the first layer. The third layer interacts strongly with the second layer, more weakly with the first layer, and usually only slightly with the solid.

 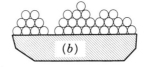

Figure 3.2 An illustration of (a) monolayer adsorption, and (b) multilayer adsorption.

People usually distinguish between two fundamentally different types of adsorption: **multilayer adsorption,** where there are several layers of adsorbate on the surface of the adsorbent, and **monolayer adsorption** where all the adsorbate is held in close proximity to the solid. Both types of adsorption are illustrated in Figure 3.2. Multilayer adsorption is basically a condensation process; attractive interactions between adsorbate molecules cause gases to condense into a liquidlike film on top of the molecules in the first monolayer. Generally, one only gets multilayer adsorption when the adsorbate/adsorbate interactions are large compared to kT. Typically, multilayer adsorption occurs when one is working at temperatures that are close to the boiling point of a fluid at the pressure of interest. Pore condensation effects and a process called "layering" also play an important role in multilayer adsorption. These effects are discussed in Section 4.12. Monolayer adsorption, on the other hand, is usually dominated by the direct interactions between the adsorbate and the surface. Monolayer adsorption can be achieved at temperatures hundreds of degrees above the boiling point of the adsorbate. For example, at a pressure of 10^{-7} torr (760 torr = 1 atm) CO adsorbs onto a supported platinum catalyst at 400 K. However, multilayers occur on the surface only after cooling to 80 K at a pressure of about 1 atm.

Adsorbate layers typically have densities between those of a gas and those of a liquid. Typically, at a monolayer coverage, the density is close to that in a liquid, which means that adsorbate molecules are in close proximity to one another. At those densities, it does not take very many molecules to form a monolayer. Consider adsorption of water in a layer 3 Å thick on a surface. The coverage of the water in molecules/cm^2, N_a, is given by

$$N_a = \left(1 \frac{\text{g}}{\text{cm}^3}\right) (3 \times 10^{-8} \text{ cm}) \left(\frac{6 \times 10^{23} \text{ molecules}}{10 \text{ g}}\right) = 1 \times 10^{15} \text{ molecules/cm}^2$$

Therefore a monolayer of water would contain about 1×10^{15} molecules/cm^2. This is a typical result. In monolayer adsorption, adsorbate coverages never exceed $1-2 \times 10^{15}$ molecules/cm^2, so very few molecules are needed to form a monolayer. Note that at a coverage of 1×10^{15} molecules/cm^2 one needs 6×10^8 cm^2 of surface to hold a mole of gas. An area of 6×10^8 cm^2 corresponds to about 12 football fields. Hence, the solid needs to have a considerable surface area before the solid can adsorb a significant quantity of gas.

Fortunately, there are materials with tremendous surface areas. Generally, these are highly porous materials; there is a tremendous surface area within the pores. Charcoal and activated carbon can have surface areas as high as 1000 M^2/g, i.e., one gram of material will have a surface area of 1000 M^2. A Linde molecular sieve can have a surface area as high as 600 M^2/g, while typical catalyst supports have surface areas on the order of 100 M^2/g. An area of 1000 M^2/g corresponds to compressing the surface area of a football field into a few cubic centimeters of material. Figure 3.3 shows how little of these adsorbents one needs in order to get a surface area similar to that of a football field. These adsorbents make it possible to adsorb a reasonable quantity of gas into a limited amount of adsorbent.

The nature of bonding between adsorbates and surfaces is still subject to some interpretation. Taylor [1931] suggested that it was important to distinguish between two radically different types of adsorption: **Chemisorption,** where there is a direct chemical bond between the adsorbate and the surface, and **physisorption,** where there is no direct bond. Instead the

Figure 3.3 The amount of Linde molecular sieve, activated carbon, and γ-alumina catalyst support needed to have 5000 M^2 of surface, i.e., about the area of a football field.

adsorbate is held by physical (i.e., van der Waals) forces. On a more fundamental level, when a molecule is chemisorbed, the electrons are shared between the adsorbate and the surface. As a result, the adsorbate's electronic structure is significantly perturbed. The surface's electronic structure is perturbed to a lesser extent. In contrast, physisorption is governed by polarization (i.e., van der Waals) forces. The surface does not share electrons with the adsorbate. As a result, the electronic structure of the adsorbate is perturbed to a much lesser extent. Therefore, a more direct test of whether a molecule is physisorbed or chemisorbed is:

> A molecule is **physisorbed** if it adsorbs without undergoing a *significant* change in electronic structure.

> A molecule is **chemisorbed** if the molecule's electronic structure is *significantly* perturbed upon adsorption.

Typical chemisorption energies are 15–100 kcal/mole for simple molecules. That compares to 2–10 kcal/mole for physisorption.

Note, however, that even though we often try to distinguish between physisorption and chemisorption, it is not possible to discriminate between them in every case. Molecules are often distorted when they adsorb on solid surfaces, as illustrated in Figure 3.4. Sometimes the molecules stay intact during the adsorption process, while at other times bonds break. Often, bond lengths change significantly. For example, Figure 3.4 shows that hydrogen dissociates upon adsorption on Pt(111). In contrast, while benzene is significantly distorted upon adsorption, no bond scission occurs. Ethylene is distorted upon adsorption at 100 K, then bonds break at 300 K. Taylor [1931] proposed that molecules are chemisorbed when bond scission occurs, and physisorbed otherwise. However, a more modern interpretation is that a molecule is chemisorbed when the electronic structure of the molecule on the surface is significantly different than in the gas phase. For example, Steiniger, Ibach, and Lehwald [1982] found that when ethylene adsorbs onto a 100 K Pt(111) sample, no bonds break. However, the carbon–carbon bond in the ethylene stretches by 0.15 Å, and the hybridization of the carbon atoms changes from sp^2 to sp^3. Hence, most modern workers would say that ethylene chemisorbs on Pt(111); CO chemisorbs on Pt(111) too. Crossley and King [1980] showed that the geometry of the adsorbed CO is similar to that in the gas phase. However, the vibrational frequency of the CO bond shifts by 200 cm^{-1} upon adsorption. Hence, CO is said to be chemisorbed on Pt(111).

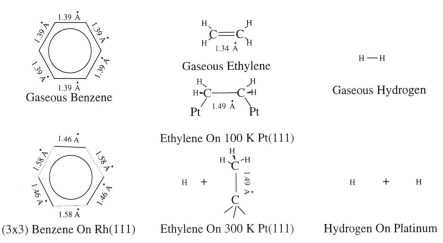

Figure 3.4   A comparison of the structure of various molecules in the gas phase and on a solid surface. (Geometric data from Lin et al. [1987] and Farkis [1935].)

There are a series of other examples where the results are not as clear. For example, Fisher [1991] found that ammonia adsorbs molecularly on Pt(111) with the lone pair on the nitrogen pointing down toward the surface. There are small changes in the ultraviolet photoemission spectrum (UPS) of the adsorbed ammonia, suggesting that the electronic structure of the ammonia changed during adsorption. Hence, one would initially conclude that ammonia is chemisorbed. However, Scheffler et al. [1979] found that the UPS spectrum of xenon changes slightly when xenon adsorbs on Pd(110), implying that the electronic structure of the xenon changed upon adsorption. Chen et al. [1984] expanded these findings to many other materials. Engel and Gomer [1970] and Ehrlich and Hudda [1959] found interesting changes in work function when xenon adsorbs on a tungsten field emitter. It is hard to imagine any sort of chemical bonds between a xenon atom and a surface. As a result, xenon is said to physisorb even though xenon's electronic structure changes slightly during the adsorption process. The changes in the electronic structure of the xenon are attributed to polarization of the xenon as described in Section 3.3.

In a more general way the question of whether something is physisorbed or chemisorbed can be a matter of degree. An adsorbing gas is usually assumed to be chemisorbed when the adsorbate's bond energy to the surface is more than 10 kcal/mole, and physisorbed if its bond energy is 10 kcal/mole or less. However, the precise distinction between physisorption and chemisorption is somewhat arbitrary.

Another subtlety is that a given molecule can usually physisorb and chemisorb on the same surface. Often, a molecule first physisorbs and then is converted into a chemisorbed state. At equilibrium, there will be a mixture of physisorbed and chemisorbed molecules on the surface. As a result, one often discusses adsorption systems as though all of the adsorbed molecules are chemisorbed. However, in reality, there are always some physisorbed molecules on the surface, and they can contribute to the overall surface reaction process.

3.2   CHEMISORPTION OF GASES ON SURFACES

From the 1930s to the 1950s there were many studies of the chemisorption process. In a summary of the key findings from this work, Trapnell and Hayward [1971] show that one can group the elements in the periodic table according to the reactivity of each of the elements. For example the surfaces of calcium, strontium, and barium show similar reactivity. However,

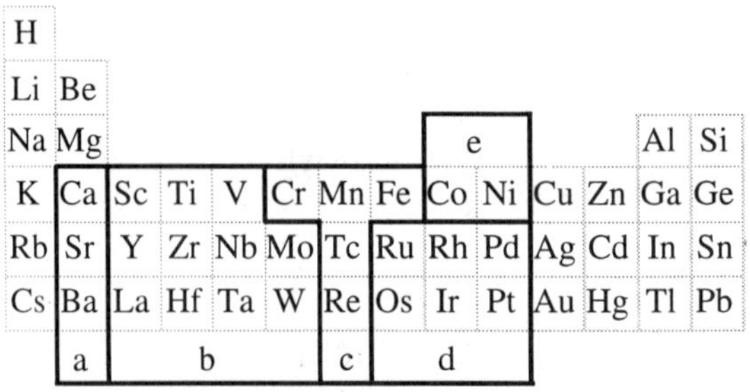

```
H
Li Be
Na Mg                                    e           Al Si
K  Ca Sc Ti V  Cr Mn Fe Co Ni Cu Zn Ga Ge
Rb Sr Y  Zr Nb Mo Tc Ru Rh Pd Ag Cd In Sn
Cs Ba La Hf Ta W  Re Os Ir Pt Au Hg Tl Pb
   a      b       c     d
```

Figure 3.5 Classification of metals and semiconductors according to the chemical reactivity of their surfaces. (After Trapnell and Hayward [1971].)

the surfaces of scandium, ytridium, lanthanium, and titanium show different reactivity than calcium, strontium, and barium. Figure 3.5 summarizes Trapnell and Hayward's groupings. Basically groups of adjacent elements show similar reactivity for simple molecules. Elements outside of the group show different reactivity.

Table 3.2 summarizes the reactivity of a number of gases with a variety of surfaces in ultrahigh vacuum (UHV). The table includes all of the metals considered by Trapnell and Hayward, miscellaneous other metals, a group of semiconductors (Si, Ge, InP) and semi-

TABLE 3.2. An Updated Version of Trapnell and Hayward's [1971] Table of the Reactivity of a Series of Gases with the Metals in Figure 3.5 and Miscellaneous Other Substances

Metal	H_2	O_2	N_2	CO	C_2H_2 C_2H_4	CH_4 C_2H_6	CH_3OH	H_2O
Group a	2 or 3	3	2	3	3	?	?	3
Group b	3	3	3	3	3	2	3	?
Group c	3	3	2	3	3	?	3	1
Group d	3	3	2	3	3	2	3	1
Group e	3	3	2	3	3	2	3	1
Cu	2	3	2	1	1 or 3	0	1	1
Ag	0	2 or 3	0	0	1	0	1	1
Au	0	0	0	3	3	0	1	1
Al	0	def-3	0	3	2	?	3	3
K, Na, Li	0	3	0	0	3 C_2H_2 0 C_2H_4	?	?	3
Si, Ge	0 or 2	3	0	0	3	2	3	3
InP	0	3	0	0	?	?	?	?
NiO	0	1			3 C_2H_2		3	3
ZnO	def-3	def-3	?	1	? C_2H_4	?		
MgO						0 CH_4		
Al_2O_3	0	1	1	1	?	1 CH_4 50 K	3	3
SiO_2	def-2	def-3				1 C_2H_6 def-2		
NaCl	0	?	1	0–100 K	1	0–100 K	0	3
LiF				1–40 K		1–45 K		

Note: 3—A rapid uptake of gas at 300 K and 10^{-4} torr pressure; 2—a slower possibly activated uptake of gas at 3–500 K and 10^{-4} torr pressure; 1—detectable uptake at 100 K in UHV but not at 300 K; 0—no uptake at 100 or 300 K in UHV; ?—no data in UHV; *def*—results on surfaces with defects such as metal or oxygen vacancies.

conducting oxides (NiO, ZnO), and then a group of insulators (NaCl, LiF) and insulating oxides (Al_2O_3, SiO_2, MgO). The table indicates that the ability of a material to adsorb gas varies with whether the material is a metal, semiconductor, or insulator, and the position of the material within the periodic table. Some gases adsorb strongly on transition metals, while other gases adsorb strongly on insulators or semiconductors. Generally, one finds that metals in the middle of the periodic table are most reactive for adsorption of some gases while metals at the edges of the portion of the periodic table in Figure 3.5 are most reactive for adsorption of other gases. One interesting feature is that very reactive metals such as potassium are not good adsorbents for some substances even though less reactive metals such as platinum are good adsorbents for those substances. As a result, the trends in Table 3.2 are different than one might expect.

The trends in Table 3.2 are important enough that someone working in surface chemistry should know the results in the table in detail. However, the reader could skip to the next section without loss of continuity.

Let's start with oxygen. The table indicates that oxygen reacts with all of the metals in Table 3.2 except gold implying that oxygen is very reactive. In contrast, hydrogen reacts quite easily with all of the metals in groups a through e in Figure 3.5, but it only reacts weakly with copper, silver, gold, silicon, and germanium. More recent data show that hydrogen atoms will react with all of these substances. However, the adsorption of H_2 has a large activation barrier on germanium and silicon so that the adsorption rate at 300–500 K is negligible. Hydrogen adsorption on copper, silver, and gold is different in that the adsorption process is activated, and the heat of adsorption of hydrogen on these three metals is small. As a result, higher temperatures are needed to get enough energy to cross the activation barrier. At those temperatures one needs pressures in the order of several atmospheres or more to get a significant amount of hydrogen to adsorb.

Carbon monoxide adsorbs readily on all of the surfaces in groups a through e. CO can adsorb on copper at 77 K, but it desorbs again upon heating to 300 K. CO basically does not adsorb on silver, gold, silicon, and germanium. Similarly, alkenes and alkynes react readily with all of the metals in groups a to e, and with silicon and germanium. The adsorption is reversible on copper, silver, and gold, and it happens that the extent of sorption varies with conditions. Hayward claims that ethylene and acetylene do adsorb at 300 K and 10^{-4} torr on copper and gold, but not silver. However, more recent data show that while both molecules will adsorb on copper, silver, and gold at 280 K, they desorb again at slightly higher temperatures. As a result, it is not obvious whether to say that ethylene and acetylene bind on copper, silver, and gold at 300 K.

The behavior of methane and ethane is more complex. Methane and ethane do not adsorb on any metal surface at 300 K. However, Brass and Ehrlich [1987], Ceyer [1988], and McMaster and Madix [1993] found that one can get them to adsorb if one heats the methane and ethane to high temperature or accelerates the methane and ethane with a molecular beam. Interestingly, Weinberg [1992] found that higher paraffins will adsorb on iridium and platinum at 300 K, but the process is activated so less adsorption is seen at 100 K.

Nitrogen and water show different trends than all of the rest of the gases. Water only physisorbs on most clean transition metal surfaces. Yet it chemisorbs on most metal oxides, group Ia and IIa metals, and on silicon and germanium. Hence the adsorption of water seems to follow a different trend than all of the other gases in Table 3.2.

Nitrogen also shows some different trends. Nitrogen readily adsorbs on polycrystalline samples of group b metals. The adsorption process is activated on group a and c metals. However, interestingly the sticking probability varies significantly with surface structure. Ehrlich and Hudda [1959] exposed a tungsten field emitter tip to nitrogen, and found that they could image nitrogen atoms on the parts of the tip corresponding to the planes indicated as light areas in Figure 3.6. However, even after repeated exposure to nitrogen, no nitrogen atoms could be detected on the parts of the tip corresponding to the planes indicated by shaded regions in Figure 3.6. Therefore, Ehrlich and Hudda asserted that the rate of nitrogen ad-

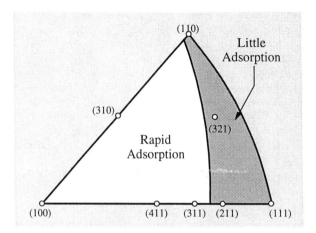

Figure 3.6 The rate of adsorption of nitrogen on tungsten as a function of the position of the plane within the stereographic triangle. (Data of Ehrlich and Hudda [1963], Delchar and Ehrlich [1965], and Adams and Germer [1971].)

sorption was greater on the parts of the tip corresponding to the light areas in Figure 3.6 than on the parts of the tip corresponding to the shaded regions in Figure 3.6.

Delchar and Ehrlich [1965] and Adams and Germer [1971] verified this conclusion on extended single crystals. Little nitrogen adsorption was seen on W(110), W(211), W(111), and W(321), while nitrogen readily adsorbs on W(100), W(310), W(411), W(210), and W(311) single crystals. As a result, Adams and Germer concluded that the rate of nitrogen adsorption was highly dependent on surface structure, as indicated in Figure 3.6.

More recent data show that nitrogen atoms can adsorb on all of the faces of most transition metals. However, there is a significant activation barrier to adsorption on many of the faces. For example, Cosser et al. [1980] found that the activation energy for N_2 adsorption is 4.2 kcal/mole on W(110). The adsorption rate is negligible at 300 K in UHV. However, rapid adsorption is seen if the N_2 is dissociated into nitrogen atoms prior to adsorption, or if the N_2 is accelerated so that it has enough energy to surmount the activation barrier of adsorption. This example illustrates that although the rate of many adsorption processes varies strongly with surface structure, often the surface structural variation is associated with a change in the barrier to adsorption rather than a change in the bonding of the adsorbate to the surface.

3.2.1 The Nature of the Adsorbed Layer

The nature of the adsorbed layer is not obvious from the previous discussion. Faraday [1834] asserted that molecules would interact with surfaces, but the molecules would not necessarily change. However, Langmuir [1912] and Freeman [1913] found that when H_2 adsorbs onto a white-hot platinum or tungsten filament, hydrogen atoms desorb from the surface at high temperatures. Langmuir proposed that the hydrogen dissociated even at room temperature. It took several years to verify this hypothesis. However, in 1929 Bonhoeffer and Harteck [1929] showed that ortho-hydrogen (i.e., H_2 with the nuclear spins pointed in the same direction on each hydrogen atom) is converted to para-hydrogen upon adsorption/desorption. These results suggested that the H–H bond breaks when hydrogen adsorbs. Later, after deuterium was discovered, Farkis and Farkis [1933, 1934] and Farkis [1935] showed that there was complete isotopic scrambling in H_2/D_2 mixtures when hydrogen adsorbed on platinum. Ortho–para conversion and isotopic scrambling are thought to occur by a bond scission process. Farkis showed that both occur instantaneously upon adsorption on platinum wire. As a result Farkis

[1935] concluded that hydrogen adsorbs dissociatively on platinum metals. Similar conclusions were also obtained for all other platinum group metals.

Langmuir found it was convenient to distinguish between two different types of adsorption: **nondissociative adsorption** when the adsorbing molecules stay intact during adsorption, and **dissociative adsorption** processes, where bonds break when molecules adsorb. Nondissociative adsorption processes are sometimes called **molecular** adsorption processes.

At first people did isotope exchange experiments to try to understand when gases will adsorb molecularly or dissociatively. However, in the late 1950s, developments in surface spectroscopy allowed people to examine the state of the adsorbate directly. One of the things that was discovered is that a given gas can adsorb molecularly on one metal and dissociatively on another metal. Brodén et al. [1976] examined the data for the dissociation of a series of molecules on transition metals surfaces and found that they could identify a line in the periodic table for each molecule such that the molecule would adsorb dissociatively on all of the metals to the left of the line and nondissociatively on all of the molecules to the right of the line. For example, Figure 3.7 shows data for CO, NO, N_2, and O_2 adsorption on a number of transition metals. At 300 K CO adsorbs dissociatively on molybdenum and tungsten and nondissociatively on ruthenium and nickel. Note, however, that the position of the boundaries in Figure 3.7 are highly dependent on conditions. For example, in UHV, one does not usually observe significant dissociation of CO except on tungsten and molybdenum. However, Goodman [1984] examined CO dissociation on Ni(111) and Ni(100). They found that the CO dissociates at 673 K and 1 atm pressure even though little CO dissociation is seen on Ni(111) or Ni(100) in UHV. Other results show that CO can dissociate on all of the transition metals

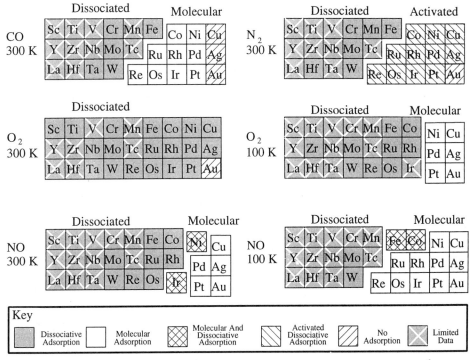

Figure 3.7 Part of the periodic table showing which metals dissociate various gases at 10^{-6} torr and 100 or 300 K, and which do not. (This is an updated version of a figure presented by Brodén et al. [1976].)

except silver and gold. However, the dissociation process has a large activation barrier on some metals. In such a case, a molecule desorbs before it dissociates in UHV. As a result only molecular adsorption is seen in UHV even though the CO can dissociate under other conditions.

This is a general result. If one goes to high enough temperatures and pressures, most molecules will dissociate on most transition metal surfaces. However, some molecules dissociate easily, while others are more difficult to dissociate. One observes dissociative adsorption only when the temperatures and pressures are high enough that the adsorbate can cross the activation barrier to dissociation before desorption occurs.

Another difficulty is that the rate of dissociation varies with surface structure. For example, at 300 K, NO adsorbs molecularly on most faces of platinum. However, Banholzer et al. [1983] found that NO adsorbs dissociatively at 300 K on Pt(410). Sugai et al. [1993] found that NO also dissociates on Pt(310), and there is some dissociation on defect sites on Pt(100). Thus, while there is a general correlation between the ability of a metal to break bonds and the position of the metal within the periodic table, the correlation does not work in detail because certain faces of metals are better able to break bonds than other faces of the same metal.

Another complication is that there is very little data on the adsorption of gases on the metals on the left of the VIA metals (i.e., to the left of tungsten, molybdenum, and chromium) in the periodic table. In the literature, it has been assumed that most species will dissociate on the metals on the left side of the periodic table. However, Trapnell and Hayward [1971] report that H_2, N_2, and CO will not dissociate on an evaporated potassium or cesium film. Many investigators have shown that H_2 does not adsorb on beryllium, sodium, magnesium, potassium, and calcium. As a result, it seems that the metals on the left side of the periodic table are less reactive than the metals in the center of the table. The reduced reactivity needs further documentation, however.

Lennard-Jones [1932] proposed a simple model to try to put the distinction between molecular and dissociative adsorption on a more fundamental basis. Consider the interaction of a B_2 molecule with a flat surface. One can imagine that as the molecule approaches the surface, the molecule interacts via a Lennard-Jones potential such as the one indicated by the line that is labeled molecular state in Figure 3.8. However, following Lennard-Jones it is assumed that the interaction energy is only a function of Z, the distance of the molecule from the surface, and not on the molecule's other two coordinates X and Y. This is, of course, an approximation, but it is a useful approximation. Now consider the potential for the interaction of two separate B atoms with the same surface. Note that in the gas phase, it takes some energy, $E_D^{B_2}$ to dissociate the B_2 molecule. As a result, at long distances from the surface the energy of two separate B atoms will be $E_D^{B_2}$ higher than the energy of a B_2 molecule. Now when the two B atoms interact with the surface, they can form bonds. As a result, the energy of the two separate B atoms will be lowered. Lennard-Jones assumed that the energy of the two separate B atoms would follow the one-dimensional potential labeled "dissociated state" in Figure 3.8. Again it is assumed that the interaction of the atoms with the surface is a function of the distance of the atoms from the surface, but not a function of the other two coordinates X and Y.

Notice that the energy of the molecular state and the energy of the dissociated state cross at some point above the surface in Figure 3.8. Lennard-Jones noted that from quantum mechanics if there are two states of the system, a molecular state and a dissociated state, the two states will interact to avoid the curve crossing. The ground state of the system will follow the dark line in Figure 3.8, where the molecule starts out in the molecular state, but as the molecule comes close to the surface the molecule can cross an activation barrier and enter the dissociated state. Therefore, Lennard-Jones asserted that one of the main distinctions between molecular and dissociative adsorption is associated with the height of the barrier between the molecular and dissociative states.

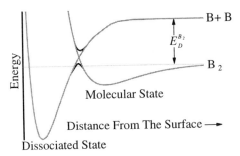

Figure 3.8 A plot of a one-dimensional potential surface for the adsorption of a B_2 molecule on a surface. (After Lennard-Jones [1932].)

Lennard-Jones extended these ideas by considering the three cases shown in Figure 3.9. The first case is one where the molecular state is more strongly bound (i.e., has a lower energy) than the dissociated state. In that case molecular adsorption is the preferred bonding configuration. However, one can still get to a dissociative state by dissociating the B_2 before it adsorbs or by starting with hot molecules. Nevertheless, the dissociated state would be an unstable configuration, so under most circumstances one should see mainly molecular adsorption.

The second case in Figure 3.9 is one where the dissociative state has a lower energy than the nondissociated state, but there is a finite activation barrier to get into the dissociative state. An incoming molecule could dissociate in this case, but if the activation barrier is high, one would still see mainly molecular adsorption.

The third case in Figure 3.9 is more complex. In this case there is a molecular well and a dissociated well and an activation barrier between them. However, the barrier lies below the energy of the B_2 molecule. As a result, an incoming B_2 molecule will always have enough energy to surmount the barrier and dissociate. Dissociative adsorption should be the norm in this situation. Note, however, that it is possible for the incoming molecule to lose some of its energy before the molecule dissociates. In that case, the molecule can get caught in the molecular well and have to surmount a barrier to dissociate. In such a case, the molecular

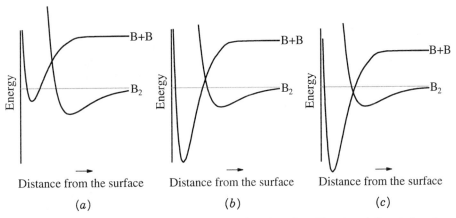

Figure 3.9 Lennard-Jones's model of (*a*) pure molecular adsorption, (*b*) activated dissociative adsorption, (*c*) unactivated dissociative adsorption.

state is called a **molecular precursor.** Dissociative adsorption through a molecular precursor is quite common on surfaces.

Lennard-Jones [1932] considered these cases and some others in detail. The main conclusions from Lennard-Jones's analysis is that there are two key issues to consider when trying to understand whether a molecule adsorbs molecularly or dissociatively. The first issue is whether the molecular state or the dissociative state has a lower energy. One usually only sees dissociative adsorption when the dissociated state has a lower energy. However, a second issue is where there is a barrier to dissociation. Usually some sort of barrier exists. If the barrier lies below the zero-point energy of the incoming molecules, then virtually all of the incoming molecules will surmount the barrier. However, if the barrier lies above the zero point energy, only a fraction of the molecules can cross the barrier and dissociate. The presence of a barrier can lead to molecular adsorption even though dissociative adsorption is thermodynamically favored.

Lennard-Jones's model can explain, at least in a qualitative way, all of the data in Figure 3.7. The depth of the atomic well in Figure 3.8 increases as one moves to the left of the periodic table. The increase in the atomic well lowers the activation barrier for dissociation, which makes it easier for a molecule to dissociate on a metal on the left side of the periodic table. However, Lennard-Jones's model fails to predict the structure sensitivity of the adsorption process seen in Figure 3.6. It also does not explain the absence of H_2 dissociation on potassium and does not adequately predict the behavior of complex molecules such as hydrocarbons.

3.2.2 More Complex Behavior

Hydrocarbons show behavior that is much more complex than one would expect from Lennard-Jones's simple models. It is usually possible to form several different kinds of molecularly adsorbed species and many different dissociatively adsorbed species. For example, Cassuto et al. [1991] found that when ethylene adsorbs onto a 40 K Pt(111) sample the ethylene goes into a weakly π-bonded state where the structure of the ethylene is not significantly perturbed from the gas phase. However, Steiniger et al. [1982] showed that on a 90 K Pt(111) surface, the ethylene forms a di-σ complex (Pt–CH_2–CH_2–Pt) with sp^3 hybridization, and an ethylene–ethylene bond that has been elongated from 1.34 Å to 1.49 Å. Backman and Masel [1990] also detected a π-bound form of ethylene on Pt(210) with a carbon–carbon length of 1.42 Å, while Yagasaki and Masel [1991] found that all three molecular forms of ethylene form during ethylene adsorption on (2x1)Pt(110). This example illustrates that there can be three molecular forms of ethylene on the same surface, so just saying that the ethylene is molecularly adsorbed does not define the state of the ethylene.

In a similar way, there are multiple forms of dissociated ethylene on platinum. For example, Kesmodel, Dubois and Somorjai [1979] showed that on Pt(111) the ethylene dissociates upon adsorption at 300 K to produce an adsorbed hydrogen atom and a $CHCH_2$ complex. The $CHCH_2$ complex is quickly converted to a species called ethylidyne (Pt$_3\equiv CHH_3$) (see Figure 3.10). The ethylene undergoes further dehydrogenation to CCH_2, CCH species at higher temperatures. At temperatures above 600 K, only hydrogen and adsorbed carbon are seen. This example shows that one can also form multiple adsorbed forms of dissociated ethylene on platinum. This is a general result. When a molecule adsorbed on a surface, the molecule often dissociates to yield a variety of different species.

One can imagine modeling the adsorption of ethylene via a Lennard-Jones-type model. The ethylene first adsorbs in a molecular state that Lennard-Jones would call a molecular precursor state. The ethylene then has to go over a small barrier to molecularly chemisorb, and a second barrier to dissociatively adsorb to form ethylidyne. The molecule would then go over a series of other barriers before it reaches its ground state, i.e., adsorbed carbon atoms and hydrogen. Note, however, that this case is fundamentally different from the case in Figure 3.9 in that there are local bonds between the adsorbate and the surface. The

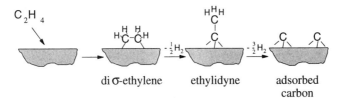

Figure 3.10 The mechanism of ethylene decomposition on Pt(111). (Proposed by Kesmodel et al. [1979] and confirmed by Ibach and Lehwald [1979].)

orientation of the incoming molecule relative to the surface atoms matters. As a result, the potential energy surface for the reaction cannot be represented via a one-dimensional barrier like that in Figure 3.9.

One of the things one finds is that although the final state of adsorbed ethylene is the same on most faces of transition metals (i.e., adsorbed carbon and hydrogen), the pathway that the reaction follows can depend on the surface structure. For example, the decomposition of ethylene follows the mechanism in Figure 3.10 on all of the hexagonal faces of group VIII metals examined so far. However, interestingly, entirely different mechanisms are seen on other surface structures. For example, Hatzikos and Masel [1987] found that on the square (1x1) reconstruction of Pt(100), the reaction followed the mechanism shown in Figure 3.11 where the ethylene decomposed to form di-σ vinylidene at 300 K. These findings are discussed in greater detail in Chapters 6 and 10. The thing to remember for now, however, is that hydrogens can undergo complex rearrangements when they adsorb on transition metals, and the nature of the surface intermediates can vary with surface structure.

The results for ethylene are typical of a wide class of rearrangements that occur when species adsorb on metal surfaces. Often, a given species can bind in a variety of different configurations, and the preferred configuration varies with both metal and face. The species then rearranges to form a variety of different intermediates, many of which bear little relationship to the original species.

It happens that simple one-dimensional models like those in Figure 3.9 cannot represent these interactions. Figure 3.9 assumes that the interaction energy can be written as a function of a single variable, the distance of the atom from the surface. However, in the case of hydrocarbons, there are several variables such as bond lengths and bonding positions. The observation that the decomposition varies with surface structure suggests that the details of the coordination of the molecule with the surface affects what intermediates form. One cannot represent the interaction with the simple one-dimensional Lennard-Jones-type models because these models ignore how the details of the adsorbate surface bonding affect the reaction pathway. However, one can model the system by examining how the reaction pathway is modified as the geometry of the surface atoms in the vicinity of the adsorbate changes. These ideas are discussed in Chapters 6 and 10.

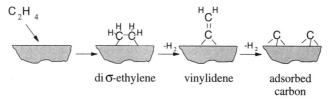

Figure 3.11 The mechanism of ethylene decomposition on (1x1)Pt(100). Proposed by Hatzikos and Masel [1987] and confirmed by Sheppard [1988].)

Another complexity that is not considered in Lennard-Jones's model is that the surface structure can change during adsorption. CO adsorption on (2x1)Pt(110) is a typical example. When a clean Pt(110) sample is heated in vacuum the surface atoms assume what is called a (2x1) reconstruction. See Figures 2.51 and 2.63. Ferrer and Bonzel [1982] found that when CO adsorbs onto (2x1)Pt(110), the surface atoms move around to produce a (1x1) reconstruction. Surprisingly, the surface stays in a (1x1) reconstruction even after the CO desorbs. Yagasaki and Masel [1991] and Wang and Masel [1991] showed that (1x1)Pt(110) is chemically different than (2x1)Pt(110).

Years ago, it was thought that surface reconstructions during adsorption were uncommon. However, recent data reviewed by Somorjai and Van Hove [1988] shows that reconstructions occur quite often. For example, Wander, Van Hove, and Somorjai [1991] found that when the ethylidyne species forms during ethylene adsorption on Rh(111), the three rhodium atoms that bond to the ethylidyne move together so that they can better bind the ethylidyne. These results suggest that adsorption is a very complicated process. Bonds in the adsorbate break; surface atoms move. Hence, one needs a complex model to understand adsorption in detail.

The thing that makes the analysis of adsorption even more difficult is that in many systems different things can happen at once. For example, Wang and Masel [1994] studied acetylene adsorption on (2x1)Pt(110). They found that when the acetylene adsorbs at 100 K, the surface structure does not change upon adsorption. In that case, the acetylene adsorbs molecularly at 100 K, and dehydrogenates upon heating to eventually form a carbonaceous overlayer. Interestingly, if one adsorbs the acetylene at 300 K rather than 100 K, the surface structure changes during adsorption and an entirely different set of surface intermediates is seen upon heating. As this example illustrates, adsorption can be a very complex process; the adsorbing gas can change the surface structure, which in turn changes what happens when additional gas adsorbs. As a result, one can get very different results when one adsorbs at low temperature and heats than when one adsorbs at some reaction temperature.

3.2.3 The Geometry of the Adsorbed Layer

Over the years, there have been many attempts to understand the geometry of the adsorbed layer. Key early work was done by Langmuir [1912, 1913, 1915, 1918], who showed that he could model the adsorption process by assuming that the adsorbate was held on a series of distinct sites on a solid surface, as illustrated in Figure 3.12. Later, Taylor [1927] noted that on a rough surface like that in Figure 2.5 there are many different types of sites to adsorb gas. As a result, multiple sites could contribute to the adsorption process.

It took many years and the advent of surface spectroscopy before people began to realize

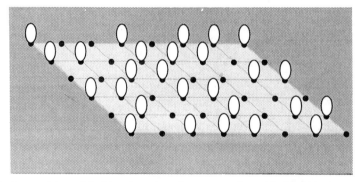

Figure 3.12 Langmuir's model of the adsorption of gases on surfaces. The black dots represent possible adsorption sites, while the white ovals represent adsorbed molecules.

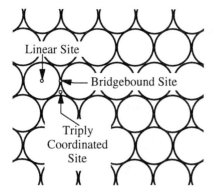

Figure 3.13 Illustration of the linear, bridgebound, and triply coordinated sites on Pt(111).

that multiple sites are important even during adsorption on single crystals. For example, Crossley and King [1980] examined carbon monoxide adsorption on a Pt(111) single crystal using infrared (IR), and found that the CO can adsorb in one of two sites, a so-called **on-top** or **linear site** directly above a surface atom, and a **bridgebound site** between two adjacent surface atoms. Subsequent investigators have also suggested that the CO can adsorb on **triply coordinated site** above a three-fold hollow. All three sites are illustrated in Figure 3.13. Crossley and King [1980] proposed that at low coverages the CO stands up on an on-top site in a so-called $(\sqrt{3} \times \sqrt{3})$ R 30° structure, illustrated in Figure 3.14, where the $(\sqrt{3} \times \sqrt{3})$ R 30° notation was described in Chapter 2. The axis of the C–O bond is perpendicular to the surface and the carbon atom in the CO is pointing toward the surface. The $(\sqrt{3} \times \sqrt{3})$ structure fills up at a coverage of 0.33×10^{15} molecules/cm^2, i.e. at the coverage shown in Figure 3.14. However, at moderate temperatures additional COs can be added to the layer. The additional COs are accommodated by pushing some of the on-top COs in the $(\sqrt{3} \times \sqrt{3})$ structure to a bridgebound site, and moving around the other COs, to free up bridgebound sites for further adsorption. That produces the C(4x2) structure in Figure 3.14. More recently, Persson, Tüshaus, and Bradshaw [1990] showed that at even higher coverages CO can form a complex **domain wall** structure on Pt(111). These structures are also called **uniaxial-commensurate** structures. A typical domain wall structure is shown in Figure 3.15. The domain wall structure repeats in both the x and y directions in the figure. However, the structure is not periodic in the x direction in the figure. Instead there are finite domains with a C(4x2)-like structure and domain walls with higher concentrations of CO. Persson et al. have shown that the average width of the C(4x2) domains slowly decreases as more CO is added to the layer, and that there is a corresponding increase in the number of domain walls so the average coverage increases. Still, there is always a distribution of domain widths. As

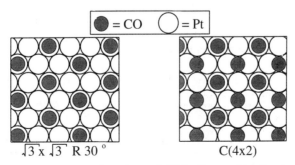

Figure 3.14 The binding sites for CO adsorption on Pt(111). (Proposed by Crossley and King [1980].)

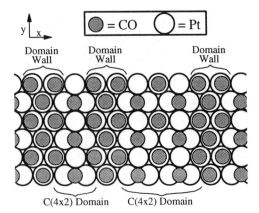

Figure 3.15   The domain wall structure of CO on Pt(111). (Proposed by Persson et al. [1990].)

a result, at moderate to high coverages the layer does not have true two-dimensional order, although locally the structure behaves as though it is ordered.

Blyholder [1964] proposed a simple model to explain how CO can bind to a metal surface. He viewed the bonding of CO as a donor/acceptor complex, where the CO transfers electrons from its bonding orbitals into the metal while the metal donates electrons back into the antibonding orbitals of the CO. This proposal has since been verified by a number of techniques.

Carbon monoxide adsorption on most group VIII metal surfaces is much like that on Pt(111).* The CO first adsorbs linearly, but at higher coverages one observes a mixture of bonding sites. Most often the CO stands up on the surface. However, Fisher et al. [1991] have speculated that CO can lie down on Rh(100). Another complication is that at very high coverages CO can go into what is called **compression layer.** The compression layer is not well defined in the literature. It generally refers to a structure where the adsorbate densities are so high that adsorbate molecules are pushed away from their preferred bonding positions. Hence, it is said that the layer is compressed. Often a compression layer will have a domain wall structure like that in Figure 3.15. However, Williams and Weinberg [1979] found that when CO adsorbs on Ru(001) the layer seems to form the **incommensurate** layer illustrated in Figure 3.16. The COs are closely packed together in an ordered hexagonal overlayer, but the lattice dimension of the overlayer is not simply related to the lattice dimension of the substrate so that the COs do not sit over any preferred site. It has been found that the CO layer maintains a hexagonal arrangement of atoms over a reasonable range of coverages. However, the lattice dimension of the CO overlayer is compressed as one adsorbs gas. Further, Pfnür et al. [1980] used reflection infrared spectroscopy to show that all of the COs have an equivalent bonding, i.e., even though there are apparently multiple binding sites, the bonding of the CO is equivalent on all of the sites. DeAngelis, Glines, and Anton [1992] measured the binding energy of the CO in the compression layer. They found that the bond energy of the CO in the compression layer decreases from 27 to 25 kcal/mole with increasing

*Carbon monoxide adsorption on tungsten and molybdenum is even more complex than carbon monoxide adsorption on group VIII metals. On the closed packed faces of tungsten and molybdenum (i.e., W(110) and Mo(110)) the CO adsorbs molecularly at 90 K. However, the CO dissociates upon heating to 120 K. The dissociated CO layer is fairly stable. Interestingly, if one does temperature programmed desorption (TPD) on the CO layer, one finds that the dissociation process is reversible: The C and O recombine to CO that desorbs in a number of peaks between 900 and 1400 K. Similar chemistry is also seen on all of the other faces of tungsten and molybdenum that have been examined so far, except that there is evidence that some of the CO dissociates upon adsorption at 100 K. Another interesting observation is that Jaeger and Menzel [1980] have suggested that the molecular axis of the CO tilts away from the surface normal on W(100).

Figure 3.16 An illustration of the incommensurate compression layer that was proposed for CO adsorption on Ru(001).

coverage. This is quite a strong bond considering that the COs are not associated with an individual surface atom. At this point, there is the possibility that CO actually adsorbs in some as yet undiscovered domain wall structure on Ru(001). However, the best evidence is that it forms an incommensurate layer. If so, these results would imply that gases do not necessarily bind to individual sites on a surface. Rather, they can form strong bonds without binding to individual atoms.

So far, we have only discussed CO in this section. However, the results here are fairly general. Most adsorbates can adsorb on multiple sites. Sometimes one observes **random site filling** where the adsorbate bonds to a series of distinct sites on the surface of the solid, but the sites do not fill up in any particular order, as illustrated in Figure 3.17. This is in contrast to **ordered adsorption,** where the sites fill in such a way that the adsorbate forms an ordered.

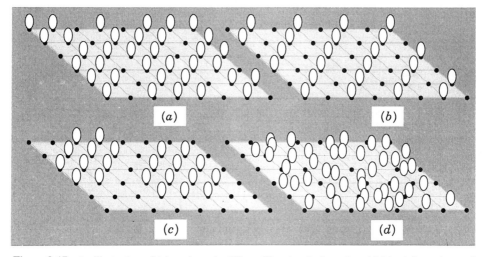

Figure 3.17 An illustration of (a) random site filling, (b) ordered adsorption, (c) island formation, and (d) random incommensurate adsorption.

structure on the surface at moderate coverages. Generally, random site filling occurs when the interaction between adsorbate molecules on adjacent sites is small with respect to kT, so there is no preference for one site over another. Ordered adsorption, on the other hand, occurs when the adsorbate/adsorbate interactions are larger than kT. In such a case the sites that are in close proximity to those that are already occupied by adsorbate fill with either an enhanced or a reduced probability according to whether the adsorbate/adsorbate interaction is attractive or repulsive. One generally gets ordered adsorption when there is a mixture of attractive and repulsive forces so that some sites are filled with enhanced probability while others are filled with reduced probability. This idea is discussed in more detail in Sections 4.6 and 4.7.

If instead the adsorbate/adsorbate interactions are just attractive and strong, the adsorbate molecules will have a tendency to cluster together in small patches called **islands.** If the adsorbate is mobile, additional adsorbate molecules will be drawn into small patches so that the rest of the surface will be cleared of adsorbate. The result is a very complex situation where small patches of the surface are covered by adsorbate but the rest of the surface is almost adsorbate free.

Normally one only observes random site filling at low coverages. At moderate coverages one usually observes either a weakly ordered structure, a strongly ordered structure, or islands.

There also is the possibility that the adsorbate will not bind to specific lattice sites, as was illustrated for the case of CO on Ru(001). We call this **random adsorption** since the adsorbate binds randomly on the lattice rather than on fixed lattice sites. It is also sometimes called **incommensurate adsorption** since the overlayer can form a periodic structure that is not commensurate with the underlying lattice.

3.3 MODELING MOLECULE-SURFACE BINDING: PHYSISORPTION

Over the years, there have been many attempts to understand why molecules bind in the way that they do. Early investigators concentrated on understanding physisorption. There is an excellent description of many models of physisorption in Kreuzer and Gortel [1986]. The material that follows is a summary of that work.

London [1930a,b] provided the first quantitative model that explained how physical (i.e., van der Waals) forces could lead to adsorption. London assumed that the adsorbate would interact with each of the atoms in a lattice via a pairwise additive potential:

$$E_{MS}(\vec{r}_M) = \sum_s E_{Ma}^s(\vec{r}_M - \vec{r}_s) \qquad (3.2)$$

where $E_{MS}(\vec{r}_M)$ is the potential energy of the molecule as it interacts with the surface as a function of $\vec{r}_M$, the position of the molecule, E_{Ma}^s is the interaction of the molecule with an atom in the surface, and $\vec{r}_s$ is the position of atoms.

London then noted that when a nonpolar gas (i.e., a gas without a permanent dipole) interacts with a nonpolar, nonmetallic substrate, each interaction could be represented by a Lennard-Jones 6–12 potential:

$$E_{Ma}^s = C_{Lond}\left(\frac{0.5\,(r_e^s)^6}{|\vec{r}_M - \vec{r}_s|^{12}} - \frac{1}{|\vec{r}_M - \vec{r}_s|^6}\right) \qquad (3.3)$$

where r_e^s is the equilibrium distance when the incoming molecule interacts with an isolated surface atom and C_{Lond} is a constant. Note that for true pairwise additive potentials, one would have to consider the fact that each surface atom will pull the incoming atom in a slightly different direction. For example, the shaded surface atom in Figure 3.18 will pull the adsorbate

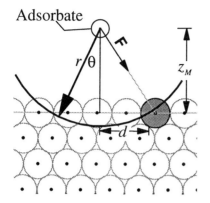

Figure 3.18 London's geometric construction to simplify the sums in Equation 3.5.

atom in the direction labeled **F** in the figure. However, London ignored that effect in his analysis.

London used what is now called the London equation to estimate C_{Lond}

$$C_{Lond} = 1.5\alpha_M\alpha_s \frac{I_M I_s}{I_M + I_s} \tag{3.4}$$

where α_M and α_s are the polarizabilities of the metal and the surface atoms and I_M and I_s are their ionization potentials. Combining Equations 3.2 and 3.3,

$$E_{Ms}(\vec{r}_M) = \sum_s C_{Lond} \left(\frac{0.5\,(r_e^s)^6}{|\vec{r}_M - \vec{r}_s|^{12}} - \frac{1}{|\vec{r}_M - \vec{r}_s|^6} \right) \tag{3.5}$$

London then simplified this sum using the geometric construction in Figure 3.18. Basically, London assumed that atoms were all identical and uniformly distributed over the solid, so he could replace the sum in Equation 3.5 by an integral over the solid:

$$E_{Ms}(z_M) = \rho_s \int_{z_M}^{\infty} \int_0^{\arctan(d/r)} \int_0^{2\pi} E_{MA}^s(r) r^2 \sin\theta \, d\phi \, d\theta \, dr \tag{3.6}$$

where z_M is the distance from the molecule to the surface, ρ_s is the density of atoms in the surface, and r, θ, and ϕ are spherical integration coordinates. Substituting Equation 3.3 into Equation 3.6 and integrating, yields

$$E_{Ms}(Z) = \rho_s \pi C_{Lond} \left(\frac{0.2\,(r_e^s)^6}{z_M^9} - \frac{1}{z_M^3} \right) \tag{3.7}$$

Figure 3.19 shows a plot of Equation 3.7 for an argon atom adsorbing on a graphite surface. Note that the potential shows a minimum at about 2 Å away from the surface; the energy at that point is 2.2 kcal/mole. By comparison Beebe [1954] measured a heat of adsorption of 2.7 ± 1 kcal/mole, so in effect London's 1930 calculation is as good as the data for this system.

Over the years there have been several corrections proposed to improve Equation 3.7. Crowell [1954, 1957] used the Kirkwood–Miller formula, Equation 3.8, to get a better value

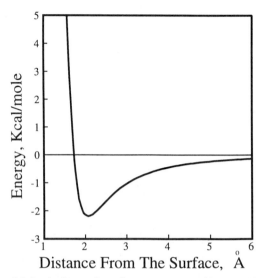

Figure 3.19 The potential for the interaction of an argon atom with a graphite surface, as calculated from Equation 3.7.

for C_{Lond}:

$$C_{Lond} = 6m_e c^2 \, \frac{\alpha_s \alpha_M}{\left(\dfrac{\alpha_s}{SM_s}\right) + \left(\dfrac{\alpha_M}{SM_M}\right)} \tag{3.8}$$

where SM_s and SM_M are the magnetic susceptibilities of the surface and the adsorbate and c is the speed of light. The Kirkwood–Miller formula seems to be a small improvement over London's original formula. Kiselev [1957] and Steele [1974] directly computed the sums in Equation 3.5. Again, this was a minor improvement. Kiselev [1957] also added in an extra energy E_{MS}^d to account for the extra interaction that occurs when a polarizable molecule interacts with the permanent dipole in an ionic solid:

$$E_{MS}^d(r) = \tfrac{1}{2}\alpha_M (F_E^s)^2 \tag{3.9}$$

where F_E^s is the electric field from the solid at the position of the incoming molecule and α_M is the polarizability of the molecule. Note the definition of the polarizability is that α_M satisfy

$$\alpha_M = \frac{E_M}{\tfrac{1}{2}F_E^2} \tag{3.10}$$

for any constant electric field, so in effect Equation 3.9 is assuming that the electric field is constant over the length scale of the molecule.

   It ends up that these old calculations did a reasonable job of predicting the strength of the binding of a nonpolar gas on nonmetallic substrates. For example, Table 3.3 compares the values of the heats of adsorption calculated from Equations 3.5, 3.8, and 3.9 to measured values for the adsorption of a number of gases on a number of surfaces. Notice that the calculations for the nonmetallic surfaces agree reasonably well with the measurements. Less

TABLE 3.3. A Comparison of Heats of Adsorption Calculated from Equations 3.5, 3.8, 3.9, and 3.15 to Measured Values

Gas	Surface	Calculated Heat of Adsorption (cal/mole)	Measured Heat of Adsorption (cal/mole)
Nonmetals			
He	Graphite	357	340
Ne	Graphite	830	830
Ar	Graphite	2220	2700
Kr	Graphite	2950	3900
C_5H_{12}	Graphite	9300	10000
C_6H_{14}	MgO	9200	9900
C_7H_{16}	MgO	10600	11400
C_6H_{14}	Graphite	11000	11800
Ar	NaCl	1978	2180
Ar	KBr	2325	2440
Metals			
He	Pt	230	265
Ne	Pt	330	370
Ar	Pt	940	1320
Kr	Pt	1240	2110
Ar	Zn	1100	1570
Ar	Cu	1090	2090

Source: Data extracted from London [1932], Ponec, Knor and Černý [1974] and Clark [1970].

good agreement is seen with metal surfaces. There are corrections because of interactions between adjacent gas molecules and the heterogeneity of the surface. However, the amazing thing is that these calculations from the 1930s to the 1950s reproduce heats of physisorption on *nonmetallic* surfaces about as accurately as they have been measured.

Of course, the methods work less well for metal surfaces. The derivation on the last few pages does not really apply to metal surfaces. A metal is quite different than an ionic solid in that many of the electrons that are available for bonding are not localized on any one atom. Rather, the electrons are free to move throughout the bulk of the metal. These electrons, called **conduction band** electrons, do most of the bonding at long distances. The analysis so far in this section assumed that the interaction of a gas molecule with a surface could be written as a pairwise sum of atomic potentials. (See Equation 3.2.) However, this approximation does not make sense in the case of metals because there are important interactions with electrons that are not associated with any one atom.

Nevertheless, Lennard-Jones [1936] showed that one can derive an approximation for the physisorption of a gas on a metal surface, which is quite similar to Equation 3.7. Lennard-Jones modeled the metal as an infinitely conducting medium. He then calculated the interactions using a modification of the derivation that follows.

Consider the interaction of a dipole with an infinitely conducting surface, as indicated in Figure 3.20. A free dipole will have field lines as indicated on the left of Figure 3.20. Now consider moving the dipole up to the surface in several different configurations. If the metal is completely conducting, there cannot be any electric field parallel to the metal's surface. As a result the field lines need to be perpendicular to the surface as indicated in the middle of Figure 3.20. It is interesting to compare two cases: the exact result where the field lines all lie perpendicular to the surface and a second case where one has replaced the surface by a second dipole that is a mirror image of the original dipole, as shown in Figure 3.20. Notice that the field lines are identical in both cases. The interaction of the dipole with the field is

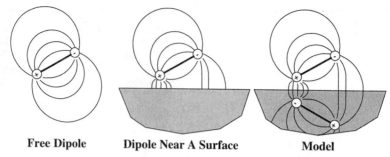

Free Dipole   **Dipole Near A Surface**   **Model**

Figure 3.20   The electric field created when a dipole is brought close to an infinitely conducting surface.

only a function of the field lines. Therefore, the interactions are identical in both systems. As a result, one can calculate the interaction of a dipole with an infinitely conducting surface by replacing the surface with a mirror image of the original dipole. The mirror image is called the **image dipole.** A subtle point is that while the surface charge does rearrange in response to the dipole, the charge distribution in the surface is not a mirror image of the dipole. Rather the dipole interacts with the surface as though there were a mirror image of the dipole in the surface even though the real charge distribution is different.

Now consider the interaction of a hydrogen atom with an infinitely conducting metal surface. The hydrogen atom consists of an electron and a proton. If the electron is at some position $\vec{r}_e$ relative to the nucleus, as indicated in Figure 3.21, the proton–electron pair will form a net dipole. That dipole will be attracted to the surface. With an infinitely conductive surface, the attractive potential, E_{attr}, is given by

$$E_{attr}(\vec{r}_e, z_M) = -\frac{e^2}{2z_M} - \frac{e^2}{2(z_M - r_z)} + \frac{2e^2}{\sqrt{(2z_M - r_z)^2 + r_x^2}} \qquad (3.11)$$

where $\vec{r}_e$, r_z, and r_x are defined in Figure 3.21 and e is the charge on an electron. The first term in Equation 3.11 is the interaction of the nucleus with its image, the second term is the

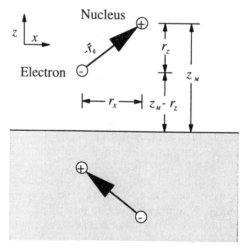

Figure 3.21   A diagram of the instantaneous interaction between an atom and a metal surface. The surface lies at the boundary between the shaded and unshaded area.

interaction of the electron with its image, and the third term is the interaction of the electron with the image of the nucleus plus the interaction of the nucleus with the image of the electron.

Equation 3.11 gives the instantaneous force between a hydrogen atom and a surface, given the position of the hydrogen nucleus and the hydrogen's electron. In reality the electron does not have a fixed position. As a result, one needs to average Equation 3.11 over all of the possible configurations of the electron to calculate a net force:

$$E_{attr}(z_M) = \int \psi_e^*(\vec{r}_e) U_{attr}(\vec{r}_e, z_M) \psi_e(\vec{r}_e) \, d\vec{r}_e \tag{3.12}$$

where $\psi_e(\vec{r}_e)$ is the wavefunction for the electron and ψ_e^* is the complex conjugate of the wavefunction.

Lennard-Jones substituted the Bohr model approximation to the wavefunction for a 1s orbital into Equation 3.12, and then integrated the equation retaining only the leading term in $1/z_M$. The result is

$$E_{attr}(z_m) = \frac{C_{LJ}}{z_M^3} \tag{3.13}$$

where C_{LJ} is a constant.

Lennard-Jones then suggested that in addition to the interaction in Equation 3.13 the incoming atom also experiences a core–core repulsion. If one assumes that the core repulsions are proportional to $1/r^{12}$, then the earlier derivation shows

$$E_{repulsive} = -\frac{C_{LJ}^1}{z_M^9} \tag{3.14}$$

where $E_{repulsive}$ is the repulsive potential and C_{LJ}^1 is a second constant. Combining Equations 3.13 and 3.14 yields

$$E_{attr}(z_m) = -\frac{C_{LJ}}{z_M^3} + \frac{C_{LJ}^1}{z_M^9} \tag{3.15}$$

Notice that Equation 3.15 has the same form as Equation 3.7. Hence, Lennard-Jones suggested that Equation 3.7 would also apply to physisorption of gases on metal surfaces. He also asserted that the equations should apply to a wide range of substances even though the equations were derived with 1s orbitals.

Unfortunately, experimentally, Equations 3.7 and 3.15 do not work well for physisorption on metal surfaces. Table 3.3 provides some example calculations. Generally, Lennard-Jones's model underestimates heats of physisorption, although the results are usually within 25% of the exact results.

There are three reasons that Lennard-Jones's model underestimates heats of physisorption on metal surfaces: no real metal is infinitely conductive, there are polarizations of the incoming molecule, and there are extra repulsive interactions due to overlap of the metal orbitals with the closed shells of the incoming molecule.

The influence of a finite surface conductivity has been considered in detail. When the surface has a finite conductivity, the field lines created when an atom approaches a metal surface are not quite perpendicular to the surface. Also, the dipole moment of the approaching atom rapidly fluctuates. Often, the surface cannot keep up with these changes. These effects reduce the net attractive potential.

There is a second effect, however, that enhances the attractive interactions. Notice that the dipole interaction in Figure 3.21 is enhanced when the electron is pointing toward the

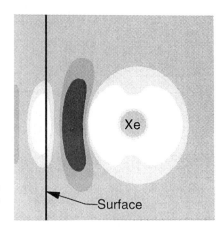

Figure 3.22   The changes in electron density that occur when a xenon atom interacts with a metal surface. (After Lang and Williams [1982].)

surface and attenuated when the electron points away from the surface. As a result, when an atom approaches a metal surface, the probability that the electron lies on the side of the atom that faces the metal is slightly larger than the probability that the electron will face the opposite direction. On average, then, there will be a net polarization when a gas atom approaches a metal surface. The electrons in the metal will be polarized too. This leads to an enhanced attractive interaction.

For example, Lang and Williams [1982] calculated the changes in electron density that occur when a xenon atom approaches a metal surface. Figure 3.22 shows a plot of some of their results. The dark areas in the figure indicate places where the electron density is enhanced. The light areas indicate places where the electron density is reduced. Notice that on average electrons flow from the zenon atom and the surface, to a position between the xenon atom and the surface. This produces a net dipole on the xenon atom. The surface responds to that dipole by creating a dipole of its own, and the two dipoles attract. The electron flows are much smaller than those in a typical chemical bond, but they are large enough to be detected in UPS. Such electron flows were not considered in Lennard-Jones's 1935 model. As a result there is an extra attraction that is not considered in Lennard-Jones's model.

There is an extra repulsion, too, which comes about because the conduction bands in the metal are not confined to the metal. Rather, the electrons spill out from the surface to form an electron cloud. According to calculations there is a reasonable electron density as much as 5 Å away from the surface. Now consider what happens when a molecule enters that electron cloud. Note that if the molecule has all closed shells, there will be electron–electron repulsions that distort the electron cloud. That leads to repulsions. These repulsions are in addition to the core–core repulsions, so they are not considered in Lennard-Jones's 1935 model. Zaremba and Kohn [1977] show that the electron–electron repulsions are much larger than the core–core repulsions. As a result, these repulsions need to be considered in the analysis.

Experimentally, Lennard-Jones's model seems to work only at large distances for metals and not at short distances. The long distance results are quite good. For example, Shih and Parsegian [1977] examined interaction of potassium atoms with a gold surface by measuring the deflection of a beam of atoms when the atoms are close to a gold cylinder. Their results are within a factor of 2 of the Lennard-Jones predictions at the range of distances used by Shih and Parsegian ($z_M > 200$ Å). Surprisingly, Mehl and Schaich [1977, 1980] found that all of the calculations done between 1935 and 1980 did not do significantly better than Lennard-Jones's 1935 model.

Nevertheless, Lennard-Jones's model does not seem to work at the distances characteristic of physisorption, that is, 2–5 Å. Generally, Lennard-Jones's model predicts much shorter bond distances and somewhat smaller heats of physisorption than are seen experimentally. For example, the heat of adsorption of zenon on tungsten is 9.5–10 kcal/mole, but Lennard-Jones's model predicts a heat of adsorption of 7 kcal/mole. Analysis of the data indicates that when the xenon atoms are in the physisorbed state, the polarizations of the xenon and the rearrangements of the electron cloud cause major perturbations to the xenon/surface interaction. Scheffler et al. [1979] found that the UPS spectrum of xenon changes slightly when xenon adsorbs on Pd(110) implying that the electronic structure of the xenon changed upon adsorption. These results are consistent with the polarizations calculated by Lang and Williams [1982]. Engel and Gomer [1970] and Ehrlich and Hudda [1959] found interesting changes in work-function when xenon adsorbs on a tungsten field emitter. These changes suggest that the outer electron cloud of the tungsten rearranges when xenon adsorbs. Similar results have since been obtained for xenon adsorption on many other surfaces. See Chen et al. [1984] for details. This appears to be a general result. Zaremba and Kohn [1977] and Lang's [1981] calculations indicate that the polarization effects and the rearrangements of the metal's electron cloud provide the largest contribution to the energy at distances characteristic of physisorption. As a result, we need a fairly sophisticated calculation to do a better job of predicting heats of physisorption on metal surfaces. Still, useful computational techniques have been developed. They are described in Sections 3.9.2 and 3.9.3. At this point, physisorption of simple gases on metal surfaces is largely well understood. There are still some questions about the dynamics of the process, however. See Kreuzer and Gortel [1986] for details.

3.4 SIMPLE MODELS OF CHEMISORPTION

Models of chemisorption, however, are not as well-developed. Years ago people used simple models of adsorbate surface bonds, based on ideas such as electronegativity differences and bond counting. Such models were able to predict properties such as heats of adsorption fairly well. However, the models were not well justified theoretically. More recently, the emphasis has been on quantum calculations, which are more easily justified theoretically, but often are not as accurate as those from the simple methods. In the next several sections, we review many of the ideas. We start with the simple models that work pretty well. We then work up to the more complex (quantum) models. We include many details, because we want the book to be a useful resource for our students. However, the reader could skip the details without loss of generality. To review the history behind our current understanding, the first models of chemisorption were proposed in the 1950s. To set the stage for the work Pauling [1939, 1960] had proposed that one could use a simple two-electron model to calculate the energy of the bond between two species

$$D(A\text{–}B) = \frac{D(A\text{–}A) + D(B\text{–}B)}{2} + \gamma_{Pa}(\chi_A - \chi_B)^2 \qquad (3.16)$$

where $D(A\text{–}A)$, $D(B\text{–}B)$, $D(A\text{–}B)$ are the energies of the $A\text{–}A$, $B\text{–}B$, and $A\text{–}B$ bonds, χ_A and χ_B are quantities called the electronegativities of A and B, and γ_{Pa} is a constant, 1/eV. Pauling fit Equation 3.16 to measured values of bond energies to calculate the values of the electronegativity. Equation 3.16 is a definition of the **electronegativity.**

Equation 3.16 is quite important to the discussion later in this chapter. It is useful therefore to derive the equation before proceeding. Equation 3.16 comes from a one-electron picture of the $A\text{–}B$ bond. In order to derive the equation consider two atoms A and B, each of which has a single electron available for bonding. Now consider taking an A atom and moving it close to a B atom. Pauling asserted that one can write the energy of atom A in its bonding configuration as a Taylor series in a quantity called the bond order, n_A^1, where the bond order is 1 if the electron density of atom A is as it would be in an $A\text{–}A$ bond and E_A^1, χ_A^1, and η_A

are constants that will be related to the electronegativity in Section 3.4.1:

$$E_A = E_A^1 + \chi_A^1 n_A^1 + \eta_A \frac{(n_A^1)^2}{2} \tag{3.17}$$

For the purposes of derivation, it is useful to define a new varible n_A via

$$n_A = n_A^1 - 1 \tag{3.18}$$

Physically, n_A will be the fraction of an electron transferred when a bond forms. Combining Equations 3.17 and 3.18 yields

$$E_A = E_A^0 + \chi_A n_A + \eta_A \frac{n_A^2}{2} \tag{3.19}$$

with

$$E_A^0 = E_A^1 + \chi_A^1 + \frac{\eta_A}{2} \tag{3.20}$$

$$\chi_A = \chi_A^1 + \eta_A \tag{3.21}$$

Similarly for B

$$E_B = E_B^0 + \chi_B n_B + \eta_B \frac{n_B^2}{2} \tag{3.22}$$

The total bond energy of the system is

$$-D(A\text{–}B) = E_A + E_B$$

$$= E_A^0 + E_B^0 + \chi_B n_B + \chi_A n_A + \frac{\eta_A n_A^2}{2} + \frac{\eta_B n_B^2}{2} \tag{3.23}$$

Pauling then assumed that there were two electrons to form bonds so that at equilibrium

$$n_a + n_B = 0 \tag{3.24}$$

Combining Equations 3.23 and 3.24 yields

$$-D(A\text{–}B) = E_A^0 + E_B^0 + (\chi_A - \chi_B)n_A + (\eta_A + \eta_B)\frac{n_A^2}{2} \tag{3.25}$$

The energy is minimized when

$$\frac{dE_T}{dn_A} = 0 \tag{3.26}$$

Substituting Equation 3.25 into Equation 3.26 and solving for n_A yields

$$n_A = \frac{\chi_B - \chi_A}{\eta_A + \eta_A} \tag{3.27}$$

Substituting Equation 3.27 into Equation 3.23 yields

$$-D(A\text{--}B) = E_A^0 + E_B^0 - \frac{(\chi_B - \chi_A)^2}{2(\eta_A + \eta_B)} \tag{3.28}$$

Similarly, if an A atom interacts with another A atom, one finds

$$-D(A\text{--}A) = E_A^0 + E_A^0 - \frac{(\chi_A - \chi_A)^2}{2(\eta_A + \eta_A)} = 2E_A^0 \tag{3.29}$$

$$-D(B\text{--}B) = E_B^0 + E_B^0 - \frac{(\chi_B - \chi_B)^2}{2(\eta_B + \eta_B)} = 2E_B^0 \tag{3.30}$$

Substituting E_A^0 and E_B^0 from Equations 3.29 and 3.30 into Equation 3.28 yields

$$D(A\text{--}B) = \frac{1}{2}\left(D(A\text{--}A) + D(B\text{--}B)\right) + \frac{1}{2(\eta_A + \eta_B)}(\chi_A - \chi_B)^2 \tag{3.31}$$

Pauling assumed that $1/[2(\eta_A + \eta_B)]$ was a universal constant γ_{pa} to derive Equation 3.16.

Eley [1950] extended this formalism to adsorption. He estimated the difference in electronegativities from measurements of something called the work function. He then used those estimates to see if he could calculate adsorbate surface bond energies from Equation 3.16.

The work function is a quantity that people measured when they were designing filaments for vacuum tubes. The emission from a filament varies with temperature and applied voltage. The derivative with respect to temperature is proportional to the work function difference between the filament and the collector (i.e., the device used to collect the electrons). Physically, the work function is the free-energy change when one removes an electron from deep inside the bulk of a metal and places the electron a long distance from the metal, as described in Zangwill [1988]. To a reasonable approximation, the work function, WF, is given by

$$WF = \frac{dE_s}{dn_s} \tag{3.32}$$

where E_s is the energy of electrons in the surface, and n_s is the number of electrons.

Eley supposed that the conduction electrons in the surface could contribute to bonding in the same way that electrons in a molecule would contribute to bonding. Generally, the electrons in the neighborhood localize around the adsorbate to form a bond, so in effect one is moving electrons from the surface onto the adsorbate or vice versa. Eley modeled that process by assuming that moving an electron from the surface onto the adsorbate is roughly equivalent to removing an electron from the metal and placing the electron at infinity. Under that assumption E_s in Equation 3.32 would be equivalent to the energy change in bonding from Equation 3.22. As a result, Eley proposed that one can use Equation 3.22 to relate the work function of a surface to the electronegativity of the surface.

Substituting E_B from Equation 3.22 for E_s in Equation 3.32 and then assuming that $\eta_s = 0$ yields

$$(0.355/\text{eV})\,WF = \chi_s \tag{3.33}$$

where χ_s is the electronegativity of the surface and the constant 0.355/eV is to convert from Millikin to Pauling electronegativity. Therefore to some approximation, the energy of a molecule-surface bond $D(M\text{--}S)$ is given by

$$D(M\text{--}S) = E_M^0 + E_s^0 + \gamma_{PA}(\chi_M - \chi_s)^2 \tag{3.34}$$

where E_M^0 and E_s^0 are the constants in Equation 3.19 appropriate for the molecule and the surface, and χ_M and χ_s are the corresponding electronegativities. At the time Eley's calculations were done people knew how to measure work function differences, but not the absolute value of the work function. As a result, Eley assumed that the work function changes that occur when gases adsorb were equal to the difference between the work function of the adsorbate and the surface. Eley further assumed that E_s^0 (i.e., twice the metal–metal bond energy) was given by

$$E_s^0 = \tfrac{1}{12} \, \Delta H_{subl}^M \tag{3.35}$$

where ΔH_{subl}^M was the heat of sublimation of the metal. Eley also assumed that E_M^0 was equal to

$$E_M^0 = \tfrac{1}{2} D(M\text{-}M) \tag{3.36}$$

The factor of 12 in Equation 3.35 arises because there are 12 nearest neighbors in the bulk. Table 3.4 shows how well Eley's calculations work. In general, Eley's formalism gave the right order of magnitude for the energy of the metal/surface bond, but it failed in detail.

In the 1950s other workers proposed several extensions of Eley's formalism. Issues that were considered were what is the proper way to calculate the difference in electronegativity in Equation 3.16 and whether Pauling's approximations were appropriate for surfaces. However, in 1950 no one had a good way to estimate the parameters in Equation 3.16 for reactions on surfaces. As a result, progress was limited.

Brennan, Hayward, and Trapnell [1960] compared all of these models to their data for oxygen adsorption on transition metals, and found that in all cases the agreement was poor. As a result, Eley's model was largely abandoned until the early 1980s.

Brennan et al. suggested an alternative model. They noted that there was a correlation between their measured heats of adsorption and the heat of formation of bulk oxides (see Figure 3.23). This led Brennan et al. to suggest that one could model the chemisorption of oxygen as though the oxygen were forming a chemical bond to the surface which is similar to the bonds in bulk oxides.

In the early 1960s many papers appeared that expanded these correlations. Roberts [1960], for example, proposed that the heat of adsorption of oxygen could be correlated with the heat of formation of a single metal oxygen bond, while the heat of adsorption of hydrogen could be correlated to the heat of formation of bulk hydrides. At the time this work was done, there was not enough data on heats of formation of bulk compounds to test this hypothesis in detail.

TABLE 3.4. A Comparison of Eley's [1950] Calculations of Heats of Adsorption to Measured Values

Gas	Surface	Measured Heat Of Adsorption (kcal per mole of H_2)	Eley's Calculation (kcal/mole)	Calculation With Flores' [1981] Values of the Parameters (kcal/mole)
H_2	Cu(111)	-10	-13.6	-12
H_2	Ni(111)	-23	-17.2	-19
H_2	W(110)	-33	-43.6	-35
H_2	Fe(110)	-27	-19.5	-28
H_2	Pt(111)	-18	-19.2	-22

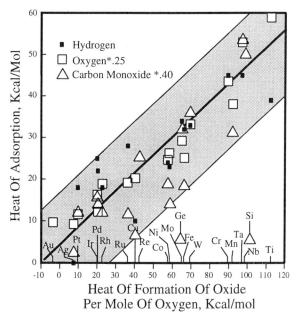

Figure 3.23   The heats of adsorption of CO ($\times 0.4$), H_2, and O_2 ($\times 0.25$) on a series of polycrystalline metal films as a function of the heat of formation of the bulk oxide per mole of oxygen.

However, the general idea that adsorbate surface bonds were similar to bulk bonds was quickly accepted in the literature.

Tanaka and Tamaru [1963] found an interesting way to expand this model. They noted that in general some metals are reactive while others are unreactive. Reactive metals tend to form strong bonds to many different adsorbates, while unreactive metals tend to only form weak bonds. Thus, if a metal forms a strong bond to an oxygen atom, it will also tend to form a strong bond to a hydroxide, nitrate, or sulfate. Tanaka and Tamaru verified this correlation for hydroxides, nitrates, sulfates, and formates of group Ia and IIa metals, i.e., Li, Na, K, Mg, Ca, Ba, etc. More recent data show that a similar correlation also works for the nitrates, sulfates, nitrides, and carbides of transition metals, although the correlation is not as good as with group I and IIa metal (see Figure 3.24). Tanaka and Tamaru then asserted that if heats of adsorption of some gas were to correlate with the heats of formation of, for example, a hydroxide or carbide, one could just as well correlate the data to the heat of formation of bulk oxides. Therefore, Tanaka and Tamaru proposed that heats of adsorption of most simple gases should correlate with heats of formation of bulk oxides.

Figure 3.23 shows how well the correlations work. Notice that the data on transition metals all fall within a band. There is a pretty good correlation between the heat of formation of bulk oxides and the heat of adsorption of oxygen on all the transition metals. However, the hydrogen data do not fit quite as well, and there are significant deviations for carbon monoxide. Still, Tanaka and Tamaru's simple correlation seems to give reasonable qualitative, if not quantitative results.

The implication of Figure 3.23 is that one can divide up the elements in the periodic table in terms of their ability to form bonds to adsorbates. Some elements are able to form strong bonds to a wide class of adsorbates, while others only bond a wide class of adsorbates weakly. As a result, if one knows that a given surface will bind a certain adsorbate strongly, one can often safely assume that the same surface will also bind other adsorbates strongly.

Table 3.5 shows a list of metal-adsorbate bond strengths calculated by Benziger [1991].

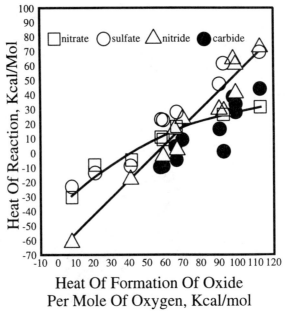

Figure 3.24 The heat of reaction of nitric, sulfuric, and formic acid, and carbon and nitrogen with a number of metals plotted as a function of the heat of formation of the corresponding oxide.

In some cases he used data for the values while other values were calculated assuming that the heats of adsorption were proportional to the heats of formation of bulk carbides, nitrides, and oxides, respectively. Notice that there is a general trend that elements to the left of the periodic table bind adsorbates more strongly than elements to the right of the periodic table. With the group VIII and IB metals the bond strengths decrease moving down the periodic

TABLE 3.5. Atom Surface Energies in kcal/mole Suggested by Benziger [1991]*

Element	Group							
	IVA	VA	VIA	VIIA	VIII	VIII	VIII	IB
	Ti	V	Cr	Mn	Fe	Co	Ni	Cu
$D(M-C)$	(216)	(196)	(181)	(171)	(166)	(162)	169	[135]
$D(M-N)$	(329)	(281)	(232)	(206)	140	(167)	138	135
$D(M-O)$	(172)	(145)	(150)	(122)	(125)	(116)	112	90
	Zr	Nb	Mo	Tc	Ru	Rh	Pd	Ag
$D(M-C)$	(220)	(205)	(183)	[165]	[140]	[130]	[130]	[70]
$D(M-N)$	(200)	(169)	155	[125]	[120]	[120]	[120]	(39)
$D(M-O)$	(191)	(151)	118	(97)	96	88	84	80
	Hf	Ta	W	Re	Os	Ir	Pt	Au
$D(M-C)$	(232)	(294)	(260)	[180]	[145]	[130]	[130]	[50]
$D(M-N)$	(201)	(178)	(137)	144	[120]	[120]	[120]	[35]
$D(M-O)$	(196)	(157)	130	(102)	(83)	92	84	79

*The values without parentheses are data; the other values are interpolations.

table. However, with IVB and VB metals the bond strengths increase moving down the periodic table. These trends are very significant, and often form a basis for catalyst design.

In the early 1960s people took the success of the Tamaru–Tanaka correlations and the apparent failure of the Eley correlations as evidence that the bonding of molecules to metal surfaces was really a local phenomenon that is much akin to the bonding in bulk oxides. However, Flores, Gabbay, and March [1981] showed that this interpretation is incorrect. Instead they noted that the data in Figure 3.23 follow all of the trends expected from Eley's model. Surfaces that form a strong bond to one molecule, oxygen, also form a strong bond to other molecules. Thus, if one adjusts E_s^0 in Equation 3.34, one can fit the data in Figure 3.23 exactly. Flores et al. also noted that they could calculate E_s^0 from bulk band structure calculations, and one gets qualitative agreement with experiment. Thus, they proposed that Eley's model may be a fundamental way to think about the binding of molecules with surfaces.

In actual practice, Flores et al. used a modified version of Equation 3.34:

$$D(M\text{-}S) = \frac{D(M\text{-}M)}{2} + E_s^0(1 - C_{FL}) + \frac{(\chi_M - \chi_s)^2}{2(\eta_M + \eta_s)} \tag{3.37}$$

where the C_{FL} term in Equation 3.37 comes about because when a molecule surface bond forms, one simultaneously breaks a metal–metal bond. Flores et al. assumed that C_{FL} followed

$$C_{FL} = 2\left(1 - \frac{V_{FL}}{N^A}\right) \tag{3.38}$$

where V_{FL} is the valency of the metal and N^A is a constant 12 for oxygen and nitrogen, and carbon, but 4 for hydrogen. They then used bulk band structures to estimate the various parameters in Equations 3.37 and 3.38. Updated values of the parameters are given in Table 3.6. Flores et al. also assumed that the electronegativity of the adsorbates were equal to Pauling's values, and that the electronegativities of the surfaces were given by the surfaces' work functions. The $1/[2(\eta_s + \eta_M)]$ term in Equation 3.37 was replaced by Pauling's universal constant $\gamma_{Paul} = 1.0/\text{eV}$ to yield

$$D(M\text{-}S) = \frac{D(M\text{-}M)}{2} + E_s^0(1 - C_{FL}) + \gamma_{Paul}(\chi_M - 0.335/\text{eV } WF_s)^2 \tag{3.39}$$

TABLE 3.6. Updated Values of E_s^0 (kcal/mol) and V_{FL} Calculated Using the Method of Flores et al. [1981] Using the Data in Benziger [1991]

Element	Group							
	IVA	VA	VIA	VIIA	VIII	VIII	VIII	IB
	Ti	V	Cr	Mn	Fe	Co	Ni	Cu
E_s^0	33.3	34.6	26.0	19.1	29.6	32.4	35.7	31.1
V_{FL}	4	5	6	5.5	5	4.5	4	3.5
	Zr	Nb	Mo	Tc	Ru	Rh	Pd	Ag
E_s^0	36.9	41.5	35.3	35.4	38.0	36.6	29.4	25.8
V_{FL}	4	5	6	5.5	5	4.5	4.0	3.5
	Hf	Ta	W	Re	Os	Ir	Pt	Au
E_s^0	36.2	44.5	44.7	43.4	46.2	43.8	44.0	32.8
V_{FL}	4	5	6	5.5	5	4.5	4	3.5

It was also assumed that oxygen forms two σ-bonds to the surface while nitrogen is held with three σ-bonds so some of the terms in Equation 3.37 needed to be corrected.

Tables 3.4 and 3.7 show the results of the calculations. It is useful to compare the calculations in Table 3.7 to the experimental results in Table 3.5. Notice that the calculations· agree with the *experimental* values (i.e., the values without parentheses or brackets) to within about 5 kcal/mole for all of the metals except copper, silver, and gold. Less good agreement is seen with the extrapolated values in parentheses in Table 3.5. The copper, silver, and gold results agree if a valence of 1.0 is used. The hydrogen results in Table 3.4 show similar agreement.

Figure 3.25 compares the heats of adsorption of hydrogen and oxygen calculated from Equation 3.39 to measured values for polycrystalline samples. Here, the agreement is not quite as good as in Table 3.7. In particular, all of the calculated heats of adsorption of oxygen are lower than the measured values. Admittedly, the data in Figure 3.25 were taken on rough films. Generally, heats of adsorptions of oxygen are larger on rough surfaces than smooth ones. Still, the errors in Equation 3.39 are larger than one can account for via the surface roughness. Therefore, while it appears that one can start to model the binding of atoms to metal surfaces with a very simple analysis, a more sophisticated model is needed to fit the data in detail.

Such results, however, are very dependent on the choice of parameters. For example, if we treat E_s^0 as a fitting parameter, one can fit the data to almost within experimental errors via Equation 3.34. Therefore, the preceding analysis seems to have some validity, even if the choice of parameters does not.

In a more fundamental way the general feature predicted by Equation 3.34 is that one can use the values of the inner potential, E_s^0, and work function to order the bonding power of surfaces. Metals with the smallest inner potential and work function will form the strongest bonds, irrespective of the species that is actually binding. It is this ordering that accounts for the general correlations in Figure 3.23 and Figure 3.24. The success of Flores et al.'s results also indicates that the binding energy of an atom to a **metal** surface is largely determined by the general ability of the metal to form bonds, and the propensity of the metal to share electrons. This is a key result that will be exploited in greater detail later in this chapter.

At present, no one has applied Flores et al.'s calculations to examine molecular adsorption and so we do not know how well it works. However, Benziger [1991] shows that if one knows the strength of a metal–carbon bond, one can get a good approximation to the bond energies of species attached to the surface with a carbon–metal single bond. Species with

TABLE 3.7. Bond Energies for Nitrogen and Oxygen on the Closed Packed Faces of Metals as Calculated via the Method of Flores et al. [1981]*

Element	Group							
	IVA	VA	VIA	VIIA	VIII	VIII	VIII	IB
	Ti	V	Cr	Mn	Fe	Co	Ni	Cu
$D(M\text{-}N)$	144	149	151	158	142	133	127	131 (122)
$D(M\text{-}O)$	110	116	116	125	106	96	89	99 (86)
	Zr	Nb	Mo	Tc	Ru	Rh	Pd	Ag
$D(M\text{-}N)$	148	148	149	—	140	133	130	135 (125)
$D(M\text{-}O)$	116	114	114	—	104	85	92	99 (89)
	Hf	Ta	W	Re	Os	Ir	Pt	Au
$D(M\text{-}N)$	152	148	152	139	137	126	117	123 (109)
$D(M\text{-}O)$	120	115	129	101	100	87	77	84 (71)

*An average of the Pauling and Millikin electronegativities was used to do the calculations. Values in parentheses were calculated using a valence of 1.0.

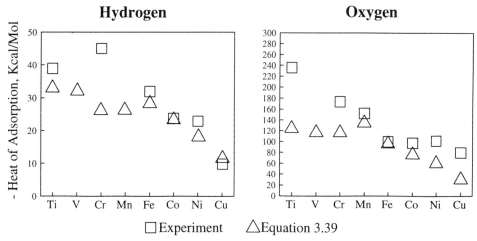

Figure 3.25 A comparison of heats of adsorption calculated from Equations 3.39 to experimental values for polycrystalline samples.

double or triple bonds are less well represented with the model. For example, the adsorption of carbon monoxide (CO) has been studied in great detail.

Blyholder [1964] showed that when CO binds to a metal surface, the CO donates electrons to the metal, and the metal back donates electrons into the CO. The back donation is not considered in Equation 3.39. As a result, the preceding analysis does less well for CO.

Another source of errors comes when one tries to extend the arguments to nontransition metal surfaces. For example, silicon and germanium both form strong oxides, implying that E_S^0 is large. Based on the correlation in Figure 3.23 one would expect CO to have a binding energy of about 60 kcal/mole on silicon and about 40 kcal/mole on germanium. Yet CO does not stick on germanium or silicon. Its binding energy is less than 7 kcal/mole. Interestingly, Asakawa et al. [1991] found that a nickel-covered silicon surface binds CO very strongly, much more strongly than nickel alone. These results indicate that the idea that one can tell whether a molecule bonds strongly or weakly to a surface based only on the magnitude of χ_s and E_S^0 does not work for adsorption on silicon or germanium.

Another problem is that there are a few species that do not follow the general correlations. Water is a typical example. Water interacts only weakly with most transition metal surfaces. Theil and Madey [1990] note that water generally adsorbs molecularly on most transition metal surfaces at 90 K and desorbs again upon heating to perhaps 200 K. Heats of adsorption are generally less than 15 kcal/mole. However, water dissociates when it interacts with nickel oxide. Matsuda et al. [1992] found that the heat of adsorption of water is 22 kcal/mole on NiO(100). Notice that water binds more strongly to nickel oxide than to nickel even though the nickel oxide is less able to bind oxygen than bulk nickel so that E_S^0 is smaller on nickel oxide than on nickel. It is not obvious, but the binding energy of hydrogen is lower on nickel oxide than on nickel as well. A similar effect occurs with many oxides. Consequently, the binding energy of water to metal surfaces does not seem to follow the trends in Figure 3.23. Water also dissociates on silicon and germanium. However, Hung, Schwartz, and Bernasek [1993] found that water adsorbs molecularly on a 100 K Fe(100) surface, and does not dissociate until the sample is heated to 250 K even though germanium has a lower E_S^0 than iron. Thus, while there is a general correlation between the ability of a metal to form bonds, and how easily species bind, the correlation does not work on nontransition metal surfaces, and it does not work for some species (i.e., CO, water).

In a more fundamental way, the material later in this chapter shows that the binding of molecules on nontransition metal surfaces is considerably different from the binding of molecules on transition and metal surfaces. Neither the simple thermodynamic correlations of the type proposed by Tanaka and Tamaru, nor Flores' model account for these differences, so neither correlation works with nontransition metals. Thus, some care must be taken when applying Flores' work or the Tanaka–Tamaru correlations to nontransition metals, and to adsorption of certain species.

3.4.1 Some Results from Density Functional Theory

The inability of Flores et al.'s methods and the Tamaru–Tanaka correlations to work for certain key species and certain key surfaces is a real limitation. If one starts with a new molecule or a new surface, one has no idea whether the correlations will work or not unless one does actually measure the quantities one is using the correlations to predict. As a result, the correlations have limited predictive value, even though they are quite good at correlating some data.

On a more fundamental basis, the correlations are asserted rather than derived. They do not have testable assumptions. As a result, one does not know whether they are going to work for any specific example. Recently, however, Parr and Yang [1989] have shown that a modification of Flores et al.'s results, Equation 3.37, can be derived from first principles for the interaction of two molecular species. The same arguments can be applied to surface reactions. The advantage of Parr and Yang's analysis is that one can identify classes of molecules and surfaces where correlations of the type proposed by Flores et al. and by Tanaka and Tamaru will work, and other classes of molecules where they will not work. As a result, Parr and Yang's analysis is quite useful.

Parr and Young's analysis [1989] started with something called density functional theory. To review the arguments in conventional quantum mechanics, one calculates a wavefunction as a function of the atomic core positions and uses the wavefunction to calculate bond energies. However, Kohn and Sham [1965] showed that one can just as well do calculations in terms of the electron density ρ_e where ρ_e is given by

$$\rho_e(\vec{r}) = \psi^*(\vec{r})\psi(\vec{r}) \tag{3.40}$$

In density functional theory one treats the total energy of the system E_T as a functional of the electron density, ρ_e, and $\Omega(r)$ the potentials associated with the interaction of the atomic cores with both the electrons and the other atomic cores in the system.

Now consider starting with a molecule and a surface that are separated by a long distance and imagine moving them together. The bond energy of the system is a state property that can be calculated via any thermodynamic pathway, so consider first moving the molecule and surface together without distorting the electronic structure of either the molecule or the surface, and then allowing the electronic structure of each species to change by, for example, transferring n_a electrons and allowing the various atomic clouds to rearrange. One can write the energy change of the system in such a situation as

$$\Delta E[n_a, \Delta\Omega] = \int \left(\frac{\delta E_T}{\delta\Omega}\right)_{n_a=0} d\Omega + \int \left(\frac{\partial E_T}{\partial n_a}\right)_\Omega dn_a \tag{3.41}$$

The first term on the right of Equation 3.41 is the energy to bring the two species together, while the second term is energy associated with bonding. The notation $(\delta E/\delta\Omega)$ is used to indicate that the first term is a functional derivative rather than a regular derivative. That distinction is important for the discussion in this section, although it is relevant in Appendix B.

The first term in Equation 3.41 is associated with a process where a gas approaches a surface without transferring electrons. The process is somewhat akin to physisorption. With a metal the first term is going to be small, and so one can easily approximate it by the heat of physisorption. Of course, with an ion on an ionic surface the first term in Equation 3.41 will be large. In that case one will have to do a better job of the analysis.

Now let's concentrate on the second term in Equation 3.41. The second term in Equation 3.41 represents the energy change when a covalent bond forms between the metal and the surface. Let's consider the energy change of the molecule and the surface separately. To some approximation one can expand the energy change of the molecules during bond formation in a Taylor series in the bond order to obtain

$$\Delta E_M = \chi'_M n_M + \eta'_M \frac{n_M^2}{2} \tag{3.42}$$

$$\Delta E_s = \chi'_s n_s + \eta'_s \frac{n_s^2}{2} \tag{3.43}$$

where χ'_M and χ'_s are the electronegativity of the adsorbate and surface, and η'_M and η'_s are constants called the **hardness** of the adsorbate and the surface. The superscripts on the electronegativity and hardness in Equations 3.42 and 3.43 indicate that the electronegativity and hardness in Equations 3.42 and 3.43 are the electronegativity and hardness when the adsorbate is close to the surface. The hardness is discussed on the next several pages, but basically it is a measure of the ability of a molecule to share electrons. According to Equation 3.27, when the hardness is small n_a, the number of electrons that are shared will be large, while when the hardness is large, n_a will be small. Thus, the hardness is inversely related to the ability of a molecule to share electrons.

Continuing the derivation one can write the total energy as

$$E_T = E_M^0 + E_s^0 + \Delta E_M + \Delta E_s$$

$$= E_M^0 + E_s^0 + n_m(\chi'_M - \chi'_s) + (\eta'_M + \eta'_s)\frac{n_M^2}{2} + \int \left(\frac{\delta E}{\delta \Omega}\right)_{n_a = 0} d\Omega \tag{3.44}$$

where we have noted that $n_m = -n_s$. Equation 3.44 is identical to Equation 3.25 except that there is an extra term that does not depend on n_m. If one plugs Equation 3.44 into the derivation on page 134, one obtains

$$D(M\text{–}S) = \frac{D(M\text{–}M)}{2} + E_s^0 + \frac{(\chi'_M - \chi'_s)^2}{2(\eta'_M + \eta'_s)} + \int \left(\frac{\delta E}{\delta \Omega}\right)_{n_a = 0} d\Omega \tag{3.45}$$

Equation 3.45 is similar to Equation 3.37, but there are some important differences. First, Equation 3.45 was derived from Equation 3.41, which is a rigorous quantum mechanical result. The only approximation made in deriving Equation 3.45 is that the energy was assumed to follow a Taylor series expansion in n_a. Parr and Yang [1989] show that this is an excellent approximation for molecules. Further, the third-order term in the Taylor expansion is always negligible. No one has tested this equation for adsorption. However, the arguments in Parr and Yang, can be applied directly to adsorption. Thus, Equation 3.45 should be essentially an exact result for adsorption even though it has not yet been tested for that purpose.

A second important difference between Equation 3.37 and Equation 3.45 is that the electronegativity and hardness in Equation 3.45 are the electronegativity and hardness when the adsorbate is held at the position of the chemisorbed bond. That may be somewhat different than the electronegativity and hardness when the adsorbate and surface are separated. Rodriguez, Campbell, and Goodman [1992] show that such changes can be quite important. See their papers for details.

Parr and Yang [1989] note that one can always calculate the electronegativity and hardness via a thermodynamic pathway. They start with the values of the electronegativity and hardness when the molecule and surface are separated, then calculate the changes in the electronegativity and hardness that occur when the molecule and surface come together. The result is

$$\chi'_M = \chi_M + \int \frac{\delta \chi_M}{\delta \Omega(\vec{r})} \, d\Omega(\vec{r}) \qquad (3.46)$$

$$\eta'_s = \eta_s + \int \frac{\delta \eta}{\delta \Omega(r)} \, d\Omega(\vec{r}) \qquad (3.47)$$

The second term in Equations 3.46 and 3.47 is called a **Fukui function.** The Fukui functions in Equations 3.46 and 3.47 account for the fact that the ability of a molecule or a surface to share electrons changes as the molecule approaches the surface. Where that is really important is in $\pi*$ back-bonding. As noted earlier, when CO bonds with the surface, electrons are transferred from the σ orbitals in the CO to the metal. In addition, there is back donation of electrons from the metal into the $\pi*$ orbitals of the CO. The derivation of Equation 3.28 only allowed electron flow in one direction. However, Equation 3.46 allows electron flow in both directions. The idea is that the effective electronegativity of the CO can change in the presence of the metal. That change increases the electronegativity of electrons in the π state, which allows back-bonding to occur.

The other key feature of Equation 3.45 is that it allows one to distinguish between the behavior of a metal and an insulator. A metal gives up electrons easily. That translates into a small value of η'_s and therefore a strong covalent interaction. An insulator, on the other hand, does not give up electrons easily. That translates into a large value of η'_s and therefore weak covalent bonding, but strong ionic bonding. Therefore, the implication of Equation 3.45 is that the reactivity of metals and insulators could be quite different. That is as observed, (see below).

More importantly, Equation 3.45 may allow one to predict what kinds of species interact strongly with surfaces. Pearson [1968] and Parr and Yang [1989] examined a similar question for the binding of ligands to inorganic complexes. Pearson noted that ligands are held to inorganic molecules by a mixture of "ionic" and "covalent" bonds. The covalent interactions are associated with the second and third terms on the right of Equation 3.45, while ionic interactions are associated with the fourth term on the right of the same equation. Each of these terms is independent. Therefore, there is no reason to suppose that molecules that form strong ionic bonds will also form strong covalent bonds, or vice versa. Pearson [1963, 1968] proposed that one could divide ligands and metal ions into three separate classes: those species that prefer to be held by mainly ionic forces, those species that prefer to bind by mainly covalent forces, and those species that are held by an equal mixture of ionic and covalent forces. Generally, species that form strong ionic bonds will have a large value of $(\delta E/\delta \Omega)$, while species form strong covalent bonds will have large values of E_s^0 and small values of η. Species that interact mainly by covalent forces are called **soft** acids and bases; species that interact mainly by "ionic" forces are called **hard** acids and bases; and species that interact by a mixture of "ionic" and "covalent" forces are called **borderline** acids and bases. Pearson [1968] shows that generally, hard acids interact strongly with hard bases, while soft acids interact strongly with soft bases. However, the interaction between a hard acid and a soft base or a soft acid and a hard base is relatively weak.

Table 3.8 shows a selection of the acids and bases considered by Pearson. Note that fairly ionic species are hard bases, while less ionic species are soft bases. Metal surfaces are soft acids or bases, semiconducting oxides (e.g., NiO, ZnO) would be borderline acid/bases, while most insulating oxides (e.g., Al_2O_3, SiO_2) would be hard acids or bases. According to

TABLE 3.8. A Selection of Hard, Soft, and Borderline Acids and Bases Discussed by Pearson [1968] with Additions Due to Parr and Yang (1989)

Hard acids	H^+, Li^+, Mg^{2+}, Cr^{3+}, Co^{3+}, Fe^{3+}, Al^{3+}, $Al(CH_3)_3$, bulk silicon
Soft acids	Cu^+, Ag^+, Pd^{2+}, Pt^{2+}, $Ga(CH_3)_3$, O, Cl, N, RO, ROO, metallic atoms, metal clusters
Borderline acids	Mn^{2+}, Fe^{2+}, Co^{2+}, Ni^{2+}, Cu^{2+}, Ru^{2+}, Os^{2+}, Ir^{3+}, Rh^{3+}
Hard bases	F^-, Cl^-, H_2O, NH_3, OH^-, $CH_3\,COO^-$, RO^-, ROH, O^-
Soft bases	I^-, CO, C_2H_4, $P(C_6H_5)_3$, C_6H_6, H^-, H_2S, metal surfaces
Borderline bases	C_5H_5N, NO_2^-, SO_3^{2-}, $C_6H_5NH_2$

Pearson's classifications bulk silicon would be a borderline acid/base. However, later in this chapter we note that silicon has something called a "surface state," which is soft.

Klopman [1968, 1974] used perturbation theory to quantify these ideas for inorganic molecules. His analysis used a simplified model, where it was assumed that the main interaction between two species could be modeled as an electron transfer between a Lewis acid and a Lewis base. The unoccupied orbitals localized on an atom in the acid interact with the occupied orbitals localized on an atom in the base to form a bond. Klopman further assumed that none of the orbitals are distorted on the acid or base during the reaction and that there was little charge transfer. Klopman then used perturbation theory to derive an expression for the strength of the interaction between the Lewis acid, A, and the Lewis base, B. The result was

$$\Delta E_r = \left[\frac{-Z_A Z_B}{R_{AB}\epsilon} \right] + 2 \sum_{\substack{empty\,orbitals \\ on\,the \\ acceptor\,a_{mo}}} \sum_{\substack{filled\,orbitals \\ on\,the \\ donor\,b_{mo}}} \frac{(C_A^{a_{mo}} C_B^{b_{mo}} \Delta\beta_{AB})^2}{E_{b_{mo}} - E_{a_{mo}}} \qquad (3.48)$$

where ΔE_r is the interaction energy between A and B; Z_A and Z_B are the charges on atoms A and B before reaction; R_{AB} is the distance between A and B; ϵ is the dielectric constant of the media; $C_A^{a_{mo}}$ is the coefficient of the atomic orbitals of atom A in molecular orbital a_{mo}; $C_B^{b_{mo}}$ is the coefficient of the atomic orbitals of atom B in molecular orbital b_{mo}; $E_{a_{mo}}$ and $E_{b_{mo}}$ are the energies of the a_{mo} and b_{mo} molecular orbitals, and $\Delta\beta_{AB}$ is the strength of the interaction between atoms A and B. Klopman calls β_{AB} a "resonance energy" and notes that it is related to the overlap of the orbitals on atoms A and B and the polarizability of the atoms. The factor of 2 arises because of the way β is defined. The first term in Equation 3.48 represents the electrostatic interaction between the molecules, while the second term represents the covalent interaction.

Klopman noted that the second term dominates when the energy difference in the denominator of Equation 3.48 is small and $\Delta\beta_{AB}$ is large, while the first term dominates under the opposite circumstances.

Klopman then noted that some acids and bases have properties that maximize the ionic interactions and some that maximize the covalent interactions. He therefore proposed that the former group could be classified as hard acids and bases, while the latter group could be classified as soft acids and bases. Physically soft acids and bases are good electron acceptors and donors, while hard acids and bases are weak electron acceptors and donors.

Klopman found that in general hard acids, hard bases, soft acids, and soft bases have the following characteristics:

Hard acid: An acceptor with no low-lying unoccupied orbitals so that it has a small affinity for electrons and remains positively charged during a reaction. Such species will have a small $\Delta\beta_{AB}$ and a very negative $E_{b_{mo}}$. Examples include solvated ions of Al^{3+}, Mg^{2+}, H^+, and surfaces such as alumina or silica.

Hard base: A donor with no high-lying donor orbitals, so that it has little capacity to donate electrons and a small value of $\Delta\beta_{AB}$. Examples include F^-, OH^-, H_2O, amines, and surfaces such as MgO or TiO_2.

Soft acid: A species that easily accepts charge. Generally, the species will have a high affinity for electrons, and a high polarizability (i.e., large $\Delta\beta_{AB}$) so that it can easily form covalent bonds. Examples include Hg^{2+}, Ag^+, Pt^+, and most small metal clusters.

Soft base: A species that easily gives up charge. Generally, the species will have a low electronegativity and a high polarizability (i.e., large $\Delta\beta_{AB}$) so that it can easily form covalent bonds. Examples include I^-, RS^-, and H^-, and most metal surfaces.

At present these ideas have not been extensively applied to adsorption on surfaces. However, they are potentially quite useful. Notice that H_2S and CO, which are soft acids, bind strongly to metals and only weakly to insulating oxides. In contrast, water, which is hard, only interacts weakly with most clean metal surfaces, but it interacts strongly with silicon and most insulating metal oxides. Thus it appears that Pearson's classifications will be useful for surface reactions, even though the classifications have not yet been extensively applied for that purpose.

Further, Parr and Yang's [1989] results show that one can quantify the ideas. For example, water has a much larger value of η than most other species, which explains why water only binds weakly on transition metals. In contrast H_2S has a small value of η, so it can form a strong covalent bond to a metal.

One can explain many of the trends in Table 3.2 using these ideas. For example, each of the groupings of metals in Figure 3.5 corresponds to metals with similar electronegativities and hardnesses. It is easy to understand why all of the metals in each group behave the same. Similarly, O_2, H_2, and N_2 behave quite similarly, and quite differently than H_2O, which is expected, since H_2O is much harder than all of the other species.

Of course, there are some exceptions to these rules. Copper is less reactive than platinum even though the hardness and electronegativity of the surfaces are similar. CO and H_2 do not readily adsorb on potassium even though the electronegativity of potassium is favorable for bond scission. Thus, there are a few cases that are not explained by these simple models.

Generally, the problem is that Pearson's simple models were proposed for simple inorganic molecules, and do not consider the role of the d-electrons in the chemistry. The absence of d-interactions leads to the exceptions cited in the last paragraph.

The d-electrons are especially important for CO. CO has an intermediate value of η that allows a moderate covalent interaction with most species. However, CO also interacts via π-back-bonding. Later in this chapter we note that π-back-bonding is strongly attenuated in the absence of d-electrons, which explains why CO only interacts weakly with silicon. At present, Parr and Yang's ideas are well known in the density functional literature. However, they have not been applied to adsorption in a detailed way. Still, the ideas are mathematically rigorous in the absence of d-electrons and do seem to offer some interesting insights into the nature of adsorption in many cases.

In a more general way, Parr and Yang's analysis tells us when the correlations of Tanaka and Tamaru or Flores et al. should work and when they should fail. Flores et al.'s correlation basically assumes that a soft species such as an atom is reacting with a soft surface like a transition metal. Thus, these correlations would be expected to work only with soft species. The Tanaka–Tamaru correlations are more general in that one gets Tanaka–Tamaru relations for both hard–hard and soft–soft interactions. However, the correlations would not be expected to work in the hard–soft case.

3.4.2 Electronegativity Equalization

Another idea that comes out of Parr's work is that the principle of electronegativity equalization can be used to say something about the nature of the species that bind to the surface. The easiest way to derive the principle of electronegativity comes from thermodynamics. Consider two species A and B that are brought together. One can define μ_A as the zero K limit of the chemical potential of an electron on species A by

$$\mu_A = \frac{\partial E_A}{\partial n'_A} = \frac{\partial E_A}{\partial n_A} = \chi_A \qquad (3.49)$$

where again E_A is the energy of species A; n'_A is the total number of electrons on A; and n_A is the number of electrons that A transfers; and χ_A is the electronegativity of A. A similar definition also applies for B

$$\mu_B = \frac{\partial E_B}{\partial n_B} \qquad (3.50)$$

At equilibrium, the chemical potentials of the electrons on both species must be equal. If one substitutes Equations 3.42 and 3.43 into Equations 3.49 and 3.50 and then sets μ_A and μ_B equal and solves for n_a, the number of electrons that are transferred, one arrives at Equation 3.27. This is an important result, because it says that if two species have different electron chemical potentials, then at equilibrium electrons will flow from one species to the other to equalize the chemical potential.

Now where that becomes really important is that one can use the principle of electronegativity equalization to learn what types of species can bind to surfaces. Consider the binding of a methoxy species (CH_3O) to a metal surface. One could conceivably form a CH_3O^+ species, a CH_3O^- species, or a neutral covalently bonded CH_3O species. In the gas phase CH_3O^+ has an electron affinity of 8.3 eV, while a typical metal surface has a work function (electron affinity) of 4–6 eV, as indicated in Table 3.9. Now consider bringing a CH_3O^+ ion up to a metal surface. In the gas phase the CH_3O^+ ion has a much larger electron affinity than a typical metal. The electron affinity of the CH_3O^+ ion could change slightly near a metal, but that effect is usually less than 0.2 eV. As a result, an adsorbed CH_3O^+ ion will have a much larger electron affinity than a typical metal. Therefore, the principle of electronegativity equalization implies that at equilibrium an electron will flow from the metal to the CH_3O^+ ion, producing a neutral species. Similarly, a gas phase CH_3O^- species has an ionization potential of 1.57 eV, while a typical metal has an electron affinity of 4–6 eV. The

TABLE 3.9. Work Functions of a Variety of Surfaces Extracted from the *CRC Handbook*

Surface	Clean Surface Work Function (eV)	Surface	Clean Surface Work Function (eV)
Ag(111)	4.74	Ir(210)	5.00
Ag(110)	4.52	Mo(110)	4.95
Au(111)	4.24	Mo(111)	4.55
Cu(111)	4.94	Mo(332)	4.55
Cu(110)	4.48	Mo(112)	4.36
Cu(100)	4.59	Pt(111)	5.70
Fe(100)	4.67	Ni(111)	5.35
Ir(111)	5.76	Ge(111)	4.80
Ir(100)	5.67	Si(111)(p)	4.60

electron affinity of a CH_3O^- will not change more than perhaps 0.3 eV upon adsorption. Therefore, the principle of electronegativity equalization implies that if a metal would bind to a CH_3O^- species, an electron would flow from the CH_3O^- to the metal, again producing a species that is almost neutral. A more detailed analysis indicates that the final species is not quite neutral. Instead there is a small residual net negative charge in the metal/oxygen bond. However, the implication of the principle of electronegativity equalization is that the final species must be nearly neutral.

One can generalize these results using the data in Table 3.9 and Table 3.10. Generally, transition metal surfaces are good donors and acceptors of electrons. As a result, ions tend to be neutralized when they approach a metal surface. A more detailed analysis indicates that most clean metal surfaces are weak Lewis bases. Metal surfaces donate electrons easier than they accept them. As a result, most of the intermediates one observes when species bind to metal surfaces are neutral or slightly negative. Positively charged species are rare.

There are exceptions to that general rule, however. For example, potassium metal is such a strong Lewis base that potassium can donate electrons to most transition metal surfaces: i.e., the electron affinity of potassium is so low that K^+ is stable on many metal surfaces. Experimentally, Taylor and Langmuir [1933] found that when they cover a tungsten surface with cesium or potassium, Cs^+ and K^+ ions desorb from the surface upon heating. Thus, it is clear that Cs^+ and K^+ can form on a metal surface. In general, most species are nearly neutral when they adsorb on a metal surface. However, a very electropositive or very electronegative species will retain its net charge.

One additional complication is that the presence of adsorbate can change the character of surfaces to allow ions to form. The case of hydronium ions (H_3O^+) is an example of this. In the gase phase, hydronium is a weak acid. The data in Table 3.10 indicates that bare hydronium ions have a higher electron affinity than most metal surfaces, so a bare hydronium ion should be neutralized when it adsorbs on a clean metal surface. However, a hydrated hydronium ion is a stronger acid. The data in Table 3.10 indicate that hydrated hydronium ions have an electron affinity somewhere between that of Pt(111) and Cu(110) surfaces. As a result, hydrated hydronium ions could be stable on a Pt(111) surface but not a Cu(110) surface. In recent studies, Sass et al. [1991], Lackey et al. [1991], and Kizhakevariam and Stuve [1992] have examined the coadsorption of water and hydrogen on Pt(111) and Cu(110). At low water coverages the hydrogen and water adsorb separately on Pt(111). However, as one raises the water coverage above one monolayer, hydronium ions form on Pt(111). In contrast, no hydronium ions are detected on Cu(110). Note that the principle of electronegativity equalization implies that a hydrated hydronium ion could be stable on Pt(111) but not on Cu(110). Thus, it is not surprising that a H_3O^+ ion forms on Pt(111) but not on Cu(110). In general, metal surfaces will neutralize ions. However, one can stabilize an ion

TABLE 3.10.   The Electron Affinity and Ionization Potentials of a Series of Gas Phase Species

Species	Electron Affinity (eV)	Species	Ionization Potential (eV)
CH_3O^+	8.6	CH_3O^-	1.57
H^+	13.65	H^-	0.754
CO^+	14.01	CO^-	1.4
OH^+	13.00	OH^-	1.825
K^+	4.341	Cl^-	3.617
$HCOO^+$	12.8	$HCOO^-$	3.23
H_3O^+	6.11	O^-	1.461
Hydrated H_3O^+	4.86	F^-	3.399

Source: Extracted from the tables of Hotop et al. [1985], Rosenstock et al. [1977], Levin and Lias [1981], Lias et al. [1988]; the data of Gomer [1977], and calculations of Talbi [1988].

in the presence of water, or oxygen (i.e., metal oxide), or other hard species. Ions can form in the presence of other hard species.

The preceding arguments are very general. One can use Lias's tables of electron affinities to predict whether a species will be an ion or neutral when it adsorbs on surfaces. One can use the electromotive series in the *CRC Handbook* to get values for hydrated ions. Gomer and Trison [1977] showed that one has to subtract 4.86 eV from the liquid values to keep everything on the same energy scale. Lackey et al. [1987, 1991] recommend instead subtracting 4.4 eV to account for the heat of physisorption of water. Generally, one would expect most species on metal surfaces to be nearly neutral. The principle of electronegativity equalization implies that one can get charged species only when the adsorbate is a strong Lewis acid, or Lewis base, or stabilized by other species. More precisely, the equilibrium between the charged and neutral species is given by the Nernst equation

$$\Delta G_{ionization} = kT \ln \left(\frac{N_{charged}}{N_{neutral}} \right) \tag{3.51}$$

where $\Delta G_{ionization}$ is the difference in the electron affinities of the charged and neutral species; $N_{charged}$ and $N_{neutral}$ are their concentrations; k is Boltzmann's constant, and T is temperature. If we plug numbers into Equation 3.51, one finds that generally the fraction of charged species should be small. This prediction agrees with experiment.

Interestingly, if one applies the same analysis to an insulating oxide (e.g., Al_2O_3, SiO_2), one gets almost the opposite result. Alumina and silica are typically completely ionic. The strongest Lewis acid sites on alumina have an electron affinity of 7.8 eV, while an aluminum cation has an electron affinity of 24 eV. Therefore, one would expect ionic species to interact most strongly with alumina. Again, this prediction agrees with experiment.

At present, the idea of electronegativity equalization has been used to discuss adsorption on oxide surfaces. It has also been extensively applied to electrochemical systems. However, it really has not been applied in detail for adsorption in UHV, although it would seem to be very useful for that purpose.

Another observation that comes from Parr's analysis is that the Fukui terms (i.e., the second term in Equations 3.46 and 3.47), can have an important effect on how molecules bind. The second term in Equation 3.46 measures how the chemical potential of the molecule and surface change as the molecule approaches the surface. Physically, when the molecule approaches the surface, the presence of empty orbitals on the molecule will make it easier for the surface to lose charge, while the presence of empty orbitals on the surface will make it easier for the molecule to lose charge. Fukui [1975] analyzed these ideas in detail for the interaction of two species, A and B, and noted that the main stabilization occurs when the highest occupied molecular orbitals (HOMO) in A interact with the lowest unoccupied molecular orbitals (LUMO) in B, and vice versa. These ideas are developed in greater detail in Chapter 10.

The fact that a few orbitals lead to the main stabilization leads to some interesting orientational specificity because one only gets strong interactions when the key orbitals overlap. For example, Figure 3.26 shows the HOMO and LUMO for CO. Both show a higher electron density on the carbon end of the molecule. Therefore, CO will tend to bind with the carbon atom toward the surface. These ideas have not been extensively applied to surface reactions, but again they should be quite useful.

In a more general way, Parr's analysis reduces to Flores et al.'s model in the case where one only has soft–soft interactions. A similar result is obtained for the case of hard–hard interactions, as is described later in this chapter. The net result is that we can predict fairly accurate heats of adsorption in many systems with relatively simple calculations.

Of course, these simple calculations fail in some circumstances. The interaction of a hard surface with a soft adsorbate or the interaction of a soft adsorbate with a hard surface is not

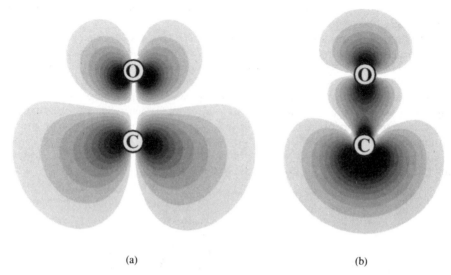

(a)   (b)

Figure 3.26   The LUMO (*a*) and HOMO (*b*) for CO.

well represented by the model. On metals, one usually cannot predict accurate adsorbate positions. Nor can one predict the effects of coadsorbants on the adsorption process. Reactive transition states are not at all well represented and the role of the d-bands is ignored. As a result, while the methods so far in this chapter are quite useful, they have important limitations.

3.5 QUANTUM MECHANICAL METHODS

In principle, quantum mechanical methods similar to those used for molecules or solids should allow one to overcome these limitations. The idea is to start with the Schrödinger equation

$$\mathcal{H}_T(\hat{r})\psi_e(\hat{r}) = E_T\psi_e(\hat{r}) \tag{3.52}$$

and then make some approximations to calculate the energy of the system E_T and the wavefunction for the electrons $\psi_e(\hat{r})$ as a function of the Hamiltonian for the system $\mathcal{H}_T(\hat{r})$. If the approximations are good enough, one can in principle calculate the properties of any adsorbate/surface system. In actual practice, one often has to make some severe approximations to actually do the calculations with a reasonable amount of computer time. As a result, many quantum mechanical methods are not as accurate as one would initially suppose. Still, quantum mechanical methods have the aura of being mathematically rigorous, and they have well-defined assumptions. Thus, quantum calculations are well respected in the literature, even though if one examines them objectively (e.g., do they agree with the experiment?), one does not find that quantum methods are superior to the methods we have discussed so far in this chapter.

In the next several sections, an attempt is made to use quantum mechanics and density functional techniques to quantify the ideas in Section 3.4. Because of length requirements, our discussion of quantum mechanical methods is necessarily brief. However, a more detailed discussion can be found in Holloway and Nørskov [1991], Van Santen [1991], Reutte [1992], Pacchioni et al. [1992], Shustorovich [1991], and Feibelman [1982, 1989]. Our discussion assumes that the reader is already familiar with the use of the Hartree–Fock approximation and density functional methods to calculate the properties of molecules. We also assume that the reader is familiar with the band structure of solids. When we actually teach our surface science course, we find that many of our students are not familiar with that material. As a result, we give lectures on the key ideas. However, this material cannot be put into this chapter. It appears in Appendices B and C. The reader may want to carefully study those appendices before proceeding here. These appendices are available directly from the author.

Appendix B is a review of quantum mechanics and molecular orbital (MO) theory. The most important part of Appendix B is Section B.2 in which the interactions between atoms and molecules are described. There are long-range attractions due to van der Waals forces (these are also called **correlation forces**), short-range attractions due to bond formation, short-range **core–core** repulsions, and repulsions due to the overlap of filled molecular orbitals. The latter repulsions are also called **Pauli repulsions.**

The attractions due to bond formation arise because when unfilled orbitals overlap there is an exchange interaction which allows the electrons in the orbitals to pair up to form bonds. The exchange interaction produces an attractive force. The core–core repulsions arise because the atomic cores are both positively charged, and positive charges repel. The Pauli repulsions are more complicated. When filled orbitals overlap, the electrons in the orbitals repel. The electron clouds deform to reduce the electron–electron repulsion. Pauli showed that it costs energy to deform the wavefunctions. As a result at short distances, filled orbitals on adjacent atoms repel. At long range, however, filled orbitals on adjacent atoms attract. The details of the attractions are rather complex and cannot be explained in detail in this chapter. However, qualitatively, at long range the electrons in A are attracted to the core of B and repelled by the electrons on B, and vice versa. The net force would be zero if the electrons in A and B would move independently. However, if the electrons from atom A move away from atom B whenever the electrons from atom B move close to A, there is a net attraction which lowers the energy of the system. As a result, at long range atoms with filled bands attract. More details of these ideas can be found in Appendix B.

Appendix C is a discussion of the band structure of solids. The most important part of Appendix C for the discussion in this chapter is Section C.1 in which we review the terminology used to discuss the band structure of solids. Recall that when a series of atoms come together to form a solid, the energy levels interact to form a **band.** One usually designates an individual state in the band by the ***k*-vector** of the state, where the k-vector is related to two π over the wavelength of the state in the band. Generally, there are two bands that will be most important for the discussion in this chapter, the **valence band** and the **conduction band.** The valence band is the highest fully occupied band in the pure material (i.e., the fully occupied band with the highest energy), while the conduction band is the band with the next highest energy. In a semiconductor, the valence band is filled while the conduction band is empty. In contrast, in a metal, the conduction band is partially occupied. One can show that if a band is partially occupied in a material, the material can conduct electricity.

People normally discuss band structures of solids in terms of two key quantities: The **Fermi level** and **density of states** (DOS). The Fermi level, E_f, is the chemical potential of the electrons in the band. Qualitatively all of the electron states with energies below E_f are occupied (i.e., filled with electrons) most of the time while all of the states with energies above E_f are empty most of the time. More precisely, the probability that a state with energy E_k is occupied is given by the Fermi function defined in Appendix C. The density of states is defined as the number of states per unit energy in the band. Generally, there are many energy levels in the band, and the levels are not equally distributed over the energy spectrum of the band. The density of states accounts for that. Generally one sees two kinds of states in a density of states plot: a broad distribution of states for states delocalized over the material, and a series of sharp states for states localized on individual atoms. All of this material is described in detail in Appendix C.

3.6 ADSORPTION ON METALS

Our discussion will start with adsorption on metals because that is where there has been the most work. Adsorption on semiconductors and insulators is treated later in this chapter. Researchers began to use quantum mechanical methods in the 1930s to study adsorption on metal surfaces. At the time the electronic structure of molecules and solids was just beginning to be understood. One could do an approximate calculation on a small molecule using an approximation to molecular orbital (MO) theory. Metals were being treated in the context of a free-electron or nearly free-electron model (see Appendix C). Many new computational techniques were developed, and an attempt was made to also apply the methods to surfaces.

3.6.1 Cluster Models of Adsorption on Surfaces

One approach was to treat the adsorbate/surface complex as a molecule. At the time this work was going on, people were just learning how to do MO calculations on molecules. One could do calculations on perhaps 5–10 atoms, with perhaps 30 electrons. Therefore, an attempt was made to find a way to model the adsorbate/surface as a molecule with 5–10 atoms. The way people did that was to treat the surface as a small isolated cluster of atoms that has the same symmetry as the bulk surface, as indicated in Figure 3.27. One then calculates the strength of the bond between the adsorbate and the cluster by solving the Schrödinger equation using either the exact methods described in Appendix B or some approximate methods that are described below.

The earliest adsorbate/surface cluster calculations used the Hückel method. In the Hückel method, one starts with the matrix version of the Hartree–Fock or Kohn–Sham equations (see Appendix B) and then approximates the various terms in the resultant eigenvalue expression. A common approximation is to assume that $\langle \psi_i | \mathcal{H} | \psi_j \rangle$, the strength of the interaction between two molecular orbitals in the adsorbate/surface cluster is given by

$$\langle \psi_i | \mathcal{H} | \psi_j \rangle = \begin{cases} \epsilon_i & \text{for } i = j \\ \epsilon_{xc}^{ij} \langle \psi_i | \psi_j \rangle & \text{for } i \neq j \end{cases} \tag{3.53}$$

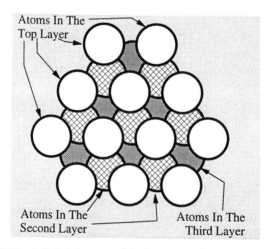

Atoms In The Top Layer

Atoms In The Second Layer

Atoms In The Third Layer

Figure 3.27 A 12,7,6 cluster model of a (111) surface of an FCC metal.

where ϵ_i and ϵ_{xc} are a series of constants that are fit to data on bulk compounds. Generally, the initial results with this method looked quite promising. Hückel calculations were usually able to predict fairly accurate heats of adsorption and adsorbate geometries for atomic adsorbates. However, they work less well for molecular adsorbates, particularly species with π-bonds. Later, people did calculations with the CNDO and MINDO method. CNDO and MINDO are variants of the Hückel methods where better approximations are used for $\langle \psi_i | \mathcal{H} | \psi_j \rangle$, as indicated in Table 3.11. CNDO and MINDO are better techniques for organic molecules, but they did not improve the predictions of the Hückel calculations.

The failure of Hückel calculations for molecular adsorbates was not totally unexpected because Hückel calculations are not particularly accurate for π-bonded species in the gas phase. However, the surprising finding was that the use of more accurate computational methods did not improve the agreement between the calculations and experimental heats of adsorption. Instead the calculations and experiments diverged. At first it was thought that this was a result of inaccuracies in the computational methods. Years ago people thought that there was a "cluster-surface analogy," where reactions on metal surfaces look quite similar to reactions on clusters of 5–20 metal atoms. For example, Plummer, Sanek, and Miller [1978] showed that the UPS spectrum of CO on an Ir(100) surface is very similar to the UPS spectrum of CO in an adsorbed $Ir_4(CO)_{12}$ cluster. However, more recent work described below, has shown that this system is an exception. Usually, a small metal cluster has a very different electronic structure than an extended metal surface. The differences in electronic structure perturb the adsorbate binding energies and reaction pathways so that the binding energies and reaction pathways are quite different on a cluster than on an extended metal surface. As a result, small metal clusters are usually not good models of extended metal surfaces.

To a chemist, this should not be a surprising result. Recall that the reactivity of a benzene molecule is strongly affected by substituents that are placed in the para position 3 Å away from the active site. Physically, the substituent induces what is called a Friedel oscillation in electron density, which induces the reactivity differences. Metals are much more polarizable than benzene. As a result, the Friedel oscillations have a longer range in metals than in benzene. A change 15–20 Å away from a given site can have a major influence on the reactivity of that site. As a result, a cluster that extends 20–30 Å in all directions from the adsorption site is needed to adequately represent adsorption on a metal surface. That means cluster sizes must be in the order of 500 atoms before the cluster can represent a surface. Such a cluster is too large to handle with 1995 computational tools.

Surprisingly, however, cluster calculations remain very popular in the literature. Therefore, it is important to get a picture of the magnitude of the errors. In the remainder of this section we will review the errors. However, one can skip to the next section without loss of generality.

TABLE 3.11. Some of the Approximate Methods Used to Do Cluster Calculations

Method	Overlap Approximation	Approximation to Coulomb Integral	Approximation to Exchange Integral
Extended Hückel	Exact (fixed atomic orbitals)	Equation 3.53 with $\epsilon_{xc}^{ij} = (\epsilon_i + \epsilon_j)$	
CNDO	$S_{ij} = \delta_{ij}$	Calculated self consistently for spherically symmetric orbitals	$\beta_i \delta_{ij}$ where the β_i are a set of parameters fit to data
MINDO	Exact (fixed atomic orbitals)	Calculated self consistently for spherically symmetric orbitals	Equation 3.53 for orbitals on the same atom, zero elsewhere

Calculations have shown that when one creates a cluster one confines the wavefunction to a small volume. The situation is analogous to a particle in a box. Recall that the wavefunctions for a particle in a box has many nodes. The nodes arise because the wavefunction is confined by the walls of the box. In the same way, when one confines the wavefunction to a small cluster, one creates a series of nodes in the wavefunction. Those nodes dominate the properties of the cluster.

It happens that the properties of the cluster only converge very slowly with cluster size. For example, Messmer [1977] showed that one can calculate the properties of cubic clusters analytically using the Hückel approximation with one orbital on each atom. Table 3.12 shows how the electronegativity and hardness of the cluster vary with cluster size for an example with $\epsilon_s = 1$ eV. Notice that the 125 atom cluster still has an electronegativity that is 0.41 eV above that of the 10^6 atom cluster, while the hardness is off by a factor of 17. Messmer [1977] theorized that this result was an artifact of the Hückel calculations. Wood [1980], however, did more accurate calculations, which showed that the workfunction of a 300 atom cluster would often be as much as 0.5 eV lower than the workfunction of a surface. Bauschlichler did CI calculations, which showed that the workfunction oscillates wildly with cluster size. The material in Section 3.4 shows that heats of adsorption often scales with the workfunction of the surface. Thus, based on the ideas in Section 3.4, one would not expect a small cluster to be a good representation of a surface.

Experiments on the properties of small clusters in molecular beams agree with this conclusion. For example, Smalley and co-workers (see Geusic et al. [1985]) and Whetten et al. [1985] found that the ionization potential of small iron, cobalt, niobium, and palladium clusters varies markedly with cluster size as indicated in Figure 3.28. In all cases the ionization potentials of the clusters are far different from those of bulk metals. The reactivity of the clusters also shows large discrepancies. For example, Whetten et al. [1985] found that an eight-atom cobalt cluster barely adsorbs hydrogen even though hydrogen rapidly chemisorbs on bulk cobalt. Whetten et al. [1986] also found that they could correlate the variations in the ionization potential of the clusters with the variation in the reactivity of the clusters. This is just as one would expect from Equation 3.45. Whetten et al.'s and Smalley et al.'s experiments show that heats of adsorption on a small metal cluster are quite different from heats of adsorption on extended metal surfaces. Thus, at first sight one would *not* expect a small cluster to be a good model for heats of adsorption on a metal surface.

Another experimental observation is that mechanisms of reactions on small metal clusters can be quite different from the mechanisms of reactions on extended surfaces. For example, the decomposition of ethylene has been examined by many investigators. Much of the data are reviewed by Yagasaki and Masel [1994]. When ethylene decomposes on the closed packed faces of platinum, palladium, rhodium, ruthenium, osmium, iridium, and cobalt, an intermediate called triply coordinated ethylidyne forms. A diagram of the ethylidyne is shown in

TABLE 3.12.   The Variation in the Properties of a Series of Cubic Clusters

Number of Atoms on a Side	Total Number of Atoms	χ_s, eV (work function)	η_s, eV
2	8	1.50	1.50
3	27	2.12	1.41
4	64	2.42	1.21
5	125	2.59	1.03
8	512	2.82	0.71
10	1000	2.87	0.58
50	1.26×10^5	2.994	0.12
100	10^6	3.00	0.06

Source: Calculated from the Hückel calculations of Messmer [1977].

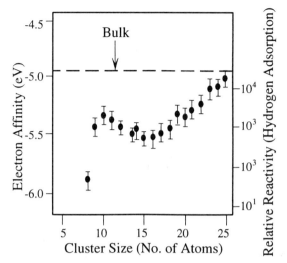

Figure 3.28 The variation in the ionization energy and reactivity of small cobalt clusters as a function of cluster size. (Data of Whetten et al. [1985].)

Figure 3.29. However, triply coordinated ethylidyne formation has never been detected during ethylene decomposition on small metal clusters in the absence of excess hydrogen. Instead the main decomposition products are vinyl, vinylidene, and ethan-1-*yl*-2-ylidyne. Slichter and co-workers [1983, 1992] examined the size dependence in detail, and found that vinylidene is formed on small supported platinum clusters, while ethylidyne is formed on clusters with thousands of atoms. Deeming [1974, 1975] found that ethylidyne can be synthesized on a small metal cluster, via hydrogenation of ethan-1-*yl*-2-ylidyne. However, ethylidene is not a direct product of ethylene decomposition. These results show that the pathway of a simple decomposition reaction can be quite different on a small metal cluster than on an extended surface. As a result, there is no reason that one should expect there to be an analogy between a reaction on a small metal cluster and a reaction on an extended surface.

Still, cluster calculations can be done exactly, and commercial codes are available to do the calculations. As a result, there has been considerable effort devoted to adapting cluster calculations so that they do a reasonable job of predicting chemisorption energies and reaction mechanisms.

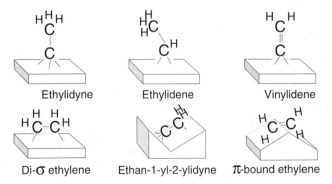

Figure 3.29 Some of the intermediates seen during ethylene decomposition on transition metals.

TABLE 3.13.   The Heat of Adsorption of H_2 on a Threefold Hollow on a Series of Nickel Clusters Designed to Simulate Ni(111)*

Cluster	Heat of Adsorption (kcal/mole)	Cluster	Heat of Adsorption (kcal/mole)	Cluster	Heat Of Adsorption (kcal/mole)
$Ni_4(3,1)$	−8.3	$Ni_{10}(3,7)$	24.7	$Ni_{13}(12,1)$	−23.9
$Ni_{17}(3,7,7)$	−22.5	$Ni_{19}(12,7)$	48.5	$Ni_{20}(3,7,7,3)$	−8.7
$Ni_{22}(12,7,3)$	27.9	$Ni_{25}(12,7,6)$	18.7	$Ni_{28}(12,7,6,3)$	1.9
$Ni_{40}(21,13,6)$	10.5			Experiment Ni(111)	−23

Source: The calculations are from Panas, Siegbahn, and Wahlgren [1987]. Experiment from Christmann et al. [1973].
*In the table a positive heat of adsorption implies that the adsorption process is endothermic (i.e., no hydrogen adsorbs).

Table 3.13 shows what state-of-the-art cluster calculations were like when this book was being written. This is a calculation of the heat of adsorption of hydrogen on a nickel cluster designed to simulate a Ni(111) surface. To keep this case in perspective note that experimentally Riley [1992] finds that nickel clusters show one of the smallest variations in properties with cluster size (see also Zhu et al. [1993]). As a result, in effect, one would expect adsorption of hydrogen on nickel to be the easiest case to examine using the cluster model.

Table 3.13 lists several clusters. All of the clusters have the same symmetry as Ni(111), but a different number of atoms. For example, a $Ni_{25}(12,7,6)$ cluster would be a 25-atom nickel cluster made of three stacked (111) planes. The surface plane has 12 atoms, the second plane has 7 atoms, while the third plane has 6 atoms. Figure 3.27 shows a diagram of a Ni(12,7,6) cluster. Additional cluster diagrams are given in Panas, Siegbahn, and Wahlgren [1987].

The calculations in Table 3.13 were done to the CI level of accuracy and are essentially exact, but only consider adsorption on the (111) sites on the cluster.

Notice that the calculated heats of adsorption fluctuate wildly with cluster size. The heats of adsorption on the 19-, 22-, 25-, 28-, and 40-atom clusters are positive (i.e., the adsorption process is endothermic). In Chapter 4 we show that the heat of adsorption needs to be negative for significant adsorption to occur. Hence, the results in Table 3.13 imply that hydrogen will not adsorb on the 19-, 22-, 25-, 28-, or 40-atom clusters. The heat of adsorption on the 20-atom cluster is negative. However, the binding energy of the hydrogen is such that all of the hydrogen should desorb below 150 K. By comparison, experimentally, hydrogen sticks with near unity probability at 300 K on Ni(111) and does not desorb to 400 K.

Again we emphasize that the results in Table 3.13 are essentially exact. Hydrogen cannot adsorb on the (111) sites on the 19-, 22-, 25-, 28-, or 40-atom clusters. Therefore the implication of the results in Table 3.13 is that essentially exact calculations of adsorption of gases on 20- to 40-atom clusters are not good representations of what happens on surfaces in even the simplest case. As a result, one needs to be exceedingly cautious about using cluster calculations to make predictions about surfaces.

In spite of the failure of exact cluster calculations to reproduce adsorbate behavior in even the simplest case and the large experimental differences between the behavior of clusters and the behavior of extended surfaces, cluster calculations have remained one of the more popular ways to model metal–surface bonds. The approach that is usually taken is to find a way to adjust the cluster calculations so they fit the experimental data in a few simple cases, and then hope that the calculations will have some significance for other cases. At present, there is no evidence that such an approximation works, but nevertheless the approximation has become very popular in the literature.

Panas et al. did extensive work to try to understand why cluster calculations do not fit data for adsorption on extended surfaces. They found that one of the reasons that many of the clusters in Table 3.13 do not adequately represent a surface is that the electrons in the cluster are already paired up into bonds, so there are no dangling bonds available for reaction. As a result, one has to break metal–metal bonds to get adsorption to occur. In contrast, on a Ni(111) surface there are many partially filled orbitals available for bonding. Therefore, Panas et al. asserted that one of the reasons that the clusters give poor energies is that one needs to modify the electronic structure of the clusters before bond formation can occur, while much smaller distortions of the electronic structure are needed on a nickel surface.

Panas et al. then did a series of calculations to see whether "preparing the cluster for bonding" would allow better energies. Physically, preparing a cluster for bonding is equivalent to calculating an exact bond energy and then adding on a correction factor to account for the energy it takes to convert the cluster from its ground state to a state that in some way approximates that of a metal surface. The idea is that while the absolute energies calculated via cluster calculations are substantially in error, the main effect is to produce an energy offset that is a function of the size and shape of the cluster, but not necessarily the specific adsorbate. If one could find the correction factor and if it worked out that the correction factor is not a function of the adsorbate, then one could still predict useful things with a cluster calculation. Panas et al. tried a number of ways to prepare the surface for bonding, and eventually settled on a scheme where the correction factor was the energy it takes to promote the cluster from its ground state to a state that has the same orbital symmetry as the Ni(111) surface. Unfortunately, the excited cluster does not have donor and acceptor orbitals at the correct energies. Further, the final state of the adsorbate on the cluster may or may not represent the correct ground state of the adsorbate because in reality the correction factor depends on the electronegativity and hardness of the adsorbate. Thus, the utility of simple cluster methods is limited.

In my view, so far exact cluster calculations have not proved to be useful models of surfaces. Experimentally, UPS spectra of adsorbed species usually show broad features, characteristic of species that are bound with electrons that are rapidly exchanging with the bulk (see Section 3.8). The exchange process can provide a source or sink for electrons. This is important because when an incoming molecule approaches a surface, the electrons in the surface can easily move out of the way. As a result, the Pauli repulsions discussed in Section 3.2 are reduced. According to Lennard-Jones's model in Section 3.2, such a change will result in a reduction in the activation barrier for dissociation.

Unfortunately, the exchange is difficult to include in a simple cluster model. A bare cluster has very different properties than a surface. As a result, a cluster does not seem to be a useful model of a surface.

Just to quantify the errors, Table 3.14 shows the best results that Panas et al. have obtained for hydrogen adsorption on Ni(111) after preparing a cluster for bonding. These are probably the best CI cluster calculations in the literature. In general, all of the clusters now bind hydrogen. However, in some cases the calculated heats of adsorption are still off by more

TABLE 3.14.   The Corrected Heat of Adsorption of H_2 on a Threefold Hollow on a Series of Nickel Clusters Designed to Simulate Ni(111)

Cluster	Corrected Heat of Adsorption (kcal/mole)	Cluster	Corrected Heat of Adsorption (kcal/mole)	Cluster	Corrected Heat of Adsorption (kcal/mole)
$Ni_4(3,1)$	−8.3	$Ni_{10}(3,7)$	−35.5	$Ni_{13}(12,1)$	−23.9
$Ni_{17}(3,7,7)$	−22.5	$Ni_{19}(12,7)$	−27.9	$Ni_{20}(3,7,7,3)$	−8.7
$Ni_{22}(12,7,3)$	−13.1	$Ni_{25}(12,7,6)$	−19.9	$Ni_{28}(12,7,6,3)$	−28.9
$Ni_{40}(21,13,6)$	−10.9	Experiment Ni(111), Christman et al. [1973]		−23

Source: The calculations are from Panas et al. [1989].

than a factor of 2. Further, the calculated heats of adsorption do not seem to converge with increasing cluster size. Panas et al. note that if one averages all of the numbers ignoring the 4-atom cluster, one finds that the average energy −21 kcal/mol then agrees reasonably well with the experimental value for a Ni(111) surface, −23 kcal/mole. However, this result is very dependent on the clusters one includes in the average. For example, if one only includes the five largest clusters in the calculations, the average heat of adsorption works out to be −16 kcal/mole. By comparison Eley's 1952 calculation gave −17 kcal/mole and the experiments give −23 kcal/mole. Panas et al.'s results show that accurate CI cluster calculations are not good models for adsorption on metal surfaces, even though such calculations continue to appear in the literature. One might anticipate that in the future people will be able to do calculations on larger clusters. According to Wood [1980] a 1000-atom cluster should be a good approximation of a metal surface. However, a 40-atom cluster is not. Thus, one needs to be exceedingly cautious about believing the cluster calculations in the current literature.

At the time this book was being written, cluster calculations had not been extensively applied to bonding on semiconductors or insulators. However, the electron states are much more localized on semiconductors and insulators than on metals. As a result, one might anticipate that cluster calculations will work much better on semiconductors and insulators than they do on metals.

3.7   THE SURFACE ELECTRONIC STRUCTURE

The reason that the cluster calculations failed is that they did not adequately model the electronic structure of the surface. Recall that in the bulk, the electron states of a solid can be represented by a band. The electronic structure of a metal surface is much like bulk electronic structure. There are bands of states similar to those in the bulk plus some additional states that arise because of perturbations due to the surface. The bulk-like states are called **projected bulk states,** while the other states are called **surface states.** A description of both types of states is given in Appendix C. The cluster model fails because it does not allow bands to form. Instead, the electrons pair into bonds. It costs energy to break the bonds when a species adsorbs on the cluster. As a result, the energetics of adsorption on small metal clusters can be quite different from the energetics of adsorption on surfaces.

The way to get around this difficulty is to try to put some information about the electronic structure of the surface into a quantum calculation. Calculations that attempted to do this started in the 1930s. At the time, people were treating the electronic structure of metals via what is called the free-electron model. Therefore, it was useful to try to treat a surface via a similar model.

Just to keep the discussion in perspective, note that there are two kinds of bands in a typical metal, localized bands and delocalized bands. The delocalized bands are also called conduction bands or sp bands. They extend through the metal and carry electricity. The localized bands are also called d-bands. They are more closely held to the atoms in the system. The conduction bands are well represented by the free-electron model but the d-bands are ignored.

3.7.1   The Jellium Model of Metal Surfaces

It ends up that when gases approach a metal surface, most of the bonding is with the conduction band. On a simple (i.e., nontransition) metal, the bond to the conduction band accounts for perhaps $99^+\%$ of the bond energy. On a transition metal, about 90–95% of the bond energy is associated with an interaction with the conduction band, while the other 5–10% is associated with an interaction with the d-band. Thus, if one can model the interaction of the adsorbate with the conduction band, one can understand the main features of binding of adsorbates to surfaces.

Figure 3.30 shows a plot of the wavefunctions of the sp bands near a closed packed face of a simple metal. The wavefunction has two parts, a real part and a complex part. In the bulk, both parts show sinusoidal oscillations. The sinusoidal variation is shown on the left part of Figure 3.31. Now when one cuts a surface, the wavefunction changes. There are no atoms above the surface to conduct electricity. As a result, the wavefunction shows a sinusoidal decay (see Appendix C for details).

Scanning tunneling microscopy (STM) indicates that the sp wavefunction also shows small variations parallel to the surface. However, such variations are not important to the binding

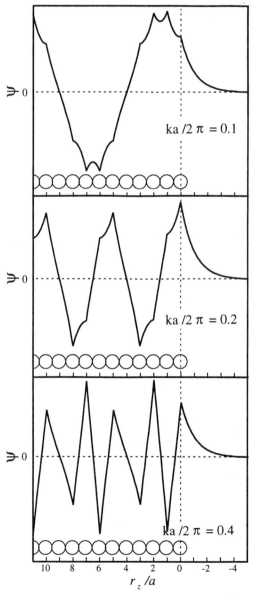

Figure 3.30 The wavefunctions for the s-bands in a metal as a function of r_z, the distance perpendicular to the surface calculated as described in Appendix C.

of molecules to flat surfaces. They are, however, important to the binding of molecules to stepped surfaces.

3.7.2 Quantification of the Jellium Model

If one wants to calculate the binding of the molecule with a metal surface, one could, in principle, solve the Schrödinger equation to calculate the properties of the system exactly. However, that requires a tremendous amount of computer time, so such calculations are rare. A much more common approach is to fit the surface wavefunction to a simplified model, and then use that simplified model to calculate properties.

The simplest simplified model is called the **jellium** model of metal surfaces. The idea in the jellium model is to represent the surface wavefunction, $\psi_{\vec{k}}$, via a free-electron wavefunction below the surface and a sinusoidal decay above the surface, i.e.,

$$\psi_{\vec{k}} = \begin{cases} C_{\vec{k}} \exp\left(-i(k_x r_x + k_s r_y)\right) \\ \qquad \cdot \exp\left(-\left(\dfrac{2m_e E_s^0}{\hbar^2} - k_z^2\right)^{1/2} z\right) & \text{for } z > 0 \\[2ex] C_{\vec{k}} \exp\left(-i(k_x r_x + k_y r_y)\right) \\ \qquad \cdot \left(\cos(k_z z) - \left(\dfrac{2m_e E_s^0}{\hbar^2} - k_z^2\right)^{1/2} \sin(k_z z)\right) & \text{for } z < 0 \end{cases} \tag{3.54}$$

where k_x, k_y, and k_z are the x, y, and z components of the k-vector for the state, r_x, r_y, and z are the x, y, and z components of the position vector, m_e is the mass of the electron, $C_{\vec{k}}$ is a normalization constant, and E_s^0 is a constant. Physically, E_s^0 is the average potential that the conduction electrons see when the electrons are far below the surface. Variations in electron density parallel to the surface are ignored.

One can show that the jellium wavefunction is the eigenstate of the Hamiltonian, $\mathcal{H}_{\text{jell}}$, given by

$$\mathcal{H}_{\text{jell}} = \begin{cases} -\dfrac{\hbar^2}{2m_e} \nabla^2 & \text{for } z > 0 \\[2ex] -\dfrac{\hbar^2}{2m_e} \nabla^2 - E_s^0 & \text{for } z < 0 \end{cases} \tag{3.55}$$

where the surface is presumed to lie at $z = 0$.

In the problem set we ask the reader to verify that the jellium wavefunction equals a free-particle wavefunction for $z < 0$, and an exponential decay for $z > 0$.

Figure 3.31 shows a plot of a jellium wavefunction calculated with Equation 3.54. Notice that the wavefunction looks very much like the s-band wavefunction in Figure 3.30 for $k_a/2\pi < 0.3$. To keep this in perspective, for metals to the left of lead in the periodic table, all of the occupied states have $k_a/2\pi < 0.3$. Thus the jellium model can be a reasonable model for the surface s-band of a metal. The d-band requires additional states, however.

When people use the jellium model, they fit it to bulk band structures. In Section 3.5 we found that in order to fit surface properties, one needs to reproduce the electronegativity and hardness of the surface. Electronegativity and hardness do not directly appear in the jellium model. However, if one adjusts the Fermi-level in the model to the experimental workfunction, one ends up with a surface that has the right electronegativity. Similarly, hardness is

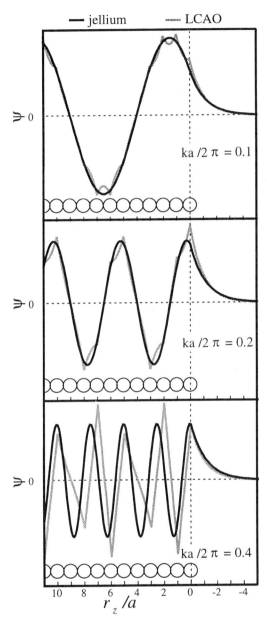

Figure 3.31   A comparison of the LCAO wavefunction for an s-band metal to the jellium wavefunction calculated from Equation 3.54.

not given directly in the jellium model. However, one does have the parameter E_s^0. As one varies E_s^0, one varies the average value of the electron density in the surface. It is easy to show that hardness is related to electron density. Therefore, if one chooses E_s^0 so that the average electron density in the jellium, ρ_{av}, matches the average electron density in the spaces between the atoms in the metal, the hardness of the surface will equal the hardness of the jellium. ρ_{av} is called the **interstitial electron density.** Morruzzi, Janak, and Williams et al.

[1978] provide accurate calculations of the interstitial electron density for the 3d and 4d metals. Experimental values are given in DeBoer et al. [1988] and in Table 3.15.

In some older calculations people did not actually fit the Fermi level in the jellium wavefunction to experimental values. Instead, they set the electron density to be correct, and then calculated the workfunction, assuming that the potential E_s^0 is created as a uniform positive background, and then chose the density of the positive background to be equal to the interstitial electron density and thereby maintained neutrality. This is called a charge-compensated jellium surface. Such an approximation has been widely adopted in the literature. However, it does not do that good a job of predicting the chemical properties of surfaces because, in effect, one is arbitrarily setting the Fermi level of the surface to a value determined by averaged bulk properties.

In the solid-state physics literature it is also common to report a value of a parameter r_S rather than the interstitial electron density, where r_S is defined by

$$\frac{4}{3} \pi r_S^3 = \frac{1}{\rho_{av}}$$

(3.56)

where ρ_{av} is the interstitial electron density in electrons/bohr^3. Experimentally r_S varies from about 1.5 bohr for iron, chromium, and molybdenum (1 bohr = 0.52 Å) to 5 bohr for cesium as indicated in Figure 3.32. The interstitial electron density reaches a maximum (i.e., r_s reaches a minimum) in the middle of the periodic table. Later in this book we find that those metals are also the most reactive for most bond-scission processes.

Figure 3.33 shows a diagram of the electron density outside of a charge-compensated jellium surface for an r_S of 2 and 5. The average value of the electron density is much larger for $r_S = 2$ than for $r_S = 5$. However, there are electron density oscillations below the surface in both cases. When one scales the values, one finds that the amplitude of the oscillations is larger for $r_S = 5$ than for $r_S = 2$, which means that in general metals with a low average

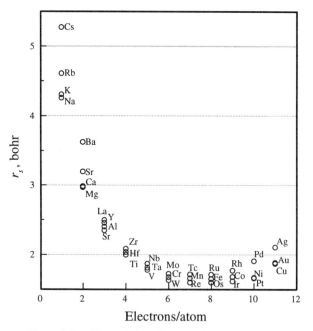

Figure 3.32 The variation of r_s over the periodic table.

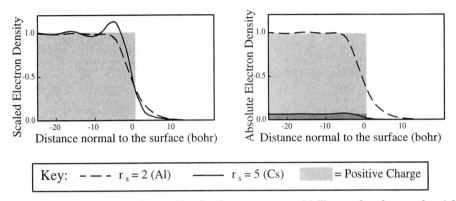

Figure 3.33 and key:

Key: – – – $r_s = 2$ (Al)   —— $r_s = 5$ (Cs)   ▓ = Positive Charge

Figure 3.33 The electron density outside of a charge compensated jellium surface for $r_s = 2$ and 5, after Halloway and Nørskov, [1991]. (*a*) Actual electron density, (*b*) scaled electron density.

electron density will have larger electron density oscillations than metals with a higher electron density.

Another interesting feature in Figure 3.33 is that the electron density spills out about 3 Å above the surface (i.e., the top of the outer electron cloud of the surface atoms) for most metals. These electrons are available for bonding to adsorbates. The electron density above the surface is proportional to the bulk interstitial electron density. As a result, the bulk interstitial electron density should be a key parameter in determining the reactivity of metal surfaces.

3.8 ADSORPTION ON JELLIUM

The advantage of the jellium model is that it allows one to use a relatively simple model of the surface and still calculate some reasonable properties of adsorbates on surfaces. Gurney [1935] proposed a simple formalism to do the calculations. He assumed that an orbital in the adsorbate was interacting with a band of levels in the metal. He was then able to derive some useful results for adsorbate surface binding. A synopsis of the analysis follows.

Consider the interaction of a single adsorbate atom with a surface. For simplicity assume that the energy levels of the surface in the absence of the adsorbate obey

$$\mathcal{H}_s \phi_{\vec{k}} = e_{\vec{k}} \phi_{\vec{k}} \tag{3.57}$$

where $\mathcal{H}_s$ is the surface Hamiltonian, $\phi_{\vec{k}}$ is the wavefunction for a state in the surface band, and $e_{\vec{k}}$ is a constant.

Similarly assume that the adsorbate obeys

$$\mathcal{H}_a \phi_a = e_a \phi_a \tag{3.58}$$

where $\mathcal{H}_a$ is the Hamiltonian for an isolated adsorbate molecule, ϕ_a is the wavefunction for an eigenstate of the Hamiltonian, and e_a is the energy of the eigenstate. When the adsorbate moves close to the surface, the Hamiltonian becomes

$$\mathcal{H}_{eff} = \mathcal{H}_s + \mathcal{H}_a + V_{as} \tag{3.59}$$

where V_{as} is the potential associated with the adsorbate surface interaction. For the material that follows, it is useful to define a set of ψ_l as the eigenstates of $\mathcal{H}_{eff}$. Gurney proposed that one could expand the ψ_l's as a linear combination of the energy levels of the surface, and the adsorbate, i.e.,

$$\psi_l = \sum_{\vec{k}} c_{l,\vec{k}} \phi_{\vec{k}} + \sum_a c_{l,a} \phi_a \tag{3.60}$$

where the ϕ_a and $\phi_{\vec{k}}$ are the wavefunctions for the energy levels of the adsorbate and surface, respectively, and the $c_{l,\vec{k}}$ and $c_{l,a}$ are a series of constants.

Gurney showed that when the adsorbate interacts with a surface the energy levels of the adsorbate are broadened due to interactions with the surface. According to second-order perturbation theory, the individual orbitals are split into a series of bands with a density of states given by

$$\text{DOS}(\epsilon) = \left[\frac{1}{VOL_{zone}} \int\limits_{\substack{Brillouin \\ zone}} |\langle \phi_a | V_{as} | \phi_{\vec{k}} \rangle|^2 \, \delta(\epsilon - \epsilon_{\vec{k}}) \, d\vec{k} \right]^{-1} = \Delta(\epsilon)^{-1} \tag{3.61}$$

where VOL_{zone} is the volume of the first Brillouin zone in k-space.

3.8.1 The Newns–Anderson Model

In actual practice, Gurney did not do quantitative calculations. However, Newns [1969] showed that Gurney's model arose naturally from a formalism developed by Anderson [1961] for defects in solids and that calculations could be done analytically for adsorption on a jellium surface. This work is called the **Newns–Anderson** model of adsorption. The details of the derivation are not very important. The key result, however, is that one can derive the density of states for adsorption on a jellium surface from a modification of Equation 3.61:

$$\text{DOS}(\epsilon) = \frac{\Delta(\epsilon)}{\left(\epsilon - \epsilon_a + \dfrac{\Lambda(\epsilon)}{\pi} \right)^2 + \Delta(\epsilon)^2} \tag{3.62}$$

where $\Delta(\epsilon)$ is given by Equation 3.61 and Λ is given by

$$\Lambda(\epsilon) = \int \frac{\Delta(\epsilon')}{\epsilon - \epsilon'} \, d\epsilon' \tag{3.63}$$

Figure 3.34 shows a schematic of the density of states calculated from Equation 3.62 for two limiting cases: a metal with a broad band and a metal with a narrow band similar to the d-band in a metal. The adsorbate starts out with a series of sharp energy levels. When the adsorbate interacts with the surface, each of those sharp energy levels is split into two bands, a band of bonding levels and a band of antibonding levels. If the surface has a narrow distribution of states, then the two bands will be separate, and one can describe the bonding via a conventional molecular orbital picture. However, the normal situation on a metal or semiconductor surface is that the bonding and antibonding bands are wide enough that they overlap. The result is a combination band centered near the original energy levels of the adsorbate.

The most important application of the Newns–Anderson model has been in the interpretation of ultraviolet photoemission spectroscopy (UPS) data. In UPS one directs an ultraviolet

Surface   Adsorbed   Adsorbate
Complex

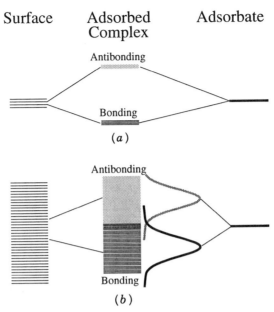

Figure 3.34 A schematic of the density of states calculated via Equation 3.62 for the interaction of an adsorbate with a surface with (a) a narrow band and (b) a wide band.

photon at a surface, and measures the number of electrons that leave the surface as a function of energy. That energy is related to the initial energy, E, of the electrons by

$$E = E_{photo} - 2\pi\hbar\nu \tag{3.64}$$

where E_{photo} is the measured energy and $\hbar$ is Planck's constant and ν is the frequency of the incident photos. To some approximation, the number of electrons leaving the surface at any initial energy is proportional to the density of states at that energy. Thus, to some approximation one can use UPS to measure the density of states of an adsorbate. Figure 3.35 compares the UPS spectrum of N_2O adsorbed on W(110) to the UPS spectrum of N_2O in the gas phase. In the gas phase one observes a series of sharp lines. However, in adsorbed N_2O those lines expand into broad resonances. The results are in qualitative agreement with predictions of Equation 3.62. There is instrumental broadening in the data in Figure 3.35 that complicates the analysis. Also, several peaks are overlapping. Still, the implication of the results in Figure 3.35 are that adsorbate energy levels do broaden upon adsorption, in qualitative agreement with the predictions of Equation 3.62.

Holloway and Nørskov [1991] provide an interesting model to explain why the peaks are broadened. Consider starting with an electron in the adsorbate energy level, and contemplate what happens when the adsorbate moves close to the surface. As the adsorbate approaches the surface, there is some probability that the electron will jump from the adsorbate to the surface and a different electron will jump from the surface to the adsorbate. From the uncertainty principle, resonances that have a finite lifetime, Δt, also have a finite width, ΔE, where

$$(\Delta E)(\Delta t) = 2\pi\hbar \tag{3.65}$$

Therefore, if the electron has a finite lifetime on the adsorbate, the adsorbate levels will be broadened by an amount given by Equation 3.65. Holloway and Nørskov [1991] quantified

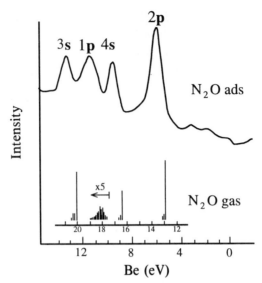

Figure 3.35 A comparison of the UPS spectrum of N_2O adsorbed on a W(110) surface to the UPS spectrum of N_2O in the gas phase. (Data of Masel et al. [1978].)

these ideas by solving the time-dependent Schrödinger equation to show that the probability that the initial electron stays in the adsorbate $P_{init}(t)$ is given by

$$P_{init}(t) = \exp\left(\frac{-2\Delta_0 t}{\hbar}\right) \qquad (3.66)$$

where t is time and Δ_0 is an average value of $\Delta(\epsilon)$. A comparison of Equations 3.65 and 3.66 shows that Δ_0/π is equal to the half-width of the adsorbate energy level. Therefore, the implication of Equation 3.65 is that the adsorbate energy levels are broadened because the electrons spend a finite lifetime on the adsorbate.

Another way to use Holloway and Nørskov's model is to note that the model allows one to use UPS to learn something about how an adsorbate bonds to the surface. One can distinguish between two cases, one where the electron is rapidly exchanging with the bulk and one where the electron is localized on the atom. The lifetime of the electron on the adsorbate will be much shorter if the electron is rapidly exchanging with the bulk. That will lead to a broad resonance. In contrast, if the adsorbate forms a local bond with the surface, only a sharp resonance will be seen. A sharp resonance will also be seen if an ionic bond forms, because an ion does not rapidly exchange electrons with the bulk. A broad resonance will be seen only when there is rapid exchange of electrons with the bulk of the solid. Therefore, one can use the width of UPS peaks to learn something about the bonding of molecules to surfaces.

Experimentally, one can observe both broad and narrow resonances. Typically, when a molecule adsorbs on an insulator, only sharp resonances are seen. Resonances are broader when molecules adsorb on semiconductors. However, it is still often possible to resolve some of the fine structure in the UPS peaks. In contrast on metals one usually observes fairly broad resonances, such as those shown in Figure 3.35, where all of the fine structure is washed out. The implication of the width of the resonance in Figure 3.35 is that the molecules are held on the surface via a fairly delocalized bond where electrons are exchanging with the bulk. On a transition metal surface, one often also observes a few fairly sharp resonances in angular resolved UPS due to interactions with the d-bands. However, the broad resonances usually dominate. Therefore, it seems that delocalized bonding is quite important when adsorbates bond to metal surfaces.

3.8.2 Lang–Williams Theory

Over the years, there have been many attempts to model the delocalized bonds in detail. One of the most important papers was the work of Lang and Williams [1978], who found a way to calculate the energy levels exactly for the interaction of a spherically symmetric atom with jellium. The derivation comes from the Lippman–Schwinger equation, which is described in many standard quantum mechanics books. Consider the interaction of a spherically symmetric atom with jellium. The wavefunctions for the states of the system are eigenstates of an effective Hamiltonian $\mathcal{H}_{eff}$, where $\mathcal{H}_{eff}$ is given by

$$\mathcal{H}_{eff} = \mathcal{H}_{jell} + V_{sph} \tag{3.67}$$

where $\mathcal{H}_{jell}$ is the effective Hamiltonian for the jellium model Equation 3.55 and V_{sph} is the extra spherical potential associated with the adsorbate. The wavefunction ψ_l for any state of the system must satisfy

$$\mathcal{H}_{eff}\psi_l = e_l\psi_l \tag{3.68}$$

Substituting Equation 3.67 into Equation 3.68 and rearranging yields

$$\psi_l = \psi_0 + (\mathcal{H}_{jell} - e_l)^{-1}V_{sph}\psi_l \tag{3.69}$$

where ψ_0 is a particular solution of the Schrödinger equation satisfying

$$(\mathcal{H}_{jell} - e_l)\psi_0 = 0 \tag{3.70}$$

and the $(\mathcal{H}_{jell} - e_l)^{-1}$ term in Equation 3.69 is an operator. Physically, ψ_0 is the surface wavefunction before adsorption. Generally, many k-states satisfy Equation 3.70, and they all contribute to bonding. One needs to include all of them (i.e., put the actual surface wavefunction) in Equation 3.69. Following Lippman and Schwinger, it is useful to define a Green's function $G(\vec{r}, \vec{r}_1)$ to be an operator that satisfies

$$(\mathcal{H}_{jell} - e_l)^{-1}\mathcal{F}(\vec{r}) = \int G(\vec{r}, \vec{r}_1)\mathcal{F}(\vec{r}_1)\, d\vec{r}_1 \tag{3.71}$$

for any arbitrary function $\mathcal{F}(\vec{r}_1)$. Combining Equations 3.69 and 3.71 yields

$$\psi_l(\vec{r}) = \psi_0(\vec{r}) + \int G(\vec{r}, \vec{r}_1)V_{sph}(\vec{r}_1)\psi_l(\vec{r}_1)\, d\vec{r}_1 \tag{3.72}$$

Next, let's show

$$G(\vec{r}, \vec{r}_1) = \int \frac{\psi_{\vec{k}}(\vec{r})\psi_{\vec{k}}^*(\vec{r}_1)\, d\vec{k}}{e_{\vec{k}} - e_l + i\delta} \tag{3.73}$$

where the integral in Equation 3.73 goes over all of the states of the surface, $\psi_{\vec{k}}$ is given by Equation 3.54, $e_{\vec{k}}$ is the energy of $\psi_{\vec{k}}$, and $\delta = 0^+$. According to Equation 3.71, $G(\vec{r}, \vec{r}_1)$ must satisfy

$$\mathcal{F}(\vec{r}) = (\mathcal{H}_{jell} - e_l)\int G(\vec{r}, \vec{r}_1)\mathcal{F}(\vec{r}_1)\, d\vec{r}_1 \tag{3.74}$$

for any arbitrary function $\mathfrak{F}(\vec{r})$, the set of all $\psi_{\vec{k}}$ forms a complete set. Therefore, we can always write any function $\mathfrak{F}(\vec{r})$ as a sum of the $\psi_{\vec{k}}$:

$$\mathfrak{F}(\vec{r}) = \int \mathfrak{F}_{\vec{k}_1}(\vec{k}_1)\psi_{\vec{k}_1}(\vec{r}) \, d\vec{k}_1 \tag{3.75}$$

where the $\mathfrak{F}_{\vec{k}}(\vec{k}_1)$'s are a set of coefficients that still need to be determined.

Let's define $\mathfrak{F}_1(\vec{r})$ via

$$\mathfrak{F}_1(\vec{r}) = (\mathcal{K}_{jell} - e_l) \int G(\vec{r}, \vec{r}_1)\mathfrak{F}(\vec{r}_1) \, d\vec{r}_1 \tag{3.76}$$

Next let's show that if we choose $G(\vec{r}, \vec{r}_1)$ via Equation 3.73, $\mathfrak{F}_1(\vec{r}) = \mathfrak{F}(\vec{r})$. Combining Equations 3.73, 3.74, and 3.75 yields

$$\mathfrak{F}_1(\vec{r}) = (\mathcal{K}_{jell} - e_l) \int \int \int \frac{F_{\vec{k}}(\vec{k}_1)\psi_{\vec{k}}(\vec{r})\psi_{\vec{k}}^*(\vec{r}_1)\psi_{\vec{k}_1}(\vec{r}_1)}{e_{\vec{k}} - e_l + i\delta} \, d\vec{k} \, d\vec{k}_1 \, d\vec{r}_1 \tag{3.77}$$

performing the integral over $\vec{r}_1$ using Equation 3.164 yields

$$\mathfrak{F}_1(\vec{r}) = (\mathcal{K}_{jell} - e_l) \int \int \frac{\mathfrak{F}_{\vec{k}}(\vec{k}_1)\psi_{\vec{k}}(\vec{r}) \, \delta(\vec{k}_1 - \vec{k}) \, d\vec{k} \, d\vec{k}_l}{e_{\vec{k}} - e_l + i\delta} \tag{3.78}$$

Integrating over $\vec{k}_1$ and moving $(\mathcal{K}_{jell} - e_l)$ inside the integral yields

$$\mathfrak{F}_1(\vec{r}) = \int \frac{\mathfrak{F}_{\vec{k}}(\vec{k})(\mathcal{K}_{jell} - e_l)\psi_{\vec{k}}(\vec{r})}{e_{\vec{k}} - e_l + i\delta} \, d\vec{k} \tag{3.79}$$

but

$$\mathcal{K}_{jell}\psi_{\vec{k}} = e_{\vec{k}}\psi_{\vec{k}} \tag{3.80}$$

combining Equations 3.79 and 3.80 and noting $\delta = 0^+$

$$\mathfrak{F}_1(\vec{r}) = \int \mathfrak{F}_{\vec{k}}(\vec{k})\psi_{\vec{k}}(\vec{r}) \, d\vec{k} = \mathfrak{F}(\vec{r}) \tag{3.81}$$

Equality follows from Equation 3.75. The implication of Equation 3.81 is that if we choose the Green's function as in Equation 3.73, then Equation 3.74 will be satisfied for any random function $\mathfrak{F}$. As a result, Equation 3.72 will work as well. An explicit expression for $G(\vec{r}, \vec{r}_1)$ for the jellium model is given in most quantum mechanics texts. For example, see Shiff [1968]. A Green's function for a nearly free-electron surface is given in Masel, Miller, and Miller [1975]. As a result, the Green's function in Equation 3.72 can be looked up for a jellium or nearly free-electron surface.

Equation 3.72 is the key result in this section. Lippman and Schwinger showed that one can use it iteratively. One starts with a guess for ψ_l, which we will call ψ_l^{Guess}. Substituting ψ_l^{Guess} into the right side of Equation 3.72 yields

$$\psi_l(\vec{r}) = \psi_0(\vec{r}) + \int G(\vec{r}, \vec{r}_1)V_{sph}(\vec{r}_1)\psi_l^{Guess}(\vec{r}_1) \, d\vec{r}_1 \tag{3.82}$$

Lippman and Schwinger showed that if one initially sets $\psi_l^{Guess} = \psi_0$, one can plug into Equation 3.82 to calculate a reasonable value for ψ_l. Substitution of that value of ψ_l for ψ_l^{Guess} in Equation 3.82 gives a more accurate value for ψ_l. Repeating further increases the accuracy of the calculations. Under most circumstances, the results eventually converge to the exact wavefunction. As a result, one can calculate eigenstates for the interaction of a spherically symmetric atomic potential with a jellium surface to arbitrary accuracy by iteratively solving Equation 3.82.

In principle, that is all that one has to do to solve Lang and Williams's model. However, that takes considerable computer time. As a result, Lang and Williams actually used a different algorithm to do the computations. They noted that a modification of Equation 3.82 also arises in low energy electron diffraction (LEED). In LEED one shoots electrons at a surface and those electrons scatter from the atoms in the surface. One can compute the scattering rate by starting with an incident wave $\psi_0 = \psi_l^{Guess}$ and using the equivalent of Equation 3.82 to calculate the scattering from the atom. Iteration corresponds to multiple scattering.

In a similar way, the solutions of Equation 3.82 correspond to a scattering problem where an electron from the bulk of the metal goes up to the surface and interacts with the adsorbate atom and scatters. This produces a continuous flow (i.e., exchange) of electrons from the bulk to the adsorbate and back to the bulk of the metal. However, at any instant there is a finite probability of finding an electron in a bond between the adsorbate and the surface. This electron binds the adsorbate to the surface.

Beeby [1967] found a way to solve the scattering problem to arbitrary accuracy for LEED, and Lang and Williams [1978] adapted this formalism to adsorption. In effect, the procedure is to expand the wavefunction near the atomic core via an LCAO basis set, i.e.,

$$\psi_l = \sum_a c_{l,a}\phi_a \quad \text{for} \quad |\vec{r}| < r_{max} \tag{3.83}$$

and then use Equation 3.82 to yield an eigenvalue expression. The details are not important to the discussion here. However, the result is a computationally efficient model of the adsorption process, where the only parameters are the potential, ϕ_a for the adsorbate, and the jellium Hamiltonian for the surface.

Lang and Williams used their model to examine the adsorption of a series of atomic adsorbates on a jellium model of an aluminum surface. They calculated the change in the density of states of the aluminum due to adsorption, and the electron density around each atom as a function of energy. Density of states curves and electron density plots are widely used in the literature, so it is useful to go through an example to see what information can be obtained from the various plots.

Figure 3.36 shows the density of states curves for lithium, chlorine, and silicon adsorption on aluminum. The lithium curve shows a broad peak with a single resonance at the 2s level of lithium. The peak lies above the Fermi level (E_f) of the surface. Note that states above E_f are generally unoccupied. Therefore, the fact that the centroid of the lithium energy level is above the Fermi level implies that the lithium level is unoccupied in the adsorbed state, i.e., electrons will flow from the adsorbate (lithium) to the surface during adsorption. Physically, lithium has a smaller electronegativity than aluminum, so when lithium adsorbs on aluminum, the lithium will give up some charge to equalize the electronegativity of the system. Note, however, that the electronegativities are such that lithium only needs to give up about 2/3 of an electron to equalize the electronegativity of the system. Therefore, there needs to be about 1/3 of an electron on the lithium. More precisely, the electron needs to spend about 1/3 of its time on the lithium. The electron hops back and forth between the lithium and the surface, which leads to a broad resonance. The position of the lithium curve in Figure 3.36 gives information about how many electrons are transferred. The center of the lithium curve lies above the Fermi level. An isolated lithium atom will have all of the states up to the center of the lithium filled, but once the lithium adsorbs only states up to the Fermi level are filled

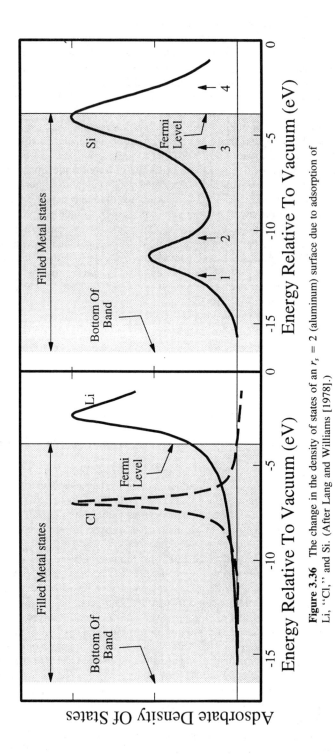

Figure 3.36 The change in the density of states of an $r_s = 2$ (aluminum) surface due to adsorption of Li, "Cl," and Si. (After Lang and Williams [1978].)

(at 0 K). Therefore, the fact that the center of the lithium curve is above the Fermi level implies that lithium gives up electrons upon adsorption.

Note, however, that the Li curve extends below the Fermi level. States below the Fermi level are filled at 0 K. Therefore, the fact that the lithium curve extends below the Fermi level implies that the electrons do spend some time on the lithium. One can calculate the fractional change on the lithium at 0 K by integrating the area under the curve up to the Fermi level. When one does that, one finds that lithium has about 1/3 of an electron, in agreement with the expectations of electronegativity equalization.

Another way to view Figure 3.36 is that the lithium is forming a covalent bond with the surface. When a covalent bond forms, the electron spends part of its time on the lithium and part of its time on the surface. That leads to a finite width resonance, as described in Section 3.8.1.

Figure 3.36 also shows a density of states curve calculated for "chlorine." In this case, there is a single sharp resonance at about 7 eV that Lang and Williams associate with the $3p$ level of chlorine. One has to be a little careful with this plot, because Lias et al. [1988] find that Cl^- has an electron affinity of 3.617 eV, which means that *in the gas phase* the 3p level is only 3.617 eV below the vacuum zero not 7 eV. One would expect the 3p level of Cl^- to shift upon adsorption. However, the broadening of the peak is small, so the shift should be small, too. Therfore, a shift to 7 eV is unlikely. As a result, Lang and Williams's calculations may not really represent chlorine. Still, the calculations show that a negative ion with an electronegativity of about 7 eV will form a bond with a sharp resonance with the surface band.

It is useful to refer back to Equation 3.65 in order to try to interpret this result. According to the equation a sharp resonance implies that the electrons spend a long time on the adsorbate before hopping back to the surface. Therefore, the sharp resonance for "chlorine" in Figure 3.36 implies that the electron spends a long time on the "chlorine" before jumping back to the surface. Note that if electrons were being shared between the metal surface and the adsorbate, one would expect a resonance whose half-width is a reasonable fraction of the metal band width. Thus, the fact that the "chlorine" resonance is so sharp implies that electrons are not being shared. Therefore, the metal chlorine bond must be fairly ionic. It is not completely ionic, because there is still a tail in the energy distribution above E_f. Lang and Williams also show that the average negative charge on the chlorine is also less than 1.

Electron density maps shown in Figure 3.37 reinforce the idea that the "chlorine"–metal bond is mainly ionic. The easiest way to see that is to look at the "total minus superstition" curves in Figure 3.37, which are plots of the changes in total electron density that occur when each of the species adsorb. In the "chlorine" case, the "chlorine" gains electron density upon adsorption while the surface loses electron density. In contrast, in the lithium case the lithium and the surface both lose electron density to form a bond between the surface and the lithium. There is also structure near the core of the lithium. That structure arises because in an isolated lithium atom, the 2s and 2p wavefunctions have a node (i.e., a sign change) at r $\approx$ 1 bohr. Chlorine shows structure near its core, too. However, that structure was artificially suppressed in Figure 3.37.

The implication of Figure 3.37 is that the surface transfers electrons to the "chlorine" when a bond forms, but there is no back donation, which suggests that the bond is completely ionic. In contrast in the lithium case, the adsorbate and the surface both donate electrons to form the bond. That is further evidence that the chlorine–surface bond is mainly ionic, while the lithium–surface bond is mainly covalent.

The silicon case in Figure 3.37 is more complex. The electron density is depleted from the silicon atom and the surface to form a bond between the silicon and the surface. However, unlike the lithium case, the enhanced electron density has a banana shape. The lobe does not look like a sigma bond. Also the electron density is enhanced on the silicon orbitals that are pointing away from the surface. The fact that some of the electron density is enhanced moving

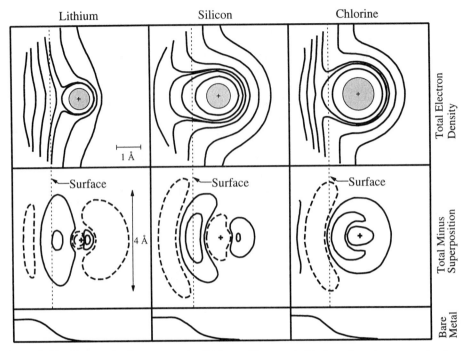

Figure 3.37 Electron density contours for lithium, silicon, and chlorine adsorption on an aluminum surface. The dashed and solid lines indicate places where electron density has been depleted or enhanced, respectively. (After Lang and Williams [1978].)

away from the silicon–surface bond implies that there is a partial antibonding interaction between the silicon and the surface.

Lang and Williams explain this electron density profile with the aid of Figure 3.36 and Figure 3.38. The density of states profile in Figure 3.36 shows two peaks: one corresponding to the 3s level of silicon, and a second corresponding to the 3p level of silicon. Both levels are broadened significantly upon adsorption. which implies that silicon forms a covalent bond

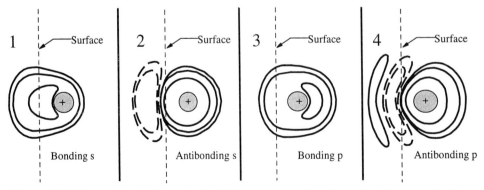

Figure 3.38 Energy selected electron density difference contours for silicon adsorption on an aluminum surface. The dashed lines indicate places where electron density has been depleted. (After Lang and Williams [1978].)

with the surface. The resonances overlap, which implies some sp coupling. However, there are separate resonances for the 3s and 3p levels, which implies that the silicon is not forming an sp^3 hybrid upon adsorption. The 3s level starts out filled, and it remains filled. It broadens, which means that it bonds. However, note that since the 3s level starts out filled there must also be an antibonding interaction.

Detailed examination of the electron density profiles reenforce this view. Figure 3.38 shows the electron density changes on the silicon at four different energies, one below the center of the 3s level, one above the center of the 3s level, one below the center of the 3p level, and one above the center of the 3p level as indicated in Figure 3.36. The electron density at the first position shows a simple enhancement of the electron density around the two atoms. Such an enhancement in electron density is characteristic of a bonding interaction. However, the second case shows a node between the atom and the surface. A node is characteristic of an antibonding interaction. Similarly, the third case represents a bonding interaction while the fourth case represents an antibonding interaction. One again has to be careful with the silicon results in Figure 3.36, Figure 3.37, and Figure 3.38 because the experimental s–p splitting in silicon is much smaller than in Lang and Williams calculations, so in reality sp^x hybrids will form. However, the implication of Figure 3.38 is that the bonding of adsorbates to surfaces can be quite complex, with both bonding and antibonding interactions contributing to the net binding energy.

Lang and Williams also estimated the heats of adsorption. However, the calculations only showed fair agreement with experiment. At the time the calculations were done, people knew how to do accurate density functional calculations for metals, but not molecules. In particular, people did not have accurate atomic core potentials for atoms or molecules. The errors in the core potentials produced inaccuracies in the calculations.

3.8.3 Qualitative Results

However, the real strength of Lang and Williams' calculations is that they allow one to understand what the bonds between an adsorbate and a metal surface are like. For example, Figure 3.37 shows how the electron density changes during adsorption. Notice that the electron density is only strongly perturbed in the region of the adsorbate. The fact that the changes in electron density occur only in the region of the adsorbate has been taken to imply that chemisorption is a fairly localized phenomenon. However, according to Equation 3.82 there is a continuous exchange of electrons between the adsorbate and the bulk. Therefore, it is incorrect to view the bond as a purely localized phenomenon. Rather the adsorbate forms a bond to the entire surface band, and not just to an individual atom. Still the interaction is effectively screened so that most of the changes in the electron density occur close to the adsorbate.

Good evidence that delocalized bonding occurs comes from the UPS data in Figure 3.35. Note that if one were bonding to a single atom, each of the bonds between the adsorbate and the surface would have a fixed energy. Hence, one would expect to observe a series of sharp resonances in UPS. However, the experiments show that instead there are a series of broad resonances. These broad resonances are associated with exchange with the bulk. As a result, while the adsorbate is only producing a local disturbance in the electron density of the surface, the adsorbate is, in effect, bonding with the entire conduction band. One has to be a little careful with that statement, because if one includes d-bands in the calculation, one will get sharp resonances characteristic of localized states. However, the key feature of Lang and Williams's model is that adsorbates can be held to surfaces with delocalized bonds to the conduction band. On simple metals the delocalized bonds dominate. As a result, the bonding in a metal tends to be quite different than the bonding on a small molecule.

Another key prediction from Lang and Williams's model is that the strength of the bond of an adsorbate to a *simple metal* surface is largely determined by a few key properties: the jellium Hamiltonian for the surface and the adsorbate core potential and energy levels as given by Equations 3.72 and 3.73. In Section 3.8.2 we noted that the jellium Hamiltonian is normally fit to the Fermi level (i.e., electronegativity) and interstitial electron density of the metal. In density functional calculations one fits the adsorbate core potential to the hardness and electronegativity of the adsorbate. That produces a model that in many cases correlates quite well with experiment.

For example, in our discussion of Flores et al.'s model in Section 3.4, we noted that one can estimate the heats of adsorption of hydrogen, oxygen, and carbon if one knows the interpotential and workfunction of the surfaces, and the electronegativity of the adsorbates. The fact that such a correlation works follows directly from Lang and Williams's model.

Another thing that has been done in the literature is to correlate the reactivity of surfaces to the parameters in Lang and Williams's model, i.e., the interstitial electron density and workfunction of the surface. For example, Figure 3.39 shows a plot of the rate of electrolysis of water on a series of metals and the workfunction of the metal. In general, there is a fairly good correlation, which implies that the workfunction of the surfaces tells a considerable amount about the reactivity of the surface. There are two lines, however, so the workfunction is not the only predictor of the electrochemical rate.

More recently, Yagasaki and Masel [1994] used these ideas to examine ethylene adsorption on transition metals. When ethylene adsorbs on a metal surface it can form anything from a physisorbed complex with a carbon–carbon bond order of 2.0 to a so-called di-σ complex. (M–CH$_2$–CH$_2$–M) with a carbon–carbon bond order of 1.0. There is also a π-bound complex with a bond order of 1.4 to 1.6. Figure 3.40 shows a correlation between the bonding mode of ethylene and the bulk interstitial electron density. No adsorption is seen on those metals with a low interstitial electron density. Pi-bonding is seen on those metals with an intermediate interstitial electron density, and di-σ bonding is seen on those metals with a high interstitial electron density.

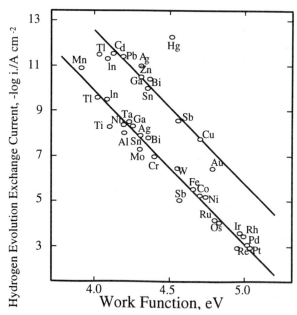

Figure 3.39 A correlation between the rate of electrolysis of water on a series of metals and the electrochemical potential (i.e., workfunction in solution) of the metal. (After Bockris [1954] and Trasatti [1972].)

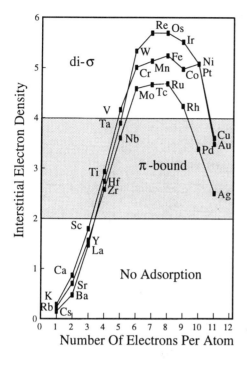

Figure 3.40 A correlation between the bonding mode of ethylene on various closed packed metal surfaces at 100 K and the interstitial electron density of the bulk metal. (After Yagasaki and Masel [1994].)

One can further quantify these ideas. Stuve and Madix [1989] and Bent et al. [1988] showed that one can calculate the carbon–carbon bond order from vibrational data. This calculation gives a bond order of 2.0 for ethylene and 1.0 for ethane. Figure 3.41 shows a plot of the bond order of the carbon–carbon bond adsorbed ethylene on a series of 100 K closed packed metal surfaces against the interstitial electron density of the solid. The interstitial electron density was taken from calculations of Morruzzi et al. [1978] and tables of DeBoer et al. [1988]. Notice that there is a good correlation between the bond order and the bulk interstitial electron density. In fact, the correlation fits the data as well as the data have been measured. Another interesting observation is that the correlation in Figure 3.40 predicts that the bond order in chemisorbed ethylene will be greater than 2 for ethylene adsorption on a potassium surface (interstitial electron density = 0.003 electrons/bohr^3) or on a calcium surface (interstitial electron density = 0.09 electrons/bohr^3). It is impossible for a metal–H$_2$CCH$_2$ complex to have a carbon–carbon bond order greater than 2. As a result, the implication of the correlation in Figure 3.40 is that ethylene should not *chemisorb* on a potassium or calcium surface. By comparison, Hayward [1964] reports that ethylene does not adsorb at all at 100 K on an evaporated calcium or potassium film.

Of course, that does not necessarily imply that no ethylene can adsorb on potassium or calcium at any conditions. One would expect that ethylene would condense (i.e., physisorb) on a potassium or calcium surface at low enough temperatures. However, there are no data that show that yet. Instead, no adsorption is seen even at 100 K. Thus, there is a clear correlation between the bonding mode of ethylene and the interstitial electron density. The form of the correlation is as predicted by Lang and Williams's model.

If one examines the Lang–Williams model more carefully, one finds that there are two key limits: a low surface electron density limit and a high surface electron density limit. In the low electron density limit, the ability of charge to localize on the adsorbate limits bonding. In that case, the interstitial electron density seems to correlate quite well with the bonding. In contrast, in the high surface electron density limit the adsorbate can move in or out to find

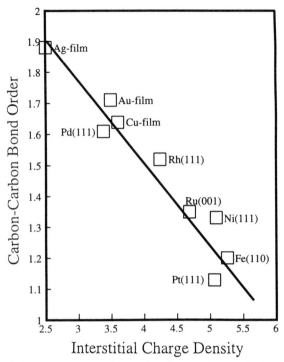

Figure 3.41 A correlation between the carbon–carbon bond order on adsorbed ethylene on various closed packed metal surfaces at 100 K and the interstitial electron density of the bulk metal. (After Yagasaki and Masel [1994].)

its ideal electron density. In that case, one would expect the bonding to be controlled by the difference between the electronegativity of the adsorbate and the surface. At present this prediction seems to work, at least for atomic adsorbates.

There are limitations to Lang and Williams's model for molecules, however. Molecules are more complex than atoms in that multiple bonds form between the adsorbate and the surface. For example, if one adsorbs ethylene onto a 100 K metal surface, and then heats to about 300 K a variety of intermediates such as ethylidyne (CH_3–$C\equiv$), vinylidene ($CH_2=C=$), or ethan-1-yl-2-yldyne result (–CH_2–$C\equiv$) (see Figure 3.29). These species all have very similar interactions with jellium, so one cannot use Lang–Williams' model to distinguish between them. The problem is that the jellium model ignores the spatial variations in electron density. The variations matter to the adsorption of molecules. For example, ethylidyne ($\equiv C$–CH_3) stands up on the surface. It can form easily on a flat surface. However, ethan-1-yl-2-ylidyne ($\equiv C$–CH_2–) needs bonds on both sides of the molecule. Such bonds are difficult to form on a flat surface, but they can form on a stepped surface. The jellium model ignores the difference between a flat and stepped surface, so it cannot reproduce this chemistry.

Hatzikos and Masel [1987] showed that the positions of d-bands also play a role in determining whether ethylidyne, vinylidene, or ethan-1-yl-2-ylidyne forms. Such complexities are difficult to consider with the Lang–Williams model.

Another difficulty is in predicting diffusion energies. The Lang–Williams model ignores the core–core interactions or variations in potential parallel to the surface. As a result, it predicts that there is no barrier to surface diffusion on metals. Experimentally, however, the diffusion process is activated, although the activation energy is usually 10% or less of the heat of adsorption. Generally, the Lang–Williams-type models get binding energies right to within 10% or so. That is good enough for many problems. However, if you want to consider a problem like surface diffusion where 10% variations in bond energy with position are significant or a problem like ethylene adsorption where different intermediates have energies that are within 10% of each other, one needs to do a more sophisticated calculation.

3.9 THE NEED FOR AN IMPROVED MODEL

The reason that Lang and Williams's model does not get the last 10% of the binding energy is that the model only considered the interaction of the adsorbate with a jellium surface. The jellium model is a fair representation of the s-band of a metal. However, the d-band is not well represented by the model, and core–core repulsions are ignored. Also, the jellium model ignores variations in electron density parallel to the surface. Such variations are not important on a flat surface. However, with a stepped surface, the electron density varies over the step, as shown in Figure 3.42. The electron density looks flat over the terrace, but then there is a large change in slope at the step. Such variations are difficult to represent with a jellium model.

Note, however, that when Equation 3.82 was derived it was not necessary to assume that the electronic structure of the surface followed a jellium model. The jellium model comes into the calculations only when one calculates an explicit form for the surface Green's function from Equation 3.73. If the calculations had used a Green's function appropriate for a real surface, the results would be exact. Therefore, there is no reason, in principle, that the Lang-Williams model should have failed for the ethylene case. Rather the failure comes only when one approximates the surface Green's function with a jellium Green's function.

It is easy to show that in principle, Equation 3.73 can be used to calculate the Green's function for a real surface. Note, that there was nothing in our derivation of Equation 3.73 that required the surface band to follow a jellium model. If we would have used the real surface Hamiltonian in Equation 3.74, then all of the derivation would have applied as before. The only change would be that we would have to include all of the eigenstates of the surface Hamiltonian in Equation 3.74, not just the jellium eigenstates, i.e.,

$$G(\vec{r}, \vec{r}_1) = \int_{\substack{core \\ levels}} \frac{\psi_{\vec{k}}(\vec{r})\psi_{\vec{k}}^*(\vec{r}_1)\, d\vec{k}}{e_{\vec{k}} - e_l + i\delta} + \int_{s\text{-}band} \frac{\psi_{\vec{k}}(\vec{r})\psi_{\vec{k}}^*(\vec{r}_1)\, d\vec{k}}{e_{\vec{k}} - e_l + i\delta}$$

$$+ \int_{d\text{-}band} \frac{\psi_{\vec{k}}(\vec{r})\psi_{\vec{k}}^*(\vec{r}_1)\, d\vec{k}}{e_{\vec{k}} - e_l + i\delta} \tag{3.84}$$

Note that the wavefunctions for the s-band and d-band will have mixed s-d character due to the presence of s-d coupling.

It happens that for a simple metal (i.e., a metal without d-electrons) the exact Green's function is not very different than the jellium Green's function. Appendix C shows that the

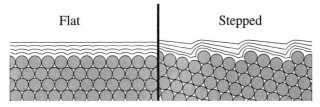

Flat   Stepped

Figure 3.42 A comparison of typical electron density contours on a flat and stepped surface.

free-electron (i.e., jellium) wavefunctions are similar to the nearly free-electron and LCAO wavefunctions except at positions near the cores of the surface atoms, and k-vectors near the band edge. With a simple metal, the corrections for k-vectors near the band edge are not very important to bonding. Initially, the states near the band edge start out empty, and those states usually remain empty during the adsorption process. One does not observe any change in these states so in effect, the k-states near the band edge do not contribute significantly to bonding. Thus, for simple metals little error is introduced by incorrectly representing the k-states near the band edge.

The errors in the jellium wavefunction near the atomic cores can be important. Generally, the electron densities are much higher near the atomic cores than expected from the jellium model. The high electron density leads to a short-range core–core repulsion that is ignored in the jellium model. That effect is very important when the atomic cores overlap. However, the core–core repulsions are usually fairly short range. As a result, while core–core interactions can be important when molecules collide with surfaces, the core–core repulsions are small at normal chemisorption distances. As a result, the core interactions only give a small correction to the wavefunction. In Section 3.10, we show that this correction is easily handled.

A more serious correction comes for transition metals. There are s-bands and d-bands in a transition metal, and neither is that well represented by a jellium model. The d-electrons are described in Appendix C. Basically the d-electrons are localized around the atoms. Their wavefunctions do not look anything like a free electron. As a result, one cannot use a jellium model to represent the d-band of a metal.

The s-band of a transition metal also requires correction. Many books say that the s-band is well represented by the free-electron model. However, there is s-d coupling, which complicates the analysis. Also if one looks carefully at the band structure of transition metals, one usually finds that there are band edges a few volts below E_f. For example, the nickel band structure in Appendix C shows a band edge at the X point which is 4.5 eV below E_f. The wavefunctions near such band edges are not well represented by a free electron (i.e., jellium) model, and they can be quite important to adsorbate/surface bonding.

In the early 1980s, Nørskov et al. [1981, 1985] did a number of calculations where they put the real wavefunctions for simple metals into the Lang and Williams formalism, and used those wavefunctions to calculate properties. Generally, they found reasonable agreement with heats of adsorption of hydrogen on simple metals (e.g., Al, Mg). However, the approach has been limited because it is difficult to calculate the surface Green's function for anything other than a simple metal. As a result, Lang and Williams's techniques have not been used for many examples.

3.10   THE EFFECTIVE MEDIUM MODEL

Fortunately, Nördlanger, Holloway, and Nørskov [1984], and Nørskov [1990] derived an approximation, called the **effective medium model** which allows one to include the d-bands in the calculations. The idea of the effective medium model is to start with Lang and Williams's formulation of the wavefunction (Equation 3.82), plug in the exact Green's function (Equation 3.84), and then approximate various terms in the resultant expression. Combining Equations 3.82 and 3.84 yields

$$\psi_l = \psi_0 + \int G_{core}(\vec{r}, \vec{r}_1) V_{core}(\vec{r}_1) \psi_l^{Guess}(\vec{r}_1) \, d\vec{r}_1$$
$$+ \int G_{s\text{-}band}(\vec{r}, \vec{r}_1) V_{sph}(\vec{r}_1) \psi_l^{Guess}(\vec{r}_1) \, d\vec{r}_1 \qquad (3.85)$$
$$+ \int G_{d\text{-}band}(\vec{r}, \vec{r}_1) V_{sph}(\vec{r}_1) \psi_l^{Guess}(\vec{r}_1) \, d\vec{r}_1$$

where G_{core}, $G_{s\text{-}band}$, and $G_{d\text{-}band}$ are the Green's functions for interactions with the core, the s-band, and the d-band, respectively.

The first term in Equation 3.85 is the core–core repulsion, the second term is the interaction of the adsorbate with the s-band, and the third term is the interaction of the adsorbate with the d-band. Nördlanger et al. showed that the first term in Equation 3.85 could be very accurately approximated by a core–core potential, while the second term can be approximated by an adsorbate-jellium potential. The third term requires a direct calculation, however.

Most of Nördlanger et al.'s work concentrated on devising a way to calculate the second term in Equation 3.85. They proposed what they call an **effective medium** approximation to do so. In the effective medium approximation, one replaces the exact interaction of the adsorbate with the s-band with the interaction of the adsorbate with a jellium surface. One makes the further assumption that the electron density is only varying slowly moving away from the surface so that when the adsorbate is attached to the surface the adsorbate is in effect surrounded by (i.e., embedded in) a uniform electron gas.

Consider the interaction of a hydrogen atom with a uniform electron gas. The hydrogen starts out with one filled and one empty 1s orbital. According to the material in Appendix B the hydrogen will have three main interactions with the uniform electron gas, a core/electron gas attraction, a core-filled orbital repulsion, and an attractive exchange–correlation interaction. In Appendix B we note that the exchange/correlation interaction goes as the electron density to the 1/3 power, while the other two interactions go as the electron density to the first power. Therefore, at low electron gas densities, the exchange–correlations will dominate. This leads to a lowering of the energy of the system when the hydrogen approaches the surface (see Figure 3.43). Physically, what is happening is that the hydrogen is bonding with the electron gas and thereby lowering the energy of the system.

Now consider what happens as the density of the electron gas increases. At first one gets additional bonding. However, eventually all of the orbitals will fill up. At that point, additional electron density will not produce any more bonding. However, it will lead to additional Pauli

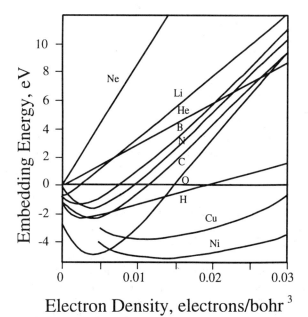

Figure 3.43 The effective medium potential for the interaction of a number of atoms with a homogeneous electron gas. (After Nørskov, [1981].)

repulsions, as described in Appendix B. As a result, the energy of the system will move up again as shown in Figure 3.43.

Now, if one looks at a more electronegative atom such as oxygen, the minimum in the potential occurs at a higher electron density. That produces a larger lowering of the energy of the system and therefore, a somewhat stronger surface adsorbate bond.

Now consider the interaction of a hydrogen or oxygen atom with a surface. Figure 3.43 shows that there is an optimal electron density for binding hydrogen or oxygen. Therefore, if the surface could provide that optimal electron density, it would form a strong bond to the adsorbate. Figures 3.36 and 3.37 show that the electron density over the surface varies over a wide range. The adsorbate can always move in close enough to the surface atoms to find its optimal electron density.

Note, however, that the electron density around the surface atom also changes as the adsorbate approaches the surface. Each given surface atom will also have an optimal electron density as indicated in Figure 3.43. Therefore, at equilibrium, one has to balance the electron density needs of the adsorbate and the electron density needs of the surface.

The net effect of all of this is that one can write the interaction of an adsorbate with the s-band on a surface as a sum of two terms, the change in the energy of the adsorbate due to the overlap of the adsorbate with the electron cloud of the surface, and the change in the energy of each of the surface atoms due to the overlap of the electron cloud of the adsorbate. Nördlanger et al. showed that each of these terms can be represented as an individual potential, $V_{EM}^{A}(\rho_e^A)$, where the potential is a function of ρ_e^A, the electron density on atom A due to all of the other atoms in the system. The constant $V_{EM}^{A}(\rho_e^A)$ is called the embedding energy of atom A. Nørskov [1990] points out that in addition there will be a local interaction between each of the atoms in the system due to overlap with the electron core. Therefore, the total energy of the system, E_T will be given by

$$E_T = \tfrac{1}{2} \sum_A \sum_B V_{cc}^{AB} + \sum_A V_{EM}^A(\rho_e^A) + \textit{d-band corrections} \qquad (3.86)$$

where V_{cc}^{AB} is the core interaction between atoms A and B. The factor of 1/2 in Equation 3.86 comes about to avoid double counting the core interaction. In principle, the embedding energy $V_{EM}^A(\rho_e^A)$ should be a function of the electron density and work function of the surface. However, people often assume that the embedding energy is only a function of the electron density, and then calculate a work function by choosing a value of E_s^0 in the jellium Hamiltonian, 3.54, to maintain charge neutrality (in effect this approximation assumes that there is no buildup of charge on cores in the metal).

In actual practice, Nördlanger et al. [1984] ignored the d-band corrections. They then attempted to calculate the potentials using an early version of density functional theory, with only modest success. For example, Figure 3.44 compares their calculated heat of adsorption of hydrogen and oxygen on a variety of transition metals to experimental values. Generally, the heats of adsorption of hydrogen are reasonable, but the heats of adsorption of oxygen are a factor of 2 to 4 too high.

The reason that the calculations failed is that they used an early version of density functional theory that overestimated the electron affinity of oxygen. Also, the third term in Equation 3.85 (i.e., the interaction with the d-bands) was ignored. However, Jacobson, Nørskov, and Pushka [1987] and Kress and DePristo [1988] derived potentials that used more modern versions of density functional theory. These potentials seem to do a better job of fitting real data than Nørdlanger et al.'s original formulation. The details of these alternative descriptions are well described in Jacobson and Nørskov [1988] and Raeker and DePristo [1991] and will not be repeated here. However, the basic idea is to add corrections to Equation 3.86 to better account for the various interactions. For example, embedding functions have been adjusted to do a better job of describing the metal surface. Recall that there are two parameters in the

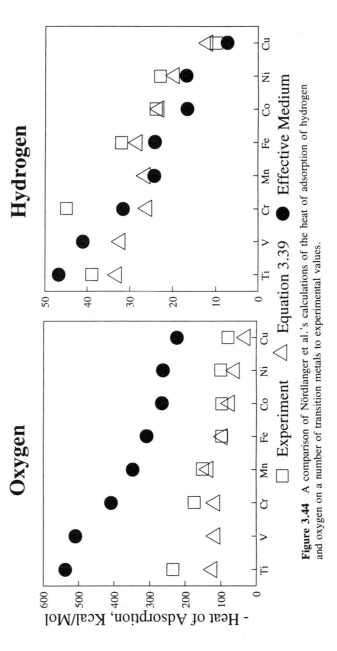

Figure 3.44 A comparison of Nördlanger et al.'s calculations of the heat of adsorption of hydrogen and oxygen on a number of transition metals to experimental values.

jellium model: the interstitial electron density and the work function of the surface. However, Nördlanger et al.'s original formulation, only considered electron density variations, and fixed the work function to maintain charge neutrality (assuming a uniform positive background). That assumption is relaxed in the more modern formulations. Also, people find it useful to calculate some of the terms in Equation 3.85 explicitly: specifically the interactions with the d-bands and some of the exchange interactions. The result is a formulation that looks quite similar to Nördlanger et al.'s original formulation. However, one can calculate reasonably accurate heats of adsorption with the modified formalism.

It works out that the bonds with the d-electrons look like simple covalent bonds. They are usually treated in the same way the covalent bond is treated in a solid: A single d-orbital in the metal interacts with orbitals in the adsorbate to form a bond. There are some corrections because the d's are part of a band. However, people often ignore that and still calculate useful properties.

3.10.1 Summary of Predictions of the Effective Medium Model for the Adsorption on Metal Surfaces

Qualitatively, the effective medium models are quite similar to Lang and Williams's model. According to the effective medium model, the main bonding is with the conduction band. However, there are also local interactions due to the presence of the d-bands and surface states.

Another prediction of the model is that there are a few key parameters that can be used to characterize the surface; i.e., the interstitial electron density over the surface, the surface's work function, the local interactions due to the d-bands, and the steric effects due to the atomic cores. Forgetting about the computational details, the key prediction of the effective medium model is that one can use these few parameters to learn about chemistry.

The interstitial electron density works out to be a key parameter. Figure 3.45 shows a plot of the interstitial electron density of a number of metals from Moruzzi et al.'s [1978] very accurate APW calculations and DeBoer et al.'s [1988] fits to bulk data. DeBoer et al's values are systematically larger than Morruzzi et al.'s so we have multiplied DeBoer et al.'s values by 0.9 in the figure. Numerical values are given on Table 3.15. The interstitial electron density starts out low at gold, silver, and copper, then increases, moving to the left in the periodic table, reaching a maximum at rhenium and osmium. The interstitial electron density then decays again, moving further to the left in the periodic table.

Interestingly, there seems to be an approximate correlation between the reactivity of surfaces for simple bond scission processes and the interstitial electron density. The data in Table 3.2 show that gold, silver, and copper are virtually inactive for H_2, CO, NO, or N_2 dissociation. Generally, the reactivity of metals increases, moving to the left in the periodic table, in a way that approximately tracks the interstitial electron density. There are very few data on the dissociation of molecules on metals to the left of tungsten (i.e., where the interstitial electron density goes down again). However, calcium does not readily adsorb hydrogen or nitrogen. Evaporated potassium films are fairly inactive for most dissociation processes even though potassium forms quite strong bonds with the dissociation products. Thus, there seems to be a qualitative but not quantitative correlation between the interstitial electron density and the ability of metals to break bonds. Such trends are as expected from the effective medium model.

Another thing that seems to correlate with the interstitial electron density is the bonding mode of the adsorbed species. Earlier in this chapter we noted that when ethylene adsorbs on metal surfaces at 100 K, the carbon–carbon bond order of the ethylene is reduced from its gas phase value of 2.0 to a value of 1.0 to 1.9. The carbon–carbon bond order varies significantly from metal to metal, as indicated in Figure 3.41. Years ago people discussed such variations in terms of a change in the bonding mode of the ethylene. In inorganic clusters,

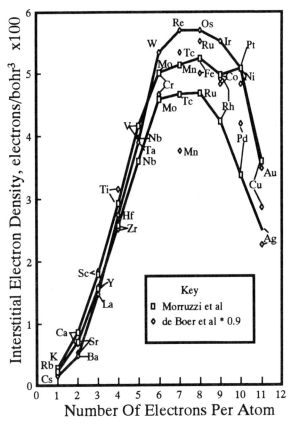

Figure 3.45 The interstitial electron density for several metals.

TABLE 3.15 The Interstitial Electron Density of a Number of Metals in Hundredths of an Electron/bohr^3*

Metal	Morruzzi	DeBoer	Metal	Morruzzi	DeBoer	Metal	DeBoer
K	0.29	0.27	Rb	0.23	0.22	Cs	0.17
Ca	0.87	0.75	Sr	0.70	0.59	Ba	0.53
Sc	1.80	2.05	Y	1.57	1.77	La	1.64
Ti	2.93	3.51	Zr	2.58	2.80	Hf	3.05
V	4.17	4.41	Nb	3.60	4.41	Ta	4.33
Cr	5.01	5.18	Mo	4.59	5.55	W	5.93
Mn	5.14	4.17	Tc	4.67	5.93	Re	6.33
Fe	5.24	5.55	Ru	4.68	6.13	Os	6.33
Co	4.98	5.36	Rh	4.24	5.45	Ir	6.13
Ni	5.08	5.36	Pd	3.38	4.66	Pt	5.64
Cu	3.60	3.18	Ag	2.50	2.52	Au	3.87
Na	0.49	0.55	Mg	1.44	1.60	Al[†]	3.73

Source: Calculated by Morruzzi et al. [1978] and as fit to data by DeBoer [1988].
*The values from DeBoer should be multiplied by 0.9 to make them compatible with Morruzzi's values.
[†]Morruzzi's value.

ethylene usually shows a bond order of 1.4–1.8 when the ethylene is bound to a single metal atom, while the ethylene shows a bond order 1–1.2 when the ethylene is bound to two metal atoms. Years ago people thought that a similar process was occurring when ethylene adsorbs on metal surfaces; i.e., the ethylene is bound to a single metal atom when the ethylene is in a π-bound state (i.e., bond order 1.4–1.9), while the ethylene is bound to two metal atoms when the ethylene is in a di-σ state (i.e., bond order 1–1.3). People now know, however, that such changes in the bonding configuration do not occur. Rather different metals have a different ability to donate electrons into the carbon–carbon bond on the ethylene. Metals that donate electrons easily form di-σ complexes, while metals that do not donate electrons as easily form π-bound complexes. Experimentally, the ability to donate electrons into the carbon–carbon bond on ethylene correlates quite well with the interstitial electron density of the metal, as indicated in Figures 3.40 and 3.41. This is as expected from the effective medium model.

At this point, no one knows whether this is a general effect. One often observes a change in the bonding mode of species with changing metal. However, the changes are usually more complex than in the ethylene case. For example, when formaldehyde ($H_2C{=}O$) adsorbs on transition metals, one can form a so-called η_1 species with a carbon–oxygen bond close to 2.0 or a so-called η_2 species with a bond order close to 1.0. Generally, it is thought that the η_2 species lies down on the surface so that the carbon and oxygen in the formaldehyde can both bond to the surface. In contrast, the η_1 species stands up on the surface; there is a single bond to the oxygen. This case is fundamentally different from the ethylene case because the orientation of the molecule is changing the orbitals that contribute to the bond change. As a result, this case is more difficult than the ethylene case. Experimentally, there is the general correlation that the η_1 species forms on metals with a low interstitial electron density, while the η_2 species forms on metals with higher interstitial electron density. That is generally what would be expected from the effective medium model, although the details still need to be clarified.

The effective medium model is also useful in understanding promoter chemistry. When one makes a supported metal catalyst, one often adds an alloying agent called a **promoter** to modify the metal in the catalyst. The promoter forms an alloy with the metal in the catalyst and thereby changes the properties of the metal. Generally, one can modify both the electronic structure of the metal and the shape of the metal particles.

Consider for the moment what would happen if one added an alloying agent that had a high electronegativity. When an alloy forms, the electronegativity of the system must equalize. If the alloying agent is more electronegative than the metal, electrons will flow from the metal to the alloying agent. That will reduce the interstitial electron density of the metal. Notice that according to Figure 3.41 the interactions with the metal would be reduced. Similarly if one adds an alloying agent that added electrons one would strengthen the interactions with the surface. In order to work out the details, one has to do a calculation to see how the interstitial electron density of the system changes as an alloy agent is added. However, one can make some useful predictions using the principle of electronegativity equalization described in Section 3.4.2.

The effective medium model has also proved to be a good way to quantitatively model adsorption data. For example, Holloway and Nørskov [1991] consider the adsorption of hydrogen on a variety of faces of nickel. According to Nördlanger et al.'s [1984] calculations, the embedding energy of hydrogen shows a shallow minimum at an electron density of about 0.01 electron/bohr^3, as shown in Figure 3.43. Now consider the adsorption of hydrogen on Ni(111) and Ni(100). According to Table 3.15 bulk nickel has an interstitial electron density of 5.08 electrons/bohr^3. This interstitial electron density is high enough that a hydrogen atom can always find its ideal electron density on any face of nickel. However, if hydrogen would adsorb on, for example, a threefold hollow on Ni(111), that electron density would need to be provided by three nickel atoms, while the hydrogen can find sites where four atoms are available for bonding on Ni(100). Therefore, on the average hydrogen would need to share less electron density with each surface atom on Ni(100) than on Ni(111). According to Figure 3.45 the electron density decays exponentially away from each surface atom. Therefore, one way for the hydrogen to find its ideal electron density would be for the hydrogen to move closer to the individual surface atoms on Ni(111) than on Ni(100). One has to be careful with the arguments, because the geometries of the two surfaces are different. However, Jacobson et al. [1987] was able to show that the hydrogen should sit farther away from the Ni(111) surface than from the Ni(100) surface. This agrees with the experimental results given in Table 3.16. Similarly, Nørskov showed that the vibrational frequency of the hydrogen should increase as it moves in closer to the surface atoms because the core potential is stiffer there. Again that prediction agrees with experiment. A similar effect is also seen with oxygen, as indicated in Table 3.16. Carbon monoxide is different because it can bind at a linear site or a bridgebound site on the same surface, as indicated in Figure 3.13. However, the adsorbate-surface vibrational frequency is usually higher on the linear site because the CO is bound to fewer atoms on the linear site than on the bridgebound site. There are exceptions, however, (see Mapledoram, Wander, and King [1993]).

This is a general result. Linearly bound molecules are usually held more closely than bridgebound molecules, which in turn are held more closely than multiply coordinated molecules. Those changes produce changes in the vibrational spectra of adsorbates.

This effect has been studied the most for CO adsorption on transition metals. Generally, CO can bind to linear, bridgebound or triply coordinated sites, as indicated in Figure 3.13. CO bound to a linear site shows a vibrational frequency between about 2000 and 2200 cm^{-1}, while CO bound to a bridgebound site shows a vibrational frequency between about 1850 and 2000 cm^{-1}. CO bound to a triply or quadruply coordinated site shows a vibrational frequency below 1850 cm^{-1}.

The CO vibrational frequency also shifts from one metal to the next. Based on the preceding discussion, one might expect there to be some correlation between the vibrational frequency of the CO and the interstitial electron density of the metal. However, Figure 3.46 shows a plot of the vibrational frequency of CO adsorbed on a number of closed packed metal surfaces as a function of the interstitial electron density of the metal. Notice that there is *not* a good correlation between the vibrational frequency of the CO and the bulk interstitial electron density. Interestingly, Figure 3.47 shows how the vibrational frequency of ethylene adsorbed on transition metals correlates with the interstitial electron density. With ethylene one finds an excellent fit between the interstitial electron density and the vibrational data.

So far, no one has really explained in detail why the vibrational frequency of ethylene correlates with the interstitial electron density while the vibrational frequency of CO does not. One usually models the binding of both ethylene and CO via what is called a *Dewar–Chatt–Duncanson model*, i.e., the adsorbate donates electrons into the metal and the metal

TABLE 3.16   **The Bond Length and Vibrational Frequency of Hydrogen and Oxygen Atoms and Carbon Monoxide Adsorbed on Ni(111) and Ni(100)**

	Hydrogen		Oxygen		Carbon Monoxide	
Surface	Distance from Surface (Å)	Adsorbate–Surface Vibrational Frequency (cm^{-1})	Distance from Surface (Å)	Adsorbate–Surface Vibrational Frequency (cm^{-1})	Form	Adsorbate–Surface Vibrational Frequency (cm^{-1})
Ni(111)	1.65	1115	1.88	580	Linear	480
Ni(100)	1.83	614	1.96	350	Bridged	359

Source: After Jacobson et al. [1987] with additional data from Willis, Lucas and Mahan [1983].

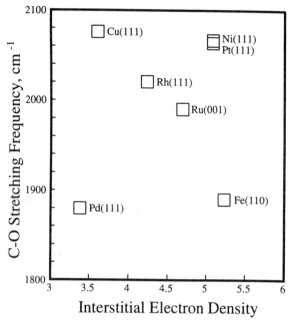

Figure 3.46   A correlation between the low coverage limit of the vibrational frequency of CO adsorbed on a series of closed packed metal surfaces and the interstitial electron density of the metal.

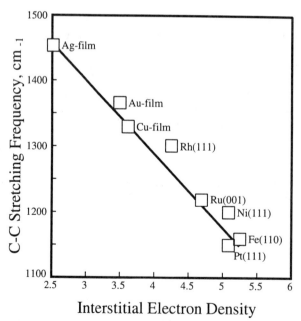

Figure 3.47   A correlation between the vibrational frequency of the C–C stretch in C_2D_4 adsorbed on a series of closed packed metal surfaces at 100 K and the interstitial electron density of the metal.

donates electrons into the antibonding orbital of the adsorbate. Electron transfer from the adsorbate to the metal tends to increase the vibrational frequency of the adsorbate, while back donation decreases it. There is an important difference between ethylene and CO, however. The lowest unoccupied molecular orbitals (LUMO) in ethylene are concentrated along ethylene's molecular axis. According to the analysis in Section 3.6, the ethylene would like to maximize overlap between the LUMO and the surface, which means that ethylene will lie with its molecular axis parallel to the surface. In contrast, the LUMO of CO is concentrated on the carbon, which means that the CO would like to stand up on the surface.

In Chapter 10 we show that this difference in geometry causes the interactions to change. Ethylene interacts mainly with the s-band, while the antibonding orbitals in CO have a significant interaction with the d's. The success of the correlation in Figure 3.47 plus the failure of the correlation in Figure 3.46 suggests that the strength of the interaction with the s-electrons is largely determined by the geometry of the adsorption site and the interstitial electron density of the solid. In contrast, the interaction with the d-electrons does not scale as the interstitial electron density.

Nørskov [1990] tried to correlate the CO frequency to the local interactions with the d-bands as measured by something called the percent d-filling ignoring the interactions with the s-band. However, the correlations were not any better than the ones in Figure 3.46. Physically, the CO vibrational frequency is affected by both s and d electrons. As a result, one must include both effects to get good correlations.

A somewhat more complex analysis is needed to predict the adsorption sites and bond energy. Consider the bond energy. Earlier we noted that hydrogen can always find its ideal electron density somewhere on a nickel surface. Now consider what happens on iron. All of the arguments we applied to nickel also work for iron. The hydrogen can find its ideal electron density over any face of iron. Therefore, the electron density on the hydrogen is about the same no matter whether the hydrogen adsorbs on Ni(111) or Fe(110).

Experimentally, however, the heat of adsorption of H_2 is about -23 kcal/mole on Ni(111) and -27 kcal/mole on Fe(110). The heat of adsorption of atomic hydrogen is -64 kcal/mole on Ni(111) and -66 kcal/mole on Fe(110). Therefore, the binding energy of hydrogen is not quite the same on Fe(110) and Ni(111) even though the hydrogen can find the same total electron density on both surfaces.

There are two main differences between iron and nickel: a difference in the local interaction of the hydrogen with the d-bands of the metal, and a difference in the electron affinity of the two metals, leading to a difference in the energy of the electron transfer process. Holloway, Lundqvist, and Nørskov [1984] considered the role of the d-bands in detail. In a case of a simple adsorbate, such as hydrogen, the results are simple. According to Lang and Williams's analysis in Section 3.8.2 and Equation 3.86, the energy level of the adsorbate will be lowered by an amount Δe_l given by

$$\Delta e_l = \int\int \psi_l^*(\vec{r})V_{sph}(\vec{r}_1)G_{d\text{-}band}(\vec{r},\ \vec{r}_1)\psi_l(\vec{r}_1)\,d\vec{r}\,d\vec{r}_1 \qquad (3.87)$$

where $\psi_l(\vec{r})$ is the adsorbate wavefunction, V_{sph} is the adsorbate potential, and G is the d-band Green's function (i.e., the last term in Equations 3.84 and 3.85).

If the adsorbate energy levels are below the d-band, one can simplify this expression by substituting Equation 3.73 for the Green's function into Equation 3.87 and then further make the assumption that the d-bands can be approximated as a single "group orbital" ψ_k (i.e., average orbital) with energy $E_{d\text{-}band}$.

After considerable algebra one obtains

$$\Delta e_l = \frac{2(1-f_d)}{E_{d\text{-}band}-\epsilon_l^0}\left|\int \psi_l^*(\vec{r})V_{sph}(\vec{r})\psi_k(\vec{r})\,d\vec{r}\right|^2 \qquad (3.88)$$

where ϵ_l^0 is the energy of orbital l in the absence of the interaction with the d-bands, and f_d is the fractional filling of the d-band. The factor of 2 arises in Equation 3.88 because there are two electrons in each orbital. The factor of f_d arises because only empty orbitals in the d-bands can interact with filled orbitals on the adsorbate.

Figure 3.48 shows the results of Holloway et al.'s calculations of the interaction energy as a function of the d-band filling. Two different curves are shown, one for an adsorbate with an energy level 1.5 eV below the d-band, and one with an energy in the center of the d-band. When the energy level is far below the d-band, Δe_l increases monotonically with the d-band filling, as one would expect from Equation 3.88. However, the situation is more complex if the energy level of the adsorbate lies at the Fermi level. In that case, electrons can be transferred from the adsorbate to the d-band, and from the d-band to the adsorbate. The rate of transfer from the adsorbate to the d-band is proportional to $(1 - f_d)$, while the transfer in the other direction is proportional to f_d. As a result, the net interaction energy is approximately proportional to $f_d(1 - f_d)$, which has a maximum at $f_d = 0.5$.

Figure 3.49 shows how the d-band filling of the transition metals varies over the periodic table. Generally, the d-bands are empty for the alkali metals. In the gas phase, the d-orbitals are empty in Ca, Sc, and Ba. However, Morruzzi et al.'s [1978] band structures show that in solid Sc and Ca the d's are partially filled. (No data are given for barium.) The d's fill up moving to the right from potassium in the periodic table. The d's are completely filled at the Ib metals (i.e., Cu, Au, Ag). There also is the observation that the d-band filling is slightly higher with the 4d metals than with the 3d or 5d metals.

It is useful to consider what Equation 3.88 predicts for the interaction of the d-bands. According to Equation 3.88 the interaction with the d-bands scales with $(1 - f_d)$. Consequently, one would expect the interaction with the d-bands to increase moving to the left on the periodic table from gold, silver, and copper because f_d in Equation 3.88 decreases in that direction. The d-band filling only changes slightly moving down the periodic table (i.e., from Ni to Pd to Pt). However, the integral in Equation 3.88 changes greatly. One can show that the integral is approximately proportional to the overlap between orbitals in the metal and the orbitals in the adsorbate. The orbital overlap is determined by the size of the orbitals (i.e., one over the orbital exponents), and the distance between the adsorbate and the metal atoms. One can calculate the orbital overlap from the data in Harrison [1980]. Harrison fit bulk band structures to an LCAO model, where each of the individual wavefunctions, ψ_l, was assumed

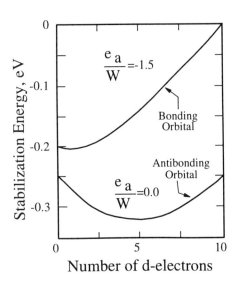

Figure 3.48 The energy stabilization due to the d-bands as a fraction of the d-band filling. (After Holloway et al. [1984].)

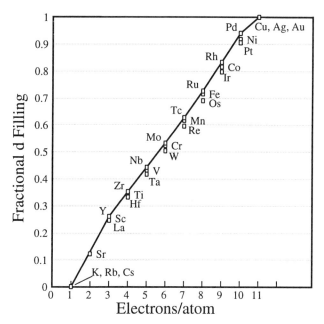

Figure 3.49   The fractional d filling for the transition metals reported by Harrison [1980]. The Sr point comes from a fit to Morruzzi et al.'s [1978] band structure.

to be of the form

$$\psi_l(\vec{r}) = C_{3.89} L^l_{m+l}[\zeta r](r\zeta)^2 Y^m_l[\Theta(\vec{r})] \exp[-\zeta r] \qquad (3.89)$$

where L^n_m is a Laguerre polynomial; Y^n_m is a spherical harmonic; Θ is an angular coordinate; ζ is the orbital exponent; and $C_{3.89}$ is a constant. Physically, the orbital exponent is proportional to the size of the orbital. Figure 3.50 shows how one over the orbital exponents of the s and d-orbitals varies moving around the periodic table. Notice that the s-orbitals are always larger than the d's. However, the difference between the size of the s's and d's decreases moving down the periodic table. To put these results in perspective, Harrison shows that the covalent radius of a metal is determined by the size of the s-orbitals. Experimentally, a simple adsorbate like hydrogen will stay at 1 Å or so outside of the s-orbital. Harrison also shows that the interaction of an s-orbital with a d-orbital goes as

$$\psi^*_l(\vec{r}) V_{sph}(\vec{r}) \psi_k(\vec{r})\, d\vec{r} = V_{sd}(\zeta r_0)^{-(3/2)} \exp(-\zeta r_0) \qquad (3.90)$$

where r_0 is the distance between the surface atom and an adsorbate atom and V_{sd} is a constant. Plugging numbers into Equation 3.90 shows that the squared integral in Equation 3.88 is 20 times larger on platinum than on nickel. Consequently, the implication of Figure 3.50 is that with the 3d metals, the d-bands barely overlap the bonding orbital of an adsorbate-like atomic hydrogen. In contrast, with platinum there is much larger d-band overlap. As a result one would expect the interaction with the d-bands to substantially increase moving down the periodic table.

Experimentally, the heat of adsorption of most atomic adsorbates increases moving to the left across the periodic table from gold (see Table 3.5). This is as one would expect due to the interaction with the d-bands. However, in most cases the heat of adsorption decreases moving down the periodic table, even though the interaction with the d-band increases sub-

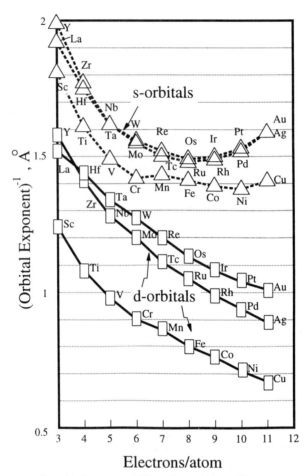

Figure 3.50 One over the orbital exponent. The values of the orbital exponent were estimated by Harrison [1980].

stantially in that direction. This is not as expected if the variations in heats of adsorption were caused mainly by an interaction with the d-bands. Also, the variations in heats of adsorption across the 3d metals (i.e., Cu, Ni, Co, Fe, . . .) are similar to those seen in the 5d metals (i.e., Au, Pt, Ir, Os, . . .) even though the interactions with the d-bands are much larger with the 5d metals than with the 3d metals. Therefore, experimentally the heats of adsorption do not vary in the way one would expect if the changes were caused by variations in the interactions with the d-band.

Another point is that there is no evidence for a strong difference between the variations of heats of adsorption of gases on d-band metals and on metals where no d's are present for bonding. For example, in Section 3.4 we discussed the Tanaka–Tamaru correlations, which say that there is a general correlation between heats of adsorption of a variety of gases and the heats of formation of the bulk oxide as given in Figure 3.23. Figure 3.23 includes data for the adsorption of a variety of gases on silicon, germanium, copper, silver, and gold. Notice that the heats of adsorption of hydrogen and oxygen on silicon, germanium, copper, silver, and gold follow just the same trends as all of the transition metals even though there are no d's available for bonding on copper, silver, gold, silicon, and germanium. Therefore, experimentally it does not seem that the variations in heats of adsorption of atomic adsorbates are mainly associated with variations in the interactions with the d-bands, even though some calculations say otherwise. Rather, there must be something else that either modifies or swamps the d-band effects.

A likely possibility is a difference in the electron affinity of various surfaces. Recall from Section 2.6 that when one creates a surface, one lowers the coordination numbers (i.e., the number of nearest neighbors) of the atoms in the surface. That, in turn, leads to a reduction in the electron density of the atoms in the surface to a value below the value preferred by the surface atoms (see the dicussion in Section 2.6). This induces a Tamm state to form on the surface. Now when a molecule adsorbs, the molecule shares electrons with the surface. The surface atoms gain electron density, which lowers the energy of the surface by shifting the Tamm state. The changes in the surface energy are different with different metals, because each metal has its own electron affinity. Experimentally, the variations in heats of adsorption of atomic adsorbates seem to correlate much better with the variations in electron affinities than with variations in the properties of the d-band.

One can quantify these ideas using the effective medium model. Recall, that in the effective medium model the embedding energy of the surface atoms tells how the energy of the surface atoms change as the surface atoms share electrons with their neighbors. Therefore, one should be able to get some insight into the variations in the electron affinities of the surface atoms by examining the properties of the embedding energies.

For the purposes of derivation it is useful to fit the embedding energies to a Taylor series in the electron density:

$$E_{EM} = E_{EM}^0 + \chi_A^0(\rho_s - \rho_s^0)c + \eta_A^0(\rho_s - \rho_s^0)^2 c^2/2 \qquad (3.91)$$

where ρ_s^0 is the electron density on the surface atoms before adsorption; ρ_s is the electron density on the surface atoms after adsorption; and E_{EM}^0, χ_A^0, η_A^0, and c are all constants. Notice the close similarity between Equation 3.91 and Equation 3.44. In fact, if we fix the constant c so that the quantity $(\rho_s - \rho_s^0)c$ is unity when a single adsorbate surface bond forms, the two formulations would be identical. Therefore, all of the results from Section 3.5 still apply. In particular, the concepts of hard and soft bases described in Section 3.4.1 should directly carry over to metal surfaces. In particular, if everything else is equal, soft species (e.g., CO, H_2S, C_2H_4) should bind more strongly than hard species (e.g., H_2O, HCl). This is generally as observed. There are some exceptions to this general rule, because of dissociation reactions. For example, methanol, which is a borderline, only binds weakly on most transition metal surfaces. However, the methanol can dissociate to form $H_{(ad)}$ and $CH_3O_{(ad)}$, both of which bind strongly. Still, there is the general finding that if everything else is equal, soft acids and bases bind more strongly to metal surfaces than hard acids and bases.

Another implication of Equation 3.91 is that variations in heats of adsorption of atomic adsorbates occur because the effective electronegativity of the surface to form bonds varies from one metal to the next. Such an implication is not well established in the surface science literature, but it is well established for alloys (see DeBoer et al. [1989]). Generally, one would expect electronegativity differences to drive adsorption reactions. A subtle point is that the electronegativities in Equation 3.91 are the electronegativities for sharing electrons with an adsorbate rather than the energy change in taking an electron and putting it in an adsorbate. This electronegativity can be quite different than the surface workfunction (i.e., the energy to remove an electron from the surface). Rodriquez et al. [1991] show that such effects are quite important in interpreting adsorption results. Therefore, the key implications of Equation 3.91 is not that the surface workfunction determines the ability of a metal to form bonds. Rather, Equation 3.91 implies that there are a series of constants to characterize the basic bonding power of metals. Metals that form a strong bond with one species will also tend to form a strong bond with other species independent of the species that adsorb.

At present, there is no consensus as to what values one should use for the constants. In

Jacobson and Nørskov's [1987] formulation of the effective medium model for the embedding energy was assumed to vary nearly parabolically with the electron density as shown in Figure 3.43. However, Kress and DePristo [1988] assumed a near linear form for the embedding energy, as suggested by Puska [1987], and added an extra core interaction to compensate. Raeker and DePristo [1991] also showed that the embedding energy should vary slightly with the work function of the surface. However, these corrections are fairly small. In practice one can use the effective medium model to make simple calculations of heats of adsorption. The calculations can be done by hand, and the results are almost as accurate as the best calculations in the literature.

Unfortunately, such simple calculations are not able to predict adsorbate geometries. Consider the adsorption of CO on Pt(111) and Pd(111). On either surface the CO can adsorb on a linear site, a bridgebound site or a triply coordinated site, where the sites are shown in Figure 3.16. Experimentally, the CO first adsorbs on a linear site on Pt(111) and only adsorbs on a bridgebound site at higher coverages. In contrast, on Pd(111), the bridge site is occupied first, and one only observes linear binding at higher coverages.

The differences between linear and bridge bonding is very subtle. The CO can find its ideal electron density on either site. The electron exchange with the bulk is identical on both sites. However, on the linear site, the main local interactions are with a single surface atom, while in the bridgebound case there are strong local interactions with three surface atoms. If the energy of the surface atoms varied linearly with electron density, then both cases would be exactly equivalent. In that case, the CO would not have a preferred bonding configuration. In actual practice, the energy of the surface atoms varies nonlinearly with electron density and core repulsions and d-band overlaps vary from one site to another. That allows one site to be preferred over another. However, the effect is small so that theoretically adsorbates should not show any large preference for one binding site over another.

Experimentally, the results are unclear. LEED indicates that most adsorbates are held in ordered arrays with fixed binding sites at low to moderate temperatures. However, the preferred binding site often changes with coverage. Also, the activation energies in moving from one site to another are usually only a small fraction (i.e., perhaps 10%) of the total binding energy of the adsorbate. Therefore, while experimentally some sites are favored over others, the effect is often associated with a small change in energy. All of these results are a further indication that the main bonding of an adsorbate to a metal surface is through the conduction band and the Tamm states and that the local interactions with the d-bands only have a secondary effect.

The d-bands are thought to play an important role in dissociative adsorption, however. One case that has been considered in great detail in the literature is the dissociative adsorption of H_2 on transition metals. Hydrogen adsorbs dissociatively on most transition metals. The heat of adsorption varies from one metal to the next, so for example the heat of adsorption is about 9 kcal/mole on Cu(110), 13 kcal/mole on Pt(110), and 21 kcal/mole on Ni(110). The activation energy for dissociative adsorption is too small to measure on Ni(110) and Pt(110), 1–2 kcal/mole on Pt(111) and Ni(111), and about 10 kcal/mole on Cu(110) and Cu(111).

It is difficult to explain why the activation energy for H_2 dissociation on Cu(110) is so much larger than the activation energy for H_2 dissociation on Pt(110) based on the Lennard-Jones model (Figure 3.16). The heats of molecular adsorption are almost the same on both surfaces, while the heat of dissociative adsorption is only 4 kcal/mole different on Pt(110) and Cu(110). Yet there is a 10-kcal/mole difference in the activation energy for dissociation.

There are two main differences between copper and nickel or platinum: the interstitial electron density and the d-band filling. Copper has an unusually low interstitial electron density, and copper's d-band is completely filled and unavailable for bonding. In contrast, platinum and nickel have a higher interstitial electron density and the d-bands are available for bonding. Both factors will influence the ability of the surface to break bonds. However, so far, most theoretical work has concentrated on the role of the d-band filling on the bond scission process. Basically there are two models; a model based on the cluster calculations of Harris and Anderson [1985], and a model based on the effective medium results of Nørskov [1990].

The best model comes from the work of Jacobson et al. [1987]. In Chapters 9 and 10 we show that when a bond in the adsorbate breaks, electrons usually flow from the surface into antibonding orbitals in the adsorbate. The electron flow weakens the adsorbate–adsorbate bonds and allows them to break. It happens that in the adsorbed state, the antibonding levels of many simple molecules (e.g., CO, NO, H_2, O_2) lie close to the Fermi level of the metal. Physically, the electronegativity of a molecule lies somewhere between the bonding and antibonding levels in the molecule. When a molecule adsorbs, the electronegativity of the molecule equilibrates with the electronegativity of the surface. That brings the antibonding levels in the adsorbate down so they overlap the Fermi level of the surface. The d-bands on platinum, nickel, and the other group VIII metals lie close to the Fermi level of the surface so the d-bands can interact strongly with the antibonding orbitals in the adsorbate. The result, according to Equation 3.88, is a strong interaction between the antibonding levels in the adsorbate and the d-bands in the metal. Such interactions make it easy to break bonds in the adsorbate. In contrast, the d-bands in copper lie 3 eV below the Fermi level (see Figure 3.39). The energy levels of the d-bands lie far below the antibonding orbital of hydrogen. Equation 3.88 predicts a much weaker interaction. As a result, it is harder to break a hydrogen–hydrogen bond on copper than nickel.

One can extend these arguments to many other systems. For example, unlike the H_2 case the lowest unoccupied antibonding orbital in O_2 is at low enough energy to strongly interact with the d-bands on copper, which partially explains why oxygen dissociates at 300–400 K on copper, but hydrogen does not. We discuss this model in greater detail in Chapter 10. One does not want to push this model too far, because the integral in Equation 3.88 is the interaction energy. However, the thing to remember for now is that while the d-bands only play a small role in determining heats of adsorption, they are quite important in promoting bond scission. Hence, the d's are quite important to the chemistry.

There is another model of the role of the d-bands that is well regarded in the theoretical literature even though the model does not seem to have a strong correlation with the experiments. Recall that when a H_2 molecule approaches a surface, the H_2 experiences a Pauli repulsion as discussed in Appendix B. Harris and Andersson [1985] considered how the Pauli repulsion would be modified by the presence of the d-bands. Note that the Pauli repulsions arise because electrons are displaced as the H_2 approaches the surface. Those electrons can flow into the bulk. However, the electrons can instead move into the d-orbital. The d-orbital has a large density of states near the Fermi level, so it does not cost much energy to put an electron into the d-bands. As a result, the d-bands can lower the Pauli repulsions.

At present, however, it is unclear whether this is a substantial effect on metal surfaces. Harris and Andersson [1985] based their analysis on calculations of 2-atom clusters. With a 2-atom cluster there is no bulk to accept charge, so the role of the d-bands in lowering Pauli repulsions is quite important. However, with a bulk metal surface electrons can also flow into the bulk. Therefore, one would expect the d-bands to play a much smaller role in lowering the Pauli repulsions on a metal surface than on a 2-atom cluster.

Experimentally, one can get a measure of the difference in the Pauli repulsions on nickel, copper, and platinum from molecular beam scattering. In Appendix E we note that when a helium atom bounces from a surface, the main interactions are Pauli repulsions. If the Pauli repulsions are much stronger on copper than on platinum or nickel, one should find that the surface potential is much stiffer on copper than on platinum or nickel. Recently, Rettner and co-workers [1992, 1993] examined hydrogen scattering from Cu(111), and found that the molecular scattering potential is *not* significantly stiffer on copper than on nickel or platinum. Thus, it does not seem that the Pauli repulsions are significantly stronger on copper than on nickel or platinum.

It is my view that with a metal the d-bands lower the barriers to dissociation mainly by interacting with the antibonding oribital as suggested by Nørskov. In Chapter 10, we show that one can often correlate rate data to that model, and so it is fairly useful. Still, the d-bands do provide a good source or sink for electrons. Therefore they can reduce Pauli repulsions. So far that does not appear to be a substantial effect on metals, because the bulk also provides a convenient source or sink for electrons. However, the effect is probably quite important in other systems such as the metal oxides, which we discuss in Section 3.13.2. Thus, Harris and Andersson's model is probably quite useful, although not for metals.

3.11   MORE ADVANCED COMPUTATIONAL METHODS

There are some effects that are not included in the methods described so far in this chapter. These methods so far in this chapter have viewed the surface as a cloud of electrons, and has not explicitly considered the effects of the presence of individual surface atoms on the interactions of adsorbates with surfaces. For example, the jellium model views the surface wavefunction via a constant potential. There are no atomic cores or local variations in the potential. The effective medium model puts in some variations, but only in an ad hoc way. Thus, there is some key physics missing for certain systems. The presence of surface atoms leads to two key effects: the atomic cores in the surface repel the atomic cores of the adsorbate, which shifts the adsorbate's position; and local bonds can form between the adsorbate and the local orbitals (e.g., s-d hybrids) on the surface atoms.

In actual practice, the presence of core–core repulsions does not seem to make much difference in the behavior of molecules on flat metal surfaces. A typical adsorbate might sit 2–3 Å above a surface. At that distance, the core–core repulsion is effectively screened. As a result, the Pauli repulsions (i.e., repulsions associated with the interactions between the valence electrons) are much larger than the core–core repulsions. However, core–core repulsions can occasionally be important to coadsorption systems. For example, when potassium is adsorbed onto a transition metal surface, the potassium gives up its valence electrons to the surface. That produces a bare core on the surface that can interact with other adsorbates. Core–core repulsions are also quite important to the interactions of energetic particles with surfaces since at higher energies the atomic cores can overlap. Core–core repulsions are much more important on stepped surfaces, and on semiconductors and insulators with a stepped surface, the atomic cores of the step atoms can be close enough to interact with an adsorbate. That produces some interesting steric effects that are discussed in Chapter 10. Therefore, one would not expect the models so far in this chapter to work as well on stepped surfaces as on close packed planes.

Another serious limitation is the absence of specific, localized interactions in the models. When one has an atomic adsorbate, or a molecule that forms a single bond to the surface, these effects are not very important. However, a specific interaction becomes quite important when species form multiple bonds to a surface. For example, at room temperature, adsorbed ethylene converts to ethylidyne ($\equiv$C–CH$_3$) on some metal surfaces, vinylidene ($>$C$=$CH$_2$) on other metal surfaces, and ethan-1-yl-2-ylidyne ($\equiv$C–CH$_2$–) on a third group of surfaces. Localized bonds control which of these species form. Those local bonds are not considered in the effective medium bond. The situation is less clear when the adsorbed species forms a double or triple bond to the surface. However, there is some experimental evidence that local bonds contribute in those cases as well.

3.11.1   Slab Calculations

One needs to do a complete quantum calculation to consider the effects of core–core repulsions and local bonds on adsorption. In Appendix B we note that one can use GVB or CI programs to do essentially exact calculations of the properties of small molecules or periodic solids. Generally, one can easily do calculations with up to 30 transition metal atoms per unit cell and calculations with 200 atoms per unit cell are feasible. As a result, if one could find a way to approximate an adsorbate/surface system as a periodic solid with 30–200 atoms per unit cell, one could calculate the properties of the adsorbate/surface system exactly.

Slab calculations attempt to do that. The idea in the slab calculations is to consider a finite slab of atoms like that shown in Figure 3.51. The slab is infinite and periodic in the directions parallel to the surface, but finite in the direction perpendicular to the surface. Now consider covering the slab with a periodic overlayer of adsorbate, like that in Figure 3.17b. The net system will be just a two-dimensional solid, with a lattice dimension equal to the lattice dimension of the adsorbate. There are 20 or 50 atoms in the unit cell of the two-dimensional solid. One can do essentially exact density functional calculations for such a slab. Appendix B shows that the wavefunctions for the s-band from a finite line of atoms closely approximates the s-band of a semi-infinite line of atoms when there are about 10 atoms in the line. In the same way, the wavefunctions for the s-band from a slab 6–10 atoms thick is virtually identical to the s-band of a semi-infinite solid. In copper, Freeman, Wang, and Krakaver [1981] found that the d-band was well approximated by a slab with a half-width of 6 atoms. However, on titanium Feibelman [1989] found that a slab with a half-width of 11 atom thick was needed. So far slab calculations have considered mainly atomic adsorbates. However, the available evidence is that one can use slab calculations to calculate heats of adsorption of atoms and photoemission spectra of atomic adsorbates about as well as they have been measured. Unfortunately, it requires a very large amount of computer time to do so, so the number of calculations is limited.

3.11.2 Embedded Cluster Method

An alternate approach is called the **embedded cluster method.** The embedded cluster method was first proposed by Grimley [1974]. The idea is that the main problem with the jellium model is that it ignores the effect of atomic cores. Thus, if one adds atomic cores to the calculation, one should be able to reproduce all of the effects. In actual practice, one does a CI or density functional calculation on the cluster, and then uses a Green's function technique similar to that discussed in Section 3.8.2 to embed the cluster in jellium. One usually considers a small cluster of atoms that directly interact with the adsorbate, plus a few surrounding atoms. Usually a 20-atom cluster is sufficient.

An embedded cluster has the same workfunction and density of states (hardness) as a real surface. Further, it can exchange electrons with the bulk. As a result one would think that

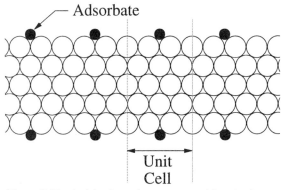

Figure 3.51 A slab of metal atoms covered by adsorbate.

the embedded cluster is a reasonable representation of a surface. However, it works out that there are some extra nodes in the wavefunction due to the small size of the clusters. Further, the orientation of the d-orbital tends to be wrong since local atom–atom interactions along the edge of the clusters are important in pinning the positions of the d-orbital. Thus the original embedded cluster calculations did not work that well.

Recently, however, Whitten [1992] developed a scheme to eliminate the extra nodes and to pin the d-orbitals. Wang and Whitten [1993] applied the method to hydrogen adsorption on Ni(111), and found that they could correctly predict the observed adsorbate geometry, vibrational frequency, and bond energy. Further, they were able to reproduce the experimentally observed activation energy of adsorption. Admittedly, hydrogen adsorption on nickel is the easiest case. The wavefunctions are fairly localized and the d-electrons do not have a significant contribution to the metal/hydrogen bond. It is unclear whether the methods will work in more difficult cases. However, the fact that the calculations worked at all is exciting because, after all, this is a case where simple cluster calculations have failed.

3.11.3 Predictions of the Advanced Computational Methods

I would have liked to have written a section telling the reader about all the new and exciting predictions from advanced computational methods. However, at present, I cannot say much. There are a few very nice calculations on individual systems, as reviewed by Fiebelman [1982, 1989]. However, the calculations are so expensive that when this book was being written, there was not enough work from which to draw any broad conclusions using the more advanced computational methods.

The one thing that has been examined in detail is the role of coadsorbates on the interaction of molecules on surfaces, e.g., how does adsorbed potassium affect the adsorption of CO on nickel? Clearly the coadsorbate affects the electronegativity of the surface. However, there is another effect, which is that the adsorbate introduces what is called a Friedel oscillation in the surface electron density. A positively charged adsorbate will attract electron density from its surrounding, producing a region that has a higher electron density and a surrounding region of lower electron density. The region of lower electron density is positively charged so it attracts additional electrons. The net result is the sinusoidal variation in electron density shown in Figure 3.52. Now, when a second molecule comes down on the surface, it interacts with the Friedel oscillations. The result is that the bonding of the second molecule is perturbed. The consequences of the perturbations are discussed in Section 4.7. At present, there are

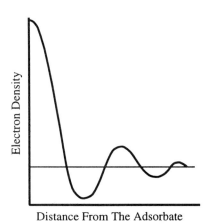

Figure 3.52 A schematic of the Friedel oscillation created when a molecule adsorbs on a jellium surface.

Electron Density

Distance From The Adsorbate

many similar details that in principle could be explored by the advanced computational methods, but they have not been explored to date.

3.12 SUMMARY OF ADSORPTION ON METALS

Still, it is interesting to summarize what we know about adsorption on metals today. There are several key observations that come out of the analysis described in this chapter. First, it is important to note that Pearson classifies metal surfaces as soft species. Metal surfaces interact strongly with other soft species and much more weakly with hard species. The main bonding is to the conduction band. The bond is localized near the surface. However, UPS shows that the electrons in the bond are rapidly exchanging with the bulk. The strength of the bonding is determined by electronegativity (i.e., workfunction) differences and measures that Parr and Yang would call the hardness of the surface.

Lang and Williams's results show that one does not really want to use the hardness, per se, to characterize surfaces. Rather, the key variable is the interstitial electron density, which in some way is related to the hardness of the surface. The effective medium model replaces the interstitial electron density with the electron density at the position of the adsorbate, but that electron density is proportional to the interstitial electron density.

There is a key feature that comes out of the calculations, which is that the electronegativity and hardness of a surface are universal properties of the surface. They do not vary with the adsorbate. Thus, a surface that binds one species strongly also binds many other species strongly. This result accounts for the general correlations shown in Figure 3.23. There is an important limitation to the correlations, however, in that they only apply to the interactions of soft species with metal surfaces. They do not work for the interaction of a hard species with a metal or a soft species with an insulator. The correlations also ignore the effects of local bonds, which can be quite important in the bonding of reactive intermediates.

At the time this book was being written, local bonding was not well understood. Slab calculations and embedded cluster calculations can in principle allow one to model the local interactions. However, when this book was being written, the computations were very time-consuming and there were no general guidelines. Bare cluster calculations have been done. However, bare cluster calculations show little correlation with experiment. Still, at this point we largely understand the binding of atoms and simple molecules to metal surfaces. The binding of more complex species such as reactive intermediates is still the subject of active investigation.

3.13 ADSORPTION ON SEMICONDUCTOR SURFACES

At this point, we leave our discussion of adsorption on metal surfaces and consider adsorption on semiconductor surfaces. All of the computational methods described in the last several sections for adsorption on metals can also be used to understand adsorption on semiconductor surfaces. Slab calculations have proved very useful for semiconductors. A slight modification of the Hückel method, called the **tight bonding approximation,** seems to also work quite well. Therefore, there is no reason to change computational methods when one changes from metals to semiconductors.

However, the calculations show that adsorption on a semiconductor surface is much different from adsorption on a metal surface. Electrons are much less mobile in the bulk of a semiconductor than in the bulk of a metal. As a result, the exchange with the bulk is suppressed. Instead, the bonding on the surface is dominated by local bonds between the adsorbate and the atoms in the surface of the semiconductor. Further, the electronic structure of a semiconductor surface is far different from that of a metal. Thus, the idea of representing the surface via jellium does not work.

3.13.1   The Surface Electronic Structure of Semiconductors

Recall that in the bulk, the bonding in a semiconductor looks like the bonding in a molecule. There are s–p–d hybrids that form local bonds. There are dangling bonds on the surface. The dangling bonds in silicon were already mentioned in Section 2.6. Basically, when one cuts the surface of a semiconductor, one leaves dangling bonds, as indicated in Figures 2.46 and 2.55. The dangling bonds like to pair up, but pairing causes lattice strain, so there are usually some unpaired bonds in the surface. For example, Figure 2.47 shows an STM image of the Si(111) surface. Notice, that if one adjusts the tip voltage properly, one can image the dangling bonds. The dangling bonds appear as bright spots in Figure 2.47. Generally, the surface chemistry of a semiconductor surface is controlled by the properties of the dangling bonds, and of the dimer pairs that are formed when the bonds pair up. Of course, there are electrons in the spaces between the dimer pairs and dangling bonds. It is not completely clear what those electrons are like, but one model is that those electrons look like hard spheres.

Lannoo and Friedel [1991] describe the surface electronic structure of silicon in some detail. Basically, the surface structure is unexpectedly simple, given the complexity of the surface reconstructions. There are dangling bond states associated with the unpaired electrons and surface states associated with the dimer pairs (see Figure 2.55 and 2.56). However, basically, one can think about the surface electron structure in what is called the tight binding model. One assumes that each surface atom forms an sp^3 hybrid, and then works out all of the electronic states associated with the lattice strain.

The key feature is the dangling bonds. A typical silicon surface has about $1-5 \times 10^{13}$ dangling bonds/cm^2 with one electron and one hole associated with each dangling bond. By comparison, a doped silicon sample only has perhaps 10^{14} to 10^{18} electrons or holes per *cubic* centimeter. Now consider the first 100 Å of the surface. There are $2-10 \times 10^{13}$/cm^2 states associated with the dangling bonds, and only 10^8 to 10^{10} states associated with the dopants. As a result, the properties of the dangling bonds swamp the properties of the dopants. In the solid-state literature, people say that the dangling bond states **pin the Fermi level** in silicon. What they mean by such a statement is that the density of states associated with the dangling bonds is so high that the dangling bond states overwhelm the effect of dopants on the surface electronic structure of silicon.

The dangling bonds are also very important chemically. In a fairly recent paper, Avouris and Cahill [1992] examined the properties of the dangling bond states with STM, and concluded that the dangling bonds are "metallic." They have a high density of states, so they can form soft bonds even though bulk silicon usually forms only hard bonds.

The situation is slightly more complex with a compound semiconductor like GaAs. When one creates a surface in a compound semiconductor the surface atoms reconstruct to pair up the dangling bonds, and to eliminate the surface dipole, as discussed in Section 2.6.5. That leaves atoms on the surface (e.g., Ga and As) with very different electronegativities. Electrons flow in response to that electronegativity difference, as described in Section 3.4.2. Some of the gallium atoms end up with nearly empty dangling bonds, while the arsenics end up with filled dangling bonds. The result is that the density of unpaired electrons and holes is much smaller on GaAs than on the surface of silicon. Consequently, the surface states do not pin the fermi level in GaAs even though the surface states pin the fermi level in silicon. Instead the surface is mainly ionic, although there are still some residual dangling bonds.

Semiconductor oxides (e.g., NiO, TiO$_2$) also are mainly ionic, although again there are still some dangling bonds. At present it is thought that the dangling bonds are associated with oxygen vacancies in the surface. Years ago, Kasowski and Tait [1979] calculated the electronic properties of the (100) and (110) surfaces of a TiO$_2$ sample that was presumed to have

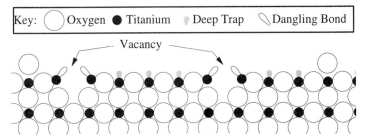

Figure 3.53 Some of the dangling bond states in the surface of TiO_2. (The figure is a schematic that is based on the analysis of Elian and Hoffman [1975] and Van Santen [1991].)

an ideal bulk terminated (rutile) structure with many oxygen vacancies. Relaxations around the oxygen vacancies were ignored. Kasowski and Tait found that there are a series of dangling bonds in all of the surfaces of TiO_2. The dangling bonds are centered on the titanium atoms in oxygen vacancies in the TiO_2 structure (see Figure 3.53). Van Santen [1991] suggests that Elian and Hoffmann's [1975] calculations imply similar things, although Elian and Hoffmann avoid reaching that conclusion. There are some limitations to these old calculations, because it is now known that TiO_2 does not form a bulk terminated structure. Rather there are strong relaxations around the oxygen vacancies, and the titanium atoms at the vacancies rehybridize because of the influence of the strong electric fields caused by the charges near the vacancy. However, the idea that there are dangling bonds on oxygen vacancies does seem to explain many observations. For example, Qain, Vohs, and Bonnell [1992] examined the properties of $TiO_2(110)$ using Scanning Tunnelling Microscopy. Generally, Qian, Vohs and Bonnell found that the surface was mainly almost ionic, but there were a number of oxygen vacancies in the surface. Tit, Halley, and Michalewitcz [1992] examined the electronic structure around the oxygen vacancies using a simple model where they considered the interaction of the d-levels on the titanium with the 2s and 2p levels on the oxygen. Tit, Halley and Michalewitcz found that they could model the STM results if they assumed that the electronic structure around the oxygen vacancy was perturbed in the way one would expect if one of the d-orbitals in the titanium atoms around the vacancy formed a dangling bond, albeit the dangling bonds had a different orientation that those in Figure 3.53. These results are new enough that they still need to be confirmed. However, the general picture that emerges from the current work is that the surfaces of semiconducting oxides are fairly ionic. Nevertheless, one does get dangling bonds and other electronic structure perturbations around oxygen vacancies. The properties of the oxygen vacancies seem to control the chemistry of the oxide.

3.13.2 Bonding to Semiconductor Surfaces

The easiest way to think about the bonding on a semiconductor surface is to consider the bonding to be just like the bonding in an organic molecule. For example, hydrogen forms a bond to a silicon surface that looks just like the σ-bond in a typical hydrocarbon. There are some extra complications with a compound semiconductor or a transition metal oxide due to the ionicity of the surface and the presence of d-electrons. However, to a first approximation, one can view the bonds between adsorbates and semiconductor surfaces as being simple σ-bonds, which are just like the bonds in a simple organic or inorganic molecule. Bond energies are also similar to the bond energies of the σ bonds in silicon compounds.

It happens that Parr and Yang [1989] have shown that Equations 3.41 to 3.50 (i.e., electronegativity equalization) work just as well in organic molecules as in other systems. Therefore, one would expect that all of the results from Section 3.6, 3.6.1, and 3.7 still apply. Silicon, for example is a hard acid/base. One would expect it to interact strongly with hard acids and bases such as HCl and H_2O, and more weakly with soft acids and bases such as CO and H_2. Experimentally, H_2O and HCl dissociate at about 100 K on Si(100)(2x1), while CO and H_2 do not adsorb. The strength of the binding should be approximated by Equation 3.27. Again that is as observed. In general, all of the principles from the last several sections also apply to the adsorption on semiconductor surfaces. The main difference is that the nature of the bonding is different, so the parameters in Equation 3.88 change.

Experimentally, most of the adsorption properties of elemental semiconductors (i.e., Se, Ge) are controlled by the dangling bonds in the surface of the semiconductor. For example, Figure 3.54 shows how the UPS spectrum of a Si(111) surface changes when trimethyl indium adsorbs. The UPS spectrum of a clean silicon sample shows a dangling bond state at 0.8 below E_f. Figure 3.54 shows how the UPS spectrum changes when trimethyl indium adsorbs. Notice that the dangling bond state disappears during the adsorption process. This is a common observation during adsorption on silicon surfaces. Physically, what is happening is that the trimethyl indiums are reacting with the dangling bonds on the Si(111) surface to form stable sp^3 bonded complexes. This example illustrates one very common bonding mode on silicon; i.e., the formation of an sp^3 hybrid via an interaction with a dangling bond.

The dangling bonds on silicon are quite important because they are the only soft centers on the surface. Therefore, they are the point of attachment for all soft species. Hard species can interact in a more general way, however. One common bonding mode is of an addition reaction over a dimer pair. Addition reactions are especially favorable on Si(100). Recall that one cuts a Si(100) surface, one leaves two dangling bonds on each silicon atom. The dangling bonds on adjacent pairs of atoms then react to form the (2x1) structure shown in Figure 2.55. The binding in the dimer pair is similar to the binding in the double bond in an olefin. The electronic structure looks somewhat like an sp^2 hybrid. Therefore, one would expect the surface to show some of the same reactions one observes in olefins. Experimentally, addition reactions of hard acids and bases such as alkyl-halides and water are quite facile. The dimer

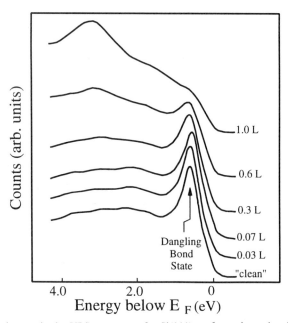

Figure 3.54 The changes in the UPS spectrum of a Si(111) surface when trimethyl indium adsorbs, after Lin et al. [1993].

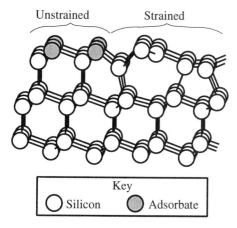

Unstrained   Strained

Key
○ Silicon   ● Adsorbate

Figure 3.55 A diagram showing how an adsorbate can relieve the lattice strain in a semiconductor surface.

pair also seems to be a strong Lewis base that can bind hard Lewis acids like trimethylaluminum or aluminum atoms. One finds, however, that soft species (CO, H_2) do not dissociatively adsorb, even though the dissociation process is thermodynamically favorable. Of course, in the gas phase olefins do not react rapidly with CO or H_2 either. The dimer pairs are much like ethylene, so perhaps the lack of reactivity with CO or H_2 is not surprising.

Insertion reactions of hard acids and bases are also fairly common on silicon surfaces. Recall that a considerable amount of lattice strain occurs when the dangling bonds on a clean silicon surface pair up to form bonds. Now consider inserting an adsorbate atom into the strained bond. Notice that the adsorbate can bridge between the structure, as shown in Figure 3.55, and thereby reduce the lattice strain. As a result, bonding like that in Figure 3.55 is quite favorable on silicon surfaces.

Another effect of the insertion reactions is that silicon surfaces often rearrange (i.e., reconstruct) when gases adsorb. Recall that clean silicon surfaces show complex reconstructions due to a competition between pairing of dangling bonds and lattice strain. The dangling bonds are saturated when gases adsorb and the strain is relieved via insertion reactions. As a result, the silicon surface structure usually changes when gases adsorb. The most common reconstruction is a simple relaxation back to the bulk terminated structure. However, very complex reconstructions are also seen. An example is given in Figure 3.56.

There is one other key process that happens with silicon, which is a direct displacement reaction: an adsorbate atom replaces a silicon atom in the silicon structure, and that silicon atom interacts with a second adsorbate atom to form an overlayer. So far, replacement reactions have been observed during oxygen, germanium, and aluminum adsorption on silicon. They seem to be particularly important at higher temperatures (>800 K). However, there has been speculation that they also occur at room temperature.

There has been much less work on adsorption on compound semiconductors (e.g., GaAs) than on adsorption on silicon. Generally, it is thought that the bonding on a compound semiconductor surface is similar to that on silicon. However, there are extra complications due to the ionicity of the surface. In particular, it is thought that the buildup of surface stresses described in Section 2.6.1 plays an important role in the adsorption of gases on compound semiconductors, although the details are missing at present.

Experimentally, GaAs is much less reactive than silicon. Recall that when a GaAs surface forms, electrons flow from the dangling bonds on the gallium to the dangling bonds on the arsenic to equalize the electronegativity of the surface. The result is a surface with a stable empty dangling bond on the gallium and a stable filled dangling bond on the arsenic. Unlike silicon, there are no half-filled orbitals available for bonding. As a result, GaAs is much less reactive than silicon, and InP is virtually inert.

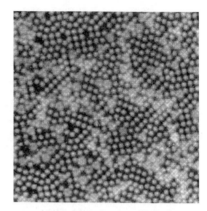

4% Monolayer Cobalt

Clean

Figure 3.56 A comparison of the STM image of clean Si(111) and Si(111) covered by 4% of a monolayer of cobalt. (Courtesy of D. Cahill.)

The adsorption of gases on semiconducting oxides (e.g., TiO_2, Ti_2O_3, and NiO) also needs much more work. McKay and Heinrich [1985] and Heinrich [1989] examined H_2, O_2, CO, SO_2, and H_2O adsorption on NiO(100) and TiO_2(110). They found that none of these gases adsorbed on a pristine freshly cleaned NiO(100) surface, while only H_2O adsorbed on a pristine freshly cleaned TiO_2(110) surface. The H_2O desorbed again upon heating. Stepped surfaces also show low reactivity. In contrast, Heinrich [1989] found that surfaces that had been sputtered or annealed to produce defects such as oxygen vacancies readily adsorb all of these gases. Further, the UPS features associated with the defect states disappear when gases adsorb. Therefore, it is thought that the adsorption process is dominated by the defects.

At this point, it is unclear how the adsorbate binds at these defect sites. Older calculations indicate that most of the binding occurs through the dangling bonds at the oxygen defects shown in Figure 3.53. However, it is now known that the electric fields in the neighborhood of the defects perturb the interactions. As a result, it is not known whether one can view the bonding as a simple interaction with a dangling bond. Still the best available evidence is that the binding occurs through the defects, although the details are unclear.

One of the more interesting things about semiconducting metal oxides is that one can dope them and thereby change their properties. In the semiconductor industry, one adds dopants to change the electrical properties of a solid. For example, if one adds a small amount of gallium to silicon, the gallium will replace one of the silicon atoms in the lattice. The gallium assumes as sp^3 configuration. However, gallium has one less valence electron than silicon. As a result, when one dopes with gallium one creates a solid with some unoccupied states. A similar analysis shows that if one adds arsenic to silicon, one gets a system with a few extra electrons in the normally unoccupied states in the conduction band. Substances like gallium that create empty levels are called p-dopants, while substances like arsenic that create extra filled states are called n-dopants.

Now consider what happens to the Fermi level when the silicon is doped. Notice that if one adds a p-dopant, the probability of a state below the band gap being empty increases. From the Fermi equation, when the probability of a state being empty increases, the Fermi level in the bulk decreases. Similarly, if one adds an n-dopant the probability that a state above the band gap increases. Again according to the Fermi equation, the Fermi level in the silicon increases too. Another way to look at this process is to use the principle of electronegativity equalization to predict what happens. The p-dopants are more electropositive than silicon, while the n-dopants are more electronegative than silicon. Now consider what happens when one adds a p-dopant to silicon. According to the principle of electronegativity equalization electrons have to flow from silicon to the dopant to equalize the electronegativity of the system. Silicon is a hard base (it is also a hard acid). Therefore, according to Equations 3.42 and 3.49, a small change in the electron density of the silicon will produce a large change in the electronegativity of the bulk silicon.

It happens that with silicon, the surface Fermi level does not change because of Fermi level pinning (i.e., the surface is a soft acid so its Fermi level does not shift). However, it is thought that the surface Fermi level does shift with most other semiconductors. When the Fermi level changes, the workfunction and hence the electronegativity of the surface changes, too. According to Equation 3.28 bond strengths should also change. Volkenshtein [1991] cites many examples of this effect from the literature before 1950. No one has documented similar changes in the modern literature, but the older literature indicates that the dopant effect is quite important.

Still, when one designs a commercial metal oxide catalyst, one adds dopants to modify the properties of the oxide. The dopants are thought to work just like the dopants in silica. They add or subtract electrons from the oxide, which changes the electronegativity of the oxide, and therefore modifies its ability to form bonds. This promoter chemistry has not been studied in detail. However, commercially, it seems to be quite important.

3.13.3 A Two-State Model for Adsorption on Semiconductors

Quantitative calculations of interactions of molecules with silicon surfaces use three different approaches: a two-level approximation, a tight binding/slab calculation, and a full abinitio/slab or embedded cluster calculation. The two-level approximation assumes that a single orbital on the adsorbate, $\phi_a(\vec{r})$ interacts with a dangling bond on the surface $\phi_s(\vec{r})$. Let's define $\mathcal{H}_s(\vec{r}_s)$ to be the Hamiltonian for the surface dangling bond and $\mathcal{H}_a(\vec{r}_a)$ to be the Hamiltonian for the adsorbate. It is useful to assume that the Hamiltonian for the system $\mathcal{H}_e$ is given by

$$\mathcal{H}_e = \mathcal{H}_a(\vec{r}) + \mathcal{H}_s(\vec{r}) + v_{as}(\vec{r}) \tag{3.92}$$

where v_{as} is the interaction between the adsorbate and the surface and $\vec{r}$ is the position of the electron. For the purpose of derivation, it is useful to assume that the wavefunction ψ_{as} for an electron the adsorbate surface bond is given by

$$\psi_{as} = c_a\phi_a(\vec{r}) + c_s\phi_s^1(\vec{r}) \tag{3.93}$$

where c_a and c_s are constants and

$$\phi_s^1 = \frac{\phi_s - \phi_a\langle\phi_a|\phi_s\rangle}{|\phi_s - \phi_a\langle\phi_a|\phi_s\rangle|} \tag{3.94}$$

Note ϕ_a and ϕ_s^1 are orthogonal to each other. ψ_{as} must satisfy the Schrödinger equation:

$$\mathfrak{IC}_e \, \psi_{as} = E\psi_{as} \tag{3.95}$$

Substituting Equations 3.92 and 3.93 into Equation 3.95, multiplying by ϕ_a and ϕ_s, and integrating yields

$$\begin{bmatrix} E_s - E & V_{as} \\ V_{as} & E_a - E \end{bmatrix} \begin{bmatrix} c_s \\ c_a \end{bmatrix} = \begin{bmatrix} 0 \\ 0 \end{bmatrix} \tag{3.96}$$

where

$$E_s = \langle\phi_s^1|\mathfrak{IC}_e|\phi_s^1\rangle \tag{3.97}$$

$$E_a = \langle\phi_a|\mathfrak{IC}_e|\phi_a\rangle \tag{3.98}$$

$$V_{as} = \langle\phi_a|\mathfrak{IC}_e|\phi_s^1\rangle \tag{3.99}$$

Equation 3.96 has two eigenvalues:

$$E_{Bonding} = \left(\frac{E_s + E_a}{2}\right) - \sqrt{\left(\frac{E_s - E_a}{2}\right)^2 + V_{as}^2} \tag{3.100}$$

$$E_{Antibonding} = \left(\frac{E_s + E_a}{2}\right) + \sqrt{\left(\frac{E_s - E_a}{2}\right)^2 + V_{as}^2} \tag{3.101}$$

The first eigenvalue corresponds to the bonding state of the adsorbate, while the second eigenvalue corresponds to the antibonding state of the adsorbate.

Fischer [1972] and Harrison [1980] did extensive calculations to determine the E_a, E_s, and V's for covalent solids. Tables of values are given in Harrison [1980] and Mönsh [1993] and in the supplemental material for this chapter. Generally, one finds that E_a, the energy level for the adsorbate, depends on the adsorbate but not on the surface, so one can use atomic values for E_a. Figure 3.57 shows a plot of the E_a for several atomic adsorbates. Generally, E_a varies linearly with the electronegativity of the atom for all of the elements except hydrogen.

Table 3.17 shows a table of E_s for each of the dangling states in Si, Ge, GaAs, and InP. The E_s are for reconstructed surfaces. They don't vary significantly when gases adsorb, unless the surface structures change. In silicon and germanium there is a single dangling bond state, while in GaAs and InP there are two dangling bond states: an empty acceptor state on the metal atom (i.e., Ge and In) and a filled donor state on the Group IV component (i.e., As or P).

One also needs an approximation for V_{as} to do any calculations. Harrison [1980] used extensive pseudopotential calculations to determine values of V_{as} for bulk semiconductors. Harrison showed that one can fit all of the data to within 10% with the simple formula:

$$V_{as} = C_{Harrison}\left(\frac{\hbar^2}{m_0 \, d_{as}^2}\right) \tag{3.102}$$

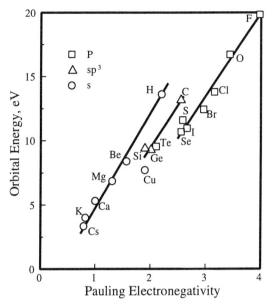

Figure 3.57 Values of E_a for a series of isolated atoms as determined by Hartree-Fock calculations. After Mönch [1993].

where m_0 is the rest mass of the electron, d_{as} is the distance between the adsorbate and the surface atom, $C_{Harrison}$ is a constant that depends on the orbitals involved in the bond, and $\hbar$ is Planck's constant; $C_{Harrison}$ equals 1.89 for the σ interaction of an s-orbital in an absorbate with a dangling bond in a semiconductor, while $C_{Harrison}$ equals 2.63 for the σ interaction of a p-orbital in an adsorbate with a dangling bond in a semiconductor. Mönch [1993] suggests the following approximation for d_{as}:

$$d_{as} = r_a + r_s + C_{M\ddot{o}}|\chi_a - \chi_s| \tag{3.103}$$

where r_a and r_s are the covalent radii of the adsorbate and the surface atom, respectively, χ_a and χ_s are their electronegativities, and $C_{M\ddot{o}}$ is a constant that varies between 0.02 and 0.06 Å/eV.

Mönich [1993] shows that Equation 3.100 fits data for the donor levels of adsorbates on semiconductor surfaces under conditions where the main interactions are with dangling bonds (i.e., at low coverages). For example, Figure 3.58 shows how the core levels of a silicon

TABLE 3.17 The Energies of the Dangling Bond States in a Series of Semiconductors, Extracted from the More Extensive List of Mönch [1993]

Semiconductor	Valance Band Top (eV)	Dangling Bond States		
		Acceptor State (eV)		Donor State (eV)
Si	−9.35		−9.38	
Ge	−8.97		−9.29	
GaAs	−9.64	−7.14		−11.46
InP	−10.03	−6.56		−11.96

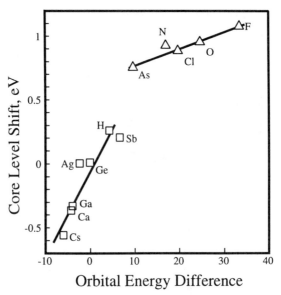

Figure 3.58 The shifts in the Si(2p) levels in the Si(111) surface due to the adsorption of gas. (Data tabulated in Mönch [1993].)

substrate change when various gases adsorb. Generally, the energy of the core levels vary linearly with the difference in E_a and E_s. This is as one would expect if the bond energy follows a linear expansion of Equation 3.100. There are two lines in the figure because $C_{Harrison}$ changes according to whether an s or p level is involved in the bonding. Therefore, it seems that Equation 3.100 can be used to model the changes in the energy levels of silicon when molecules adsorb.

Mönch [1993] also implies that the heat of adsorption will also be accurately predicted via Equation 3.100, provided the conditions are such that the main interaction is with the dangling bonds. At present, however, there is a scarcity of data on heats of adsorption on semiconductor surfaces. As a result it is difficult to be sure whether this claim is correct. Equation 3.100 certainly gives an approximation to heats of adsorption on semiconductors at low coverages. However, one does not know how accurate the approximation is.

Equation 3.100 fails when there are significant reconstructions of the surface during adsorption. When the surface reconstructs, E_s changes in a complex way. The two-state model ignores that physics, so it fails when the surface reconstructs. One needs to do a fairly large calculation to understand the effects of the reconstructions on the adsorbate/surface bonds. Generally, people do full abinitio slab or embedded cluster calculations to model these effects. The results generally reproduce the photoemission data on changes of the adsorbate density of states almost as well as they can be measured. It is also thought that the slab calculations can predict accurate heats of adsorption. However, accurate measurements of heats of adsorption on semiconductor surfaces are scarce. As a result, further confirmation of the accuracy of the slab calculations is needed.

3.14 ADSORPTION ON IONIC SURFACES

There is much less information about the binding of molecules on insulating surfaces than about the binding of molecules on metals or semiconducting surfaces. Generally, people use cluster methods similar to those described for metals to model insulators and claim some success. One does have to worry about how to terminate the cluster without creating an unphysical electric field due to the presence of unpaired ions. At present, those details of how to terminate the cluster are not well understood. Also, it has been difficult to verify the models, because there are very little data taken under pristine, well-defined conditions.

3.14.1 The Surface Electronic Structure of Insulating Oxides

There only has been a limited amount of work on the surface electronic structures of insulators. Years ago, there were calculations that suggested that the surfaces of insulating oxides (e.g., MgO, Al_2O_3) were mainly metallic. However, more recent experiments indicate that the surfaces of insulators are mainly ionic. A purely ionic surface is fairly inert, but on a real surface, there are usually a series of defect structures that dominate the chemistry. For example, with silica/aluminas there are two key types of defect structures that are important to the chemistry: Lewis acid sites and Brønsted acid sites. Lewis acid sites contain an aluminum atom in threefold sp_2 hybrid with an empty dangling orbital available to accept charge. Brønsted acid sites contain an OH, with a readily transferred proton. Generally, the chemistry of silica/aluminas is controlled by the properties of the defects. One can dope the oxide (i.e., add impurities) to change the electron negativity of the defect. That completely changes the chemistry. At present, however, many of the details are not well understood. See Tsukada, Adachi, and Satoko [1983], Moffat [1991], or Heinrich and Cox [1994] for details.

Recently, there has been some interest in the surface electronic structure of alkali halides (e.g., NaCl, LiF). EELS indicates that there is an intrinsic surface state in the alkali halides, but the nature of this surface state is unclear. See Saiki et al. [1993] for details.

3.14.2 Adsorption on Insulating Surfaces

Experiments show that pristine/freshly cleaved insulating oxides are fairly inert. For example, Heinrich [1989] finds that CO, H_2, O_2, and SO_2 do not adsorb on a MgO(100) surface that had been cleaved in UHV. However, once the surface is sputtered and annealed to create some defects, all of the gases adsorb. H_2O adsorbs on even a freshly cleaved sample. There is less UHV data for adsorption on silica/aluminas. However, it is thought that they show similar behavior.

At present, there are few data on the adsorption of gases on pristine insulator surfaces. However, there are considerable data on the adsorption of gases on surfaces with high concentrations of defects. Generally, the binding process is fairly complex. The adsorbate usually first reacts with a defect, as illustrated in Figure 3.59, to create an ion. Then the ion migrates out over the defect free regions of the surface.

The heat of adsorption on the defect-free regions of the surface is determined by ionic and

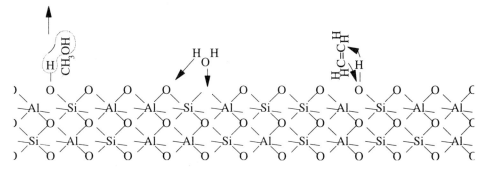

Figure 3.59 Some typical binding processes on silica/aluminas.

polarization forces. Neutral nonpolar species usually interact fairly weakly with an insulating oxide. Generally, a nonpolar species is polarized when it interacts with an ionic surface, so there is a small binding energy as given in Equation 3.9. However, the total binding energy is usually less than 10 kcal/mole. Species with a strong dipole (e.g., H_2O) can interact more strongly with an ionic surface. A typical binding energy might be 15 kcal/mole. By contrast, an ionic species can be bound by 25–100 kcal/mole. As a result, ionic species tend to be the main adsorbates on insulating oxides.

At present, there are no data that show how, for example, the binding energy of molecules varies systematically with the composition on pristine surfaces of the insulating oxides. Generally, we would expect the binding energy of neutral species or a defect-free surface to be proportional to the surface electric field, and the polarizability of the molecules. Generally, one would expect a strong interaction with hard acids or bases. However, it has been difficult to locate data to illustrate that. There is a good discussion of our current understanding of adsorption on oxides in Heinrich and Cox [1993]. The reader should refer to this reference for details.

There is considerable work on the adsorption on zeolites. Zeolites are silica/aluminas with strong Brønsted and Lewis acid sites. Generally, the acid sites are hard centers. They interact strongly with hard acids and bases, and more weakly with soft acids and bases.

People usually discuss the properties of zeolites in terms of the electronegativity of the surface and its hardness. All of the ideas are identical to those in Sections 3.4.1 and 3.4.2. For example, one can change the acidity of the zeolite by doping it to change the average electronegativity of the substrate. Electrons flow to equalize the electronegativity of the zeolite, but that changes the properties of the acid sites in the zeolite. The result is that one can change the acidity of zeolites via doping. Figure 3.60 shows how the acid strength, as measured via infrared varies over a number of zeolites. Notice that there is a good correlation between the acid strength of the zeolite and the electronegativity of the zeolite, as expected from Equation 3.45. As a result, one can tailor the properties of zeolites in the same way that one can tailor the properties of metals or metal oxides.

Mortier and co-workers [1986, 1990] have done extensive tests of electronegativity equalization for adsorption and reactions in zeolites. They developed a code called electronegativity equalization method (EEM) to calculate the interaction of a molecule with a zeolite. The basic idea is to assume that the main bond between the adsorbate and the zeolite is an ionic bond. The presence of the ions in the zeolite polarizes the atoms in the adsorbate. One calculates the net polarization by adding up all of the electrostatic forces on the atoms as

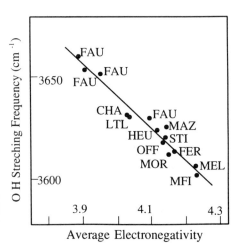

Figure 3.60 The OH stretching frequency in a number of ion-exchanged zeolites as a function of the average electronegativity of the zeolite. (After Jacobs and Mortier [1982].)

TABLE 3.18 The Electronegativities and Hardnesses of a Series of Species in Zeolites.

Species	χ_M^* (eV/electron)	η_M^* (eV/electron^2)
H($\delta+$)	4.4	13.77
H($\delta - 1$)	3.17	9.92
C	5.68	9.05
N	10.60	13.2
O	8.5	11.08
F	32.4	90.00
Al	-2.24	7.76
Si	1.33	6.49
P	2.90	6.29

Source: Calculated by Jacobs and Mortier [1982].

illustrated in Example 3.C. The calculation starts with a table of the effective electronegativities and hardnesses of all of the atoms in a molecule. The molecule is then placed near an acid site in a simulated zeolite, and then the charges on all of the atoms are allowed to reorganize in response to both direct charge transfer to the acid sites in the zeolite and the electric field created by the ions in the zeolite. Generally, it is found that the electric fields in the zeolite cause the charges on a polarizable molecule to reorganize. The reorganization creates hard centers that can interact strongly with the acid sites in the zeolite. Mortier shows that all of the effects can be calculated from Equation 3.45. Note that, unlike metals, the last term in Equation 3.45 (i.e., the electrostatic interaction) is quite important in zeolites. Usually, the electrostatic (i.e., ionic) forces dominate the other forces, i.e., most adsorbates are held to ionic surface with ionic bonds. A subtle point is that just as with metals and semiconductors, the electronegativity of a species in a zeolite is different from its electronegativity in the gas phase. Table 3.18 summarizes some of the values of the electronegativities in zeolites. See Mortier [1990] for further details.

3.15 SYNOPSIS AND PLANS FOR THE NEXT SEVERAL CHAPTERS

In summary, this chapter has provided an overview of the binding of molecules on solid surfaces. We started with a general synopsis of adsorbate–surface bonding, and pointed out the important differences between *ab*sorption, multilayer *ad*sorption, and monolayer *ad*sorption. We then focused on monolayer adsorption and described how molecules bind to surfaces. We noted that molecules form chemical bonds to surfaces. However, the nature of the adsorbate–surface bond varies significantly with the substrate. On metals, the largest contribution to bonding comes from an interaction with the conduction band. The conduction band shares electrons with the adsorbate. The electron density changes only in the neighborhood of the adsorbate. However, there is a rapid exchange of electrons with the bulk. With a transition metal, there is an additional interaction with the d-bands. The interaction is more like a local bond. In actual practice the d's only contribute a small part (i.e., less than 10%) of the total bond energy. However, they are quite important to understanding reactions on transition metals.

The interaction of adsorbates with semiconductors is much different than on metals. Generally, one can view the bonds as simple covalent interactions, although there are some corrections due to lattice strain and the strong electric field at the surface of a compound semiconductor. At low coverages, the main bonding is with dangling bonds in the adsorbate.

However, at higher coverages, one also gets insertion reactions into strained bonds in the surface.

On insulating oxides, the main bonding is ionic. One can calculate the bond energy by adding up the electrostatic interactions of the adsorbate with all of the ions in the surface, provided one includes the polarization of all of the molecules, as indicated in Example 3.C. Note, however, that there can also be covalently bonded species at defects on insulating oxides. Often the covalently bonded species dominate the chemistry.

At present, people largely understand what determines the strength of bonding of adsorbates to surfaces. Years ago Eley [1950] proposed that the strength of the bond was largely determined by the general ability of the adsorbate and surface to form bonds, and the difference in the electronegativities of the adsorbate and the surface. Such a model arises naturally from density functional theory and the effective medium model of adsorbate–surface binding, and it does fit data on the adsorption of atomic species on metal surfaces. A modified version of the theory is also used to discuss adsorption on the defect sites of semiconductor and oxide surfaces. The model only considers about 90% of the bond energy when a molecule interacts with a transition metal surface, so it misses some subtle effects. However, the overriding principle is that the strength of adsorbate–surface bonds is determined by the basic ability of the surface to form bonds, and the electronegativity and hardness of the various species, independent of whether the surface is a metal, semiconductor, or insulator.

At this point, the adsorption of gases on surfaces is still an active research area. Molecular adsorption on metals still needs more work, as does adsorption of gases on semiconducting and insulating surfaces. However, after over 100 years of work we are now starting to understand adsorption on surfaces.

3.16 SOLVED EXAMPLES

Example 3.A Estimation of Heats of Adsorption Using Bond Counting

Benziger [1991] discusses a simple idea to estimate heats of adsorption of reactive fragments. The idea is to simply add up the energies of all of the bonds in the molecule to estimate a heat of adsorption. The method is analogous to the methods described by Benson [1976] for gas phase molecules.

Consider an adsorbed fragment RX_{ad}. Benziger [1993] shows that to a first approximation, one can estimate, $\Delta H_f(RX_{ad})$, the heat of formation of the RX_{ad} species via

$$\Delta H_f(RX_{ad}) = \Delta H_f(RX\cdot) + \frac{n_{sx}}{n_x} D(MX) \qquad (3.104)$$

where $\Delta H_f(RX\cdot)$ is the heat of formation of the $RX\cdot$ radical in the gas phase, $D(MX)$ is the strength of the MX bond from Table 3.5, n_x is the total number of bonds to atom X, and n_{sx} is the number of bonds between atom X and the surface.

As an example of this analysis consider forming a methoxy (CH_3O_{ad}) on platinum. The heat of formation will be

$$\Delta H_f(CH_3O_{ad}) = +3.5 \text{ kcal/mole} - \tfrac{1}{2}(84 \text{ kcal/mole}) = -38.5 \text{ kcal/mole} \qquad (3.105)$$

where 3.5 kcal/mole is the heat of formation of methoxy in the gas phase.

Calculations like this are used quite often in the literature. They are not particularly accurate (errors are as large as 0.5 eV). However, often they are as good as can be done when the electronegativity or hardness of the various species, is not known.

Example 3.B   An Example of a Two-State Model Calculation

In Section 3.13.3 we presented the two-state model of adsorption on semiconductor surfaces. The objective of this example is to illustrate the use of the method. Consider the interaction of a chlorine atom with a dangling bond in a silicon surface. According to Table 3.17, the surface has a donor level of -9.38 eV, while according to Table 3.20 chlorine has an acceptor level at -13.78 eV. When the chlorine comes up to the surface, the electron from the dangling bond reacts with the chlorine to form a new energy level, as given by Equation 3.100

$$E_{Bonding} = \left(\frac{(-9.38) + (-13.78)}{2} \right) - \sqrt{\left(\frac{9.38 - 13.78}{2} \right)^2 + V_{as}^2} \qquad (3.106)$$

where V_{as} is given by Equation 3.102. Following Equation 3.103

$$d_{as} = 1.17 \text{ Å} + 0.99 \text{ Å} - \frac{0.04 \text{ Å}}{\text{eV}} |3.16 - 1.67| = 2.1 \text{ Å} \qquad (3.107)$$

where 1.17 Å is the covalent radius of silicon, 0.99 Å is the covalent radius of chlorine, 3.16 is the electronegativity of an isolated chlorine molecule, and 1.67 is the electronegativity of bulk silicon. According to Equation 3.102

$$V_{as} = 2.63 \, (7.62 \text{ eV-Å}) \left(\frac{1}{2.11 \text{ Å}} \right)^2 = 4.5 \text{ eV} \qquad (3.108)$$

where 2.63 is $C_{Harrison}$ and 7.62 eV-Å^2 is $\hbar^2/m_0$. Plugging V_{as} into Equation 3.106 yields

$$E_{Bonding} = -16.6 \text{ eV} \qquad (3.109)$$

The heat of adsorption of chlorine per mole of atomic chlorine is

$$\Delta H_{ad} = 16.6 \text{ eV} - (-9.38 \text{ eV} - 1.23 \text{ eV}) = 6 \text{ eV} \qquad (3.110)$$

where 1.23 eV is the standard heat of formation of atomic chlorine and -9.38 eV is the initial energy of the electron before a bond forms.

Example 3.C   EEM Calculation (Mortier's Method)

In Section 3.14 we noted that Mortier [1990] has shown that the EEM can be used to calculate the interaction of molecules with zeolites. The objective of this example is to illustrate the idea. The problem will be to calculate the change in the charge distribution of water due to the presence of another nearby water molecule.

Consider a molecule or a cluster of molecules. In the EEM model one computes the total energy of the cluster, E_{total}, as a sum over all of the atoms in the cluster:

$$E_{total} = \sum_i \left(E_i^0 - \chi_1^* n_i^1 + \frac{\eta_i^*}{2} (n_i^1)^2 - N_i^e \sum_{j \neq i} \frac{Z_j}{r_{ij}} + \frac{N_i^e}{2} \sum_{j \neq i} \frac{N_j^e}{r_{ij}} + \frac{1}{2} \sum_{j \neq i} \frac{Z_i Z_j}{r_{ij}} \right) \qquad (3.111)$$

where i and j are indices of all of the atoms in the cluster; E_i^0 is the energy of an isolated i atom; χ_i^* and η_i^* are the electronegativity and hardness of an isolated i atom; r_{ij} is the distance

TABLE 3.19 Electronegativities of the Elements*

H 2.20																	He
Li 0.98 1.01	Be 1.57 1.49											B 2.04 1.69	C 2.55 2.20	N 3.04 2.48	O 3.44	F 3.98	Ne
Na 0.93 0.96	Mg 1.31 1.22											Al 1.61 1.50	Si 1.90 1.67	P 2.19 1.68	S 2.58	Cl 3.16	Ar
K 0.82 0.80	Ca 1.00 0.90	Sc 1.36 1.15	Ti 1.54 1.30	V 1.63 1.51	Cr 1.66 1.65	Mn 1.55 1.58	Fe 1.83 1.75	Co 1.88 1.81	Ni 1.91 1.85	Cu 1.90 1.62	Zn 1.65 1.45	Ga 1.81 1.46	Ge 2.01 1.61	As 2.18 1.70	Se 2.55	Br 2.96	Kr
Rb 0.82 0.74	Sr 0.95 0.85	Y 1.22 1.14	Zr 1.33 1.21	Nb 1.60 1.42	Mo 2.16 1.65	Tc 1.90 1.88	Ru 2.20 1.92	Rh 2.28 1.92	Pd 2.20 1.94	Ag 1.93 1.57	Cd 1.69 1.44	In 1.78 1.38	Sn 1.96 1.47	Sb 2.05 1.56	Te 2.10	I 2.66	Xe
Cs 0.79 0.69	Ba 0.89 0.82	La 1.10 1.08	Hf 1.30 1.26	Ta 1.50 1.44	W 2.36 1.70	Re 1.90 1.92	Os 2.20 1.92	Ir 2.20 1.97	Pt 2.28 2.00	Au 2.54 1.83	Hg 2.00 1.49	Tl 2.04 1.38	Pb 2.33 1.45	Bi 2.02 1.47	Po 2.00	At 2.20	Rn
Fr 0.70	Ra 0.90	Ac 1.10															

*The top value is Pauling's [1960] value for isolated elements. The bottom value is Miedema, dChâtel, and Deboer's [1980] for covalent solids. Miedema's values have been multiplied by 0.355 to put them on the Pauling scale.

TABLE 3.20 Values of $-E_a$ for the Elements*

1	2	3	4	5	6	7	8	9	10	11	12	13	14	15	16	17	18
H 13.6																	He 25.0
Li 5.34	Be 8.41 5.79											B 13.46 8.43	C 19.37 11.07	N 26.22 13.84	O 34.02 16.72	F 42.78 19.86	Ne 52.5 23.1
Na 4.94	Mg 6.88 3.84											Al 10.70 5.71	Si 14.79 7.58	P 19.22 9.54	S 24.01 11.60	Cl 29.19 13.78	Ar 34.7 16.1
K 4.01	Ca 5.32	Sc 5.63 8.59	Ti 5.89 10.98	V 6.11 12.55	Cr 6.32 13.84	Mn 6.50 15.37	Fe 6.71 15.05	Co 6.90 15.28	Ni 7.08 15.54	Cu 7.27 2.37 15.8	Zn 7.96 4.02 19.8	Ga 11.55 5.67	Ge 15.15 7.33	As 18.91 8.98	Se 22.86 10.68	Br 27.00 12.43	Kr 31.4 16.1
Rb 3.75	Sr 4.85	Y 5.18 6.62	Zr 5.55 9.04	Nb 5.82 10.84	Mo 6.03 12.55	Tc 6.20 14.51	Ru 6.43 15.14	Rh 6.61 16.23	Pd 6.78 17.44	Ag 6.91 2.61 18.52	Cd 7.21 3.99 20.2	In 10.14 5.37	Sn 13.04 6.76	Sb 16.02 8.14	Te 19.12 9.54	I 22.34 10.97	Xe 25.7 12.4
Cs 3.36	Ba 4.29	La 4.34 9.49 4f 8.04	Hf 5.50 9.84	Ta 5.77 11.54	W 6.02 13.44	Re 6.19 13.44	Os 6.41 14.03	Ir 6.63 15.14	Pt 6.80 15.30	Au 6.48 2.38 17.98	Hg 7.10 3.95 19.02	Tl 9.82 5.23	Pb 12.48 6.53	Bi 15.19 7.79	Po 17.96 9.05	At 20.82 10.33	Rn 23.8 11.6

*The top value is ϵ_a for an s electron; the second number is ϵ_a for a p electron. The transition metals have a third number, which is ϵ_a for a d electron. Data from Fischer [1972], Huzinga [1984], and Mönch [1993]. All values are measured in eV.

between atom i and atom j in the molecular cluster; N_i^e and N_j^e are the number of electrons on atoms i and j in the stable cluster; Z_i and Z_j are the charges on the atomic cores of atoms i and j; and n_i^1 is the number of electrons that atom i transfers when the molecular cluster is assembled from its atoms. For the purposes of derivation it is useful to note that

$$N_i^e = N_i^0 + n_i^1$$

$$N_j^e = N_j^0 + n_j^1 \tag{3.112}$$

where N_i^0 and N_j^0 are the total number of electrons in an isolated i or j atom, respectively.

The first three terms in Equation 3.111 are analogous to the first four terms in Equation 3.44, the fifth term is the electrostatic interaction between the electrons on atom i and the cores of the other atoms in the system, the sixth term is the electrostatic interaction between the charges on the various atoms in the molecular cluster, and the seventh term is the core–core repulsions between the atoms in the cluster.

The objective of the EEM calculation is to calculate the charges on each of the atoms, by noting that at equilibrium the electronegativity of each of the atoms in the cluster must be equal. One could use the analysis in Section 3.4 to do that. Note, however, that the presence of the electrostatic interaction changes the electronegativity of each of the atoms in the cluster. One can calculate a new electronegativity, μ_i from Equation 3.49

$$\mu_i = \frac{\partial E_{total}}{\partial n_i^1} \tag{3.113}$$

Substituting Equations 3.111 and 3.112 into Equation 3.113 yields

$$\mu_i = \chi_i^* + \eta_i^* \, n_i^1 + \sum_{j \neq i} \frac{n_j^1}{r_{ij}} + \frac{1}{2} \sum_{j \neq i} \frac{(N_j^0 - Z_j)}{r_{ij}} \tag{3.114}$$

The principle of electronegativity equalization implies that all of the μ_i should be equal. Let's define μ_0 to be the as yet unknown value of μ_i. Further, assume that we know the geometry of the molecular cluster and that there are N_a atoms in the cluster. Notice that if we substitute $\mu_i = \mu_0$ into Equation 3.114 for each value of i, we will get N_a linear equations, one for each value of i. There are $N_a + 1$ unknowns, that is, all of the n_i^1 and μ_0. We can get another equation by noting that electrons are conserved when a molecular cluster forms, i.e.,

$$\sum_i n_i^1 = 0 \tag{3.115}$$

Equations 3.114 and 3.115 represent $N_a + 1$ linear equations in $N_a + 1$ unknowns. Therefore they can be solved to calculate the charge distributions in the molecular cluster. Those change distributions can be substituted back into equation 3.111 to calculate the total energy of the system.

Uytterhoeven and Mortier [1992] did extensive tests of the method, and Figure 3.61 shows some of their results. In the case presented in the figure, Uytterhoeven and Mortier calculated the charge distribution in a water cluster assuming that the water does not distort. Generally, they found that the calculations give approximately correct charge distributions, although there is a small systematic error since they did not allow the water to distort. In a similar way there is a small systematic error in the bond energy. Still, the calculations seem to be quite good, especially considering their simplicity of methods.

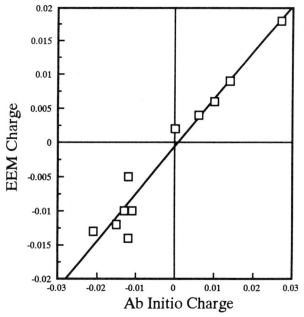

Figure 3.61 The changes in the charge on the oxygen atoms in a water molecule due to the presence of other water molecules. (Calculations of Uytterhoeven and Mortier [1992].)

PROBLEMS

3.1 Define the following in three sentences or less.

(1) Adsorption	(19) Ordered adsorption
(2) Desorption	(20) Dissociative adsorption
(3) Physisorb	(21) Nondissociative adsorption
(4) Absorption	(22) Molecular adsorption
(5) Adsorbate	(23) Molecular precursor
(6) Chemisorb	(24) Islands
(7) Multilayer adsorption	(25) Domain wall structure
(8) Monolayer	(26) Image dipole
(9) Monolayer adsorption	(27) Jellium
(10) Linear CO	(28) Projected bulk bands
(11) Bridgebound CO	(29) Surface state
(12) Threefold hollow	(30) Hard acid
(13) Triply coordinated CO	(31) Hard base
(14) Bridging site	(32) Soft acid
(15) On-top site	(33) Soft base
(16) Incommensurate adsorption	(34) Electronegativity
(17) Compression layer	(35) Hardness (Pearson's)
(18) Random adsorption	(36) Electronegativity equalization method

3.2 Summarize the adsorption of gases on metal surfaces.

(a) What are the bonds like?

(b) What are the key factors that determine the bond strength?

(c) What do the d-electrons do?

3.3   Summarize the adsorption of gases on semiconductor surfaces.

(a) What are the bonds like?

(b) What are the key factors that determine the bond strength?

(c) What is the role of dangling bonds?

(d) How do the dimer pairs affect the binding process?

(e) How does lattice strain affect the binding?

(f) How is the binding of hard spheres different from soft species on semiconductors?

(g) Why are elemental semiconductors (Si, Ge) less able to dissociate molecules than transition metals?

(h) Why doesn't the argument in (g) apply to a defected nickel oxide surface?

3.4   Use Benziger's analysis in Example 3.A to estimate the heat of reaction for the following two reactions on platinum, palladium, and nickel:

(a) $CH_3OH_{(ad)} \rightarrow CH_3O_{(ad)} + H_{(ad)}$

(b) $CH_3OH_{(ad)} \rightarrow CH_{3(ad)} + HO_{(ad)}$

3.5   Summarize the adsorption of gases on oxide surfaces.

(a) What are the bonds like on a pristine oxide surface?

(b) How are the bonds different on the defects?

(c) What determines the strength of the bonding on the pristine regions of the oxides?

(d) According to Mortier what determines the strength of the bonding on the defects?

3.6   In Section 3.1 we said that a saturated monolayer roughly corresponds to 1×10^{15} molecules/cm^2 of gas. The objective of this problem is to calculate a more accurate monolayer coverage for a variety of molecules on Ni(111). Consider the adsorption of (1) carbon monoxide, (2) xenon, (3) methane, (4) benzene, (5) Freon 113, (6) hydrogen atoms (7) trimethyl gallium, (8) trimethylamine. Assume that at low temperatures each of these molecules binds molecularly on Ni(111).

(a) Calculate the saturation density, assuming that the molecules bind to a surface in a hexagonal close packed layer (see Chapter 2) with as tight a packing configuration as possible. Assume that the nonspherical molecules adsorb in a flat configuration. Use the molecular dimensions you learned in organic chemistry to estimate the effective molecular sizes accounting for the van der Waals radius of the various atoms. How many molecules can you fit into 1 cm^2 of the metal's surface?

(b) Assume that instead the molecules bind so that they are centered over the threefold hollow sites on the Ni(111) surface. How many threefold hollow sites will each molecule block? Calculate the saturation coverage in this case.

3.7   In Section 3.1 we noted that Kayser found that when he adsorbed ammonia onto a glass rod, he could not detect any adsorption. However, when he drew the rod into a fiber 0.0261 cm in diameter and 22325 m long, 8.26 cc-stp of ammonia adsorbed. The question is "how much ammonia was he putting on the glass fiber?"

(a) What ammonia coverage did Kayser obtain in molecules/cm^2, assuming that all of the ammonia adsorbed on the external surface of the fiber (i.e., that there were no pores)?

(b) Use analysis like that in Problem 3.6a to estimate the saturation coverage of ammonia in the first monolayer.

(c) How many layers of ammonia would have to account for the difference?

(d) Is it reasonable for that many layers of ammonia to form at room temperature? How different will a thick layer of ammonia be than liquid ammonia? What is the vapor pressure of liquid ammonia at room temperature?

3.8   Figure 3.2 shows that one can squeeze 5000 M^2 of surface area (i.e., about the area of a football field) into less than 10 cm^3 of Linde molecular sieve. Consider starting with a thin sheet the size of a football field.

   (a) How thick would the sheet have to be to squeeze 5000 M^2 into a volume of 10 cm^3? Do not forget to count the area on both sides of the sheet. Express your answer in Å.

   (b) In fact, when you roll up the sheet you will have to leave spaces between the layers in the sheet to admit gas. Assume that only 33% of the final sieve is solid. How thick would the sheet have to be?

   (c) Would the sheet be self-supporting? How could you construct a self-supporting structure? (*Hint:* What do people do to strengthen a building or a bridge?)

3.9   Earlier in this chapter we said that molecules can be held to surfaces with a variety of different types of bonds. However, years ago it was thought that molecules were held with only local bonds. Assume that heats of adsorption are proportional to the number of bonds a molecule can make with the solid. Consider a Pt(111), (1x1) Pt(110), and (2x1) Pt(110) surface, and consider adsorption of CO on the linear, bridgebound, and threefold coordinative sites on each surface. (a) Estimate heat of adsorption via bond counting, assuming that each CO–metal bond has an energy of 10 kcal/mole. (*Hint:* How many bonds form at each site?) (b) How would you check to see if the bond-counting procedure allows you to accurately estimate heats of adsorption?

3.10 Qualitatively, why does CO bind more strongly to platinum than H_2O, while H_2O binds more strongly to silicon than CO?

3.11 Consider the following gases. Would they bind more strongly to platinum or to silicon? (a) H_2S, (b) H_2O, (c) CO, (d) HCOOH, (e) NH_3, (f) HCl, (g) Cl_2. (*Hint:* Look at Table 3.8 and Figure 3.23.)

3.12 Repeat Problem 3.11 comparing

   (a) platinum and nickel

   (b) platinum and tungsten

   (c) nickel and copper

   (d) platinum and germanium

   (e) platinum and a gold–silicon alloy (assume the properties of the alloy are halfway between gold and silicon)

3.13 Use the two-states model to predict the heat of adsorption of (a) chlorine, (b) cesium, (c) arsenic on GaAs(110).

3.14 List the assumptions in the following models, and describe how well the assumptions agree with experiment.

   (a) Lang-Williams model

   (b) Cluster model

   (c) Slab model

   (d) Effective medium model

   (e) What key features are missing from each of the models?

3.15 Qualitatively, why are the cluster models generally not accurate for adsorption on surfaces?

   (a) Why is key physics missing? How are the properties of small clusters different from the properties of extended surfaces?

   (b) One of the differences between clusters on surfaces is the absence of exchange with the bulk. How will exchange with the bulk affect the Pauli repulsions that occur when gases impinge on surfaces?

(c) According to Lennard-Jones's model, how will those changes affect the activation energy for adsorption?

(d) Use your results in (b) and (c) to explain why an 8-atom iron or cobalt cluster will not adsorb hydrogen.

3.16 Figure 3.22 shows the changes in electron density that occur when a zenon atom approaches a metal surface.

(a) Compare the shape of the plot to the shapes of the plots in Figure 3.37. Is there a significant difference?

(b) Why, then, is xenon said to physisorb? What is the key difference between the bonding of xenon and the bonding of the other species?

3.17 In Section 3.8.3 we noted that the principle of electronegativity equalization implies that there will be considerable flow of electrons from a gallium atom to an arsenic atom on a GaAs surface. (a) Look up the gas phase electron affinity of Ga^+ and As^-. Use Equation 3.51 for gallium and for arsenic to calculate the charge transfer, assuming that the electron affinities on the surface are equal to the gas phase values. (b) How do the electron affinities compare to those in Table 3.17?

3.18 Consider the adsorption of cesium and chlorine on a GaAs(110) surface.

(a) Would the cesium rather bind to the gallium or the arsenic?

(b) Would the chlorine rather bind to the gallium or the arsenic?

(c) Estimate the bond energies in each case.

3.19 Use the data in Table 3.17 to compare the reactivity of Si, GaAs, and InP.

(a) Which surface is a better electron donor?

(b) Which surface is a better electron acceptor?

(c) Based on your results in (a) and (b), which surface would you expect to be the most reactive?

3.20 Section 3.4.1 discusses the principle of electronegativity equalization, which states that when two species come together, charge flows to equalize the electronegativity of the system.

(a) Consider the first four species in Table 3.9. Will each species be slightly positively charged or slightly negatively charged when it adsorbs on a Cu(110) surface? How about a Pt(111) surface? For the purposes of this problem assume that the electron affinity of the species when they adsorb is equal to the gas phase electron affinity.

(b) Use Equation 3.51 to estimate the net charge flow at 300 K.

(c) Compare to the value from Equation 3.27, assuming

$$\frac{1}{(\eta_A + \eta_B)} = \frac{2 \text{ electrons}^2}{eV} \text{ (i.e., Pauling's value)}$$

(d) What do you conclude from this?

3.21 Discuss the use of electronegativity equalization to calculate the binding of molecules to surfaces.

(a) What are the key ideas?

(b) According to the model, what determines bond strengths?

(c) According to the model, how are metals, semiconductors, and insulators different?

(d) Why are the electronegativity and hardness different on a metal, a semiconductor, and an insulator surface? (*Hint:* Relate the difference to a difference in the binding process.)

(e) How are those differences accounted for in Mortier's electronegativity equalization method?

3.22 Discuss Mortier's electronegativity equalization method as a way of calculating the strengths of the bonds between two species.

(a) What are the key ideas?

(b) According to the model, what determines bond strengths? (*Hint:* Which terms in Equation (3.111) are largest?)

(c) Based on the model, would you expect H_2S or H_2O to adsorb more strongly on alumina?

3.23 Figure 3.37 shows the changes in electron density that occur when gases adsorb on aluminum.

(a) Describe in your own words what the plots tell you.

(b) Explain how you can tell that chlorine is negatively charged from the plot.

(c) Make a xerox of the plot and identify

(1) The lithium–surface bond

(2) The silicon–surface bond

(3) The antibonding silicon–surface orbital

(d) Look at the size of the lithium–surface bond. How wide is it? How far does it extend into the bulk of the aluminum? How do your results compare to the aluminum–aluminum distance?

(e) Now look at the total electron density in the figure. Why is there so little difference in the total electron density in the three cases?

3.24 Figure 3.36 shows the change in the density of states that occurs when "chlorine," "lithium," and silicon adsorb on aluminum.

(a) What is the net charge on the "lithium" and the "chlorine"? What would the density of states curve look like if the net charge was zero?

(b) What is the net charge on the silicon?

(c) What is the significance of there being two maxima in the silicon density of states?

(d) What would the silicon density of states curve be like if the silicon formed an sp^3 hybrid with all three sp^3 hybrids being equivalent?

(e) How would inequivalent sp^3 hybrids change your answer in (d)?

3.25 In Section 3.14.2 we noted that one can dope a zeolite to change its electronegativity.

(a) Consider starting with silica/alumina, and using an ion-exchange process to replace some of the silicons with aluminums. How will the electronegativity of the zeolite change?

(b) How would that change affect the binding energy of hydrogen? Will the zeolite be a stronger or a weaker acid? Try to be quantitative.

(c) Repeat for phosphorus.

3.26 Benesh and Liyanage, *Surface Sci.* **261**, 207 [1992] use a slab calculation to examine the binding of oxygen on (1x1)Pt(110). Figure P3.26a shows a density of states (DOS) curve from their paper. What can you tell about the binding from the DOS curve?

(a) Is the oxygen positively charged, negatively charged, or nearly neutral?

(b) Is the bond mainly covalent or ionic?

(c) Is the main bonding localized or delocalized?

(d) There is a negative peak in the DOS–difference spectrum. What is the significance of the loss of intensity? (*Hint:* Look at Figure 3.54.)

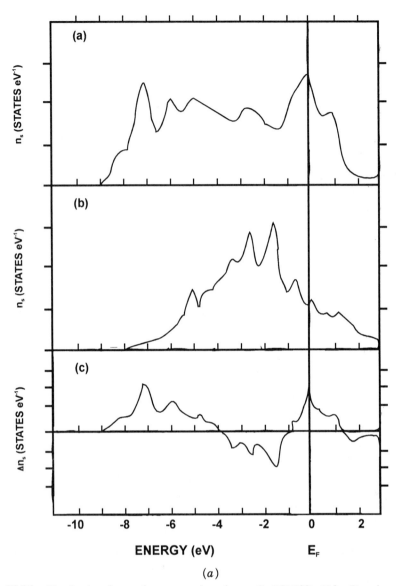

(a)

Figure P3.26a The density of states for oxygen adsorption on (1x1)Pt(100). (After Benesh and Liyanage [1992].) (a) Oxygen and surface; (b) clean surface; (c) difference.

(e) Figure P3.26b shows the changes in electron density that occur when oxygen adsorbs on the surface. Are the electron density plots consistent with your findings in (a) to (d)?

(f) What is the significance of the fact that the bonds on the oxygen do not point toward any of the surface atoms? When you answer this question, refer back to your answer in (c).

(g) Why do the electron states on the metal atoms point in the way they do? (Hint: Look at Figures C.21 and C.22.)

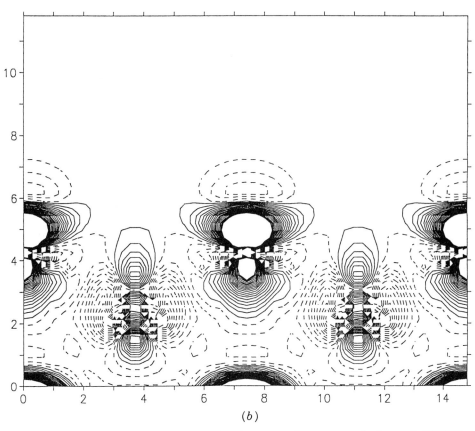

(b)

Figure P3.26b The changes in electron density that occur when oxygen adsorbs on (1x1)Pt(100). (After Benesh and Liyanage [1992].)

3.27 Consider Lennard-Jones's model described in Section 3.2.

    (a) Describe the key features of the model.

    (b) How does the dissociation process change as the well depth of the molecular and dissociative wells change?

    (c) How does the dissociation process change as the stiffness of the potential (i.e., the derivative of the potential moving up from the bottom of the well) changes.

    (d) How does the dissociation process change as $E_D^{B_2}$ changes?

3.28 The objective of this problem is to try to quantify Lennard-Jones' model from Section 3.2, page 119. Molecular hydrogen adsorbs weakly on most transition metals. Assume that the molecular well in Figure 3.9 is given by a Morse potential:

$$E_{molec} = \frac{3 \text{ kcal}}{\text{mol}} \left\{ (1 - \exp\left[-St\ (z - 5\ \text{Å})\right])^2 - 1 \right\}$$

where St is the stiffness of the potential. Also assume that the atomic well is given by

$$E_{atomic} = (104 \text{ kcal/mol} - \Delta H_{ad}) \left\{ (1 - \exp\left[-1.0/\text{Å}\ (z - 3\ \text{Å})\right])^2 \right\} + \Delta H_{ad}$$

where ΔH_{ad} is the heat of dissociative adsorption of H_2 (*note:* ΔH_{ad} is negative), z is the distance above the surface, and the 104 kcal/mole is the bond energy of an H_2 bond. Assume $St = 2.0/\text{Å}$.

(a) Verify that the curve looks like the one in Figure 3.9.

(b) Calculate the activation energy for dissociation as a function of ΔH_{ad}.

(c) Use the data in Table 3.4 to predict which surfaces will dissociate H_2, and whether the process is activated. Compare your results to Table 3.2.

3.29 Hydrogen atoms bind quite strongly to a Si(111)(7x7) surface. However, H_2 does not stick. It is thought that the adsorption process has an activation barrier of over 30 kcal/mole, although so far the barrier has not been measured on a defect-free surface. The question is "why does the barrier exist on Si(111)(7x1) when no barrier exists on Pt(110)?"

(a) First consider Lennard-Jones's model. According to the results in Problem 3.28 if one has a deep atomic well, one will always get gas to adsorb. However, if one changes the stiffness of the potential one can get a barrier. How stiff does the potential have to be to get a barrier of 30 kcal/mole? Use Figure 3.23 to estimate the heat of adsorption of hydrogen.

(b) Does it make sense that the barrier will be stiffer on silicon than on platinum? How is the binding different on silicon than on platinum? How will that difference in binding lead to a difference in the stiffness of the potential? (*Hint:* Look at Pauli repulsions.)

(c) Are the results in (b) sufficient to account for the difference in the stiffness? Compare the stiffness in (a) to the stiffness of a typical covalent bond ($\sim 1\text{-}3/\text{Å}$). Is it reasonable that the molecular potential will be much stiffer than a covalent bond? (*Hint:* What will the binding of H_2 on Si(111)(7x7) be like?)

(d) Another reason that the adsorption of H_2 is inhibited on Si(111)(7x7) is that when H_2 dissociates, one needs 2 adjacent sites to hold the adsorbed gas. What do you think the preferred binding site for hydrogen atoms will be on Si(111)(7x7)? Locate that preferred binding site on the picture of the Si(111)(7x7) surface in Figure 2.47.

(e) Calculate the distance between adjacent hydrogen sites. Note that the atomic coordinates of a Si(111)(7x7) lattice can be found in S. Y. Tong, H. Huang, C. M. Wei, W. E. Packard, K. F. Men, G. Glander, M. B. Webb, *J. Vac. Sci. Tech.,* A **6,** 615 [1988]. Compare to the distance between two threefold hollows on Pt(111).

(f) Is there something else that makes it harder for H_2 to adsorb on silicon than on platinum? (Hint: See Chapter 10.)

3.30 Consider the adsorption of a F^- ion on a platinum surface.

(a) Derive an equivalent of Equation 3.11 for this situation.

(b) Plug in numbers to calculate the heat of adsorption, assuming that the fluorine sits at its covalent radius above the platinum surface.

(c) Now consider the adsorption of F_2. Assume that F_2 will dissociate to form two F^- ions upon adsorption. Estimate the heat of adsorption of F_2.

3.31 The objective of this problem is to relate the effective medium model to the concepts of electronegativity and hardness.

(a) What would the effective medium potential for a soft species be like? How about a hard one?

(b) Study Figure 3.43 carefully. Rank each of the species in terms of the hardness or softness. (*Hint:* Use Equation 3.91.) Ignore He and Ne.

(c) How would changes in electronegativity be reflected in Figure 3.43? (*Note:* This is a subtle question; think it through carefully using Equation 3.91.)

(d) According to the analysis in Section 3.4, when two species come together, electrons flow to equalize the electronegativity of the system. How would such a change be reflected in Figure 3.43?

(e) What additional information do you know to determine which of the species in Figure 3.43 had the highest electronegativity?

3.32 Use Equation 3.100 to estimate the heat of adsorption of Ga, Al, and As on a Si(111)(7x7) surface.

(a) Assume that there are three bonds to p-orbitals in each case.

(b) How will your answer in (a) change if instead an sp^2 hybrid forms? (*Hint:* The wavefunction for an sp^2 orbital is $(\phi_s + \sqrt{2}\,\phi_p)/\sqrt{3}$.)

(c) How about an sp^3 hybrid?

3.33 In Section 3.8.1 we noted that one can use the width of UPS peaks to estimate the lifetime of the electrons in the adsorbate–surface bond.

(a) Consider the N_2O data in Figure 3.35. Measure the half-widths of each of the peaks associated with N_2O, and then use Equation 3.63 to estimate the average lifetime of the electrons in each of the orbitals in the N_2O.

(b) Assume that the lifetime of the electrons was (1) 10^{-6} sec., (2) 10^{-9} sec., and (3) 10^{-12} sec. How broad should the peaks be?

(c) What do you conclude from your results?

3.34 Table 3.2 summarizes the reactivity of a number of gases on a number of surfaces.

(a) Focusing on the metals, what general trends do you observe? Which metals are the most able to adsorb gases? Which metals are least able to adsorb gases?

(b) Is there a correlation between your results in (a) and the position of the metal in the periodic table? What parts of the periodic table are the most reactive?

(c) How well do the results correlate with the Tanaka–Tamaru correlations? Are the metals that form the strongest metal oxides also the most reactive during adsorption? Are there any metals that form strong oxides, and yet are relatively unreactive?

(d) How well do the results correlate with the variations in interstitial electron density? Are there any metals with a high interstitial electron density and yet are relatively unreactive? Are there any metals with a low interstitial electron density, and yet are relatively reactive? Do metals with similar interstitial electron density show similar reactivity?

(e) What do you conclude from this?

3.35 Kellog [1994]; Clark and Seebauer [1995]; and Gomer [1990] review the available data for diffusion of atoms on surfaces. The diffusion process can often be viewed via a hopping model where an atom detaches itself from one site on a surface and jumps to an adjacent site. Generally, diffusion barriers on metals are small; perhaps 10% of the bond energy. However, the barriers are larger on semiconductor surfaces; they are typically between 40% and 90% of the bond energy. The objective of this problem is to try to relate these differences in behavior to differences in bond energy.

(a) Qualitatively, how is the binding of atoms on semiconductor surfaces different from that on metals?

(b) Try to relate the differences in (a) to differences in the diffusion process. (*Hint:* What bonds break when atoms move laterally on a surface?)

3.36 Heitzinger, Gebhard, and Koel, *J. Phys. Chem* **97,** 5327 [1993], examined the adsorption of ethylene on a palladium monolayer on Mo (100).

(a) Consider the adsorption of Pd on Mo(110). According to Pearson [1968] an isolated palladium atom has an electronegativity of 4.45 eV. Now consider what happens when palladium atoms adsorb on Mo(110). Based on the data in Table 3.9 and the principle of electronegativity equalization, what direction should the electrons flow when palladium adsorbs on Mo(110).

(b) J. Rodriguez, R. A. Campbell, P. W. Goodman, *J. Phys. Chem* **95,** 5716 [1991], *Surface Sci.* **240,** 71 [1990] verify that electrons flow from the palladium to the surface. How should this electron flow affect the binding of ethylene on the Pd overlayer? (*Hint:* Look at Figure 3.65.)

(c) How do your predictions compare to the findings of Heitzinger, Gebhard, and Koel?

3.37 The objective of this problem is to use the Fukui functions to predict adsorbate geometries.

(a) Consider the adsorption of benzene on Pt(110). Based on Pearson's [1968] hardness and electronegativity of benzene (χ = 4.1 eV, η = 5.3 eV), would you expect the benzene to mainly donate electrons, mainly accept electrons, or form a donor–acceptor complex with the Pt(111) surface? (*Hint:* If you use the Pauling electronegativity, be sure to multiply Pearson's electronegativity values by 0.355 to convert them to the Pauling scale.)

(b) Based on your results in (a), would you look at F_j^0 , F_j^+ or F_j^- to understand the binding of benzene on Pt(111)?

(c) Look up the LUMO and HOMO for benzene. A suitable reference is L. Salem and W. L. Jorgensen, *The Organic Chemist's Book of Orbitals*, Academic Press, NY [1973]. Based on the shapes of the molecular orbitals, would you expect the benzene to lay flat or stand up on the surface?

(d) Now consider the adsorption of pyridine (χ = 4.4, η = 5). What is different about the MOs of pyridine and benzene?

(e) How would you expect those differences to affect the bonding of pyridine?

(f) Show that pyridine could stand up on a suitable surface due to an interaction with the MOs centered on the nitrogen end of the molecule.

(g) Would you expect the pyridine to stand up or lie flat on Pt(111)?

(h) How would your answer change on Cu(111)?

3.38 Walsh, Davis, Muryn, Thorton, Dhanak, and Prince, *Phys. Rev. B* **48,** 147 [1993] examined the orientation of benzene and pyridine on ZnO(1010). They found that the benzene lies flat on the surface while the pyridine adopts an upright geometry with the plane of the pyridyne parallel to the (1120) plane. Use the ideas from Problem 3.37 to explain these results.

3.39 In Section 3.2.1 we noted that CO can bind in a linear site, a bridgebound site, or a triply coordinated site on Ni(111), while hydrogen adsorbs in a triply coordinated site. The objective of this problem is to understand how the bonding of the CO and the hydrogen changes from one site to another.

(a) First consider the interactions of hydrogen with the s-band. How should the interaction with the s-band change from one site to another? (*Hint:* Look at Figure C.23 in Appendix C.)

(b) Now consider the interactions of hydrogen with the d-band. How should the interactions with the d-band change from one site to another? (*Hint:* Look at Figure C.25 in Appendix C.)

(c) Now consider the energy gain of the nickel atoms when the hydrogen adsorbs. Look carefully at Figure 3.46. Is it better for one nickel atom to gain all of the electron density or for the electron density to be shared between three nickel atoms? (*Hint:* How does the slope of the embedding energy of the nickel change with increasing electron density?)

(d) Summarize your results in (a), (b), and (c) to predict where hydrogen will adsorb.

(e) How is the binding of CO different from that of hydrogen? (*Hint:* Look at the Blyholder model in Section 3.2.1.)

(f) How could the differences in (e) lead to a different binding site for CO?

3.40 Roberts, Madix, and Crew, *J. Catalysis* **141**, 300 [1993] and Davis and Barteau *Surface Sci.* **268**, 11 [1992] examined the role of oxygen on reactions of ethylene on silver and formaldehyde on palladium. Oxygen has a variety of different effects on the chemistry. However, in this problem we focus on one aspect of the chemistry, which is the ability of oxygen to modify the properties of the palladium or silver.

(a) Based on the analysis in Section 3.4.2, would you expect oxygen to be a net electron donor or a net electron acceptor?

(b) How long range will the effect be? (*Hint:* Look carefully at Figure 3.37.)

(c) How will the changes in (a) affect the binding of ethylene? (*Hint:* Look at Figure 3.40 and Figure 3.41.)

(d) By analogy to your results in (c), how would the oxygen affect the binding of formaldehyde?

3.41 The objective of this problem is to highlight the difference between Pauling's version of electronegativity equalization, Parr and Yang's version, and the effective medium model. Consider the Pt(210) surface discussed in Figure 2.9 and Table 2.4. Table 2.4 indicates that when the (210) surface forms, the topmost atom has a lower electron density than the remaining atoms.

(a) How does the electronegativity of the atoms vary from the topmost layer to the bulk? (*Hint:* Consider the principle of electronegativity equalization.)

(b) Look at Figure 3.42. According to the figure, is it possible for the atoms to have the same electronegativity and different electron densities? What is going wrong? (*Hint:* The last part is a subtle question, think about it now, then go to part c.)

(c) What prevents the topmost layer of atoms on the (210) surface from moving in far enough to have the same electron density as the other atoms? Assume that the other atoms could rearrange to keep their electron density constant.

(d) Where is the extra term in Equation 3.86?

(e) Now go back to question (b). Note that when one moves the atoms to change them, the electron density and the core–core repulsions also change. Therefore the core–core potential will influence the electronegativity of the system. Use Equations 3.86 and 3.91 to derive an equation for the influence of the core–core repulsions on the electronegativity of the system.

(f) Relate your answer in (e) to Equations 3.44 and 3.46. What do your results tell you about the last term in Equations 3.44 and 3.46?

3.42 Consider the interaction of a sodium ion (Na^+) with a water molecule. Assume that the water molecule binds as shown in Figure P3.42.

(a) Use the EEM model to calculate the charges on the oxygen. Assume that there is no change transfer to the lithium ion, and that the charges are identical on both hydrogen atoms.

(b) Now remove the sodium ion and calculate again how the charges differ.

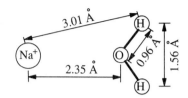

Figure P3.42 Geometry for Problem 3.42.

(c) Consider the binding of a water molecule to the sodium ion. Calculate the changes in each of the terms in Equation 3.11. Which terms dominate?

(d) Use Equation 3.103 to estimate the change in the length of the O–H bond in the water when the water approaches a sodium.

(e) Look up Badger's rule from your course in spectroscopy. How much should the vibrational frequency of the O–H change due to the changes in (c)?

More Advanced Problems

3.43 Read Appendix B, Section B.3, and then describe in your own words why there is a barrier to the reaction $H + D_2 \rightarrow HD + D$.

(a) How do Pauli repulsions contribute to the barrier?

(b) Now apply the ideas to surfaces. Why could there be a barrier to the dissociative adsorption of hydrogen?

(c) Relate your findings in (b) to the Lennard-Jones model—If you reduced the Pauli repulsions, how would the shape of the molecular and atomic wells change?

(d) What does Lennard-Jones's model tell one about what one can do to change the barrier in addition to reducing the Pauli repulsions?

(e) Are the effects large enough to account for the difference between the barrier for hydrogen adsorption on Pt(110) and Cu(110)?

3.44 The objective of this problem is to see how well the Lennard-Jones model reproduces the trends in Figure 3.7.

(a) Figure 3.7 suggests that molecules will dissociate more easily moving to the right across the periodic table. When one moves to the right in the periodic table, one changes the bond energy for dissociative adsorption as shown in Figure 3.23 and Tables 3.5 and 3.6. How well do the data follow the trends expected from the Lennard-Jones model?

(b) Use Tables 3.5 and 3.7 to make plots of the heat of dissociative adsorption of each of the species on a variety of metals as a function of the position, i.e., column, of the metal in the periodic table. Identify the line separating dissociative and nondissociative adsorption. According to the Lennard-Jones model, the line should be the same for each row in the periodic table. How well do the data fit the Lennard-Jones model?

(c) Now consider the difference from one adsorbate to the next. Are the lines for each adsorbate at the same absolute positions or at different positions? Can the trends be explained by the Lennard-Jones model?

(d) Try to extend your reasoning to hydrogen adsorption on potassium, aluminum, or silicon. (The adsorption of hydrogen is activated on those surfaces.) What would your plots in (b) and (c) predict for this? Why did the Lennard-Jones model fail?

3.45 In Section 3.2 we noted that the adsorption of nitrogen on tungsten is structure sensitive. Some of the faces of tungsten rapidly adsorb nitrogen, while others do not, as summarized in Figure 3.6. The objective of this problem is to see if those results can be explained by the effective medium model.

(a) First examine the structure of the (100), (411), (311), (310), and (521) planes of tungsten (a BCC metal). All of these planes dissociate nitrogen. What do the planes have in common? Look particularly at the coordination number of the atoms. The program SURFSTRU (available from Professor Masel) and the tables in Nicholas [1965] *An Atlas of Models of Crystal Surfaces* (Gordon and Breach, NY) would be particularly helpful for this purpose.

(b) Now examine the (211), (111), (321), and (110) faces of tungsten. These planes do not dissociate nitrogen. How are the planes that dissociate nitrogen different from the planes that do not dissociate nitrogen? (*Hint:* Nitrogen atoms adsorb on a fourfold hollow on W(100).)

(c) Now compare your results to the effective medium model. The effective medium model relates changes in reactivity to changes in d-band filling, changes in the metal electron affinity, and changes in the surface electron density. The fractional d-filling of all of the faces of tungsten are almost the same as is the electron affinity. However, the interstitial electron density of the surface atoms does vary. To a first approximation assume that the electron density ρ_e at a surface atom is given by

$$\rho_e = \frac{\rho_e^{AV}(CN)}{M}$$

where ρ_e^{AV} is the bulk interstitial electron density, CN is the coordination number of the atoms (i.e., the number of nearest neighbors), and M is constant. $M = 8$ for BCC metals and $M = 12$ for FCC and HCP metals. Also assume that the nitrogen sees the average interstitial electron density of its nearest neighbors. How well do the data correlate to the effective medium model?

(d) Use the results in (c) to make predictions about other metals. What will happen if you go to chromium or iron? (*Hint:* How do the interstitial electron density and the % d-filling change?)

3.46 Figure 3.7 shows data on the dissociation of CO, N_2, O_2, and NO on transition metals. In the literature people often say that there is a correlation between the ability of a surface to break bonds and the heat of reaction for the dissociation process.

(a) Consider the dissociation of CO. The CO starts out with perhaps $\frac{1}{4}$ of a metal–carbon bond, and forms full metal–carbon and metal–oxygen bonds when the carbon dissociates. Use the data in Table 3.5 to estimate the heat of reaction for the reaction $CO_{(ad)} \rightarrow C_{ad} + O_{ad}$ on all of the metals in Figure 3.7. Is there a correlation between the heat of reaction and the ability to dissociate CO?

(b) Repeat the calculation for oxygen and nitrogen. Is there a correlation for each of the metals?

(c) Now compare the results for oxygen, nitrogen, and CO. Can you find one critical heat of reaction where dissociation occurs, or does the critical heat of reaction vary from molecule to molecule? (*Note:* In Chapter 10 we show that the dissociation rate varies strongly with surface structure even though the heat of reaction is almost constant, so clearly the heat of reaction is not the whole story in determining rates of reaction.)

3.47 Rodriguez, Campbell, and Goodmann, *J. Phys. Chem* **95**, 5716 [1991] examined the electron flows when copper or palladium atoms adsorb on metal surfaces. They found that under some circumstances, the electron density flowed in the opposite direction than one would expect based on an examination of the gas phase electronegativities.

(a) Consider the adsorption of palladium or copper atoms on a W(110), Ir(111), Mo(110), Pt(111) surface. In the gas phase isolated palladium atoms have an electronegativity 4.45 eV, while isolated copper atoms have an electronegativity of 4.48 eV. Based on the gas phase electronegativities, which way should electrons flow when palladium or copper adsorbs on Ir(110), Ir(111), Mo(110), and Pt(111).

(b) According to Rodriguez et al., which way do the electrons actually flow?

(c) How can you account for these differences based on Equations 3.45, 3.46, and 3.47?

3.48 Figure 3.20 shows a diagram of the electric field that is created when a dipole is brought close to a surface. Note the formation of the "image dipole" in the right diagram in the figure. This image dipole is quite important to surface infrared spectroscopy. In infrared, the light interacts with the changes in the surface dipole that occur when molecules vibrate.

(a) Consider a molecule with a permanent dipole oriented perpendicular to the surface. How will the image dipole be oriented? Will the image dipole reinforce or cancel the net surface dipole? How will it affect the changes in the dipole when the molecule vibrates?

(b) How will the answer to (a) change for a molecule with a permanent dipole oriented parallel to the surface.

(c) What do these results tell you about the ability of IR to detect molecules with dynamic dipoles perpendicular and parallel to the surface?

3.49 Section 3.8.2 describes the jellium wavefunctions for a number of metals.

(a) How are the aluminum–jellium wavefunctions different than the palladium–jellium wavefunctions?

(b) How does the workfunction of aluminum compare to that of palladium? How should that affect the reactivity of aluminum versus Pd for hydrogen adsorption?

(c) In reality, hydrogen adsorption on Pd(110) is unactivated, while Rendulic, *Surface Sci.* **272**, 34 [1992] reports a 13-kcal/mole activation barrier for hydrogen adsorption on Al(110). How can you account for that difference?

3.50 In the recent literature, there has been considerable interest in creating "hydrogen-passivated" Si(111) and Si(100) surfaces.

(a) The simplest passivation procedure is to adsorb about 5% of monolayer of hydrogen to tie up the dangling bonds. How will this affect the reactivity of the silicon? Are there still sites that show unusual reactivity? Think about the strained bonds. How will the reactivity of the strained bonds compare to the reactivity of dangling bonds?

(b) Now consider adding even more hydrogen so that all of the strained bonds are saturated, and the silicon reverts back to a (1x1) structure. How will the chemistry change?

(c) Compare the changes in (b) to the changes in reactivity that occur when an olefin is hydrogenated to a paraffin. Describe the analogy in detail.

(d) Will the surface in (b) be completely unreactive? (*Hint:* Is a paraffin completely unreactive?)

3.51 Section 3.11 noted that a pristine, freshly cleaved MgO(100) surface is fairly inert.

(a) Draw the surface structure of the MgO(100) surface assuming that MgO has a simple face-centered cubic (FCC) lattice with an oxygen atom at each of the lattice points and a magnesium between each oxygen. Be sure to include the sizes of the atoms in your picture.

(b) Now look at the geometry perpendicular to the surface. Show that the geometry is equivalent to a Taskers type III lattice described in Section 2.6.4.

(c) Consider adding a series of hydroxyl (i.e., OH^-) ions to the surface. Would the hydroxyls prefer to bond to the O^- or Mg^+ ions on the surface?

(d) Calculations indicate that if an OH^- would bind, the MgO surface would rumple. The magnesium ions would move toward hydroxyls. Show that such a process produces a net dipole on the surface.

(e) Use an analysis like that in Section 2.6.4 to show that the surface stress is infinite in (d). Therefore, the surface must reconstruct or facet when OH^- adsorbs.

(f) How do the results in (e) affect the ability of a pristine MgO surface to adsorb gas?

3.52 In our discussion of the two-state model in Section 3.13.3 we assumed that E_s does not change in the presence of the adsorbate, which in effect implies that

$$\langle \phi_s | \mathcal{H}_e | \phi_s \rangle = \langle \phi_s^1 | \mathcal{H}_e | \phi_s^1 \rangle$$

The objective of this problem is to see when this approximation is correct.

(a) Substitute Equation 3.94 into Equation 3.97 to get an expression for $\langle \phi_s^1 | \mathcal{H}_e | \phi_s^1 \rangle$ in terms of $\langle \phi_s | \mathcal{H}_e | \phi_s \rangle$ and V_{as}.

(b) Plug in some typical numbers (e.g., choose hydrogen on silicon) and see how large a correction is needed. (*Note:* In practice, Harrison [1980] adjusted V_{as} to keep E_s constant.)

REFERENCES

Adams, D. L., and L. H. Germer, *Surface Sci.* **27**, 21 (1971).

Anderson, P. W., *Phys. Rev.* **124**, 41 (1961).

Anderson, P. W., *Concepts in Solids*, Benjamin-Cummings, Menlo Park, CA (1963).

Asakawa, T., K. Tanaka, and I. Toyoshima, *J. Phys. Chem.* **95**, 727 (1991).

Avouris, P., D. G. Cahill, *Ultramicrosc.* **42**, 838 (1992).

Backman, A. L., and R. I. Masel, *J. Phys. Chem.* **94**, 5300 (1990).

Bancroft, W. D., *J. Phys. Chem.* **21**, 441 (1917); **21**, 573 (1917); **21**, 644 (1917); **21**, 734 (1917); **22**, 22 (1918); **22**, 345 (1918); **22**, 433 (1918).

Banholzer, W. E., Y. O. Park, K. Mak, and R. I. Masel, *Surface Sci.* **128**, 176 (1983).

Bartmess, J. E., T. A. Scott, and R. T. McIver, *J. Am. Chem. Soc.* **101**, 6047 (1979).

Bauschlicher, C. W., S. R. Langhoff, and P. R. Taylor, *Accurate Quantum Chemical Calculations*, NASA (1989).

Beebe, R. A., and D. M. Young, *J. Phys. Chem.* **58**, 93 (1954).

Beeby, J. L., *J. Phys. C*, **1**, 82 (1968).

Benson, S. W., *Thermochemical Kinetics*, Wiley, NY (1976).

Bent, B. E., C. M. Mate, C.-T. Kao, A. J. Slavin, and G. A. Somorjai, *J. Phys. Chem.* **92**, 4720 (1988).

Benziger, J. B., in E. Shustorovich, ed., *Metal-Surface Reaction Energetics*, VCH, NY (1991).

Berzelius, J. J., *Edinburg New Philos. J.* **21**, 243 (1836).

Blyholder, G., *J. Phys. Chem.* **68**, 2772 (1964).

Bockris, J. M., in J. M. Bockris, ed., *Modern Aspects of Electrochemistry*, **1**, 180 (1954).

Bonhoeffer, K. F., and P. Hartek, *Naturwiss.* **17**, 1821 (1929).

Brass, S. G., and G. Ehrlich, *J. Chem. Phys.* **87**, 4285 (1987).

Brennan, D., D. O. Hayward, and B. M. W. Trapnell, *Proc. Royal Soc. A* **256**, 81 (1960).

Brodén, G., T. N. Rhodin, C. F. Bruker, R. Benbow, and Z. Hurych, *Surface Sci.* **59**, 593 (1976).

Bu, Y., J. C. S. Chu, and M. C. Lin, *Surface Sci.* **285**, 243 (1993).

Bu, Y., L. Ma, and M. C. Lin, *J. Vac. Sci. Tech. A* **11**, 2931 (1993).

Cassuto, A., M. Mane, and J. Jupille, *Surface Sci.* **249**, 8 (1991).

Ceyer, S. T., *Ann. Rev. Phys. Chem.* **39**, 479 (1988).

Chappuis, P., *Wiedermann's Annalen Der Physik und Chemie*, **12**, 161 (1881).

Chen, L. Y., and S. C. Ying, *Phys. Rev. Lett.* **73**, 700 (1994).

Chen, L. Y., and S. C. Ying, *Mod. Phys. Lett. B* **9**, 307 (1995).

Chen, X. C., J. E. Cunninham, and C. P. Flynn, *Phys. Rev. B* **30**, 7317 (1984).

Clark, A., *The Theory of Adsorption and Catalysis*, Academic Press, NY (1970).

Cosser, R. C., S. R. Bare, S. M. Francis, and D. A. King, *Vacuum* **31**, 503 (1980).

Cox, P. A., *Transition Metal Oxides*, Clarendon Press, Oxford (1992).

Crossley, A., and D. A. King, *Surface Sci.* **95**, 131 (1980).

Crowell, A. D., *J. Chem. Phys.* **22**, 1397 (1954); **26**, 1407 (1957).

Davey, H., *Philos. Trans. Phyl. Trans.* **23**, 115 (1816), **24**, 77 (1817).

DeAngelis, M. A., A. M. Glines, and A. B. Anton, *J. Chem. Phys.* **96**, 8582 (1992). *J. Vac. Sci. Tech.*, **10**, 3507 (1992).

DeBoer, F. R., et al., *Cohesion in Metals: Transition Metal Alloys*, North Holland, Amsterdam (1988).

Deeming, A. J., and M. Underhill, *J. Chem. Society Dalton Trans.* 1415 (1974).

Deeming, A. J., S. Hasso, and M. Underhill, *J. Chem. Society Dalton Trans.* 1614 (1975).

Delchar, T. A., and G. Ehrlich, *J. Chem. Phys.* **42**, 2686 (1965).

Döbereiner, J. W., *Annales de Chimie et de Physique*, **24**, 91 (1823).

Döbereiner, J. W., *Archiv. Natwl.* **2**, 225 (1824).

Döbereiner, J. W., *J. Der Physik*, **42**, 60 (1824).

Döbereiner, J. W., *Philos. Mag.* **65**, 150 (1825).

Döbereiner, J. W., *J. Physik,* **31**, 512 (1834).

Dubois, L. H., D. G. Castner, and G. A. Somorjai, *J. Chem. Phys.* **72**, 5234 (1981).

DuLong, P. L., and L. J. Theonard, *Annales de Chimie* **23**, 440 (1822); **24**, 330 (1823).

Ehrlich, G., and F. G. Hudda, *J. Chem. Phys.* **30**, 493 (1959).

Eley, D. D., *Trans. Faraday Soc.* **8**, 34 (1950).

Elian, M., and R. Hoffmann, *Inorg. Chem.* **14**, 1058 (1975).

Engel, T., and R. Gomer, *J. Chem. Phys.* **52**, 5572 (1970).

Euceda, A., D. M. Bylander, and L. Kleinman, *Phys. Rev. B* **28**, 528 (1983).

Faraday, M., *Quarterly Journal of Sci.* **7**, 106 (1819).

Faraday, M., *Philos. Trans. Royal Soc. London* **124**, 55 (1834).

Farkis, A., *Orthohydrogen, Parahydrogen and Heavy Hydrogen*, Cambridge University Press, Cambridge (1935).

Farkis, A., and L. Farkis, *Nature* **132**, 894 (1993); *Proc. Royal Soc. A* **144**, 467 (1934).

Feibelman, P. J., *Prog. Surf. Sci.* **12**, 287 (1982).

Feibelman, P. J., *Ann. Rev. Phys. Chem.* **40**, 261 (1989).

Feibelman, D. R. Hamann, and F. T. Himpsel, *Phys. Rev. B* **22**, 1734 (1980).

Ferrer, S., and H. Bonzel, *Surface Sci.* **119**, 234 (1982).

Fischer, C. F., *Atomic Data* **4**, 301 (1972).

Fisher, G., *Chem. Phys. Lett.* **79**, 452 (1991).

Flores, F., I. Gabbay, and N. H. March, *Chem. Phys.* **63**, 391 (1981).

Fontana, F., *Recherches physics sur la nature de l'air nitreux et de l'air dephlogistiqué*, Paris (1776). (This reference is cited in Priestly, 1777, and in the British Museum Catalog, but is not generally available.)

Fontana, F., *Societie Italiana della Scienze, Rome, Memorie di Mathematica e di fis.* **1**, 648 (1782). (This journal is in the University of Michigan collection but only limited copying is permitted.)

Freeman, A. E., *J. Am. Chem. Soc.* **35**, 927 (1913).

Freeman, A. J., C. L. Fu, and E. Wimmer, *J. Vac. Sci. Tech. A* **4**, 1265 (1986); *Phys. Rev. Lett.* **55**, 2618 (1985).

Freeman, A. J., D. S. Wang, and H. Krakaver, *Phys. Rev. B* **24**, 3094 (1981).

Freundlich, H., *Kapillarchemie*, Academische Verlag, Leipzig (1909).

Fukui, K., *Theory of Orientation and Stereoselection*, Springer-Verlag, Berlin (1975).

Geusic, M. E., M. D. Mores, J. R. Heath, and R. E. Smalley, *J. Chem. Phys.* **83**, 2293 (1985).

Gomer, R., Rep. *Prog. Phys.* **53**, 917 (1990).

Gomer, R., and G. Tryson, *J. Chem. Phys.* **66**, 4413 (1977).

Goodman, D. W., *Acc. Chem. Res.* **17**, 194 (1984).

Grimley, T. B., *Adv. Catal.* **12**, 24 (1960).

Grimley, T. B., in D. O. Goodman, ed., *Dynamic Aspects of Surface Physics*, Editrice Composori, Bologna (1974).

Grimley, T. B., and C. Pisani, *J. Phys. C* **7**, 2831 (1974).

Grimley, T. R., *J. Vac. Sci.* **8**, 31 (1971).

Gurney, R. W., *Phys. Rev.* **47**, 479 (1935).

Hansen, W. N., and G. J. Hansen, *Phys. Rev. Lett.* **58**, 955 (1986).

Hansen, W. N., and G. J. Hansen, *Phys. Rev. A* **36**, 1396 (1987).

Hansen, W. N., and D. M. Koub, *J. Electron. Chem.* **100**, 493 (1979).

Harris, J., and S. Andersson, *Phys. Rev. Lett.* **55**, 1583 (1985).

Harris, J., C. Holmberg, and S. Andersson, *Phys. Scripta T* **13**, 155 (1986).

Harrison, W. A., *Electronic Structure and the Properties of Solids*, Freeman, San Francisco (1980).

Hatzikos, G. H., and R. I. Masel, *Surf. Sci.* **185**, 479 (1987).

Hatzikos, G. H., and R. I. Masel, in J. Ward, ed., *Catalysis 1987*, p. 883, Elsevier, Amsterdam (1988).

Hayward, D. O., B. M. W. Trapnell, *Chemisorption*, Butterworths (1964).

Henrich, V. E., in *Surfaces Interfaces Ceramic Materials*, p. 173, NATO ASI Series E, (1989).

Henrich, V. E., *Rep. Prog. Phys.* **48**, 1481 (1985).

Henrich, V. E., and P. A. Cox, *App. Surf. Sci.* **72**, 277 (1993).

Henrich, V. E., and P. A. Cox, *The Surface Science of Metal Oxides*, Cambridge University Press, NY (1994).

Henry, W. D., *Philos. Trans.* **114**, 266 (1824).

Henry, W. D., *Ann. Philos.* **25**, 422 (1825).

Holloway, S., B. I. Lundqvist, and J. K. Norskøv, *Proc. 4th Int. Congr. Catal.*, 85 (1984).

Holloway, S., and J. Norskøv, *Surface Science Lecture Notes*, vol. 1, Liverpool University Press (1991).

Hotop, H., and W. C. Weinberger, *J. Chem. Phys. Ref. Data* **14**, 731 (1985).

Hung, W. H., I. Schwartz, and S. K. Bernasek, *Surf. Sci.* **294**, 21 (1993).

Huzinga, S., *Gaussian Basic Sets for Molecular Calculations*, Elsevier, NY (1984).

Jacobs, P. A., and W. J. Mortier, *Zeolites* **2**, 226 (1982).

Jacobson, K. W., and J. K. Nørskov, *Phys. Rev. Lett.* **59**, 2764 (1987); **60**, 2496 (1988).

Jacobson, K. W., J. K. Nørskov, M. J. Pushka, *Phys. Rev. B* **35**, 7423 (1987).

Jaeger, R. M., and O. Menzel, *Surface Sci.* **93**, 71 (1980).

Kaswoski, R. V., and R. H. Tait, *Phys. Rev. B* **20**, 5168 (1979).

Kayser, H., *Wiederman's Ann. Phys. Chem.* **14**, 451 (1881).

Kellog, G. L., *Surface Sci. Rep.* **21**, 1 (1994).

Kesmodel, L. L., L. H. Dubois, and G. A. Somorjai, *Chem. Phys. Lett.* **56**, 267 (1978); *J. Chem. Phys.* **70**, 2180 (1979).

Kiselev, A. V., *2nd Int. Congr. Surfacity Activity*, **2**, 168, Butterworths, NY (1957).

Kiselev, A. V., *Quat. Rev.* **25**, 99 (1961).

Kiselev, V. F., and O. V. Krylov, *Adsorption and Catalysis on Transition Metals and Their Oxides*, Springer-Verlag, Berlin (1989).

Kizhakevariam, N., and E. Stuve, *Surface Sci.* **275**, 223 (1992).

Klopman, G., *J. Am. Chem. Soc.* **90**, 233 (1968).

Klopman, G., *Chemical Reactivity and Reaction Paths*, Wiley, NY (1974).

Kohn, W., and L. J. Sham, *Phys. Rev.* **140**, 1133 (1965).

Krakaver, H., M. Posternak, and A. J. Freeman, *Phys. Rev. B* **19**, 1706 (1979).

Kress, J. D., and A. E. Depristo, *J. Chem. Phys.* **87**, 4700 (1987); **88**, 2596 (1988).

Kreuzer, H. J., and Z. W. Gortel, *Physisorption Kinetics*, Springer-Verlag, Berlin (1986).

Lackey, D., J. Schott, and J. K. Sass, *J. Electron. Spec. Relat. Phenom.* **54/55**, 649 (1990).

Lackey, D., J. Schott, J. K. Sass, S. I. Woo, and F. T. Wagner, *Chem. Phys. Lett.* **184**, 277 (1991).

Lang, N. D., *Phys. Rev. Lett.* **46**, 842 (1981).

Lang, N. D., and A. R. Williams, *Phys. Rev. B* **18**, 616 (1978).

Lang, N. D., and A. R. Williams, *Phys. Rev. B* **25**, 2940 (1982).

Langmuir, I., *J. Am. Chem. Soc.* **34**, 1310 (1912); **35**, 105 (1913); **37**, 1139 (1915); **40**, 1361 (1918).

Langmuir, I., *Trans. Faraday Soc.* **17**, 607 (1922).

Langmuir, I., and K. H. Kingdon, *Science* **57**, 58 (1923); *Proc. Roy. Soc. A* **107**, 61 (1925).

Lannoo, M., and P. Friedel, *Atomic and Electronic Structure of Surfaces, Theoretical Foundations*, Springer-Verlag, NY (1991).

de Laplace, P. S., *Mechanique Celeste*, Supplement to Book 10, Chez J. M. B. Duprat, Paris (1806).

Lennard-Jones, J. E., *Trans. Faraday Soc.* **28**, 333 (1932).

Lennard-Jones, J. E., and A. F. Devonshire, *Nature* **137**, 1069 (1936).

Levin, R. D., and S. G. Lias, *Ionization Potential and Appearance Potential Measurements*, NBS (1981).

Lias, S. G., J. F. Liebman, and R. D. Levin, *J. Phys. Chem. Ref. Data* **13**, 695 (1984).

Lias, S. G., J. E. Bartmess, J. F. Liebman, J. L. Holmes, R. D. Leom, and W. G. Mallard, *Gas Phase Ion and Neutral Thermochemistry*, *J. Chem. Phys. Ref. Data* **V 17 Suppl** (1988).

Lin, R. F., G. S. Blackman, M. A. Van Hove, and G. A. Somorjai, *Acta Cryst., Sect. B* **43**, 368 (1987).

London, F., *Z. Physik*, **63**, 245 (1930).

London, F., *Z. Physik Chem. B* **11**, 222 (1930).

London, F., *Trans. Faraday Soc.* **33**, 8 (1937).

Magnus, A., *Z. Phys. Chem. A* **142**, 401 (1929).

Magnus, G., *Poggendorf's Ann.* **3**, 81 (1825).

Magnus, G., *Annal de Chemie*, **39**, 344 (1853); *Philos. Mag.* **6**, 334 (1853); *Annal. der Physik*, **89**, 604 (1853).

Mapledoram, L. D., A. Wander, and D. A. King, *Chem. Phys. Lett.* **208**, 409 (1993).

Masel, R. I., R. P. Miller, and W. H. Miller, *Phys. Rev. B* **12**, 5545 (1975).

Matsuda, T., H. Taguch, and M. Nagao, *J. Therm. Anal.* **38**, 1835 (1992).

McKay, J. M., and V. E. Henrich, *Phys. Rev. B*, **32**, 6764 (1985).

McMaster, M. C., and R. J. Madix, *J. Chem. Phys.* **98**, 9963 (1993).

McMaster, M. C., C. R. Arumainayagam, and R. J. Madix, *Chem. Phys.* **177**, 461 (1993).

McMaster, M. C., S. L. M. Schroeder, and R. J. Madix, *Surface Sci.* **297**, 253 (1993).

Mehl, J. J., and W. L. Schaich, *Phys. Rev. A*, **16**, 921 (1977); **21**, 1174 (1980).

Messmer, R. P., *Phys. Rev. B* **15**, 1811 (1977).

Miedema, A. R., P. F. dChâtel, and F. R. Deboer, *Physica* **100B**, 1 (1980).

Moffat, J., ed., *Theoretical Aspects of Heterogeneous Catalysis*, Van Nostrand Reinhold, NY (1991).

Mönch, W., *Semiconductor Surfaces and Interfaces*, Springer-Verlag, NY (1993).

Morozzo, C. L., *Obs. Phys. Hist. Nat. Arts* **23**, 362 (1783).

Morozzo, C. L., *J. Phys. Chim. Hist. Nat. Arts* **58**, 374 (1803).

Morruzzi, V. L., J. F. Janak, A. R. Williams, *Calculated Electronic Properties of Metals*, Pergamon, NY (1978).

Mortier, W. J., in J. Moffat, ed., *Theoretical Aspects of Heterogeneous Catalysis*, p. 135, Van Nostrand Reinhold, NY (1991).

Mortier, W. J., S. K. Ghosh, and S. Shankar, *J. Am. Chem. Soc.* **108**, 4315 (1986).

Moulijn, J. A., R. A. Van Santen, and P. W. N. M. Van Leeven, *Catalysis, An Integrated Approach to Homogeneous, Heterogeneous and Industrial Catalysis*, Elsevier, Amsterdam (1993).

Müller, K., *Prog. Surf. Sci.* **42**, 245 (1993).

Newns, D. M., *Phys. Rev.* **178**, 1123 (1969).

Nördlanger, P., Holloway, S., and J. N. Nørskov, *Surface Sci.* **136**, 59 (1984).

Nørskov, J. K., *Phys. Rev. B* **20**, 446 (1979).

Nørskov, J. K., *J. Vac. Sci. Tech.* **18**, 420 (1981).

Nørskov, J. K., *Rep. Prog. Phys.* **50**, 1253 (1990).

Nørskov, J. K., S. Holloway, and N. D. Lang, *Surface Sci.* **137**, 65 (1985); **150**, 24 (1985).

Nørskov, J. K., A. Houmøller, P. K. Johansson, and B. I. Lundqvist, *Phys. Rev. Lett.* **46**, 257 (1981).

Oswald, W., *Lehrbook Der Allgemeinen Chemie*, p. 778, Leipzig, (1875); 2nd ed. (1885).

Pacchioni, G., F. Parmigiani, and P. S. Bagos, *Cluster Models for Surface and Bulk Phenomena*, Plenum, NY (1992).

Panas, I., P. E. M. Siegbahn, and U. Wahlgren, *Theor. Chim. Acta* **74**, 167 (1988); *Chem. Phys.* **112**, 325 (1987).

Parr, R. G., and W. Yang, *Density-Functional Theory of Atoms and Molecules*, Oxford University Press, NY (1989).

Pauling, L., *The Nature of the Chemical Bond*, Cornell University Press, Ithaca, NY (1939); 3d ed. (1960).

Pearson, R. G., *J. Am. Chem. Soc.* **85**, 3533 (1963).

Pearson, R. G., *J. Chem. Educ.* **45**, 643 (1968).

Persson, B. N. J., M. Tüshaus, and A. M. Bradshaw, *J. Chem. Phys.* **92**, 5034 (1990).

Pfnür, H., D. Menzel, F. M. Hoffman, A. Ortiga, and A. M. Bradshaw, *Surface Sci.* **93**, 431 (1980).

Plummer, E. W., W. R. Sanek, and J. S. Miller, *Phys. Rev. B* **18**, 1673 (1978).

Ponec, V., T. Knor, and S. Ĉerney, *Adsorption on Solids*, CRC Press, Cleveland (1974).

Priestley, J., *Experiments on Different Kinds of Air*, J. Johnson, London, vol. I (1775). vol. II (1777), vol. III (1779).

Puska, M. J., R. M. Nieminen, and P. Jena, *Phys. Rev. B* **35**, 6059 (1987).

Qain, Z., J. M. Vohs, and O. A. Bonnell, *Surface Sci.* **274**, 35 (1992).

Raeker, T. J., and A. E. DePristo, *Int. Rev. Phys. Chem.* **10**, 1 (1991).

Rettner, C. T., H. A. Michaelson, and D. J. Averbach, *Phys. Rev. Lett.* **68**, 1164 (1992); *Faraday Disc.* **96**, 17 (1993).

Riley, S. J., *Ber. Bunsen Ges. Physik. Chem.* **96**, 1104 (1992).

Roberts, M. W., *Nature (London)*, **188**, 1020 (1960).

Rodrizuez, J. A., R. A. Campbell, and D. W. Goodman, *Phys. Rev. B* **46**, 7077 (1992); *Science* **257**, 897 (1992); **260**, 1527 (1993); *J. Vac. Sci. Tech.* **10**, 2540 (1992).

Rosenstock, H. M., K. Draxl, B. T. Steiner, and J. T. Herron, *Energetics of Gaseous Ions*, NBS (1977).

Ruette, F., *Quantum Chemistry Approaches to Chemisorption and Heterogeneous Catalysis*, Kluwer Press, (1992).

Sabatier, P., *Catalysis in Organic Chemistry* Van Nostrand, NY (1906), 2d ed. (1918).

Saiki, K., W. Rui, and A. Koma, *Surface Sci.* **287**, 644 (1993).

Sass, J. K., D. Lackey, J. Schott, and B. Straehler, *Surf. Sci.* **247**, 239 (1991).

Scheele, C. W., *Chemische Ubhandung von der Luft und dem Feuer*, p. 96, Bergmann Press, Upsalla, (1777). (An English translation appears in L. Dobbin, ed., *The Collected Papers of Carl Wilhelm Scheele*, p. 173, G. Bell and Sons, Ltd, 1931.)

Scheele, C. W., March 1, 1773 letter cited in C. C. Gillespie, ed., *Dictionary Of Scientific Biography*, vol. 12, p. 143 (1974).

Scheffler, H., K. Horn, A. M. Bradshaw, and K. Kambe, *Surface Sci.* **80**, 69 (1979).

Schönhummer, K., and O. Gunnarson, *Phys. Rev. B* **22**, 1629 (1980).

Seebauer, E. G. and C. E. Allen, *Prog. Surface Sci.* **49**, 1 (1995).

Shih, A., and V. A. Parsegian, *Phys. Rev. A* **12**, 835 (1977).

Shustorovich, E., *Metal Surface Reaction Energetics*, VCH, NY (1991).

Slichter, C. P., P. K. Wang, and J. H. Sinfelt, *Phys. Rev. Lett.* **53**, 82 (1984).

Slichter, C. P., C. A. Klug, and J. H. Sinfelt, *Israel J. Chem.* **32**, 185 (1992).

Somorjai, G. A., and M. A. van Hove, *Prog. Surface Sci.* **30**, 201 (1988).

Steele, W. A., *The Interaction of Gases with Solid Surfaces*, Pergamon Press, Oxford (1974).

Steiniger, H., H. Ibachs, and S. Lehwald, *Surface Sci.* **115**, 685 (1982).

Stevenson, D. P., *J. Chem. Phys.* **23**, 203 (1955).

Sugai, S., K. Shimizu, H. Watanabi, H. Miki, K. Kawasaki, and T. Kioka, *Surface Sci.* **287**, 455 (1993).

Sugai, S., K. Takeuchi, T. Ban, H. Miki, K. Kawasaki, and T. Tioka, *Surface Sci.* **282**, 671 (1993).

Talbi, D., F. Pauzat, and V. Ellinger, *Chem. Phys.* **126**, 291 (1988).

Tanaka, K., and K. Tamaru, *J. Catal.* **2**, 366 (1963).

Taylor, H. S., *Proc. Royal Soc. London A* **108**, 105 (1925).

Taylor, H. S., *J. Am. Chem. Soc.* **53**, 578 (1931).

Taylor, J. B., and I. Langmuir, *Phys. Rev.* **44**, 423 (1933).

Taylor, H. S., G. B. Kistiakowsky, *Z. Physik Chem.* **125**, 341 (1927).

Taylor, H. S., G. B. Kistiakowsky, and E. W. Flosdorf, *J. Am. Chem. Soc.* **49**, 2220 (1927).

Thiel, P., *Acc. Chem. Res.* **24**, 31 (1991).

Thiel, P., and T. Madey, *Surface Sci. Rep.* **7**, 211 (1987).

Tit, N., J. W. Halley, and M. T. Michalewicz, *Surface Interface Anal.* **18**, 87 (1992).

Trasatti, S., in B. E. Conway, and J. M. Bockris, *Modern Aspects Electrochem.* **13**, 81 (1979).

Tsukada, M., H. Adachi, and C. Satoko, *Prog. Surface Sci.* **14**, 113 (1983).

Uytterhoeven, L., and W. J. Mortier, *J. Chem. Soc. Faraday Trans.* **88**, 2747 (1992).

Van Santen, R. A., *Theoretical Heterogeneous Catalysis*, World Scientific (1991).

Volkenshtein, F. F., *Electronic Processes on Semiconductor Surfaces*, Consultants Bureau, NY (1991).

Wander, A., M. Van Hove, and G. A. Somorjai, *Phys. Rev. Lett.* **67**, 626 (1991).

Wagner, F. T., and T. E. Moylan, *Surface Sci.* **206**, 187 (1988).

Wang, H., and J. Whitten, *J. Chem. Phys.* **89**, 5329 (1989); **91**, 126 (1989); **96**, 5529 (1992); **98**, 5039 (1993).

Wang, J., and R. I. Masel, *J. Am. Chem. Soc.* **113**, 5850 (1991).

Wang, J., and R. I. Masel, unpublished results 1994.

Weinberg, W. H., *J. Vac. Sci. Tech. A.* **10**, 2271 (1992).

Whetten, R. L., D. M. Cox, D. F. Trevor, and A. Kaldar, *Phys. Rev. Lett.* **54**, 1494 (1985).

Whitten, J. L., *Chem. Phys.* **177**, 387 (1993).

Williams, E. D., W. H. Weinberg, *Surf. Sci.* **82**, 93 (1979).

Willis, R. F., A. A. Lucas, and G. D. Mahan, in D. A. King and D. P. Woodruff, *Chemical Physics of Solid Surfaces and Heterogeneous Catalysts*, vol. 2, p. 59 Elsevier, Amsterdam (1983).

Wood, D. M., *Phys. Rev. Lett.* **46**, 749 (1980).

Yagasaki, E., and R. I. Masel, *Surface Sci.*, **222**, 430 (1991); **226**, 51 (1991).

Yagasaki, E., and R. I. Masel, *Spec. Rep. Catal.* **11**, 165 (1994).

Young, D. M., A. D. Crowell, *Physical Adsorption of Gases*, Butterworth, NY (1962).

Young, T., in J. Peacock, ed., *Miscellaneous Works*, vol. 1, p. 418 (1855).

Zangwill, A., *Physics at Surfaces*, Cambridge Univesity Press, Cambridge (1988).

Zaremba, E., and W. Kohn, *Phys. Rev. B* **13**, 2270 (1976); **15**, 1769 (1977).

Zhu, L., J. Ho, E. K. Paks, and S. J. Riley, *J. Chem. Phys.* **98**, 2798 (1993); *Z. Phys. D* **26**, 313 (1993).

4 Adsorption II: Adsorption Isotherms

PRÉCIS

In the previous chapter we provided a description of the binding of molecules to surfaces. However, in order to model reactions on surfaces, one needs to know more information about the adsorption process. In particular, one needs to know how much of each reactant adsorbs and how quickly the reactant adsorbs. The objective of this chapter is to provide a description of how to quantify how much gas adsorbs. We concentrate on equilibrium adsorption, i.e., adsorption under conditions where the adsorption is reversible and the adsorbed layer is in equilibrium with the adsorbing gas. We emphasize understanding and modeling adsorption isotherms which are plots of the amount of gas that adsorbs on a surface as a function of the pressure of the gas at constant temperature. The approach is to first provide an overview of adsorption and adsorption isotherms. We then derive equations for adsorption isotherms using a variety of idealized models. We find that these idealized models give a good qualitative picture of the adsorption process, although they are not able to quantitatively describe some of the details. We then outline some of the techniques that are now beginning to be used to model these details. The latter discussion is limited because modeling efforts are still evolving. However, an attempt will be made to indicate where the analysis is heading.

4.1 INTRODUCTION

As noted in the Chapter 3, adsorption is a process where molecules from the gas phase or from a solution bind in a condensed layer on a solid or liquid surface. The first quantitative measurements of adsorption were made by Morozzo [1783, 1803] and Rouppe and Norden [1799]. They found that a number of gases will adsorb on charcoal and that the quantity of gas that adsorbs depends on the composition of the gas. Von Saussure [1814] examined the adsorption process in more detail and discovered that the amount of gas that adsorbs depends on the temperature, pressure, and composition of the gas. He made a number of plots of the amount of gas that adsorbs as a function of the pressure of the gas. Today, these plots would be called **adsorption isotherms.**

Adsorption isotherms were studied heavily in the nineteenth century, and the early part of the twentieth century. Over the years thousands of adsorption isotherms have been measured. Much of the data have been reviewed in Brunauer [1945] or Valenzuela and Meyers [1989]. One should refer to their books for details.

Figure 4.1 and Figure 4.2 show some typical adsorption isotherms. Figure 4.1 shows isotherms for CO adsorption on Pd(111) at a variety of temperatures. At 300 K one observes significant adsorption even at very low pressures. At higher temperatures one needs to go to progressively higher pressures to get significant gas to adsorb. Still, the isotherm saturates at half a monolayer in all cases. Kohn, Szany, and Goodman [1992] have found that at these temperatures there is little additional adsorption when one raises the pressure to one atmosphere. The data in Figure 4.1 are typical of an isotherm for adsorption of a noncondensable gas on a clean metal surface under conditions where the adsorbate forms a strong bond to

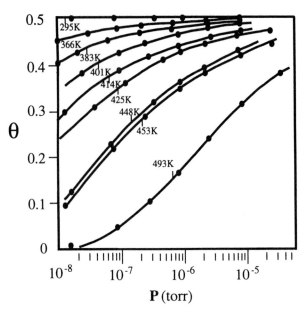

Figure 4.1   Adsorption isotherms for CO on Pd(111). (Data of Ertl and Koch [1970].)

the surface. One often gets a monolayer of gas to adsorb at low pressures, but little additional gas adsorbs at higher pressures. If one takes a more weakly bound species (e.g., CO_2, H_2O, C_2H_6), moderate pressures have to be used to get significant gas to adsorb. However, again the adsorption isotherm saturates at one monolayer. Condensable gases show a different effect. The first monolayer often adsorbs at low pressures, but when the partial pressure of the adsorbate approaches the adsorbate's bulk vapor pressure, additional gas adsorbs. Usually the second monolayer condenses at pressures that are slightly below the bulk vapor pressure of the adsorbate, but one does not get more than about three layers until the partial pressure of the adsorbate equals its bulk vapor pressure. At that point a thick liquid film condenses.

Adsorption isotherms on highly porous materials are somewhat different than isotherms on bulk metals because of the capillary condensation effects mentioned in Section 3.1. Figure 4.2 shows a series of adsorption isotherms for krypton, argon, and ammonia adsorption on graphitized carbon black. These are typical of the kinds of adsorption isotherms observed on porous materials. Notice that each of the isotherms shows different behavior. At 94.72 K (i.e., krypton's boiling point at 1 atm pressure) krypton condenses at low pressures to form a single monolayer at 0.1 torr. The amount of adsorption is almost constant at krypton pressures between 0.1 torr and 2 torr. However, then one begins to adsorb a small amount of gas into the second monolayer. The second monolayer fills up suddenly at krypton pressures between 17 and 22 torr. Then there is a second plateau. At krypton pressures above 30 torr, the amount of adsorption increases continuously until a bulk liquid is formed at a krypton pressure of 760 torr.

Figure 4.3 shows a blowup of the low-pressure submonolayer part of the krypton isotherm shown in Figure 4.2. Notice that initially little krypton adsorbs. However, once about a third of a monolayer has been reached, the adsorption process accelerates. The amount of adsorption then levels off at about one monolayer.

Argon at 91.3 K shows behavior similar to that of krypton. Initially the argon adsorbs slowly, but there is a rapid increase in the curve at about $\frac{1}{3}$ and about $\frac{3}{4}$ of a monolayer. The argon saturates at about a monolayer. There is more argon than krypton in the first monolayer

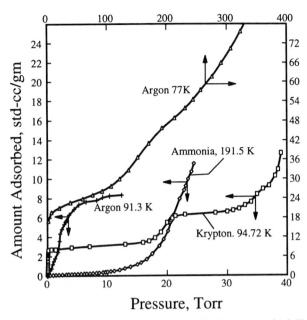

Figure 4.2 Isotherms for adsorption of krypton, argon at 77 K, argon at 91.3 K, and ammonia on graphitized carbon black. (Data of Putnam and Fart [1975], Ross and Winkler [1955], Basset, Boucher, and Zettlemoyer [1968], and Bomchil et al. [1979], respectively.)

because the argon data were taken on a sample with a higher surface area than the krypton data, and because argon is smaller than krypton.

Figure 4.2 also shows argon data taken at 77 K (i.e., the boiling point of argon at 1 atm pressure). Again, there is some interesting behavior at low pressures that cannot be clearly discerned in the figure, but the key thing to recognize is that there is no clear plateau in the figure at a monolayer coverage. Instead there is a continuous increase in the amount of gas that adsorbs. This continuous increase occurs because of multilayer adsorption and pore

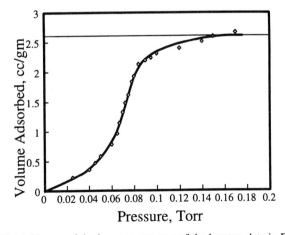

Figure 4.3 A blowup of the low-pressure part of the krypton data in Figure 4.2.

condensation. At 77 K the interactions between adjacent argon molecules is comparable to kT where k is Boltzmann's constant and T is the temperature. That is why liquid argon forms at 77 K. In Section 3.1 we noted that multilayers form under such conditions. The pore condensation effects described in Section 3.1 then come into play. According to the Kelvin equation, Equation 3.1, the vapor pressure of argon in the smallest pores in graphitized carbon black is less than 100 torr (760 torr = 1 atm). As a result, once the argon pressure exceeds 100 torr those pores will fill up with argon. At higher pressures, larger pores begin to fill until all of the pores are filled at a pressure of 760 torr.

The 77 K argon data illustrate a general phenomenon. When a substance is condensed on a solid at temperatures close to the substance's boiling point, multilayers can form at pressures considerably below the vapor pressure of the bulk fluid. These multilayers substantially change the nature of the adsorption process, as is described in Section 4.12.

The ammonia data in Figure 4.2 illustrate another interesting phenomenon. Ammonia does not strongly interact with graphite, so initially little ammonia adsorbs, but once the ammonia begins to adsorb, additional ammonia hydrogen bonds to the ammonia that first adsorbed, so the amount of adsorption increases rapidly. The result is very nonlinear behavior where there are patches (i.e., islands) on the surface which are covered by ammonia and other patches that are adsorbate free.

Figure 4.2 also exemplifies another important point: the adsorption of one gas on a given adsorbant does not allow one to say much about the adsorption of a different gas even when the gases are quite similar. Krypton, for example, shows distinct breaks in the adsorption curve at 0.5, 1, and 2 monolayers, while argon shows breaks at $\frac{1}{3}$, $\frac{3}{4}$ monolayers and continuous behavior above a monolayer.

In 1945, Brunauer proposed that one can classify the different kinds of behavior seen during the adsorption of gases on solids into one of the five general forms shown in Figure 4.4. With a flat surface one usually observes one of two types of adsorption isotherms: **Type I** isotherms where the amount of gas increases with increasing pressure and then saturates at about a monolayer coverage, and **type II** isotherms where the amount adsorbed increases with increasing pressure, starts to level off, and then starts to grow again at higher pressures. Type I isotherms are characteristic of monolayer adsorption. Similar isotherms also arise if the adsorbant has very small pores that all fill up at once. Type II isotherms are characteristic of multilayer adsorption. In Figure 4.2, argon follows a type I adsorption isotherm at 91.3 K and a type II isotherm at 77 K. Occasionally one observes a so-called **type III** adsorption isotherm where initially there is very little adsorption. However, once a small droplet or island

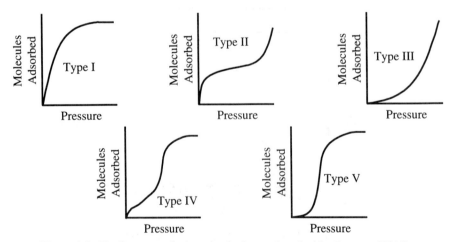

Figure 4.4   The five types of adsorption isotherms described by Brunauer [1945].

of adsorbate nucleates on the surface, additional adsorption occurs more easily because of strong adsorbate–adsorbate interactions. Ammonia adsorption on graphitized carbon is an example of a type III adsorption isotherm. Brunauer [1945] also defines a **type IV** and a **type V** adsorption isotherm. They are illustrated in Figure 4.4. Type IV and type V isotherms usually occur when multilayers of gas adsorb onto the surface of the pores in a porous solid. Initially, the adsorption looks like a type II or a type III adsorption, but eventually the adsorbed layer gets so thick that it fills up the pores. As a result, no more gas can adsorb and the isotherm saturates. The krypton data in Figure 4.2 do not fit any of these classifications. Krypton shows a stepwise adsorption profile where the first monolayer fills up with gas, then at some higher pressure the second monolayer fills up with gas. These stepwise profiles are associated with so-called layering transitions where ordered phases are formed in the second monolayer. See Gregg and Sing [1967] for details.

4.2 ANALYTICAL MODELS FOR REVERSIBLE MONOLAYER ADSORPTION

People have been trying to model adsorption since the work of Van Saussure [1814]. In the material later in this book, the most important class of adsorption is reversible monolayer adsorption where adsorbing molecules reversibly attach themselves to the surface of the solid. The first paper on reversible adsorption of which this author is aware is Scheele's 1777 paper, which showed that gases can be removed from coal by heating. The gases readsorb when the coal cools. Up until 1909, all of the models of adsorption were empirical. Then, in a series of papers between 1913 and 1918, Langmuir derived a simple model for an adsorption isotherm which is still in use today. Since Langmuir's day, there have been many improvements to his model. However, we are still in the position where we can understand only the simplest adsorption systems. In the following section, we review some of the most important models of *reversible* adsorption of monolayers of gas. Multilayer adsorption is discussed in Section 4.12. Other adsorption systems are discussed in Chapters 5, 6, and 10.

4.3 LANGMUIR ADSORPTION MODEL

As just noted, the most important model of monolayer adsorption came from the work of Langmuir [1913, 1915, 1918]. Langmuir considered adsorption of an ideal gas onto the idealized surface. The gas was presumed to bind at a series of distinct sites on the surface of the solid, as indicated in Figure 4.5, and the adsorption process was treated as a reaction

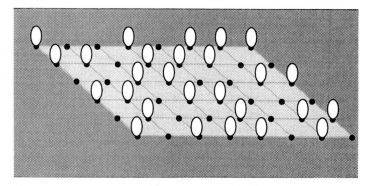

Figure 4.5 Langmuir's model of the structure of the adsorbed layer. The black dots represent possible adsorption sites, while the white ovals represent adsorbed molecules.

where a gas molecule A_g reacts with an empty site, S, to yield and adsorbed complex A_{ad}:

$$A_g + S \rightleftharpoons A_{ad} \qquad (4.1)$$

In the material that follows we derive an expression for the Langmuir adsorption isotherm for a number of examples. The derivation starts by deriving an equilibrium expression for the adsorption of a gas. Then, a mass balance is used to calculate an adsorption isotherm.

4.3.1 Kinetic Derivation of the Langmuir Adsorption Isotherm

Langmuir's original derivation assumed that adsorption and desorption were elementary processes where r_{ad} the rate of adsorption and r_d the rate of desorption are given by

$$r_{ad} = k_{ad} P_A [S] \qquad (4.2)$$

$$r_d = k_d [A_{ad}] \qquad (4.3)$$

where P_A is the partial pressure of A over the surface, $[S]$ is the concentration of bare sites in number/cm^2, $[A_{ad}]$ is the surface concentration of A in molecules/cm^2, and k_{ad} and k_d are constants.

At equilibrium, the rate of adsorption equals the rate of desorption. Setting $r_{ad} = r_d$ and rearranging yields

$$\frac{[A_{ad}]}{P_A [S]} = \frac{k_{ad}}{k_d} = K^A_{equ} \qquad (4.4)$$

where we have defined K^A_{equ} to be the ratio of k_{ad} and k_d.

For the analysis later in this chapter it is also useful to define an *equilibrium* enthalpy, entropy, and free energy of adsorption ΔH_{ad}, ΔS_{ad}, ΔG_{ad}, respectively, by

$$\frac{[A_{ad}]}{P_a [S]} = K^A_{equ} = \frac{1}{P_r} e^{-\Delta G_{ad}/RT} = \frac{1}{P_r} e^{\Delta S_{ad}/R} e^{-\Delta H_{ad}/RT} \qquad (4.5)$$

Note that Adamson [1960, 1990] also defines several other types of heats of adsorption, including an isosteric heat of adsorption and a calorimetric heat of adsorption. They are similar to the heat of adsorption in Equation 4.5, but they have a slightly different numerical value. Often when people report heats of adsorption, they do not say which heat of adsorption they are reporting. Hence, some care must be taken when one is using Equation 4.5 to make sure one is using the right heat of adsorption.

4.3.1.1 Single Adsorbate Case. In the next part of this chapter we use Equation 4.4 to derive an expression for the amount of gas that adsorbs onto a solid surface as a function of the partial pressure of the gas and the equilibrium constant of adsorption. The general approach is to combine Equation 4.4 with a site balance to calculate an equilibrium coverage. Inherent in this derivation is the assumption that we have a finite number of sites to hold gas, so that as we raise the pressure there will be fewer and fewer sites available to hold additional adsorbate. We start with the simplest case: the adsorption of a single adsorbate onto a series of equivalent sites on the surface of the solid. For the purposes of the derivation we assume that (1) the adsorbing gas adsorbs into an immobile state, (2) all sites are equivalent, (3) each site can hold at most one molecule of A, and (4) there are no interactions between adsorbate molecules on adjacent sites so that K^A_{equ} will be independent of the coverage of the adsorbed species, $[A_{ad}]$. From the analysis in Section 4.3.1:

$$\frac{[A_{ad}]}{P_a[S]} = K_{equ}^A \tag{4.6}$$

One also needs a site balance to derive the isotherm

$$[S] = [S_0] - [A_{ad}] \tag{4.7}$$

with $[S_0] = S_0/a$. Note that $[S_0]$ is the concentration of sites in number/cm^2.
Substituting the expression for $[S]$ from Equation 4.6 into Equation 4.7 yields

$$\frac{[A_{ad}]}{P_A K_{equ}} + [A_{ad}] = [S_0] \tag{4.8}$$

Rearranging and defining θ_A, the fraction of the surface sites covered with A via

$$\theta_A = \frac{[A_{ad}]}{[S_0]} \tag{4.9}$$

yields

$$\theta_A = \frac{K_{equ}^A P_A}{1 + K_{equ}^A P_A} \tag{4.10}$$

Equation 4.10 is the **Langmuir adsorption isotherm** for noncompetitive, nondissociative adsorption.

Equation 4.10 was one of Langmuir's key results. It predicts that adsorption of a gas on a surface follows a type I adsorption isotherm (see Figure 4.4). At low pressures the coverage varies linearly with pressure. However, the coverage saturates with increasing pressure. In the problem set, we ask the reader to show that there is a close analogy between the Langmuir adsorption isotherm and a modified van der Waals equation of state.

$$P_g(v - b_w) = \mathfrak{F}(kT)$$

where v is the molar volume, b_w is a constant related to the volume of a molecule, and $\mathfrak{F}(kT)$ is a function of kT. The van der Waals equation predicts that at low pressure the density varies linearly with pressure. However, at high pressure the density saturates because each molecule has a finite volume. In the same way the Langmuir adsorption isotherm, Equation 4.10, predicts that at low pressure the density of molecules on the surface varies linearly with coverage. However, at high coverages the density of molecules on the surface saturates since each molecule blocks a finite region on the surface.

Since Langmuir's time, Equation 4.10 has been found to fit a wide variety of adsorption systems. Equation 4.10 often works under conditions where Equations 4.2 and 4.3 fail. Therefore, the Langmuir adsorption isotherm is a very general result for chemisorption systems.

Volmer and Mahnert [1925] and Fowler [1935] wrote several papers where they tried to explain how Equation 4.10 could work under conditions where Equations 4.2 and 4.3 fail. They showed that one can derive Equation 4.10 without making many of the assumptions in Section 4.3.1. Thus, the Langmuir adsorption isotherm is more general than it would initially appear. Below, we use Volmer and Fowler's approach to rederive Equation 4.10.

The derivation will start by assuming that gas was adsorbing on a fixed array of sites as in Figure 4.5, and that there are no interactions between adjacent gas molecules. They show that under this assumption sites fill randomly and the Langmuir adsorption isotherm applies. This kind of adsorption system has come to be called an **ideal lattice gas,** since the gas molecules are bound to a lattice and the interactions between gas molecules are similar to those in an ideal gas.

In the material that follows, we reproduce their derivation. We first use statistical mechanics to an expression for the equilibrium constant for reaction 4.1. The result is Equation 4.21. We then do a site balance to rederive the Langmuir adsorption isotherm, Equation 4.10. The derivation requires some knowledge of classical statistical mechanics. The reader may wish to review the material in Section 1 of Appendix A before proceeding with the derivation.

Consider adsorption of N identical molecules of A onto a series of S_0 localized sites on a solid surface. From statistical mechanics, the Helmholtz free energy of the adsorbed molecules, A_s, can be calculated from

$$A_s = -kT \ln (Q^N_{canon}) \tag{4.11}$$

where

k = Boltzmann's constant
T = absolute temperature
Q^N_{canon} = the canonical partition function, i.e., the partition function for N adsorbate molecules

If all of the sites are equivalent, the molecules on each site do not interact, and all of the adsorbed molecules are immobile, then we show in Appendix A that the canonical partition function for the system becomes:

$$Q^N_{canon} = \frac{S_0!}{(S_0 - N)!\, N!}\, q_s^N \tag{4.12}$$

where

q_s = the partition function for an individual adsorbed molecule
$\dfrac{S_0!}{(S_0 - N)!\, N!}$ = the number of ways of placing N molecules on S_0 sites

Combining Equations 4.11 and 4.12, and using Stirling's approximation, that is,

$$\ln(X!) = X \ln (X) - X$$

yields

$$A_s = -kT \ln (Q^N_{canon})$$
$$= kT\,[-S_0 \ln (S_0) + N \ln (N) + (S_0 - N) \ln (S_0 - N)$$
$$- N \ln q_s + S_0 - N - (S_0 - N)] \tag{4.13}$$

Canceling like terms,

$$A_s = kT\,[-S_0 \ln (S_0) + N \ln (N) + (S_0 - N) \ln (S_0 - N) - N \ln q_s] \tag{4.14}$$

Thermodynamics defines the chemical potential of the adsorbate layer by

$$\mu_s = \left(\frac{dA_s}{dN}\right)_{T, S_0} \tag{4.15}$$

Combining Equations 4.14 and 4.15 yields

$$\mu_s = kT \{0 + [1 + \ln(N)] - [1 + \ln(S_0 - N)] - \ln(q_s)\} \tag{4.16}$$

Deleting like terms

$$\mu_s = kT \ln\left(\frac{N}{S_0 - N}\right) - kT \ln(q_s) \tag{4.17}$$

At equilibrium, the chemical potential of the adsorbed layer, μ_s, must equal the chemical potential of the gas over the layer, μ_g. The chemical potential of an ideal gas is given by

$$\mu_g = kT \ln\left(\frac{P_A}{P_r}\right) - kT \ln(q_g) \tag{4.18}$$

with

P_A = the partial pressure of A over the surface
P_r = the reference pressure, 1 atm
q_g = the partition function of a gas molecule at the reference pressure

Combining Equations 4.17 and 4.18 and noting that at equilibrium $\mu_g = \mu_s$ yields

$$kT \ln\left(\frac{N}{(S_0 - N)}\right) = kT \ln\left(\frac{P_A}{P_r}\right) - kT \ln\left(\frac{q_g}{q_s}\right) \tag{4.19}$$

Rearranging and defining a constant K_{equ}^A by

$$\frac{N}{(S_0 - N)P_A} = \frac{1}{P_r}\left(\frac{q_s}{q_g}\right) = K_{equ}^A \tag{4.20}$$

where K_{equ}^A is called the equilibrium constant for adsorption.

Dividing the top and bottom of Equation 4.20 by the surface area of the solid, a, yields

$$\frac{[A]}{[S] P_A} = K_{equ}^A \tag{4.21}$$

with

$[A] = N/a$ = the surface concentration of A
$[S] = S_0/a - [A]$ = the surface concentration of bare sites.

The implication of Equation 4.21 is that one can indeed treat the adsorption process as a reaction between gas molecules and free sites, as indicated in reaction 4.1. Note that Equation 4.21 is identical to Equation 4.4. Therefore Equation 4.10 applies to any lattice gas, provided that there is only one species adsorbing on the surface, the species adsorbs nondissociatively, and that there are no interactions between species on adjacent sites.

4.3.1.2 Competitive Adsorption.   Still, Equation 4.10 assumes that there is only one species adsorbing onto the surface. However, during a reaction there often will be two or more species adsorbing simultaneously. Those two species often compete for the same sites. As a result, a different adsorption isotherm is needed for the competitive adsorption of two different gases.

Consider two species A and B that compete for the same adsorption sites. Assume that (1) all sites are equivalent; (2) each site can hold at most one molecule of A or one molecule of B, but not both; and (3) there are no interactions between adsorbate molecules on adjacent sites. One can use a derivation similar to that in Section 4.3.1.1 to show

$$\frac{[A_{ad}]}{P_A[S]} = K_{equ}^A \quad \frac{[B_{ad}]}{P_B[S]} = K_{equ}^B \tag{4.22}$$

We also need a site balance:

$$[S] = [S_0] - [A_{ad}] - [B_{ad}] \tag{4.23}$$

Combining Equations 4.22 and 4.23 and rearranging yields

$$\theta_A = \frac{K_{equ}^A P_A}{1 + K_{equ}^A P_A + K_{equ}^B P_B} \tag{4.24}$$

$$\theta_B = \frac{K_{equ}^B P_B}{1 + K_{equ}^A P_A + K_{equ}^B P_B} \tag{4.25}$$

Equations 4.24 and 4.25 are the Langmuir adsorption isotherm for a **competitive nondissociative** adsorption.

4.3.1.3 Dissociative Adsorption.   The one other case of special importance is when a molecule D_2 dissociates into two atoms upon adsorption. The Langmuir adsorption isotherm for dissociative adsorption will be derived by assuming that (1) D_2 completely dissociates to two molecules of D upon adsorption; (2) the D atoms adsorb onto distinct sites on the surface of the solid and then move around and equilibrate; (3) all sites are equivalent; (4) each site can hold at most one atom of D, and (5) there are no interactions between adsorbate molecules on adjacent sites. One can use a derivation similar to the one in Section 4.3.1.1 to show

$$\frac{[D_{ad}]}{P_{D_2}^{1/2}[S]} = K_{equ}^D \tag{4.26}$$

The factor $P_{D_2}^{1/2}$ comes about because one gas phase molecule produces two adsorbed species. Combining Equation 4.26 with a site balance like that in Equation 4.7 yields

$$\theta_D = \frac{K_{equ}^D P_2^{1/2}}{1 + K_{equ}^D P_{D_2}^{1/2}} \tag{4.27}$$

where θ_D is the fraction of the surface sites covered by atoms of D. Equation 4.27 is the Langmuir adsorption isotherm for a **dissociative** adsorption process.

Similarly a gas that adsorbs molecularly, blocking two sites obeys

$$\frac{[E_{ad}]}{P_E[S]^2} = K_{equ}^E \qquad (4.28)$$

A derivation like that in the previous section shows

$$\theta_E = 1 - \frac{1}{2P_E K_{equ}^E S_0}(1 - \sqrt{1 + 4P_E K_{equ}^E S_0}) \qquad (4.29)$$

Equations 4.10, 4.24, 4.25, and 4.27 were Langmuir's key results. They revolutionized the way people thought about adsorption. Notice that as P_A increases, the coverage of A increases. However, the total amount of adsorption saturates at modest pressures. As a result, one often finds that the coverage of a component is independent of the pressure of the component in the gas phase. In contrast, if one is absorbing gases into a liquid, the amount of gas that *ab*sorbs into the liquid usually does not saturate as the pressure of A increases. Hence, *ad*sorption of a gas onto a surface is fundamentally different than *ab*sorption of the gas into a liquid.

Another interesting effect occurs during the coadsorption of two gases, A and B. Notice that according to Equation 4.24 when the partial pressure of A increases the coverage of B decreases. The reduction occurs even though A and B do not interact on a surface. This is in contrast to the situation in a liquid where a reduction occurs only when there is a direct or indirect interaction between A and B. Hence, Equations 4.10, 4.24, 4.25 and 4.27 show that *ab*sorption into a liquid is fundamentally different than *ad*sorption onto a surface in that *ab*sorption and *ad*sorption follow much different rate laws.

4.3.2 Comparison to Data

Equations 4.10, 4.24, 4.25, and 4.27 are still used to model adsorption systems today. According to Equation 4.27 if a dissociative adsorption process follows a Langmuir adsorption isotherm, then a plot of θ^{-1}, versus $P_A^{-1/2}$ should be linear.

$$\frac{1}{\theta} = \frac{1}{K_{equ}^A P_A^{1/2}} + 1 \qquad (4.30)$$

Figure 4.6 shows several of these plots for the dissociative adsorption of hydrogen on Pt(111) at 213, 241, and 281 K. Notice that the data are linear over a reasonable range of coverages, but deviates from the line at high and low coverages. This is typical of what is observed on single crystal surfaces; larger deviations are seen on rough surfaces. As a result, while the Langmuir adsorption isotherm can be a useful approximation, it does not fully explain what is observed.

The deviations from the Langmuir adsorption isotherm occur because of two effects. In many cases there are several different sites on a given surface that can hold the adsorbate. Taylor [1925] and Constable [1925] showed that the binding energy of a given molecule can vary significantly from site to site. As a result, corrections are needed to the Langmuir adsorption isotherm to account for multiple inequivalent sites.

Another correction is needed because of adsorbate/adsorbate interactions. As noted in Section 3.1 adsorbate densities are similar to those of a liquid. In a liquid, there are usually strong molecule/molecule interactions that scale with the density. If one extrapolates the liquid data to the densities in an adsorbed layer, one would expect there to be significant adsorbate/

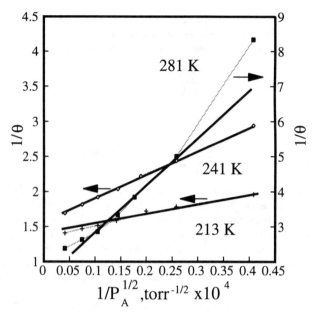

Figure 4.6   A plot of a series of adsorption isotherms for hydrogen adsorption on Pt(111). Dotted lines and symbols: data. Solid Lines: fits to the Langmuir adsorption isotherm. (Data of Ertl, Neuman, and Streit [1977].)

adsorbate interactions at coverages approaching one monolayer. Hence, adsorbate/adsorbate interactions need to be accounted for in order to model adsorption in detail.

4.4   MODIFICATIONS OF THE LANGMUIR ADSORPTION ISOTHERM: INEQUIVALENT SITES

Over the years, there have been many attempts to propose modified adsorption isotherms that account for the deviations. Table 4.1 lists many of these isotherms, along with their advantages and disadvantages. In the following sections we describe each of these isotherms in detail. First, we describe a series of analytical methods that can be used to reproduce adsorption isotherms, but not the detailed structure of the adsorbed layer. Then we move on to lattice gas models that can represent the structure of the adsorbed layer. An attempt was made to include the most important isotherms, but there are others; see Trapnell and Hayward [1964] or Doraiswamy [1991] for others.

The simplest deviations from the Langmuir adsorption isotherm come during adsorption on rough inhomogeneous surfaces. On a rough surface, there are multiple sites available for adsorption; the heat of adsorption varies from site to site. Hence major modifications to the Langmuir adsorption isotherm are needed in such systems. Much of the work on inhomogeneous surfaces have been reviewed by Jaroneic and Madey [1988]. Readers should refer to their books for details.

Generally, it has been presumed that when there are multiple sites, each site acts independently during the adsorption process and each site follows a Langmuir adsorption isotherm. It is usually convenient to define a distribution function $D(E)$, where $D(E)$ is the number of sites with binding energy between E and $E + dE$, and $D(E)$ is normalized via

TABLE 4.1 A Comparison of the Advantages and Disadvantages of Several Adsorption Isotherms

Isotherm	Advantages	Disadvantages
Langmuir	Best one parameter isotherm	Ignores adsorbate/adsorbate interactions
Freundlich, Toth	Two parameters	No physical basis for equation
Multisite	Many parameters	Good for inhomogeneous surfaces. Wrong physics for single crystals
Tempkin Fowler Slygin–Frumkin	Account for adsorbate/adsorbate interactions in an average sense	Does not consider how the adsorbate layer is arranged
Lattice gas	Complete description of adsorbate/adsorbate interactions for commensurate layers	Requires a computer to calculate isotherm
		Assumes commensurate adsorption
	Predicts arrangement of adsorbed layer	Parameters used in the model are difficult to determine

$$\int_0^\infty D(E)\,dE = 1 \qquad (4.31)$$

If it is assumed that since the coverage on each site follows Equation 4.10, then at any gas pressure the fraction of the total surface area covered by an adsorbate, A, is given by

$$\theta_A = \int_0^\infty \left(\frac{K_{equ}^A(E)P_A}{1 + K_{equ}^A(E)P_A} \right) D(E)\,dE \qquad (4.32)$$

The main advantage of Equation 4.32 is that it provides a way to artificially add additional parameters (i.e., $D(E)$,) into the modeling equation so that one can more closely reproduce experimental adsorption data. Sips [1948] showed that in fact one can fit any adsorption isotherm exactly if one treats $D(E)$ as a function to be fit to the isotherm. Several investigators have used Sips methods to estimate $D(E)$. However, the physical significance of $D(E)$ is not clear since the model has ignored adsorbate/adsorbate interactions.

4.4.1 The Freundlich Adsorption Isotherm

The most important multisite adsorption isotherm for rough surfaces is the Freundlich adsorption isotherm:

$$\theta_A = \alpha_F P^{C_F} \qquad (4.33)$$

where α_F and C_F are fitting parameters. Equation 4.33 implies that if one makes a log-log plot of adsorption data, the data will fit a straight line. Equation 4.33 has two parameters, while Langmuir's equations only has one. As a result, Equation 4.33 often fits adsorption data on rough surfaces better than the Langmuir's equations. In the older literature, Equation

4.33 was used extensively. However, it was found that equation 4.33 usually only fits adsorption data taken over a small pressure range and the equation has little predictive value. As a result, Equation 4.33 is now rarely used.

A related equation is the Toth equation. Note, that one can rearrange the Langmuir adsorption isotherm, Equation 4.10, to yield:

$$\theta = \frac{P_A}{\dfrac{1}{K_{equ}^A} + P_A} \tag{4.34}$$

Toth [1971] modified Equation 4.34 by adding two extra parameters α_{T_o} and c_{T_o} as follows:

$$\theta^{c_{T_o}} = \frac{\alpha_{T_o} P_A^{c_{T_o}}}{\dfrac{1}{K_{equ}^A} + P_A^{c_{T_o}}} \tag{4.35}$$

Freundlich [1909] and Toth [1962] actually originally discussed these equations as empirical expressions to correlate data. However, Zeldowitch [1935] and Sips [1948] worked backwards from the Freundlich equation and showed that if one assumed that $K_{equ}^A(E)$ and $D(E)$ are given by

$$K_{equ}^A(E) = K_{equ}^{A,0} e^{+E/kT} \tag{4.36}$$

$$D(E) = \frac{\alpha_F}{kT(K_{equ}^{A,0})^{c_F}} \frac{\sin \pi C_F}{\pi} e^{-E/kT} \tag{4.37}$$

one can derive equation Equation 4.33 theoretically. In the literature, there has been considerable discussion to whether there is any physical significance to Equation 4.37. However, it does not appear that Equation 4.37 is physically significant.

4.4.2 The Multisite Model

The Freundlich adsorption isotherm is used mainly for rough inhomogeneous surfaces. With single crystals, it has been more common to assume that the adsorbate can adsorb on a finite number of sites, each of which follows a Langmuir adsorption isotherm. Following Langmuir [1922], one then calculates the total coverage of an A, θ_A by summing over the individual types of sites:

$$\theta_A = \sum_i \frac{\chi_i K_{equ}^i P_A}{1 + K_{equ}^i P_A} \tag{4.38}$$

where χ_i = the fraction of sites of type i.

It happens that χ_i and K_{equ}^i can be measured directly from adsorption–desorption measurements as outlined in Chapters 5 and 6. As a result, Equation 4.38 should in principle be quite useful. In fact, years ago, Equation 4.38 was used quite regularly even though it had not been rigorously verified experimentally. In recent years lattice gas calculations have largely replaced the multisite model. However, the multisite model is still used occasionally.

4.5 ADSORBATE/ADSORBATE INTERACTIONS

One of the difficulties with Equation 4.38 is that it ignores adsorbate/adsorbate interactions. In Section 3.1 we noted that at saturation, the adsorbed layer has a density like that of a liquid. There are strong interactions between adjacent molecules in a liquid that scale as the density. An extrapolation of the liquid data to densities characteristic of adsorbed layers shows that there should be strong interactions between adjacent adsorbate molecules in an adsorbed layer at near monolayer coverages.

Experimentally, there is clear evidence for adsorbate/adsorbate interactions in heat of adsorption data. For example, Figure 4.7 shows how the heat of adsorption of CO, NO, and H_2 on Pt(111) varies with coverage. Notice that all three heats of adsorption vary with coverage. With H_2 and NO there are smooth variations, while CO shows distinct jumps in the heat of adsorption with changing coverage. There is a problem in interpreting the data in the literature in that not all of the data were measured under equilibrium conditions. For example, Vasquez, Muscat, and Madix [1994] found that their measured heat of adsorption for CO on Ni(100) varied substantially with coverage even at low coverage when they froze CO onto the surface at 100 K (see Figure 4.8). However, there was little variation in the heat of adsorption with coverage up to coverages of 0.5 L when the adsorbate layer was annealed at 400 K to allow the adsorbed molecules to equilibrate. Hence one has to be careful in interpreting the heat of adsorption data in the literature. Still, there is a good evidence that the heat of adsorption does vary substantially with coverage in many systems, and this variation needs to be modeled.

A variation in the heat of adsorption with coverage implies that the binding of the adsorbate to the surface varies with coverage. Such a variation could only occur if there were strong adsorbate/adsorbate interactions.

There are two kinds of adsorbate/adsorbate interactions: direct interaction between adjacent adsorbed molecules and indirect interaction where the adsorbate changes the surface, which in turn affects the adsorption of other adsorbate molecules.

Generally the direct interactions are similar to those in a liquid. There are van der Waals attractions and hard-core repulsions. There can also be hydrogen bonding.

The indirect interactions are more complex. These interactions arise because when the adsorbate bonds to the surface, electrons are donated from the surface to form the adsorbate/

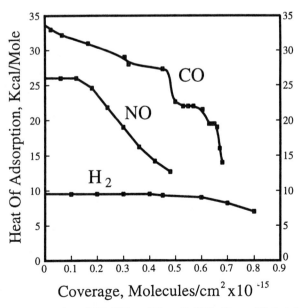

Figure 4.7 The heat of adsorption of CO, NO, and H_2 on Pt(111). (Data of Ertl et al. [1977], Seebauer, Kong, and Schmidt [1986] and Christmann, Ertl, and Pignet [1976].)

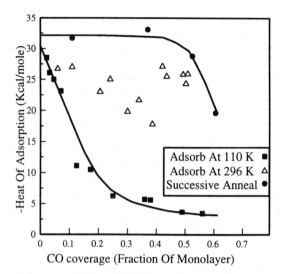

Figure 4.8 The heat of adsorption of CO on Ni(100) as a function of coverage measured by adsorbing at 296 K, or successively annealing. (Data of Vasquez et al. [1994].)

surface bond. The donation of electrons changes the electronic properties of the surface, which in turn affects the bonding of all of the other adsorbate molecules. As a result, the bonding of all the molecules in the layer can change with coverage.

Experimentally, both indirect and direct adsorbate/adsorbate interactions affect heats of adsorption in similar ways. As a result, it is difficult to distinguish between a direct and an indirect interaction from an adsorption isotherm. However, it is easy to distinguish between the two spectroscopically. For example, Figure 4.9 shows some of Olsen and Masel's [1988] data for the position of linear-^{12}CO infrared band seen during adsorption of ^{12}CO on Pt(111). The data are plotted as a function of coverage during experiments where the ^{12}CO on the surface was diluted with (a) ^{12}CO, and (b) ^{13}CO. In work on supported catalysts Hammaker, Fransis, and Gischens [1965] showed that when a ^{12}CO is added to a ^{12}CO layer, the position of the linear CO band is affected by both direct and indirect adsorbate/adsorbate interactions. However, when ^{13}CO is added to the ^{12}CO layer, there is a quantum effect so that the band position is affected only by the indirect interactions. Figure 4.9, shows that the band shifts when ^{13}CO is added to the layer, implying that there are indirect interactions, i.e., a change in the ability of the surface to donate electrons to the adsorbed ^{12}CO with changing coverage. However, different results are obtained when ^{12}CO is added to the layer. The only difference between diluting with ^{13}CO and ^{12}CO is that there are direct interactions in the latter case. Thus, the difference in behavior observed when the layer is diluted with ^{12}CO rather than ^{13}CO implies that there are also direct interactions between the adsorbed molecules.

4.5.1 The Tempkin Adsorption Isotherm

Tempkin [1940] considered the effects of indirect adsorbate/adsorbate interactions on adsorption isotherms. He noted that experimentally heats of adsorption would more often decrease than increase with increasing coverage. Often the results could be linearized provided we were not working over too large a range of coverage. Therefore, Tempkin derived a model assuming that as he loaded up the surface with adsorbate, the heat of adsorption of all the molecules in the layer would decrease linearly with coverage due to adsorbate/adsorbate interactions:

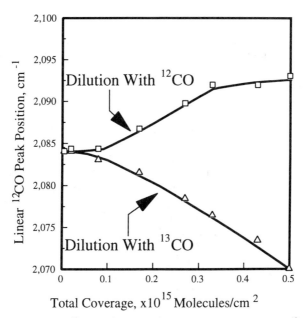

Figure 4.9 The position of the ^{12}CO band as a function of coverage when (a) ^{12}CO or (b) ^{13}CO is added to the layer. (Data of Olsen and Masel [1988].)

$$\Delta H_{ad} = \Delta H^0_{ad}(1 - \alpha_T\theta) \tag{4.39}$$

where α_T is a fitting parameter. Temkin then assumed that the Langmuir adsorption isotherm still applied to the adsorbed layer, except that K^A_{equ} varies with coverage, as follows:

$$K^A_{equ} = K^{A,0}_{equ}e^{-(\Delta H^0_{ad}\alpha_T\theta/kT)} \tag{4.40}$$

Rearranging Equation 4.10 yields

$$K^A_{equ}P = \frac{\theta}{1 - \theta} \tag{4.41}$$

Substituting the expression for K^A_{equ} from Equation 4.40 into Equation 4.41 yields

$$\ln{(K^{A,0}_{equ}P_A)} = \frac{\Delta H^0_{ad}\alpha_T\theta}{kT} + \ln{\left(\frac{\theta}{1 - \theta}\right)} \tag{4.42}$$

Tempkin did most of his work at coverages (θ) of about 0.5. As a result he assumed that the second term on the right of Equation 4.42 was negligible. In that case,

$$\ln{(K^{A,0}_{equ}P_A)} = \frac{\Delta H^0_{ad}\alpha_T\theta}{kT} \tag{4.43}$$

Equation 4.43 is called the Tempkin adsorption isotherm. It is also called the Tempkin–Frumkin isotherm, since it is equivalent to a form suggested by Slygin and Frumkin [1935].

4.5.2 The Fowler Adsorption Isotherm

Fowler [1935] derived Equation 4.42 using a somewhat different formalism. His approach was to assume that there were direct rather than indirect adsorbate/adsorbate interactions. If there was a molecule on one site, the heat of adsorption on an adjacent site would be reduced by an amount $\Delta H_{ad}^{0}\alpha/2$ where α is positive for repulsive interactions and negative for attractive interactions. Fowler then noted that since there are attractive or repulsive interactions between molecules on adjacent sites, the probability of filling a given site should depend on whether the sites adjacent to the given site are filled or empty. However, he could not account for that variation computationally. As a result, he made what is called a mean field approximation, i.e., he assumed that the main effect of the repulsive interaction will be to vary the heat of adsorption of an average molecule in the layer. This approach is described in greater detail in Section 4.9. However, qualitatively, if a molecule on site 1 reduces the heat of adsorption on site 2 by $\Delta H_{ad}^{0}\alpha/2$, then if there is a molecule on site 2, the binding on site 1 will also be reduced by $\Delta H_{ad}^{0}\alpha/2$. Therefore, the total heat of adsorption would be reduced by $\Delta H_{ad}^{0}\alpha$. The probability that a site is occupied is θ. Therefore, if the coverage is θ, the heat of adsorption of an incoming molecule will, on average, be reduced by $\Delta H_{ad}^{0}\alpha\theta$. Notice that according to Equation 4.39 the heat of adsorption will be reduced by $\Delta H_{ad}^{0}\alpha\theta$. Hence, Equation 4.39 also applies to Fowler's derivation. Equations 4.40 and 4.41 apply as well, under the assumption that the probability of a given site being occupied is independent of whether adjacent sites are occupied. Therefore, Equation 4.42 should apply too. Equation 4.42 has come to be called the Fowler adsorption isotherm even though Fowler [1939] called it a crude approximation.

4.5.3 Other Analytical Adsorption Isotherms

There have been many other attempts to put adsorbate/adsorbate interactions into Langmuir-like models. For example, Halsey and Taylor [1947] rederived the Freundlich adsorption isotherm by assuming that the heat of adsorption varied according to

$$\Delta H_{ad} = \Delta H_{ad}^{0} \ln (\theta_A) \qquad (4.44)$$

See Tompkins [1978] for details. Brunauer, Love, and Keenan [1942] and Sips [1948] derived the Slygin–Frumkin [1935] adsorption isotherm

$$\theta_A = \frac{1}{K_{sf}} \ln (\alpha_{sf} P_A) \qquad (4.45)$$

by assuming

$$D(E) = \begin{cases} D & \text{if } E_{\min} < E < E_{\max} \\ 0 & \text{otherwise} \end{cases} \qquad (4.46)$$

and eliminating some of the terms in the resulting expression. Brunauer [1945] and Trapnell and Hayward [1964] describe several other adsorption isotherms. However, none of them fits data better than the ones described so far in this chapter.

4.5.4 Summary of Analytical Models

Equations 4.33, 4.35, 4.38, and 4.43 have been used extensively in the literature. Their main advantage is that they contain more parameters than the Langmuir adsorption isotherm. Hence, they can fit an adsorption isotherm more accurately than the Langmuir adsorption isotherm. Also, on a rough porous material like that in Figure 2.5, there really is reason to believe that there are many different types of sites to hold gas. Hence, if one is trying to fit data taken on a rough surface, there are some physically reasonable arguments why one should use equations 4.33, 4.35, 4.38, and 4.43.

The problem, however, is that Equations 4.33, 4.35, 4.38, and 4.43 do not work as well for adsorption on a highly ordered surface. For example, it is useful to look back at the data for argon and krypton adsorption on graphitized carbon black shown in Figures 4.2 and 4.3. Notice that the adsorption isotherm for krypton in Figure 4.3 shows a distinct change in slope at $\frac{1}{2}$ a monolayer, while the adsorption isotherm for argon at 91.3 K in Figure 4.2 shows changes in slope at $\frac{1}{3}$ and about $\frac{3}{4}$ of a monolayer. Figure 4.10 shows a plot of some typical isotherms calculated from Equations 4.33, 4.35, 4.38, and 4.43. Note that the changes in slope seen in Figures 4.2 and 4.3 are not reproduced by Equations 4.33, 4.35, 4.38, and 4.43. One can fit the changes in slope by modifying Equation 4.38, assuming that there are multiple sites on the graphitized carbon surface to hold gas with complex kinetics. However, Dash [1975] has shown that the changes in slope in Figure 4.3 occur under conditions where the graphite is completely ordered, and the krypton always adsorbs on the same site. Thus, there are some important features of adsorption isotherms on highly ordered surfaces that are not reproduced by any of the adsorption isotherms that were discussed so far in this chapter.

The ammonia data in Figure 4.2 are also not reproduced in detail by any of the equations so far in this chapter. As noted previously, ammonia does not interact strongly with graphite, and so initially there is little adsorption. However, once one gets a small patch of ammonia, additional ammonia can hydrogen bond to the patch, and so the patch grows. The result is that the adsorbate overlayer looks like that in Figure 4.11, where some patches of the surface are completely saturated with adsorbate, while other patches are almost adsorbate free. The patches of adsorbate are called islands. One can try to fit the data in Figure 4.2 without considering the patches by, for example, assuming that the adsorbate actually is adsorbing in small pores. However, when ammonia adsorbs on a flat surface the isotherm still looks like

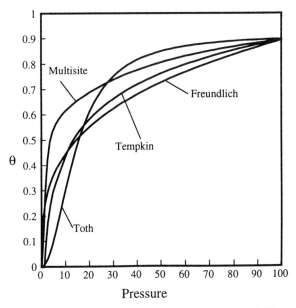

Figure 4.10 A plot of some typical isotherms calculated from Equations 4.33, 4.35, 4.38, and 4.43.

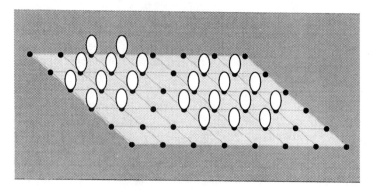

Figure 4.11 A schematic of island formation on a square surface.

that in Figure 4.2. Again that is behavior that is not reproduced by any of the models that have been discussed so far in this chapter.

Years ago, none of these issues would have been a problem because people were working with such rough surfaces that the effects of disorder swamped all effects discussed in this section. However, in the last 20 years, adsorption on ordered surfaces has become increasingly important. Highly ordered zeolites and graphitized carbons are now used as selective adsorbants. Film growth reactions are run on single-crystal semiconductors. The technology is shifting to highly ordered adsorbants. As a result, the inability of Equations 4.33, 4.35, 4.38, and 4.43 to reproduce data on highly oriented surfaces is a serious limitation.

The reason that Equations 4.33, 4.35, 4.38, and 4.43 fail is that they do not consider how the interactions between various molecules of adsorbate affect the arrangement of the adsorbed layer. Equations 4.33, 4.35, 4.38, and 4.43 assume that the adsorbate randomly populates the available adsorption sites. That assumption is good when there is a random adsorption process as shown in Figure 3.17. However, in reality one often observes ordered adsorption or islands as shown in Figure 3.17. Those ordered structures are not considered in the derivation of Equations 4.33, 4.35, 4.38, and 4.43.

Physically, when there are attractive adsorbate/adsorbate interactions the adsorbing molecules will tend to cluster together. The adsorbate clusters will attract extra adsorbate, which leads to extra adsorption. Similarly, if there are repulsive adsorbate/adsorbate interactions, the molecules will tend to separate. That leads to modifications of the adsorption isotherm. In either case, we get behavior that is not reproduced by the models discussed up to this point in this chapter.

4.6 ADSORBATE PHASE BEHAVIOR

The tendency for molecules to cluster and/or separate also has some other important consequences. In Section 3.2 we discussed CO adsorption on Pt(111). We noted that Crossley and King [1980] found that at low coverages the CO goes to an on-top site in a so-called $(\sqrt{3} \times \sqrt{3}\ R\ 30°)$ structure illustrated in Figure 3.14. The $(\sqrt{3} \times \sqrt{3})$ structure fills up at a coverage of 0.33×10^{15} molecules/cm^2. However, at moderate temperatures additional COs can be added to the layer. The additional COs are accommodated by pushing some of the on-top COs in the $(\sqrt{3} \times \sqrt{3})$ structure to a bridgebound site and moving around the COs to free up other bridge bound sites for further adsorption. Crossley and King suggested that at a coverage of 0.5×10^{15} molecules/cm^2 the system forms an ordered C(4x2) overlayer, as illustrated in Figure 3.14. At higher coverages CO Pt(111) also shows a complex adsorbate arrangement that was originally assumed to be a compression layer. However, CO is now thought to form the so-called domain wall structure shown in Figure 3.15.

Most atomic (e.g., O_{ad}, N_{ad}, S_{ad}, Se_{ad}, Au_{ad}) and diatomic adsorbates (e.g., O_2, NO) behave similarly to CO. When the adsorbate first adsorbs, it adsorbs randomly. However, as the coverage increases one begins to see ordered overlayers, at least at low temperatures. The structure changes as more gas adsorbs. However, at least at low temperatures, most simple adsorbates show ordered atomic arrangements. Hence ordering of the adsorbate is basically a universal phenomenon that does not depend on the nature of the adsorbate.

It is useful to consider a simple example to try to understand why the adsorbate tends to form an ordered arrangement. Consider adsorption of a gas onto the square lattice shown in Figure 4.12. Assume that the gas molecules bind to the fixed lattice sites indicated as small black circles in Figure 4.12. Further, assume that the interaction between the gas molecules is such that adjacent gas molecules can get no closer than the gray circles in Figure 4.12.

Figure 4.12 illustrates how the adsorbate arrangement changes as the adsorbate coverage increases. At low coverage, the adsorbate can adsorb randomly, as shown in Figure 4.12a. However, once gas adsorbs on a given site, there will not be room to adsorb gas on the sites that are directly to the left or right or up or down from the given site. Thus, those adjacent sites are blocked. There is room for gas to adsorb on the adjacent sites that are at a 45° angle from the given site. As a result, once a given site is filled, one can only add further gas to the sites that are at a 45° angle from the given site. Now consider adding gas randomly to the 45° sites. At first there is no problem. However, eventually one gets to a situation like that in Figure 4.12b where there are gaps in the adsorbate layer. However, one can verify that all of the vacant sites are blocked. At that stage, there are patches of gas on the surface that have some local order, but the coverage is less than saturation. In order to add more gas, one has to find ways to consolidate the gaps. That can happen by adsorbate molecules hopping from the edge of one ordered patch on the surface to the next one. Eventually the surface is completely ordered, as shown in Figure 4.12c.

This example shows that if large molecules that bind to specific sites are adsorbed, and the molecules block adjacent sites, the adsorbate layer will order at moderate coverages.

One can extend these ideas to molecules that can get closer than those in Figure 4.12. Consider, for example, the adsorption of an adsorbate that is small enough to fit on a series of adjacent sites, but the adsorbate/adsorbate interactions are repulsive so that the adsorbate molecules would prefer to not adsorb on adjacent sites. In that case, adsorbate molecules would still lower their energy if they adsorb on the sites on the 45° positions. At low temperatures, the adsorbate layer will go to these low-energy configurations. As a result, at low temperatures we would still get ordered adsorbate overlayers.

Of course, at higher temperatures, other configurations will contribute. As a result, the adsorbate phase behavior is temperature dependent. We quantify these ideas in the next section. Nevertheless, the thing to remember for now is that interactions between adsorbate

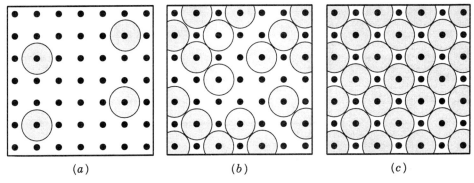

(a) (b) (c)

Figure 4.12 Schematic of the adsorbate arrangement that forms when a gas adsorbs onto a square lattice. The small dark circles are possible lattice sites. The large dotted circles are adsorbate molecules.

molecules on adjacent sites can cause the adsorbate layers to order and that the phase behavior is temperature dependent.

Experimentally, adsorbate ordering is a fairly universal phenomenon for adsorption of simple gases on smooth highly ordered surfaces at moderate temperatures. Ordering occurs whenever molecules on adjacent sites interact. Most adsorbates show suitable interactions. Adsorbate ordering is less important for microscopically rough surfaces, however, because the roughness tends to disrupt the local ordering of the adsorbate.

Over the years many investigators have tried to represent adsorbate/adsorbate phase behavior on a phase diagram where the structure of the adsorbate overlayer as a function of temperature and pressure, as indicated in Figure 4.13. One usually designates individual phases on the phase diagram using Wood's notation defined in Section 2.7 and Example 2.C. Recall that Wood's notation uses two indices (ixj), where i and j are used to specify the ratio of the lengths of the lattice vectors for the adsorbate unit cell and those for the unit cell of the underlying lattice. For example, a (3x2) overlayer would have a unit cell that is three times as long and two times as wide as the unit cell of the underlying (i.e., metal surface) lattice. Wood's notation also often includes the letter p or c to indicate whether the unit cell for an overlayer is a primitive unit cell or a conventional unit cell. For example, the overlayer in Figure 4.12c has a c(2x2) structure. Several other examples of ordered overlayers for square lattices are given in Figure 4.22. (See Example 2.C for additional examples.)

Figure 4.13 shows a phase diagram for oxygen adsorption on W(110), which was adapted from a phase diagram presented by Lagally, Wu, and Yang [1980]. Notice that one observes a series of different phases as one changes the coverage and temperature. At low temperatures the oxygen forms a P(2x1) structure at half coverage and a P(1x1) structure at full coverage. At coverages below 0.5 one observes a layer with some P(2x1) character, but it is very disordered. Similarly, at coverages between 0.5 and 1.0 the layer has a mixture of P(2x1) and P(1x1) domains. There is also some evidence for some local P(2x2) structure near a coverage of 0.75. It happens that the adsorbate arrangement changes as the temperature increases. Eventually, the phase domains disappear. In their paper, Lagally et al. represented this information as a phase diagram with a series of sharp phase boundaries to represent the idea that they have separate phases on the surface.

It happens that while this particular system does exhibit individual phase domains at

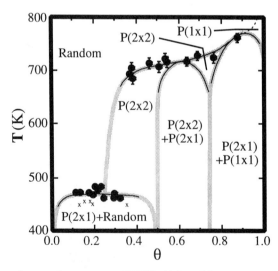

Figure 4.13 A phase diagram for oxygen on W(110). (Adapted from one presented by Lagally et al. [1980].)

intermediate coverages, it is an exception. Experimentally, one hardly ever observes sharp phase transitions. Rather, one usually observes mixed behavior where one kind of phase grows in and another decays. Calculations show that if one examines the surface at some intermediate stage, one does not usually see individual phases. Rather, one observes a random distribution of molecules where some of the adjacent sites on the (1x1) lattice are occupied and some are empty. However, there is some probability that the local configuration around a given surface atom is (1x1), (2x1), or (2x2). It is useful to discuss these properties in terms of an order parameter, which is a quantitative measure of how well the local arrangement of the molecules on the surface is represented by P(2x1), P(1x1), or P(2x2) symmetry. A precise definition of the order parameter is given in Example 4.F. It happens that people often represent the information in these more complex systems with a phase diagram, but the notation is often confusing since separate phases do not usually occur on the surface. Still, people use these diagrams, so one needed to be included in this manuscript. Hence, one of the few examples that shows real multiphase behavior was included in Figure 4.13. The figure shows shaded lines instead of sharp phase boundaries, because even in this system one does not usually observe sharp phase transitions.

In a more global picture, one can show that the changes in the structure of the adsorbed layer are analogous to solid–solid phase transitions. For example, when iron is heated to 1060°C it undergoes a first-order phase transition from a body-centered cubic (BCC) structure to a face-centered cubic (FCC) structure. Such a change is very analogous to the change from a P(2x1) to P(2x2) structure, which occurs when one puts a third of a monolayer of oxygen on W(110) and then heats. There is one fundamental difference between the phase behavior of iron and the phase behavior of an adsorbate: there is a sudden change in the structure of the iron when it is heated (i.e., there is a first-order phase transition), while a continuous, second-order transition is seen with oxygen on W(110). However, qualitatively both changes are phase transitions, and so we can get some useful information about the structural rearrangements from what is known about solid–solid phase transitions.

Experimentally, the phase transitions substantially alter adsorption isotherms. During a phase transition the coverage changes rapidly with changing pressure. Much slower changes in coverage with pressure are seen at coverages just above or below the phase transition. The change from a slow variation to a rapid one is quite important if one wants to model the adsorption process in detail. It is responsible, for example, for the changes in slope in Figure 4.3. Hence, an understanding of the phase transitions is quite important to modeling adsorption isotherms.

The phase transitions are also important to reactions. When we are running a reaction, the reaction rate will be quite different when the reactants are condensed into distinct islands than when there is a random distribution of molecules on the surface. During a phase transition, the ordering of the molecules on the surface changes. Hence, there is a large change in the reaction rate. None of the analytical models described earlier can reproduce the phase behavior. As a result, none of the models in section Figure 4.5 do a good job of reproducing some important features of actual adsorption systems.

4.7 LATTICE GAS MODELS

Lattice gas models were developed to try to model the phase behavior of the adsorbed layer. The lattice gas model starts with the same model used for the derivation of the Langmuir adsorption isotherm. It is assumed that there is a series of sites on a solid surface that is available to adsorb gas, and that adsorption occurs when a gas molecule A_g reacts with a bare site, S, to yield an adsorbed complex

$$A_g + S \rightleftharpoons A_{ad} \qquad (4.1)$$

However, in the lattice gas model one looks at each possible arrangement of the adsorbate, and adds up the effects of the local interactions between adjacent adsorbate molecules. Methods from statistical mechanics are then used to calculate the adsorption isotherm and the structure of the adsorbed phase. Physically, the lattice gas model differs from the models in Section 4.5 in that while the models in Section 4.5 treat the adsorbate/adsorbate interactions as though they are only a function of coverage, the lattice gas model considers the fact that even at a given coverage, different arrangements of surface atoms have different numbers of nearest neighbor interactions, and therefore different energies. This allows the lattice gas models to consider the influence of the adsorbate arrangement on the adsorption isotherm. Thus the lattice gas models are better able to model adsorbate/adsorbate phase behavior than the models in Section 4.5.

Fowler [1935] provided an initial description of the properties of a lattice gas, and Hill [1952] expanded the results. Much of the work was qualitative until about 1980, when improvements in computers allowed lattice gas models to be used on a routine basis. Tovbin [1991] and Persson [1992] review much of the work. The purpose of the next section is to derive equations for the adsorption behavior of a lattice gas. More details are given in Hill [1960], Binder [1984], and Tovbin [1991]. There is an equivalent problem called the Ising problem. It is described quite well by Feynman [1972], and Plische and Bergeren [1989].

4.7.1 Derivation of the Basic Equations

In this section we use statistical mechanics to derive the basic equations for the phase behavior of a lattice gas. The reader may want to review the material in Appendix A before proceeding.

First we need to introduce some notation. In Sections 4.3.1 and 4.3.1.1 we calculated the isotherm from Q_{canon}^N, the canonical partition function; i.e., the partition function for N molecules arranged on the layer. However, for the derivation to follow it will be more convenient to work with Q_{grand} the grand-canonical partition function where the grand-canonical partition function is the partition function for all possible states of the layer including all possible numbers of adsorbed molecules. Q_{grand} differs from Q_{canon}^N in that Q_{grand} includes all possible arrangements of any number of adsorbed molecules, while Q_{canon}^N only considers the arrangements of N adsorbed molecules. Q_{grand} is given by

$$Q_{grand} = \sum_m e^{-\beta E_m} \tag{4.47}$$

where m is an index of all possible states of the adsorbed layer; E_m is the energy of that state; and $\beta = 1/kT$.

In principle the index m should go over all of the states of the system (i.e., all the arrangements of the adsorbed molecules and all of the internal modes of the molecules). However, if we assume that the adsorbed molecules are identical (i.e., that there is only one type of adsorption site), then we can separate the sum in Equation 4.47 into two terms: a sum over n, all of the arrangements of the adsorbed layer, and a sum over I, the index of all possible internal states of the adsorbed molecules as indicated in Appendix A. One can then define the partition function for all the arrangements of the adsorbed layer, Q_S, by

$$Q_S = \sum_n g_n e^{-\beta E_n} \tag{4.48}$$

where E_n = the expectation value of the free energy when the adsorbate layer is in arrangement n.

In the derivation to follow it is useful to number all of the sites on the surface by a single index i and define an occupancy number ε_i where $\varepsilon_i = 0$, if the site is empty, and $\varepsilon_i = 1$, if the site is occupied. If one assumes that the adsorbed molecules interact via a pairwise additive potential, with two-body but no three-body interactions, and only direct interactions between adsorbate molecules, then one can compute the energy, E_n, of any state of the system, n, by

$$-E_n = H_0 + H_1 \sum_i \varepsilon_i + \sum_i \sum_{j>i} h_{ij} \varepsilon_i \varepsilon_j \tag{4.49}$$

where

H_0 = the free energy of the surface in the absence of the adsorbate
H_1 = the free energy of an adsorbed molecule
h_{ij} = the interaction energy between a molecule on site i and a molecule on site j; a positive h_{ij} corresponds to an attractive interaction, while a negative h_{ij} corresponds to a repulsive interaction

One can show from statistical mechanics that if the adsorbate adsorbs nondissociatively

$$H_1 = \mu_S + kT \ln (q_s) = kT \ln (K_{equ}^{A=0} P_A) \tag{4.50}$$

Note that Equation 4.49 is an approximation that has ignored three-body and four-body interactions, (i.e., terms in $\varepsilon_i \varepsilon_j \varepsilon_k$ and $\varepsilon_i \varepsilon_j \varepsilon_k \varepsilon_l$), and so while it usually works quite well, it is not exact.

One can show, $h_{ij} = h_{ji}$. Therefore

$$-E_n = H_0 + H_1 \sum_i \varepsilon_i + \tfrac{1}{2} \sum_i \sum_j h_{ij} \varepsilon_i \varepsilon_j \tag{4.51}$$

Equations 4.49 and 4.51 are the key equations for adsorption of a gas onto a two-dimensional lattice. According to Equations 4.49 and 4.51 the total energy has three terms, a term for the bare lattice, a term proportional to $\sum_i \varepsilon_i$, and some cross terms. Note that

$$\sum_i \varepsilon_i = S_0 \theta \tag{4.52}$$

Therefore, the second term in Equations 4.49 and 4.51 accounts for the variation in the free energy with coverage. The third term in Equations 4.49 and 4.51 is more complex. In the discussion that follows, we show that h_{ij} is different if sites i and j are adjacent to one another rather than if sites i and j are separate. As a result, the last term in Equation 4.51 varies with the arrangement of the adsorbed layer (see Example 4.E for further details). The variation in the free energy of the adsorbed layer with changing adsorbate arrangement causes certain adsorbate arrangements to have lower free energy than others. The lowering of the surface free energy allows ordered adsorbate arrangements to form.

One can calculate the adsorbate arrangement at 0 K, using a simple analysis described in Example 4.E. However, an elaborate analysis is needed at higher temperatures. We usually teach the analysis in our graduate thermodynamics and statistical mechanics course, not in our surface science course. The material, however, does not appear in a coherent way in any textbook. As a result, we present the material here. Still, the reader can skip to Section 4.8 without any loss of coherence.

To proceed with the analysis one needs to consider which interactions to include in the calculations. First, it is useful to develop some notation. We are going to want to calculate an isotherm for adsorption on a two-dimensional lattice, so consider an atom adsorbed on the site labeled i on the square lattice shown in Figure 4.14. The atom on site i interacts most

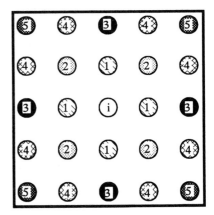

Figure 4.14 The (1) nearest neighbors, (2) next nearest neighbors, (3) third nearest neighbors, etc., for atom i on a square lattice.

strongly with the atoms on the sites that are labeled 1. Those sites are called first nearest neighbor sites. Often there are also interactions with the atoms labeled 2. Those sites are called the next nearest neighbor sites. In a similar way, there can be interactions with the third, fourth, and fifth nearest neighbor sites. In principle, there can also be interactions with molecules farther away from i. However, these interactions are usually ignored.

There are two kinds of adsorbate/adsorbate interactions that need to be considered: direct adsorbate/adsorbate interactions and indirect adsorbate/adsorbate interactions. Direct interactions are usually van der Waals-type interactions. Molecules repel if they are packed together so their van der Waals radii overlap. The molecules attract at larger distances. Generally, the direct attractive forces are rather weak, although they can still be important at low temperatures.

The indirect interactions are more complex. Indirect interactions arise because the presence of an adsorbate perturbs the surface, which in turn perturbs the adsorption of further adsorbate molecules. For example, consider the adsorption of an adsorbate on a jellium surface. For the moment assume that the adsorbate attracts electrons to form a local bond with the surface. When the adsorbate attracts electrons, the electrons will be depleted for some distance around the adsorbate. That produces a net positive charge, which then attracts additional electrons. The effect is that one then gets an oscillation in electron density, as indicated in Figure 4.15. The oscillation in electron density is called a *Friedel oscillation*.

Now consider adsorbing a second molecule onto the surface. If the second molecule adsorbs at one of the peaks in the electron density, the Friedel oscillation from the second

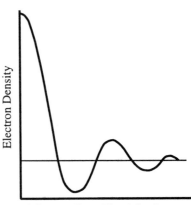

Figure 4.15 A schematic of the Friedel oscillation created when a molecule adsorbs on a jellium surface.

molecule will reinforce the Friedel oscillation from the first molecule. The second molecule will see an enhanced interstitial electron density. According to the analysis in Section 3.10.1 the heat of adsorption of an electron withdrawing species will be enhanced when the interstitial electron density around the atom increases. In effect, then, the second molecule will be attracted to sites where there are local maxima in the Friedel oscillations so h_{ij} will be positive. Another effect is that the presence of the second molecule will enhance the Friedel oscillations on the first molecule so the heat of adsorption of the first molecule will be enhanced. That also produces a positive h_{ij}. In contrast, if the second molecule adsorbs into the valleys in electron density caused by the Friedel oscillations on the first molecule the opposite arguments will apply. The Friedel oscillations will conflict, which produces a negative h_{ij}.

Notice that one gets hills and valleys in the Friedel oscillations. One gets positive h_{ij} at the hills, and negative h_{ij} at the valleys. As a result, h_{ij} does not show simple monotonic behavior moving away from the adsorption site.

There is complication in real systems because the strength of the Friedel oscillation varies with crystallographic direction. For example, Figure 4.16 shows how the electron density of a Mo(100) surface varies with direction when sulfur adsorbs on the Mo(100) surface. Notice that the Friedel oscillation is stronger along the [100] direction than along the [110] direction. This leads to anisotropies in the h_{ij}. Additional examples showing such anisotropies can be found in Feibelman [1989] and White and Akhter [1988].

In the case of a real surface, there is an additional complication because surface atoms can move during the adsorption process. A classic example is CO adsorption on Pt(100) where the surface changes from a (5x20) to (1x1) reconstruction when CO adsorbs. The reconstruction occurs because the heat of adsorption of CO is much larger on the (1x1) regions of the surface. As a result, the COs tend to cluster onto (1x1) islands on a (5x20) surface, so the net adsorbate/adsorbate interaction is extremely positive.

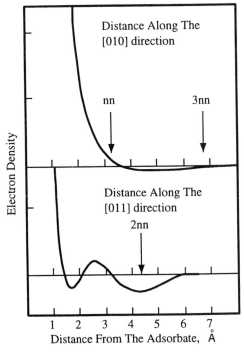

Figure 4.16 The Friedel oscillations seen when sulfur adsorbs on Mo(100). (Adapted from Gellman et al. [1987].)

So far, most of the data showing attractive adsorbate/adsorbate interactions due to surface reconstructions has been taken on metal surfaces. However, one usually observes surface reconstructions when gases adsorb on semiconductors. Calculations indicate that the reconstructions should affect the heat of adsorption on adjacent sites (see Rockett [1990] or Harris [1992] for details). Hence, calculations suggest that the indirect interactions due to surface reconstructions are going to be quite important during adsorption on semiconductors. There is not yet any data to check the calculations, however.

4.7.2 Conversion to the Ising Model

The next several sections outline the methods that are used to calculate an adsorption isotherm, assuming that all of the h_{ij} are known. Experimentally, they are rarely known. However, we will find that if we know the adsorbate phase behavior, we can back calculate the h_{ij}.

First, it is useful to change notation slightly. It happens that there is a problem, called the *Ising problem* whose solution appears in the literature. The Ising problem has the same Hamiltonian as that in Equation 4.51 except that the occupancy number is defined by

$$\xi_i = 2\varepsilon_i - 1 \tag{4.53}$$

Hence, if we substitute Equation 4.53 into Equation 4.51, we convert the solution of the problem to one that has already been solved, and we can just look up the solution. In this notation, $\xi_i = 1$ corresponds to a filled site, $\xi_i = -1$ corresponds to an empty site, and $\Sigma \xi_i = 0$ corresponds to a half-filled surface. Making the substitution,

$$-E_n - H_0 + H_1 \sum_i \left(\frac{\xi_i + 1}{2}\right) + \frac{1}{2} \sum_i \sum_j \frac{(\xi_i + 1)(\xi_j + 1)}{4} h_{ij} \tag{4.54}$$

Expanding out the sums

$$-E_n = H_0 + H_1 \sum_i \left(\tfrac{1}{2}\right) + \tfrac{1}{8} \sum_i \sum_j h_{ij} + \tfrac{1}{2} \sum_i \xi_i \left(H_1 + \tfrac{1}{2} \sum_j h_{ij}\right) + \tfrac{1}{8} \sum_i \sum_j h_{ij} \xi_i \xi_j \tag{4.55}$$

In the material that follows it is demonstrated that in this representation

$$\sum_n g_n e^{-\beta E_n} = \sum_{\xi_1 = \pm 1} \sum_{\xi_2 = \pm 1} \cdots \sum_{\xi_{S_0} = \pm 1} e^{-\beta E_n} \tag{4.56}$$

If all sites are equivalent, the *i*th site will be equivalent to the first site. Therefore, $\sum h_{ij} = \sum_j h_{1j}$. In the work to follow it will be useful to define a constant h_0, by

$$\sum_j h_{ij} = 4h_0 \tag{4.57}$$

The factor of 4 was added to simplify algebra later in this section. Physically, h_0 is a measure of the strength of the adsorbate/adsorbate interactions.

Substituting Equation 4.57 into Equation 4.56 and noting that $\sum_i 1 = S_0$ yields

$$-E_n = H_0 + \frac{S_0}{2}(H_1 + h_0) + \frac{(H_1 + 2h_0)}{2} \sum_i \xi_i + \frac{1}{4} \sum_i \sum_{j>i} \xi_i \xi_j h_{ij} \tag{4.58}$$

where, as before, S_0 is the number of sites available to adsorb gas. We can rewrite Equation 4.58 as

$$-E_n = H_{S_0} + F \sum_i \xi_i + \frac{1}{4} \sum_i \sum_{j>i} h_{ij} \xi_i \xi_j$$

$$H_{S_0} = H_0 + \frac{S_0}{2}(H_1 + h_0) \qquad (4.59)$$

with

$$F = \frac{(H_1 + 2h_0)}{2}$$

Note H_{S_0} is the average energy of a randomly populated half-filled surface, $2F$ is the extra energy gained if molecules are randomly added to a half-filled surface, and h_{ij} is the interaction energy between molecules on site i and j. One can show from statistical mechanics that

$$2\beta F = \ln (K_{equ}^{\theta=0.5} P_A) \qquad (4.60)$$

Equation 4.59 is a standard Hamiltonian (i.e., energy expression) in statistical mechanics. It was first discussed by Ising [1925] for magnetic materials. In a ferromagnet, there is a lattice of spins that can interact with an external magnetic field. The spins can be oriented in either the positive or the negative direction with respect to the magnetic field. One can show that the Hamiltonian for the magnet is the same as that for a lattice gas. One can also derive similar Hamiltonians for the composition in an alloy, or the configuration of a polymer chain (see Hill [1960] or Feynman [1972] for details).

The Ising problem has been studied extensively in the physics literature. It was solved exactly in one dimension as outlined below. Onsanger [1944] solved the problem exactly in two dimensions when $\theta = 0.5$ or $F = 0$. In the remainder of this section we present a solution of the Ising problem using the methods of Feynman [1972], Schultz, Mattis and Lieb [1964], Pilschke and Bergersen [1989], and Binder [1984]. The reader should refer to these references for details.

4.7.3 Overview of the Solution Methods

In the next three sections we discuss several methods that have been used to calculate adsorption isotherms for lattice gases. First, we describe matrix methods, and show that we can compute the partition function for the lattice gas as the trace, Tr, of a matrix $\hat{V}^N$.

$$Q_{Grand} = Tr[\hat{V}^N] \qquad (4.61)$$

We also show that the coverage can be written as the trace of a second matrix, $\hat{V}_1 \hat{V}^{n-1}$:

$$\theta = \langle \epsilon_1 \rangle = \frac{1}{Q_{Grand}} Tr[\hat{V}_1 \hat{V}^{N-1}] \qquad (4.62)$$

Detailed expressions for the various matrices are given in Sections 4.7.5 and 4.7.6, and Example 4.H. We then use the matrix methods to calculate the properties of adsorption isotherms.

We find that the matrix methods are very useful for step surfaces, but less useful for flat surfaces. We then describe some Monte Carlo methods that are very useful for flat surfaces. The Monte Carlo method is used to provide our key results.

Finally, we describe some approximation methods, called "the Bragg–Williams mean field method," "the Bethe–Peirls cluster methods," and "the quasichemical method" and see when they are useful.

The following sections include detailed derivations of all of the key equations, so the material will require some patience. Nevertheless, this material is worth studying in detail. Matrix methods are widely used in statistical mechanics, and the partition function is often expressed as a trace of a matrix. Hence, it is useful to know how the key equations arise. Monte Carlo methods are also used very heavily throughout the statistical mechanics literature. Hence, they are very useful to understand. Mean field/cluster methods are becoming less important as computers improve. However, much of the older work was done with mean field/cluster methods, and cluster methods often allow one to derive useful approximations. Hence, they are included for completeness.

4.7.4 Exact Solution for the One-Dimensional Ising Model: $F = 0$

In order to illustrate the methodology used to solve the Ising problem, we are going to solve a simpler problem first, which is the adsorption of a gas onto a series of sites arranged into a one-dimensional lattice. The one-dimensional lattice can be a series of sites on the ridges of a stepped surface, or the binding sites on a long-chain polymer or protein. To simplify the algebra, we also assume that $h_{ij} = h$ if the sites are directly adjacent to each other and 0 elsewhere, that is,

$$h_{ij} = \begin{cases} h & \text{if } j = i \pm 1 \\ 0 & \text{if } j \neq i \pm 1 \end{cases} \tag{4.63}$$

Note $h_0 = 0.25\, n_n h$, where n_n is the number of sites adjacent to i ($n_n = 2$ for a line of sites). Substituting h_{ij} from Equation 4.63 into Equation 4.59 yields

$$-E_n = H_{S_0} + F \sum_i \xi_i + \frac{h}{4} \sum_i \xi_i \xi_{i+1} \tag{4.64}$$

First consider the case for $F = 0$ or $\theta = 0.5$. If $F = 0$, the second term in Equation 4.59 is zero. If $\theta = 0.5$, there are an equal number of filled and empty sites. Therefore

$$\sum_i \xi_i = 0 \tag{4.65}$$

In either case the second term on the right of Equation 4.59 vanishes. Equation 4.59 becomes

$$-E_n - H_{S_0} + \frac{h}{4} \sum_i \xi_i \xi_{i+1} \tag{4.66}$$

Note that in the $\theta = 0.5$ case the term in F dropped out of the equation because the energy of the half-filled surface is already in H_{S_0}.

In the next section we compute the partition function for the layer. According to Equation 4.48, the partition function can be computed as a sum over all of the arrangements of molecules in the layer, n. It is convenient to change variables from n, the arrangements of the layer, to $\xi_1, \xi_2, \ldots, \xi_{S_0}$, the occupancy of each site. Notice that for any arrangement of the adsorbed layer, values of $\xi_1, \xi_2, \ldots, \xi_{S_0}$ are fixed. If a given site is empty, ξ_i is -1, and $+1$ if the site is filled. Therefore, each arrangement of the surface layer will give a unique value of ξ_1, ξ_2, . . . , ξ_{S_0}. One can also go the other way, and use the values of ξ_1, ξ_2, . . . , ξ_{S_0} to describe the arrangement of all of the molecules within the layer. Therefore, if one wants to sum over all of the arrangements of the adsorbed layer, one can simply sum over all values of ξ_1, ξ_2, . . . , ξ_{S_0}. Equation 4.48 becomes

$$Q = \sum_{\xi_1 = \pm 1} \sum_{\xi_2 = \pm 1} \cdots \sum_{\xi_{S_0} = \pm 1} \exp(-\beta E_n) \qquad (4.67)$$

Substituting Equation 4.66 into Equation 4.67 yields

$$Q = \exp(\beta H_{S_0}) \sum_{\xi_1 = \pm 1} \sum_{\xi = \pm 2} \cdots \sum_{\xi_{S_0} = \pm 1} \exp\left(\frac{\beta h}{4} \sum_{i=1}^{S_0 - 1} \xi_i \xi_{i+1}\right) \qquad (4.68)$$

Note, the last sum only goes to $S_0 - 1$ because there is no $S_0 + 1$ site. Now we have to find a way to do the sum. It happens that the sum can be done directly. However, for the purposes of discussion, it is useful to do the sum using a recursion relationship. We will define a series of partition functions Q_1, Q_2, . . . , Q_{S_0} via

$$Q_j = \exp(\beta H_{S_0}) \sum_{\xi_1 = \pm 1} \sum_{\xi_2 = \pm 1} \cdots \sum_{\xi_j = \pm 1} \exp\left(0.25\ \beta h \sum_{i=1}^{j-1} \xi_i \xi_{i+1}\right) \qquad (4.69)$$

Note that

$$Q_{j+1} = \exp(\beta H_{S_0}) \sum_{\xi_1 = \pm 1} \sum_{\xi_2 = \pm 1} \cdots \sum_{\xi_j = \pm 1} \sum_{\xi_{j+1} = \pm 1} \exp\left(0.25\ \beta h \sum_{i=1}^{j} \xi_i \xi_{i+1}\right) \qquad (4.70)$$

Equation 4.70 has a term $\xi_i \xi_{i+1}$ rather than $\xi_{i+1} \xi_{i+2}$ since the sum in Equation 4.69 only goes to $j - 1$.
Rearranging

$$Q_{j+1} = \exp(\beta H_{S_0}) \sum_{\xi_1 = \pm 1} \sum_{\xi_2 = \pm 1} \cdots \sum_{\xi_j = \pm 1} \exp\left(0.25\ \beta\ h \sum_{i=1}^{j-1} \xi_i \xi_{i+1}\right)$$
$$\cdot \left\{ \sum_{\xi_{j+1} = \pm 1} \exp(0.25\ \beta h\ \xi_j \xi_{j+1}) \right\} \qquad (4.71)$$

Now consider the term in braces in Equation 4.71. If $\xi_j = 1$, we can write the two terms in the sum as

$$\sum_{\xi_{j+1} = \pm 1} \exp(0.25\ \beta h \xi_j \xi_{j+1}) = \exp[0.25\ \beta h(1)(1)] + \exp[0.25\ \beta h(1)(-1)]$$
$$= 2 \cosh(0.25\ \beta h)$$

If $\xi_j = -1$, we can write the two terms in the sum as

$$\sum_{\xi_{j+1} = \pm 1} \exp(0.25\ \beta h \xi_j \xi_{j+1}) = \exp[0.25\ \beta h(-1)(1)] + \exp[0.25\ \beta h(-1)(-1)]$$
$$= 2 \cosh(0.25\ \beta h)$$

Therefore, independent of whether $\xi_j = +1$ or -1

$$Q_{j+1} = 2 \cosh (0.25 \beta h) \left[\exp(\beta H_0) \sum_{\xi_1 = +-1} \sum_{\xi_2 = \pm 1} \cdots \sum_{\xi_j = \pm 1} \exp \left(0.25 \beta h \sum_{i=1}^{j-1} \xi_i \xi_{i+1} \right) \right]$$

(4.72)

The term in square brackets in Equation 4.72 is equal to Q_j. Therefore

$$Q_{j+1} = Q_j [2 \cosh (0.25 \beta h)]$$ (4.73)

Equation 4.73 is a recursion relationship that allows us to compute Q_{j+1} from Q_j. We use it by starting with Q_1, and using the recursion relationship to sequentially compute Q_2, then Q_3, $Q_4 \ldots Q_{S_0}$. The result is

$$Q_{S_0} = Q_1 [2 \cosh (0.25 \beta h)]^{S_0 - 1}$$ (4.74)

Substituting $j = 1$ into Equation 4.69 gives

$$Q_1 = 2 \exp (\beta H_{S_0})$$ (4.75)

Therefore

$$Q_{S_0} = 2 \exp (\beta H_{S_0}) [2 \cosh (0.25 \beta h)]^{S_0 - 1}$$ (4.76)

Equation 4.76 is the partition function for adsorption of gas onto a line of sites with $F = 0$, where F is the free energy of adsorption onto a half-filled surface.

The partition function defines all of the properties of the layer. For example, in Section 4.7.5 we show that the average number of adsorbed molecules in the layer, $\langle N \rangle$, is given by

$$\frac{1}{\beta} \left(\frac{\partial \ln (Q_{S_0})}{\partial H_1} \right)_{\beta, H_0, h_{ij}} = \langle N \rangle$$ (4.113)

where S_0 is the number of sites. Hence, one can calculate the adsorption isotherm once one knows the partition function. Substituting Equation 4.76 into Equation 4.113 shows that $\theta = 0.5$, independent of h for $F = 0$.

4.7.5 Exact Solution for the One-Dimensional Ising Model: $F \neq 0$

Conceptually, it is easy to generalize these results to other systems. For example, if one wants to calculate an adsorption isotherm for a two-dimensional lattice with complex adsorbate/adsorbate interactions, one can define a Hamiltonian like that in Equation 4.51. One can then use methods from statistical mechanics to calculate the partition function as a function of the partial pressure of the adsorbate. The coverage of the adsorbate can be calculated from Equation 4.113. A plot of the coverage versus adsorbate pressure is the adsorption isotherm.

The one difficulty with this procedure is that for all of the cases except a one-dimensional lattice with $F = 0$, the mathematics is more difficult. For example, in the derivation below we show that with a one-dimensional lattice with $F \neq 0$, we get a term like

$$\exp (\beta \xi_j [F + 0.25 h \xi_{j-1}])$$

which has a different value according to whether the $j - 1$ site is occupied or empty. Hence, one cannot compute the necessary sums as described in Section 4.7.4. In the next three sections we describe the principal methods one uses to calculate the partition functions for these more complex examples. The derivations in the next two sections require some familiarity with statistical mechanics, so the reader may want to review the material in Appendix A before proceeding. However, the final algorithm, described in Section 4.8 requires less background. The derivation is more difficult than the derivations in the previous sections. The methods described in these sections, however, are also applicable to a wide variety of other problems, including the phase behavior of alloys, colloids, polymers, liquid crystals, and solutions, the configurations of proteins, and the binding of biological molecules. Hence, the methods are worth learning. Still, the reader could skip down to Equation 4.115 without loss of continuity.

The derivation will consider adsorption of a gas onto a chain of S_0 sites where eventually we are going to let S_0 go to infinity. In order to simplify the algebra, we assume that the chain obeys periodic boundary conditions where the chain repeats every S_0 sites and the S_0th site is adjacent to the first site:

It is useful to rewrite the Hamiltonian as

$$-E_n = H_{S_0} + F \sum_i \left(\frac{\xi_i + \xi_{i+1}}{2} \right) + \frac{h}{4} \sum_{i=1}^{S_0} \xi_i \xi_{i+1} \tag{4.77}$$

Note that the last sum now goes to S_0 because when there are periodic boundary conditions the $S_0 + 1$ site can be occupied, while it could not have been occupied if there are only S_0 sites.

Substituting Equation 4.77 into Equation 4.67 yields

$$Q_{S_0} = \exp(\beta H_{S_0}) \sum_{\xi_1 = \pm 1} \sum_{\xi_2 = \pm 1} \cdots \sum_{\xi_{S_0} = \pm 1} \exp\left(\beta \sum_{i=1}^{S_0} \left[\left(\frac{h}{4}\right) \xi_i \xi_{i+1} + \frac{F}{2}(\xi_i + \xi_{i+1}) \right] \right)$$

$$\tag{4.78}$$

The first term in the Equation 4.78 is easy to evaluate. However, the sum requires more work. It is useful to concentrate on that term. Defining a modified partition function Z_{S_0} by

$$Z_{S_0} = \sum_{\xi_1 = \pm 1} \sum_{\xi_2 = \pm 1} \cdots \sum_{\xi_{S_0} = \pm 1} \exp\left(\beta \sum_{i=1}^{S_0} \left[0.25\, h\xi_i \xi_{i+1} + \frac{F}{2}(\xi_i + \xi_{i+1}) \right] \right) \tag{4.79}$$

Rearranging

$$Z_{S_0} = \sum_{\xi_1 = \pm 1} \sum_{\xi_2 = \pm 1} \cdots \sum_{\xi_{S_0} = \pm 1} \prod_{i=1}^{S_0} \exp\left(\beta \left[0.25\, h\xi_i \xi_{i+1} + \frac{F}{2}(\xi_i + \xi_{i+1}) \right] \right) \tag{4.80}$$

In order to evaluate this partition function, one needs to use a mathematical trick. The approach will be to find a matrix representation of the partition function and then use matrix algebra to evaluate the terms. Consider an individual term, $V_{\xi_i \xi_{i+1}}$, in the product above, where

$$V_{\xi_i \xi_{i+1}} = \exp\left(\beta\left[0.25\, h\xi_i \xi_{i+1} + \frac{F}{2}(\xi_i + \xi_{i+1})\right]\right) \tag{4.81}$$

One can define a transfer matrix $\hat{V}$ and four constants $V_{++}\, V_{+-}\, V_{-+}$ and V_{--} by:

$$\hat{V} = \begin{pmatrix} V(\xi_i = 1,\, \xi_{i+1} = 1) & V(\xi_i = 1,\, \xi_{i+1} = -1) \\ V(\xi_i = -1,\, \xi_{i+1} = 1) & V(\xi_i = -1,\, \xi_{i+1} = -1) \end{pmatrix} = \begin{pmatrix} V_{++} & V_{+-} \\ V_{-+} & V_{--} \end{pmatrix} \tag{4.82}$$

Combining Equations 4.81 and 4.82 yields

$$\hat{V} = \begin{pmatrix} \exp[\beta(0.25\, h + F)] & \exp[-0.25\,\beta h] \\ \exp[-0.25\,\beta h] & \exp[\beta(0.25\, h - F)] \end{pmatrix} \tag{4.83}$$

In the derivation that follows, the partition function for the adsorbed layer is computed via a recursion relation similar to that in Section 4.7.4. The analysis starts with a chain with only one site that interacts with itself via periodic boundary conditions and then calculates a partition function for that case. Next a chain containing two sites is considered and it is assumed that the second site interacts with the first via periodic boundary conditions. A partition function for that case is derived as well. A recursion relationship is then used to calculate a partition function for S_0 sites. That partition function is used to calculate an adsorption isotherm.

Consider the case of a chain with only one site that interacts with itself via periodic boundary conditions. In that case,

$$\xi_i = \xi_{i+1} = \xi_1 \tag{4.84}$$

The modified partition function becomes

$$Z_1 = \sum_{\xi_{i+1} = \xi_i = \pm 1} \exp\left(\beta\left[0.25\, h\xi_i \xi_{i+1} + \frac{F}{2}(\xi_i + \xi_{i+1})\right]\right) \tag{4.85}$$

Note that there are only two terms in the sum. The $\xi_i = \xi_{i+1} = +1$ and the $\xi_i = \xi_{i+1} = -1$ term. The $\xi_i = \xi_{i+1} = +1$ term is V_{++}, while the $\xi_i = \xi_{i+1} = -1$ term is V_{--}. Therefore,

$$Z_1 = V_{++} + V_{--} \tag{4.86}$$

Mathematicians call the sum of the diagonal elements of a matrix the trace (Tr) of the matrix. Notice that

$$Z_1 = \text{Tr}\,[\hat{V}] \tag{4.87}$$

Now consider having two elements in the chain, ξ_j and ξ_k, that interact back on themselves via periodic boundary conditions. In that case, the partition function Z_2 is given by

$$Z_2 = \sum_{\xi_j = \pm 1} \sum_{\xi_k = \pm 1} \exp\left(\beta\left[0.25\, h\xi_j \xi_k + \frac{F}{2}(\xi_j + \xi_k)\right]\right)$$
$$\cdot \exp\left(\beta\left[0.25\, h\xi_k \xi_{k+1} + \frac{F}{2}(\xi_k + \xi_{k+1})\right]\right)$$

with

$$\xi_{k+1} = \xi_j \tag{4.88}$$

In this case, there are four terms in the sum. One can show that they are

$$Z_2 = V_{++}V_{++} + V_{+-}V_{-+} + V_{-+}V_{+-} + V_{--}V_{--} \tag{4.89}$$

One can relate the expression in Equation 4.89 to the trace of $\hat{V}^2$. Matrix algebra shows that

$$\hat{V}^2 = \begin{pmatrix} V_{++}V_{++} + V_{+-}V_{-+} & V_{++}V_{+-} + V_{+-}V_{--} \\ V_{-+}V_{++} + V_{--}V_{-+} & V_{--}V_{--} + V_{-+}V_{+-} \end{pmatrix} \tag{4.90}$$

Notice that the four terms in Equation 4.89 all appear on the diagonal of $\hat{V}^2$. Therefore

$$Z_2 = \text{Tr}\,[\hat{V}^2] \tag{4.91}$$

One can generalize this result by noting that

$\hat{V}^{S_0}$

$$= \begin{pmatrix} \sum_{k=\pm1}\sum_{l=\pm1}\cdots\sum_{M=\pm1} V_{+k}V_{kl}V_{lm}\cdots V_{M+} & \sum_{k=\pm1}\sum_{l=\pm1}\cdots\sum_{M=\pm1} V_{+k}V_{kl}V_{lm}\ldots V_{M-} \\ \sum_{k=\pm1}\sum_{l=\pm1}\cdots\sum_{M=\pm1} V_{-k}V_{kl}V_{lm}\ldots V_{M+} & \sum_{k=\pm1}\sum_{l=\pm1}\cdots\sum_{M=\pm1} V_{-k}V_{kl}V_{lm}\ldots V_{M-} \end{pmatrix} \tag{4.92}$$

where M is the index of the S_0 site. According to Equation 4.79 the modified partition function is the sum of all of the $V_{12}V_{23}\ldots V_{M1}$ terms that have the same index at both sides of the product. Notice that all of the appropriate terms appear in the diagonal elements in Equation 4.92 and that there are no extra terms. Therefore,

$$Z_{S_0} = \text{Tr}\,[\hat{V}^{S_0}] \tag{4.93}$$

In order to do calculations with Equation 4.93 one needs an expression for the trace of $\hat{V}^{S_0}$. It can be obtained as follows. Assume that $\hat{V}$ has eigenvalues λ_1 and λ_2 and corresponding orthonormal eigenvectors:

$$\vec{\phi}_1 = \begin{pmatrix} \phi_{11} \\ \phi_{12} \end{pmatrix} \qquad \vec{\phi}_2 = \begin{pmatrix} \phi_{21} \\ \phi_{22} \end{pmatrix} \tag{4.94}$$

with $\vec{\phi}_1 \cdot \vec{\phi}_2 = 0$. Now define a conjugate matrix $\hat{\phi}$ by

$$\hat{\phi} = \begin{pmatrix} \phi_{11} & \phi_{21} \\ \phi_{12} & \phi_{22} \end{pmatrix} \tag{4.95}$$

Note that since the eigenvectors are orthonormal

$$\hat{\phi}\hat{\phi}^T = \begin{bmatrix} 1 & 0 \\ 0 & 1 \end{bmatrix} \qquad \hat{\phi}^T\hat{V}\hat{\phi} = \begin{bmatrix} \lambda_1 & 0 \\ 0 & \lambda_2 \end{bmatrix} \tag{4.96}$$

where $\hat{\phi}^T$ is the transpose of $\hat{\phi}$. As a result

$$\hat{V}^{S_0} = VV \dots V = (\hat{\phi}\hat{\phi}^T)V(\hat{\phi}\hat{\phi}^T)V(\hat{\phi}\hat{\phi}^T) \dots (\hat{\phi}\hat{\phi}^T)V(\hat{\phi}\hat{\phi}^T) \tag{4.97}$$

or

$$\hat{V}^{S_0} = \hat{\phi}(\hat{\phi}^T V \hat{\phi})(\hat{\phi}^T V \hat{\phi})(\hat{\phi}^T \dots \hat{\phi})(\hat{\phi}^T V \hat{\phi})\hat{\phi}^T \tag{4.98}$$

Combining Equations 4.96 and 4.98 yields

$$\hat{V}^{S_0} = \hat{\phi} \begin{pmatrix} \lambda_1 & 0 \\ 0 & \lambda_2 \end{pmatrix}^{S_0} \hat{\phi}^T = \hat{\phi} \begin{pmatrix} \lambda_1^{S_0} & 0 \\ 0 & \lambda_2^{S_0} \end{pmatrix} \hat{\phi}^T \tag{4.99}$$

But for any matrices $\hat{A}$, $\hat{B}$, and $\hat{C}$.

$$\mathrm{Tr}\,(\hat{A}\hat{B}\hat{C}) = \mathrm{Tr}\,(\hat{B}\hat{C}\hat{A}) \tag{4.100}$$

Therefore

$$Z_{S_0} = \mathrm{Tr}\,(\hat{V}^{S_0}) = \mathrm{Tr}\left[\hat{\phi} \begin{pmatrix} \lambda_1^{S_0} & 0 \\ 0 & \lambda_2^{S_0} \end{pmatrix} \hat{\phi}^T \right] = \mathrm{Tr}\left[\begin{pmatrix} \lambda_1^{S_0} & 0 \\ 0 & \lambda_2^{S_0} \end{pmatrix} \hat{\phi}^T \hat{\phi} \right] = \lambda_1^{S_0} + \lambda_2^{S_0} \tag{4.101}$$

One can determine the eigenvalues of $\hat{V}$ from

$$\det \begin{Vmatrix} V_{++} - \lambda & V_{+-} \\ V_{-+} & V_{--} - \lambda \end{Vmatrix} = 0 \tag{4.102}$$

Giving

$$\lambda_1 = \exp\left(\frac{\beta h}{4}\right) \cosh(\beta F) + \sqrt{\exp(0.5\,\beta h) \sinh^2(\beta F) + \exp(-0.5\,\beta h)} \tag{4.103}$$

$$\lambda_2 = \exp\left(\frac{\beta h}{2}\right) \cosh(\beta F) - \sqrt{\exp(0.5\,\beta h) \sinh^2(\beta F) + \exp(-0.5\,\beta h)} \tag{4.104}$$

Note that $|\lambda_1| > |\lambda_2|$. Therefore, in the limit that S_0 goes to infinity $\lambda_1^{S_0} \gg \lambda_2^{S_0}$. Therefore,

$$Z_{S_0} = \lambda_1^{S_0} \tag{4.105}$$

Substituting Equation 4.105 into Equation 4.78 yields

$$Q_{S_0} = \exp(\beta H_{S_0})\,\lambda_1^{S_0} \tag{4.106}$$

Equation 4.106 is the partition function for adsorption of a gas on a chain of S_0 sites.

The next stage is to derive an expression for the number of the molecules on the surface. Before we proceed, it is useful to note that we can rewrite Equation 4.51 as

$$-E_n = H_0 + H_1 N + \tfrac{1}{2} \sum_i \sum_j h_{ij}\varepsilon_i\varepsilon_j \tag{4.107}$$

where N is the number of adsorbed molecules. The expectation value of N, $\langle N \rangle$, is given by

$$\langle N \rangle = \frac{\sum_n N \exp^{-\beta E_n}}{Q_{S_0}} \qquad (4.108)$$

Combining Equations 4.107 and 4.108 yields

$$\langle N \rangle = \frac{\sum_n N \exp^{\beta(H_0 + H_1 N + \Sigma_i \Sigma_{j>i} h_{ij} \xi_i \xi_j)]}}{Q_{S_0}} \qquad (4.109)$$

Note that

$$Q_{S_0} = \sum_n \exp^{([\beta(H_0 + H_1 N + \Sigma_i \Sigma_{j>i} h_{ij} \varepsilon_i \varepsilon_j)]} \qquad (4.110)$$

Now consider

$$\frac{1}{\beta}\left(\frac{\partial \ln (Q_{S_0})}{\partial H_1}\right)_{\beta, H_0, h_{ij}} = \frac{1}{\beta Q_{S_0}}\frac{\partial Q_{S_0}}{\partial H_1} \qquad (4.111)$$

Combining Equations 4.110 and 4.111 yields

$$\frac{1}{\beta}\left(\frac{\partial \ln (Q_{S_0})}{\partial H_1}\right)_{\beta, H_0, h_{ij}} = \frac{\sum_n \frac{1}{\beta}\frac{\partial}{\partial H_1} \exp^{[\beta(H_0 + H_1 N + \Sigma_i \Sigma_{j>i} h_{ij} \varepsilon_i \varepsilon_j)]}}{Q_{S_0}} \qquad (4.112)$$

Note that for any atomic arrangement n, the values of N, and all of the ε_i are fixed. Therefore

$$\frac{1}{\beta}\left(\frac{\partial \ln (Q_{S_0})}{\partial H_1}\right)_{\beta, H_0, h_{ij}} = \frac{\sum_n N \exp^{[\beta(H_0 + H_1 N + \Sigma_i \Sigma_{j>i} h_{ij} \varepsilon_i \varepsilon_j)]}}{Q_{S_0}} = \langle N \rangle \qquad (4.113)$$

Substituting Q_{S_0} from Equation 4.106 into Equation 4.113 yields

$$\langle N \rangle = \frac{S_0}{2} + \frac{S_0}{2}\frac{1}{\lambda_1}\frac{\partial \lambda_1}{\partial \beta F} \qquad (4.114)$$

Substituting λ_1 from Equation 4.103 into Equation 4.114 and performing the algebra yields

$$\langle N \rangle = \frac{S_0}{2} + \frac{S_0}{2}\frac{\sinh (\beta F)}{\sqrt{\sinh^2 (\beta F) + \exp\left(-\frac{\beta h}{4}\right)}} \qquad (4.115)$$

Equation 4.115 is the isotherm for adsorption of a gas onto a one-dimensional line of sites with first nearest neighbor interactions. It is a simple exercise to show that Equation 4.115

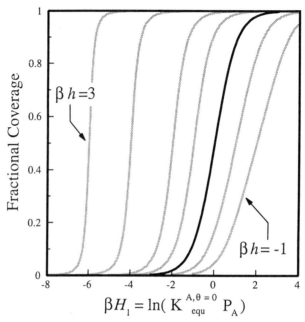

Figure 4.17 A series of isotherms calculated from Equation 4.115 for $\beta h = 3, 2, 1, 0.5, 0, -0.5,$ -1. The Langmuir line ($\beta h = 0$) is shaded.

goes to Equation 4.10 (i.e., the Langmuir adsorption isotherm) when $h = 0$. In the literature, it is common to plot the isotherms calculated by lattice gas methods as a function of either F, H_1, or μ_s, rather than partial pressure of the adsorbate over the surface. However, note that F, H_1, and μ_s are proportional to the log of the pressure (see Equation 4.50).

Figures 4.17 and 4.18 show plots of the isotherm calculated from Equation 4.115 for a variety of values of βh. The curves all look qualitatively the same. We have plotted the curves both versus βH_1 (i.e., chemical potential) and versus $\exp (\beta H_1 + \beta h)$ (i.e., dimensionless pressure). When we plot the data versus chemical potential, we observe S-shaped curves for all values of βh. When H_1 is negative, the pressure is small so the coverage is small, while if H_1 is positive, the pressure is high so the coverage approaches a monolayer. When $\beta h < 0$, the adsorbed molecules repel each other, so that at a given chemical potential, the amount of gas that adsorbs is less than that predicted by the Langmuir adsorption isotherm (i.e., $\beta h = 0$). When $\beta h > 0$, the adsorbed molecules attract each other, so that at a given chemical potential, the amount of gas that adsorbs is larger than that predicted by the Langmuir adsorption isotherm. However, the general shape of the adsorption isotherm when plotted against chemical potential is independent of h, at least in the one-dimensional limit.

If one replots the data versus dimensionless pressure as shown in Figure 4.18 one finds that there are some qualitative differences in the curves. At low pressures, all of the isotherms predict that the coverage varies linearly with pressure. For $\beta h = 0$ (i.e., the Langmuir case) the adsorption isotherm varies linearly with coverage and then saturates. When $\beta h < 0$ the curves look similar to the Langmuir isotherm. However, they approach saturation more gradually than the Langmuir adsorption isotherm. In contrast, the curves saturate more quickly than Langmuir's isotherm. In addition for $\beta h > 1.75$, there is an inflection point in the curve at low pressure. As one raises the pressure, the density of adsorbed molecules (i.e., the coverage) rises very quickly with pressure. The behavior looks quite like that which one would expect for a first-order phase transition. In fact, it is not a true first-order phase transition. In Example 4.G we demonstrate that a true first-order phase transition cannot occur

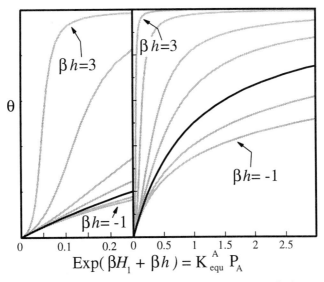

Figure 4.18 A replot of the data from Figure 4.17 versus dimensionless pressure.

in one dimension. However, for large values of βh one can get something that looks qualitatively like a first-order phase transition even though in reality the order is always greater than 1.

4.7.6 Exact Solution: Adsorption on an Infinite Strip

One can get first-order phase transitions with a two-dimensional lattice, however. The phase transitions represent important physics that cannot occur in one dimension. Therefore, we need to treat the two-dimensional case separately from the one-dimensional case. There are two ways to calculate adsorption on a two-dimensional surface. One can either treat the surface as a semifinite strip ℕ atoms wide by M atoms long and let M and ℕ go to infinity. Alternately, one can do an MC calculation on a similar system. In the next two sections we consider both types of calculations.

First, consider adsorption on a strip with two rows of M = $S_0/2$ identical sites where M will eventually go to infinity. Such a strip could be representative of adsorption on a step-surface where the terraces are just wide enough to hold two molecules of adsorbate. In the next two pages a derivation of the equations is provided for the adsorption isotherm for adsorption on a strip two sites wide. As before periodic boundary conditions are used so that the strip repeats after M sites. For the purposes of the derivation, each adsorption site will be numbered with two indices, i and l, where i goes from 1 to M and $l = 1$ or 2.

The solution of this problem is similar to the solution in the last section. We define an occupancy vector $\vec{\varepsilon}$ by

$$\vec{\varepsilon}_i = \begin{pmatrix} \varepsilon_{i,1} \\ \varepsilon_{i,2} \end{pmatrix} \tag{4.116}$$

with

$$\varepsilon_{i,l} = \begin{cases} 1 & \text{if the } (i, l) \text{ site is occupied} \\ 0 & \text{if the } (i, l) \text{ site is empty} \end{cases} \tag{4.117}$$

As earlier, the energy of any state of the system is

$$-E_n = H_0 + H_1 \sum_{(i,l)} \varepsilon_{i,l} + \sum_{(i,l)} \sum_{(j,m)>(i,l)} h_{ijlm} \varepsilon_{i,l} \varepsilon_{j,m} \qquad (4.118)$$

If it is assumed that there are only first nearest neighbor interactions

$$E_n = H_0 + H_1 \sum_{(i,l)} \varepsilon_{il} + h \sum_i (\varepsilon_{i,1}\varepsilon_{i+1,1} + \varepsilon_{i,2}\varepsilon_{i+1,2} + \varepsilon_{i,1}\varepsilon_{i,2}) \qquad (4.119)$$

Following the solution in the last section, it is useful to define a function V by

$$V\begin{bmatrix} \varepsilon_{i,1} & \varepsilon_{i,2} \\ \varepsilon_{i+1,1} & \varepsilon_{i+1,2} \end{bmatrix} = \exp(\beta H_0 + \beta H_1 (\varepsilon_{i,1} + \varepsilon_{i,2})$$

$$+ \beta h(\varepsilon_{i,1}\varepsilon_{i,2} + \varepsilon_{i,1}\varepsilon_{i+1,1} + \varepsilon_{i,2}\varepsilon_{i+1,2})) \qquad (4.120)$$

It is also useful to define a matrix $\hat{V}$ by

$$\hat{V} = \begin{bmatrix}
V\begin{bmatrix}11\\11\end{bmatrix} & V\begin{bmatrix}11\\10\end{bmatrix} & V\begin{bmatrix}11\\01\end{bmatrix} & V\begin{bmatrix}11\\00\end{bmatrix} \\
V\begin{bmatrix}10\\11\end{bmatrix} & V\begin{bmatrix}10\\10\end{bmatrix} & V\begin{bmatrix}10\\01\end{bmatrix} & V\begin{bmatrix}10\\00\end{bmatrix} \\
V\begin{bmatrix}01\\11\end{bmatrix} & V\begin{bmatrix}01\\10\end{bmatrix} & V\begin{bmatrix}01\\01\end{bmatrix} & V\begin{bmatrix}01\\00\end{bmatrix} \\
V\begin{bmatrix}00\\11\end{bmatrix} & V\begin{bmatrix}00\\10\end{bmatrix} & V\begin{bmatrix}00\\01\end{bmatrix} & V\begin{bmatrix}00\\00\end{bmatrix}
\end{bmatrix} \qquad (4.121)$$

A derivation like that in the last section yields

$$Q = \text{trace}\ [\hat{V}^M] \qquad (4.122)$$

Hence, one could use methods like those described in the last section to calculate an adsorption isotherm. Such a calculation is given in Example 4.H at the end of this chapter. It happens that if the $\varepsilon_{i,1}\varepsilon_{i,2}$ coupling in Equation 4.120 were zero, the strip for $l = 1$ would be independent of the strip for $l = 2$. As a result, $\hat{V}$ would have two independent solutions. If there is $\varepsilon_{i,1}\varepsilon_{i,2}$ coupling, these solutions can split, making a pseudo-two-phase region possible. Example 4.G shows that for any finite strip, one will not have a true phase transition. Rather, one observes a pseudo-phase transition where there is a rapid change in the state of the system for a small change in conditions. Rikvold et al. [1986] calculated adsorption isotherms using a 10-site-wide strip, and found what appears to be two phase regions for certain critical values of the parameters.

Strip calculations are very useful for adsorption on stepped surfaces. The molecules on the same terrace are much closer together than molecules on different terraces (see Figure 4.19). As a result, the gas molecules interact strongly with other gas molecules on the same terrace. However, the interactions between molecules on adjacent terraces are much weaker. As a result adsorption on a stepped surface can often be simulated by a strip a few sites wide (see Example 4.F).

Figure 4.19   A schematic of adsorption on a stepped surface.

There is a problem with using a strip to simulate adsorption on a flat surface, however. Typically, one needs a fairly wide strip to simulate the phase behavior seen on a flat surface. The $\hat{V}$ matrix for a strip l sites wide has 4^l elements. From a practical standpoint, if the strip is wider than 10 or 12 elements, one cannot fit the $\hat{V}$ matrix onto most computers, much less calculate its eigenvalues. Yet one often needs a strip 20–60 sites wide to adequately simulate the actual phase changes that are seen during adsorption on low index faces of transition metals. Hence, strip calculations have not been that useful for simulating adsorption on flat surfaces.

4.8 MONTE CARLO SOLUTION

The Monte Carlo/Metropolis algorithms outlined by Binder [1984, 1988] work much better for such surfaces. The idea behind Monte Carlo algorithms is to calculate the partition function in Equation 4.48 by randomly choosing states of the system, calculating the energy of the state, and summing each state with a probability of $\exp(-\beta E_n)$ as indicated in Equation 4.48. If it is assumed that $g_n = 1$,

$$Q_s = \sum_n e^{-\beta E_n} \tag{4.123}$$

A given property of the layer, F_L, can be computed by calculating an expectation value of the property:

$$\langle F_L \rangle = \frac{\sum_n F_n e^{-\beta E_n}}{Q_s} \tag{4.124}$$

where F_n is the value of the property F_L when the system is in configuration n. For example, if one substitutes N for F_n in Equation 4.124, one finds that the expectation value of the number of adsorbed molecules is as given in Equation 4.108.

In most cases, it is inefficient to choose states completely randomly and then weighing them with $\exp(-\beta E_n)$ because the main contributions to Equation 4.124 come from a small fraction of the states. However, a modification of the Monte Carlo algorithm, due to Metropolis et al. [1953] allows one to calculate reasonable adsorption isotherms with a minimum of computational effort. The procedure is to choose a distribution of states of the system so that each state has a probability $\exp(-\beta E_n)$ of being included in the distribution. States with low energy are counted multiple times in the distribution, while states with a high energy are counted once or less. It works out that the properties of the system become simple averages over all of the states in the distribution (see Equation 4.125).

The Metropolis algorithm starts with some initial configuration, and then takes "a Monte Carlo step" where one makes a random change in the configuration of the system. If the energy of the system goes down, then the system is assumed to move to the new configuration. If the energy of the system goes up by an amount ΔE, the move is made if $\exp(-\beta \Delta E) >$ rand, where rand is a random number between 0 and 1. One then calculates the properties of some quantity F_L by computing the average value of F_L at the end of each step, being sure to include all of the steps, independent of whether a move is made or not. Metropolis et al. [1953] showed that if one chooses states in this way, one will choose the low energy and high energy states just the right number of times so that the expectation value of any quantity F_L is given by

$$\langle F_L \rangle = \frac{1}{n_s} \sum_{steps, s} F_s \qquad (4.125)$$

where n_s is the number of steps and F_s is the value of F_L at the end of the Monte Carlo step. Note if no move is made one should calculate F_s at the final configuration (which equals the initial configuration).

In actual practice, one uses the following algorithm to compute the isotherm:

(1) One starts with some initial configuration of the system, for example, a clean surface, and fixes a value of β, H_1, and h in Equation 4.51.

(2) One then takes a step where one picks a site, and "flips" the configuration of the site. If the site is occupied, the adsorbed molecule is presumed to desorb. If the site is empty, a molecule is presumed to adsorb. Physically, one might want to pick a site at random. However, Binder [1984] shows that the same isotherms are obtained whether the sites are flipped randomly or systematically. The algorithm converges more quickly near a phase transition, however, if one alternates flipping of random sites with stepping through the lattice and flipping sites.

(3) One then calculates the energy change associated with flipping the site. If the energy of the system goes down, the flip is accepted. If the energy of the system goes up by an amount ΔE, the flip is accepted with a probability of $\exp(-\beta \Delta E)$. One usually does the calculations using periodic boundary conditions where the surface is presumed to loop back onto itself. However, this is not essential for the calculations.

Generally, one steps through the lattice a few hundred or thousand times to get the algorithm going. One then calculates a coverage, by summing the coverage periodically (e.g., once every 100 times through the lattice) over the next several thousand times through the lattice. The equilibrium coverage is then calculated via a simple average. Binder and Landau

(1981) found that a 40×40 lattice is needed to get reasonably accurate results. They were able to get reasonable isotherms by stepping through the lattice a few hundred times. In 1991 our IBM RISC station took less than a second to do such a calculation. Hence it was quite viable. We have found that the Monte Carlo calculation needs many more steps (~ 10,000) to converge near a phase transition. Our RISC station takes 25 seconds to step through a 40×40 lattice 10,000 times. A copy of the computer program used to do the Monte Carlo calculations is included in the supplemental material at the end of the chapter.

4.8.1 Qualitative Results

Figure 4.20 shows a series of adsorption isotherms that we have calculated using Binder's algorithm. We have chosen a few different values of βh to illustrate the effects. When $\beta h = 0$, the Monte Carlo results are identical to those for the Langmuir adsorption isotherm. As βh increases, there are deviations from the Langmuir adsorption isotherm. Initially the calculated isotherm looks qualitatively like Equation 4.115 (i.e., the one-dimensional result). However, Equation 4.115 does not fit quantitatively.

Figure 4.20 is a key result in this chapter, and so it is useful to describe it in detail. Let's start at low pressure. According to Equation 4.50, at low enough pressure βH_1 will be negative, and so from Figure 4.20 the coverage will be negligible. However, as the pressure increases, βH_1 increases, and so the coverage increases. Notice, that independent of βh most of the adsorption takes place between a βH_1 of -5 and a βH_1 of 4. Hence, if βH_1 were less than about -5, little gas would adsorb. Conversely, if βH_1 were above about 4, little gas would desorb. Therefore, Figure 4.20 implies that there are fundamental constraints to adsorption. If one wants to adsorb a gas, one has to work at conditions where βH_1 is greater than -5. In most systems βH_1 increases (i.e., becomes less negative) with decreasing surface temperature. As a result, the implication of Figure 4.20 is that one has to go to a low enough temperature so that βH_1 is greater than -5 for there to be significant adsorption. Similarly,

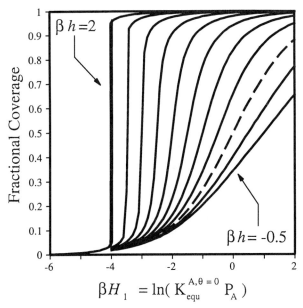

Figure 4.20 The coverage of adsorbate vs. βH_1 calculated using a Metropolis/Monte Carlo algorithm on a (40 × 40) lattice for $\beta h = -0.5, -0.25, 0, \ldots, 2.0$. The dashed line is for $\beta h = 0$ (i.e., the Langmuir adsorption isotherm).

one needs to go to high enough temperatures that βH_1 is less than 4 for there to be significant desorption.

Now consider starting at low temperature, and slowly raising the pressure. Let us first assume the adsorbate/adsorbate interactions are attractive, and that we choose a temperature where βh is 2.0. At a βh of 2, significant quantities of gas begin to adsorb at a βH_1 of -4.5. First, the amount of adsorption increases slowly with increasing pressure. However, at a βH_1 of -4.0 there is a sudden change in the coverage with increasing gas pressure. These sudden changes in coverage are associated with a phase transition in the two-dimensional adsorbate layer. When the adsorbate/adsorbate interactions are attractive, the adsorbate molecules tend to attract each other. At low temperature one observes first-order phase transitions where the coverage jumps suddenly when there is a small change in H_1 (i.e., a small change in the external gas pressure) due to condensation of adsorbate molecules into islands. In the actual calculations, one finds that the coverage changes drastically but continuously over a small range of H_1. However, Schick [1981] shows that with attractive interactions, one gets a sharp first-order phase transition with an infinite two-dimensional lattice.

Now consider raising the temperature and repeating the same experiment. Note that β equals $1/kT$; therefore, as you raise the temperature, βh goes down. Figure 4.21 shows a replot of the low-pressure portion of the data. Note that the sharp phase transitions disappear as you raise βh. As a result, one cannot get island formation unless βh is sufficiently large. Example 4.G demonstrates that islands form only when $\beta h > 1.7627$. When $1.7627 > \beta h \geq 0$, the surface layer is disordered. The point for $\beta h = 1.7627$, $\theta = 0.5$ represents a critical point that is much like the critical point seen in fluid systems. There are two stable phases at temperatures below the critical point: a low-density gas-like phase and a high-density solid-like phase. The individual phases are no longer stable at temperatures above the critical point. Instead, one observes a continuous transformation from gas-like behavior to solid-like behavior increasing pressure.

Interestingly, repulsive interactions give phase transitions too. When the adsorbate/adsorbate interactions are repulsive, an adsorbate molecule would rather be next to an open site than another adsorbate. Hence the adsorbate molecule will, in effect, attract empty sites. In Examples 4.E and 4.F we show that there will also be a series of phase transitions when the adsorbate/adsorbate interactions are strongly repulsive. Note, however, that there are no

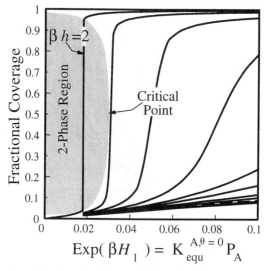

Figure 4.21   A replot of the data from Figure 4.20 as a function of dimensionless pressure.

discontinuities in Figure 4.20 when βh is repulsive. This is because repulsive interactions lead to continuous (e.g., second-order) transitions not first order transitions. Binder and Landau [1981, 1988] discuss a series of other cases. For example, if there are repulsive first nearest neighbors and attractive second nearest neighbor interactions, one gets a series of different ordered phases as the coverage increases.

In general, one can usually reproduce any observed adsorbate phase behavior with a lattice calculation. For example, if an experiment shows that the adsorbate goes into the P(2x2) overlayer shown in Figure 4.22a with sites 1, 2, and 4 empty and 3 and 5 occupied, one can force the calculation to reproduce the experimental structure if one assumes that the interactions between site i and sites 1 and 2 are repulsive and that the interactions between i and sites 3 and 5 are attractive. If one observes phase changes with increasing coverage or heating, one may have to carefully adjust the parameters to fit the data. However, so far all of the examples that have been examined in detail could be fit with a *generalized version* of the lattice gas model presented in Section 4.7.

The reason that one needs to consider a generalized version of the lattice gas model is that in the specific formulation of the lattice gas model presented in Section 4.7 the Hamiltonian has been simplified in a way that limits the range of phase behavior that can be reproduced. For example, Figure 4.23 is a replot of the data from Figure 4.20 as a function of $2\beta F$ rather than as a function of βH_1. The shapes of each of the individual curves is identical in both figures because $2F$ is equal to $H_1 + 2h$. However, the figures look different: H_1 is the energy change when populating a site on an empty surface. Hence, the curves in Figure 4.20 collapse upon themselves at low coverage. In contrast, $2F$ is the energy change upon populating a site on a half-filled surface. As a result, the curves in Figure 4.23 collapse upon each other at half coverage.

Figure 4.23 only applies to a square lattice. However, Figure 4.24 shows a similar calculation for a hexagonal lattice. The hexagonal lattice results are similar to the square lattice

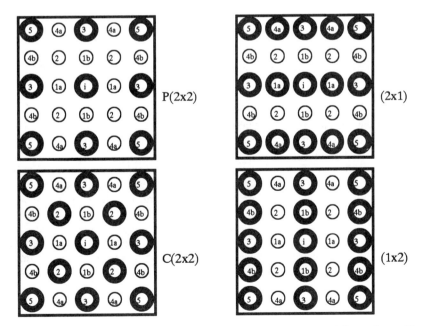

Figure 4.22 The adsorption of molecules in a P(2x2), C(2x2), (2x1), and (1x2) overlayer. The dark circles represent filled sites, while the open circles represent empty sites.

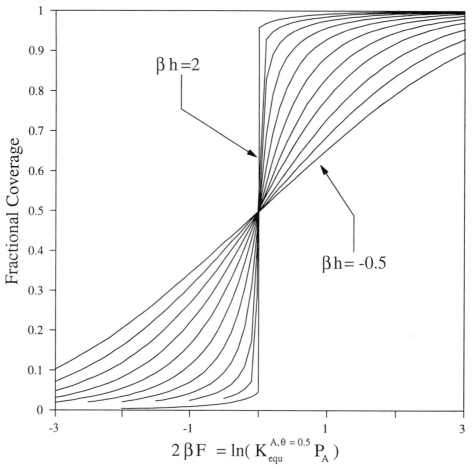

Figure 4.23 A series of adsorption isotherms calculated via the lattice gas method for adsorption on a square lattice with first nearest neighbor interactions. Curves are shown for $\beta h = -0.50, -0.25, 0, \ldots, 2.0$.

results, except that the phase transition occurs at a slightly lower value of βh. There are more neighbors to interact with on a hexagonal lattice, so each individual neighbor needs to only provide a smaller interaction.

Notice that all the isotherms in Figure 4.23 are symmetric about the point $F = 0$, $\theta = 0.5$, independent of h. This symmetry is inherent to the Hamiltonian in Equation 4.59 and is not dependent on the fact that we have limited our derivations to first nearest neighbor interactions. Notice that if we replace ξ_i by $-\xi_i$ and F by $-F$ in Equation 4.59 the energy does not change. Replacing ξ_i by $-\xi_i$ is equivalent to removing the adsorbate from all of the filled sites and putting adsorbate on all of the empty sites. Since such a change does not change the Hamiltonian, the phase behavior must be the same after the replacement. Replacing ξ_i by $-\xi_i$ also changes θ to $(1 - \theta)$. As a result, the adsorption isotherm must be symmetric about the point $\theta = (1 - \theta)$, i.e., $\theta = 0.5$ whenever the Hamiltonian in Equation 4.59 applies.

Experimentally, one often finds that the isotherm is approximately symmetric about $\theta = 0.5$. However, it rarely is exactly symmetric. Rikvold et al. [1983], Binder and Landan [1981], and Binder [1984] show that some more complex adsorption isotherms can be fitted

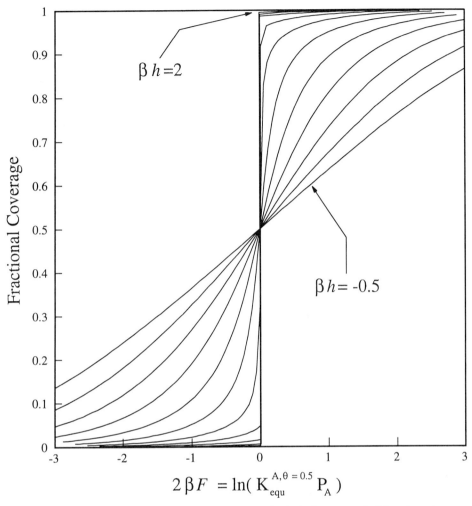

$$2\beta F = \ln(K_{equ}^{A,\theta = 0.5} P_A)$$

Figure 4.24 A series of adsorption isotherms calculated via the lattice gas method for adsorption on a hexagonal lattice with first nearest neighbor interactions. Curves are shown for $\beta h = -0.50$, -0.25, $0, \ldots , 2.0$.

if a three-body term is added to the Hamiltonian:

$$-E_n = H_{S_0} + F \sum_i \xi_i + \tfrac{1}{4} \sum_i \sum_{j>i} h_{ij}\xi_i\xi_j + \tfrac{1}{8} \sum_i \sum_{j>i} \sum_{l>j} h_{ijl}\xi_i\xi_j\xi_l \qquad (4.126)$$

where h_{ijl} is the three body interaction energy. Binder [1984] also adds a four-body term. The one difficulty with adding three- and four-body terms is that one gets so many parameters in the calculation that the results lose physical significance. However, it is possible to fit any phase behavior observed so far with a generalized lattice gas model, provided the adsorbate binds to distinct lattice sites and there are no indirect adsorbate/adsorbate interactions.

It happens that one does not necessarily have to do calculations to model the phase behavior of simple adsorbates. As noted previously, Ising models have already been extensively studied in the statistical mechanical literature. There are many examples that have already been solved. Usually these examples are solved in the magnetic formalism. However, if one is working on an example that has a Hamiltonian that has already been solved, one can simply look up the phase diagram and not do any calculations.

Physicists catalog the solutions in the literature by something called a **universality class.** Chemists and chemical engineers are most familiar with the concept of universality in terms of corresponding states relationships. The idea is that a series of phase diagrams can be put on a universal plot, provided one defines suitable dimensionless variables. In the same way, adsorbate phase diagrams can be put on universal plots where the dimensionless variables are the coverage, βh and βF. There are several different universal plots corresponding to different kinds of interactions. Physicists call each group of solutions a universality class. Generally, solutions of the same universality class have the same dimensionality (i.e., two-dimensional or three-dimensional), symmetry, and degeneracy. It happens that so far people have only found five universality classes that matter to adsorption of a single adsorbate on infinite, flat, periodic surfaces: the Ising universality class, the three-dimensional Potts universality class, four-dimensional Potts universality class, the XY universality class, and the Heisenberg universality class. The properties of the individual universality classes are described by Schick [1981]. The phase behavior of each of these universality classes has already been solved. Hence, if one is working with a problem that fit one of these universality classes, we can often simply look up the phase behavior without doing any calculations.

For example, Figure 4.23 is a phase diagram for Ising universality class. In Example 4.G we show that $\beta h = 1.7626\ T_c/T$ and $K_{equ}^{A,\theta=0.5}\ P = P/P_c$ at T_c where T_c and P_c are the critical temperature and pressure of the system. As a result Figure 4.23 is a corresponding states curve that is much like the corresponding states curves from thermodynamics. The form of the curve in Figure 4.23 is more useful because it is plotted in a way that allows one to use it even under conditions where P_c and T_c are unknown. Example 4.A at the end of this chapter illustrates this approach for hydrogen adsorption on Ni(110). One should refer to the example for details.

The universality concept also is useful for the description of surface phase transitions. For example, consider adsorption of a gas onto the square lattice shown in Figure 4.22. Rottman [1981] and Schick [1981] show that if there are only first nearest neighbor interactions, then there are two possible phase diagrams for the system. The simplest phase diagram comes when the first nearest neighbor interactions are attractive. At high temperatures, the lattice will simply fill randomly. At lower temperatures, however, there can be a first-order phase transition where because of attractive interactions, some of the adsorbate condenses into islands while the remaining part of the surface is nearly cleared of adsorbate. This phase transition is very similar to a gas–liquid phase transition. See Schick [1981] for details.

Entirely different phase behavior is seen if the adsorbate/adsorbate interactions are repulsive. In this case, a molecule on site i in Figure 4.22 repels molecules at the 1a and 1b sites, but does not affect molecules at the 2 sites. As a result, the system would like to form a C(2x2) overlayer where the 1a and 1b sites are empty and the 2 sites are occupied. In fact, Schnick [1981] shows that the layer does condense into a C(2x2) overlayer at moderate temperatures and pressures. If the interaction is not too repulsive, then at higher pressure more adsorbate molecules can squeeze into the gaps in the C(2x2) overlayer to eventually form a (1x1) overlayer. Systems that form (1x1) or C(2x2) overlayers are called members of "Ising Universality class," and it happens that all of the phase behavior of molecules in the Ising Universality class has already been worked out.

If there are second nearest neighbor interactions, then the phase behavior can change. A description of the phase behavior can be found in Examples 4.E, 4.F, and 4.G at the end of this chapter. The simplest case is where there are attractive first and second nearest neighbor interactions. In that case, the system still shows the gas–liquid-type transitions described earlier. If the first nearest neighbor interactions are repulsive and the second nearest neighbor interactions are either attractive or weakly repulsive, then the system still has a tendency to form a C(2x2) overlayer. Again, the system happens to follow the Ising universality class, whose behavior is known.

Different behavior is seen when the second nearest neighbor interactions are repulsive and strong enough to compete with the first nearest neighbor interactions. There are two cases, one where the first nearest neighbor interactions are attractive and a second where the second nearest neighbor interactions are repulsive. If the first nearest neighbor interactions are repulsive, then a molecule on site i in Figure 4.22 will repel molecules at sites 1a, 1b, and 2. As a result, the system would have a tendency to go into a P(2x2) overlayer where the 1a, 1b, and 2 sites are empty. However, if the first nearest neighbor interactions are attractive, there would be nothing that stops a molecule from adsorbing on one of the sites labeled 1a in Figure 4.22. One could also get adsorption on site 1a if the first nearest neighbors were repulsive but the pressure was high enough to overcome the repulsions. Once the molecule is adsorbed, on site 1a, the second nearest neighbor interactions would tend to suppress adsorption on the sites labeled 1b. However, there would be nothing to prevent molecules from adsorbing on the other 1a sites or the 4a sites. As a result, if the second nearest neighbor repulsions are stronger than the first nearest neighbor attractions, gas would preferentially adsorb on the 1a and 4a sites. This produces lines of molecules on the surface. If all of the interactions are right, then one can form a (2x1) overlayer on the surface. By symmetry one can also form a (1x2) overlayer. We call systems that can form a (1x2), (2x1), or (2x2) overlayer members of the XY universality class. It happens that there is an anomaly with XY universality class in that it does not always show universal phase behavior because of complications with the intermixing of (2x1) and (1x2) domains. However, there are many examples of XY behavior that have been solved, and one can often use the solutions directly.

Schick [1980, 1981] also considers several other types of phase behavior. Hexagonal surfaces show behavior characteristic of the Ising universality class if all directions and all sites are equivalent. If instead the system shows threefold degeneracy, it follows three-state Potts universality class. Four-state Potts universality class comes about on a hexagonal surface with two inequivalent sites arranged in a threefold symmetric pattern. The reader is asked to work out the phase behavior for hexagonal surface in Problem 4.37. Schick [1980] also solves this example. The linear site on a hexagonal lattice can also show behavior characteristic of Heisenberg universality class (see Schick [1981]). Again the phase behavior of each of these universality classes is well known, so if one has an example one can simply look up the phase behavior (see Schick [1980, 1981]).

Experimentally, the idea of universality has some important implications. For example, Barber [1983] shows that ^4He adsorbed on graphite should show the behavior characteristic of Potts three-state universality class with a continuous phase transition at $\frac{1}{3}$ coverage. However, if the graphite is covered by a closed packed krypton layer, the system should shift to Ising Universality class. The phase transition is at $\frac{1}{2}$ coverage in Ising Universality class. Tejwana et al. [1980] examined the phase behavior of both systems, and found that on a clean surface the phase transition in the ^4He layer is at $\frac{1}{3}$ coverage, but the phase transition moves to $\frac{1}{2}$ coverage when the substrate is covered with krypton. All of this occurs even though the changes in the intermolecular forces are small. These results demonstrate the power of universality in determining phase behavior.

The one limitation to the universality concept at present is that in general the phase behavior of the lattice gas model has been worked out only for a perfect infinite lattice. However, real surfaces have defects. Einstein, Bartlet, and Roelofs [1988] considered the role of defects on adsorbate phase behavior and found that the behavior of the system near a phase transition was affected by defects even when there is only one defect every 40 unit cells. The changes are subtle, and may not be important for every application. However, one does need to consider these effects if one wants to model the behavior of the system near a phase transition (see Barber [1983] for details).

There is one other implication of the universality concept that has not yet been tested experimentally. It was stated earlier in this section that the universality class of a given adsorption system is determined by the symmetry and degeneracy of the adsorbed phase. Notice that if one goes from a single adsorbate to a coadsorption system, one adds an extra degree of freedom to the system. This in turn changes the universality class. No one has examined the implications of such a change in universality class in detail. However, it would seem that the concept of universality would imply that at temperatures below the critical temperature, mixtures should show quite different phase behavior than the individual components in the mixture.

4.9   APPROXIMATION METHODS:
THE BRAGG–WILLIAMS APPROXIMATION

The one difficulty with the Monte Carlo solutions is that for a general type of interaction one does not get an analytical expression for the adsorption isotherm. As a result, it takes a considerable amount of computer time to fit a given experimental data set. From a practical standpoint, one often wants an analytical expression for the adsorption isotherm. One has to use an approximation to find it. In this section we will review the key approximation methods used today.

The simplest approximation is called the *Bragg–Williams* or *mean field approximation*. Bragg and Williams [1934] proposed the approximation to calculate the structure of an alloy. However, Fowler [1935] extended it for use on the surface. The idea behind the Bragg–Williams approximation is to treat the adsorbate/adsorbate interactions in an average sense. One ignores the fact the surface may order, and instead calculate the effects of the adsorbate/adsorbate interactions by replacing the occupancy of the nearest neighbor sites, ε_j in Equation 4.51, by an average value of the occupancy for the whole layer. This approximation ignores the fact that an adsorbate on one site will affect the probability that an adjacent site is filled. Equation 4.51 becomes

$$-E_n = H_0 + H_1 \sum_i \varepsilon_i + \frac{1}{2} \sum_{i,j} \frac{h_{ij}}{2} \varepsilon_i \frac{\langle N \rangle}{S_0} \qquad (4.127)$$

The factor of $\frac{1}{2}$ in the sum is needed to avoid double counting: if an atom on site i reduces the heat of adsorption on site j by $h/2$, and the atom on site j reduces the heat of adsorption on site i by $h/2$, the net result is that the total energy is reduced by h.

If it is assumed that $h_{ij} = h$ for nearest neighbor sites and zero elsewhere, one obtains

$$-E_n = H_0 + \sum_i \varepsilon_i \left(H_1 + 0.25 \, hn_n \frac{\langle N \rangle}{S_0} \right) \qquad (4.128)$$

where n_n is the number of nearest neighbors of a given site. The partition function becomes

$$Q_{S_0} = \exp\{\beta H_0\} \sum_{\varepsilon_1 = 0,1} \sum_{\varepsilon_2 = 0,1} \cdots \sum_{\varepsilon_{S_0} = 0,1} \prod_i \exp\left[\beta \varepsilon_i \left(H_1 + 0.25 \, hn_n \frac{\langle N \rangle}{S_0} \right) \right] \qquad (4.129)$$

Performing the sums using a recursion analysis like that in Section 4.7.4 yields

$$Q_{S_0} = \exp(\beta H_0) \left[1 + \exp\left(\beta H_1 + 0.25 \, hn_n \frac{\langle N \rangle}{S_0} \right) \right]^{S_0} \qquad (4.130)$$

Substituting the expression for $\langle N \rangle$ from Equation 4.130 into Equation 4.113 yields

$$\theta = \frac{\langle N \rangle}{S_0} = \frac{\exp (\beta H_1 + 0.25 \ \beta h n_n \theta)}{1 + \exp (\beta H_1 + 0.25 \ \beta h n_n \ \theta)} \qquad (4.131)$$

Rearranging,

$$\frac{\theta}{1 - \theta} \exp (-0.25 \ \beta h n_n \theta) = \exp (\beta H_1) \qquad (4.132)$$

It is easy to show that Equation 4.132 is equivalent to the Fowler adsorption isotherm, Equation 4.42.

Equation 4.132 has some interesting properties. For example, Figure 4.25 is a plot of the coverage as a function of H_1 calculated from Equation 4.132. Note that when h is sufficiently large, the coverage/chemical potential plot shows an S-shaped curve where different values of θ give the same values of H_1. This implies that there are multiple stable phases of the adsorbed layer. The S-shaped curve is called a *van der Waals loop*; it is seen whenever there are two stable phases. If one is on the top or bottom of the loop, only one phase is stable. However, two phases are stable in the middle of the loop, one can calculate how much of each phase that should be present via the lever rule (Maxwell construction). Qualitatively, the phase diagram calculated via the Bragg–Williams approximation is very similar to the phase diagrams in Figure 4.20. However, there is a finite two-phase region with the Bragg–Williams isotherm, while there is a sharp first-order phase transition in the exact solution. Still, the qualitative effect that there can be multiple adsorbed phases is reproduced with the Bragg–Williams approximation, even though the Bragg–Williams result does not fit the exact phase behavior for a two-dimensional lattice gas exactly.

4.9.1 Bethe–Peierls Approximate Solution

The reason that the Bragg–Williams approximation does not fit exactly is that it ignores the local ordering of the atoms on the surface. As noted in Section 4.6, molecules often order on surfaces and so the probability of one site being occupied is not independent of the probability of an adjacent site being occupied. This changes the isotherm.

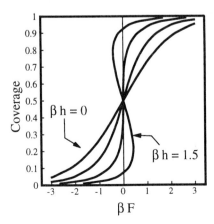

Figure 4.25 A plot of the coverage as a function of H_1 calculated from the Bragg–Williams isotherm Equation 4.132.

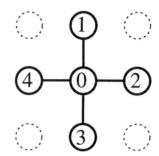

Figure 4.26 The Bethe–Peierls cluster for a square lattice.

Bethe [1935] and Peierls [1936] developed an approximation that to first order accounts for some of the ordering. Their idea was to consider a cluster of a few atoms. They then treat the interactions within the cluster exactly, and treat the other interactions with a mean field approximation.

Consider the cluster shown in Figure 4.26 with a central atom and its nearest neighbors. If it is assumed that there are only first nearest neighbor interactions, the central atom (i.e., atom 0) will interact with all of the other atoms in the cluster, while the remaining atoms n_n atoms will also interact with the atoms outside of the cluster. As a result, one can calculate the energy of the cluster, E_C, from

$$E_C = H_1 \sum_{j=0}^{n_n} \varepsilon_j + h\varepsilon_0 \sum_{j=1}^{n_n} \varepsilon_j + \sum_{j=1}^{n_n} \varepsilon_j (2g) \qquad (4.133)$$

where the parameter g is used to represent the interactions of the edge atoms in the cluster with all of the atoms outside of the cluster. The parameter g will be calculated self-consistently. Note that the definition of g in Equation 4.133 is slightly different than the definition of g in Bethe's papers. The two are related by:

$$g_{Bethe} = g - 0.25 \, (n_n - 1)h \qquad (4.134)$$

The partition function of the cluster is

$$Q_c = \sum_{\varepsilon_0 = 0, 1} \sum_{\varepsilon_1 = 0, 1} \cdots \sum_{\varepsilon_{n_n} = 0, 1} \exp{(-\beta E_C)} \qquad (4.135)$$

One can perform the sum by considering the terms for $\varepsilon_0 =$ and $\varepsilon_0 = 1$ separately and using the recursion relationship in Section 4.7.4. The result is

$$Q_C = [1 + \exp{\{\beta(H_1 + 2g)\}}]^{n_n} \\ + \exp{\{\beta H_1\}} [1 + \exp{\{\beta(H_1 + 2g + h)\}}]^{n_n} \qquad (4.136)$$

One still needs a value for the parameter g. It can be found by noting that the cluster approximation gives two different expectation values for the coverage; θ_0 the expectation value of the coverage on the central atom and θ_1 the expectation value of the coverage on an edge atom. The derivation requires some algebra, but the result is

$$\theta_0 = \langle \varepsilon_0 \rangle \qquad (4.137)$$

$$= \frac{\exp{\{\beta H_1\}} [1 + \exp{\{\beta(H_1 + 2g + h)\}}]^{n_n}}{([1 + \exp{\{\beta(H_1 + 2g)\}}]^{n_n} + \exp{\{\beta H_1\}} [1 + \exp{\{\beta(H_1 + 2g + h)\}}]^{n_n})}$$

$$\theta_1 = \langle \varepsilon_1 \rangle$$

$$= \frac{\begin{aligned}&\exp\{\beta(H_1 + 2g)\}\,[1 + \exp\{\beta(H_1 + 2g)\}]^{n_n - 1}\\ &+ \exp\{\beta(2H_1 + 2g + h)\}\,[1 + \exp\{\beta(H_1 + 2g + h)\}]^{n_n - 1}\end{aligned}}{\begin{aligned}&([1 + \exp\{\beta(H_1 + 2g)\}]^{n_n}\\ &+ \exp\{\beta H_1\}\,[1 + \exp\{\beta(H_1 + 2g + h)\}]^{n_n})\end{aligned}} \tag{4.138}$$

Note the expectation values in Equations 4.137 and 4.138 were calculated by using the recursion relationship in Section 4.7.4 and the relation

$$\langle \varepsilon_j \rangle = \left(\frac{1}{Q_C}\right) \sum_{\varepsilon_0 = 0, 1} \sum_{\varepsilon_1 = 0, 1} \cdots \sum_{\varepsilon_{n_n} = 0, 1} \varepsilon_j \exp(-\beta E_C) \tag{4.139}$$

Since θ_0 should equal θ_1, one can solve Equations 4.137 and 4.138 simultaneously to obtain values of g and $\theta = \theta_1 = \theta_2$. Equations 4.137 and 4.138 are the Bethe–Peirls adsorption isotherm.

In actual practice, Equations 4.137 and 4.138 have n_n roots. However, the only physically meaningful roots come when g is real and between 0 and $(n_n - 1)h/2$. One normally calculates g via a search procedure. A suitable program can be found in the supplemental material at the end of the chapter.

It happens that it is a little easier to use the Bethe approximation in reverse, i.e., pick a value of θ, and calculate a value of βH_1. For the purposes of derivation it is useful to define a new variable G_{Bethe} by

$$G_{Bethe} = \exp\{\beta(H_1 + 2g)\} \tag{4.140}$$

Substituting Equation 4.140 into all the terms in g in Equations 4.137 and 4.138 and solving the two equations simultaneously assuming that there is only one phase on the surface yields

$$\exp(\beta H_1) = \frac{G_{Bethe}\,[1 + G_{Bethe}]^{n_n - 1}}{[1 + \exp(\beta h)\,G_{Bethe}]^{n_n - 1}} \tag{4.141}$$

Substituting Equations 4.140 and 4.141 into Equation 4.137 and canceling the like terms yields

$$G_{Bethe}^2 \exp(\beta h)(1 - \theta) + G_{Bethe}(1 - 2\theta) - \theta = 0 \tag{4.142}$$

Note that from Equation 4.140, G_{Bethe} must be positive. Therefore, G_{Bethe} is the larger root of Equation 4.142. If one knows θ, one can calculate G_{Bethe} analytically from Equation 4.142. One can then substitute the resultant expression into Equation 4.141 to yield an analytical expression for βH_1. Therefore, if one knows θ, one can calculate βH_1 analytically.

Typically one finds that the coverages predicted by the Bethe approximation is within 5% of the Monte Carlo results for $\beta h \le 4 \ln(n_n)/n_n$. The Bethe approximation fails for $\beta h > 4 \ln(n_n)/n_n$ in that it predicts S-shaped curves similar to those in Figure 4.25 with multiple roots for each value of βH_1, while the exact solution shows a sharp two-phase transition at $\beta F = 0$. Thus, the Bethe approximation makes a sizable error in the two-phase region.

4.9.2 A Modified Bethe Approximation

Fortunately, it is easy to define a modified Bethe approximation that is closer to the exact behavior. The modified Bethe approximation is derived by starting with the simple Bethe approximation, and forcing it to have a first-order phase transition at $F = 0$. Specifically, one makes the following approximation:

(1) One assumes that the modified Bethe approximation follows the Bethe results when $\beta h \leq 4 \ln (n_n)/n_n$ (i.e., when only one phase is present).

(2) In all other cases, one assumes that the isotherm follows the lower branch of the S-shaped curve for $\beta F < 0$, and the upper branch of the S-shaped curve for $\beta F > 0$. That produces a sharp phase transition at $\beta F = 0$.

Figure 4.27 shows a plot of the isotherm calculated via the modified Bethe approximation for various values of βh. The figure is essentially indistinguishable from the exact results except when $1.8 < \beta h < \ln (4)$. The modified Bethe approximation predicts a sharp phase transition for $\beta h < \ln (4) = 1.38$ while the exact solution predicts that two phases will be stable when $\beta h < \sinh^{-1} (4) = 1.76$. However, the actual isotherm is not that much different. The biggest error occurs at $\beta h = 1.75$, (i.e., just at the critical point). The modified Bethe approximation predicts a first-order phase transition from $\theta = 0.12$ to $\theta = 0.88$ at $\beta F = 0$, while the exact result shows the same coverage change when βF goes from -0.03 to $+0.03$. Much smaller errors are seen at other values of βh. Thus while the modified Bethe approximation is not exact, it has the advantage of offering analytical solutions, and it is within a few percent of the exact results over the entire coverage range.

For the material later in this book, it will be useful to obtain an analytical expression for the phase boundary at $F = 0$ in the modified Bethe approximation. Note that $2\beta F = \beta H_1 + n_n \beta h/2$. Therefore, when $F = 0$, $1\beta H_1 = -n_n \beta h/2$. Substituting $(\beta H_1 = -n_n \beta h/2)$ into Equation 4.141 with $n_n = 4$ yields

$$[\exp (\beta h)\, G_{Bethe}^2 - 1]\, [\exp (\beta h)\, G_{Bethe}^2 + \{3 \exp (\beta h) - \exp (2\beta h)\}\, G_{Bethe} + 1] = 0$$

$$(4.143)$$

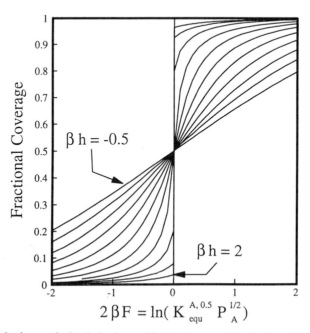

$$2\beta F = \ln(K_{equ}^{A,\, 0.5}\, P_A^{1/2})$$

Figure 4.27 An isotherm calculated via the modified Bethe approximation for adsorption on a square lattice.

while making the same substitution with $n_n = 6$ yields

$$[\exp{(\beta h)}\, G_{Bethe}^2 - 1]\, [\exp{(\beta h)}\, G_{Bethe}^2 - A_{Bethe}\, G_{Bethe} + 1]$$
$$\cdot\, [\exp{(\beta h)}\, G_{Bethe}^2 - B_{Bethe} + 1] = 0 \tag{4.144}$$

where A_{Bethe} is the larger root of

$$A_{Bethe}^2 + A_{Bethe}\,[5 - \exp{(2\beta h)}] + 10\exp{(2\beta h)} - 5\exp{(3\beta h)} - \exp{(\beta h)} = 0 \tag{4.145}$$

and B_{Bethe} is the smaller root of the same equation. One can show that Equation 4.143 has three positive roots and one negative root for $\beta h > \ln{(4)}$, while Equation 4.144 has five positive roots and one negative root for $\beta h > \ln{(6)}$.

The first factor in Equations 4.143 and 4.144 has a positive root and a negative root. The positive root is the solution corresponding to the $\theta = 0.5$ point in the S-shaped curve in Figure 4.25, while the negative root is spurious. The second factor in Equations 4.143 and 4.144 has two roots corresponding to the upper and lower phase boundaries in Figure 4.21. There are two additional roots of Equation 4.144. Again, they are not physically meaningful.

Notice that Equations 4.143, 4.144, and 4.145 can be decomposed into a series of second-order equations in the variable G_{Bethe}. Therefore, the equations can be solved analytically to yield G_{Bethe} as a function of βH_1. That result can be substituted into Equation 4.142 to calculate a phase envelope.

For the purposes of derivation, it is useful to define G_{Bethe}^{Lower} as the smaller root of

$$\exp{(\beta h)}\, G_{Bethe}^{2}{}_{Lower} + [3\exp{(\beta h)} - \exp{(2\beta h)}]\, G_{Bethe}^{Lower} + 1 = 0 \tag{4.146}$$

It is easy to show that G_{Bethe}^{Lower} is the value of G_{Bethe} corresponding to the lower phase boundary at $F = 0$. Substituting Equation 4.146 into Equation 4.142 yields

$$\theta_{\substack{Lower \\ Boundary}} = \frac{G_{Bethe}^{Lower} + \exp{(\beta h)}\, G_{Bethe}^{2}{}_{Lower}}{1 + 2\, G_{Bethe}^{Lower} + \exp{(\beta h)}\, G_{Bethe}^{2}{}_{Lower}} \tag{4.147}$$

Equation 4.146 is not particularly accurate for $1.8 < \beta h < \ln{(4)} = 1.38$. With $n_n = 4$ Equation 4.147 predicts a critical point at $\beta h = 1.38 \ldots$, while the exact critical point comes at $\beta h = 1.7627 \ldots$. Thus, the Bethe approximation is far from exact at the critical point.

Interestingly, by the time $\beta h = 2$, the errors are substantially reduced. For example, with $n_n = 4$, the Bethe approximation gives a lower phase boundary at $\theta = 0.0357$ at $\beta h = 2$, while the exact phase boundary is at $\theta = 0.0416$. To put this result in perspective, at $\beta h = 1.948$ the Bethe approximation predicts a phase boundary at $\theta = 0.0416$. Thus, it seems that the phase boundary with the Bethe approximation is almost correct, except near the critical point.

4.9.3 The Kukuchi Cluster Approximation

The main limitation of the modified Bethe approximation is that it ignores second nearest neighbor interactions. As a result the modified Bethe approximation fails when second nearest neighbor interactions are important. However, there are a series of other approximations that are simply related to the Bethe–Peirls approximation and can include second and higher nearest neighbor interactions. Kukuchi [1951] extended Bethe's work to include clusters of variable size. If one includes second nearest neighbors into the Bethe cluster, one can easily treat the second nearest neighbor interactions. The results again work out to be within 5% of the exact results of temperatures well above the critical temperatures for island formation, although errors arise when islands form. In practice, if one only has first nearest neighbor interactions, the simple Bethe approximation is very close to the exact solution of the lattice gas model except right at the phase transitions. Hence, the use of large clusters does not significantly improve the predictions of the Bethe model when there are only first nearest neighbor interactions. However, work on the Ising problem indicates that the Kukuchi method gives significantly better results than the Bethe model when there are second nearest neighbor or higher interactions (see Barley [1972] for details). No one has considered the use of the Kukuchi method to model adsorption. However, one can anticipate that it could be quite useful for that purpose.

In fact, the Kukuchi cluster approximation may be a better representation of a real adsorption system than the lattice gas model. After all, the lattice gas model assumes that the surface shows long-range periodicity. Experimentally, real surfaces show periodicity, but they are only periodic for a finite number of sites. Hence, at present it is not known whether the exact solution of the lattice gas model or a cluster approximation will better represent adsorption on a real surface.

4.9.4 Quasichemical Approximation

There is another approximation called the *quasichemical approximation*, which tries to consider the same physics as the Bethe approximation, i.e., the local ordering of the lattice. However, the quasichemical approximation considers a cluster containing a pair of sites and ignores the fact that the sites are confined to a lattice. One then derives a partition function from the equilibrium expression for the pairs. The result is a quadratic equation that can be solved in closed form. The quasichemical approximation was very popular in the 1940s, 1950s, 1960s, and early 1970s, but has become less important today. It is now known that the quasichemical approximation is much less accurate than the Bethe approximation for adsorption on a lattice (the quasichemical approximation is thought to be reasonable for a liquid). Still there are important calculations from the 1970s that used the quasichemical approximation, and a few groups have continued to use the approximation today. Therefore, the quasichemical approximation is reviewed below. The reader should refer to Hill [1960], Fowler and Gugenheim [1960], Steele [1974], or Lombardo and Bell [1991] for a more complete discussion of the quasichemical approximation.

The quasichemical approximation is derived by considering all the individual pairs of adjacent sites and then making the ad hoc assumption that the occupancy of each individual pair of sites is independent of all the other pairs. For example, there are four pairs of sites in the Bethe cluster shown in Figure 4.26, one containing sites 0 and 1, another containing sites 0 and 2, a third containing sites 0 and 3, and a fourth containing sites 0 and 4. The quasichemical approximation assumes that the occupancy in these pairs of sites is independent so that the 01 pair can be filled with two adsorbate molecules at the same time that the 02 pair is empty even though the 01 and 02 pair both contain atom 0. This assumption works in a liquid where pairs of atoms are independent. However, it is not correct on a lattice. Still, Gugenheim [1952] has claimed that a reasonable approximation to a lattice gas can be obtained by making this assumption and that the results can be obtained in closed form. Until recently, no one knew that this was not an accurate approximation. Following Lombardo and Bell [1991], it is useful to define three quantities: P_{oo}, the probability that both sites in a given pair are occupied, P_{ov}, the probability that one site in a given pair is occupied and one is vacant, and P_{vv}, the probability that both sites are empty. One can show that

$$P_{oo} + P_{ov} + P_{vv} = 1 \qquad (4.148)$$

$$P_{oo} + \tfrac{1}{2} P_{ov} = \theta \tag{4.149}$$

Fowler and Gugenheim [1960] show that if the occupancy of the various pairs of sites is independent

$$\frac{P_{oo} P_{vv}}{\left(\dfrac{P_{ov}}{2}\right)^2} = \exp(-\beta h) \tag{4.150}$$

The 2 on the left of the equation comes about because there are two ways of placing one atom in two sites.

One can solve Equations 4.147, 4.149, and 4.150 simultaneously for P_{oo}, P_{vv}, and P_{ov}. The result is

$$P_{oo} = \theta + \frac{1 - X_{Qc}}{2[\exp(-\beta h) - 1]} \tag{4.151}$$

$$P_{ov} = \frac{1 - X_{Qc}}{\exp(-\beta h) - 1)} \tag{4.152}$$

$$P_{vv} = 1 - \theta - \frac{1 - X_{Qc}}{\exp(-\beta h) - 1} \tag{4.153}$$

with

$$X_{Qc} = \sqrt{1 + 4\,\theta(1 - \theta)\,[\exp(-\beta h) - 1]} \tag{4.154}$$

Fowler and Gugenheim [1960] and Steele [1974] derive adsorption isotherms from these equations. None of the derivations are mathematically rigorous, so only the final result is presented in this chapter:

$$K^0_{equ,A} P_A \exp(-2n_n\beta h) = \frac{\theta}{1 - \theta}\left[\frac{(X_{Qc} - 1 + 2\theta)}{(X_{Qc} + 1 - 2\theta)}\left(\frac{1 - \theta}{\theta}\right)\right]^{n_n/2} \tag{4.155}$$

Equation 4.155 is the quasichemical approximation to the isotherm for a two-dimensional lattice gas. Equation 4.155 is much less accurate than Equation 4.141, so there is no reason to use Equation 4.155. Still, some investigators continue to do so.

4.10 INCOMMENSURATE ADSORPTION

The one major limitation of the lattice gas models and associated cluster approximations is that they assume that the adsorbate goes onto an ordered array of distinct lattice sites on the surface of the solid. Up until a few years ago, most people had thought that there were indeed distinct lattice sites for adsorption, and hence the lattice gas models were quite well accepted. However, recent data call into question whether adsorbates overlayer are highly ordered at reasonable temperatures.

Most of the information about ordering of adsorbate overlayers comes from two techniques: low-energy electron diffraction (LEED) and scanning tunneling microscopy (STM). In LEED one measures a diffraction pattern of the surface. One only observes a sharp diffraction pattern from the regions of the surface that are highly ordered. Hence, a simple use of LEED where one only looks at the diffraction spots tends to bias the measurement toward the ordered regions of overlayer. One can use LEED to assess order more carefully if one measures the sharpness of the LEED spots and the amount of diffuse background in the LEED pattern. If one had a perfectly ordered overlayer, then the LEED pattern should be extremely sharp with no diffuse background. Sharp LEED patterns are often seen with atomic adsorbates. However, it is unusual to observe a sharp LEED pattern from a molecular adsorbate. More often one observes a fuzzy LEED pattern with a diffuse background. A fuzzy LEED pattern suggests that there may be some disorder in the surface layer or that the adsorbed molecules do not stick at distinct lattice sites. Hence, there is need to consider the possibility of a random adsorption where some of the adsorbate molecules do not bind to specific lattice sites.

STM also shows that at reasonable temperatures, actual overlayers may not always be so highly ordered. Recently, there have been several attempts to do STM of adsorbed molecules. Occasionally one observes ordered molecular overlayers at 300 K. However, in many cases no images of ordered overlayers can be obtained with STM at 300 K. Land [1991] has found many cases where images of ordered overlayers are observed at very low temperatures (i.e., ~ 40 K) but not at 300 K. These results suggest that adsorbate molecules may not be fixed at specific lattice sites at 300 K. Rather, the adsorbate may be in some sort of random configuration with a high degree of mobility.

This brings up the issue of when the adsorbate layer is mobile. In the material so far in this chapter, it was assumed that the adsorbate was bound to fixed lattice sites, and not moving. Such adsorbate behavior is called *immobile adsorption*. More precisely, immobile adsorption is when the adsorbate bonds to specific sites on the surface of the solid and stays there for a long time in comparison to vibrational times. It is possible to identify a specific binding site in these situations, e.g., directly on top of the surface atoms or in a threefold hollow. However, there is another type of adsorption called *mobile adsorption*. Mobile adsorption occurs when the adsorbate bonds strongly to the surface, but not to a specific site. Rather the adsorbate is free to move over the surface of the solid.

Over the years there has been some controversy whether mobile adsorption is real (see Averbach and Rettner [1987] or Weinberg [1987]). The issue is complex because a given molecule can be mobile on a time scale of seconds, and immobile on a time scale of femtoseconds. In Example 4.D we show that if the adsorbate diffuses at a rate of 50,000 Å/sec or more, the mobility has an important effect on the properties of the adsorbed layer. Therefore we discuss mobile and immobile adsorption by assuming that the adsorbate is mobile if it moves at about 50,000 Å/sec, and immobile if it is stationary on that time scale.

Recently, STM work of Land et al. [1991] has provided strong evidence for mobile and immobile adsorption. Land et al. exposed a clean Pt(111) sample to ethylene at 300 K, then cooled the sample to various temperatures. When the ethylene adsorbs at 300 K, the molecule loses a hydrogen and rearranges to form a species called ethylidyne ($Pt_3 \equiv CCH_3$). The ethylidyne stands upright on the surface. At 160 K, one observes an ordered pattern in the STM image. Land et al. [1991] showed that each white blob in the picture corresponds to an individual ethylidyne molecule. Land et al. found that at 160 K, the white blobs form an ordered pattern. Hence, it is clear that the ethylidyne layer is immobile at 160 K. However, at 300 K, the STM pattern smears out; one cannot detect individual ethylidynes at fixed sites in the lattice. Instead one observes a diffuse background in the picture. This suggests that at 300 K, the ethylidynes are no longer fixed at distinct lattice sites, at least in the presence of the STM. Instead they can move along the surface. If the adsorbate is free to move over the surface, we say that the gas adsorbs into a **mobile state.**

According to calculations, mobile adsorption occurs because the bonding of the adsorbate to a surface often is more complex than the bonding in a simple isolated molecule. In a simple molecule we can often describe the bonding by assuming that individual bonds form between adjacent atoms in the molecule. When a gas adsorbs on a surface, it can form a local bond to the surface atoms. However, in Sections 3.8.3 and 3.10 we noted that the adsorbate can also form a more delocalized bond. For example, in a metal there are conduction band electrons that move freely through the metal. They are not localized on any one atom and hence can carry electricity. One can show that an adsorbate can form a bond with the conduction band electrons. The bonding electrons will be localized around the adsorbate. However, the electrons that form the bond do not come from any one specific atom. Rather, the electrons come from a band that is being shared by all of the atoms in the metal. As a result, one cannot think of the bonding as being composed of local bonds between the adsorbate and one or more surface atoms. Rather, the adsorbate is held by a delocalized bond between the adsorbate and all of the atoms in the solid.

In Section 3.10 it was shown that molecules are held to surfaces by polarization (i.e., van der Waals) forces and a mixture of localized and delocalized bonds. The delocalized bonds can lead to behavior that is not seen in simple molecules.

For example, consider displacing a molecule that is held onto a metal surface by a delocalized bond to the conduction band of the metal. Notice that the molecule would be hard to detach from the surface since in order to detach the molecule, one would have to break the adsorbate-surface bond. However, if the adsorbate molecule were moved laterally along the surface, the bonding electrons could move along with the adsorbate. As a result, the bonding between the adsorbate and the surface would not change significantly when the adsorbate moves laterally.

In any real system, the actual bonding involves a mixture of localized and delocalized bonds. If the delocalized bonding dominates, then the barrier to surface diffusion, E_B, will be small. Often the barrier is less than kT, where k is Boltzmann's constant and T is the absolute temperature so that the adsorbate can move freely over the surface. Similarly, when the localized bonding dominates, bonds break when the adsorbate moves laterally. Hence, E_B will be much greater than kT. As a result, the adsorbate will be largely fixed in one spot. Note, however, that even if a molecule is adsorbed into an immobile state, it will still slowly diffuse over the surface. However, the time constant of the motion will be much longer than vibrational times so that the molecule will be largely fixed while it reacts.

Experimentally, most gases can adsorb into either a mobile or an immobile state. Generally, the heat of adsorption into an immobile state is usually assumed to be larger than that into an mobile state. In mobile adsorption the adsorbate is held only with delocalized bonds, while in immobile adsorption there can be a mixture of localized and delocalized bonds. Thus, there often is more total bonding in an immobile state. As a result, immobile adsorption is often thought to be thermodynamically favored at low temperatures. Still, incoming molecules can initially be adsorbed into a mobile state. Hence, both mobile and immobile adsorption are important during reactions on solid surfaces.

4.10.1 Two-Dimensional Equations of State: Ideal Lattice Gas

There is not a large body of literature on the consequences of mobile adsorption on adsorption isotherms. Many years ago Volmer [1925] proposed that one could calculate an adsorption isotherm for a mobile layer via an equation of state. This idea is to treat the adsorbed layer as a two-dimensional gas that is free to translate over the surface of the solid, but is prevented from leaving the surface. Adamson [1960, 1990] notes that such an approach works well for adsorption on a liquid surface and it seems to work for adsorption of argon on graphite at moderate temperatures. Hence, it also be applicable to other adsorption systems.

Adamson derived an equation for an ideal two-dimensional gas by assuming that a series of molecules adsorbed to form a two-dimensional gas. Each molecule was assumed to have a surface area σ (i.e., each molecule blocks a surface area σ) and move with a Maxwell–Boltzmann velocity distribution in two dimensions. Adamson asserts that if N molecules of adsorbate are adsorbing onto a surface with an area, a, and the molecules have a two-dimensional Maxwell–Boltzmann velocity distribution, then in the absence of adsorbate/adsorbate interactions one can write an equation of state for the layer as follows:

$$\Pi(a - N\sigma) = NkT \tag{4.156}$$

where Π is called the film pressure or spreading pressure. It is defined as the force per unit length that a group of molecules would produce if the molecules were confined to a small box on the surface of the solid. The spreading pressure is a two-dimensional analog of a pressure in a three-dimensional gas. See Adamson [1960, pp. 124 and 145] or Adamson [1990, pp. 87–99] for details.

Note that $(a - N\sigma)$ is the surface area available for motion of the adsorbate. Hence, Equation 4.156 is the surface equivalent of an ideal gas law $(PV = NkT)$ with an excluded volume.

Adamson shows that one can calculate the adsorption isotherm from the equation of state. The derivation is as follows. First, it is useful to define a coverage by

$$\theta = N\frac{\sigma}{a} \tag{4.157}$$

Substituting Equation 4.157 into the left side of Equation 4.156 yields

$$(a - N\sigma) = a(1 - \theta) \tag{4.158}$$

In the gas phase, the chemical potential, μ_g, follows:

$$\left(\frac{\partial \mu_g}{\partial P}\right)_T = \frac{V}{N} \tag{4.159}$$

Adamson [1960, 1990] demonstrates that one can calculate the chemical potential of the adsorbed layer, μ_s, from

$$\left(\frac{\partial \mu_s}{\partial \Pi}\right)_T = \left(\frac{a}{N}\right) \tag{4.160}$$

Combining Equations 4.156, 4.157, and 4.158 yields

$$\Pi = \frac{NkT}{a(1 - \theta)} = \frac{\theta kT}{\sigma(1 - \theta)} \tag{4.161}$$

Substituting the value of Π from Equation 4.161 and the value of N from Equation 4.157 into Equation 4.160 yields

$$\frac{d\mu_s}{d\theta} = kT\left[\frac{1}{\theta} + \frac{1}{(1 - \theta)} + \frac{1}{(1 - \theta)^2}\right] \tag{4.162}$$

Integrating

$$\mu_s = kT \ln\left[\frac{\theta}{1 - \theta}\right] + \frac{kT}{1 - \theta} = C_1 \tag{4.163}$$

where C_1 is a constant. Equation 4.163 should approach Equation 4.17 in the limit of zero coverage. That will happen if the constant, C_1, is chosen so that

$$C_1 = kT\left[1 + \ln(q_s)\right] \tag{4.164}$$

Combining Equations 4.163 and 4.164 yields

$$\mu_s = kT \ln \left(\frac{\theta}{1 - \theta} \right) + \frac{kT \, \theta}{1 - \theta} - kT \ln (q_s) \tag{4.165}$$

at equilibrium $\mu_g = \mu_s$. Equating the value of μ_g from Equation 4.18 with μ_s from Equation 4.165 yields

$$\ln \left(\frac{q_s P_A}{q_g P_r} \right) = \ln \left(\frac{\theta}{1 - \theta} \right) + \frac{\theta}{1 - \theta} \tag{4.166}$$

Substituting for q_s from Equation 4.20 into Equation 4.166 and taking the exponential of both sides yields

$$K_{equ}^A P_A = \frac{\theta}{1 - \theta} \, e^{(\theta/1 - \theta)} \tag{4.167}$$

Equation 4.167 is the isotherm for adsorption of a gas into a random mobile layer on a solid surface. The derivation here assumed that the adsorbate had a velocity distribution like that of an ideal gas and there were no adsorbate/adsorbate interactions (except for the excluded volume). However, one would have obtained the same equation if one had assumed some other velocity distribution. Hence Equation 4.167 is a fairly general result for mobile adsorption. It does not apply to cases where there are adsorbate/adsorbate interactions, however.

4.10.2   Other Two-Dimensional Equations of State

Adamson [1990] outlines several other equations of state. However, none of these other equations of state seems to explain data better than Equation 4.167. There are other equations of state in the literature and it happens that a two-dimensional hard sphere fluid can on occasion, be a reasonable approximation for adsorption. The isotherm for a two-dimensional hard sphere fluid is derived by assuming that the adsorbed molecules interact via a series of hard-sphere repulsions, i.e., the interaction is assumed to be infinitely repulsive if the adsorbing molecules are within a radius, R_0, of each other and zero otherwise.

Schaaf and Talbot [1989] analyze the hard sphere adsorbate in detail. They consider two limiting cases, the mobile case, where the adsorbate is free to move around to find holes in the lattice, and the immobile case, where the adsorbate sticks and blocks some sites forever. The immobile case is called **random sequential adsorption** in the literature. Widom [1963] showed that if the adsorbate is immobile, the adsorbate layer never reaches equilibrium. Physically, one can imagine that incoming molecules can adsorb and block sites, and never reach equilibrium because the adsorbate molecules cannot move about to create spaces for other molecules. The effect is that molecules are forced to adsorb at low-energy sites. A detailed analysis of this phenomenon is delayed until the next chapter where we discuss the kinetics of adsorption processes. However, it is important to recognize that experimentally, one can get a situation where the adsorption process does not reach equilibrium. That is why the variation in the heat of adsorption with coverage in Figure 4.8 changed according to how the layer was prepared.

4.10.3   Monte Carlo Solution

Schaaf and Talbot [1989] also consider the case where the adsorbate layer is mobile. In that case the isotherm is calculated using a variation of the Metropolis algorithm described in Section 4.8. One starts with some initial configuration, and move an adsorbate molecule by an amount $\vec{r}$. If the energy goes down, one accepts the step. If the energy goes up one accepts the step with a probability exp $(-kT)$. One then calculates an equation of state using methods outlined by Widom [1963], Adams [1974], and Speedy [1977].

To summarize the derivation, Widom [1963] defines a configuration partition function Q^N, such that the total partition function for N adsorbed molecules is

$$Q_{canon} = q_s^N Q^N \qquad (4.168)$$

One can show that

$$Q^N = \frac{1}{N!} \sum_t \exp(-\beta E_t) \qquad (4.169)$$

where E_t is the energy of the system when the N adsorbed molecules are in arrangement t. The chemical potential of the layer becomes

$$\mu_s = -kT \frac{\partial \ln(Q_{canon})}{\partial N} = -kT \ln(q_s) - kT \frac{\partial \ln(Q^N)}{\partial N} \qquad (4.170)$$

But N is an integer. For large values of N

$$\frac{\partial \ln(Q^N)}{\partial N} = \frac{\ln(Q^N) - \ln(Q^{N-1})}{1} = \ln\left(\frac{Q^N}{Q^{N-1}}\right) = \ln\left(\frac{Q^{N+1}}{Q^N}\right) \qquad (4.171)$$

Following Widom [1963], it is useful to define a function $\psi(\vec{p})$ as the energy change that occurs when a single atom is added to the layer at point $\vec{p}$. Widom [1963] shows that one can rearrange Equation 4.169 to give

$$\frac{Q^{N+1}}{Q^N} = \frac{1}{a} \left\langle \int \exp[-\beta\psi(\vec{p})] \, d\vec{p} \right\rangle \qquad (4.172)$$

where a is the area of the layer. Notice that if one knows all of the forces between adsorbed molecules, one can calculate the expectation value of ψ as a function of coverage using the Metropolis algorithm. One can then plug the expectation value into Equations 4.170 and 4.171 to calculate the chemical potential of the layer as a function of coverage. The adsorption isotherm is just a plot of the coverage versus chemical potential. Hence, it can be calculated very easily.

It happens that most of the people who do these types of Monte Carlo calculations are interested in liquids and not surfaces and so they do not actually report adsorption isotherms. Instead, they usually report equations of state. Widom [1963] shows that one can calculate the equation of state for a two-dimensional fluid from:

$$\frac{\Pi a}{NkT} = 1 - \frac{1}{4\,kT} \frac{\left\langle \int \Phi(\vec{p}) \exp[-\beta\psi(\vec{p}) \, d\vec{p} \right\rangle}{\left\langle \int \exp[-\beta\psi(\vec{p})] \, d\vec{p} \right\rangle} \qquad (4.173)$$

where

$$\Phi(\vec{p}) = \sum_j (\vec{p} - \vec{r}_j) \frac{dh_{pj}}{d\vec{r}_j} \qquad (4.174)$$

where $\vec{r}_j$ is the position of atom j, and h_{pj} is the interaction between an adsorbate molecule at $\vec{p}$ and a second adsorbate molecule at $\vec{r}_j$ (see Ree and Hoover [1964], Adams [1974], or Gray and Gubbins [1984] for details).

Metropolis et al. [1953] and Ree and Hoover [1964] simulated a hard sphere fluid. They found that in two dimensions the fluid obeys

$$\frac{\Pi a}{NkT} = 1 + \theta + \left(\frac{4}{3} - \frac{\sqrt{3}}{\pi}\right)\theta^2 + (0.5327)\theta^3 + (0.3338)\theta^4 + (0.1992)\theta^5 \quad (4.175)$$

Henderson [1975] showed that Equation 4.175 can be approximated by

$$\frac{\Pi a}{NkT} = \frac{1 + \theta^2/8}{(1 - \theta)^2} \quad (4.176)$$

One can then use a derivation like that in Section 4.10.1 to calculate an adsorption isotherm. Schaff and Talbot [1989] obtain the following result:

$$K^A_{equ}P_A = \theta \exp\left[2 - \frac{2}{(1 - \theta)^2} + \frac{7}{8}\frac{\theta}{(1 - \theta)^2} + \frac{7}{8}\ln(1 - \theta)\right] \quad (4.177)$$

Equation 4.177 is the isotherm for mobile adsorption of a hard sphere fluid.

Over the years these results have been generalized to many other types of interactions. For a review see Gray and Gubbins [1984] or Levesque [1987]. Most of the effort has concentrated on understanding two-dimensional liquids. However, very recently there have been some attempts to model adsorption systems. The fundamental difference between an adsorbed phase and a two-dimensional liquid is that in the adsorbed phase, there are periodic adsorbate/surface interactions that modify the phase behavior of the adsorbed species. These are quite important. Unfortunately, at present there is not a good enough model for the forces between molecules and the effects of adjacent molecules on the adsorbate/surface interaction to do the calculations accurately. As a result, the work so far has been limited to fitting data via a highly parameterized model. That has not yielded any important new insights. However, one can anticipate the situation improving over the next several years, and eventually the Monte Carlo calculations will be the preferred way to model the behavior of adsorption systems in cases where one needs to know the details of the phase behavior.

4.11 PERSPECTIVE

Of course, one will not always need to model all of the details. In such a case, a less sophisticated method will suffice. For example, if one were simply trying to use the model to fit an adsorption isotherm in order to design an absorber, one would not need to know the detailed phase behavior of the adsorbates. Rather, one would only need to know the coverage as a function of the pressure. One can fit coverage versus pressure data with any of the analytical adsorption isotherms described in Section 4.4. Generally, if one is working over a small range of coverages, the Langmuir adsorption isotherm will be sufficient, while if one works over a wider range of conditions, a different model will be needed. The Freundlich or Toth adsorption isotherm is often used in the latter case. Valenzeula and Meyers [1989] show that in fact the Toth equation can be used to fit a wide range of adsorption data.

Over the years, there has been considerable effort devoted toward understanding which analytical model works best. For example, Figure 4.8 shows that heats of adsorption often vary significantly with coverage. If one wanted to reproduce heat of adsorption data, one would have to use a complex model. However, if one is trying to fit an adsorption isotherm over a limited coverage range, one could instead use a Langmuir adsorption isotherm, even though the Langmuir adsorption isotherm assumes that the heat of adsorption is constant.

In fact, in many cases the differences between the adsorption isotherms predicted by the various methods are subtle. For example, Figure 4.28 was calculated by starting with the exact results for adsorption on a square lattice with $\beta h = -0.5$, and then fitting that isotherm with some of the other equations in this chapter. The Bethe approximation is essentially exact over the entire range. The Toth equation also fits the exact results quite well provided one sets $\alpha_{TO} = 0.75$. Interestingly it is also possible to fit the exact results with a Langmuir adsorption isotherm if one assumes that each adsorbate molecule takes up two sites not one, and then adjust the saturation coverage appropriately. Notice that the differences between the various results in Figure 4.28 are often not significant. While one can devise plots that distinguish between the various models as shown in Problem 4.7, if one has adsorption data over a limited range of conditions, one can often fit the data equally well with several different adsorption isotherms.

The problem, of course, is that even though one fits the data with a model, there is no assurance that the model is correctly representing what is happening on the surface. The Langmuir adsorption isotherm only closely fits the data in Figure 4.28 when one makes the

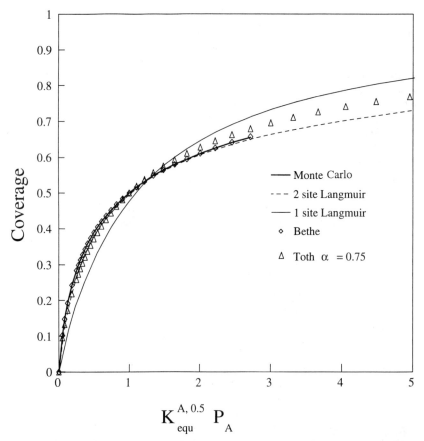

Figure 4.28 A comparison of the exact solution for the adsorption of a gas on a square lattice with βh = -0.5 to (a) the Bethe approximation, (b) the Toth equation with $\alpha_{TO} = 0.75$, and (c) the two-site Langmuir isotherm.

incorrect assumption that the adsorbate bonds to two sites. There is some connection between the two-site model and the Monte Carlo calculations in that in a system with repulsive interactions, adsorption of a gas on one site reduces the probability that an adjacent site will be filled with adsorbate. However, the parameters in Figure 4.28 were chosen so that there is no site blocking; all of the sites fill at high coverages. Yet the two-site Langmuir adsorption isotherm fits the Monte Carlo calculations better than the one-site Langmuir model even though the surface phase behavior is closer to that of the one-site Langmuir model than the two-site Langmuir model. This example illustrates the danger of using the adsorption isotherm to learn about the adsorbate behavior. There is no assurance that a model that adequately describes the adsorption isotherm will be correctly representing the behavior of the molecules on the surface and in fact there are examples where a very poor model fits better than a model that has some but not all of the correct physics. A real example is given in Example 4.B at the end of this chapter.

On a more fundamental level, the Langmuir adsorption isotherm, and all its analytical extensions discussed in Sections 4.4.1, 4.4.2, 4.4.3., 4.4.4, and 4.4.5, are not able to reproduce the detailed structure (i.e., the adsorbate arrangement) in the adsorbed phase. As noted in Section 3.1, one often observes a series of distinct phase transitions when gases adsorb onto surfaces. Often these phase transitions only have subtle effects on the adsorption isotherm. However, they have much larger effects on other properties of the adsorbed layer.

Later in this book, the findings in this chapter are used to make predictions about the kinetics of reactions between the adsorbate. Notice that with a simple reaction between two adsorbed molecules, A and B, the reaction will occur much more frequently when A and B are well mixed on the surface than when A and B separate into distinct islands. As a result, the detailed structure of the surface phase can have an important affect on the reactivity of adsorbed species. Hence, a model that ignores the possibility of island formation will not do a good job of reproducing the kinetics of surface reactions.

None of the analytical models described in this chapter reproduce the phase behavior of the adsorbed layer in detail. As a result, even if a given analytical model is able to reproduce a measured adsorption isotherm, there is no guarantee that the model will be useful for kinetics calculations.

The Monte Carlo methods described in Sections 4.8 and 4.10.3 are the best way to model the phase behavior of the adsorbate. At this point, if the surface atoms are fixed, and we know all of the interactions between the adsorbed molecules, we can calculate the phase behavior of any mixture of adsorbates exactly. Often the Bethe approximation is sufficient for many purposes. At present, we do not have good models for adsorbate/adsorbate interactions or the effects of adsorbates on the adsorbate/surface interactions. However, there is now considerable effort devoted toward understanding these parameters (see Fiebelman [1989] for details). We can anticipate having reasonable models for these interactions in the next several years.

4.12 MULTILAYER ADSORPTION

All of the material so far in this chapter only applies to monolayer adsorption. In most reacting systems, the reaction is confined to a single monolayer, and so one just has to consider adsorption in the first monolayer to understand reactions. However, for completeness, we also briefly discuss multilayer adsorption. The reader is referred to Gregg and Sing [1982] for a more complete description of multilayers.

4.12.1 The BET Adsorption Isotherm

Multilayer adsorption is fundamentally different from monolayer adsorption in that while adsorbate/surface interactions control monolayer adsorption, adsorbate/adsorbate interactions

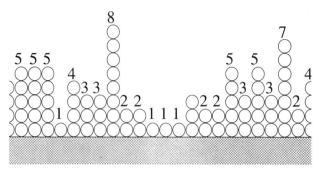

Figure 4.29 Brunauer's model of multilayer adsorption; that is, a random distribution of sites covered by one, two, three, etc., adsorbate molecules.

control multilayer adsorption. In a groundbreaking paper, Brunauer, Emmett, and Teller [1938] derived the first isotherm for multilayer adsorption. This isotherm is now called the BET equation. The derivation starts by dividing up the surface into the number of sites that are empty, the number of sites that are covered by one monolayer, the number of sites that are covered by two monolayers, etc., and then assuming a random distribution of the various types of sites as illustrated in Figure 4.29. One then defines a series of equilibrium constants $K_1, K_2 \ldots$, where K_i is the equilibrium constant between the gas phase and the number of sites covered by i adsorbed layers; K_i is defined by

$$K_i = \frac{[A]_i}{P_A [A]_{i-1}} \tag{4.178}$$

where $[A]_0$ is the number of bare sites, and $[A]_i$ is the number of surface sites covered by i molecules. Normally one assumes

$$K_2 = K_3 = K_4 \ldots = K_m \tag{4.179}$$

Combining Equations 4.178 and 4.179 yields

$$[A]_1 = K_1 P_A [A]_0 \tag{4.180}$$

$$[A]_2 = K_2 P_A [A]_1 = K_1 K_m P_A^2 [A]_0 \tag{4.181}$$

$$[A]_i = K_i P_A [A]_{i-1} = K_1 K_m^{i-1} P_A^i [A]_0 \tag{4.182}$$

The total number of sites on the surface, S_0, is given by

$$S_0 = \sum_{i=0}^{\infty} [A]_i = [A]_0 + \sum_{i=1}^{\infty} K_1 K_m^{i-1} P_a^i [A]_0 \tag{4.183}$$

Performing the sum yields

$$S_0 = [A]_0 + \frac{K_1 P_A [A]_0}{1 - P_A K_m} \tag{4.184}$$

The total concentration of molecules on the surface $[A]$ is given by

$$[A] = \sum_{i=1}^{\infty} i[A]_i = \sum_{i=1}^{\infty} iK_1 K_m^{i-1} P_A^i [A]_0 \tag{4.185}$$

Performing the sum

$$[A] = \frac{K_1 P_A [A]_0}{(1 - K_m P_A)^2} \tag{4.186}$$

Dividing Equation 4.186 by Equation 4.184 yields

$$\frac{[A]}{S_0} = \frac{\dfrac{K_1 P_A}{(1 - K_m P_A)^2}}{\left[1 + \dfrac{K_1 P_A}{1 - K_m P_a}\right]} \tag{4.187}$$

Defining two quantities x_B and c_B by

$$x_B = P_A K_m \qquad c_B = \frac{K_1}{K_m} \tag{4.188}$$

and rearranging yields

$$\boxed{\frac{[A]}{S_0} = \frac{c_B x_B}{(1 - x_B)\,[1 + (c_B - 1)x_B]} \tag{4.189}}$$

Equation 4.189 is called the BET equation. Note that the coverage goes to infinity when $x = 1$ so that the gas begins to condense. Brunauer, Emmett, and Teller [1938] showed that condensation occurs on a flat surface when $P_A = P_A^{sat}$, where P_A^{sat} is the vapor pressure of liquid A at the temperature where the isotherm was measured. Substituting $x = 1$ and $P_A = P_A^{sat}$ into Equation 4.188, solving for K_m, and substituting the resulting expression back into Equation 4.188 yields

$$x_B = \frac{P_A}{P_A^{sat}} \tag{4.190}$$

Figure 4.30 shows a plot of the BET Equation 4.189, for various values of c_B. Notice that the BET equation goes from type III behavior to type I behavior as c_B increases. Hence, the BET equation can reproduce the qualitative isotherms shown in Figure 4.4. The BET equation is the only two-parameter model proposed so far that reproduces this qualitative behavior. Thus, the BET equation was an important advance.

Still, the BET equation rarely fits all of the subtle details of an adsorption isotherm on a modern adsorbant. For example, the argon and krypton data in Figure 4.2 show a series of changes in slope. They are not reproduced by the BET equation. A detailed comparison of the krypton data in Figure 4.2 and Figure 4.3 to the BET equation is given in Example 4.I. At low coverage the experimental data show a phase transition due to attractive adsorbate/adsorbate interactions. The BET equation ignores these interactions so it misses the phase transition. The BET equation also does not fit the multilayer data very well, because it ignores surface tension. Similarly, the small bumps in the argon data, and the details of the ammonia data are not well represented by the model. Nevertheless, the BET equation does fit the data at coverages near a monolayer. Hence, the BET equation is a useful way to estimate how much gas adsorbs in a monolayer.

Brunauer [1961] notes that these deviations are caused by the approximations used to derive Equation 4.178. According to Equation 4.178 the equilibrium constant for adsorption

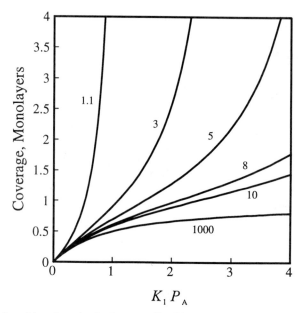

Figure 4.30   A plot of the adsorption isotherm predicted by the BET equation for various values of c_B.

of gas into the lth layer is independent of coverage. However, the results in Section 4.6 show the equilibrium constant for adsorption of a gas into a two-dimensional layer is independent of coverage only when the adsorbate/adsorbate interactions are negligible. If adsorbate/adsorbate interactions were negligible, one would not get multilayers to form. Hence, there is an inherent contradiction in the BET equation in that one is simultaneously assuming that there are no adsorbate/adsorbate interactions and that multilayers form.

In addition, Equations 4.178 and 4.190 have ignored the effects of surface tension. In Section 3.1 it was noted that the vapor pressure of a fluid in a small pore is less than the vapor pressure of the bulk fluid. Equation 4.190 ignores that effect. The vapor pressure of a curved boundary is greater than the vapor pressure of a flat surface. That effect is also ignored in Equation 4.190. The result is that there are significant deviations from the BET equation when the pores in a porous material start to fill. Brunauer [1961] defended these approximations by noting that if one tries to do better one will invariably need more parameters in the model. However, one does need to do better if one wants to fit adsorption isotherms in detail.

One of the main applications of the BET equation is in the measurement of surface areas. The approach is to measure an adsorption isotherm for nitrogen or krypton. One then fits the data to Equation 4.189 to get a value of S_0, the number of molecules that could adsorb in a single layer. One then multiplies by an area of a molecule, e.g., 16.2 Å^2 for nitrogen, to get a number for the surface area of a porous solid. Example 4.I shows how the technique works. Most measurements of surface areas are made in this way. Hence, the BET equation forms a basis for the primary coverage calibration in most supported catalyst systems.

4.12.2   Wetting, Capillary Condensation, and Other Advanced Topics

Over the years, there have been many attempts to improve the BET equation. Early investigators tried to use the Kelvin equation, Equation 3.1, to modify Equation 4.190. Typically, one would assume a pore size distribution, and use the distribution to correct the BET equation. At one point people even thought that one could do such a calculation in reverse and calculate the pore size distribution of a porous solid. However, people now know that the Kelvin equation does not work when the pore size gets below 10 molecular diameters. As a result, such an approach is limited.

The problem with using a more sophisticated model is that the computations are very complex; one needs to know the pore-size distribution and how the pores are connected. Such data are rarely available for real adsorbants. As a result, most of the calculations in the literature have been done on highly idealized surfaces. The calculations are not far enough along that they can be used to routinely model adsorption. As a result these methods will not be considered in this monograph. The reader should refer to Evans [1987], Dietrich [1988], or Evans and Parr [1990] for a description of some of the calculations. There is a good description of multilayer adsorption in Gregg and Sing [1982] and Nicholson and Parsonage [1982]. The reader should refer to their work for further information about this topic.

4.13 SOLVED EXAMPLES

Earlier in this chapter a number of equations were derived to model adsorption. In this section we solve a few examples to illustrate some of the properties of the equations and to show how the equations can be used to fit data.

Example 4.A Fitting Adsorption Data

The objective of this problem is to show how to fit experimental adsorption isotherms using the equations in Sections 4.3 to 4.9.

Christmann et al. [1974] measured very accurate adsorption isotherms for hydrogen adsorption on Ni(110). The data in Table 4.2 were obtained at 89°C. How well do these data fit (a) the Langmuir adsorption isotherm, (b) the Lattice gas model, the (c) Bethe approximation, the (d) Fowler adsorption isotherm (i.e., Bragg–Williams approximation)?

TABLE 4.2 The Amount of Hydrogen that Adsorbs on an 89°C Ni(110) Sample as a Function of the Hydrogen Pressure*

Pressure (torr)	Coverage (monolayers)
3×10^{-8}	0.003
1×10^{-7}	0.021
2×10^{-7}	0.041
3×10^{-7}	0.059
4×10^{-7}	0.171
5×10^{-7}	0.482
6×10^{-7}	0.667
1×10^{-6}	0.781
2×10^{-6}	0.845
3×10^{-6}	0.877
4×10^{-6}	0.890
1.25×10^{-5}	0.959
1.5×10^{-5}	0.975
2×10^{-5}	0.984

*Data of Christmann et al. [1974].

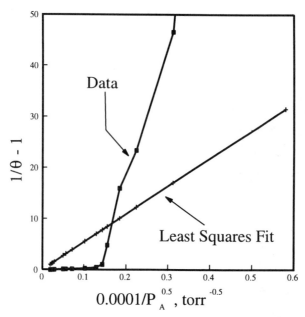

Figure 4.31   A plot of Christmann's data in the form suggested by Equation 4.30.

SOLUTION

The Langmuir Adsorption Isotherm.   Hydrogen dissociates on Ni(110), therefore the appropriate Langmuir adsorption isotherm will be Equation 4.27. From Equation 4.30, a plot of $(1/\theta - 1)$ versus $1/P_A^{0.5}$ should be linear. However, Figure 4.31 shows the plot and it is very nonlinear. As a result, one does not get a very good fit to the data with the Langmuir adsorption isotherm.

To illustrate how bad the fit is with the Langmuir adsorption isotherm, K_{equ}^A was calculated in two separate ways. First a least squares regression was used to fit the data to Equation 4.30. The regression assumed that $(1/\theta - 1)$ approach 0 at high pressure. The result is the bottom line in Figure 4.32. We have also adjusted K_{equ}^A to get something that which is qualitatively correct. However, neither of these regression schemes reproduce the data accurately. When people fit data to the Langmuir adsorption isotherm, they often do not assume that the adsorbate dissociates. Instead, they fit the data to the nondissociative form of the isotherm, the dissociative form for the isotherm, and the isotherms for multiple sites and try to see which one fits best. It happens that one can fit Christmann's data can be fitted better with

$$\theta = \frac{K_{equ}^A P_A^{1.5}}{1 + K_{equ}^A P_A^{1.5}} \tag{4.191}$$

than with Equation 4.27. Equation 4.191 is the Langmuir adsorption isotherm for adsorption of a species that forms three-atom clusters on the surface. Occasionally you read a paper that assumes that if the data fit the Langmuir adsorption isotherm for some special adsorbate arrangements, e.g., clusters, then it implies that the adsorbate really has that arrangement. However, in my view, this is nonsense. In the case of hydrogen adsorption on Ni(110) it is clear that no clusters actually form. Instead, the H_2 dissociates. The phase behavior that one observes is associated with attractive interactions in the adsorbed layer, not cluster formation.

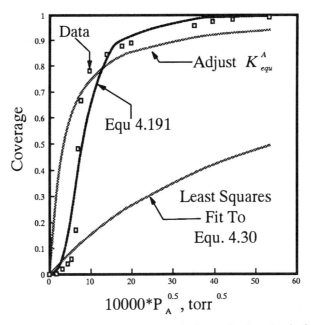

Figure 4.32 Fits of Christmann's data to the langmuir adsorption isotherm.

This example illustrates one of the dangers with using the adsorption isotherm to learn about the state of the adsorbate. There are many examples where a reasonable fit to an adsorption isotherm comes with an equation derived with entirely wrong physics.

Lattice Gas Model. Section 4.8 provided a solution of the lattice gas model for a square lattice. Ni(110) is a rectangular lattice. However, if one assumes that strength of the nearest neighbor interactions are the same in the [001] and [1$\overline{1}$0] directions, then the solution will be the same for a rectangular lattice as for a square lattice. One can then fit the data to the gas model replotting the data on the same scale as Figure 4.21.

One does need to consider that H_2 dissociates upon adsorption. Note that for a dissociative adsorption process

$$2\beta F = \ln\ (K_{equ}^{A,\theta=0.5}\ P_A^{1/2}) \qquad (4.192)$$

Therefore, one can fit the data to a lattice gas model by plotting θ versus $\ln(P_A^{1/2})$ and shifting the curve left and right until one obtains a good fit. Figure 4.33 shows such a plot. Notice that for $\theta < 0.5$ the data seem to fall on the line for $\beta h = 1.5$, while the data are closer to the $\beta h = 1.0$ line at higher coverages. Hence one can get a reasonable fit to the data with a lattice gas model with $\beta h = 1$ to 1.5 (see Figure 4.34).

Note, however, that again even though the adsorption isotherm can be fit with the lattice gas model, there is no guarantee that the lattice gas model is correctly representing the adsorbate phase behavior. Binder and Landau [1981] show that the first nearest neighbor interaction is strongly repulsive during hydrogen adsorption on Ni(100). And yet Figure 4.34 shows that the adsorption isotherm can be accurately reproduced by assuming that the first nearest neighbor interaction was attractive. Hence, a good fit was obtained with an incorrect model. What is happening physically is that while the first nearest neighbor interaction for hydrogen adsorption on nickel is repulsive, the second nearest neighbor interaction is attractive. The analysis in Section 4.8 shows that the adsorbate will form a C(2x2) layer if there

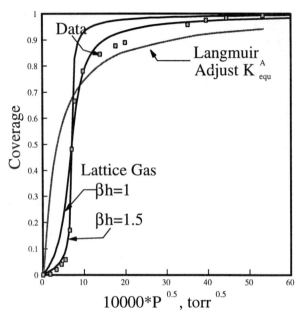

Figure 4.33   A replot of the data from Figure 4.34 on the same scale as the data in Figure 4.32.

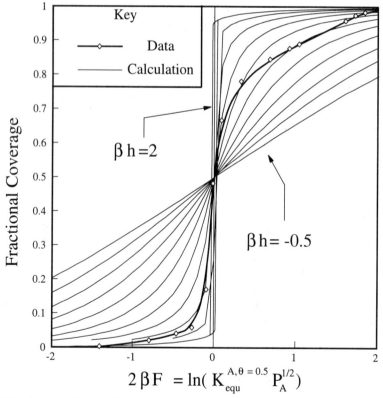

Figure 4.34   A comparison of Christmann's data to the predictions of the lattice gas model with symmetric first nearest neighbor interactions.

are strongly repulsive first nearest neighbor repulsions and attractive second nearest neighbor interactions. It is easy to show that if there are strongly repulsive first nearest neighbor interactions and attractive second nearest neighbor interactions, the adsorption of the gas on the C(2x2) sites will closely approximate the results in Figure 4.23, with h being the strength of the interaction between molecules on the C(2x2) sites (i.e., the second nearest neighbor interaction). Hence, Figure 4.23 is approximately correct for attractive second nearest neighbor interactions even though it was derived for first nearest neighbor interactions.

One might expect that one can get an even better fit if one allowed the interaction energy to be different in the [001] and [1$\bar{1}$0] directions. However, the Hamiltonian will still be symmetric about the point $\theta = 0.5$, $F = 0$. As a result one does not improve the fit of the data if one allows the interaction energy to be different in the [001] and [1$\bar{1}$0] directions.

Physically the variation in the effective value of βh occurs because there are indirect interactions between the adsorbed hydrogen atoms. The adsorbed hydrogen changes the electronegativity of the surface, which in turn changes the effective interaction between adjacent hydrogen atoms. Binder and Landau [1981] show that one can fit the data for hydrogen adsorption on Ni(100) precisely by including a three-body term in the Hamiltonian. The three-body term is given in Equation 4.121. While it is not immediately obvious that this approach is correct, in Problem 4.21 we ask the reader to show that one can simulate an indirect interaction by modifying the two- and three-body terms in the Hamiltonian and adding a long-range component. Hence, one can simulate an indirect interaction via a lattice gas model provided one modifies the Hamiltonian appropriately.

Bethe Approximation. One can also try to fit the data with the Bethe approximation. The Bethe approximation predicts phase behavior that is very similar to that for the lattice gas approximation. As a result one might expect the results for the Bethe approximation to be close to the results for the lattice gas. Figure 4.35 compares the predictions of the Bethe approximation to the experiment data. One can observe that, as with the lattice gas model, the Bethe approximation fits the data provided βh is allowed to vary with coverage. There

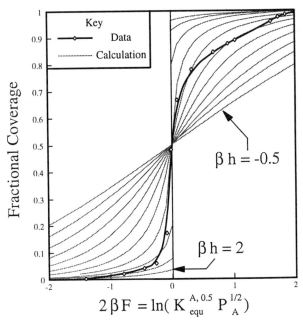

Figure 4.35 A comparison of Christmann's data to the predictions of the modified Bethe model.

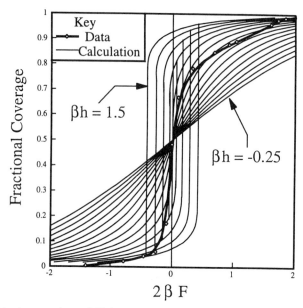

Figure 4.36 A comparison of Christmann's data to the Bragg–Williams' approximation.

are some slight differences between the values of βh estimated from Figures 3.34 and 3.35. However, the differences are well within experimental parameters under all conditions. Hence, the modified Bethe approximation is essentially as good as the data.

Fowler–Bragg–Williams Approximation. The Fowler adsorption isotherm (i.e., the Bragg–Williams approximation) does not fit the data, however. Figure 4.36 compares Christmann's data to the predictions of the Bragg–Williams model. One can fit the high coverage portion of the data with the Bragg–Williams model with $\beta h = 0.75$, i.e., a βh 50% smaller than the value needed to fit the data with a lattice gas. However, one cannot fit the low-to-moderate coverage part of the isotherm at all with the Bragg–Williams isotherm. Notice that at low coverages the data lie below the line for $\beta h = 1.5$ in Figure 4.36. However, with a βh of 1 or more, the Bragg–Williams approximation predicts that there will be an S-shaped curve similar to that in Figure 4.22. Note, however, that the data in Table 4.2 were generated by starting with a clean surface, exposing the surface to gas, and then measuring the coverage as they raised the pressure. It is easy to show that one would not measure the entire isotherm in such an experiment. Rather, one would start at the bottom of the S-shaped curve, and would ride along the bottom of the S until one got out to the maximum pressure point on the bottom of the S. One would jump to the upper branch of the S with further increases in pressure. If one then lowered the pressure, one would stay on the upper branch of the S until one got to the minimum pressure point on the upper branch. One would then jump to the lower branch. Therefore, one would not measure the same isotherm with increasing and decreasing pressure. Rather, there should be a hysteresis loop. Experimentally, one does not observe the hysteresis loop. Hence, we cannot get a meaningful fit to the data at moderate coverages with the Bragg–Williams model.

Example 4.B

The prior example showed that adsorption data can often be accurately reproduced with several different adsorption isotherms. The objective of this problem is to do a comparison to a case

TABLE 4.3   The Amount of Hydrogen that Adsorbs on a Ni(100) Surface*

$\beta F = kT$ $\ln(K_{equ,BL}^{A,\theta=1.5}\, P_A^{0.5})$	Coverage	$\beta F = kT$ $\ln(K_{equ,BL}^{A,\theta=1.5}\, P_A^{0.5})$	Coverage
−2.14	0.004	−0.98	0.289
−2.00	0.015	−0.94	0.394
−1.70	0.028	−0.90	0.443
−1.60	0.039	−0.66	0.469
−1.47	0.057	−0.53	0.487
−1.34	0.084	−0.40	0.495
−1.20	0.128	−0.27	0.497
−1.07	0.193	−0.14	0.498

*Calculated for $-\beta h = \beta h_{nn} = 10/9$.

where everything is known to see if there is any physics in the fits. Binder and Landau [1981] did an extensive Monte Carlo simulation of hydrogen adsorption on Ni(100). They calculated the isotherm given in Table 4.3. The objective of this problem is to try to fit these results with a multisite Langmuir adsorption isotherm. Assume that the saturation coverage and all of the constants in the equations are variables.

SOLUTION

In Problem 4.32 we ask the reader to show that if the hydrogen dissociates and then follows a multisite Langmuir adsorption isotherm, then

$$K_{equ}^A P_A^{0.5} = \frac{\dfrac{\theta}{\theta_s}}{\left(1 - \dfrac{\theta}{\theta_s}\right)^n} \qquad (4.193)$$

where n is the number of sites covered by adsorbate, and θ is the fractional coverage of A and θ_s is the saturation coverage. If take K_{equ} to be a variable, then a plot of the log of the right side of Equation 4.193 as a function of log of F, should match the data except for a constant. Figure 4.37 shows such a plot where K_{equ} was adjusted to fit the data. Notice that one can fit the data quite well with the multisite Langmuir adsorption isotherm if one assumes that each hydrogen atom covers one-quarter of a site at low coverages and one-eighth of a site at higher coverage! However, note that hydrogen really covers a single site, not one-quarter of one-eighth of a site. This example again illustrates the difficulty with taking parameters derived from a fit of an adsorption isotherm and using the parameters for other purposes.

Example 4.C   Estimating H_1 from First Principles

The examples in Sections 4.3 to 4.9 provide a series of plots of the amount of gas adsorbed as a function of H_1 where

$$H_1 = \mu_g + kT \ln (q_s)$$

$$= kT \ln \left[\left(\frac{q_s}{q_g}\right)\left(\frac{P_A}{P_r}\right)\right]$$

$$= kT \ln [K_{equ}^{A,0} P_A] \qquad (4.194)$$

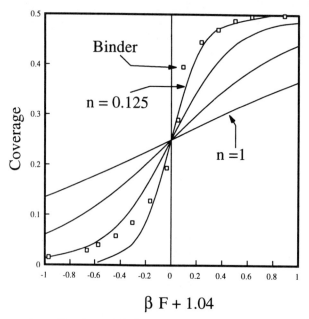

$$\beta\, F + 1.04$$

Figure 4.37 A comparison of Landau and Binder's calculations to coverages calculated via the multisite Langmuir adsorption isotherm, with a saturation coverage of 0.5 and n = 1, 0.5, 0.25, 0.125.

According to the analysis negligible adsorption should be seen when βH_1 is less than -3; the surface is usually saturated when βH_1 is greater than 2 to 6. When $\beta(H_1 + 0.5\, n_n h) = 0$, the coverage is 0.5. The objective of this example is to put some numbers into Equation 4.194 so one can get a measure of the magnitude of H_1 in a variety of circumstances.

Assume that the surface is exposed to 10^{-6} torr of a nonlinear adsorbate at 300 K. Calculate H_1 as a function of the binding energy of the adsorbate.

SOLUTION

We need expressions for q_g and q_s in order to evaluate the various terms in Equation 4.194. In the gas phase a nonlinear molecule with n_0 atoms has 3 translational modes, 3 rotational modes, and $3n_0 - 6$ vibrational modes. When the molecule adsorbs, several of the translational and rotational modes are converted into vibrational modes. The zero-point energy of the molecule also changes. If the translational, vibrational, and rotational modes of the molecule are independent, then we can express q_g and q_s as a product of the partition functions of the normal modes of the adsorbate molecule in the gas phase and on the surface:

$$q_g = q_t^3 q_r^3 q_v^{3n_0 - 6} \tag{4.195}$$

$$q_s = q_t^{n_t} q_r^{n_r} q^{3n_0 - n_t - n_r} e^{-(\Delta U_a/kT)} \tag{4.196}$$

where q_t, q_r, and q_v are the partition functions for an individual rotational, vibrational, or translational mode, respectively; n_t and n_r are the number of translational and rotational modes of the adsorbed molecules; and ΔU_a is the internal energy change upon adsorption measured from the zero point of the molecule in the gas phase and on the surface.

Table 4.4 shows the magnitude of the various partition functions. It works out that the

TABLE 4.4 **Equations for the Partition Function for Translational, Rotational, Vibrational Modes and Electronic Energy Levels**

Type of Mode	Partition Function	Approximate Value of the Partition Function for Simple Molecules
Translation of a molecule of an ideal gas in a one-dimensional box of length a_x	$q_t = \dfrac{(2\pi m_g kT)^{1/2} \, a_x}{h}$	$q_t \approx 1 - 10/\text{Å} \; a_x$
Translation of a molecule of an ideal gas at a pressure P_A and a temperature T	$\dfrac{q_t^3}{N} = \dfrac{(2\pi m_g kT)^{3/2}}{h^3}\left(\dfrac{kT}{P_A}\right)$	$q_t^3 \approx 10^6 - 10^7$
Rotation of a linear molecule with moment of inertia I	$q_r^2 = \dfrac{8\pi^2 I k T}{s_n h^2}$, where s_n is the symmetry number	$q_r^2 \approx 10^2 - 10^4$
Rotation of a nonlinear molecule with moments of inertia of I_a, I_b, I_c about three orthogonal axes	$q_r^3 = \dfrac{8\pi^2 (8\pi^3 I_a I_b I_c)^{1/2} kT^{3/2}}{\sigma h^3}$	$q_r^3 = 10^4 - 10^5$
Vibrational of a harmonic oscillator when energy levels are measured relative to the harmonic oscillator's zero-point energy	$q_v = \dfrac{1}{1 - \exp\left(-h\nu/kT\right)}$ where $\nu =$ the vibrational frequency	$q_v \approx 1 - 3$
Electronic level (assuming that the levels are widely spaced)	$q_e = \exp\left(-\dfrac{\Delta U}{kT}\right)$	$q_e = \exp\left(-\beta\Delta U\right)$

ratio of q_s/q_g varies considerably according to the state of the adsorbed molecule. If all of the rotational and translational modes of the adsorbed molecule are converted to vibrations (i.e., if the adsorbate is held in a tightly bound immobile state), then

$$\frac{q_s}{q_g} \approx 10^{-9} \exp\left(-\beta\Delta U_a\right) \qquad (4.197)$$

where ΔU_a is called the binding energy of the adsorbed molecule. It is the change in the internal energy of the molecule when it adsorbs, measured from the zero-point energy of the molecule in the gas phase to the zero point energy of the molecule when it is adsorbed.

If the molecule is strongly bound but it can still rotate around the adsorbate surface bond

$$\frac{q_s}{q_g} \approx 10^{-7} \exp\left(-\beta\Delta U_a\right) \qquad (4.198)$$

Alternately, if the molecule is free to rotate and translate in the two-dimensional plane of the surface, then at a density of 1×10^{14} molecules/cm^2,

$$\frac{q_s}{q_g} \approx 10^{-2} \exp\left(-\beta\Delta U_a\right) \qquad (4.199)$$

Hence q_s/q_g can vary by 10^7 according to the detailed state of the molecule on the surface.

In fact people often assume that most strongly bound species are immobile when they adsorb. As a result, q_s/q_g is often somewhere between the value given in Equation 4.197 and Equation 4.199. However, there are many exceptions and so data are needed if one wants to calculate an accurate value of H_1 via Equation 4.194.

At present there are only a few systems where there has been enough careful work so that the state of the adsorbate is known well enough to calculate q_s/q_g. For example, Lanzillotto, Yates et al. [1987], and Netzer and Madey [1982] have examined ammonia adsorption in Ni(111) in great detail. Figure 4.38 shows one of their ESDIAD images of the adsorbed ammonia. The ESDIAD image shows distinct molecular features that appear to be stationary. The appearance of a stationary image implies that the ammonia adsorbs into an immobile or partially immobile state. However, the part of the image associated with the hydrogens is axial symmetric about the Ni–N bond. Therefore Lonzilloto et al. concluded that the ammonia is rotating freely around the ammonia–nickel bond. Lanzillotto et al. also measured the vibrational spectrum of the adsorbed ammonia and found that the vibrational spectrum of the ammonia does not change significantly upon adsorption. One does need a rotational constant to actually calculate the partition function, and no one has measured it yet. However, if a typical value for the rotational constant is plugged in, one finds that the ratio of the partition functions follows Equation 4.198. Substituting Equation 4.198 into Equation 4.194 yields

$$-H_1 \approx 3.6.6\, kT + \Delta U_a \qquad (4.200)$$

Therefore, to get an H_1 of zero at 10^{-6} torr 300 K

$$\Delta U_a \approx -21.9\ \text{kcal/mole}$$

It works out that if $-\Delta U_a$ is less than 19 kcal/mole H_1 will be less than -3, so there will be negligible adsorption.

It happens that in most systems there are not enough data to tell whether the molecule is rotating or not. However, if we assume that the adsorbed molecule is immobile (we will test this assumption in the next problem), we conclude that if $-\Delta U_a$ is less than 18–21 kcal/mole, one will get little adsorption at 10^{-6} torr and 300 K.

One can turn this idea around and use Equation 4.200 to estimate the binding energy of a molecule from the adsorption isotherm. The idea is that since $\theta = 0.5$ comes at $2F = H_1 + 0.5\, n_n h \approx 0$, if one knows the temperature where $\theta = 0.5$, we can estimate the average

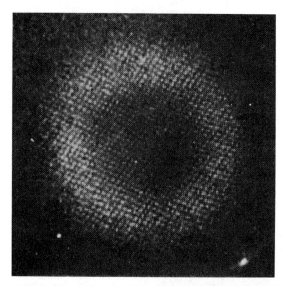

Figure 4.38   An ESDIAD image of ammonia on Ni(111). (From Netzer and Madey [1982].)

TABLE 4.5 The Average Binding Energy Needed to
Give Approximately Half a Monolayer Coverage When a
Surface Is Exposed to 10^{-6} torr of a Strongly Bound
Immobile Adsorbate at Various Temperatures

Temperature (K)	Approximate Average Binding Energy $-\Delta U_a + 0.5\, n_n h$ (kcal/mole)
100	7
300	22
500	37
750	55
1000	73

binding energy of the molecule, $-\Delta U_a + 0.5\, n_n h$, from Equation 4.200. Table 4.5 shows some results of the calculations.

Figure 4.20 shows that the coverage is near unity when $F = 3$ to 6, while the coverage is near zero when F is below about -3. As a result, if the binding energy of the gas is more than a few kcal/mole larger than the value given in Table 4.5, the system will form a saturated monolayer, while if it is more than a few kcal/mole smaller than the value in Table 4.5, hardly any molecules will stick.

Example 4.D   When Does Surface Mobility Affect Properties?

In Example 4.C the partition function for the adsorbed layer was calculated assuming that the adsorbate layer was immobile. However, in reality the adsorbate layer is held in a periodic potential, and there is a finite barrier to jumping from one site to the next. The objective of this problem is to see how the finite barrier affects the partition function for the adsorbed layer.

Consider the adsorption of a gas onto a one-dimensional surface, and assume that the adsorbate is held onto the surface by a two-dimensional sinusoidal potential $E(x) = E_D/2$ $[\sin(2\pi x/a_x) + \sin(2\pi y/a_y)]$.

Hill [1960] shows that the partition function for a particle in a sinusoidal potential can be approximated by:

$$q_{xy} = \frac{\pi E_D}{\theta kT} e^{-E_D/kT} I_0^2\left(\frac{E_D}{2\,kT}\right) \left(\frac{1}{1 - \exp\left(-\left[\dfrac{2h^2}{m_g a x^2 E_p}\right]^{1/2} \dfrac{E_D}{2\,kT}\right)} \right)^2 \qquad (4.201)$$

where m_g is the mass of the adsorbate, and I_0 is the modified bessel function of the first kind. As with Table 4.4, this partition function applies when energies are measured from the zero-point energy of a molecule in the sinusoidal potential.

(a) Compare the partition function calculated from Equation 4.201 to the partition function in Table 4.4. For $\theta = 0.25$, how large does E_D have to be before the partition function for the particle in a sinusoidal potential is within a factor of 10 of the value for a typical vibration from Table 4.4?

(b) Assume that the molecule diffuses a distance x given by $x^2 = 10^{13} \exp(-E_D/kT)$ Å/ sec t_{diff}, where t_{diff} is the diffusion time. How far will the molecule diffuse in one second at the value of E_D calculated in part (a)?

SOLUTION

Assume $T = 300$ K; $a_x = 3$ Å $= 3 \times 10^{-10}$ meters; $E_D = 10$ kcal/mole $= 6.9 \times 10^{-20}$ joules/molecule; $m_g = 28$ amu $= 4.65 \times 10^{-26}$ kgn:

$$\frac{E_D}{kT} = \frac{(6.9 \times 10^{-20} \text{ joules/molecule})}{(1.38 \times 10^{-23} \text{ joules/molecule K}) (300 \text{ K})} = 16.7$$

$$q_{xy} = \frac{\pi(16.7)}{0.25} \exp\left(-16.7\right) I_0^2 \,(8.35)$$

$$\cdot \left(\frac{1}{1 - \exp\left(-\left(\dfrac{2\,(6.63 \times 10^{-34}/\text{s})^2}{(4.65 \times 10^{-26} \text{ kg}) (3 \times 10^{-10})^2 \, 6.9 \times 10^{-20}}\right)^{1/2} 8.35\right)}\right)$$

Looking up values of I_0 from Abromowitz and Stegun, *Handbook of Mathematical Functions*, Dover, NY [1970] shows

$$q_{xy} = \frac{(1.03)}{0.25} \left(\frac{1}{1 - \exp\left(-0.459\right)}\right)^2 = 30.5$$

Similar calculations show

E_D, kcal/mole	q_{xy}
1	340
2	140
5	56
10	30.5
15	22.1
20	17.8

From Table 4.4, $q_v^2 = 1$ to 9. An average value is 5 so the mobility of the adsorbate changes the partition function for the adsorbate by an order of magnitude when $E_D \approx 5$ kcal/mole.

$$x_D^2 = (10^{13} \text{ Å}^2/\text{sec}) \exp\left[-\frac{(5 \text{ kcal/mole})}{RT}\right] (1 \text{ sec}) = 2.4 \times 10^9 \text{ Å}^2$$

$$x_D \approx 50{,}000 \text{ Å}$$

Example 4.E Landau Energies to Estimate Phase Behavior

The purpose of this problem is to develop criteria for what phase is formed during adsorption on a square lattice.

Calculate the energy/adsorbed molecule as a function of H_1 and the strength of the first and second nearest neighbor interactions for **random** adsorption of a gas onto (a) a (1x1) layer, (b) a C(2x2) layer, (c) a P(2x2) layer, (d) a (2x1) layer. (e) Use those results to calculate what structures are possible for a square lattice. Assume a square lattice in all cases.

SOLUTION

Random adsorption on a (1x1) layer: For random adsorption the probability of any site being filled is θ. E_i, the energy of a atom adsorbed at site i, becomes

$$-E_i = H_1 + 4h_{nn}\theta + 4h_{2nn}\theta \qquad (4.202)$$

The factor of 4 comes about because there are four nearest neighbors and four second nearest neighbors.

Random adsorption on a C(2x2) layer: With a C(2x2) layer there are no first nearest neighbors, but there are second nearest neighbors. If it is assumed that the C(2x2) layer is partially filled, then

$$-E_i = H_1 + 4h_{2nn}\theta \qquad (4.203)$$

Random adsorption on a P(2x2) layer: With a P(2x2) layer there are no first or second nearest neighbors

$$-E_i = H_1 \qquad (4.204)$$

Random adsorption on a (2x1) layer: With a (2x1) layer there are two first nearest neighbors and no second nearest neighbors

$$-E_i = H_1 + 2h_{nn}\theta \qquad (4.205)$$

Equations 4.202 through 4.205 are called the Landau expressions for the energy.

Implications of These Results. Notice that the (1x1) layer has the lowest energy when $E_{(1x1)} < E_{C(2x2)}$, $E_{(1x1)} < E_{P(2x2)}$, $E_{(1x1)} < E_{(2x1)}$. Substituting Equations 4.202, 4.203, 4.204, and 4.205 into this yields

$$4\,h_{nn} + 4\,h_{2nn} > 4\,h_{2nn} \qquad (4.206)$$

$$4\,h_{nn} + 4\,h_{2nn} > 0 \qquad (4.207)$$

$$4\,h_{nn} + 4\,h_{2nn} > 2\,h_{nn} \qquad (4.208)$$

Rearranging Equation 4.206

$$h_{nm} > 0 \qquad (4.209)$$

Rearranging Equation 4.208

$$h_{nn} + 2\,h_{2nn} > 0 \qquad (4.210)$$

Note that if Equations 4.209 and 4.210 are satisfied, Equation 4.207 will be satisfied too. Therefore, the (1x1) layer will have the lowest energy when

$$h_{nn} > 0 \quad h_{nn} + 2\,h_{2nn} > 0$$

A similar analysis shows that the C(2x2) layer will have the lowest energy when

$$h_{nn} < 0 \quad h_{2nn} > 0$$

The P(2x2) layer will have the lowest energy when

$$h_{2nn} < 0 \qquad h_{nn} < 0$$

The (2x1) layer will have the lowest energy when

$$h_{nn} > 0 \qquad h_{nn} + 2 h_{2nn} < 0$$

The preceding analysis used criteria based on the Landau energies of the layer and assumed that the sites fill randomly. This is equivalent to a Bragg–Williams approximation. In reality the sites do not fill randomly. However, one can show from renormalization group theory that the actual adsorbate arrangements one observes correspond to the ones one derives from the Landau energy expression (see Schick [1981] or Plischke and Bergersen [1989] for details).

The implication of the results in this problem is that if $h_{nn} > 0$ and $h_{nn} + 2h_{nn} > 0$, one could only form a (1x1) layer. On the other hand, if $h_{2nn} < 0$ and $h_{nn} < 0$, then *at low enough temperatures* one would form a P(2x2) layer at coverages below 0.25. As one raises the coverage, one could form either a C(2x2) or a (2x1) overlayer or go directly to a (1x1) overlayer. The (2x1) overlayer forms when E_i for the (2x1) overlayer is lower than for the C(2x2) and the (1x1) overlayer. After some algebra one can show that the P(2x2) layer is converted to a C(2x2) layer when $2h_{2nn} > h_{nn}$ (with $h_{nn} < 0$ and $h_{2nn} < 0$), while the (2x1) layer forms when $2h_{2nn} < h_{nn}$ (with $h_{nn} < 0$ and $h_{2nn} < 0$). One can work out other cases. If $h_{nn} < 0$ and $h_{2nn} < 0$, the adsorbate will first form a C(2x2) pattern, then go to a (1x1) disordered pattern as one raises the coverage.

Notice that all of these conclusions come from the Landau expressions, which are easy to find. Hence, one does not need an extensive calculation to work out the low-temperature limit of the adsorbate phase diagram.

Example 4.F   Order Parameters

Schick [1981] provides an alternate scheme to arrive at the same results as in Example 4.D. His derivation is complex. However, the same results can be obtained more simply by considering a quantity $\Theta(\vec{q})$, which is called the *order parameter* for an arrangement of molecules with a reciprocal lattice vector $\vec{q}$, where the reciprocal lattice vector is defined in the supplemental material for Chapter 2. $\Theta(\vec{q})$ is given by

$$\Theta(\vec{q}) = \frac{1}{QS_0} \sum_n \sum_i \sum_{q_i} \epsilon_i \exp(-j\vec{q}_i \cdot \vec{r}_i) \exp(-\beta E_n) \qquad (4.211)$$

where ϵ_i is the occupancy of site i when the system is in arrangement n; $\vec{r}_i$ is the position of site i; and j is the square root of -1. Also, $\Theta(\vec{q})$ is called the order parameter for a given arrangement of the adsorbate. Physically $\Theta(\vec{q})$ is the coverage in a given ordered state; $\Theta(\vec{q} = 0)$ is the total coverage. As an exercise, the reader may want to compute the order parameter for each of the structures shown in Figure 4.22.

To illustrate how one can use the order parameter to tell something about why molecules arrange themselves in a given configuration, assume a gas is adsorbing on a square lattice. Show that if there are only first nearest neighbor interactions, one can only get (1x1) and (2x2) structures.

Mean Field Solution. The mean field solution is derived by assuming that the occupancy θ_i of any given site i is given by

$$\theta_i = \theta(\vec{q} = 0) + \sum_{q \neq 0} \exp(j\vec{q} \cdot \vec{r}_i) \theta(q) \qquad (4.212)$$

The free energy of the system is A_f, given by

$$A_f = \langle E_N \rangle = \left\langle -\sum_i H_i \theta_i - \tfrac{1}{2} \sum_{i,J \neq i} h_{ij} \theta_i \theta_j^* \right\rangle \tag{4.213}$$

Substituting Equations 4.212 and 4.213 into Equation 4.108 and summing yields

$$A_f = -H_1 S_0 \theta (\vec{q} = 0) + \frac{1}{2} \sum_{i,j \neq i} \theta^2 (\vec{q} = 0) \, h_{ij}$$

$$+ \frac{1}{2} \sum_{q \neq 0} \sum_{i,j \neq i} h_{ij} \, |\theta(\vec{q})|^2 \exp \left[j\vec{q} \cdot (\vec{r}_i - \vec{r}_j) \right]$$

$$+ \left. kT \left(\sum_i \ln \left(\frac{\theta_i}{1 - \theta_i} \right) + \text{higher order terms} \right) \right. \tag{4.214}$$

At low-energy temperatures the last term will be negligible. Therefore, the most stable arrangement of the atom will be the one that minimizes a function $h(q)$, given by

$$h(q) = -\sum_{j \neq i} h_{ij} \exp \left[j\vec{q} \cdot (\vec{r}_i - \vec{r}_j) \right] \tag{4.215}$$

The function $h(q)$ is called the Landau energy for a given arrangement of the adsorbate. Therefore, the stable low-temperature arrangements are the ones that minimize the Landau energy.

Now consider a square lattice with lattice dimension a_l and assume that an adsorbate adsorbs that has only nearest neighbor interactions of strength h. There will be four terms in Equation 4.215 corresponding to $\vec{r}_i - \vec{r}_j = (a_l, 0), (-a_l, 0), (0, a_l)$, and $(0, -a_l)$. Performing the sum in Equation 4.215.

$$h(\vec{q}) = -2h \cos (q_x a_l) - 2h \cos (q_y a_l) \tag{4.216}$$

where q_x and q_y are the x and y components of $\vec{q}$. Note q_x, q_y must satisfy $q_x, q_y < \pi/a_l$.

First consider attractive interactions ($h > 0$) $h(\vec{q})$ are between 0 and $-2h$. Notice that the minimum value of $h(\vec{q})$ comes when $\vec{q} = 0$. According to Equation 4.212, $\vec{q} = 0$ produces a (1x1) lattice. Therefore with $h > 0$ only a (1x1) lattice is possible.

Now consider $h < 0$. When $h < 0$, $\vec{q} = \left(\dfrac{\pi}{a_l}, \dfrac{\pi}{a_l} \right)$ minimizes $h(\vec{q})$. According to Equation 4.212, $\vec{q} = \left(\dfrac{\pi}{a_l}, \dfrac{\pi}{a_l} \right)$ corresponds to a C(2x2) structure. Therefore, with $h < 0$ the system will form a C(2x2) structure at intermediate coverages.

One can continue the analysis to derive all of the results in Example 4.E. The advantage of Schick's analysis, however, is that one does not have to postulate a series of structures and see what structures occur. Instead one can calculate which structures are stable, by finding the $\vec{q}$'s that minimize Equation 4.215.

Example 4.G Limitations of the Low-Temperature Analysis

In Examples 4.E and 4.F consider the low-temperature limit, where the Landau energy determines the phase behavior. As we raise the surface temperature, the Landau energy becomes less important, and entropy starts to play a role in the adsorbate phase behavior. Eventually the surface disorders. This brings up the questions 'under what conditions is the adsorbate phase ordered or disordered? When is it appropriate to assume that the adsorbate layer is well mixed so that the local concentration equals the global concentration?' One can solve this problem by calculating the critical temperature of the system. At temperatures below the critical temperature, one can have multiple phases on the surface. However, at temperatures above the critical temperature multiple phases are unstable. Instead, the surface is well mixed. The question, though, is, how does one calculate the critical temperature?

SOLUTION

To simplify this question we are only going to treat the case where there are first nearest neighbor interactions. If there are only first nearest neighbor interactions, then from symmetry if a phase transition exists it will always be at $F = 0$. Therefore, we will only analyze the $F = 0$ case.

Domain Wall Solution. One way to calculate the critical temperature is to assume that the surface is covered by a single island, and calculate the free-energy change that occurs when the island breaks up. If the free energy of the system goes down when islands break up, islands will be unstable. Peierls [1936] used such an analysis to show that islands are always unstable in one dimension. The key arguments are given in Wannier [1966] and Pilschke and Bergersen [1989] and reproduced below.

It is easiest to solve this problem using the Hamiltonian in Equation 4.59. If $F = 0$

$$-E_n = H_{S_0} + \frac{h}{4} \sum_i \xi_i \xi_{i+1} \tag{4.217}$$

First consider the lines of sites shown in Figure 4.39. Notice that the minimum energy state of the system (i.e., the ground state) comes when the sum over i in Equation 4.217 is maximized. This occurs when all of the ξ_i's are 1 or all the ξ_i's are -1. Hence, there are two ground states of the system, one with all of the sites filled and the other with all of the sites empty. Each of these states has an energy E_0, given by

$$-E_0 = H_{S_0} + \frac{hS_0}{4} \tag{4.218}$$

Now consider the first excited state shown in Figure 4.39, where the first i sites are occupied, and the remaining $S_0 - i$ sites are empty. From Equations 4.217 and 4.218 the energy of the system, E_e, is

$$E_e = E_0 + \frac{h}{2} \tag{4.219}$$

One can calculate the entropy of the state from the equations in Appendix A. Recall that the statistical mechanical definition of the entropy, S_1, for any system is

$$S_1 = \sum_n P_n \ln P_n \tag{4.220}$$

Let's consider a microcanonical ensemble where we fix the energy of the system at E_e. Notice that the energy of the system is independent of i, the site at which we switch from filled to empty sites, for $i = 1$ to $S_0 - 1$. Therefore, there are $S_0 - 1$ equivalent states of the system, each of which is equally probable. Another factor of 2 arises because the energy is the same independent of whether the first i sites are occupied and the remaining $S_0 - 1$ are empty or vice versa. Therefore, there are a total of $\mathbb{N} = 2(S_0 - 1)$ states of the system, each of which

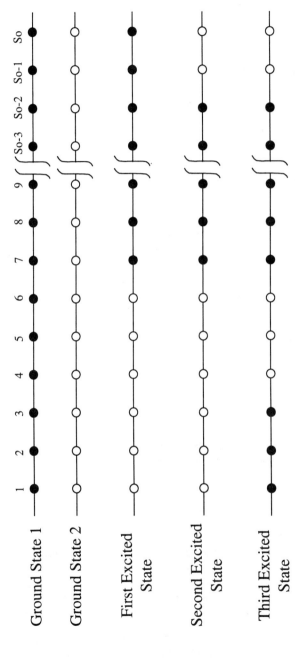

Figure 4.39 A diagram of the ground state and the first three excited states of a lines of sites obeying the Hamiltonian in Equation 4.217. The solid circles represent filled sites, while the empty circles represent empty sites.

is equally probable. The probability of each state is $1/\mathbb{N}$. Substituting $P_n = 1/\mathbb{N}$ into Equation 4.220 yields

$$S_1 = -k \sum_{n=1}^{\mathbb{N}} \frac{1}{\mathbb{N}} \ln\left(\frac{1}{\mathbb{N}}\right) = k \ln (\mathbb{N}) \qquad (4.221)$$

where $\mathbb{N} = 2(S_0 - 1)$ for the first excited state. A similar analysis shows $\mathbb{N} = 2$ for the ground state. Therefore, the entropy change in moving to the first excited state, ΔS_e

$$\Delta S_e = k \ln (2(S_0 - 1)) - k \ln 2 = k \ln (S_0 - 1) \qquad (4.222)$$

Therefore, the free-energy change, ΔG_e, associated with converting the system into the first excited state is

$$\Delta G_e = \frac{h}{2} - kT \ln (S_0 - 1) \qquad (4.223)$$

Notice that when S_0 is infinite, the free-energy change associated with the excitation is always negative at any temperature above 0 K. Therefore, excitations of the type shown in Figure 4.39 will occur at any nonzero temperature. A similar analysis shows that the free-energy change associated with converting the first excited state into the second excited state, and the excitation converting the second excited state into the third excited state will be negative. Such excitations lead to the breakup of islands at any nonzero temperature. As a result a one-dimensional lattice gas cannot phase separate into islands except at 0 K.

In two dimensions, however, the results are different. Consider adsorption of a gas onto an $(N \times N)$ square lattice. Again, there are two ground states of the system, one with all of the sites filled and one with all of the sites empty. Each of these states has an energy E_0 given by

$$E_0 = H_{S_0} - \frac{N^2 h}{4} \qquad (4.224)$$

Now consider an excitation that divides up the surface into two regions, one completely covered by adsorbate and another completely empty, as indicated in Figure 4.40. The energy E_n of the excited state is given by

$$E_n = N_B \frac{h}{2} + E_0 \qquad (4.225)$$

where N_B is the number of segments of length L_B in the boundary between the occupied and unoccupied regions of the surface.

According to equation 4.221, the entropy associated with the change is equal to $k \ln (\mathbb{N}) - k \ln 2$, where $\mathbb{N}$ is the number of equivalent states. One can calculate the number of equivalent states by computing the number of possible configurations of the boundary between the filled and occupied states in Figure 4.40.

Consider moving from the top to the bottom of Figure 4.40 along the boundary. In general, at each vertex, one can move in one of three directions: to the left, right, or down. Note, however, that since the boundary cannot loop back on itself, if in the last step the boundary moved to the left, on the next step the boundary can move down or to the left, but not to the right. Therefore, at each vertex there are two or three possibilities.

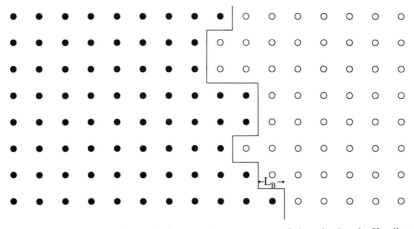

Figure 4.40 A diagram of the first excited states of a square array of sites obeying the Hamiltonian in Equation 4.217. The solid circles represent filled sites, while the empty circles represent empty sites.

According to Equations 4.221 and 4.222, entropy of the excitation, ΔS_e, is just the number of possible configurations of the bonding. Since there are two or three possibilities at each vertex, the entropy of a given chain is given by

$$\Delta S_e = k \ln [(m_e)^{N_B}] = N_B k \ln [m_e] \qquad (4.226)$$

where m_e is a number between 2 and 3. There are S_0 possible starting points for the chain. Therefore the free-energy change associated with forming the chain is

$$\Delta G_e = N_B \left[\frac{h}{2} - kT \ln (m_e) \right] - kT \ln (S_0) \qquad (4.227)$$

Notice that N_B increases at least proportionally to N. Therefore, the first and second terms in Equation 4.224 will increase proportionally to N while the third term in the equation will only increase as the log (N). In the limit that N goes to infinity the first two terms will be larger than the third. As a result, excitations of the type in Figure 4.40 will lower the free energy of the system when

$$\frac{h}{2} - kT \ln (m_e) \leq 0 \qquad (4.228)$$

A similar argument can be made for further subdividing the structure in Figure 4.40. As a result, the layer will be well mixed when Equation 4.225 is satisfied. The layer will separate into islands otherwise. If one rearranges Equation 4.225, we can show that islands can form when:

$$\beta h = \frac{h}{kT} > 2 \ln (m_e) \qquad (4.229)$$

However, m_e is between 2 and 3. Therefore,

$$1.387 \ldots = 2 \ln (2) \geq 2 \ln (m_e) \geq 2 \ln (3) = 2.197 \ldots \qquad (4.230)$$

The system will be well mixed if $\beta h \leq 2 \ln (m_e)$. As a result, $\beta h = 2 \ln m_e$ represents a critical point of the system above which two phase mixtures are impossible.

Bragg–Williams Solution. The preceding arguments considered the formation of a domain wall. However, another way to arrive at the same result is to ask when multiple adsorbed phases are stable.

Note that from Figure 4.21, if βh is large, the Bragg–Williams approximation predicts that there are multiple solutions of Equation 4.116. Note that if

$$\left.\frac{dH_1}{\partial \theta}\right|_{\theta = 0.5} \geq 0 \tag{4.231}$$

there will be only one stable phase. Otherwise, there will be two stable phases, substituting the value for H_1 from Equation 4.116 into Equation 4.230 and doing a page of algebra shows that in the Bragg–Williams approximation, multiple phases will be stable when

$$\beta h > \frac{1}{0.25 \, n_n} = 1 \text{ for a square lattice} \tag{4.232}$$

Onsanger's Solution. The Bragg–Williams solution is, of course, a poor approximation. However, Onsanger [1944] presented an exact solution for a lattice gas. The original derivation is rather difficult to follow. There is a simplified version by Schultz [1964] that starts with the matrix formulation of the partition function in Section 4.76 and letting the size of the matrix approach infinity. Unfortunately, the derivation requires many pages of algebra. Therefore, we only present the result, that multiple phases are stable when

$$\text{Sinh} \ (0.5 \ \beta h) > 1 \tag{4.233}$$

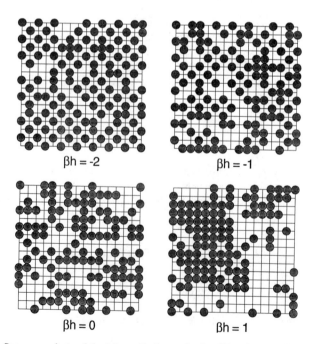

$\beta h = -2$ $\beta h = -1$

$\beta h = 0$ $\beta h = 1$

Figure 4.41 Some snapshots of the Monte Carlo results for $\beta F = 0$ and different values of βh.

or

$$\beta h > 1.7627 \dots \qquad (4.234)$$

A comparison of Equations 4.229 and 4.234 shows that the exact result is obtained in the domain wall solution if it is assumed that $m_e = 2.41$.

Monte Carlo Solution. One can also see whether ordered phases are stable by calculating an order parameter for the ordered phase from Equation 4.211 using the Monte Carlo algorithm in Section 4.8 and in the supplemental material at the end of the chapter. The calculations are identical to those in Section 4.8, and will not be repeated here. However, we ask the reader to do the calculations in Problem 4.35.

Figure 4.41 shows some of the results of the calculations. Notice that as βh increases, the local ordering increases. However, there is no long-range order. This local ordering is very important to our discussion of surface reactions in Chapter 7.

Example 4.H Strip Calculations

As noted in Section 4.7.6, strip calculations are very useful in the study of adsorption on stepped surfaces. The objective of this example is to show how one can actually do the calculations.

Consider the adsorption of a gas onto a square terrace two sites wide as illustrated in Figure 4.19. Experimentally, one usually finds that the binding energy is different near the bottom of the step than on the step edge. The objective of this example is to calculate an adsorption isotherm assuming:

(1) There are two sites on the terrace that are not equivalent

(2) There are only first nearest neighbor interactions of strength $4h$

(3) There is no adsorption on the steps

SOLUTION

Assume we have a strip M sites long and 2 sites wide with periodic boundary conditions. Following the notation in Section 4.7.6, assume that Hamiltonian is given by

$$-E_n = H_0 + \sum_{i=1}^{M} (H_{11}\varepsilon_{i,1} + H_{12}\varepsilon_{i,2}) + 4h \sum_{i=1}^{M} (\varepsilon_{i,1}\varepsilon_{i,2} + \varepsilon_{i,1}\varepsilon_{i+1,1} + \varepsilon_{i,2}\varepsilon_{i+1,2}) \qquad (4.235)$$

where $\varepsilon_{i,l}$ is zero when a given site is empty and 1 when a given site is filled. One can calculate θ_1 and θ_2 the fractional coverage on sites of type 1 and 2 from

$$\theta_1 = \langle \varepsilon_{2,1} \rangle = \frac{1}{Q} \sum_n \varepsilon_{2,1} \exp(-\beta E_n) \qquad (4.236)$$

$$\theta_2 = \langle \varepsilon_{2,2} \rangle = \frac{1}{Q} \sum_n \varepsilon_{2,2} \exp(-\beta E_n) \qquad (4.237)$$

Expanding n in Equations 4.236 and 4.237 in terms of $\varepsilon_{1,n}, \varepsilon_{1,2} \cdots$

$$\langle \varepsilon_{2,1} \rangle = \frac{\exp(\beta H_0)}{Q} \sum_{\varepsilon_{1,1}=0}^{1} \sum_{\varepsilon_{1,2}=0}^{1} \sum_{\varepsilon_{2,1}=0}^{1} \sum_{\varepsilon_{2,2}=0}^{1} \cdots \sum_{\varepsilon_{m,1}=0}^{1} \sum_{\varepsilon_{m,2}=0}^{1} \varepsilon_{2,1} \Pi_i$$

$$\cdot \exp[\beta H_{11}\varepsilon_{i,1} + \beta H_{12}\varepsilon_{i,2} + 4\beta h(\varepsilon_{i,1}\varepsilon_{i,2} + \varepsilon_{i,1}\varepsilon_{i+1,1} + \varepsilon_{i,2}\varepsilon_{i+1,2})] \qquad (4.238)$$

Following the derivation in Section 4.7.6, it will be useful to define two functions V and V_1 by

$$V\begin{bmatrix} \varepsilon_{i,1} & \varepsilon_{i,2} \\ \varepsilon_{i+1,1} & \varepsilon_{i+1,2} \end{bmatrix} = \exp\,(\beta H_{11}\varepsilon_{i,1} + \beta H_{12}\varepsilon_{i,2}$$

$$+\, 4\beta h(\varepsilon_{i,1}\varepsilon_{i,2} + \varepsilon_{i,1}\varepsilon_{i+1,1} + \varepsilon_{i,2}\varepsilon_{i+1,2})) \tag{4.239}$$

$$V_1\begin{bmatrix} \varepsilon_{i,1} & \varepsilon_{i,2} \\ \varepsilon_{i+1,1} & \varepsilon_{i+1,2} \end{bmatrix} = \varepsilon_{i+1,1}\,V\begin{bmatrix} \varepsilon_{i,1} & \varepsilon_{i,2} \\ \varepsilon_{i+1,1} & \varepsilon_{1+1,2} \end{bmatrix} \tag{4.240}$$

It will also be useful to define the matrices $\hat{V}$ and $\hat{V}_1$ by

$$\hat{V} = \begin{bmatrix} V\begin{bmatrix}11\\11\end{bmatrix} & V\begin{bmatrix}11\\10\end{bmatrix} & V\begin{bmatrix}11\\01\end{bmatrix} & V\begin{bmatrix}11\\00\end{bmatrix} \\ V\begin{bmatrix}10\\11\end{bmatrix} & V\begin{bmatrix}10\\10\end{bmatrix} & V\begin{bmatrix}10\\01\end{bmatrix} & V\begin{bmatrix}10\\00\end{bmatrix} \\ V\begin{bmatrix}01\\11\end{bmatrix} & V\begin{bmatrix}01\\10\end{bmatrix} & V\begin{bmatrix}01\\01\end{bmatrix} & V\begin{bmatrix}01\\00\end{bmatrix} \\ V\begin{bmatrix}00\\11\end{bmatrix} & V\begin{bmatrix}00\\10\end{bmatrix} & V\begin{bmatrix}00\\01\end{bmatrix} & V\begin{bmatrix}00\\00\end{bmatrix} \end{bmatrix} \tag{4.241}$$

$$\hat{V}_1 = \begin{bmatrix} V_1\begin{bmatrix}11\\11\end{bmatrix} & V_1\begin{bmatrix}11\\10\end{bmatrix} & V_1\begin{bmatrix}11\\01\end{bmatrix} & V_1\begin{bmatrix}11\\00\end{bmatrix} \\ V_1\begin{bmatrix}10\\11\end{bmatrix} & V_1\begin{bmatrix}10\\10\end{bmatrix} & V_1\begin{bmatrix}10\\01\end{bmatrix} & V_1\begin{bmatrix}10\\00\end{bmatrix} \\ V_1\begin{bmatrix}01\\11\end{bmatrix} & V_1\begin{bmatrix}01\\10\end{bmatrix} & V_1\begin{bmatrix}01\\01\end{bmatrix} & V_1\begin{bmatrix}01\\00\end{bmatrix} \\ V_1\begin{bmatrix}00\\11\end{bmatrix} & V_1\begin{bmatrix}00\\10\end{bmatrix} & V_1\begin{bmatrix}00\\01\end{bmatrix} & V_1\begin{bmatrix}00\\00\end{bmatrix} \end{bmatrix} = \begin{bmatrix} V\begin{bmatrix}11\\11\end{bmatrix} & V\begin{bmatrix}11\\10\end{bmatrix} & 0 & 0 \\ V\begin{bmatrix}10\\11\end{bmatrix} & V\begin{bmatrix}10\\10\end{bmatrix} & 0 & 0 \\ V\begin{bmatrix}01\\11\end{bmatrix} & V\begin{bmatrix}01\\10\end{bmatrix} & 0 & 0 \\ V\begin{bmatrix}00\\11\end{bmatrix} & V\begin{bmatrix}00\\10\end{bmatrix} & 0 & 0 \end{bmatrix} \tag{4.242}$$

When $\langle \varepsilon_{2,1} \rangle$ is calculated, it will be necessary to sum terms like

$$\varepsilon_{2,1}\, \Pi_i \exp\,[\beta H_{11}\varepsilon_{i,1} + \beta H_{12}\varepsilon_{i,2} + 4\beta h(\varepsilon_{i,1}\varepsilon_{i,2} + \varepsilon_{i,1}\varepsilon_{i+1,1} + \varepsilon_{i,2}\varepsilon_{i+1,2})]$$

$$= \varepsilon_{2,1}\,V\begin{bmatrix}\varepsilon_{1,1}&\varepsilon_{1,2}\\\varepsilon_{2,1}&\varepsilon_{2,2}\end{bmatrix} V\begin{bmatrix}\varepsilon_{2,1}&\varepsilon_{2,2}\\\varepsilon_{3,1}&\varepsilon_{3,2}\end{bmatrix} V\begin{bmatrix}\varepsilon_{3,1}&\varepsilon_{3,2}\\\varepsilon_{4,1}&\varepsilon_{4,2}\end{bmatrix} V\begin{bmatrix}\varepsilon_{4,1}&\varepsilon_{4,2}\\\varepsilon_{5,1}&\varepsilon_{5,2}\end{bmatrix} \cdots$$

$$= V_1\begin{bmatrix}\varepsilon_{1,1}&\varepsilon_{1,2}\\\varepsilon_{2,1}&\varepsilon_{2,2}\end{bmatrix} V\begin{bmatrix}\varepsilon_{2,1}&\varepsilon_{2,2}\\\varepsilon_{3,1}&\varepsilon_{3,2}\end{bmatrix} V\begin{bmatrix}\varepsilon_{3,1}&\varepsilon_{3,2}\\\varepsilon_{4,1}&\varepsilon_{4,2}\end{bmatrix} V\begin{bmatrix}\varepsilon_{4,1}&\varepsilon_{4,2}\\\varepsilon_{5,1}&\varepsilon_{5,2}\end{bmatrix} \cdots$$

$$\tag{4.243}$$

A derivation like that in Section 4.7.5 shows

$$Q = \exp\,(-\beta H_0)\,\mathrm{Tr}\,[\hat{V}^M] \tag{4.244}$$

$$\langle \varepsilon_{2,1} \rangle = \frac{\exp{(-\beta H_0)}}{Q} \, \text{Tr} \, [\hat{V}^{M-1}] \qquad (4.245)$$

Assume $\hat{V}$ has eigenvalues λ_1, λ_2, λ_3, λ_4, and corresponding eigenvectors $\vec{\phi}_1$, $\vec{\phi}_2$, $\vec{\phi}_3$, $\vec{\phi}_4$. If the calculation is done under conditions where there is no two-phase region, one can assume $|\lambda_1| > |\lambda_2| \geq |\lambda_3| \geq |\lambda_4|$. Defining a conjugate matrix $\hat{\phi}$ as in Section 4.7.5 shows

$$\langle \varepsilon_{2,1} \rangle = \frac{\exp{(-\beta H_0)}}{Q} \, \text{Tr} \, [\hat{V}_1 \hat{\phi} \hat{\lambda}^{M-1} \hat{\phi}^T] \qquad (4.246)$$

with

$$\hat{\lambda} = \begin{bmatrix} \lambda_1 & 0 & 0 & 0 \\ 0 & \lambda_2 & 0 & 0 \\ 0 & 0 & \lambda_3 & 0 \\ 0 & 0 & 0 & \lambda_4 \end{bmatrix} \qquad (4.247)$$

But for any matrices $\hat{A}$ and $\hat{B}$ $\text{Tr} \, [\hat{A}\hat{B}] = \text{Tr} \, [\hat{B}\hat{A}]$. Therefore

$$\langle \varepsilon_{2,1} \rangle = \frac{\exp{(-\beta H_0)}}{Q} \, \text{Tr} \, [\hat{\phi}^T \hat{V}_1 \hat{\phi} \hat{\lambda}] \qquad (4.248)$$

A derivation like that in Section 4.7.5 shows

$$\frac{q}{\exp{(-\beta H_0)}} = \lambda_1^M + \lambda_2^M + \lambda_3^M + \lambda_4^M = \lambda_1^M \left[1 + \left(\frac{\lambda_2}{\lambda_1}\right)^M + \left(\frac{\lambda_3}{\lambda_1}\right)^M + \left(\frac{\lambda_4}{\lambda_1}\right)^M \right] \qquad (4.249)$$

Substituting Q from Equation 4.249 into Equation 4.248 yields

$$\langle \varepsilon_{2,1} \rangle = \frac{\text{Tr} \, [\hat{\phi}^T \hat{V}_1 \hat{\phi} \hat{L}^2]}{\lambda_1 \left[1 + \left(\frac{\lambda_2}{\lambda_1}\right)^M + \left(\frac{\lambda_3}{\lambda_1}\right)^M + \left(\frac{\lambda_4}{\lambda_1}\right)^M \right]} \qquad (4.250)$$

where

$$\hat{L} = \begin{bmatrix} 1 & 0 & 0 & 0 \\ 0 & \left(\dfrac{\lambda_2}{\lambda_1}\right)^{M-1/2} & 0 & 0 \\ 0 & 0 & \left(\dfrac{\lambda_3}{\lambda_1}\right)^{M-1/2} & 0 \\ 0 & 0 & 0 & \left(\dfrac{\lambda_4}{\lambda_1}\right)^{M-1/2} \end{bmatrix} \qquad (4.251)$$

in the limit that $M \to \infty$

$$\langle \varepsilon_{2,1} \rangle = \frac{\text{Tr} \, [\hat{L}\hat{\phi}^T \hat{V}_1 \hat{V}_2 \hat{\phi} \hat{L}]}{\lambda_1} \qquad (4.252)$$

with

$$\hat{L} = \begin{bmatrix} 1 & 0 & 0 & 0 \\ 0 & 0 & 0 & 0 \\ 0 & 0 & 0 & 0 \\ 0 & 0 & 0 & 0 \end{bmatrix} \qquad (4.253)$$

Note

$$\hat{\phi}L = [\vec{\phi}_1|0|0|0] \qquad (4.254)$$

where $\vec{\phi}_1$ is the eigenvector corresponding to λ_1, the largest eigenvalue of $\hat{V}$.

After some algebra one obtains

$$\langle \varepsilon_{2,1} \rangle = \frac{\vec{\phi}_1^T \hat{V}_1 \vec{\phi}_1}{\lambda_1} \qquad (4.255)$$

Now consider the case where $\beta H_{11} = 0.25$, $\beta H_{12} = \beta H_{11} - \frac{1}{4}$, $\beta h = 0.5$. The $\hat{V}$ matrix becomes

$$\hat{V} = \begin{bmatrix} \exp(1.75) & \exp(1.25) & \exp(1.25) & \exp(0.75) \\ \exp(0.75) & \exp(0.75) & \exp(0.25) & \exp(0.25) \\ \exp(0.50) & \exp(0.0) & \exp(0.50) & \exp(0.0) \\ \exp(0.0) & \exp(0.0) & \exp(0.0) & \exp(0.0) \end{bmatrix} \qquad (4.256)$$

We have calculated the eigenvalues and the eigenvectors of $\hat{V}$ using the power method. The result is

$$\vec{\phi}_1^T = (0.8675 \quad 0.3647 \quad 0.2776 \quad 0.1933) \qquad \lambda_1 = 8.81 \qquad (4.257)$$

Plugging numbers into Equation 4.255 yields

$$\langle \varepsilon_{2,1} \rangle = \frac{(0.8675 \quad 0.3647 \quad 0.2776 \quad 0.1933)}{(8.81)}$$

$$\cdot \begin{bmatrix} \exp(1.75) & \exp(1.25) & 0 & 0 \\ \exp(0.75) & \exp(0.75) & 0 & 0 \\ \exp(0.50) & \exp(0.00) & 0 & 0 \\ \exp(0.00) & \exp(0.00) & 0 & 0 \end{bmatrix} \begin{pmatrix} 0.8675 \\ 0.3647 \\ 0.2776 \\ 0.1933 \end{pmatrix}$$

$$= 0.81 \qquad (4.258)$$

We have calculated an adsorption isotherm by repeating this procedure for many other values of H_1. The results are shown in Figure 4.42. For $\beta h = 0$, the results are identical to those in Figure 4.17. As βh increases the two lines collapse, and the coverage changes more strongly with pressure (i.e., F) than one would expect from Equation 4.115.

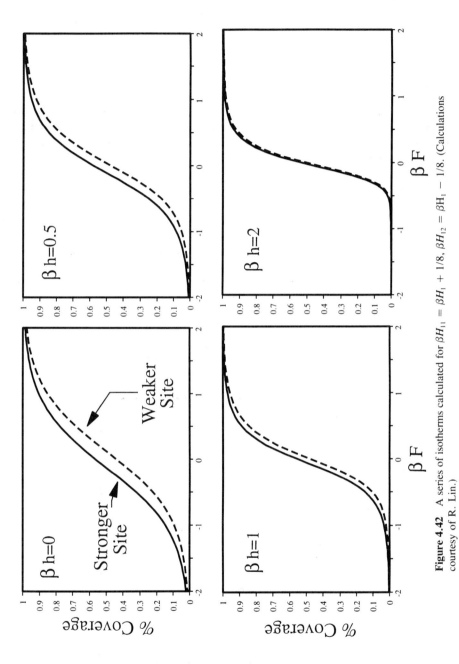

Figure 4.42 A series of isotherms calculated for $\beta H_{11} = \beta H_1 + 1/8$, $\beta H_{12} = \beta H_1 - 1/8$. (Calculations courtesy of R. Lin.)

327

Example 4.I Use of the BET Equation to Calculate Surface Areas

In Section 4.5 it was noted that the BET equation is often used to measure surface areas. The objective of this problem is to see how well it works.

Consider the krypton data in Figure 4.12. Use the BET equation to calculate the area of a monolayer. How well does the BET equation fit the data?

SOLUTION

When one calculates the area of a monolayer with the BET equation, one first calculates the quantity of gas in the monolayer, and then multiplies by a molecular area to get the area of a monolayer. In Figure 4.12, quantities of gas are measured in cubic centimeters (cc) at standard temperature and pressure (STP). Therefore, it is convenient to define a quantity, V_M, the volume krypton in a monolayer measured in cc/gm of adsorbent.

From Equation 4.189

$$\frac{V}{V_M} = \frac{[A]}{[S_0]} = \frac{c_B x_B}{(1 - x_B)\,[1 + (c_B - 1)\,x_B]} \tag{4.259}$$

where V is the volume gas adsorbed in cc/gm. Rearranging Equation 4.259 yields

$$\frac{x_B}{V(1 - x_B)} = \frac{1}{V_M c_B} + \frac{(c_B - 1)}{V_M c_B}\, x_B \tag{4.260}$$

where $x_B = P/P_{sat}$, where $P_{sat} = 42.36$ torr at 94.72 K.

The implication of Equation 4.260 is that a plot of $\dfrac{x_B}{V(1 - x_B)}$ versus x_B should be linear.

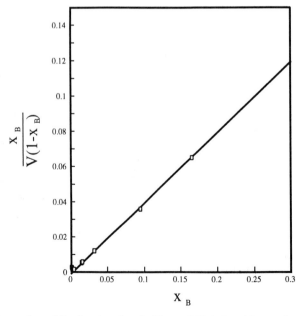

Figure 4.43 The portion of the krypton data in Figure 4.12 below 10 torr plotted as suggested by Equation 4.260.

TABLE 4.6   The Surface Area of Several Adsorbates*

N_2(77 K)	16.2 $\overset{\circ}{A}{}^2$
Ar(77 K)	13.8 $\overset{\circ}{A}{}^2$
Kr(77 K)	20.2 $\overset{\circ}{A}{}^2$
n C_4H_{10}(273 K)	44.4 $\overset{\circ}{A}{}^2$
C_6H_6(293 K)	43.0 $\overset{\circ}{A}{}^2$

Suggested by Dollimer et al. [1976].
*These adsorbates are common used for surface area measurements.

If one examines the entire range of data in Figure 4.12, one does not find that the BET equation fits the data very well. However, in order to calculate a surface area one only needs to fit the data in the region near one monolayer coverage (i.e., at pressures between 1 and 10 torr). Figure 4.43 shows a plot of $\dfrac{x_B}{V(1-x_B)}$ versus x_B for the data below 10 torr. Notice that the plot of the data taken at pressures below 10 torr is linear with a slope of 0.428 g/cc and an intercept of 1.44×10^{-3} g/cc.

Solving Equation 4.259 for V_M

$$V_M = \frac{1}{(\text{slope} + \text{intercept})} = 2.34 \text{ cc/g}$$

The surface area A becomes

$$A = 2.34 \text{ cc/g} \left(\frac{6 \times 10^{23} \text{ atoms}}{22,400 \text{ cc}}\right)\left(\frac{20.2 \ \overset{\circ}{A}{}^2}{\text{atoms}}\right) \times \frac{\text{m}^2}{10^{20} \ \overset{\circ}{A}{}^2} = 12.7 \text{ m}^2/\text{g}$$

where the area of a krypton molecule is given by Dollimer, Spooner, and Turner [1976].

Figures 4.44 and 4.45 compare the krypton data in Figure 4.12 to the fits with the BET

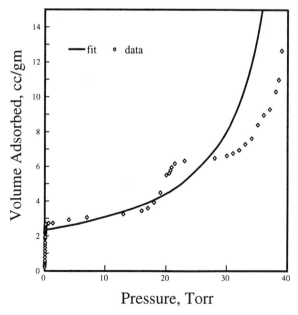

Figure 4.44   A comparison of the krypton data in Figure 4.12 to the best fit with the BET equation.

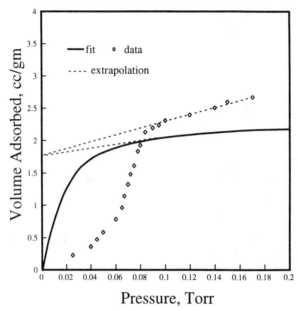

Figure 4.45   A blowup of the portion of the data in Figure 4.44 below 4 torr.

equation. The BET equation does a good job in fitting the data from 1 to 20 torr, but diverges significantly both at low and high coverages. This is typical. Generally, the BET equation fits the data well at coverages near a monolayer but nowhere else. One can still use the BET equation to estimate a surface area, however. One calculates a surface area by estimating V_M, the volume of gas at STP that adsorbs in a monolayer. One can easily show that V_M is well approximated by the extrapolation of the plateau in the adsorption isotherm to zero pressure. Figure 4.45 shows the extrapolation. Notice that the BET equation does give an accurate value of V_M even though the BET equation does not do well at either low or high coverage. Hence, the BET can be used to measure a surface area even though it is an inaccurate model of the adsorption process.

4.14   SUPPLEMENTAL MATERIAL

In the problem set, the reader is asked to use a computer program to do a Monte Carlo calculation. Therefore, it is helpful to provide a listing of a working Monte Carlo routine. The following FORTRAN routine was used to actually calculate the figures in the chapter. A C version of a simplification of the routines is also included in case the reader is unfamiliar with FORTRAN. The main program is called Isotherm and it is listed later in the listing. The program starts by setting up a 40x40 occupancy matrix, which contains 1 if a site is occupied and 0 if the site is empty. The program starts by randomly filling the occupancy matrix. It then takes 4000 Monte Carlo steps to randomize the occupancy matrix. The coverage is calculated by averaging over the next 5000 MC steps. There are two tricks in the program. We calculate a matrix of values of $\exp(-\beta \Delta E)$ in the main routine to avoid having to calculate the exponentials over and over. There also are two different Monte Carlo subroutines. One chooses sites randomly; one chooses sites in order. Both routines assume periodic boundary conditions. It is suggested that the reader run these routines on a workstation, as they are slow on a PC.

```
      subroutine step(n,o,exe,exf)
c ****************************************************
c   Subroutine which goes to each site in the lattice and
c   takes a Monte Carlo step
c   exe[N] and exf[H] hold exp(-beta Δ e) [Nn] as a function
c   of the number of nearest neighbors of a given site
c   o holds the occupancy numbers of the site
c   n is the total number of adsorbed molecules
c ****************************************************
      real exe(5),exf(5),d
      integer n, o(40,40), i,j,ip,im,jp,jm,nnij
      do 10 i=1,40
      ip=i+1
      im=i-1
      if(i.eq.40) ip=1
      if(i.eq.1)im=40
      do 10 j=1,40
      jp=j+1
      jm=j-1
      if(j.eq.40)jp=1
      if(j.eq.1)jm=40
      nnij=o(ip,j)+o(im,j)+o(i,jp)+o(i,jm)
c
c   nnij is the number of nearest neighbors of site i,j
c
      if(o(i,j).eq.1) then
      d=exf(nnij+1)
      else
      d=exe(nnij+1)
      endif
      if(d.gt.rand() ) then
         if(o(i,j).eq.0) then
         o(i,j)=1
         n=n+1
         else
         o(i,j)=0
         n=n-1
         endif
      endif

c flip it

   10 continue
      return
      end

      subroutine steprand(n,o,exe,exf)
c ****************************************************
c   Subroutine to which randomly chooses 9600 sites and takes
c   a Monte Carlo step at each site
c   exe and exf hold exp(-beta delta e) as a function of the
c   number of nearest neighbors of a given site
c   o holds the occupancy numbers of the site
c   n is the total number of adsorbed molecules
c ****************************************************
      real exe(5),ext(5),d
      integer n, o(40,40),ii, i,j,ip,im,jp,jm,nnij
      do 10 ii=1,1600
      j=1599.*rand()
      j=j/40
      i=j-40*i+1
      i=i+1
      ip=i+1
      im=i-1
      if(i.eq.40) ip=1
      if(i.eq.1)im=40
      jp=j+1
      jm=j-1
      if(j.eq.40)jp=1
      if(j.eq.1)jm=40
      nnij=o(ip,j)+o(im,j)+o(i,jp)+o(i,jm)
c
c   nnij is the number of nearest neighbors of site i,j
c
      if(o(i,j).eq.1) then
      d=exf(nnij+1)
      else
      d=exe(nnij+1)
      endif
      if(d.gt.rand() ) then

         if(o(i,j).eq.0) then
```

Main Program

```fortran
      program isotherm
      integer o(40,40), n,nst,i,j,ii,nr
      real exe(5),exf(5),an,cov,lang,t(4)
      an=rand(1000)
      an=rand()
      an=rand()
      call setup(n,o,0.01)
      do 600 kk=1,11
      bh=0.25*(kk-3)
      write(6,100) bh
100   format(/,."bh = ",f7.3,///
     1 * "H1 langmuir lattice gas Bethe",/,
     1 beta  "    coverage  coverage    Coverage"
     1 )

      do 500 ii=1,11
      bh1= -4.0 +0.01"(ii-6)
      an=0
      nst=0
      do 1 i=1,5
      exe(i)=exp(bh1+bh*(i-1) )
      ext(i)=1/exe(i)
1     continue
      do 5 i=1,4000
      call step(n,o,exe,exf)
      do 10 j=1,500
      do 15 i=1,10
15    call step(n,o,exe,exf)
      call steprand(n,o,exe,exf)
      an=an+n
      nst=nst+1
      cov=an/nst/1600.
10    continue
      lang=(1.+tanh(bh1*0.5))/2.
      call bethe(exe(1),exe(2)/exe(1),t,nr)
      write(6,101) bh1, lang, cov,(t(i),i=1,nr)
101   format(2x,f7.3,6(2x,f8.5))
500   continue
```

```fortran
      o(i,j)=1
      n=n+1
      else
      o(i,j)=0
      n=n-1
      endif
      endif
10    continue
      return
      end

      subroutine setup(n,o,p)
      integer n,o(40,40),i,j
      n=0
      do 10 i=1,40
      do 5 j=1,40
      if(p.gt.rand()) then
      o(i,j)=1
      n=n+1
      else
      o(i,j)=0
      endif
5     continue
10    continue
      return
      end

c***************
c     Compute the bethe approximation
c***************
      subroutine bethe(a,b,t,nr)
      real a,b,t(4),x,xmin,xmax,f,fold,
      real f1,step,z,za,zb,fa,fb
      integer nr,i,kount
      xmax=b**3
      xmin=xmas
      f1=1.
```

```fortran
      nr=0
      if(xmin.gt.0.95)xmin=0.95
      if(xmax.lt.1.05)xmax=1.05
      step=0.01*(xmax-xmin)
      fold=((1+a*b*xmin)**3-x*(1+a*xmin)**3
c***************** search for a root

      do 10 i=1,100
      x=xmin+step*i
      f=(1+a*b*x)**3-x*(1+a*x)**3
      if(f*f1.le.0.) then
c********************Found a root

      nr=nr+1
      za=x-step
      zb=x
      fa=fold
      fb=f
c****************Zoom in using secant method
      do 5 kount=1,5
      z=zb-fb*(za-zb)/(fa-fb)
      lf(z*za.le.0.0) then
                      zb=z
      fb=(1+a*b*z)**3-z*(1+a*z)**3
                      else
                      za=z
      fa=(1+a*b*z)**3-z*(1+a*z)**3
                      endif
      continue
5     z=zb-fb*(za-zb)/(fa-fb)
c***************** Compute the coverage
      t(nr)=a*(1+a*b*z)**4
      t(nr)=t(nr)/(t(nr)+(1+a*z)**4)
      f1=-f1
      endif
      fold=f
10    continue
      return
      end
```

```fortran
600   continue
      stop
      end
```

Start of C Program

```c
/* Montecarlo Calculation on a Lattice */
#include <stdio.h>
#include <conio.h>
#include <stdib.h>
#include <math.h>
#include <graph.h>
float rands[256]; int ira = 1;
float random(void);
float random()
/* return a random number between 0 and 1 */
{
int l, ran;
if(ira)
{for(i=0,i<256;i++) rands[i]=rand();
ira=0;
}
i=rand()/128; i=rands[i]/128;
ran=rands[i];
rands[i]=rand();
   return ( (float) ran )/32767.;
}
int n; /* n = the number of absorbed molecules */
int o[40][40]; /* o is the occupancy number of the i,j, site */
float exe[5]; /* exe[nn] is exp( β ΔE(nn) ) where ΔE(nn) is the
                  free energy change when a molecule adsorbs onto a
                  site which has nn nearest neighbors */
float ex[5]; /* exf[nn] is exp( β ΔE(nn) ) where ΔE(nn) is the
                  free energy change when a molecule desorbs from a
                  site which has nn nearest neighbors */

mclatt(void);
mclatt()
{
```

333

```
/***********************************************
c   Subroutine to visit each point [i][j] on the lattice and do a monte
c   carlo step there 0 ≥ i ≥ 39 , 0 ≥ j ≥ 39
c
***********************************************/

   float d;
   int i,j; /* i and j are the index of site [i][j] */
   int ip,im,jp,jm; /* ip is the index of the site to the left of i
                       im is the index of the site to the right of i
                       jp is the index of the site above j
                       jm is the index of the site below j
                       */
   int nnij;/* nnij is the nunmber of nearest neighbors to site i,j */
   for(i=0;i<40;i++)
   {  ip=i+1;
      im=i-1;
      if(i==39) ip=0; /* impose periodic boundary conditions */
      if(i==0) im=39; /* impose periodic boundary conditions */
      for(j=0;j<40;j++)
      {  jp=j+1;
         jm=j-1;
         if(j==39)jp=0;/* impose periodic boundary conditions
                          */
         if(j==1)jm=1;/* impose periodic boundary conditions
                          */

         nnij=o[ip][j]+o[im][j]+o[i][jp]+o[i][jm]; /* count number
                                                      of neighbors*/

         if(o[i][j]==1) d=exf[nnij];
         else d=exe[nnij];
         if(d> random() ) /* should I take the step? */
         { /* yes accept the step */
            if(o[i][j]==0) /* if site originally unoccupied */
            { o[i][j]=1; /*absorb a molecule

               n=n+1;
            }
            else
            { o[i][j]=0; /*desorb a molecule*/
               n=n-1;
```

```
   return 1;
}

main(void);
main() /* montecarlo calculation of an isotherm */
{

int nsteps;/* nsteps = the number of MC steps */
int nn; /* nn = number of nearest neighbors*/
int nsamp; /*nsamp is the number of times to sample*/
int step; /* step is an index of montecarlo steps*/
float cov; /* cov is the coverage; we will compute cov from
              cov = Σ number of adsorbed atoms at the end of a step
                    ────────────────────────────────────────────────
                    (number of mc steps)(number of sites)
                    */

float an; /* Σ number of adsorbed atoms at the end of a step */

float lang; /* langmuir's approximation to the isotherm */

float bh; /* bh = βh */
float bH1; /* bH1 = β*H1 */
FILE *out;
out = fopen("isotherm.dat",'W');
nsteps=50;
srand(1000);rand();rand(); /* start random number generator*/
__clearscreen(__GCLEARSCREEN);
setup((float) 0.01);
bh = (float)0.;
fprintf(out, "\n\n******************************\n βh = %7.3f \n\n",bh);
fprintf(out," β*H1  Montecarlo Langmuir\n");
fprintf(out,"       coverage\n");
for(bH1=(float)(-4.0-.2,*bh);bH1<(float)(4.01-2.*bh);bH1+=(float)0.1)
{ an=(float)0.0;
  for(nn=0;nn<5;nn++)
  { exe[nn]=(float)exp((double)(bH1+bh*((float)nn) ));
    exf[nn]=((float)1)/exe[nn];
  }
  for(step=1;step<40;step++)/* take 40 steps to randomize */
  { mclatt();
```

334

```c
        _settextposition(10,30);
        printf("randomizing step %5i          ",step);
    }

    for( nsamp=0;nsamp<nsteps;nsamp++)
    {

    for( step=0; step < 11; step++) /* take 11 MC steps */
    {

        mclatt();
        _settextposition(10,30);
        printf("β*H1 = %7.3f, sample = %5i,          step
                = %5i ",
                bH1, nsamp, step);
    }

    an=an+n;

    cov=an/((float)1600)/((float)nsteps);
    lang=(float)(1.+tanh(( (double)bH1)* 0.5))/2.;
    fprintf(out, " %7.3f %7.3f %7.3f\n", bH1, cov, lang);
    }
    fclose(out);
    printf("done\7\7\7\7\7\7\7\7\7\7\7\7\7\7\7\7\7\7\7\7\7\7");
    return 2;
```

```c
    }
    }
    kbhit();/* check for cntrol-c */
    return 1;

setup( float );

setup( float p)
/* subrouting to assign a random initial configuration to the
   occupancy matrix where p is the probability that a site
   is occupied */
{

    int i,j;
    n=0;
    for (i=0;i<40;i++)
    for (j=0;j<40;j++)
    { if(p>random() )
    { o[i][j]=1;
                    n=n+1;
    }
    else o[i][j]=0;
    }
}
```

PROBLEMS

4.1   Define the following in three sentences or less.

(1) Immobile adsorption
(2) Mobile adsorption
(3) Random site filling
(4) Ordered adsorption
(5) Incommensurate adsorption
(6) Monolayer
(7) Multilayer
(8) Dissociative adsorption
(9) Coadsorption
(10) Adsorption isotherm
(11) Langmuir adsorption isotherm
(12) Noncompetitive adsorption
(13) Competitive adsorption
(14) Freundlich adsorption isotherm
(15) Multisite model
(16) Lattice gas model

(17) Next nearest neighbor
(18) Freidel oscillation
(19) Monte Carlo calculation
(20) Transfer matrix
(21) Bragg–Williams mean field approximation
(22) Bethe approximation
(23) Quasichemical approximation
(24) Order parameter
(25) Critical point
(26) Direct adsorbate/adsorbate interactions
(27) Indirect adsorbate/adsorbate interactions
(28) Universality class
(29) Universality

4.2   Use the data in Table 3.1 to calculate the surface area of Chappuis's charcoal in m^2/g of charcoal, assuming that each adsorbed CO_2 molecule covers 20 $\overset{\circ}{A}\,^2$ of surface.

4.3   Outline the key assumptions used to derive the following adsorption isotherms.

(a) Langmuir adsorption isotherm for noncompetitive, nondissociative adsorption

(b) Langmuir adsorption isotherm for competitive adsorption

(c) Langmuir adsorption isotherm for dissociative adsorption

(d) Freundlich adsorption isotherm

(e) Toth equation

(f) Multisite adsorption isotherm

(g) Tempkin adsorption isotherm

(h) Fowler adsorption isotherm

(i) Slygin–Frumkin adsorption isotherm

(j) Lattice gas model

4.4   Activated charcoal is now sold as filters for air cleaners that are used in homes and restaurants. The filters are reasonably effective in removing aromatic hydrocarbons from cigarette smoke. Assume that you are designing a system for a restaurant and want to only have to replace the activated carbon once a year. Calculate the amount of activated carbon needed to remove the tar from 50,000 cigarettes/year. Assume each cigarette produces 10 mg of tar.

4.5   How many air molecules do you need to displace from the soles of your shoes when you take a step? Make any assumptions you need, but get an answer.

4.6   E. G. Seebauer, A. C. F. Kong, and L. D. Schmidt, *Surface Sci.* **176,** 134 [1986] measured isotherms for NO adsorption on Pt(111). The data in the following table were obtained at 300 K.

Coverage $\times 10^{14}$ molecules/cm^2	Pressure (torr)
0.27	4.5×10^{-8}
0.31	2.3×10^{-7}
0.36	1.1×10^{-6}
0.39	4.0×10^{-6}
0.43	1.6×10^{-5}
0.58	Saturation

How well can one fit these data with (a) the Langmuir adsorption isotherm, (b) the Toth equation, (c) the lattice gas model, (d) the Bethe approximation, (e) the Bragg–Williams approximation? (*Note:* NO adsorbs molecularly on Pt(111).)

4.7 R. Eisinger and R. C. Alkire, *J. Electrochem. Soc.* **130**, 93 [1983] measured an isotherm for adsorption of β-naphanol on glassy carbon from an aqueous solution as a function of the potential of the glassy carbon electrode. The data in the following table were obtained.

− 0.26 V		− 0.51 V		− 1.26 V	
Concentration $(10^{-5}$ M)	Amount Adsorbed 10^{10} molecules/ cm^2	Concentration	Amount Adsorbed 10^{10} molecules/ cm^2	Concentration	Amount Adsorbed 10^{10} molecules/ cm^2
0	0	0	0	0	0
0.15	0.43	0.19	0.41	0.36	0.31
0.49	0.76	0.57	0.72	0.88	0.52
0.96	0.99	1.05	0.93	1.45	0.68
1.50	1.18	1.61	1.09	2.06	0.81
2.06	1.31	2.20	1.23	2.66	0.92

(a) Try to fit the data to a Langmuir adsorption isotherm. How well does it fit?

(b) Eisinger and Alkire point out that when β-naphanol adsorbs on the electrode, n_w water molecules are displaced. Derive an adsorption isotherm for this case. You might want to look at the derivation in J. Bockris and D. S. Winkels, *J. Electrochem. Soc.* **111**, 736 (1964).

(c) How well does your isotherm from part (b) fit for the data? Assume $n_w = 5$–7.

(d) Why does the data vary with the electrode potential? Look at Example 4.C. What changes in the partition function when the electrode potential changes?

4.8 K. H. Mayo, M. Nunez, C. Burke, C. Starbuck, D. Lauffenburger, and C. R. Savage, *J. Biological Chem.* **264**, 17838 [1989] measured an isotherm for the binding of epidermal growth factor (EDF) on human fibroblast cells. The results in the following table were obtained.

EDF in Solution $\times 10^{-6}$	EDF on Cells $\times 10^{-4}$	EDF in Solution $\times 10^{-6}$	EDF on Cells $\times 10^{-4}$
0.127	0.37	2.00	3.21
0.195	0.51	2.67	4.01
0.262	0.65	3.33	4.17
0.331	0.84	4.06	3.73

EDF in Solution × 10^{-6}	EDF on Cells × 10^{-4}	EDF in Solution × 10^{-6}	EDF on Cells × 10^{-4}
0.394	1.14	4.69	4.19
0.464	1.01	5.35	4.14
0.472	1.31	6.76	5.02
0.603	1.36	13.44	6.64
0.667	0.98	20.02	8.86
1.33	2.22	62.4	10.02
		85.0	10.33

(a) How well do these data fit the (1) Langmuir adsorption isotherms, (2) Freundlich adsorption isotherm, (3) one-dimensional lattice gas model?

(b) How well do these data fit the modified isotherm discussed in Problem 4.7?

(c) Mayo et al. fit their data with a two-site model

$$EGF + S_1 \rightleftharpoons C_1 \qquad C_1 + S_2 \rightleftharpoons C_2$$

where S_1 and S_2 are two kinds of sites. Derive a Langmuir adsorption isotherm for this case. How well does this isotherm fit Mayo's data?

(d) What do you conclude from your findings?

4.9   G. Liu, Y. Li, and J. Jonas, *J. Chem. Phys.* **95**, 6892 [1991] measured an isotherm for adsorption of nitrogen on a novel silica glass at 77 K. The results in the following table were obtained.

Pressure (ATM)	Amount Adsorbed (cc-STP/g)	Pressure (ATM)	Amount Adsorbed (cc-STP/g)
0.017	19	0.209	63
0.041	31	0.319	72
0.070	39	0.381	79
0.110	48	0.440	84
0.162	56	0.530	95

Use these data to calculate the surface area of the silica.

4.10   In Section 4.3.1 adsorption isotherms were derived using the canonical partition function, while the grand canonical partition function was used in sections 4.7–4.9. (a) Outline the differences between the two partition functions. (b) How do you decide which one to use?

4.11   Describe (a) what types of systems lead to each of the adsorption isotherms in Figure 4.4 and 4.10 (b) what are the key physics that lead to each of the isotherms (e.g., strongly adsorbate/adsorbate attractions and weakly attractive adsorbate/surface attractions).

4.12   The mean field/cluster approximation is often used to model reactions on surfaces. During a reaction, the arrangement often reaches a local equilibrium, where the local arrangement of the atoms is in equilibrium but there is no long-range order. How would you use the mean field/cluster approximation to model the adsorbate arrangement in such a situation?

4.13 Write 3 to 5 pages describing the key predictions of the Ising model and explaining qualitatively why they occur. Be sure to include both general ideas and specific predictions.

4.14 In Section 4.8 and Example 4.G we found that it is impossible for a two-phase mixture to exist at a temperature above the critical temperature where the critical temperature is given by $\beta h = 1.7627$. Why does the critical point arise? Explain, in your own words, why individual phases are unstable above the critical temperature. Your answer should be less than 4 sentences and not contain any equations. (*Hint:* Think about the balance between entropy and enthalpy in the system.)

4.15 (a) In liquid–vapor equilibrium, why is there a critical temperature above which no separate condensed phases exist?

(b) Qualitatively, why is there a critical temperature in two dimensions but not in one dimension?

(c) Show mathematically that in the case of a one-dimensional Ising model it is impossible to have two phases at temperatures above 0 K.

(d) Why is the critical point at a lower temperature in the mean field approximation than in the Onsanger solution?

(e) Why is the critical point calculated by the Bethe approximation above that calculated by the mean field approximation, but below the exact result?

4.16 We showed several different adsorbate arrangements in Figures 2.71, 2.72, 2.73, and P2.22. What types of interactions lead to the adsorbate arrangements in these figures? For example, Attractive first nearest neighbors and repulsive second nearest neighbors.

4.17 Earlier in this chapter we found that when there are repulsive interactions between molecules, the system orders.

(a) Why do systems order when there are repulsive interactions between the molecules?

(b) Show that first-order phase transitions are impossible in an Ising model with repulsive interactions, i.e., show that it is impossible for the system to segregate into two separate phases.

(c) Show that if the molecules in a system show strong second nearest neighbor repulsions and weak first nearest neighbor attractions, the system can form (2x1) domains so the system follows the XY universality class.

(d) What kinds of interactions could lead to the situation in part (c)? (*Hint:* Consider a nonspherical molecule or a system with a strong Friedel oscillation.)

4.18 We showed that the partition function for the Ising problem could be written as a trace of the transfer matrix. What is the physical significance of the trace of the matrix?

4.19 In Example 4.C it was noted that if a molecule has a binding energy considerably less than the value in Table 4.5 there will be little adsorption of the molecule. The derivation assumed, however, that the gas adsorbed molecularly. The results change if the gas adsorbs dissociatively. Consider the dissociative adsorption of H_2, O_2, and assume that the H_2 and O_2 dissociate into atoms upon adsorption. Calculate the binding energy of hydrogen or oxygen *atoms* needed to get a coverage of 0.5 at 300, 500, and 700 K. Do not forget to consider the bond-dissociation energy.

4.20 In the catalysis literature it is common to use an equation like Equation 4.30 to fit data to the Langmuir adsorption isotherm. However, in the biological literature, when one is studying the binding of molecules from solution it is more common to make what is called a Scatchard [1949] plot, which is a plot of the $[A]/C_A$ versus $[A]$ where C_A

is the concentration of adsorbate A in solution. One can also get a linear form by plotting θ versus $C_A (1 - \theta)$.

(a) Rearrange Equation 4.10 to show that if the adsorption follows a Langmuir adsorption isotherm, then the Scatchard plot and the plot of θ versus $C_A (1 - \theta)$ will also be linear.

(b) Which of the three forms will show deviations from the Langmuir adsorption isotherm most clearly? Assume you have data for the amount adsorbed as a function of concentration, and that the data follow either the Fowler adsorption isotherm or Equation 4.115 with $\beta h = -0.5$. Assume that you have data at $\theta = 0, 0.1, 0.2, 0.3, \ldots, 0.9$. However, you only know the number of molecules adsorbed versus concentration, you do not actually know θ. (*Hint:* First, use the Fowler equation to calculate data, then try to fit the data to the Langmuir isotherm using a modification of Equation 4.30, the Scatchard plot, and a plot of θ versus $C_A (1 - \theta)$.)

(c) Which equation is best for fitting data? Consider the data from Example 4.A. Which form gives the best least squares regression. Be sure to consider that hydrogen dissociates.

4.21 A lattice gas/Monte Carlo program is included in the supplemental material for this chapter.

(a) Outline in words the steps the program executes to calculate the adsorption isotherm via the Monte Carlo method. (*Hint:* The first step is to randomly populate the occupancy matrix.)

(b) Identify where the program
   (1) Fills the occupancy matrix
   (2) Computes values of exp $(-\beta \Delta E)$
   (3) Does the actual Monte Carlo steps
   (4) Computes the Bethe approximation
   (5) Computes the Langmuir approximation
   (6) Prints out the results

(c) We define two matrices exe(nn) and exf(nn) via

$$\text{exe(nn)} = \exp [-\beta \Delta E_{nn}^A]$$

$$\text{exf(nn)} = \exp [-\beta \Delta E_{nn}^D]$$

where ΔE_{nn}^A is the energy change in adsorbing on a site with nn nearest neighbors and ΔE_{nn}^D is the energy change in desorbing from a site with nn nearest neighbors. Show that ΔE^A is only a function of the number of nearest neighbors when there are only first nearest neighbor interactions.

(d) Where are exe and exf computed? Where are they used? Show that the program correctly decides whether to flip the state of a site.

(e) What would you change in the program if there were second nearest neighbor interactions?

(f) How does the subroutine to compute the Bethe approximation work? Outline in words what the program does to calculate a point on the adsorption isotherm.

4.22 Figure 4.41 shows a series of surface structures calculated for $\beta F = 0$, and various values of βh.

(a) Start with $\beta h = 2$. Show that the surface structure should show a c(2x2) pattern.

(b) Make a photocopy of the figure. Identify the c(2x2) regions on the photo. Also identify the boundaries adjacent c(2x2) domains and identify any regions that are not in a c(2x2) structure. Guess why these structures arise.

(c) Now look at the picture for $\beta h = -1$. Can you still identify c(2x2) structures? How does the order compare to that in part (b)?

(d) Now go to $\beta h = 0$. Do you still see ordered structures? Why do they arise? Is there any structure? (*Hint:* If you randomly filled sites, what would the diagram look like?)

(f) Use the program LATGAS, available from Professor Masel to compare the dynamics at $\beta h = 0$ and $\beta h = 2$. What is the difference?

(e) Try simulating the $\beta h = 0$ case using the program LATGAS, available from Professor Masel. Try turning off surface diffusion. Does the $\beta h = 0$ case change?

(g) Now look at the case for $\beta h = 1$. What should the surface structure be like? Does Figure 4.41 conform to your expectations?

4.23 In Chapter 7 we will find that the rates of bimolecular chemical reactions are proportional to the number of pairs of atoms on the surface. Use the program LATGAS, available from Professor Masel to calculate the number of pairs of atoms on the surface as a function of βh for a coverage of 0.1, 0.25, 0.35, 0.65, 0.75. (*Hint:* Use Figure 4.23 to pick βF.)

4.24 A compiled version of the lattice gas program (LATGAS) is available from Professor Masel. Use the program to explore the effect of βF and βh on the structure of the adsorbed layer.

(a) Start with $\beta F = 0$, $\beta h = 0$. Run the program and watch the time evolution of the coverage and structure. Do you see any structure in the layer? How much does the coverage vary?

(b) Now vary βh keeping $\beta F = 0$. At what point do you see a distinct c(2x2) phase? At what point do distinct islands form? Compare you results to the expectations from Example 4.G.

(c) Choose $\beta h = \pm 2$ and vary βF. What do you see?

4.25 In Section 4.7.5 an equation was derived for the adsorption of gas on a line of sites.

(a) Can you think of any materials that would provide a line of adsorption sites? Polymers are one example. Find several others.

(b) How would the derivation change if there were second nearest neighbor interactions?

(c) Could you still define a transfer matrix in part (b)?

(d) Provide a qualitative argument (not a detailed derivation) to show that the partition function is still a trace of a matrix in part (b).

4.26 The objective of this problem is to compute the isotherm for a finite one dimensional chain.

(a) Consider adsorption on a line of 8 sites with periodic boundary conditions. Use the derivation like that in Section 4.7.5 to obtain an expression for the exact isotherm for adsorption on the line of sites. Compute the isotherm for $\beta h = -1$, 0, 1.

(b) Write a Monte Carlo program for adsorption on the line of sites using the Monte Carlo program in the supplemental material as a guide. Compare your results to those in part (a)

4.27 Example 4.G shows that a critical point exists in two dimensions but not in one dimension. Explain in your own words why there is a critical temperature in two dimensions, while there is no critical temperature (i.e., the critical temperature is absolute zero) in one dimension. (*Hint:* Consider an island breaking up. How does the enthalpy and entropy change during island breakup vary with the size of the island in (a) one dimension, (b) two dimensions.)

4.28 Compare the analytical models discussed in this chapter. Calculate an isotherm with the Fowler equation, Equation 4.132, with $\beta h = 0.5$ and various values of βH and fit the results with the other equations in this chapter using methods desorbed in Example 4.A. When will the coverage be a factor of 2 different when you use different models? How can you distinguish between the models?

4.29 Show that the adsorption isotherm calculated via the Bragg–Williams approximation is symmetric around $F = 0$. (*Hint:* Let $\theta \Rightarrow 1 - \theta$, $H_1 + 2h \Rightarrow -H_1 - 2h$ in Equation 4.132 and show that the resultant isotherm is the same.)

4.30 Calculate the order parameter for each of the structures shown in Figure 4.22. Show that the order parameter is unity for the ideal structure. Compute the order parameter for the (2x1) structure using the reciprocal lattice vector for the (2x2) structure. What do you conclude from this?

More Advanced Problems

4.31 We discussed several models of adsorption in this chapter, starting with Langmuir's adsorption isotherm and moving up to the lattice gas models. We gave the results in an historical perspective to outline the historical development of the field. The objective of this problem is to trace the historical development to see why the various models were proposed.

(a) In the beginning of this chapter we discussed that adsorption and absorption are fundamentally different. However, when Langmuir started his work, people thought that the two processes were closely related. Compare the absorption of a gas into a liquid and adsorption onto a solid. How is an adsorption isotherm different from the isotherm for adsorption of a gas into a liquid? Why did Langmuir have to propose his isotherm instead of assuming that adsorption on a surface followed the equations developed for absorption (e.g., Henry's law plus corrections)?

(b) Why did people propose modifications to the Langmuir model, such as Equations 4.33, 4.35, 4.38, and 4.43? What were the key experimental observations that motivated the use of these modified equations?

(c) The lattice gas models are used today. The lattice gas models were originally proposed because they could be solved using statistical methods, and not because they were important. However, now many people use lattice gas models to explain their data. What were the key experimental observations that led people to use lattice gas models as opposed to the simple models in Section 4.3?

4.32 The derivations in Section 4.3 assumed that the adsorbing molecules bind to a single site. In fact, adsorbing molecules often bind to multiple sites. Consider a molecule that binds to two sites.

(a) Use a derivation like that in Section 4.3.1.1 and Example A.1, Appendix A, to show that the molecule obeys the equilibrium expression

$$K_{equ} = \frac{[A]}{[S]^2 P_A}$$

(b) Derive a Langmuir adsorption isotherm for the molecule following the derivation in Section 4.3.

(c) Consider the coadsorption of two molecules, one of which binds to one site and a second of which binds to two sites. Derive a Langmuir adsorption isotherm for the coadsorption of the two molecules. [*Hint:* You can use your equilibrium expression from part (a) for part (c).]

4.33 Figure 4.45 shows that at low coverage there are significant differences between the BET equation and Putnam's krypton data.

(a) Show that the BET equation, Equation 4.189, reduces to the Langmuir adsorption isotherm, Equation 4.10, in the limit of low coverage. (*Hint:* Define a new variable y_B via $x_B = y_B/c_B$, and use it to eliminate x_B from Equation 4.189, then take the limit as $c_B \rightarrow \infty$ and note $y_B = K_1 P_a$.)

(b) Notice that the BET equation goes to the Langmuir result only if we eliminate x_B. If we instead assume that $x_B = P/P_{sat}$, with P_{sat} finite, the equation does not approach the Langmuir result as $c_B = \infty$. Why not?

(c) Use the parameters from Example 4.I to estimate when the second monolayer starts to form.

(d) Show that the differences between the data and the BET equation shown in Figure 4.45 are associated with deviations from ideal Langmuirian behavior.

(e) Consider the size of an atom of krypton. How does it compare to the lattice dimension of graphite. Based on these data, would you expect the first nearest neighbor krypton–krypton interaction to be attractive or repulsive? How about the second nearest neighbor interaction?

(f) Fit the data to the lattice gas model. (*Note:* Be sure to consider the hexagonal symmetry.) How well does it fit? See Schick (1981) for a detailed calculation.

(g) Look up a krypton–krypton interaction potential in the literature for liquid krypton. How well do your parameters in part (e) compare to the liquid results? How can you account for the differences?

4.34 In the discussion in Section 4.7.5 we considered the adsorption of a molecule on a line of sites. However, the general equations are applicable to many other situations.

(a) Consider the adsorption of two species, A and B, on a line of sites. Note that there are three possible values for the occupancy number of each lattice site, $\epsilon_i = 0$ (empty), $\epsilon_i = A$ (filled with A), $\epsilon_i = B$ (filled with B). Show that if there are only first nearest neighbor interactions, the partition function equals the trace of a three-by-three matrix while a $3 \times 3 \times 3$ array is needed if there are first and second nearest neighbor interactions. (*Note:* One can use a 9×9 matrix instead of a $3 \times 3 \times 3$ array.)

(b) Develop a specific equation for each elements of the three-by-three matrix in terms of the chemical potential of A and B, and the strength of the AA, AB, and BB interactions. Assume that there are only first nearest neighbor interactions.

(c) Use a computer to calculate the adsorption isotherm (i.e., θ_A and θ_B as a function of βH_{1B}) assuming $\beta H_{1A} = \beta H_{1B} + 1$; $4\beta h_{AA} = 0.25$; $4\beta h_{AB} = -0.25$; $4\beta h_{BB} = 0.125$

(d) Generalize the results in part (a) to show how if c molecules adsorb on each site, and there are only first nearest neighbor interactions, the partition function becomes the trace of a $(c + 1)$ by $(c + 1)$ matrix.

(e) Now consider adsorption of c molecules onto a strip l sites wide with first nearest neighbor interactions. Show that the partition function is equal to the trace of a $(c + 1)^l$ by $(c + 1)^l$ matrix.

4.35 The Monte Carlo/Metropolis calculations for a lattice gas were described in Section 4.8. The objective of this problem is to calculate an actual adsorption isotherm using the Metropolis algorithm described in Section 4.8 and the supplemental material.

(a) Modify the program in the supplemental material to run at a fixed value of βh. Use a matrix with periodic boundary conditions. Take 100 Monte Carlo steps to start the program, and 500 to calculate an adsorption isotherm. You only need to choose $10 \times 10 = 100$ sites in steprand rather than 1600.

(b) Use your program to calculate an adsorption isotherm for $\beta h = -1$ and $\beta h = 1.75$. Compare to the results for a (40×40) lattice in Figure 4.20. Use a workstation to do the calculations.

(c) Compare your results in (b) to the Bragg–Williams and Bethe–Peierls isotherms for the same values of the parameters.

(d) Modify the program to calculate the order parameter for the c(2x2) phase for your cases in (b). When does the order parameter equal 0? When is the c(2x2) phase stable? (*Hint:* Calculate an order parameter at the end of each call to step or steprand.)

4.36 In Section, 4.10.1, Equation 4.160 was presented but not derived. The purpose of this problem is to derive Equation 4.160. Start with the thermodynamic definition of the spreading pressure

$$dE = TdS - \pi da + \mu dN \qquad (4.260)$$

and show that the Hemholtz free energy of the layer, A, follows:

$$dA = -SdT - \pi da - \mu dN \qquad (4.261)$$

Now substitute the expression for A from Equation 4.2 into Equation 4.261 to yield

$$\left(\frac{d \ln Q}{da}\right)_{N_i T} = \frac{\pi}{kT} \qquad (4.262)$$

(a) Substitute Q for a two-dimensional ideal gas into Equation 4.262 to derive equation 4.160.

(b) How would your expression change if the velocity distribution of the adsorbed layer were different from an ideal gas so that the translational partition function changed.

(c) Substitute Q from Equation 4.12 into Equation 4.262 and note that

$$a = S_0/[S_0] \qquad \text{with } [S_0] = \text{a constant}$$

to derive an equation of state for a Langmuir adsorption isotherm. Show that the result is asymptotic to Equation 4.177 as $\theta \to 0$.

4.37 Consider the adsorption of a gas onto the hexagonal lattice shown in Figure 3.14.

(a) Work out the low-temperature limit of the phase diagram, assuming that the molecules adsorb onto the on-top sites, and that there are first and second nearest neighbor interactions. Use the analyses in Examples 4.E and 4.F.

(b) If one observes the $(\sqrt{3}x\sqrt{3})$ lattice shown in Figure 3.14, what does that tell us about the adsorbate/adsorbate interactions?

(c) Norton, Davies, and Jackman [1987] found that the CO coverage is higher in the $(\sqrt{3}x\sqrt{3})$ pattern than one would expect from Figure 3.14. One proposal is that the diagram in Figure 3.14 should be reversed. All the on-top sites that are empty should be filled, and all of the sites that should be filled should be empty. What type of pairwise interactions would give us this type of adsorbate phase diagram. To aid your thinking, start with Equation 4.59 for the case in part (b), and flip the configuration of all of the sites. What do you need to do to keep the Hamiltonian the same?

(d) How would the parameters have to be different to get each of the $(\sqrt{3}\times\sqrt{3})$ patterns?

(e) Show that there is a phase transition at one-third coverage in the phase diagram for the preceding system. This is a characteristic of the three-state Potts universality class.

(f) Another characteristic of the Potts universality class is that the adsorption system is threefold degenerate. Show that the system is threefold degenerate.

4.38 There has been considerable literature on the nature of the indirect adsorbate/adsorbate interactions on metal surfaces. Einstein [1978] and co-workers modeled the interactions by assuming that the presence of an atom at site i would cause Friedel oscillation in the surface electron density near a surface atom. Feibelman [1989] shows that it may be better to model the change by considering two terms, a short-range oscillation and a longer range shift in the Fermi level (i.e., electrochemical potential) of the surface. White and Akhter [1988] discuss several other types of adsorbate/adsorbate interactions.

The objective of this problem is to ask how to include such interactions into the Hamiltonian for a lattice gas, and to examine the consequences of the interactions on adsorbate behavior.

(a) First consider the Friedel oscillations. Use the results in Einstein's paper to derive an approximation for the first, second, third, . . . nearest neighbor interactions. Show that one can simulate the effects of the Friedel by choosing the form of the interactions appropriately. How would the Friedel oscillations affect the phase behavior of the adsorbed layer?

(b) Next consider the change in the local Fermi level. Consider the nonlinearities. Show that this leads to an effective three-body term in the Hamiltonian.

(c) A third kind of adsorbate/adsorbate interaction, which has not yet been considered theoretically, but seems to be important experimentally is associated with adsorbate-induced reconstructions. The adsorbate can perturb the local arrangement of the surface atoms, which changes the bonding of adjacent adsorbate molecules. Consider some simple examples. How can one put such interactions into a lattice gas calculation?

(d) White and Akhter discuss several other types of indirect interactions. Discuss how you would include them in a lattice gas calculation.

4.39 In Problem 4.27 you showed that the $\sqrt{3}\times\sqrt{3}$ LEED pattern in Figure 3.14 for CO on Pt(111) is associated with repulsive interactions between adjacent COs. Experimentally, the COs do show repulsive interactions as evidenced by the continuous reduction of the heat of adsorption of CO with coverage shown in Figure 4.7. The objective of this problem is to understand how the reduction is possible.

(a) Experimentally, the COs stand upright on the surface with the carbon end down and the oxygen pointing away from the surface. How far apart are the adjacent COs when the CO layer is arranged in the $\sqrt{3}\times\sqrt{3}$ overlayer shown in Figure 3.14? How does this compare with the van der Waals radius of the COs? Is it reasonable that the COs would show repulsive interactions at the distances characteristic of the $\sqrt{3}\times\sqrt{3}$ overlayer shown in Figure 3.14? Note each platinum atom is 2.77 Å in diameter.

(b) The data in Figure 4.9 suggest that there is also an indirect interaction between the adjacent COs. Adsorption of one CO molecule changes the local electronic structure of the surface, which in turn affects the bonding of other COs. Discuss how these indirect interactions can produce long-range repulsions between adjacent CO molecules following the analysis in Section 4.7.1 and P. T. Feibelman, *Adv. Chemical Physics* **40**, [1989], and T. L. Einstein, *CRC Crit. Rev. Mat. Sci.* **7**, 261 [1983].

4.40 Crossley and King [1980] proposed that at moderate coverage, CO forms the C(4x2) structure shown in Figure 3.14 when the CO adsorbs on Pt(111).

(a) Notice that the COs adsorb on two different sites. How can one change the lattice gas model described in Section 4.7.1 so that one could consider two different adsorption sites?

(b) What type of interactions can produce the c(4x2) structure?

(c) What type of interactions can produce the domain wall structure in Figure 3.11?

(d) How do your conclusions compare to those in Persson, Töshavs, and Bradshaw [1990]?

4.41 C. H. Chen , S. M. Vesecky, and A. A. Gewirth, *JACS* **114**, 451 [1992] have examined the adsorption of silver on Au(111) with an AFM. They found that the silver forms the (3x3) overlayer depicted in Figure 4.46.

(a) What type of adsorbate/adsorbate interactions could product the arrangement shown?

(b) Notice that the silver adsorbs on two different types of sites. How could one simulate adsorption on two different types of sites with a Monte Carlo calculation? What would your variables be? What interaction parameters would you choose?

(c) Write a Monte Carlo routine to simulate the adsorption of silver on a 18 atom × 18 atom section of the Au(111) surface. Pick some reasonable values of the parameters. How does the ordering of the overlayer vary with silver coverage?

Figure 4.46   The structure of a silver layer on Au(111). (As proposed by Chen, Vesecky, and Gewirth, *JACS* **114**, 451 [1992].)

(d) Compute the distances between adjacent silver atoms. Is it reasonable that the silver will show repulsive interactions at this range? If the silver atoms are attached to a sulfate ligand, would it be reasonable that they show repulsive interactions? Is sulfate large enough to produce the repulsions? Why can the sulfate not be seen with atomic force microscopy (AFM)?

4.42 In Section 4.8 we noted that if one has weakly attractive first nearest neighbor interactions and strongly repulsive second nearest neighbor interactions, the adsorbate can form (2x1) domains. The objective of this problem is to describe in words how it is possible to have interactions like this.

(a) First consider a simple direct pairwise interaction. Sketch an adsorbate/adsorbate interaction potential that would produce attractive first nearest neighbor interactions and repulsive second nearest neighbor interactions.

(b) Consider a Pt(111) surface. What are the distances between a site i on top of one atom, and the adjacent first and second nearest neighbor sites? How large does the molecule have to be before second nearest neighbor interactions are repulsive? Could the first nearest neighbor interactions be attractive?

(c) Consider an adsorbate that would like to form a two-molecule cluster on the surface. Assume each cluster covers two sites, and that the interactions are repulsive between adjacent clusters. Show that this type of interaction could be modeled using the quasichemical approximation. Will this interaction produce (2x1) structures?

(d) Next consider the Friedel oscillations described in Section 4.7.1. Show that they can lead to (2x1) structures.

(e) Experimentally, one observes (2x1) structures most often when an adsorbate forms a bridge bond between two adjacent sites. Often surface atoms move. Show that if the surface atoms are displaced perpendicular to the surface, the energy of the system will be lower if there is a continuous chain of adsorbates rather than a random adsorption. Also, show that the entropy of the system is higher if there are kinks in the chain so that at low temperatures, the system would rather form straight lines of adsorbate.

4.43 Ching, Huber, Lagally, and Wang, *Surface Sci.* **77,** 550 [1978] did a Monte Carlo calculation for oxygen adsorption on W(110).

(a) Read their paper carefully. Explain, in your own words, why they conclude that there are three-particle interactions.

(b) They also calculate the strength of the two- and three-body terms. How accurate are their estimates?

(c) Would one get the same phase diagrams if one assumed that the individual pair interactions were weaker, but there are nonnegligible pair interactions out to sixth nearest neighbors?

4.44 Carbon monoxide oxidation on Pd(111) is interesting because T. Engel and G. Ertl, *Adv. Catal.* **28,** 1 [1979] found that the adsorbed carbon monoxide and oxygen phase separate into CO-rich and oxygen-rich islands during the reaction even though no islands form when either individual reactants adsorbs. Stuve, Madix, and Brundle, *Surface Sci.* **146,** 155 [1984] observed a similar effect on Pd(100). Silverberg and Ben-Shaul, *Surface Sci.* **214,** 17 [1989] used the Monte Carlo method to simulate this reaction. Look over Silverberg and Ben-Shaul's paper. Show that for the parameters chosen in their paper, no islands form when each reactant molecule adsorbs individually. Why do islands form when carbon monoxide and oxygen adsorb simultaneously? Why do Silverberg and Ben-Shaul choose the values of the parameters for the adsorbate/

adsorbate? What would change in their results if the strength of the adsorbate/adsorbate interactions were doubled. What if it were halved?

4.45 Kohn, Szanyi, and Goodman, *Surface Sci.* **274**, L611 [1992] examined CO adsorption on Pd(111) with infrared (IR). They found that at temperatures above 500 K, and pressures near 1 atm, some new CO phases are formed that are not seen at lower temperatures. The objective of this problem is to understand how such changes are possible.

    (a) Start with the data in Figure 4.1. Fit it to the lattice gas model using an analysis like that in Example 4.A.

    (b) Now consider raising the pressure and the temperature, keeping βF constant. What pressure do you need to maintain a coverage of 0.5 at 300, 400, 500, and 600 K?

    (c) How will the phase behavior change? (*Hint:* How does βh change?)

    (d) Use your results in part (c) to show that it is possible to change the phase behavior by raising the temperature and pressure.

    (e) What kinds of interactions does one need to explain the phases Kohn et al. observe at 550 K.

    (f) Do the changes in phase behavior observed by Kohn et al. make sense given your results in part (e)?

Extension of the Methods to Other Fields

4.46 Example 4.A shows that one can fit data to our corresponding states curve, Figure 4.23, even if one does not know the critical temperature and critical pressure of the system. Generalize these results to fluid systems. How can one replot the standard corresponding states curve (i.e., Z_r vs. T_r and P_r) to get a plot that can be fit to data even when P_c and T_c are unknown.

4.47 (a) Describe the Ising model in three sentences or less.

    (b) What are the key assumptions in the model?

    (c) How can the Ising model be used to model adsorption?

    (d) How can the Ising model be used to model gases?

    (e) How can the Ising model be used to model polymers/proteins?

    (f) How can the Ising model be used to model alloys?

    (g) How can the Ising model be used to model liquid solutions? What are the key assumptions needed to use the Ising model for these systems?

    (h) Why does the simple Ising model not give a good description of the behavior of a two-component mixture near the critical point?

4.48 Vrahapoulou and McHugh, *Macromolecules* **17**, 2657 [1984] examined the phase behavior of a polymer under flow. Read their paper and give a synopsis of what they found. Your answer should be half a page or less typed.

4.49 Consider using the Ising model to model a solution of A in a solvent B.

    (a) Write an expression for the Hamiltonian in terms of the strengths of the A–A, A–B, and B–B interactions.

    (b) First consider the ideal case ($h = 0$). Show that at low concentrations the system follows Raoult's law.

    (c) Now consider the case where the A–B interactions is stronger than the A–A or the B–B interaction. Use the mean field approximation to show that Henry's law will still be valid at low pressure.

(d) Will Henry's law coefficient be larger or smaller in the case in part (b) or in the case in part (c)? That is, will less or more A adsorb at a fixed fugacity of A?

(e) Develop a criterion for when Henry's law coefficient will be larger or smaller than the case in part (b).

(f) How can one use universality to develop a criterion for when there will be significant deviations from Henry's law?

(g) Under what conditions will the system separate into two liquid phases? Assume that there are 10 nearest neighbors in the system. (*Hint:* A generalization of the material given earlier says that the critical point comes when sinh $(n_n \beta h/4)$ = $n_n/2$.)

4.50 The methods in Sections 4.3–4.8 are useful for many systems in addition to adsorption from a gas. Consider the following examples:

(a) Binding of a molecule to a protein. (*Reference:* I. M. Klotz, *Quarterly Review of Biophysics* **18**, 227 [1985].)

(b) Adsorption from solution with and without an electric field. (*Reference:* J. Bockis and Swinkels, *J. Electrochem. Soc.* **111**, 736 [1964].)

Derive analogies of adsorption isotherms for these cases and discuss how well they work.

4.51 The lattice gas methods in Sections 4.4–4.8 are applicable to many other systems in addition to adsorption. The list below gives some references to other applications of the methods. Pick one of these examples and report on the application of Ising/lattice gas methods to these problems.

(a) Folding of polypeptides: J. Applequist, *J. Chem. Phys.*, **38**, 934 (1963); D. C. Poland and H. A. Scheraga, *J. Chem. Phys.*, **43**, 2071 (1965); K. Binder and D. Stauffer, in *Montecarlo Methods in Statistical Physics*, 2 ed., K. Binder, ed., Springer-Verlag, New York, p. 326 (1979).

(b) Denaturation of DNA: R. M. Wartel and E. W. Montrol, *Adv. Chem. Phys.*, **22**, 129 (1972).

(c) Configuration of Polymers In Solution: P. J. Flory, *Statistical Mechanics of Chain Molecules*, Chaps. III and IV, especially pp. 67–71, Interscience, New York (1969); see Also T. L. Hill and Y. Kantar, *Phys. Rev. A*, **44**, 5091 (1991).

(d) Crystallization of Paraffins and Polymers: J. Naghizadeh, *Adv. Chem. Phys.*, **65**, 45 (1986).

(e) Crystal Growth: H. Müller-Krumbhaar, in *Monte Carlo Methods In Statistical Physics*, 2 ed. K. Binder, ed., Springer-Verlag, New York, p. 261 (1979).

(f) Phase Behavior of Liquid Crystals: P. A. Lebwohl and G. Lasher, *Phys. Rev. A*, **6**, 426 (1977); H. J. F. Jansen, G. Vertogen, and J. G. J. Ypma, *Mol. Crystals Liquid Crystals*, **38**, 87 (1977).

(g) Deviations from Henry's Law in Solution: T. L. Hill, p. 371 [1961].

(h) Multilayer Adsorption: K. Binder and D. P. Landau, *Phys. Rev. B* **37**, 1745 (1988), and **40**, 6971 (1989).

(i) Phase Transition on a Binary Alloy: L. D. Fosdick, *Methods Comp. Phys.* **1**, 245 (1963); L. Gutman, *J. Chem. Phys.*, **34**, 1024 (1961); P. A. Flinn and G. M. McManus, *Phys. Rev.*, **124**, 54 (1961); Piscke and Bergersen [1989].

(j) Phase Transitions in Solution: Dietrich [1989]; D. W. Sullivan and M. M. Telodagama, in *Fluid Interfacial Phenomea*, C. A. Croxton, ed., Wiley, New York (1986).

(k) Phase Transitions in Colloidal Systems: L. P. Voegtli and C. F. Zukoski, *J. Colloid. Int. Sci.*, **141**, 79 (1991), S-H. Chen and R. Rajagopalan, *Statistical*

Thermodynamics of Micellar Solutions and Microemulsions, Springer-Verlag, New York (1987); A. L. Kholodenko and C. Qian, *Phys Rev. B*, **40**, 2477 (1989).

(l) Diffusion and Flow in Porous Media: D. Stauffer, *Introduction to Percolation Theory*, Taylor & Francis, London (1985); E. M. Sevick, *Chem. Eng. Sci.*, **44**, 21 (1989).

(m) Neural Networks: *Phys. Rep.*

(n) Surface Reconstructions: V. E. Zubkus and E. E. Tornav, *Surface Sci.*, **216**, 23 (1989); I. K. Robinson, E. Vlieg and K. Kern, *Phys. Rev. Lett.*, **63**, 2578 (1989).

4.52 Gugenheim [1952] shows that the methods in Section 4.9.2 can be used to calculate the properties of mixtures of miscible liquids. Discuss the application to liquids of these methods and the other methods in Section 4.9. How well do they work?

ADSORPTION BIBLIOGRAPHY

Adamson, A. W., *Physical Chemical of Surfaces*, 5th ed., Wiley, NY (1990).

Clark, A., *The Theory of Adsorption and Catalysis*, Academic Press, NY (1970).

Dash, J. G., *Films on Solid Surfaces*, Academic Press, NY (1975).

Jaroniec, M., and R. Madey, *Physical Adsorption on Heterogeneous Solids*, Elsevier, NY (1988).

King, D. A., and D. P. Woodruff, eds., *The Chemical Physics of Solid Surfaces and Heterogeneous Catalysis*, vol. 2, *Adsorption at Solid Surfaces*, Elsevier (1983).

Ponec, V., Z. Knor, and S. Cerny, *Adsorption on Solids*, Butterworth, London (1974).

Somorjai, G. A., *Chemistry in Two Dimensions: Surfaces*, Cornell University Press, (1977).

Steele, W. A., *The Interaction of Gases with Surfaces*, Pergamon Press, NY (1974).

Tompkins, F. C., *Chemisorption of Gases on Metals*, Academic Press, NY (1978).

REFERENCES

Adams, D. J., *Molecular Physics* **28**, 1241 (1974).

Adamson, A. W., *Physical Chemistry of Surfaces*, Wiley, NY (1960).

Adamson, A. W., *Physical Chemistry Of Surfaces*, 5 ed., Wiley, NY (1990).

Averbach, D. J., and C. T. Rettner, in M. Gruze and H. J. Kreuzer, ed., *Kinetics of Interface Reactions*, p. 125, Springer-Verlag, NY (1987).

Barber, M. N., in C. Domb and J. L. Lebowidz, ed., *Phase Transitions and Critical Phenomena*, vol. 8, p. 145, Academic Press, NY (1983).

Barley, D. M., in C. Domb, ed., *Phase Transitions and Critical Phenomena*, vol. 2, p. 329, Academic Press, NY (1972).

Bassett, D. R., E. A. Boucher, and A. C. Zettlemoyer, *J. Colloid. Int. Sci.* **27**, 649 (1968).

Bethe, H. A., and W. H. Wills, *Proc. Roy. Soc. A* **150**, 552 (1935).

Binder, K., ed. *Applications of the Monte Carlo Method in Statistical Physics*, Springer-Verlag, NY (1979).

Binder, K., ed. *Applications Of The Monte Carlo Method In Statistical Physics*, Springer Verlag, NY (1984).

Binder, K., and D. W. Heermann, *Monte Carlo Simulation in Statistical Physics*, Springer-Verlag, NY (1988).

Binder, K., and D. P. Landau, *Surface Sci.* **108**, 503 (1981).

Binder, K., and D. P. Landau, *Adv. Chem. Phys.* **76**, 91 (1988).

Blyholder, G., *J. Phys. Chem.* **68**, 2772 (1964).

Bragg, W. L., and E. J. Williams, *Proc. Royal. Soc. A* **145**, 699 (1934).

Bomchil, G., G. N. Harris, M. Leslie, J. Jabony, and J. W. White, *J. Chem. Soc., Faraday Trans.* **75,** 153 (1979).

Bonhoeffer, K. F., and P. Hartek, *Nat. Wiss.* **17,** 182 (1929).

Brunauer, S., *The Adsorption of Gases and Vapors*, vol. 1, Princeton University Press, Princeton, NJ (1945).

Brunauer, S., P. H. Emmett, and E. Teller, *J. Am. Chem. Soc.* **60,** 309 (1938).

Brunauer, S., in R. F. Gould, ed., *Solid Surfaces and the Gas Solid Interface*, p. 5, American Chemical Society, Washington, DC (1961).

Brunauer, S., K. S. Love, and R. G. Keenan, *J. Am. Chem. Soc.* **64,** 751 (1942).

Cassuto, A., M. Mane, and J. Jupille, *Surface Sci.* **249,** 8 (1991).

Chappuis, P., *Wiedermann's Annalen Der Physik und Chemie*, **12,** 161 (1881).

Christmann, K., G. Ertl, and T. Pignet, *Surface Sci.*, **54,** 365 (1976).

Christmann, K., O. Schober, G. Ertl, and M. Neuman, *J. Chem. Phys.* **60,** 4528 (1974).

Constable, F. H., *Proc. Roy. Soc., London, A* **108,** 355 (1925).

Crossley, A., and D. A. King, *Surface Sci.* **95,** 131 (1980).

Dietrich, S., in C. Domb and J. L. Lebowitz, eds., *Phase Transitions and Critical Phenomena*, Vol. 12, p. 1, Academic Press, NY (1988).

Dobereiner, M., *Annales de Chimie et de Physique*, **24,** 91 (1823).

Dollimare, D., P. Spooner, and A. Turner, *Surface Tech.* **4,** 121 (1976).

Doraiswamy, L. K., *Prog. Surf. Sci.* **37,** 1 (1991).

DuLong, P. L., and L. J. Theonard, *Annales de Chimie* **23,** 440 (1822); **24,** 330 (1823).

Ehrlich, G., and F. G. Hudda, *J. Chem. Phys.* **30,** 493 (1959).

Einstein, T. L., *CRC Crit. Rev. Mater. Sci.* **7,** 261 (1983).

Einstein, T. L., N. C. Bartlet, and L. D. Roelofs, In J. F. van der Veen and M. A. Van Hove, eds., *The Structure Of Surfaces II*, p. 480 Springer-Verlag, NY (1988).

Engel, T., and R. Gomer, *J. Chem. Phys.* **52,** 5572 (1970).

Ertl, G., and T. Koch, *Z. Nat. Forsch.* **25A,** 1906 (1970).

Ertl, G., M. Neuman, and K. Streit, *Surface Sci.* **64,** 393 (1977).

Evans, R., *J. Chem. Phys.* **86,** 7138 (1987).

Evans, R., and A. O. Parr, *J. Phys. Cond. Matter* **2,** 1 (1990).

Faraday, M., *Quarterly Journal Of Sci.* **7,** 106 (1819).

Faraday, M., *Philos. Trans. Royal Soc. London* **124,** 55 (1834).

Farkis, A., *Orthohydrogen, Parahydrogen and Heavy Hydrogen*, Cambridge University Press, Cambridge (1935).

Farkis, A., and L. Farkis, *Nature* **132,** 894 (1933).

Farkis, A., and L. Farkis, *Proc. Royal Soc. A* **144,** 467 (1934).

Feibelman, P. J., *Ann. Rev. Phys. Chem.* **40,** 261 (1989).

Ferrer, S., and H. Bonzel, *Surface Sci.* **119,** 234 (1982).

Feynman, R., *Statistical Mechanics*, Addison-Wesley, Menlo Park, CA (1972).

Fisher, G., *Chem. Phys. Lett.* **79,** 452 (1981).

Fontana, F., *Recherches physics sur la nature de l'air nitreux et de l'air dephlogistiqué*, Paris (1776). (This reference is cited in Priestly, 1777, and in the British Museum Catalog, but is not generally available.)

Fontana, F., *Societie Italiana della Scienze, Rome, Memorie di Mathematica e di fisica*, **1,** 648 (1782). (This journal is in the University of Michigan collection but only limited copying is permitted.)

Fowler, R. H., *Proc. Camb. Philos. Soc.* **31,** 260 (1935).

Fowler, R. H., *Statistical Physics*, Cambridge University Press, NY (1939).

Fowler, R. H., and E. A. Guggenheim, *Statistical Mechanics*, 2 ed., Cambridge University Press, NY (1955); 3 ed., (1965).

Freeman, A. E., *J. Am. Chem. Soc.* **35,** 927 (1913).

Freundlich, H., *Kapillarchemie*, Academishe Bibliotek, Leipzig (1909).

Gellman, A., W. T. Tysoe, F. Zaera, and G. A. Somorjai, *Surface Sci.* **191,** 271 (1987).

Gland, J. L., D. A. Fischer, S. Shen, and F. Zaera, *J. Am. Chem. Soc.* **112,** 569 (1990).

Gray, G. C., and K. E. Gubbins, *Theory of Molecular Fluids*, vol. I, Oxford University Press, NY (1984).

Gregg, S. J., and K. S. W. Sing, *Adsorption, Surface Area and Porosity* 2d ed., Academic Press, London (1982).

Gugenheim, E. A., *Mixtures*, Oxford University Press, London (1952).

Halsey, G. D., *J. Chem. Phys.* **16,** 93 (1948).

Halsey, G. D., and H. S. Taylor, *J. Chem. Phys.* **15,** 624 (1947).

Hammaker, R. M., S. A. Fransis, and R. P. Eischens, *Spectrochim. Acta* **21,** 1295 (1965).

Harris, S. J., and D. N. Belton, *Thin Solid Films* **212,** 193 (1992).

Henderson, D., *Mol. Physics* **30,** 971 (1975).

Hill, T. L., *Adv. Catalysis* **4,** 211, (1952).

Hill, T. L., *Statistical Mechanics* (1956).

Hill, T. L., *Statistical Thermodynamics* (1960).

Ising, E., *Z. Phys.* **31,** 253 (1925).

Kayser, H., *Wiederman's Annalen Der Physik und Chemie*, **14,** 451 (1881).

Kikuchi, R., *Phys. Rev.* **81,** 988 (1951).

Kikuchi, R., *J. Chem. Phys.* **60,** 1071 (1974).

Kohn, W. K., J. Szany, and D. W. Goodman, *Surface Sci.* **278,** L612 (1992).

Lagally, M. G., T. M. Wu, and G. C. Yang, in Sihna, ed., *Ordering in Two Dimensions*, Springer-Verlag, NY (1980).

Land, T. A., T. Michely, R. J. Behm, J. C. Hemminger, and G. Comsa, *J. Appl. Phys.* **A53,** 414 (1991).

Langmuir, I., *J. Am. Chem. Soc.* **34,** 1310 (1912); **35,** 105 (1913), **37,** 1139 (1915), **40,** 1361 (1918).

Langmuir, I., *Trans. Faraday Soc.* **17,** 607 (1922).

Lanzillotto, A. M., M. J. Dresser, M. D. Alvey, and J. T. Yates, *Surface Sci.* **191,** 15 (1987).

Lefesque, D., J. J. Weis, and J. P. Hansen, in K. Binder, ed., *The Monte Carlo Method in Statistical Physics*, Springer-Verlag, NY (1987).

Lin, R. F., G. S. Blackman, M. A. Van Hove, and G. A. Somorjai, *Acta Crystal. Sect. B* **43,** 368 (1987).

Lombardo, S. J., and A. T. Bell, *Surface Sci. Rep.* **13,** 1 (1991).

Metropolis, N., A. W. Rosenbluth, M. N. Rosenbluth, A. H. Teller, and E. Teller, *J. Chem. Phys.* **21,** 1087 (1953).

Mitscherlich, E., *Poggendorf's Annalen De Physik und Chemie*, **59,** 94 (1843).

Morozzo, C. L., *Observations Sur La Physique Sur L Historie Naturelle Et Sur Les Arts* **23,** 362 (1783).

Morozzo, C. L., *Journal de Physique de Chimie D'Historie Naturelle et des Arts* **58,** 374 (1803).

Netzer, F. P., and T. E. Madey, *Surface Sci.* **119,** 422 (1982).

Nicholson, D., and N. G. Parsonage, *Computer Simulation and the Statistical Mechanics of Adsorption*, Academic Press, NY (1982).

Norton, P. R., J. A. Davies, and T. E. Jackman, *Surface Sci.* **121,** L595 (1982).

Ohtani, H., R. J. Wilson, S. Chaing, and C. M. Mate, *Phys. Rev. Lett.* **60,** 2398 (1988).

Olsen, C. W., and R. I. Masel, *Surface Sci.* **201,** 444 (1988).

Onsanger, L., *Phys. Rev.* **65,** 117 (1944).

Oswald, W., *Lehrbook Der Allgemeinen Chemie*, p. 778, Liepzig, (1885).

Peierls, R. E., *Proc. Camb. Philos. Soc.* **32,** 477 (1936).

Persson, B. N. J., *Surface Sci. Rep.* **15,** 1 (1992).

Pilschke, M., and B. Bergersen, *Equilibrium Statistical Physics*, Prentice-Hall, Englewood Cliffs (1989).

Priestley, J., *Experiments on Different Kinds of Air*, J. Johnson, London, vol. I (1775); vol. II (1777); vol. III (1779).

Putnam, F. A., and T. Fort, *J. Phys. Chem.* **79,** 459 (1975).

Ree, F. H., and W. G. Hoover, *J. Chem. Phys.* **40,** 939 (1964).

Rikvold, P. A., W. Kinzel, J. D. Gunton, and K. Kaski, *Phys. Rev. B* **28,** 2686 (1983).

Rockett, A., *Surface Sci.* **227,** 208 (1990).

Ross, S., and W. Winkler, *J. Colloid Sci.* **10,** 319 (1955).

Rottman, C., *Phys. Rev. B* **24,** 1482 (1981).

Rouppe, H. W., with Norden, *Annales De Chemie*, ser. 1, **32,** 3 (1799).

Scatchard, G., *Ann. N.Y. Acad. Sci.* **51,** 660 (1949).

Schaaf, P., and J. Talbot, *J. Chem. Phys.* **91,** 4401 (1989).

Scheele, C. W., *Chemische Ubhandung von der Luft und dem Feuer*, p. 96, Bergmann Press, Upsalla, (1777). (An English translation appears in L. Dobbin, ed., *The Collected Papers of Carl Wilhelm Scheele*, p. 173, G. Bell and Sons, Ltd, 1931.

Scheele, C. W., March 1, 1773 letter cited in C. C. Gillespie, ed., *Dictionary of Scientific Biography*, vol. 12, p. 143 (1974).

Scheffler, H., K. Horn, A. M. Bradshaw, and K. Kambe, *Surface Sci.* **80,** 69 (1979).

Schick, M., *Prog. Surface Sci.* **11,** 245 (1981).

Schultz, T., D. Mattis, and E. Lieb, *Rev. Mod. Phys.* **36,** 856 (1964).

Seebauer, E. G., A. C. F. Kong, and L. D. Schmidt, *Surface Sci.* **176,** 134 (1986).

Sips, R, *J. Chem. Phys.* **16,** 490 (1948).

Slygin, A., and P. Frumkin, *Acta Physicochim. URSS* **3,** 791 (1935).

Somorjai, G. A., and M. A. van Hove, *Prog. Surface Sci.* **30,** 201 (1988).

Speedy, R. J., *J. Chem. Soc. Faraday Trans. II* **73,** 714 (1977).

Steiniger, H., H. Ibach, and S. Lehwald, *Surface Sci.* **115,** 685 (1982).

Taylor, H. S., *Proc. Royal Soc. London A* **108,** 105 (1925).

Taylor, H. S., *J. Am. Chem. Soc.* **53,** 578 (1931).

Tejwana, M. J., O. Ferreira, and O. E. Vilces, *Phys. Rev. Lett.* **44,** 152 (1980).

Tempkin, M. I., and V. Pyzhev, *Acta Physiochim. URSS* **12,** 217 (1940).

Toth, J., *Acta Chim. Acad. Sci. Hung* **30,** 1 (1962); **69,** 311 (1971).

Tovin, Y. K., *Prog. Surface Sci.* **34,** 1 (1991).

Trapnell, B. M., and D. O. Yayward, *Chemisorption* Butterworths, London (1964).

Valenzuela, D. P., and A. L. Meyers, *Adsorption Equilibrium Data Book*, Prentice-Hall, Englewood Cliffs (1989).

Vasquez, N., A. Muscat, and R. J. Madix, *Surface Sci.* **301,** 83 (1994).

Volmer, M. A., and P. Mahnert, *Z. Physik. Chem.* **115,** 239 (1925a); **115,** 253 (1925b).

von Saussure, T., Gilbert's *Annalen Der Physik*, **47,** 113 (1814).

Wambach, J., G. Odörfer, H-J. Freund, H. Kuhlenbek, and M. Neumann, *Surface Sci.* **209,** 159 (1989).

Wander, A., M. Van Hove, and G. A. Somorjai, *Phys. Rev. Lett.* **67,** 626 (1991).

Wang, J., and R. I. Masel, *J. Am. Chem. Soc.* **113,** 5850 (1991).

Weinberg, W. H., in M. Gruze and H. J. Kreuzer, eds., *Kinetics of Interface Reactions*, p. 94, Springer-Verlag, NY (1987).

White, J. M., and S. Akhter, *CRC Crit. Rev. Mater. Sci.* **14,** 131 (1988).

Widom, B., *J. Chem. Phys.* **44,** 388 (1963).

Yagasaki, E., and R. I. Masel, *Surface Sci.* **222,** 430 (1991).

Yagasaki, E., and R. I. Masel, *Surface. Sci.* **226,** 51 (1991).

Zeldowitch, J., *Acta Physicochim. URSS* **1,** 961 (1934).

5 Adsorption III: Kinetics of Adsorption

PRÉCIS

In the previous chapter, we discussed adsorption under conditions where the adsorbate layer was in equilibrium with the gas phase. During a reaction, however, the adsorption process does not always have time to reach equilibrium. Instead, the adsorption process is limited by kinetics. In this chapter, the kinetics of adsorption is discussed. First, we give a general overview of the kinetics of adsorption, and in particular we define two processes called *trapping* and *sticking* and distinguish between them. We then present a variety of models for trapping and sticking. The emphasis is on classic models of trapping and sticking. We ignore the effects of state-to-state transitions (i.e., quantum effects) on adsorption. Additional information can be obtained in the reviews of Morris, Bowker, and King [1983], Goodman [1974, 1976], or Ashford and Rettner [1991].

5.1 HISTORICAL OVERVIEW

As noted in the last chapter, adsorption is a process where a molecule collides with a surface and sticks. Saussure [1814] did the first quantitative study of the adsorption process. In his 1814 paper he commented that initially the adsorption process is rapid. However, the amount of adsorption quickly saturates and so the dynamics of the process could not be measured with Saussure's techniques. In the late 1800s several people tried to measure rates of adsorption. For example, Kayser [1881] examined the adsorption of air onto charcoal and reported that the adsorption process is 90% complete in what Kayser estimated to be the first second. In 1881, it was not possible to make quantitative adsorption measurements in a time scale shorter than minutes. As a result, quantitative measurements of the adsorption rate could not be made with the techniques available to Kayser.

In 1903, however, von Giesen designed a microbalance that could be used to measure the dynamics of the adsorption process. In his 1903 paper, von Giesen measured the weight of a powdered coal sample as a function of time during adsorption of ammonia. Von Giesen found that initially the adsorption rate was rapid. However, the adsorption process quickly saturated. Von Giesen also tried to model his data as a first-order approach to equilibrium, but found that his data did not fit a first-order model.

Soon thereafter, Langmuir [1916, 1917, 1918, 1921] proposed a model for the adsorption process. His idea was that when an incoming molecule collided with a surface it could be "trapped" in a weakly bound state; the molecule would then have to find a bare site before it could "stick"; if not, the molecule would desorb. A more precise definition of trapping and sticking is given below. Langmuir [1916] tested these ideas with extensive experiments and concluded that when molecules collide with surfaces they are trapped and can stick.

Langmuir modeled his adsorption data by assuming that trapping probabilities were unity. This assumption was supported by work of Knudsen [1915], who showed that when metal vapors collided with a surface they lost all memory of their incoming state. However, some years earlier Maxwell [1860] had proposed that molecules could scatter from surfaces without being trapped. Baule [1914] showed how a finite trapping probability could come about theoretically. Ellet and Olsen [1928] and Johnson [1928] verified Maxwell's ideas by examining the scattering of alkali metals from salt crystals and found that some of the molecules simply bounced. As a result, it was concluded that there is a finite probability that a molecule will bounce from a surface without being trapped.

There was not much more progress on the analysis of trapping until the early 1960s, when improvements in vacuum technology allowed trapping and sticking to be examined directly on atomically clean surfaces. In the early 1960s Zwanzig [1960] and McCarroll and Ehrlich [1963] presented a detailed model of trapping. Soon thereafter, Logan and coworkers [1966, 1968] presented the so-called cube models of adsorption. The cube models were used for the next 20 years. In the last few years there have been several improvements to the models. In particular, several investigators have made extensive use of molecular dynamics/langevin simulations to examine adsorption processes. In the next several sections we describe several models for trapping and sticking. The reader is referred to Goodman and Wachman [1976], Ashford and Rettner [1991], Arumainayagam and Madix [1991], Morris, Bowker, and King [1984], or Tully [1980] for further details.

5.2 SCATTERING, TRAPPING, AND STICKING

Our analysis of trapping and sticking starts by considering the collision of a molecule with a solid surface. Figure 5.1 shows a schematic of several of the different processes that can occur during the collision:

- The molecule can simply bounce. When a tennis ball hits a hard surface, it simply rebounds without losing significant energy. In the same way, when a molecule collides with a undeformable surface, the molecule simply rebounds without losing translational energy. We call collisions where no energy transfer occurs **elastic scattering.**

- The molecule can lose energy, but not enough energy to stay at the surface for a long time. Note that when a tennis ball is directed toward a soft surface, the tennis ball bounces, but it does not bounce as high as it would from a hard surface. In the same way, when a molecule collides with a deformable surface, the molecule can lose energy, but not necessarily lose enough energy to stay on the surface. When a molecule collides with the surface and loses energy, but not enough to stay on the surface, we say the molecule **scatters inelastically.** In general if the molecule collides with the surface and bounces either inelastically or elastically, we say that the molecule **scatters.**

- Of course if we direct a tennis ball to a soft enough surface, the tennis ball will lose all of its perpendicular energy, and so it will not bounce from the surface. In the same way, when a molecule collides with a deformable surface, the molecule can lose enough of its translational energy so that the molecule stays on the surface. In such a case, we say that the molecule is **trapped.** The process is called **trapping.**

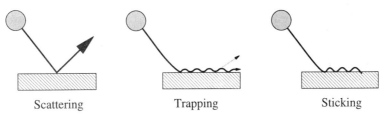

Scattering   Trapping   Sticking

Figure 5.1 A schematic of the processes that can occur when a molecule collides with a solid surface.

- Note, however, that when a molecule is first trapped it has lost enough energy so that the molecule does not immediately leave the surface. Nevertheless, the molecule is still in a weakly bound mobile state. When the molecule is in a weakly bound state, thermal motion of the surface atoms can cause the molecule to desorb. As a result, the molecule needs to be converted to a more strongly bound state if the molecule is to stay on the surface for a significant time. When a molecule collides with a surface, loses energy, and is converted into a state where the molecule remains on the surface for a reasonable time, we say the molecule **sticks.** We call the process **sticking.**

Fundamentally, trapping and sticking are quite different. Trapping rates are determined by the rate at which energy is transferred between the incoming molecule and the surface, while sticking rates are determined by the rate at which incoming molecules find sites where they can physisob or chemisorb. In general, trapping and sticking are important to all adsorption processes. However, sometimes the rate of trapping controls the adsorption rate while other times the rate of sticking controls the adsorption rate. One can tell whether trapping or sticking is controlling the adsorption rate by examining the variation in adsorption rate with changing gas temperature at fixed surface temperature. Note that trapping and sticking respond quite differently to changes in gas temperature. As the gas temperature goes up, more energy has to be removed from the gas molecules to trap them. As a result, trapping rates go down. In contrast, sticking rates are usually unaffected by gas temperature, and they can even go up with increasing gas temperature if the incoming molecules have to cross an activation barrier to stick. Hence, if one observes a decrease in the rate of adsorption with increasing gas temperature, one can conclude that the adsorption process is limited by trapping, while if the adsorption rate does not change or increases with increasing gas temperature, the adsorption process is limited by the rate at which molecules stick.

Experimentally, trapping is most important to weakly bound species (e.g., rare gases), while sticking is more important to species that are strongly bound. There is a complication, however, because a strongly bound species often first adsorbs into a weakly bound state and is then converted to a strongly bound state. Trapping is quite important to such systems. Hence, trapping is often important even with strongly bound species. In particular, trapping is important in many of the processes used to grow semiconductor films.

In the remainder of this chapter, we describe models of trapping and sticking in more detail. A variety of applications of the work are described in the examples at the end of the chapter.

5.3 TRAPPING

As noted previously, trapping is a process where a molecule loses energy so that it stays in close proximity to a surface. The best experimental evidence for trapping comes from molecular-beam experiments. When one directs a beam of hot molecules at a surface, one often can detect a small number of molecules that have spent a few hundred microseconds on the surface. These molecules have remained on the surface for a long time compared to collisional times, and yet they do not stick. Such molecules are said to have been trapped.

Unfortunately, the number of cases where people have measured good data on trapping probabilities is small. The best work has been on the trapping of rare gases on transition metals. Experimentally, one finds that the trapping probability of helium is less than neon, which is less than argon, which is less than krypton, which is less than xenon. At low temperatures, initial trapping probabilities (i.e., trapping probabilities on clean, adsorbate free surfaces) decrease with increasing temperature. However, at temperatures above 500 K, the trends are unclear. The small amount of data which exists suggests that trapping rates are smaller on stiffer lattices than softer ones. However, this trend needs further experimental verification.

In spite of the limitations in the data, the theory of trapping has been worked out in detail. There are now several models that can be used to explain trapping rates. At present, the models have not been completely verified experimentally because there is so little available data. However, the models do seem to explain the available data.

All of the models are basically the same. One calculates the probability of trapping by considering the collision of a molecule with a solid surface. Generally, one starts with a molecule that is far away from the surface and numerically integrates Newton's equations of motion for the molecule and all of the surface atoms as the molecule collides with the surface. One then averages over several different incident molecules to calculate a trapping probability.

In the remainder of this section we derive a number of equations for trapping probabilities. We assume that the reader is familiar with the idea of numerically integrating Newton's equations of motion of molecules to calculate their kinetic properties. If the reader is unfamiliar with this topic, he or she should look at Hailey [1992], Hoover [1991], or Gould and Tobochnik [1988] before proceeding. For the purposes of the derivation it is necessary to assume that the molecule impinges on a surface with some velocity v_i and that the molecule scatters as shown in Figure 5.3. The incoming molecule has a translational energy E_i given by

$$E_i = \tfrac{1}{2} m_g v_i^2 \tag{5.1}$$

where m_g is the mass of the molecule and v_i is the molecule's incoming velocity. If the molecule does not lose any translational energy, then the molecule must keep moving with the same velocity. The molecule cannot simply move along the surface forever, because eventually the molecule will collide with a surface atom or an adsorbate and desorb. Simulations indicate that desorption occurs within a few nanoseconds. As a result, the molecule must lose energy to the surface to be trapped.

One can get a measure of how probable it is for a molecule to be trapped by examining the energy flow during the collision process. Assume that a molecule scatters from a surface. One can define an energy accommodation coefficient, α, by

$$\alpha = \frac{E_i - E_o}{E_i - E_s} \tag{5.2}$$

where E_o is the average energy of the molecules that leave the surface, and E_s is the energy of a molecule that had thermally accommodated with the surface. A molecule with only translational energy has an E_s of $2kT$. If $\alpha = 0$, the molecule simple scatters, while if $\alpha = 1$, the molecule is fully accommodated with the surface. In the material which follows, we find that when $\alpha = 1$, molecules are trapped. As a result, one can model trapping by examining thermal accommodation.

Years ago people measured accommodation coefficients using a hot-wire technique. They would heat a filament in a low-pressure, low-temperature gas, and measured how much power is needed to keep the filament at some high temperature T_s. From an energy balance

$$\text{Power} = (\text{impingement rate of molecules}) \, (C_{p,g})(T_s - T_g)\alpha$$

where $C_{p,g}$ is the specific heat of the gas. We ask the reader to work out the details in Problem 5.6.

More recently, people have begun to measure accommodation coefficients using molecular beams. One directs a stream of gas at a hot surface and measures the velocity distribution of the incoming and scattered molecules using a technique called *time of flight* (i.e., the time it takes for molecules to fly some fixed distance). An example of these types of data is given in Problem 5.7. Once one knows the velocity distribution of the molecules, one can calculate the energy of the molecules, and therefore the accommodation coefficient.

It happens that the study of thermal accommodation coefficients predates Langmuir's work on trapping and sticking. Kundt and Warburg [1875] examined the viscosity of a variety of gases at low pressure using a sensitive viscometer in an attempt to verify Maxwell's [1860] prediction that the viscosity of a gas would be independent of pressure at low pressure. Kundt and Warburg found that Maxwell's relation worked well at moderate vacuums. However, at very low pressure, the apparent viscosity of the gases decreased with pressure. Kundt and Warburg proposed that the decrease was an artifact. They had measured the viscosity by calculating the force on a rotating cylinder and had assumed that a no-slip condition applied. However, Kundt and Warburg proposed that at low enough pressure, the no-slip condition no longer was valid in their viscometer. As a result, they suggested that their viscometer was no longer giving reliable results.

Goodman [1974] states that a few years later Maxwell showed that one could model the breakdown of the no-slip condition by assuming that the gases had a finite probability of thermally accommodating with the viscometer. Maxwell defined a constant that we will call f_M, which is the fraction of the molecules that thermally accommodate with a surface, and showed that if $f_M < 1$, the no-slip condition breaks down.

Maxwell's accommodation coefficients were difficult to quantify. However, Knudsen [1911, 1915] showed that if he used the definition of an accommodation coefficient given in Equation 5.2, he could measure accommodation coefficients directly by examining the heat transfer between two concentric cylinders. He also suggested that $f_M \approx \alpha$.

At about the same time Baule [1914] proposed a rather simple model for the accommodation coefficient. His idea was to keep track of what happens to a molecule as it collides with the surface and transfers energy. He computed the trajectory of the incident molecule as it made a collision with the surface, and used that information to calculate the accommodation coefficient, α. Baule's ideas have been expanded in the years since 1914. However, the general idea of calculating accommodation coefficients and trapping probabilities by keeping track of the trajectory that a molecule follows as it collides with a surfaces is still used today. The method has been generalized to examine rates of reaction, and so it is quite important.

Figure 5.2 shows what typical trajectories are like. The trajectories were calculated using the computer program in Example 5.D, which is available from Professor Masel. In the first trajectory on the left of Figure 5.2 the atom impinges onto the surface from the top left and simply bounces. In the second trajectory, however, the incident atom hits the surface, bounces a few times, and is eventually trapped. In the third trajectory the molecule bounces twice and scatters. There are several other trajectories in Figure 5.2. We recommend that if the reader

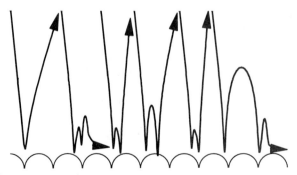

Figure 5.2 A series of trajectories seen when a molecule collides with a surface. The trajectories were calculated with the computer program in Examples 5.C and 5.D.

has access to the program in Example 5.D he or she use the program to see what atom surface collisions are like before proceeding.

One can define the trapping probability as the fraction of the trajectories where the molecules are trapped. Hence, one can calculate a trapping probability with a trajectory calculation.

In 1914, one did not have computers to do trajectory calculations, so one had to do the calculations analytically. Baule found a simple model that he could solve analytically. While Baule's model is no longer used to fit data, it does provide a framework to discuss trapping probabilities. Hence, it is quite useful.

In the next section we describe Baule's model in detail. We then use his model to calculate trapping probabilities. We find that Baule's model gives a reasonable approximation to the trapping probability. However, it does not fit data in detail. We then derive a series of other approximations that are still used to model accommodation coefficients and trapping today. A discussion of the use of the methods to examine rates of reactions appears in Chapter 7.

5.4 BAULE'S HARD SPHERE MODEL

Baule [1914] proposed the first model for the accommodation coefficient. His model assumes that accommodation coefficients are determined by the rate at which translational energy is transferred between the incoming adsorbate molecules and the surface atoms. He modeled the interaction as a hard sphere collision between an adsorbate molecule and a surface atom and assumed that the surface atom was initially stationary and isolated from other surface atoms. He then derived an expression for the accommodation coefficient as follows:

Consider the collision shown in Figure 5.3 where an incoming hard sphere molecule with velocity v_i collides with a hard sphere surface atom that is initially at rest. Assume that the incoming molecule hits the surface atom with an angle of ϕ_i with respect to the normal to the point of impact, and the molecule rebounds with a velocity v_o at an angle of ϕ_o with respect to the normal at the point of impact. With hard spheres, all of the forces of the collision are transmitted along the surface normal at the point of collision. The momentum of the adsorbate molecule is conserved along the direction parallel to the collision. Therefore,

$$m_g v_i \sin(\phi_i) = m_g v_o \sin(\phi_o) \tag{5.3}$$

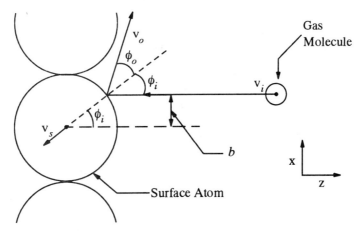

Figure 5.3 A diagram of the collision between a hard sphere adsorbate molecule and a hard sphere surface atom.

The incoming adsorbate transfers momentum to the surface atom in the direction normal to the surface atom at the point of impact, and so the surface atom rebounds along the direction normal to the impact. Conservation of the total momentum of the system in the normal direction implies

$$m_g v_i \cos(\phi_i) = m_g v_o \cos(\phi_o) + m_s v_s \tag{5.4}$$

where m_s is the mass of the surface atom and v_s is the velocity at which the surface atom rebounds after the collision. Note that in reality the surface atoms are bonded to each other so that when one surface atom moves, other surface atoms move along with it. Hence, later in this chapter we treat m_s as an effective mass of a surface atom, where the effective mass is defined as the mass of the surface atom plus some correction for the masses of the other surface atoms that move along with the surface atom.

If the surface atom (or group of surface atoms) is isolated from all other surface atoms, then the total energy of the adsorbate molecule and the surface atom will be conserved during the collision. Therefore

$$\tfrac{1}{2} m_g v_i^2 = \tfrac{1}{2} m_g v_o^2 + \tfrac{1}{2} m_s v_s^2 \tag{5.5}$$

If the initial velocity of the surface is zero, then the energy of a molecule that is fully accommodated with the surface will also be zero. As a result the thermal accommodation coefficient, α becomes

$$\alpha = \frac{\tfrac{1}{2} m_g v_i^2 - \tfrac{1}{2} m_g v_o^2}{\tfrac{1}{2} m_g v_i^2} \tag{5.6}$$

At this point there are four equations (5.3–5.6) in four unknowns, α, ϕ_o, v_o, and v_s. These equations can be solved simultaneously to derive an expression for α. The result is

$$\alpha = \frac{4 \left(\dfrac{m_g}{m_s}\right) \cos^2(\phi_i)}{\left(\dfrac{m_g}{m_s} + 1\right)^2} \tag{5.7}$$

Experimentally, one cannot control where the incoming molecule hits the surface atom. As a result, one needs to average the value of α in Equation 5.7 over some range of solid angle. If the impact is random, then one should average over the full solid angle of the hemisphere. The average value α, which will be labeled $\hat{\alpha}$, is given by

$$\hat{\alpha} = \frac{4 \left(\dfrac{m_g}{m_s}\right)}{\left(\dfrac{m_g}{m_s} + 1\right)^2} \left(\frac{\displaystyle\int_0^{\pi/2} \cos^2(\phi_i)[\sin(\phi_i)\, d\phi_i]}{\displaystyle\int_0^{\pi/2} \sin(\phi_i)\, d\phi_i} \right) = \frac{\dfrac{4}{3} \left(\dfrac{m_g}{m_s}\right)}{\left(\dfrac{m_g}{m_s} + 1\right)^2} \tag{5.8}$$

where the factor of $\sin(\phi_i)$ arises because we are integrating in spherical coordinates.

Note, however, that in a real surface the surface atoms are packed tightly together. The incoming molecule can easily make a head on collision (i.e., $\phi_i = 0$), with the surface atoms.

However, it is harder for an incoming molecule to get down the sides of the surface atom because adjacent surface atoms get in the way. As a result it is more probable for an adsorbate to hit the surface atom head on than at on obtuse angle. If one assumes that all of the collisions are head-on collisions so that $\phi_i = 0$:

$$\hat{\alpha} = \frac{4\left(\dfrac{m_g}{m_s}\right)}{\left(\dfrac{m_g}{m_s} + 1\right)^2} \tag{5.9}$$

Equation 5.8 is called Baule's expression for the accommodation coefficient, while Equation 5.9 is called Baule's equation for head-on collisions. Goodman [1967] shows that with an array of surface atoms, the collisions are not random or head on. Also, it is quite common for an incoming molecule to collide with more than one surface atom. Interestingly, Goodman [1967] finds that the equation

$$\hat{\alpha} = \frac{2.4\left(\dfrac{m_g}{m_s}\right)}{\left(\dfrac{m_g}{m_s} + 1\right)^2} \tag{5.10}$$

is an even better approximation to the accommodation coefficient.

In the 1920s and 1930s people tried to use Baule's equations to model thermal accommodation data with limited success. However, the problem was the data and not the theory. In most of the experiments in the 1920s and 1930s the surfaces were covered by monolayers of impurities. As a result the energy transfer rate was dominated by the energy transfer between the incident molecule and the impurities rather than the incident molecule and the surface atoms. It was not until the 1960s that one could reliably produce clean surfaces. In 1967, Thomas [1967] and coworkers found good qualitative agreement between the Baule model and accommodation coefficients measured on clean surfaces, provided they adjusted the effective mass of the surface to account for the fact that in general an incoming molecule will interact with more than one surface atom. Hence, Baule's models and related improvements provide a good qualitative picture of the energy transfer process.

5.4.1 The Baule–Weinberg–Merrill Approximation for Trapping Probabilities

Weinberg and Merrill [1971] used the Baule model to estimate trapping probabilities. They assumed that the interaction between the incoming gas molecule and the surface could be broken into two parts: a long-range, slowly varying attraction due to interaction of the incoming molecule with all of the atoms in the lattice, and an infinite short-range hard sphere repulsion due to the interaction of the incoming molecule with one of the surface atoms.

Consider the collision of a molecule that impinges on a surface at normal incidence. When the molecule impinges on the surface, it will experience a Lennard-Jones–like potential at long distances. At shorter distances the potential will also look like a Lennard-Jones potential. However, the potential will be different according to whether the molecule impinges directly on top of a surface atom or in the gap between two surface atoms. One cannot obtain an analytical expression for the trapping probability if one does calculations with the real potential. However, Weinberg and Merrill found that if they instead assumed that the potential is

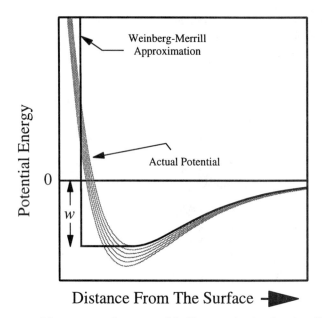

Figure 5.4   The potential energy seen by a normal incidence molecule when it collides with a solid surface. A series of lines is shown because the potential is different when the incoming atom hits at different places along the surface.

given by the solid line in Figure 5.4 where the potential looks like a Lennard-Jones potential at long distances and an infinite hard sphere repulsion at short distances, Weinberg and Merrill could obtain an analytical result.

Physically, the long-range part of the potential includes attractions from all of the atoms in the lattice. It should vary slightly with X and Y. However, Steele [1974] shows that such variations are usually not important to trapping. The use of the hard sphere repulsion is a more serious limitation. The hard sphere assumption, in effect, implies that the incoming molecule is colliding with one surface atom at a time, independent of where the surface atom hits. In reality, if the incoming molecule hits a threefold site, the incoming molecule will interact with three surface atoms simultaneously. As a result, the Baule–Weinberg–Merrill model leaves off some key physics. Still, the model can be solved analytically. Thus, the model is still quite useful.

Consider an incoming molecule with an energy E_z interacting with the Weinberg–Merrill potential. The molecule initially is accelerated by the long-range attractive part of the potential. When the molecule reaches the bottom of the attractive well, it will have an energy E_i' given by

$$E_i' = E_z + w \qquad (5.11)$$

where w is the well depth. The molecule will lose a fraction of its energy, α', during the collision with the surface. Note that α' is different from α. Here, α is the fraction of the molecule's initial energy that is lost, while α' is the energy that is lost divided by the energy of the molecule at the bottom of the attractive well. For example, consider a molecule that is incident on a cold surface, and assume $E_z = 40$ kJ/mol, $w = 60$ kJ/mol, and the molecule loses 30 kJ/mole when it collides with the surface. In such a case, $\alpha = 30/40 = 0.75$ and $\alpha' = 30/(40 + 60) = 0.3$.

From the definition of α' it is easy to show that the energy of the molecule just after it collides with the surface atom, E_o', will be given by:

$$E'_o = (1 - \alpha')E'_i + \alpha'E_s \qquad (5.12)$$

Note that the molecule can leave the surface if $E'_o > w$; if not, it will be trapped. Therefore, trapping occurs when

$$E'_o \leq w \qquad (5.13)$$

Substituting Equations 5.11 and 5.12 into Equation 5.13 shows that the molecule will be trapped whenever

$$E_z \leq \frac{\alpha'}{1 - \alpha'}(w - E_x) \qquad (5.14)$$

Weinberg and Merrill then defined a critical energy for trapping, E_{crit}, by

$$E_{crit} = \frac{\alpha'}{1 - \alpha'}(w - E_s) \qquad (5.15)$$

They then suggested that the trapping probability, P_T, is given by

$$P_T = \begin{cases} 0 & \text{if } E_z > E_{crit} \\ 1 & \text{if } E_z \leq E_{crit} \end{cases} \qquad (5.16)$$

Weinberg and Merrill noted that the derivation of Baule's model only considered a molecule that impinges on a surface with $w = 0$. In effect, therefore, Baule's model is calculating α' not α. Therefore, Weinberg and Merrill substituted Baule's approximation to $\hat{\alpha}$ for α' in Equation 5.15 to calculate E_{crit}, where $\hat{\alpha}$ was assumed to be given by Equation 5.9. Substituting $\hat{\alpha}$ from Equation 5.9 into Equation 5.15 and rearranging yields

$$E_{crit} = \frac{4\left(\dfrac{m_g}{m_s}\right)}{\left(1 - \dfrac{m_g}{m_s}\right)^2}(w - E_s) \qquad (5.17)$$

Equations 5.16 and 5.17 are the Baule–Weinberg–Merrill approximation for the sticking probability.

It works out that Equation 5.16 is only a fair approximation to reality. Equation 5.16 predicts that all molecules with energies below E_{crit} will be trapped while all molecules with energies above E_{crit} will simply scatter. Experimentally, molecules with an incident energy much less than E_{crit} will all be trapped, while those with an energy much greater than E_{crit} will all scatter (provided the energy is not so high that the atom will be buried in the surface). However, molecules with an energy close to E_{crit} will have a finite probability of being trapped. Some of the molecules with an energy less than E_{crit} will scatter, while some of the atoms with an energy more than E_{crit} will be trapped. Hence, the Baule–Weinberg–Merrill approximation is only useful in making very approximate predictions of the trapping probability.

5.4.2 An Improved Baule Hard Sphere Model: Ion Cores in Jellium

The reason that the Baule–Weinberg–Merrill model has failed is that it assumed that the accommodation coefficient would be independent of where the incident molecule hits the surface. As a result, every molecule trajectory is equivalent, and either every molecule scatters or every molecule sticks. However, note that according to Equation 5.7, α changes according to where the molecule hits the surface. If the molecule hits a surface atom head on (i.e., at $b = 0$), it transfers considerable energy. However, if the molecule hits somewhere in the middle of a surface atom, where $b > 0$, less energy is transferred. Almost no energy is transferred if the incoming atom hits the surface near the edge of a surface atom. Hence, the Baule–Weinberg–Merrill model has ignored some important physics.

Furthermore, in a real surface, the surface atoms are vibrating. If the incoming molecule hits a surface atom when the surface atom is moving up, the incoming molecule sees a different force than if the surface atom is moving down. As a result, the Weinberg–Merrill result needs to be averaged over the motion of the surface atoms.

In the remainder of this section we derive an extension of the Baule–Weinberg–Merrill approximation that includes the variation of α over the unit cell, but not the effects of the motion of the surface atoms. The effects of surface atom motion will be considered in the next two sections. The new model is called the **ion cores in jellium model.** Recall, from Chapter 3 that there are two main interactions when a rare gas atom collides with a metal surface interaction with the conduction band, and an interaction with the d-bands and atomic cores. The idea of the ion cores in jellium model is to model the interaction with the sp-bands via the jellium model, and model the interaction with the ion cores and d-bands as a hard sphere.

Figure 5.5 shows a schematic of an idealized jellium potential for the interaction of a rare gas with a close packed metal surface. As noted in Chapter 3, the jellium potential is attractive as long range, reaches a minimum at intermediate range, and then starts to be repulsive again. Note, however, that electron density in the jellium density saturates at a finite value. Therefore, the jellium potential will saturate at some finite value.

The idea in the ion cores in jellium model is to assume that the incident molecule does not hit the ion cores until the jellium potential has saturated. In that case, the energy of the incident molecule, when it collides with the ion core, will be given by Equation 5.11. The incident atom will then hit the ion core and lose energy. Then the molecules will try to escape from the attractive well. The result will be a model that is very similar to the Baule–Weinberg–Merrill model, but can be derived from first principles.

For the purposes of derivation it will be useful to define a quantity $P(b)$ as the probability that a molecule is trapped when it hits the surface at a distance b from the center of a surface atom, where b is defined in Figure 5.3. The quantity b is called the **impact parameter.** One can show that the average probability that the molecule is trapped, P_{trap}, is related to $P(b)$

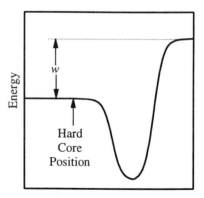

Figure 5.5 A schematic of all idealized jellium potential over a closed packed metal surface.

by

$$P_{trap} = \frac{\displaystyle\int_0^{r_{atom}} P(b)b \; db}{\displaystyle\int_0^{r_{atom}} b \; db} \tag{5.18}$$

where r_{atom} is the radius of a surface atom. The factor of b in Equation 5.18 comes about because the surface atom is cylindrically symmetric about the surface normal. Note that the ion cores in jellium model is very similar to the Baule–Weinberg–Merrill model. An atom is first accelerated and then it scatters from a hard sphere. Therefore, most of the results from Section 5.4.1 still apply. In particular, if we fix b, then we also fix α. The results from Section 5.4.1 show that for each value of α there will be a critical value of the incident energy below which all molecules will be trapped. If one calculates $P(b)$, one finds:

$$P(b) = \begin{cases} 0 & \text{if } E_z > E_{crit} \\ 1 & \text{if } E_z \leq E_{crit} \end{cases} \tag{5.19}$$

where E_{crit} is the critical energy for trapping of an incoming molecule that hits the surface atom at a distance b from the atom's center.

Combining Equations 5.18 and 5.19 yields

$$P_{trap} = \left(\frac{b_{crit}}{r_{atom}}\right)^2 \tag{5.20}$$

where b_{crit} is the value of the impact parameter b where the molecule loses just enough energy to be trapped. If $b > b_{crit}$, the incident molecule will not lose enough energy to be trapped; if $b \leq b_{crit}$, the incident molecule will be trapped.

For the derivation that follows, we will note that the accommodation coefficient, α, is a function of b. Therefore, we can rewrite α as $\alpha(b)$. Now from our definition of b_{crit}, when b equals b_{crit} the incident molecule does not quite have enough energy to escape from the surface. From the definition of E_{crit}, the molecule will not quite have enough energy when E_z equals E_{crit}. Substituting E_z for E_{crit} and $\alpha(b_{crit})$ for α in Equation 5.14 yields

$$E_z = \frac{\alpha(b_{crit})}{1 - \alpha(b_{crit})} (w - E_s) \tag{5.21}$$

We can calculate b_{crit} by substituting Equation 5.7 for α into Equation 5.21, noting that

$$\cos(\phi_l) = \frac{\sqrt{r_{atom}^2 - b^2}}{r_{atom}} \tag{5.22}$$

and solving for b_{crit}. Substituting the result into Equation 5.20 yields

$$P_{trap} = \left[1 - \frac{\left(1 + \dfrac{m_g}{m_s}\right)^2 E_z}{4\left(\dfrac{m_g}{m_s}\right)(E_z - E_s + w)} \right] \quad \text{provided } P_{trap} > 0$$

$$P_{trap} = 0 \qquad\qquad\qquad\qquad\qquad \text{otherwise} \tag{5.23}$$

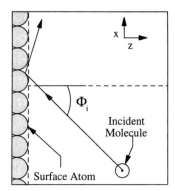

Figure 5.6 Scattering of a non-normal-incidence molecule.

So far we have only considered molecules that collide with the surface at normal incidence. In that case, the incident molecule has to lose its entire translational energy to be trapped. However, if the incoming molecule impinges at some grazing angle Φ_i, shown in Figure 5.6, then if the molecule loses its component of energy along the surface normal, the molecule will no longer be able to move away from the surface. Therefore, the molecule will be trapped. If we call z the direction of the surface normal, the component of the molecule's velocity V_z in the normal direction is given by

$$V_z = V_i \cos(\Phi_i) \tag{5.24}$$

The component of energy in the normal direction, E_z, becomes

$$E_z = \tfrac{1}{2} m_g v_z^2 = E_i \cos^2(\Phi_i) \tag{5.25}$$

Combining Equations 5.23 and 5.25 yields

$$
P_{trap} = \left[1 - \frac{\left(1 + \dfrac{m_g}{m_s}\right)^2 E_i \cos^2(\Phi_i)}{4\left(\dfrac{m_g}{m_s}\right)(E_i \cos^2(\Phi_i) - E_s + w)} \right] \quad \text{provided } P_T > 0
$$
$$
P_{trap} = 0 \qquad\qquad\qquad\qquad\qquad\qquad\qquad\qquad \text{otherwise} \tag{5.26}
$$

At the time this book was written, Equation 5.26 had not yet appeared in the literature. However, it fits measured trapping data quite well, provided w and m_s are treated as adjustable parameters. For example, Figure 5.7 compares Arumainayagam et al.'s [1990] data for argon trapping on Pt(111) to the predictions of Equation 5.26. Arumainayagam et al.'s Langevin calculations are also included in the figure. (Langevin calculations are the best calculations anyone was able to do in 1995). In the calculations we fixed m_s at 195 AMU, and varied w to obtain a good fit of the data. We find that we need a w of 8 kJ/mole to fit the data. By comparison, Arumainayagam et al. estimate a heat of adsorption of 25.9 kJ/mole.

In the literature there has been some discussion whether the w in the original Baule–Weinberg–Merrill equation should equal the heat of adsorption. However, experimentally w is normally less than the heat of adsorption because the jellium potential is not completely attractive (see Figure 5.5).

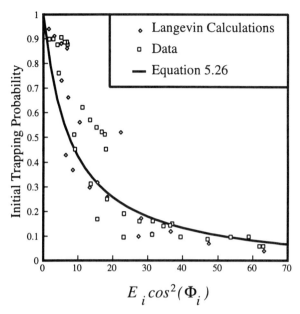

$$E_i cos^2(\Phi_i)$$

Figure 5.7 A comparison of the trapping probability for Xe on Pt(111): (a) Equation 5.26, with m_s = 195 AMU, w = 8 kJ/mole; (b) Arumainayagam et al.'s [1990] data and Langevin results.

Arumainayagam et al. took experimental data at several incidence energies and angles. According to Equation 5.26 when the data are plotted as a functon of $E_i cos^2(\Phi_i)$, all of the data should collapse onto a single line. Experimentally, however, the data do not quite collapse into a single line; a single line is obtained if one plots the data as a function of $E_i cos(\Phi_i)$. Nevertheless the difference between Equation 5.26 and the data is relatively small, provided w is adjusted appropriately.

Equation 5.26 does not show the correct Φ_i dependence because we have only considered the z momentum of the incident molecule in the derivation of Equation 5.26. The Baule–Weinberg–Merrill model assumes that whenever an incident molecule loses enough of its z momentum that the molecule cannot climb out of the attractive well, the molecule will be trapped. In the derivation of Equation 5.26, we have considered the loss of z momentum due to scattering from the lattice. However, we have not considered the possibility that the x and y momentum of the incident molecule could be converted into z momentum during the course of the collision. Note, however, that from Figure 5.2, when an incident molecule hits a surface atom at a distance b from the atom's centerline, with $b > 0$, z momentum can be converted into x momentum and vice versa. If x momentum is converted into z momentum, there will be less loss of z momentum, and hence the trapping probability will be less than that given by Equation 5.26.

A more careful analysis shows that in fact the Φ_i dependence of the trapping probability changes when x momentum is converted into z momentum. The x momentum scales as $sin(\Phi_i)$, so there is more x momentum to be transferred at grazing incidence than at normal incidence. As a result, when x momentum is converted into z momentum the trapping probability is more strongly affected at large Φ_i than at small Φ_i. This changes the angular variation of the trapping probability. We ask the reader to work out the algebra in Problem 5.17. If the trapping probabilities varies as $E_i cos^2\Phi_i$, we say the system obeys **normal energy scaling.** Normal energy scaling occurs when there is little conversion of x-y momentum into z momentum during the scattering process. Experimentally, most of the systems considered so far obey normal energy scaling reasonably well, at least at incident energies below 2–5 kcal/

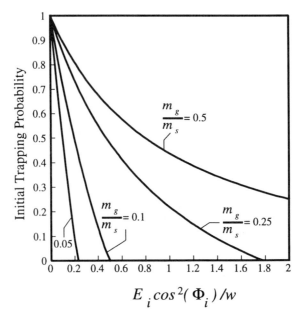

Figure 5.8 A plot of the trapping probability predicted by Equation 5.26 as a function of the incident energy of the molecule for various vales of m_g/m_s.

mole, and so Equation 5.26 is often a useful approximation. There have been exceptions, however, and those exceptions need to be considered in a careful analysis.

Equation 5.26 also reproduces most of the important qualitative trends seen by other investigators. Figure 5.8 shows a plot of the trapping probability predicted by Equation 5.26 as a function of $E_i \cos^2(\Phi_i)/w$ for several values of m_g/m_s. Equation 5.26 predicts that the trapping probability approaches unity as the well depth, w, grows. This trend agrees with the available experimental data. Equation 5.26 also predicts that when $w/E_i \cos^2(\Phi_i) > 10\text{–}20$, most incoming molecules are trapped, and it is only when the well depths are small that the trapping probabilities can be substantially less than unity. Again, that seems to agree with experimental results, although more data are needed. Equation 5.26 predicts that trapping probabilities will tend to be lower for light molecules (i.e., hydrogen and helium) than heavier species. Again this is observed. Equation 5.26 predicts that a slow-moving molecule will have a higher probability of being trapped than a fast-moving one and that trapping probabilities should be enhanced when molecules approach surfaces at grazing incidence. This is observed as well. Finally, Equation 5.26 predicts that trapping probabilities are reduced as the surface temperature increases. Physically, a hot surface can transfer energy back to the incoming molecule. As a result, the net energy transfer rate to the surface is less when the surface is hot, which in turn results in a decreased trapping probability. There are not enough data to say conclusively whether this occurs. However, the available data do seem to follow Equation 5.26.

Nonetheless, there are some deviations from Equation 5.26. Equation 5.26 predicts that the trapping probability should rapidly approach zero at high incident energy (i.e., large E_i). However, experimentally, trapping probabilities do not go to zero with increased energy as quickly as predicted by Equation 5.26. Equation 5.26 also predicts that the trapping probability will reach a maximum when m_g equals m_s and then decay to zero as illustrated in Figure 5.9. Experimentally, however, the trapping probability does not decay to zero at high mass.

Both of these deviations are mainly a result of multiple scattering where a given incident

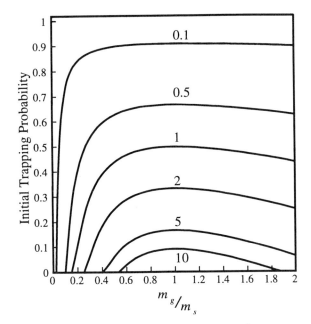

Figure 5.9 A plot of Equation 5.26 as a function of m_g/m_s for $E_i\cos^2(\Phi_i)/w = 0.1, 0.5, 1, 2, 5, 10$.

atom can collide with more than one surface atom. If a high-energy molecule hits the surface just right, it can sometimes bounce around enough times to stick. When a heavy molecule hits a surface, it loses some energy in the first collision. However, the molecule usually stays in close proximity to the surface and multiple collisions allow it to be trapped. Langevin calculations described in Section 5.7 will be able to treat these effects. However, in general, the data fit Equation 5.26 reasonably well, especially considering that Equation 5.26 is based on a 1914 model, and it only has two adjustable parameters compared to the 5-7 adjustable parameters in a typical Langevin calculation.

The one experimental trend that is not explained at all by equation 5.26 is the observation that trapping probabilities tend to be higher on a soft deformable lattice like a metal than a stiff one like a semiconductor. The experimental data are not completely clear because there are so few data. However, physically, it is reasonable to treat a metal as ion cores in jellium. In the jellium model the surface atoms can move independently so that the effective mass of the surface atom is equal to the actual mass of the surface atom. With a semiconductor, however, there are local bonds between each of the surface atoms. The surface atoms are highly coupled and it is thought that several surface atoms move together. This raises the effective mass of the surface atom. Equation 5.8 shows that when m_s increases, the rate of energy accommodation is reduced. One can still fit trapping data by adjusting the effective mass of the surface atoms, m_s, in Equation 5.26 to account for these effects. However, a more fundamental model is needed.

5.5 CUBE MODELS

Years ago the cube models were proposed as an improvement to account for the effects of the thermal motion of the surface atoms on the trapping process. The cube models are qualitatively similar to Baule's models. One models the collision of the gas with the surface as the interaction of an incoming molecule with a surface atom having an effective mass m_s.

However, Baule's model assumes that the surface atom is fixed, the cube model assumes that the surface atom is moving with some velocity distribution. The distribution of the surface atom velocities is treated as a fitting parameter in the calculations. One then solves the equations of motion using the MO method described in Appendix A to calculate an accommodation coefficient and trapping probability.

There are two versions of the cube models, the so-called "hard cube models" proposed by Logan and Stickney [1966] and the "soft cube models" proposed by Logan and Keck [1968]. The hard cube models assume that there are hard sphere collisions between the gas molecule and the surface, while the soft cube models assume that the gas/surface interaction is given by either a Morse potential or a Lennard-Jones potential. The main advantage of the hard cube model over Baule's model is that the hard cube accounts for the effect of the thermal motions of the surface atoms on the momentum transfer process. Hence, the hard cube model can better model the influence of the surface temperature on the momentum transfer during gas/surface collisions. The soft cube model does not add any new qualitative physics to the hard cube models. However, it does introduce many extra parameters into the analysis. Hence, one can fit a wider range of data with the soft cube model than the hard cube model.

The hard cube and soft cube models are hardly used today. As a result these models will not be discussed in detail in this book. An example of a cube model calculation is included in the solved examples. If the reader wants more information about either model, Goodman [1967, 1974] provides a thorough description of the hard cube and soft cube models and their applications.

5.6 ZWANZIG'S LATTICE MODEL

So-called "lattice models" have replaced the cube models in the current literature. The lattice models are similar to the cube models. One treats the trapping procss as a collision between the incoming molecule and a group of moving surface atoms. One then uses a trajectory calculation to calculate the trapping probability. However, while the cube models assume that the surface atoms are all isolated from the bulk of the solid, the lattice models assume that the surface atoms are connected to the bulk of the solid with a network of springs or anharmonic potentials. As with Baule's model, the incoming molecule is assumed to be trapped when it loses enough energy that it can no longer leave the surface. The energy transfer is presumed to occur via direct collisions with one or more of the surface atoms. The surface atoms are assumed to transfer the collisional energy to the bulk via lattice vibrations.

Zwanzig [1960] proposed the first lattice model. He represented the surface as a linear chain of atoms tied together with springs as indicated in Figure 5.10. An incoming atom is assumed to interact only with the first atom in the chain. However, the first atom interacts with the second atom in the chain etc. Hence one can get vibrational energy transfer into the lattice. Quantitatively, Zwanzig showed that one should obtain some useful results by solving

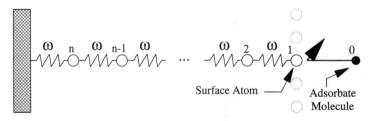

Figure 5.10 Zwangig's [1960] model of the interaction of a gas molecule with a one-dimensional chain of surface atoms.

Newton's equations of motion for the particle and the lattice. This general approach is called *molecular dynamics* (MD). If the reader is unfamiliar with MD, he or she should read Appendix A before proceeding.

In order to derive Zwanzig's results we will define x_0 as the position of the incoming molecule, x_1, x_2, x_3, x_4, x_n to be the displacement of atoms 1, 2, 3 in the chain from their equilibrium position and $F(x_1 - x_0)$ to be the force between the incoming atom and the top surface atom. Newton's laws of motion (i.e., $F = ma$) show that the trajectory of the incoming molecule should follow

$$m_g \frac{d^2 x_0}{dt^2} = F(x_1 - x_0) \tag{5.27}$$

If the top surface atom exerts a force $F(x_1 - x_0)$ on the incoming molecule, then the incoming molecule will exert an equal and opposite force on the top surface atom. Hence, one force on the top surface atom is $-F(x_1 - x_0)$. A second force arises because of the interaction of the surface atom with the lattice. If we assume that the top surface atom is connected to the second surface atom by a spring with a force constant of $m_s \omega^2$, then it is easy to show that the position of the first atom, x_1 obeys

$$m_s \frac{d^2 x_1}{dt^2} = m_s \omega^2 (x_2 - x_1) - F(x_1 - x_0) \tag{5.28}$$

Note, ω is the vibrational frequency of the top atom when the position of the second atom is fixed in space.

One can derive similar equations for the remaining atoms in the lattice. The result is

$$m_s \frac{d^2 x_2}{dt^2} = m_s \omega^2 (x_1 - x_2) - m_s \omega^2 (x_2 - x_3)$$

$$= m_s \omega^2 x_1 - 2 m_s \omega^2 x_2 + m_s \omega^2 x_3$$

$$m_s \frac{d^2 x_3}{dt^2} = m_s \omega^2 x_2 - 2 m_s \omega^2 x_3 + m_s \omega x_4$$

$$\vdots$$

$$m_s \frac{d^2 x_{n-1}}{dt^2} = m_s \omega^2 x_{n-2} - 2 m_s \omega^2 x_{n-1} + m_s \omega x_n$$

$$m_s \frac{d^2 x_n}{dt^2} = m_x \omega^2 x_{n-1} - 2 m_s \omega^2 x_n \tag{5.29}$$

Zwanzig [1960] formulated the solution of these equations as a series of Bessel functions, but he did not solve them. However, McCarroll and Ehrlich [1963] solved the equations, assuming that the force between the incoming molecule and the lattice is given by that for a truncated harmonic oscillator:

$$F(x) = K_{ME}(x - x_E^{nE}) \qquad x < x_{\max}^{ME}$$
$$F(x) = 0 \qquad\qquad x \geq x_{\max}^{ME} \tag{5.30}$$

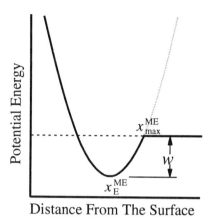

Figure 5.11 The gas/surface interaction potential used by McCarroll and Ehrlich [1963].

This is equivalent to assuming that the interaction potential is as given in Figure 5.11. McCarroll and Ehrlich also assumed that the lattice is initially at rest.

Figure 5.12 shows some of McCarroll and Ehrlich's results (we have replotted them on a different scale). In the course of their calculations McCarroll and Ehrlich made the assumption that none of the surface atoms were moving before the collision with the incident molecule. They further assumed that one could only get head-on collisions with the surface atoms, so that all trajectories are equivalent at any incident energy. If all trajectories are equal and one trajectory of the incident molecule leads to trapping, all the other trajectories of the incident molecules will lead to trapping. As a result, at any incident energy the trapping probability is predicted to be either zero or one. One can define E_{crit} as the energy where the trapping probability switches from zero to one. It is easy to show that within the McCarroll–Ehrlich

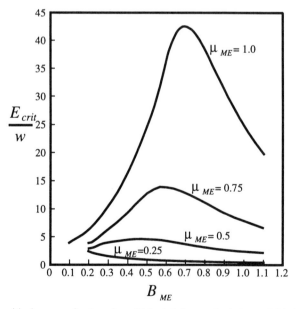

Figure 5.12 The critical energy for trapping. (Adapted from calculations of McCarroll and Ehrlich [1963].)

approximation, if the incident energy is below E_{crit}, the incident molecules will always be trapped while if the energy is above E_{crit}, the molecule will always scatter.

Figure 5.12 is a plot of E_{crit}/w as a function of B_{ME}, for various values of μ_{ME} where

$$B_{ME} = \frac{K_{ME}}{m_s\omega^2}$$

$$\mu_{ME} = \frac{m_g}{m_s} \tag{5.31}$$

and w is the well depth of the molecule/surface potential. A B_{ME} of zero corresponds to a undeformable lattice, while a B_{ME} of much more than unity corresponds to a very deformable one. Notice that when B_{ME} is zero, E_{crit} is zero, so no molecules can stick. When B_{ME} goes to infinity, the critical energy for trapping approaches the Weinberg–Merrill–Baule result for head-on collisions. Interestingly, a maximum in the trapping probability occurs at intermediate values of B_{ME}, The maximum comes about because at certain values of B_{ME} the incoming molecule moves at a velocity that is synchronized with the velocity of the surface atoms moving at the vibrational frequency of the one-dimensional lattice.* At this particular velocity, the incoming molecules can very efficiently transfer energy to the lattice.

In actual practice, one cannot use the McCarroll–Ehrlich results to fit data. Experimentally, one usually measures the trapping probability as a function of the incident angle and energy, and observes trends like those in Figures 5.7 and 5.8 where the trapping probability varies slowly with angle and energy. However, the McCarroll–Ehrlich calculation suggests that there should be a sudden change in the trapping probability from unity to zero as the incident energy goes above some critical value, E_{crit}.

Still, the McCarroll–Ehrlich results do allow one to say something about how the vibrational coupling between adjacent surface atoms effects trapping rates. One can show that when K_{ME} goes to infinity, the McCarroll–Ehrlich model approaches the Baule–Weinberg–Merrill result. Physically, a K_{ME} of infinity corresponds to a hard sphere collision. Physically, when there is a hard sphere collision, all of the energy exchange between the incident molecule and the top surface atom occurs at the instant of impact. The incident molecule is only in contact with the surface atom for an infinitesimal amount of time. During that instant, the surface atom is not displaced significantly from is equilibrium position. Notice, however, that with a spring model, the force on all of the atoms is zero when all of the atoms are at their equilibrium positions. One only gets a net force between the atoms when the atoms are displaced from their equilibrium positions. When K_{ME} is large, the gas/surface collision is done before the surface atoms have time to be displaced from their equilibrium position. As a result, when K_{ME} is large, the surface atom acts as though it is isolated from the lattice during the entire time that the surface atom is interacting with the incident molecule. Hence, for large K_{ME} the McCarroll–Ehrlich calculations approach the isolated atom limit (i.e., the Baule result).

When K_{ME} is smaller, the incident atom interacts with the lattice for a finite amount of time. According to the McCarroll–Ehrlich calculations, if we start with noninteracting surface atoms (i.e., $\omega = 0$) and slowly turn on interactions, the energy transfer rate, and hence the trapping probability, will increase because the springs carry energy away from the surface atom. However, according to the calculations, if ω increases, too much, the trapping probability will decrease again, because the surface atoms will start moving together, which raises the effective mass in the Baule–Weinberg–Merrill model.

Experimentally we do not know if all of the trends predicted by the McCarroll–Ehrlich model are correct. The experimental data are very incomplete. There is a general trend that

---

*The one-dimensional lattice actually has a band of allowed frequencies. The maximum energy accommodation occurs when the incoming molecule is synchronized with the center (i.e., maximum) of the band.

the trapping probabilities decrease on stiffer lattices. However, it is unclear whether the change is monotonic or whether the data show a maximum like that in Figure 5.12. Still all of the existing lattice calculations show these same general trends. Hence, if one believes the lattice calculations, one does have to accept the general trends seen in McCarroll and Ehrlich's calculations.

5.7 MOLECULAR DYNAMICS SIMULATIONS OF LATTICE MODELS

The McCarroll–Ehrlich solution of Zwanzig's model was used in the early 1960s. However, now molecular dynamics models (MD) have replaced the McCarroll–Ehrlich–Zwanzig results. The molecular dynamics models are three-dimensional versions of Zwanzig's model. The solid is modeled as a three-dimensional arrangement of atoms that are held together with springs or possibly anharmonic potentials. One then calculates a trapping probability by numerically integrating Newton's equations of motion for the incoming gas molecule and some group of surface atoms.

Consider an incoming molecule with an initial position $\vec{r} = (x, y, z)$ and assume that the molecule is moving toward the surface. One can calculate the trajectory of the molecule from the equations of motion of the molecule:

$$m_g \frac{d^2 \vec{r}}{dt^2} = \vec{f}(\vec{r}) \tag{5.32}$$

where $\vec{f}(\vec{r})$ is the net force on the molecule at point $\vec{r}$, and t is time. The force can be calculated from

$$\vec{f} = \left(\frac{dE_p(\vec{r})}{dx}, \frac{dE_p(\vec{r})}{dy}, \frac{dE_p(\vec{r})}{dz} \right) \tag{5.33}$$

where $E_p(\vec{r})$ is the potential energy of the molecule at point $\vec{r}$. One can then numerically integrate Equation 5.32 to see if the molecule is trapped. One then averages over several trajectories to calculate a trapping probability. The general approach is called *molecular dynamics* (MD). If the reader is unfamiliar with MD, he or she should refer to Appendix A. In all of the trapping calculations done so far, an empirical expression was used to compute the potential energy of the incoming molecule and each of the surface atoms. For example, it is common to assume that the interaction between the incoming molecule and the surface is given by a sum of Lennard-Jones potentials between the incoming molecule and each of the surface atoms. If one makes the further assumption that all of the translational energy of the incoming molecules is converted into vibrational motion of the surface atoms, then one can derive equations similar to Equations 5.32 and 5.33 for each of the surface atoms. Numerical integration of these equations allows one to calculate the trajectory of the incoming molecule. The trapping probability is calculated by randomly choosing initial conditions for the molecule, calculating trajectories, and seeing what fraction of the molecules are trapped.

There is an issue of how many surface atoms to include in the calculations. In principle, one should integrate the trajectories of all of the atoms in the solid. However, with 10^{23} atoms in the solid that cannot be done. Therefore, one normally chooses an "active ensemble" of surface atoms and only calculates the trajectories of these atoms within the active ensemble exactly. When molecular dynamics calculations were first used to model surface processes, it was common to assume that the atoms outside of the active ensemble were fixed. However, Goodman [1962] showed how to use a response function approach to treat the atoms outside of the active ensemble as a heat bath. Adelman and Doll [1974] expanded this approach to yield what is now called the generalized Langevin model. The Langevin model is still used today.

To understand how the equations for the generalized model come about, it is useful to consider the one-dimensional chain of atoms described in the last section. The chain of atoms follows the classical equations of motion given in Equations 5.27 to 5.29. In the work to follow we attempt to eliminate $x_3, x_4, \ldots, x_n$ from these equations so that we will only have to numerically integrate the equations for the incoming gas molecule and the top two atoms in the one-dimensional surface. The algebra is complex and could be skipped. However, we include it in case the reader is interested.

Our derivation will follow Adelman and Doll [1974]. Consider the one-dimensional chain of atoms shown in Figure 5.5. Taking the Laplace transform of Equation 5.29 yields

$$s^2 X_2(s) = \omega^2[X_1(s) - X_2(s) + X_3(s)] + IC_2(s)$$

$$s^2 X_3(s) = \omega^2[X_2(s) - X_3(s) + X_4(s)] + IC_3(s)$$

$$\vdots$$

$$s^2 X_{n-1}(s) = \omega^2[X_{n-2}(s) - X_{n-1}(s) + X_n(s)] + IC_{n-1}(s)$$

$$s^2 X_n(s) = \omega^2[X_{n-1}(s) - X_n(s)] + IC_n(s) \tag{5.34}$$

where s is the Laplace transform variable, $X_i(s)$ is the Laplace transform of x_i, and $IC_i(s)$ is the Laplace transform of the initial conditions

$$IC_i(s) = sx_i(t = 0) + \frac{dx_i(t = 0)}{dt} \tag{5.35}$$

Solving the bottom expression in Equation 5.34 for X_n, substituting the resultant expression in the next expression up in Equation 5.34, solving for X_{n-1}, and repeating the process yields

$$X_n(s) = \frac{\omega^2 X_{n-1}(s)}{s^2 + 2\omega^2} + \frac{IC_n}{s^2 + 2\omega^2}$$

$$X_{n-1}(s) = \frac{\omega^2 X_{n-2}(s)}{s^2 + 2\omega^2 - \dfrac{\omega^4}{s^2 + 2\omega^2}} + \frac{IC_{n-1} + \omega^2 \dfrac{IC_n}{s^2 + 2\omega^2}}{s^2 + 2\omega^2 - \dfrac{\omega^4}{s^2 + 2\omega^2}}$$

$$X_{n-2}(s) = \frac{\omega^2 X_{n-2}(s)}{s^2 + 2\omega^2 - \dfrac{\omega^4}{s^2 + 2\omega^2 - \dfrac{\omega^4}{s^2 + 2\omega^2}}} + \frac{IC_{n-2} + \omega^2 \dfrac{IC_{n-1} + \omega^2 \dfrac{IC_n}{s^2 + 2\omega^2}}{s^2 + 2\omega^2 - \dfrac{\omega^4}{s^2 + 2\omega^2}}}{s^2 + 2\omega^2 - \dfrac{\omega^4}{s^2 + 2\omega^2 - \dfrac{\omega^4}{s^2 + 2\omega^2}}}$$

$$\vdots$$

$$\tag{5.36}$$

Substituting the value of $X_3(s)$ from Equation 5.36 into the expression for $X_2(s)$ from Equation 5.34 yields

$$s^2 X_2(s) = \omega^2 X_1(s) - 2\omega^2 X_2(s) + \omega^2 R_D^n(s) X_2(s) + R_R^n(s) + IC_2(s) \tag{5.37}$$

with

$$R_D^n(s) = \cfrac{\omega^2}{s^2 + 2\omega^2 - \cfrac{\omega^4}{s^2 + 2\omega^2 - \cfrac{\omega^4}{s^2 + 2\omega^2 - \cfrac{\omega^4}{\cdots}}}} \qquad (5.38)$$

The term R_R^n equals the sum of all of the terms containing the initial conditions of atoms 3, 4, ..., n. Taking the inverse Laplace transform of Equation 5.37 by using the convolution theorem to invert the $R_D^n(s)X(s)$ term yields

$$\frac{d^2 x_2}{dt^2} = \omega^2 x_1 - 2\omega^2 x_2 + \int_0^t R_D^n(t - \tau)x_2(\tau)\, d\tau + R_R^n(t) \qquad (5.39)$$

Equation 5.39 is the Langevin equation of motion for the second layer of surface atoms. Note that equation 5.39 is very similar to the top line in Equation 5.29 except that the term in x_3 has been replaced with $R_R^n(t)$ and an integral in $R_D^n(t)$. The term $R_R^n(t)$ is called the random force; it is a force that acts on the surface atoms due to the random thermal fluctuations of the atoms in the bulk. The quantity $R_D^n(t)$ is called the memory kernel or dissipative force. It keeps the surface atom in a dynamic thermal equilibrium with the bulk.

Adelman and Doll [1976] generalized these results to an infinite three-dimensional lattice. The final equations are essentially identical to Equation 5.39, except that one needs to evaluate a more complex expression than Equation 5.38 to evaluate the dissipative force and the random force. In actual calculations, one does not usually use the computed values of $R_R^n(t)$ and $R_D^n(t)$. The quantity $R_R^n(t)$ is usually approximated as being random Gaussian noise; $R_D^n(t)$ is usually modeled approximately with a Debye function and then fit to experimental data.

When this book was being written, an attempt was made to find a series of plots which show how various parameters affect the results of the Langevin calculation. However, that material has not yet appeared in the literature.

What does appear, instead, is a series of calculations that attempt to model experimental trapping measurements. Generally, one assumes a form for the gas/surface and surface-atom/surface-atom potential with 5–7 adjustable parameters, and then numerically integrate the three-dimensional version of Equation 5.39 to calculate a trapping probability following methods outlined in Example 5.D. Langevin calculations have become the standard way to model trapping and sticking on semiconductor surfaces. They are particularly useful in modeling a process called *molecular beam epitaxy* (MBE). However, generally, one uses an active ensemble of about 150 atoms, and calculates all of the properties of the ensemble exactly. The results are generally qualitatively similar to the results of Baule's model. However, one can explicitly account for the effects of (1) conversion of lateral momentum into perpendicular momentum, (2) multiple scattering, and (3) collective motions of the surface atoms. As a result one is able to fit more of the details of a trapping measurement. The reader should see the reviews of Doll and Voter [1987] or Tully [1980] for further information.

Lattice calculations are also used to model trapping on metal surfaces. Generally, the calculations do quite a good job of fitting data, albeit the calculations have 5–7 adjustable parameters in the potential functions. Figure 5.7 shows an example of how well the calculations fit. Still, one does have to wonder whether the lattice models are correctly representing the physics of a gas/metal collision. As just noted, the lattice models assume that the energy of the incoming molecule is transferred to the vibrational motion of the surface atoms. That energy is then assumed to be transferred into the bulk of the solid via lattice vibrations. However, work in solid-state physics indicates that in a metal, heat is carried by excitations of the conduction band electrons and not by vibrations of the metal atoms. Hence, the assumption that energy is transferred vibrationally and not electronically during a gas/metal collision is questionable.

There have been a few attempts to calculate the effects of electronic excitations on the energy transfer rate. Gadzuk and Metiu [1980] showed that electronic excitations should be quite important to gas/metal energy transfer. The main effect is to change the dissipative force in Equation 5.39. Makuski [1983] and Gunnarsson and Schoenhamer [1982, 1983] described a method to calculate the effect. However, so far this work has not led to quantitative models that could be compared to experiment.

Admittedly, it is unclear whether a change in how one calculates the dissipative force will have much real effect on trapping calculations. Generally, people do not actually calculate the dissipative force; rather they fit it to data. The trapping probability works out to only be a weak function of the form of the dissipative force. As a result, the trapping probability seems to be well represented by the Langevin model, even under circumstances where the assumptions used to derive the model are questionable. Still, I believe there is a need for more work on the influence of the conduction band electrons on gas/metal energy transfer.

The influence of adsorbates on trapping rates also needs to be considered in more detail. Experimentally, trapping rates usually increase in the presence of adsorbates. For example, Roberts [1939] showed that the trapping probability of He on a tungsten ribbon changes from 0.02 on a "clean" ribbon to 0.6 if the ribbon is covered by oxygen. Equation 5.8 shows that when the incident molecule collides with a surface species which has a low effective mass, the trapping rate is larger than in the absence of the low-mass species. With metal surfaces, atoms in the adsorbate usually have a lower effective mass than the atoms in the surface. As a result, one would expect accommodation coefficients and trapping rates to increase when the surface is partially covered by adsorbate.

The adsorbate can have other effects, however. Experimentally, adsorbates make the surface look more corrugated, which enhances the conversion of tangential momentum into normal momentum. There can also be some stiffening of the lattice due to the presence of strong adsorbate/surface bonds. At present methods to model these phenomena are still being developed. An example is the work of Head-Gordon et al. [1991].

5.8 STICKING

Still, trapping is basically a well-understood process when vibrational excitation of the lattice is the dominant mode of energy transfer, coverages are low, and temperatures are high enough that quantum effects are not important. Usually, the existing models are good enough to represent trapping data. It is harder to fit scattering data because quantum effects are important to scattering. However, models exist for scattering as well.

In this section, we leave our discussion of trapping and move on to sticking. **Sticking** is a process where an incident molecule collides with a surface and then bounces around until it finds a site where it can adsorb. Sticking is fundamentally different from trapping. In trapping a molecule simply loses energy to the lattice so it no longer has enough energy to leave the surface. In sticking, however, the incident molecule not only has to lose energy, it also has to form a strong bond with the surface. Hence, while energy transfer rates determine the rate of trapping processes, sticking rates are determined by both the rate of energy transfer and the ability of the surface to form bonds.

In the literature, it is common to discuss sticking in terms of a sticking probability, $S(\theta)$, which is a function of the coverage, θ. The sticking probability is defined as

$$S(\theta) = \frac{\text{Number of molecules that stick}}{\text{Number of molecules that impinge on a surface}} \qquad (5.40)$$

The rate of adsorption, r_a, is related to the sticking probability by

$$r_a = S(\theta)\hat{I}_z \qquad (5.41)$$

where $\hat{I}_z$ is the total flux of molecules onto the surface in molecules/cm^2 sec.

Sticking probabilities are measured by exposing a surface to a fixed number of molecules and measuring how many stick. In the following we develop an equation for the sticking probability in terms of the rate of adsorption and the properties of the incident molecules. Assume that we have a stream of molecules which is moving at an angle Φ_i' with respect to the surface normal as indicated in Figure 5.6. If the molecules are moving away from the surface, they will not hit it. If the molecules are moving toward the surface, then the flux of the molecules, I_z, onto the surface will be proportional to the component of the molecules velocity in the direction of the surface normal, $\vec{n}$. The flux is also proportional to the density of molecules in the gas phase. As a result:

$$\begin{cases} I_z = (\vec{v}_i \cdot \vec{n})\, \rho_g = v_i \cos(\Phi_i')\rho_g & -\dfrac{\pi}{2} < \Phi_i < \dfrac{\pi}{2} \\ I_z = 0 & \text{otherwise} \end{cases} \qquad (5.42)$$

where ρ_g is the density of the incident molecules in the gas phase. Equation 5.42 is called Knudsen's cosine law since it was experimentally verified by Knudsen [1915].

In a modern surface science experiment, one uses a directed doser so that the incoming gases all impinge from the same direction. However, in older experiments, people would fill up their whole apparatus with gas. When one exposes a clean surface to a fixed pressure of gas, molecules impinge onto the sample from all angles. Hence one needs to average the cosine function in Equation 5.42 over all values of Φ_i' between $-\pi$ and π. The result is

$$\hat{I}_x = \left(\frac{\displaystyle\int_{-\pi}^{\pi} I_x \sin\Phi_i \, d\Phi}{\displaystyle\int_{-\pi}^{\pi} \sin\Phi_i \, d\Phi_y} \right) = \frac{v_i \rho_g}{4} \qquad (5.43)$$

The factor of $\sin \Phi_i$ in Equation 5.43 arises because we are integrating in spherical coordinates.

It is useful to use Equation 5.43 to estimate the flux in a typical experiment. Consider adsorbing carbon monoxide by filling a chamber with 10^{-6} torr of carbon monoxide at 273 K. In Appendix A, Example A.2, we show that at 10^{-6} torr and 273 K, carbon monoxide has an average molecular velocity of 4.53×10^4 cm/sec. From the ideal gas law, it has a density of 3×10^{10} molecules/cm^3. Plugging these numbers into Equation 5.43 shows that the net flux toward the surface is 3.4×10^{14} molecules/cm^2/sec. If the sticking probability were unity, then it would only take 2.9 sec to fill the surface with 1×10^{15} molecules/cm^2 of gas, (i.e., about a monolayer).

People who work in surface science often measure the number of molecules that impinge on a sample during an experiment in units of **Langmuirs,** where 1 Langmuir (L) is equal to the number of molecules that would impinge per square centimeter of surface if the surface were exposed to 10^{-6} torr of gas for 1 sec. At 273 K, a 1-L exposure of carbon monoxide corresponds to directing 3.4×10^{14} molecules/cm^2 carbon monoxide onto the surface, while

2-L exposure is equivalent to directing twice that amount. One can calculate the exposure in Langmuirs very easily by multiplying the gas pressure in torr by the exposure time in seconds and dividing by 10^{-6}. However, a difficulty arises when one measures exposures in Langmuirs. Note that molecular velocities vary with the mass and temperature of the gas, while molecular densities vary with the gas temperature. Hence, a 1-L exposure of a gas at a given temperature is not equivalent to a 1-L exposure to another gas at the same temperature or even 1 L of the same gas at a different temperature. During a 1-L exposure of carbon monoxide, 5.6×10^{14} molecules/cm^2 of carbon monoxide impinge on the surface at a gas temperature of 100 K; 3.4×10^{14} molecules/cm^2 impinge at a gas temperature of 273 K; and 3.1×10^{14} molecules/cm^2 impinge at a gas temperature of 330 K. During a 1-L exposure of hydrogen, 2.1×10^{15} molecules/cm^2 impinge on the surface at a gas temperature of 100 K, and 1.3×10^{15} molecules/cm^2 impinge at a gas temperature of 273 K. In general the exposure in molecules/cm^2, EX_A, is related to the exposure in Langmuirs, $Lang$, by

$$EX_A = \frac{2.97 \times 10^{16} \frac{\text{molecules}}{\text{cm}^2 \, \text{K}^{1/2} \, \text{AMU}^{1/2} \, \text{Langmuir}}}{\sqrt{MW \, T_g}} \, Lang \qquad (5.44)$$

where MW is the molecular weight of the adsorbate, and T_g is the *gas* temperature that may not equal the surface temperature. The variation in the conversion factor between the number of molecules and the number of Langmuirs leads to some confusion in the literature. Some workers now also report exposures in molecules/cm^2. However, one has to design a special port on one's apparatus to directly measure the number of molecules that impinge on a surface while one can usually measure a pressure without making special modifications to the apparatus. Hence, most workers still measure exposures in Langmuirs.

People use Equation 5.43 or 5.44 to measure a so-called uptake curve. In a typical experiment one exposes a surface to various amounts of adsorbate and measures the amount of gas that sticks in molecules/cm^2 [A]. The uptake curve is a plot of [A] as a function of the exposure of the gas in molecules/cm^2, EX_A. Figure 5.13 shows a typical uptake curve. Notice that the uptake of gas is initially rapid, but it quickly saturates. Most uptake curves for monolayer adsorption systems are similar to that in Figure 5.13. There is an initial rapid uptake of gas and then a slow approach to saturation. The data in Figure 5.13 do eventually saturate at a total coverage of 1 monolayer. However, no saturation is seen in systems where multilayers form.

In principle, one can use uptake data like that in Figure 5.13 to calculate a coverage-dependent sticking probability, $S(\theta)$, where $S(\theta)$ is given by

$$S(\theta) = \frac{d[A]}{dEX_A} \qquad (5.45)$$

and θ is the fractional coverage of adsorbate in percent of a monolayer. In practice, one does not get an accurate value of $S(\theta)$ from Equation 5.45. However, Morris, Bowker, and King [1984] describe a number of ways to measure $S(\theta)$ accurately.

There have been hundreds of papers that report sticking probability data. Generally, people measure an **initial sticking probability,** $S(0)$ i.e., the sticking probability in the limit of zero coverage, and the variation in the sticking probability with coverage, $S(\theta)/S(0)$. Morris et al. summarize the data up to 1983. Table 5.1 shows some of their findings as well as data from Joyce and Foxton [1984]. Measured initial sticking probabilities have varied from 10^{-7} to 1. However, the majority of the data in the literature were taken where the initial sticking probability was between 0.1 and 1. Usually, a gas with a sticking probability of less than about 0.01–0.0001 is said to not stick.

Initial sticking probabilities vary with the gas, the surface composition and orientation,

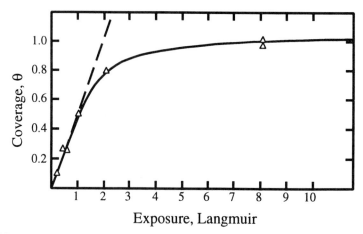

Figure 5.13 The amount of carbon monoxide that sticks on a Pt(410) surface as a function of the carbon monoxide exposure. (Data of Banholzer and Masel [1986].)

and temperature. For example, at 100 K, the sticking probability of oxygen is close to unity on most faces of platinum. However, at 300 K Ducros and Merrill [1976] found that the sticking probability of oxygen was about 0.4 on (2x1)Pt(110), while Kneringer and Netzer [1975] found that the sticking probability is less than 10^{-3} on (5x20)Pt(100). There are also some unusual features in the oxygen data, Pirig, Broden, and Bonzel [1977] found that a small number of defects will substantially raise the sticking probability of oxygen on (5x20)Pt(100). Most gases follow the same general trends seen with oxygen. The initial sticking probability varies with the gas temperature, the surface temperature, the surface structure, and the surface composition. Often, the initial sticking probability is strongly affected by a small number of defects. As a result, there can be substantial variations in the values of the initial sticking probabilities measured by various groups.

The variation of the sticking probability with coverage is easier to measure because one only needs relative rather than absolute coverages to obtain the variation in $S(\theta)/S(0)$ with coverage. Morris et al. [1984] noted that all the $S(\theta)/S(0)$ available in 1983 followed one of the six generalized curves shown in Figure 5.14. These six generalized curves can be used to classify sticking probability plots.

Curve A shows the simplest behavior: a linear drop in the sticking probability with coverage. Sticking probabilities drop with increasing coverage because the adsorbate takes up sites. If another adsorbate molecule comes in and hits the filled sites, the new adsorbate molecule cannot stick; instead it desorbs. Langmuir [1916, 1921] showed that if the adsorbate only needs one site to adsorb, and there are no adsorbate/adsorbate interactions or weakly bound states, then the system will show type A behavior. The analysis is described in Section 5.9.1. However, it happens that one rarely sees a type A sticking probability plot. The specific

TABLE 5.1 Some Typical Values of the Initial Sticking Probability at 300 K

Gas	Surface	$S(0)$	Gas	Surface	$S(0)$
H_2	Ni(100)	0.06	N_2	W(320)	0.7
H_2	Ni(111)	0.02	N_2	W(110)	$<3 \times 10^{-3}$
H_2	Pt(110)	0.2	Ga	GaAs(100)-B	1.0
H_2	Si(100)	$<10^{-4}$	As_4	GaAs(100)-B	$<10^{-3}$
CO	Pt(111)	0.67	As_4	GaAs(100)-A	0.5

Data taken from Morris et al. [1984] and Joyce and Foxton [1984].

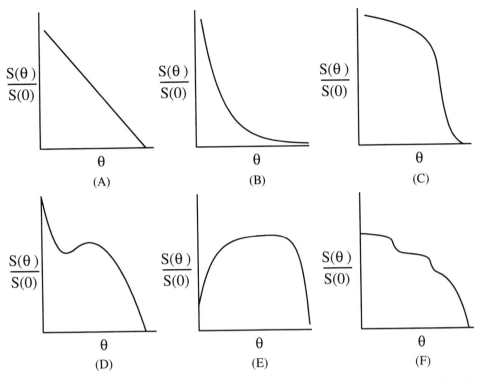

Figure 5.14 A general classification of the variation in the sticking probability with coverage. (Adapted from Morris et al. [1984].)

systems that Langmuir studied the most, the room temperature adsorption of hydrogen and oxygen on tungsten filaments of light bulbs, do follow a type A sticking probability plot up until moderate coverages. There are a few other controversial examples. However, the great majority of adsorption systems do not show such simple type A behavior.

Type B behavior is much more common than type A behavior. Type B behavior is similar to type A behavior in that there is a monotonic drop in the sticking probability with increasing coverage. However, the drop is not linear. Instead the line is curved. The curvature in the sticking probability plot can arise for several different reasons. If the adsorbate dissociatively adsorbs so it blocks two or more sites, then one will get type B behavior. One can also get type B behavior if there are strong adsorbate/adsorbate interactions or a variation in the heat of adsorption with coverage. Immobile adsorbates also show type B behavior. As a result, type B behavior is quite common.

Type C behavior occurs when the sticking probability is nearly constant up to some intermediate coverage and then drops at higher coverages. Type C behavior arises because the incoming molecules are initially trapped into a weakly bound "precursor" state. The molecules then move around the surface and find a site to adsorb. Most simple molecules follow type C behavior at low temperatures. At higher temperatures one often observes type B behavior because the weakly bound molecules desorb before they can be converted to a strongly bound state.

Type D behavior occurs when the sticking probability initially drops with increasing coverages. However, at some intermediate coverage it begins to rise again. Type D behavior usually arises in systems that show a surface reconstruction. The adsorption rate initially drops because sites are blocked. However, at higher coverages, the surface reconstructs. If the heat of adsorption of the adsorbate is higher on the reconstruction surface than on the unreconstructed one, the sticking probability will go up when the surface reconstructs. This produces type D behavior. Type D behavior is unusual on closed packed surfaces. However, it can be observed on fairly open faces of transition metals where reconstructions are common.

Type E behavior occurs when the sticking probability initially rises as one adsorbs gas. However, eventually the sticking probability drops as one fills up sites. Experimentally, type E behavior occurs mainly in trapping-dominated systems and in other systems where energy transfer plays an important role. As noted in the last section, trapping rates increase in the presence of an adsorbate because adsorbate/adsorbate energy transfer is much more efficient than adsorbate–surface energy transfer. As a result, the sticking probability initially rises as adsorbate accumulates. The sticking probability drops again at high coverages due to site blocking.

In principle one can also get type E behavior when there are strong adsorbate/adsorbate attractions so that the heat of adsorption increases as gas adsorbs. At present, however, type E behavior has only been observed in trapping-dominated systems or systems with weakly bound precursors or other systems where momentum or energy transfer plays an important role.

Type F behavior, where the sticking probability shows a pattern of plateaus and dips, is more unusual. In principle, type F behavior can arise if there is more than one adsorption site or if there is a surface phase transition during the adsorption process. However, one usually only observes type F behavior on polycrystalline samples or samples with many defects.

All of Morris, Bowker, and King's diagrams are for systems where one only can adsorb a monolayer of gas. If one does experiments where multilayers can form, the sticking probability does not go to zero as the coverage approaches a monolayer. Instead, the sticking probability levels off to a constant, nonzero value.

All of the adsorption data summarized in Morris et al. [1984] were collected by exposing a surface to a stream of gas at room temperature and measuring how much gas sticks. However, in recent years several investigators have used molecular beam systems to examine the variation in the sticking probability with incident angle and energy. Much of the work is reviewed by Ceyer [1988] and Ashford and Rettner [1991]. Figure 5.15 shows a variety of typical data. Basically, three different kinds of behavior are observed. If a given adsorption process is trapping dominated into, for example, a precursor state, then Equation 4.26 predicts that an increase in the incident energy will produce a decrease in the adsorption rate. As a result one should observe a monotonic decrease in the sticking probability with increasing beam energy. Alternatively one can get a direct adsorption into a strongly bound state. When w is high, Equation 4.26 predicts that sticking probability should be near unity at low energy and slowly decrease with increasing energy. In actual practice one does observe a slow decrease in the sticking probability with increasing incident energy, although the effect is small. However, the sticking probability is less than unity because only those molecules that hit the binding site for the molecules stick. The third kind of behavior arises when the adsorption process is activated. In such cases, the sticking probability is negligible at low energy because the incident molecules will not have enough energy to climb over the activation barrier. However, once the molecules have enough energy to cross the barrier, the sticking probability increases rapidly with increasing energy. The sticking probability continues to increase at higher energy because a larger fraction of the molecules can make it over the barrier.

Figure 5.15 illustrates several of these behaviors using data of Rendulic and Winkler [1989]. One should refer to Ceyer [1988] or Ashfold and Rettner [1991] for additional illustrations. When a beam of hydrogen adsorbs onto a Pt(110) surface at $\phi_i = 60°$, the hydrogen shows behavior characteristic of a simple precursor mechanism: the sticking probability decreases monotonically with increasing beam energy and then levels off. In contrast, on Ni(111) and Cu(111) the hydrogen shows a behavior characteristic of there being a finite activation barrier to adsorption. The sticking probability is near zero below an energy threshold. How-

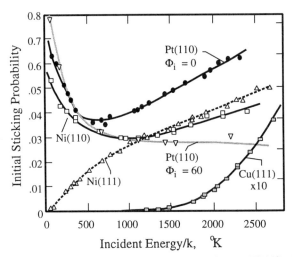

Figure 5.15 The initial sticking probability of hydrogen on a Ni(111), Ni(110), Cu(111), Pt(110) Φ_i = 10° and 60° as a function of the energy of the incident gas. (Data of Rendulic and Winkler [1989].)

ever, above the threshold the sticking probability increases with increasing energy. The data for hydrogen adsorption on Pt(110) at Φ_i = 10° and the data for hydrogen adsorption on Ni(110) show more complex behavior: the sticking probability first decreases with increasing beam energy and then rises again. Rendulic and Winkler [1989] show that this behavior arises because there are two pathways for hydrogen adsorption on Ni(110) and Pt(110), an indirect pathway where the incoming hydrogen goes into a molecular precursor before it chemisorbs, and a direct adsorption pathway where the incoming hydrogen molecules climb over an activation barrier and stick without going through the precursor state.

The adsorption of hydrogen on Pt(110) and Ni(110) illustrates an important complication in the analysis of sticking probability data. Often more than one adsorption channel (i.e., reaction pathway) affects the rate. As a result real data are hard to fit with simple models.

When this chapter was being written, an attempt was made to find an example that showed little variation in the sticking probability with changing incident energy. The search was unsuccessful, but not because of any physical reason that the sticking probability must vary strongly with incident energy. Data like that in Figure 5.15 are very difficult to take, and people do not usually bother unless they are working on a system that would be expected to show large variations in the sticking probability with incident energy. So far, no one has done extensive work on a system that would not be expected to show strong variations. As a result, while I am not aware of data showing no variation in sticking probability with incident energy, such systems probably exist. They just have not been explored.

5.9 MODELS FOR THE VARIATION IN THE STICKING PROBABILITY WITH ENERGY AND COVERAGE

One can generally model the variation of the initial sticking probability with energy and angle using either the improved Baule model described in Section 15.4 or by the molecular dynamics/Langevin (MD/L) method described in Section 5.7. The Baule model only works for trapping-dominated systems. However, MD/L works even when there is an activated process. With a trapping-dominated process, people generally assume that the trapping probability is equal to the initial sticking probability. Experimentally this seems to be a good approximation. The situation is more complex when there is an activation barrier for adsorption. If one knows the barrier, one can compute the sticking probability using the MD/L method. However, in order to really understand the trends in the data, one has to understand why activation barriers occur. We defer the latter discussion until after we have discussed activation barriers. In the remainder of this chapter we concentrate on describing models for the variation in the sticking probability with coverage.

5.9.1 Langmuir's Model

Langmuir [1921] did the first useful work on sticking probabilities. He defined the sticking probability and provided the first useful model for the variation in the sticking probability with coverage. Langmuir's model assumes that the adsorbate binds to a series of identical surface sites. When a molecule impinges on a surface and hits a bare site it will stick with probability $S(0)$. However, if it hits a filled site, the molecule will simply scatter. The sticking probability becomes $S(0)$ times the probability that an incoming molecule hits a bare site, $P_{bare}^n(\theta)$, i.e.,

$$S(\theta) = S(0) \, P_{bare}^n(\theta) \tag{5.46}$$

If the incoming molecules impinge onto sites randomly, and the molecules adsorb in a random layer where the probability that a given is occupied is independent of whether other adjacent sites are occupied, then

$$P_{bare}^n(\theta) = (1 - \theta)^n \tag{5.47}$$

where n is the number of sites that is needed to hold the adsorbate. Equation 5.47 implicitly assumes that when a molecule dissociatively adsorbs, the dissociation products move apart and randomly block sites.

Combining Equations 5.41, 5.46, and 5.47 yields

$$r_a = S(0)\hat{I}_z(1 - \theta)^n \tag{5.48}$$

Equation 5.48 is called Langmuir's adsorption law. According to Equation 5.48, when $n = 1$ the system will show type A behavior. If $n > 1$, the system will show type B behavior.

Langmuir did extensive tests of Equation 5.48 for adsorption of alkali metals on mica and adsorption of a variety of gases on tungsten and found that he could represent all of his data via Equation 5.48 with $n = 1$. Taylor and Langmuir [1933] did find that their data for the sticking of cesium on tungsten did not follow Equation 5.48. However, this system was an exception in Langmuir's work. Hence, Langmuir proposed that type A behavior was very common.

5.9.2 Relationship to the Langmuir Adsorption Isotherm

Equation 5.48 is closely related to the Langmuir adsorption isotherm. At equilibrium, the rate of adsorption, r_a, must equal the rate of desorption, r_d. If the rate of desorption is presumed to follow a first-order rate law, i.e.,

$$r_d = k_d \theta \tag{5.49}$$

then the net rate of adsorption, r_{net}, i.e., the rate of adsorption minus the rate of desorption at any coverage θ is given by

$$r_{net} = r_a - r_d = \hat{I}_x S(\theta) - k_d \theta \tag{5.50}$$

Substituting the ideal gas law into Equation 5.43 shows

$$\hat{I}_x = \frac{v_i \rho_g}{4} = \frac{v_i P_A}{4kT}$$
(5.51)

At equilibrium, $r_{net} = 0$. Substituting $\hat{I}_x$ from Equation 5.51 and $S(\theta)$ from Equation 5.48 into Equation 5.50 and solving for P_A yields

$$K^A_{equ}P_A = \frac{\theta}{(1 - \theta)^n}$$
(5.52)

with

$$K^A_{equ} = \frac{V_i S_0}{4kTk_d}$$
(5.53)

It is easy to show that Equation 5.52 is equivalent to Equation 4.10, the Langmuir adsorption isotherm for a nondissociative adsorption process. As a result, Langmuir concluded that systems that show type A behavior will follow the Langmuir adsorption isotherms. He already had shown that the Langmuir adsorption isotherm was quite common. Therefore, Langmuir suggested that type A behavior should be quite common.

5.9.3 Limitations of Langmuir's Analysis

Up until the mid-1960s most investigators assumed that type A behavior was indeed very common. However, in the late 1960s improvements in vacuum technology and surface spectroscopy made it possible to test Equation 5.48 for a wide variety of adsorbents and surfaces. It was found that equation 5.48 works quite well for adsorption of simple molecules on W(110) and adsorption of alkali's on mica (i.e., the adsorption systems examined by Langmuir). However, it happens that these systems are exceptions. Morris et al. [1984] cite 150+ measurements of sticking probabilities of simple gases on single-crystal surfaces. However, only 8 systems show type A behavior: hydrogen on Pt(211), Pd(111), and W(100); carbon monoxide, hydrogen, nitrogen and oxygen on W(110), and nitrogen on W(411). There have been several reports of type A behavior during the adsorption of gases on polycrystalline samples. However, type A behavior is unusual, at least during the adsorption of simple gases on single-crystal surfaces.

There are three key reasons that one gets deviations from Langmuir behavior. First, recall from the discussion in Chapter 3 that many species can bind to more than one site on a metal surface. However, Equation 5.47 only considers one site. Second, the pairwise interactions between adsorbed molecules modify the adsorption process. Third, a molecule that hits a filled site does not necessarily desorb. Instead, the molecule can diffuse across the surface and find another site. The role of multiple binding sites on sticking probabilities has not been extensively explored in the literature. However, the influence of pairwise forces, and the ability of impinging molecules to find bare sites has been discussed in detail.

In some of the earliest work on this topic Langmuir observed significant deviations from Equation 5.48 in his work with Taylor [1933] on the adsorption of cesium on tungsten. When cesium adsorbs on tungsten, the sticking probability is constant up to coverages near saturation. However, Equation 5.48 predicts that the sticking probability should drop continuously with increasing coverage.

Langmuir discussed this unusual behavior in his 1933 paper. Equation 5.48 assumes that the incoming molecule bounces from the surface if it impinges on a site that is occupied. However, when cesium impinges onto tungsten, the sticking probability is near unity even at high coverage. Therefore, there must be some way for an incident cesium atom to stick, even when it impinges onto an occupied site. Langmuir noted that an incident atom that impinges onto an occupied site could not chemisorb onto the site, but the atom could be trapped into a weakly bound state. The atom can then diffuse around the surface and find a site to adsorb. As a result, the sticking probability can be constant up until high coverages. This idea was amplified by Lennard-Jones [1932], Ehrlich [1955], Kisliuk [1957], Gorte and Schmidt [1978] and Cassuto and King [1981]. The weakly bound state is called a **mobile precursor,** and adsorption via a precursor state is called **precursor moderated adsorption.** Precursor moderated adsorption is fairly common, especially at low temperatures. Up until about 1980 it was unclear whether a precursor state really exists, or whether the precursor was just a convenient artifice to fit experimental sticking probability data. However, in the early 1980s several research groups examined the interaction of molecules with surfaces using molecular beams. Much of that data are reviewed in Ashford and Rettner [1991]. In most cases, one observed three different populations leaving the surface: an elastically scattered component, a component that thermally accommodates with the surface, but only stays for less than a microsecond, and a component that stays on the surface for several seconds or more. Auerbach and Rettner [1991] show that the thermally accommodated flux is associated with molecules that are trapped into a weakly bound precursor state but do not stick, while the molecules that stay on the surface for seconds or more are associated with molecules that are trapped and stick. As a result, it is now thought that the mobile precursor state is more than an artifice; rather molecules often do go into a mobile state upon adsorption.

5.9.4   Models for Precursor-Moderated Adsorption

There are two ways that one can derive kinetic expressions for precursor-modified adsorption, the successive site approach of Kisliuk [1957] and the kinetics approach of Ehrlich [1955]. Cassuto and King [1981] showed mathematically that the two approaches yield identical equations. In this section we will derive the key equations. Our derivation closely follows Cassuto and King [1981] with additions from Morris et al. [1984] and Gorte and Schmidt [1978].

To begin, it is useful to note that when a molecule collides with a surface that is partially covered by adsorbate, the adsorption process is complex. The incoming molecule first loses energy to the lattice and is trapped. The molecule then has to find a site where it can stick. At the end of Section 4.3 we noted that energy transfer rates are different when the molecule hits a bare site than when it collides with another adsorbate. Sticking rates are also different, since the molecules that hit another adsorbate and are trapped need to diffuse over the surface to find a bare site, while the molecules that are trapped at a bare site can stick directly. As a result, we need to consider two different kinds of molecules in the calculations, those that impinge on a filled site, and those that impinge on an empty site.

In principle, one would have to calculate the dynamics of all of the incident molecules to develop an accurate picture of this process. However, Kiskliuk [1957] proposed that one can model the kinetics by considering two types of precursors: an "extrinsic" precursor, which is the precursor that forms when an incoming molecule impinges an occupied site, and an "intrinsic" precursor, which forms when an incoming molecule impinges on an empty site. In the following material we designate the precursors over occupied (i.e., filled) sites with the subscript FI and precursors over unoccupied (i.e., empty) sites with the subscript MT.

Consider the kinetic scheme shown in Figure 5.16. In this scheme adsorption occurs when a gas phase molecule A_g, is trapped at a filled site to yield an extrinsic precursor, A_{FI}, or trapped at an empty site to form an intrinsic precursor, A_{MT}. The precursor is further converted into a strongly bound state A_{ad}. If we consider the possibility of desorption of the precursors, and conversion of molecules from one precursor state to another, and assume that all the reactions in Figure 5.16 are elementary, then we can derive an approximation for the rate of adsorption using an analysis as follows.

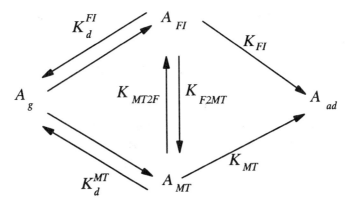

Figure 5.16 The precursor mechanism for nondissociative adsorption.

For the reaction scheme in Figure 5.16, the net rate of adsorption into the strongly bound state, r_a, is given by

$$r_a = K_{FI}[A_{FI}] (1 - \theta) + K_{MT}[A_{MT}] (1 - \theta) \tag{5.54}$$

where K_{MT} is the rate constant for the conversion of molecules in an intrinsic precursor state into a strongly bound state, while K_{FI} is a pseudo-rate constant for the direct conversion of molecules from the extrinsic precursor state to a strongly bound state. The constant K_{FI} is normally assumed to be independent of coverage even though it is much harder for the extrinsic precursor to find a bare site at high coverage than at low coverage. It is not obvious whether there should be the $(1 - \theta)$ in the last term of Equation 5.54. Some authors include it, some do not. We included it because we get a better fit to data if we do.

In order to use Equation 5.54 one has to get an expression for $[A_{FI}]$ and $[A_{MT}]$. We can do so from a mass balance. For the reaction scheme in Figure 5.16, the net rate of accumulation of extrinsic and intrinsic precursor molecules, r_{FI}, and r_{MT}, respectively, is given by

$$r_{FI} = I_z P_{trap}^{FI}\theta + K_{MT2F}[A_{MT}] - (K_d^{FI} + K_{F2MT} + K_{FI}(1 - \theta)) [A_{FI}] \tag{5.55}$$

$$r_{MT} = I_z P_{trap}^{MT}(1 - \theta) + K_{F2MT}[A_{FI}] - (K_d^{MT} + K_{MT2F} + K_{MT}(1 - \theta)) [A_{MT}] \tag{5.56}$$

where P_{trap}^{MT} and P_{trap}^{FI} are the probabilities that an incident molecule is trapped when the molecule hits an empty or filled site, respectively; K_d^{MT} and K_d^{FI} are the rate constants for the desorption of precursor molecules when they are held on empty or filled sites, respectively; and K_{F2MT} and K_{MT2F} are the pseudo-rate constants for conversion of precursor molecules from a filled site to an empty site and from an empty site to a filled site, respectively.

At first sight one might think that K_{MT2F} is negligible. In fact, however, when molecules are first trapped they are only weakly bound and still have considerable tangential kinetic energy. As a result, the molecules are very mobile. Hence, it is not unreasonable that K_{MT2F} is nonzero, and in fact experimentally K_{MT2F} can be quite large.

For the purpose of derivation we assume that all of the rate constants are independent of coverage. We also make the steady-stae approximation, $r_{FI} = 0$ and $r_{MT} = 0$. Substituting $r_{FI} = r_{MT} = 0$ into Equations 5.55 and 5.56, and solving for $[A_{MT}]$ and $[A_{FI}]$ yields

$$[A_{FI}] = \frac{I_z P_{trap}^{FI}\theta(K_d^{MT} + K_{MT2F} + K_{MT}(1 - \theta)) + K_{F2MT}I_z P_{trap}^{MT}(1 - \theta)}{(K_d^{MT} + K_{MT2F} + K_{MT}(1 - \theta)) (K_d^{FI} + K_{F2MT} + K_{FI}(1 - \theta)) - K_{F2MT}K_{F2MT}} \tag{5.57}$$

$$[A_{MT}] = \frac{K_{F2MT} I_z P_{trap}^{FI} \theta + I_z P_{trap}^{MT}(1 - \theta) (K_d^{FI} + K_{F2MT} + K_{FI}(1 - \theta))}{(K_d^{MT} + K_{MT2F} + K_{MT}(1 - \theta)) (K_d^{FI} + K_{F2MT} + K_{FI}(1 - \theta)) - K_{F2MT} K_{MT2F}} \quad (5.58)$$

Substituting Equations 5.57 and 5.58 into Equation 5.54 yields

$$\frac{r_a}{(1 - \theta)} = \left\{ \frac{\begin{aligned} & I_z P_{trap}^{FI} \theta (K_{MT} K_{F2MT} + K_{FI} K_d^{MT} + K_{FI} K_{MT}(1 - \theta) + K_{MT2F} K_{FI}) \\ & + I_z P_{trap}^{MT}(1 - \theta) (K_d^{FI} K_{MT} + K_{MT} K_{F2MT} + K_{MT} K_{FI}(1 - \theta) + K_{FI} K_{F2MT}) \end{aligned}}{(K_d^{MT} + K_{MT2F} + K_{MT}(1 - \theta)) (K_d^{FI} + K_{F2MT} + K_{FI}(1 - \theta)) - K_{F2MT} K_{MT2F}} \right\}$$

$$(5.59)$$

Equation 5.59 is the rate equation for adsorption via a mobile precursor state. Kisliuk derived a somewhat simpler expression than Equation 5.59 by assuming that K_{FI} was negligible, which is equivalent to assuming that extrinsic precursors are converted into intrinsic precursors before they strongly adsorb. If K_{FI} is negligible,

$$S(\theta) = \frac{r_A}{I_z} = (P_{trap}^{MT}) \left(\frac{K_1^P \theta + \dfrac{(1 - \theta)}{K_2^P}}{K_3^P + \dfrac{(1 - \theta)}{K_2^P}} \right) (1 - \theta) \quad (5.60)$$

$$K_1^P = P_{trap}^{FI} / P_{trap}^{MT}$$

$$K_2^P = K_{F2MT} / (K_d^{FI} + K_{F2MT})$$

$$K_3^P = \frac{[(K_d^{MT} + K_{MT2F}) (K_d^{FI} + K_{F2MT}) - K_{F2MT} K_{MT2F}]}{K_{MT} K_{F2MT}} \quad (5.61)$$

Equation 5.60 considered in detail the difference between intrinsic and extrinsic precursors. However, an alternate approach is to assume that there is rapid conversion of intrinsic to extrinsic precursors and vice versa, which is equivalent to assuming that K_{F2MT} is large so that $K_2^P = 1$. If the trapping probabilities are the same in the two states, $K_1^P = 1$. Therefore,

$$S(\theta) = P_{trap}^{MT} \left(\frac{1}{K_3 + (1 - \theta)} \right) (1 - \theta) \quad (5.62)$$

After some algebra one can show that

$$S(\theta) = \frac{S(0)}{1 + \dfrac{K_3^P \theta}{(1 + K_3^P) (1 - \theta)}} \quad (5.63)$$

with

$$S(0) = \frac{P_{trap}^{MT}}{1 + K_3^P} \quad (5.64)$$

Equation 5.63 is used less often than Equation 5.60. However, we include it for completeness. It is useful to discuss the physical significance of K_1^P, K_2^P, and K_3^P. The term K_1

is the ratio of the trapping probability on filled to that on empty sites. In Section 5.7 we noted that trapping rates tend to increase in the presence of adsorbates. As a result, K_1^P should be larger than 1. Typically, K_1^P is only slightly larger than unity. However, there are some special cases where K_1^P is as large as 15.

The term K_2^P is the probability that an extrinsic precursor diffuses to an occupied site. Physically, when an incident molecule hits an occupied site, the incident molecule can either be scattered or be trapped. If the molecule is trapped, the molecule can then either desorb again or diffuse to an empty site where it can stick. Thus $K_d^{FI} + K_{F2MT}$ is the total rate at which trapped molecules leave the extrinsic precursor state, while K_{F2MT} is the rate at which molecules leave by diffusing to empty sites. Hence, K_2^P is the ratio of the rate at which molecules leave the extrinsic precursor state by diffusing to empty sites, to the total rate at which molecules leave the extrinsic precursor state. Thus K_2^P is the probability that molecules in the extrinsic precursor state diffuse to an empty site.

Physically, K_2^P can have any value between 0 and 1. However, if K_2^P is much less than 1, precursor states are not going to be important to the adsorption process.

The term K_3^P is harder to describe. Rearranging Equation 5.61 shows

$$\left(\frac{K_3^P}{1 + K_3^P}\right) = 1 - K_2^P\left(\frac{K_{MT2F}}{K_{MT2F} + K_d}\right) \tag{5.65}$$

The last term is the rate at which intrinsic precursors diffuse to occupied sites divided by the rate at which they diffuse or desorb. Hence, the last term is the probability that if a molecule leaves the intrinsic state without sticking, the molecule leaves via diffusion. Physically, $K_3^P/(1 + K_3^P)$ can have any value between 0 and 1, which means that K_3^P can have any value between zero and infinity. A K_3^P of zero would correspond to very rapid diffusion of molecules, and hence a very mobile precursor. A K_3^P of infinity, would correspond to an immobile precursor, i.e., one that would never find an empty site before it desorbs. In the discussion below we show that the precursor model approaches Langmuir's model as K_3^P goes to infinity.

In his paper, Kisliuk [1957] also considers cases where $K_3^P/(1 + K_3^P)$ is greater than unity. However, it is unclear how $K_3^P/(1 + K_3^P)$ can be greater than unity. Furthermore, the predicted behavior has never been observed experimentally.

Figure 5.17 is a plot of Equation 5.60 for for some typical values of the parameters. When K_3^P is small, Equation 5.60 predicts that the sticking probability will be nearly constant up

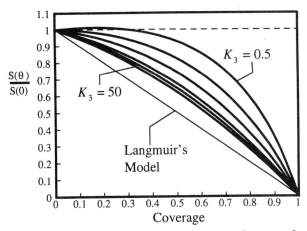

Figure 5.17 A sticking probability calculated via Equation 5.60 for $K_1^P = 1.2$, $K_2^P = 0.95$, $K_3^P = 0.5$, 1, 2, 5, 10, 50 and via Langmuir's model, Equation 4.47.

to half coverage, and then drops at higher coverages. Therefore, Equation 5.60 explains the qualitative features of type C sticking probability plots. When K_3^P is greater than unity (i.e., when the trapping probability is larger on an occupied site than an empty one), the sticking probability at moderate coverages is higher than the initial sticking probability. Hence, one can also explain type E sticking probability plots via Equation 5.60. Figure 5.17 also shows that in the limit that K_3^P grows to infinity, the precursor model approaches Langmuir's model. Hence, Equation 5.60 is useful even when the deviations from Langmuir's model are small.

In actual practice, however, Equation 5.60 provides more of a qualitative rather than a quantitative fit to sticking data. One important source of deviation from Equation 5.60 arises because precursor- and nonprecursor-moderated adsorption usually occur simultaneously. In the discussion of Figure 5.15 we noted that one observes both precursor- and nonprecursor-moderated adsorption during hydrogen adsorption on Ni(110) and Pt(110). This is a general result. Most systems that show precursor-moderated adsorption also show a direct adsorption pathway. Equation 5.60 only considers precursor-moderated adsorption. Hence, it often needs to be modified to explain real data.

Diffusion limitations also lead to deviations from Equation 5.60. As noted earlier, if the adsorbate coverage is low, an incoming molecule that happens to impinge on a filled site will not have to diffuse very far before the incoming molecule finds an empty site. However, at high coverages, diffusion lengths will be much longer. In principle, one can include the effects of diffusion in Equation 5.60 by allowing K_{FI} to be a function of coverage. However, when people use Equation 5.60, they usually assume K_{FI} is constant. As a result, one often needs additional modifications to equation 5.60 to explain real data.

A third source of error in Equation 5.60 occurs because of the effects of adsorbate/adsorbate interactions. If the adsorbate/adsorbate interaction is repulsive, K_{MT}, the rate constant for sticking of intrinsic precursors, will decrease with increasing coverage. In contrast, K_{MT} will increase with increasing coverage when the adsorbate/adsorbate interaction is attractive. The changes in K_{MT} lead to additional corrections to Equation 5.60.

Singh-Boparai, Bowker, and King [1975] did some calculations to quantify the effects of adsorbate/adsorbate interactions on sticking probability plots. Unfortunately, the calculations used the quasichemical approximation, which is now known to be inaccurate. Hence, the calculations will not be reproduced here. However, Singh-Boparai et al.'s calculations show that adsorbate/adsorbate interactions are important when one wants to model the kinetics of precursor-moderated adsorption in detail.

5.10   IMMOBILE ADSORPTION

In the previous section, we derived a correction to Langmuir's model, Equation 5.48, due to a mobile precursor. Different corrections occur if the adsorbate is immobile. The corrections arise because Equation 5.48 was derived by assuming that the probability of a site being occupied is independent of the probability that an adjacent site is occupied. In such a case the probability of having two empty sites next to each other is just the square of the probability that one site is empty. However, it happens that if the adsorbate is immobile, the probability of a site being occupied is not independent of the probability that an adjacent site is occupied. Hence, corrections are needed.

In order to understand why these corrections occur, it is useful to consider immobile adsorption of a gas, which takes up two adjacent sites. Initially, the gas molecules will adsorb randomly, and cover up sites. By chance, however, there is always some probability that a given site will be completely surrounded by adsorbate, but still be empty. If the gas requires two adjacent sites to adsorb, then the gas cannot adsorb on any site that is surrounded by adsorbate even though that particular site is empty. As a result, the gas will not be able to adsorb on some of the surface sites, even though those sites are bare. The adsorption process will stop before all of the sites are filled! In contrast, Equation 5.48 predicts that the adsorption rate process will continue until all of the sites are filled. Thus, at high coverage Equation 5.48 predicts too large a rate of adsorption for an immobile adsorbate, which requires two or more sites to adsorb. It happens that at low coverages Equation 5.48 predicts too low a rate for reasons that are discussed in the next section.

Roberts [1935] made the first experimental observations of immobile adsorption. Roberts was studying the adsorption of oxygen on a tungsten filament. He found that when he saturated the surface with oxygen at low temperature, only about 90% of the surface sites were covered by oxygen. Oxygen would adsorb on the additional 10% of the sites if the surface was heated. Roberts also found that the bare surface sites would adsorb neon at low temperatures. As a result, Roberts concluded that some sort of activated process was preventing oxygen from adsorbing on the last 10% of the sites.

At this point, Roberts noted that O_2 dissociated upon adsorption, so that two sites were needed to hold each molecule of O_2 that adsorbs. Roberts then conjectured that the adsorbed oxygen was immobile. As noted previously, in an immobile adsorption process there is a finite probability that a given empty site will be completely surrounded by adsorbate. In such a case the site is not available for adsorption even though the site is empty. Roberts supposed that the reason that he could only cover about 90% of his sample with oxygen was that the adsorbed layer was immobile over the time scale of his experiment and so some sites were left vacant during the adsorption process. It took 40 years to verify this supposition. However, it is now well established that oxygen does form an immobile layer when oxygen adsorbs on tungsten at low temperatures.

Roberts and Miller [1939] did some calculations to try to account for the effect of the immobile layer on the kinetics of the adsorption process. Roberts and Miller started with a Monte Carlo calculation. They began with a 10 × 10 lattice, and added dimers randomly to see how the adsorption process would proceed. The calculations were very tedious because all the numbers had to be computed by hand. Roberts and Miller chose a random site by drawing cards, and a second adjacent site was chosen by choosing a second card. If both sites were empty, the molecule was presumed to adsorb; if not, the molecule was presumed to scatter. At each stage of the calculation Roberts and Miller added up the fraction of the sites that were still available to adsorb dimers. In that way they were able to estimate the sticking probability as a function of coverage.

Roberts and Miller also fit their calculations to a fifth-order polynomial to yield the following semiempirical expression for the sticking probability as a function of coverage:

$$S(\theta) = [1 - 1.75 \, (\theta - 0.3215 \, \theta^2 - 0.0833 \, \theta^3 - 0.0175 \, \theta^5)] \, S(0) \qquad (5.66)$$

Today people routinely do Monte Carlo calculations to calculate sticking probabilities. The results are often fit to a polynomial expansion similar to Equation 5.66. These expansions are called **viral expansions** in the literature. In 1939, however, no one knew whether Monte Carlo calculations were acceptable methods to calculate properties of adsorbates. As a result, Equation 5.66 was not immediately accepted.

A few years after Equation 5.66 was proposed, Roberts and Miller [1939] developed an approximation to the sticking probability, which was more readily accepted. Roberts and Miller's derivation was based on a variation of the Bethe approximation discussed in Chapter 4. Their idea was to treat the filling of a small cluster of surface sites exactly, and then treat the filling of the sites around the cluster in an average sense. Consider the small cluster of sites shown as light circles in Figure 5.18a. Now consider filling up the cluster with molecules that bind on two adjacent sites as shown in Figure 5.18b–d. It is useful to define two quantities ε_0, the occupancy of the central site in the cluster (i.e., site zero) and n_m the number of molecules that overlap the cluster but do not cover the central site in the cluster. For the

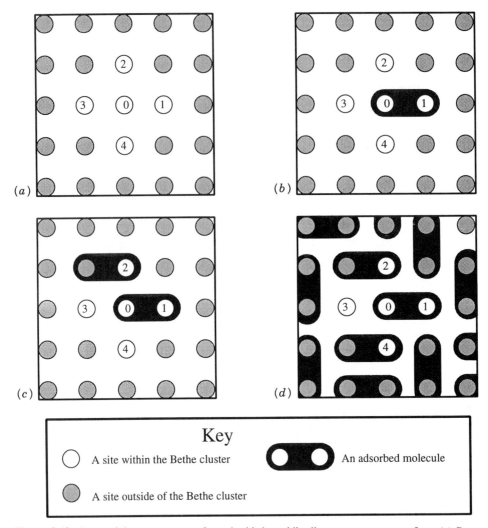

Figure 5.18 Some of the arrangements formed with immobile dimers on a square surface. (*a*) Bare surface. (*b*) One dimer. (*c*) Two dimers. (*d*) Saturated surface.

arrangement of molecules in Figure 5.18*d*, $\varepsilon_0 = 1$ and $n_m = 2$. The set of all values of ε_0, and n_m represent all of the possible arrangements of the system. In the material that follows we designate each arrangement by the pair of indices (ε_0, n_m). For example, the arrangement in Figure 5.18*d* will be designated the (1, 2) state of the system.

Table 5.2 lists each of the possible states of the system, and the degeneracies g_{ε_0, n_m} of each of these states. If a single molecule is adsorbed on the central site in the cluster as illustrated in Figure 5.18*b*, the molecule can be oriented in any of four directions. As a result, the degeneracy of the (1, 0) state of the cluster is 4. Similarly, once we fix the arrangement of the molecule covering the central site, then we can add an additional molecule on any one of three sites. Hence, the degeneracy of the (1, 1) state is $4 \cdot 3 = 12$. If we add a third molecule, it can be placed on any one of two sites. One might think that the degeneracy would be $4 \cdot 3 \cdot 2$. However, note that this is double counting arrangements, since all arrangements that switch the molecules that do not overlap the central site are equivalent. Therefore, the degeneracy of the (1, 2) site is $4 \cdot 3 \cdot 2/2 = 12$. Similarly the degeneracy of

TABLE 5.2 The Degeneracies of the Possible Arrangements of the Molecules in the Cluster Shown in Figure 5.18

ϵ_0	1	1	1	1	0	0	0	0	0
n_m	0	1	2	3	0	1	2	3	4
g_{ϵ_0,n_m}	4	12	12	4	1	4	6	4	1

the (1, 3) site is $(4 \cdot 3 \cdot 2 \cdot 1)/(3 \cdot 2 \cdot 1) = 4$. The degeneracies for the remaining cases are given in Table 5.2. Readers should check them for themselves.

The probability that any state (ϵ_0, n_m) is occupied P_{ϵ_0, n_m} is given by

$$P_{\epsilon_0, n_m} = \frac{g_{\epsilon_0,n_m} P_{\epsilon_0} P_{n_m}^{\epsilon_0}}{\sum\limits_{\epsilon_0, n_m} g_{\epsilon_0,n_m} P_{\epsilon_0} P_{n_m}^{\epsilon_0}} \qquad (5.67)$$

where P_{ϵ_0} is the probability that the central atom is in state ϵ_0, and $P_{n_m}^{\epsilon_0}$ is the probability that there are n_m molecules surrounding the central site which do not overlap the central site, given that the central site is in state ϵ_0.

It is easy to show that, in general, $P_{n_m}^{\epsilon_0}$ will be different according to whether the central site is empty or occupied. For example, if we saturate the surface with gas, we will leave some isolated individual sites but no adjacent pairs of sites. Under such circumstances it is possible for the central site in the cluster to be empty, or for one of the side sites to be empty. However, at saturation it is impossible for both the central site and a side site to be empty. (If both were empty, gas molecules could still adsorb.) Therefore, at saturation,

$$P_{n_m=0}^{\epsilon_0=0} = P_{n_m=1}^{\epsilon_0=0} = P_{n_m=2}^{\epsilon_0=0} = P_{n_m=3}^{\epsilon_0=0} = 0 \qquad P_{n_m=4}^{\epsilon_0=0} = 1$$
$$P_{n_m=0}^{\epsilon_0=1} \approx 0 \qquad P_{n_m=1}^{\epsilon_0=1} \approx 0.01 \qquad P_{n_m=2}^{\epsilon_0=1} \approx 0.1 \qquad P_{n_m=3}^{\epsilon_0=1} \approx 2.9 \qquad (5.68)$$

The top line in Equation 5.68 results from the requirement that if the central site is empty, the surrounding sites must be filled. The bottom line in Equation 5.68 comes from Roberts' Monte Carlo simulation.

Roberts and Miller did not have a way to account for the various values of $P_{n_m}^{\epsilon_0}$ exactly, and so they proposed to treat $P_{n_m}^{\epsilon_0}$ using a very rough "statistical approximation":

$$P_{n_m}^{\epsilon=1} \approx P_{n_m}^{\epsilon_0=0} \approx (P_{RM})^{n_m} \qquad (5.69)$$

where P_{RM} is a constant related to the probability that an average site is occupied. The constant P_{RM} is calculated self-consistently. Roberts and Miller supposed that Equation 5.69 would be a good approximation at low coverages, and it may not be a bad approximation at high coverages.

Roberts and Miller used Equation 5.69 to derive an approximation to the sticking probability as a function of coverage. First, the derivation notes that the probability of the central site being occupied is $P_{\epsilon_0=1}$, which must equal θ. Next, they needed an approximation for the coverage on sites 1, 2, 3, and 4. When $\epsilon_0 = 1$ and $n_m = 0$ the central site will be covered, but either site 1, 2, 3, or 4 will be covered too. Therefore, the average coverage of the site will be ϵ_0 when $n_m = 0$. Similarly, when n_m is nonzero, there will be n_m additional sites covered. Therefore, for any configuration of the system (ϵ_0, n_m) the average total coverage of adsorbate on sites 1, 2, 3, and 4 will be $\epsilon_0 + n_m$.

The average occupancy on sites 1, 2, 3, and 4, θ_s, becomes

$$\theta_s = \frac{1}{4} \sum_{\epsilon_0, n_m} P_{\epsilon_0, n_m}(\epsilon_0 + n_m) \qquad (5.70)$$

Substituting Equation 5.69 into Equation 5.67, and substituting the resultant expression into Equation 5.70, while noting that $P_{\varepsilon_0 = 1} = \theta$, $P_{\varepsilon_0 = 0} = (1 - \theta)$, and taking the the degeneracies from Table 5.2 yields

$$4\theta_s = \frac{\theta(4 + 24P_{RM} + 35P_{RM}^2 + 18P_{RM}^3) + (1 - \theta)(4P_{RM} + 12P_{RM}^2 + 12P_{RM}^3 + 4P_{RM}^4)}{\theta(4 + 12P_{RM} + 12P_{RM}^2 + 4P_{RM}^3) + (1 - \theta)(1 + 4P_{RM} + 6P_{RM}^2 + 4P_{RM}^3 + P_{RM}^4)}$$

(5.71)

Note that from self-consistency θ_s must equal θ. Therefore, one can solve Equation 5.71 self-consistently for P_{RM}.

Roberts and Miller did not actually use Equation 5.71. Instead, they noted that once they made the approximation in Equation 5.69, the average coverage did not work out to be θ unless they adjusted their values of $P_{\varepsilon_0 = 1}$ and $P_{\varepsilon_0 = 0}$ appropriately. Therefore, Roberts and Miller adjusted Equation 5.71 until it reached the correct limits. Their derivation lacks some mathematical rigor, and so it will not be reproduced here. However, Roberts and Miller suggested that one could replace Equation 5.71 with

$$4\theta_s = \theta\left(\frac{4 + 24P_{RM} + 36P_{RM}^2 + 16P_{RM}^3}{4 + 12P_{RM} + 12P_{RM}^2 + 4P_{RM}^3}\right)$$

$$+ (1 - \theta)\left(\frac{4P_{RM} + 12P_{RM}^2 + 12P_{RM}^3 + 4P_{RM}^4}{1 + 4P_{RM} + 6P_{RM}^2 + 4P_{RM}^3 + P_{RM}^4}\right)$$

(5.72)

This expression can be factored as follows

$$4\theta_s = \theta\left(\frac{(1 + 4P_{RM})(1 + 8P_{RM} + 4P_{RM}^2)}{(1 + P_{RM})(1 + 8P_{RM} + 4P_{RM}^2)}\right)$$

$$+ (1 - \theta)\left(\frac{(4P_{RM})(1 + 3P_{RM} + 3P_{RM}^2 + P_{RM}^3)}{(1 + P_{RM})(1 + 3P_{RM} + 3P_{RM}^2 + P_{RM}^3)}\right)$$

(5.73)

Canceling like terms in Equation 5.73 yields

$$4\theta_s = \frac{\theta + 4P_{RM}}{(1 + P_{RM})}$$

(5.74)

Setting $\theta_s = \theta$ in Equation 5.74 and solving for P_{RM} yields

$$P_{RM} = \frac{3\theta}{4(1 - \theta)}$$

(5.75)

Now consider a molecule that impinges onto sites 0 and 1 in Figure 5.16a. If either site 0 or 1 is filled, the incident molecule cannot stick. If both sites are empty, the incident molecule can stick. In the absence of adsorbate/adsorbate interactions the sticking probability will simply be $S(0)$. Therefore, the sticking probability becomes

$$S(\theta) = S(0)\, P_{\varepsilon_0 = 0}\delta_E$$

(5.76)

where δ_E is the probability that site 1 is empty, given that site 0 is empty. The term δ_E can have any value between 0 and 1 and is not a Dirac delta. From Equation 5.72 and Table 5.2,

$$\delta_E = \frac{1 + 3P_{RM} + 3P_{RM}^2 + P_{RM}^3}{1 + 4P_{RM} + 6P_{RM}^2 + 4P_{RM}^3 + P_{RM}^4} = \frac{1}{1 + P_{RM}} \quad (5.77)$$

Note, in deriving Equation 5.77, we used the fact that for the ($\varepsilon_0 = 0$, $n_m = 1$) configuration, there are three states with site 1 empty. Therefore, the coefficient of the ($\varepsilon_0 = 0$, $n_m = 1$) configuration on the top of Equation 5.77 is 3, etc.

Substituting the expression for P_x from Equation 5.75 into Equation 5.77, and substituting the resultant expression for δ_E into Equation 5.76, yields

$$S_{MR}(\theta) = S(0) \left(\frac{4}{4 - \theta} \right) (1 - \theta)^2 \quad (5.78)$$

where $S_{MR}(\theta)$ is the Roberts–Miller approximation for the sticking probability. Our derivation considered a square lattice with four nearest neighbors. However, when there are n_n nearest neighbors, an analogous derivation yields

$$S(\theta) = S(\theta) \left(\frac{n_n}{n_n - \theta} \right) (1 - \theta)^2 \quad (5.79)$$

Roberts and Miller compared Equation 5.79 to their Monte Carlo calculations and found that it was a good approximation for $\theta < 0.6$, but deviated significantly for $\theta > 0.7$. As a result, they concluded that Equation 5.79 was only valid at low to moderate coverage. However, Equation 5.79 was reproduced in several textbooks in the 1940s and 1950s without mentioning Roberts and Miller's provision that it only be applied for $\theta < 0.6$. As a result, Equation 5.79 became a standard approximation at even high coverages.

Unfortunately, Equation 5.79 predicts qualitatively incorrect behavior at high coverages. Note that

$$S_{MR}(\theta) \geq S(0)(1 - \theta)^2 = S_{Mobile}(\theta) \quad (5.80)$$

Therefore, the Roberts–Miller approximation predicts that the sticking probability in the immobile limit is always larger than the sticking probability in the mobile limit. At low coverages, this prediction ends up being correct. However, at high coverage $S_{immobile}(\theta) < S_{mobile}(\theta)$.

Let us consider the low coverage results first. Assume that we have a molecule that adsorbs dissociatively into two atoms, each of which covers a single site. We need to consider two cases, the case where the adsorption is immobile, so that the atoms stay on two adjacent sites, and the case where the adsorption is mobile so that the atoms can separate. Figure 5.19 shows a diagram of the two cases. If there is a dimer adsorbed on sites 0 and 1, it is not possible to adsorb a dimer on **seven** pairs of sites, i.e., the (0, 1), (0, 2), (0, 3), (0, 4), (1, 5), (1, 6), and (1, 7) pairs. However, if instead two atoms are adsorbed on sites 1 and 3, it will not be possible to adsorb a dimer on **eight** pairs of sites, i.e., the (0, 1), (1, 5), (1, 6), (1, 7), (3, 0), (3, 8), (3, 9), (3, 10) pairs. Therefore, at low coverages two isolated atoms will block more pairs of sites than an individual dimer. As a result, the sticking probability is higher with the dimer than with isolated atoms, in agreement with Equation 5.80. One can show that Equation 5.79 goes asymptotically to the correct result, in the low coverage limit.

At high coverage, though, Equation 5.79 fails. As noted previously, if the layer is immobile, there is always a chance of leaving a bare site surrounded by filled sites. If the adsorbate requires two sites to adsorb, then no gas can adsorb on a bare site that is completely surrounded with filled sites. As a result, when dimers adsorb onto an immobile layer, it is impossible to completely fill the surface with the adsorbate. Vette et al. [1974] have calculated

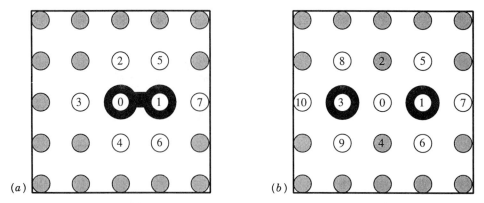

Figure 5.19   Site blocking by (*a*) two adjacent atoms, and (*b*) two atoms that separate.

the saturation coverage exactly using methods described in Section 5.10.2. Their results show that to five significant figures, the maximum coverage is $\theta_{sat} = 0.90293$ and that

$$S(\theta) = 0 \quad \text{for} \quad \theta \geq 0.90293 \tag{5.81}$$

In contrast, Equation 5.79 predicts a saturation coverage of unity. As a result, while Equation 5.79 gives qualitatively useful results at low coverages it is not useful at high coverages.

Over the years, many people have proposed improvements to Equation 5.79. The one-dimensional case was considered by Widom [1966, 1973], Flory [1939], Page [1959], MacKenzie [1962], McQuistan and Lichtman [1968], Maltz and Molta [1982], Vette et al. [1974], and Gonzalez, Hemmer, and Hoye [1974]. Vette et al. [1974] provided a general solution method that was applicable to a two-dimensional lattice. Nord and Evans [1985] and Schaaf et al. [1988] extended Vette et al.'s method to the immobile adsorption of squares, trimers, etc. Talbot and co-workers [1989] also extended the methods to random immobile adsorption of molecules that do not bind to individual lattice sites. Much of the work is beyond the scope of this monograph, so one is referred to the literature for details. However, we did want to include Vette et al.'s [1974] results so that the reader would get a general idea of how to derive a better approximation than Equation 5.79.

In the next two sections we use Vette et al.'s method to calculate an approximation for adsorption of immobile dimers on a one- and two-dimensional lattice. The algebra is complex and could be skipped. However, we include it in case the reader is interested.

5.10.1 Exact Solution for Adsorption on a Line of Sites

Consider the adsorption of dimers on the line of sites illustrated in Figure 5.20 and assume that each dimer takes up two sites. It is useful to define p_1 as the probability that if we choose

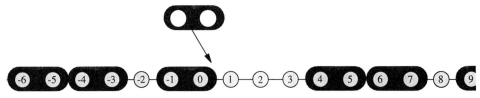

Figure 5.20   Adsorption on a line of sites.

a single site at random, that site will be vacant; p_2 as the probability that if we choose a group of two adjacent sites at random, both sites will be vacant; p_3 as the probability that if we choose a group of three adjacent sites at random, all three sites will be vacant, etc. Note that $p_1 = (1 - \theta)$. According to Equation 5.46, the sticking probability of dimers follows:

$$S(\theta) = S(0)p_2 \tag{5.82}$$

Therefore, $S(\theta)$ can be calculated by keeping track of how p_2 changes as adsorbate builds up on the surface.

Consider what happens to a random site, which we will label site 0 in the line of sites, when a dimer impinges onto the line of sites. If the dimer impinges onto sites 0 and 1 or sites -1 and 0, site zero will be covered up with probability $S(0)$. Therefore, the probability that a given site (e.g., site zero) is vacant is reduced whenever a molecule adsorbs. It is easy to show that the rate of reduction follows

$$-\frac{dp_1}{dEX} = \frac{NN_1 S(0)p_2}{N_s^1} \tag{5.83}$$

where, by definition, p_2 is the probability that a given pair of adjacent sites (i.e., sites $(0, 1)$ or sites $(-1, 0)$) is vacant, $S(0)$ is the initial sticking probability, EX is the exposure in molecules/cm, N_s^1 is the density of sites in number/cm, and NN_1 is the number of nearest neighbors to site 0. The factor of NN_1 in Equation 5.83 comes about because site 0 is covered up whenever a dimer molecule adsorbs on site $(0, 1)$ or $(-1, 0)$.

Now consider adsorption on a random pair of sites that we label the $(0, 1)$ pair. The pair of sites can be lost to further adsorption by direct adsorption on the pair. However, it will also not be possible to adsorb a dimer onto the $(0, 1)$ pair of sites if a dimer had previously adsorbed on the $(-1, 0)$ or $(1, 2)$ pairs of sites. Hence, we can block the $(0, 1)$ pair of sites to adsorption of dimers in three ways: we can adsorb on the $(0, 1)$ pair, we can adsorb on the $(1, 2)$ pair, or we can adsorb on the $(-1, 0)$ pair.

Now consider $-\dfrac{dp_2}{dEX}$, the rate at which sites are blocked due to exposure of gas. It is useful to define three quantities $B_{(0,1)}^{(0,1)}$, $B_{(-1,0)}^{(0,1)}$, $B_{(1,2)}^{(0,1)}$ as the rate at which the $(0, 1)$ pair of sites is blocked via adsorption on the $(0, 1)$, $(-1, 0)$, and $(1, 2)$ sites, respectively. It is easy to show

$$-\frac{dp_2}{dEX} = \frac{B_{(0,1)}^{(0,1)}}{N_s^1} + \frac{B_{(-1,0)}^{(0,1)}}{N_s^1} + \frac{B_{(1,2)}^{(0,1)}}{N_s^1} \tag{5.84}$$

If we assume that the sticking probability is zero on a filled site and $S(0)$ on an empty site, then

$$B_{(0,1)}^{(0,1)} = S(0) \, p_2 \tag{5.85}$$

where p_2 is again the probability that the $(0, 1)$ site is vacant.

Now consider $B_{(1,2)}^{(0,1)}$; $B_{(1,2)}^{(0,1)}$ is more subtle than $B_{(0,1)}^{(0,1)}$; $B_{(1,2)}^{(0,1)}$ is the rate at which the $(1, 0)$ pair of sites are blocked due to adsorption on the $(1, 2)$ pair of sites. If sites, 0, 1, and 2 are vacant and we adsorb a dimer onto sites 1 and 2, then the $(0, 1)$ pair of sites will be blocked for further dimer adsorption. However, if site 0 is already filled due to, for example, previous adsorption on the $(-1, 0)$ pair of sites, then the $(0, 1)$ pair is already blocked. In such a case, adsorption on the $(1, 2)$ pair will not change the ability of the $(0, 1)$ pair to adsorb dimers. Therefore, adsorption on the $(1, 2)$ pair will change in the ability of the $(0, 1)$ pair of sites to adsorb dimers only if the (0, 1, 2) triad started out vacant. That occurs with a probability of p_3. Therefore,

$$B_{(1,2)}^{(0,1)} = S(0) \, p_3 \tag{5.86}$$

A similar argument shows

$$B_{(-1,0)}^{(0,1)} = S(0) \, p_3 \tag{5.87}$$

Substituting Equations 5.86 and 5.87 into Equation 5.84 yields

$$-\frac{dp_2}{dEX} = \frac{S(0)}{N_s^!} p_2 + 2 \frac{S(0)}{N_s^!} p_3 \tag{5.88}$$

A similar argument holds for loss of triads. The (0, 1, 2) triad is lost (i.e., it is no longer completely vacant) by adsorption of a dimer on the triad itself or by adsorption on the (-1, 0) or (2, 3) pairs of sites. The latter two cases only affect the triad if the 0, 1, 2, 3 sites or the (-1, 0, 1, 2) sites all started out vacant. Furthermore, there are two ways to adsorb a dimer on the triad. Therefore, the loss of vacant triads due to the adsorption of dimers follows the following equations:

$$-\frac{dp_3}{dEX} = \frac{2S(0)}{N_s^!} p_3 + \frac{2S(0)}{N_s^!} p_4 \tag{5.89}$$

We ask the reader to derive a similar equation for adsorption of trimers in Problem 5.25.

One can generalize Equation 5.89 to consider an arbitrary line of n sites. The loss of vacant groups of n adjacent sites is given by

$$-\frac{dp_n}{dEX} = NA_n \frac{S(0)}{N_s^!} p_n + NN_n \frac{S(0)}{N_s^!} p_{n+1} \tag{5.90}$$

where NA_n is the number of ways a dimer can adsorb onto a given group of n adjacent sites, and NN_n is the number of nearest neighbors to a group of n clusters.

Note from Equation 5.83 that

$$\frac{d\theta}{dEX} = -\frac{dp_1}{dEX} = \frac{NN_1 \, S(0)}{N_s^!} p_2 \tag{5.91}$$

Therefore

$$\frac{dp_n}{d\theta} = \frac{\left(\dfrac{dp_N}{dEX}\right)}{\left(\dfrac{d\theta}{dEX}\right)} = \frac{\dfrac{dp_n}{dEX}}{\dfrac{NN_1 \, S(0)}{N_s^!} p_2} \tag{5.92}$$

Rearranging Equation 5.92

$$NN_1 p_2 \frac{dp_n}{d\theta} = \frac{N_s^!}{S(0)} \frac{dp_n}{dEX} \tag{5.93}$$

Substituting Equation 5.90 into 5.93 yields

$$-NN_1 p_2 \frac{dp_n}{d\theta} = NA_n p_n + NN_n P_{n+1} \tag{5.94}$$

It is useful to define a quantity δ_n by

$$\delta_n = \frac{p_n}{p_{n-1}} \tag{5.95}$$

Physically δ_n is the probability that if a group of $n-1$ sites is empty, a given site adjacent to the group of $n-1$ sites will also be empty.

Now consider the quantity $-NN_1 p_2 \dfrac{d \ln(\delta_n)}{d\theta}$ where $\ln(X)$ is the log of X.

$$
\begin{aligned}
-NN_1 p_2 \frac{d \ln \delta_n}{d\theta} &= -\frac{NN_1 p_2}{\delta_n} \frac{d\delta}{d\theta} = -\frac{NN_1 p_2}{\delta_n} \frac{d}{d\theta} \left(\frac{p_n}{p_{n-1}} \right) \\
&= -\frac{NN_1 p_2}{\delta_n p_{n-1}} \left(\frac{dp_n}{d\theta} \right) + \frac{NN_1 p_2 p_n}{\delta_n p_{n-1}^2} \frac{dp_{n-1}}{d\theta} \\
&= -\frac{NN_1 p_2}{p_n} \frac{dp_n}{d\theta} + \frac{NN_1 p_2}{p_{n-1}} \frac{dp_{n-1}}{d\theta}
\end{aligned} \tag{5.96}
$$

Substituting the expressions for $\dfrac{dp_n}{d\theta}$ and $\dfrac{dp_{n-1}}{d\theta}$ from Equation 5.96 into Equation 5.94 yields

$$-NN_1 p_2 \frac{d \ln \delta_n}{d\theta} = \left[\frac{NA_n p_n + NN_n P_{n+1}}{p_n} - \frac{NA_{n-1} p_{n-1} + NN_{n-1} p_n}{p_{n-1}} \right] \tag{5.97}$$

Performing the algebra

$$-NN_1 p_2 \frac{d \ln(\delta_n)}{d\theta} = (NA_n - NA_{n-1}) + NN_n \delta_{n+1} - NN_{n-1} \delta_n \tag{5.98}$$

For a line of sites

$$NA_n = n - 1$$
$$NN_n = 2 \tag{5.99}$$

for all n. Therefore

$$-2p_2 \frac{d \ln(\delta_n)}{d\theta} = 1 + 2\delta_{n+1} - 2\delta_n \tag{5.100}$$

For $n = 2$, this becomes

$$2p_2 \frac{d \ln(\delta_2)}{d\theta} = 1 + 2\delta_3 - 2\delta_2 \tag{5.101}$$

Dividing Equation 5.100 by Equation 5.101 shows

$$\frac{d \ln(\delta_n)}{d \ln(\delta_2)} = \frac{1 + 2\delta_{n+1} - 2\delta_n}{1 + 2\delta_3 - 2\delta_2} \tag{5.102}$$

The solution of Equation 5.102 is

$$\delta_n = \delta_{n-1} = \cdots = \delta_2 = \delta \tag{5.103}$$

Therefore, if we have a line of sites, the probability of finding a bare site adjacent to a line of n bare sites is independent of n provided n is 1 or more. The relation does not work for $n = 0$, however. The probability of finding a single bare site is different than the probability of finding a single bare site adjacent to one or more bare sites.

Next we develop an expression for δ as a function of coverage. From Equation 5.100, with $NN_1 = 2$

$$-2p_2 \frac{d \ln \delta}{d\theta} = 1 \tag{5.104}$$

but

$$p_2 = \delta_2 p_1 = \delta(1 - \theta) \tag{5.105}$$

Substituting Equation 5.105 into Equation 5.104 yields

$$1 = -2\delta(1 - \theta) \frac{d \ln \delta}{d\theta} = -2\delta(1 - \theta) \left(\frac{1}{\delta} \frac{d\delta}{d\theta} \right) \tag{5.106}$$

Canceling like terms and rearranging yields

$$\frac{d\delta}{d\theta} = -\frac{1}{2(1 - \theta)} \tag{5.107}$$

Integrating, and noting $\delta = 1$ at $\theta = 0$ yields

$$\delta = 1 + \tfrac{1}{2} \ln(1 - \theta) \tag{5.108}$$

Now it is useful to go back to Equation 5.93. From Equation 5.93

$$p_n = \delta_n p_{n-1} = \delta p_{n-1} \tag{5.109}$$

Applying Equation 5.109 iteratively yields

$$p_n = \delta^{n-1} p_1 = \delta^{n-1}(1 - \theta)$$

$$= (1 - \theta) (1 + \tfrac{1}{2} \ln(1 - \theta))^{n-1} \tag{5.110}$$

Substituting Equation 5.110 into Equation 5.83 yields

$$-\frac{d(1 - \theta)}{dEX} = \frac{2S(0)}{N_s^!} (1 - \theta) \left[1 + \frac{1}{2} \ln(1 - \theta) \right] \tag{5.111}$$

The solution of this equation is

$$\theta = 1 - \exp\left[2\left\{\exp\left(-\frac{EX\,S(0)}{N_s^1}\right) - 1\right\}\right] \tag{5.112}$$

Equation 5.112 is the exact solution for adsorption of dimers on a line of sites. One can calculate the sticking probability by substituting p_2 from Equation 5.110 into Equation 5.82. The result is

$$S(\theta) = S(0)\,(1 - \theta)\,[1 + \tfrac{1}{2}\ln(1 - \theta)] \tag{5.113}$$

Note that the adsorption process stops when $S(\theta) = 0$. Therefore, the value of θ that gives an $S(\theta)$ of zero is the maximum coverage that can be achieved. Substituting $S(\theta) = 0$ into Equation 5.113 and solving for θ yields

$$\theta = 1 - e^{-2} = 0.864 \cdots \tag{5.114}$$

Therefore, when gas adsorbs onto a row of sites at saturation, only 86.4% of the sites will be covered by gas.

5.10.2 Series Solution: Adsorption on a Two-Dimensional Array of Sites

In two dimensions, it is possible to get to higher coverages. Imagine filling up a two-dimensional square lattice. One could first populate the rows of sites to get 86.4% coverage, leaving 13.6% of the sites vacant. Note, however, that there is some probability that an empty site in one row is adjacent to an empty site in an adjacent row. We could add another molecule to those sites. As a result, if gas is randomly adsorbed onto a two-dimensional structure, a saturation coverage greater than 0.864 can be achieved.

Vette et al. [1974] developed a scheme to calculate the saturation coverage in two dimensions. They noted that Equations 5.113 and 5.114 only apply to a line of sites. With a two-dimensional lattice there are several different arrangements of 3, 4, 5, ... sites, each of which has different values of NA_n and NN_n. The solution of 5.21 adsorption on a two-dimensional lattice of sites is more difficult than in one dimension (see Figure 5.21). In principle, these groups of sites should have different values of p_n and δ_n. However, Vette et al. [1974] have shown that one can obtain a reasonable approximation to the two-dimensional case by assuming that all the p_3 are equal, all the p_4 are equal, etc. If so, all the δ_n will also equal δ. In such a case, Equations 5.82 to 5.98 will still apply with N_s^1 replaced by N_s, the density of sites in number/cm^2.

For a *line of sites* on a two-dimensional lattice

$$NA_n = (n - 1) \text{ for a line of sites}$$

$$NN_n = 2 \text{ for a line of sites in one dimension}$$

$$NN_n = 3 + (n - 1) \text{ for a line of sites on a triangular lattice}$$

$$NN_n = 4 + 2(n - 1) \text{ for a line of sites on a square lattice}$$

$$NN_n = 6 + 2(n - 1) \text{ for a line of sites on a hexagonal lattice} \tag{5.115}$$

Note that Equation 5.115 only applies to a line of sites. Figure 5.21 illustrates several different arrangements of four sites on a square lattice. The linear arrangement has ten nearest

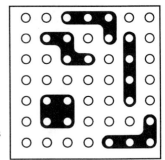

Figure 5.21 Some of the arrangements of four nearest neighbor sites on a square lattice.

neighbors, in agreement with Equation 5.115. However, the square arrangement only has eight nearest neighbors.

Vette et al. [1974] showed that one can get a reasonable approximation, by first assuming that all of the arrangements have $4 + 2(n - 1)$ nearest neighbors, and then adding in the effects of arrangements with less than $4 + 2(n - 1)$ nearest neighbors. The approach in this book is similar to of Vette et al.'s. First we assume that all of the arrangements of n adjacent sites follow Equation 5.115. Then we mention how one corrects the final equations to consider the effects of arrangements of molecules that do not follow Equation 5.115.

In the derivation to follow, we only consider adsorption of dimers on a square lattice. Different results are obtained for a hexagonal, rectangular and triangular lattice. One should refer to Vette et al. [1974] for these other results.

Consider the situation where Equation 5.115 is assumed to be exact. In such a case, a derivation similar to that in the previous section shows that all of the arrangements of sites will have the same value of δ. Substituting Equation 5.115d into Equation 5.98, and noting $p_2 = (1 - \theta)\delta$, yields

$$-4\delta(1 - \theta) \frac{d \ln(\delta)}{d\theta} = 1 + 2\delta \qquad (5.116)$$

One can show that the solution of Equation 5.116 is

$$\delta = \tfrac{1}{2} [3\sqrt{(1 - \theta)} - 1] \qquad (5.117)$$

The sticking probability becomes

$$S(\theta) = S(0) (1 - \theta) \delta = \frac{S(0)}{2} (1 - \theta) [3\sqrt{(1 - \theta)} - 1] \qquad (5.118)$$

Note that $S(\theta) = 0$ at $\theta_{sat} = 8/9$. Therefore, if one assumes that all groups of n adjacent sites have $4 + 2(n - 1)$ nearest neighbors, the saturation coverage will be 8/9.

Equation 5.118 is, of course, an approximation because it assumes that all groups of adjacent sites follow Equation 5.115. However, that is not true. Vette et al. [1974] classify the groups of sites that show derivations from Equation 5.115 as those with loops and kinks. If there is a loop or kink, a given site can be the nearest neighbor to two different sites in the original group of n sites (see Figure 5.22). As a result, the number of nearest neighbors can be less than the value given in Equation 5.115.

Vette et al. [1974] propose a hierarchy, where first one accounts for the simple kinks or loops, and then the more complex structures. One increases the number of structures until everything converges. In practice one needs to consider structures that have 1, 2, or 3 less nearest neighbors than the number given by Equation 5.115 to obtain results accurate to 4 or

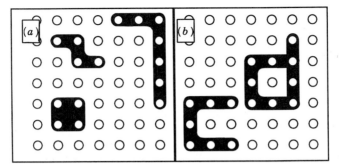

Figure 5.22 Groups of sites with (a) simple kinks or loops (b) complex kinks or loops.

5 significant figures. By considering all of these possibilities, Vette et al. obtained a system of 46 differential equations that had to be solved simultaneously. The details of the solution are beyond the scope of this monograph. However, it is interesting to note that Vette et al. find Equation 5.118 to be almost exact up to $\theta = 0.8$. There are some deviations above $\theta = 0.8$, and the saturation coverage works out to be 0.90293 rather than the 0.88889 predicted by Equation 5.118.

A number of other papers have expanded on Vette et al.'s results. Nord and Evans [1985] generalized Vette et al.'s results to the adsorption of trimers, tetramers, etc. These results are useful for the adsorption of normal alkanes. Schaaf et al. [1988] found a particularly convenient solution for square particles that is very useful for adsorption of larger molecules. None of these results are qualitatively different from those just cited, and so the derivations are not reviewed here. The reader is referred to Nord and Evans [1985] or Schaaf et al. [1988] for details.

The results on the last few pages assumed that the immobile adsorbate must adsorb on a series of lattice sites. However, with larger molecules such as polymers or proteins, or with cells, the adsorbing particles are not confined to bind to specific lattice sites. Instead they can bind randomly. However, since adjacent particles cannot overlap, the system does not follow a simple rate law. The binding of large immobile particles is called *random sequential adsorption* in the literature. Many people have worked on it. See Talbot and Schaff [1989], Brosilow, Ziff, and Vigil [1991] or Dickman, Wang, and Jensen [1991] for some recent results. Random sequential adsorption is not reviewed here because the literature is extensive and is only useful for large molecules or cells. The reader is referred to Talbot and Schaff [1989] for further details.

One of the reasons that more of these details were not included in this monograph is that as of 1995, Equation 5.118 and its improvements had not made it into the mainstream of surface science. Roberts and Miller's result, Equation 5.79, while not as accurate as Equation 5.118, was used to fit data in the 1930s and 1940s. It became the standard result for immobile adsorption, even though Roberts and Miller [1939] noted that it is only accurate at coverages below 0.6. Vette et al.'s results seem to have been ignored. Yet, when one adsorbs a diatomic gas at moderate temperatures, one often finds that it takes much longer than one would expect to saturate the surface. One reason that it takes a long time to saturate the surface is that molecules need to move around to create vacancies for adsorption and the diffusion of molecules takes a finite time at moderate temperatures. Systems that take a long time to equilibrate should be well approximated by Vette et al.'s model.

At very low temperatures, one observes somewhat different behavior. Initially, the adsorption process behaves much like that predicted by Roberts and Miller or Vette et al. However, the adsorption process does not stop when only a few isolated bare sites are left on the surface. Instead, molecules bind to the individual isolated sites in a weakly bound configuration. These molecules are converted to a strongly bound state upon annealing. One can sometimes observe these weakly bound intermediates spectroscopically (i.e., with electron energy loss spectroscopy (EELS) or ultraviolet photoemission spectroscopy (UPS), but, the species are difficult to see with temperature programmed desorption (TPD). During TPD, one heats the surface and watches what desorbs. However, in the case of immobile adsorption, the weakly bound molecules are converted to a strongly bound state before they desorb. Hence, the weakly bound molecules can be hard to detect. Still, there have been a number of examples where the heat of adsorption varies according to how the layer is prepared. Such systems are characteristic of immobile adsorption processes. Hence, the view of this author is that immobile adorption may be fairly common at low temperatures even though immobile adsorption has been largely ignored in the surface science literature up to 1995.

5.11 THE ROLE OF ACTIVE CENTERS IN ADSORPTION

Another case that is rarely considered in the literature is the case where adsorption is not occurring uniformly over the surface of the catalyst. Faraday [1835] first mentioned nonuniform adsorption in 1835. In 1925, Pease and Stewart [1925] provided conclusive evidence to show that it exists. Pease and Stewart were studying the adsorption of hydrogen on copper. They found that if they adsorbed mercury on the copper, the amount of hydrogen that adsorbed decreased by a factor of 20, but the rate of adsorption decreased by a factor of 200. If the adsorption process were occurring uniformly over the surface of the copper, the decrease in rate would be the same as the decrease in the amount adsorbed. However, it was not. As a result, Pease and Stewart [1925] proposed that adsorption was not occurring uniformly over the solid surface. Rather, it was occurring on a series of distinct active centers on the surface of the copper.

Taylor [1925, 1926] expanded these ideas substantially. Taylor's idea was that if one needed special sites for adsorption, one would also need special sites for reaction. Taylor showed experimentally that it was possible to poison a reaction without poisoning the adsorption process. As a result, Taylor suggested that most surface reactions occur only on special sites, which he termed "active sites." Dissociative adsorption is one case of a surface reaction. Hence, Taylor proposed that dissociative adsorption could only occur on special active sites.

For many years people thought that Taylor's ideas only applied to very rough surfaces. However, it is now clear that the results also apply to single crystals. When one cuts a single crystal, one leaves a sizable number of defects. It happens that one often finds that the adsorption process is affected by the defects.

For example, Figure 5.23 compares the sticking of H_2 on a well-annealed Ni(111) surface to that on an Ni(111) sample that has been sputtered so that about 1% of the surface is covered by defects. The initial sticking probability is 0.02 on the annealed surface and 0.07 on the defected surface. When people first measured data like these, they assumed that the change in the sticking probability was associated with a change in the binding of the first few molecules that adsorb. These molecules were presumed to be strongly bound to the defect sites. Note, however, that the data in Figure 5.23 show that the difference in sticking probability persists to high coverage. Hence, it appears that the hydrogen molecules can first adsorb onto defects on the Ni(111) surface, and then diffuse out onto the Ni(111) plane. It happens that one can partially block the adsorption process on the defects by adding a small amount of oxygen or sulfur to the surface. Interestingly, larger amounts of oxygen enhance the sticking probability again. Hence, the role of defects and impurities can be very complex.

When this book was being written no one knew how extensive a role defects play in real adsorption processes. Rendlic and Winkler's [1989] result implies that defect moderated adsorption may be fairly common. However, more data are needed. At present there are no

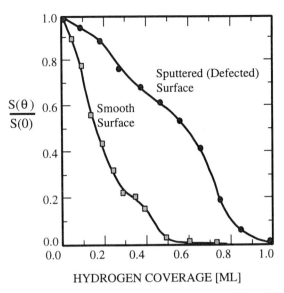

Figure 5.23 The sticking probability of H_2 on a smooth and sputtered (i.e., highly defected) Ni(111) surface. $S(0) = 0.02$ on the smooth surface and 0.07 on the defected surface. (Data of Rendulic and Winkler [1989].)

quantitative models of the effects of defects on the rate of adsorption. Serri, Cardillo, and Becker [1982] have done some work on the effects of defects on desorption. However, nothing on defect-moderated adsorption has yet appeared. Still, the available data suggest that defects may be more important to adsorption/desorption processes than was generally recognized in the surface science literature up to 1991.

5.12 MACROSCOPIC REVERSIBILITY AND TIME-REVERSAL SYMMETRY

Before closing the chapter, let us mention that there is a connection between the kinetics of adsorption and the thermodynamics of adsorption, which was first pointed out by Langmuir [1916]. At equilibrium the rate of adsorption, r_a, must equal the rate of desorption, r_d. In Section 5.9.2 we showed that one can derive the Langmuir adsorption isotherm, by deriving expressions for the rate of adsorption and desorption and setting the two expressions equal. Many investigators have extended the idea that adsorption/desorption measurements are related to equilibrium properties. One idea that appears in the literature is a principle called **detailed balancing.** Consider a surface that is held in a box of molecules. A certain fraction of the molecules will stick when they hit the surface. In order to maintain equilibrium, therefore, the same number of molecules must desorb. Furthermore, if we examine the sub-population of molecules that stick, they will be depleted from the gas phase, unless the properties of the molecules that desorb are the same as the properties of the molecules that stick. The principle of detailed balancing says that at equilibrium the rate that molecules in state m are lost from the gas phase via adsorption must equal the rate that molecules in state m are returned to the gas phase via desorption. In other words, at equilibrium the rate of adsorption of molecules in state m must equal to the rate of desorption of molecules in state m. When detailed balancing holds, equilibrium is maintained throughout the system.

In the literature, there has been considerable discussion whether detailed balancing works when the system is far from equilibrium. For example, if we measure the rate that molecules in state m are adsorbed at some conditions and the rate that molecules desorb at some other conditions, will the rate of adsorption and desorption follow the same rate law? At first one might initially say that the answer is obviously no. In Chapter 3 we found that the adsorbate undergoes a series of phase transitions as we change conditions. Hence, there is no obvious reason why adsorption and desorption should be the same when the adsorption process is done under much different conditions from the desorption process.

The argument that is made, however, is that adsorption and desorption are, in fact, closely related processes. Consider a molecule in state m that collides with a surface. In Section 5.7 we noted that we can calculate whether the molecule sticks or not by calculating the trajectory of the molecule when it is in the region of the surface. Sometimes the trajectory leads to sticking, while other times the molecule scatters. Now consider a molecule that had adsorbed on the surface. That molecule must have gotten into the adsorbed state by following a trajectory that led to sticking. Notice that if we could somehow run the trajectory in reverse, the adsorbed molecule would desorb. It is not obvious how to run the trajectory in reverse. However, note that if all of the forces on the molecule are the same during adsorption and desorption, then the results in Section 5.7 show that all of the equations for the trajectories of the molecules will be the same during adsorption and desorption. Under such a circumstance, the trajectory for the adsorbing molecule will be related to the trajectory of a desorbing molecule via time reversal, i.e., by letting $t \Rightarrow -t$ in all of the equations. If time reversal does not change anything, i.e., if the system obeys **time reversal symmetry**, then detailed balancing will hold even far from equilibrium. Brenig [1987] provides a more detailed discussion of this point.

In fact, however, only a small number of systems have been definitely shown to obey detailed balancing. Palmer et al. [1970] and Cardillo, Balooch, and Stickney [1975] examined the energy and angular dependence of the initial (i.e., zero coverage) sticking probability of hydrogen on Ni(111) and the energy and angular dependence of the desorption flux at the same temperature, and found that the two distributions were the same, implying that detailed balancing holds. Steinrück, Rondulie, and Winkier [1985] extended the measurements to moderate coverages and still found reasonable agreement. However, all of these measurements were done at temperatures close to the desorption temperature of the gas. Hence, they are really measurements that are being done at conditions close to equilibrium. Thus from the available data it is unclear whether detailed balancing works at conditions far from equilibrium.

For example, a common experiment is to adsorb a gas at low temperature and desorb it again at high temperature. If the state of the adsorbate and the surface were exactly the same during the adsorption and desorption, then all of the forces on the molecules should be the same during adsorption and desorption. As a result, detailed balancing should hold provided some criteria described by Brenig et al. [1987] also apply. However, note that in Chapter 3 we found that the state of the adsorbate is rarely the same if we adsorb at low temperatures as when we desorb at high temperatures. For example, in the discussion of Figure 3.13 we noted that it is easy to freeze molecules into a weakly bound state during adsorption. The molecules are converted to a more strongly bound state upon heating. Later in Chapter 3, we found that the arrangement of the adsorbate on the surface changes with both coverage and temperature. If the state of the adsorbate is in any way different during adsorption and desorption, the forces on the molecules will be different during adsorption and desorption. Therefore, detailed balancing will not hold. Unfortunately, when one does a nonequilibrium experiment, one often finds that something changes between when the molecules adsorb and when they desorb. Hence, it is often unclear whether one can apply detailed balancing in a nonequilibrium situation. Still, detailed balancing is used by some investigators even under nonequilibrium conditions. Hence, some care must be taken in interpreting discussions in the literature that are on detailed balancing.

5.13 PERSPECTIVE

At this point, let us summarize all of the results in this chapter, and try to put them in perspective. Trapping and sticking are complex phenomena. A variety of different kinetic processes, including energy transfer to the lattice and surface transport all play a role in the adsorption rate. We have models for most of the individual steps in the process. However, in a real example it is often difficult to know which model to apply and which effects are going to be important.

At present, what we understand the least is the role of surface transport on sticking rates. So far in this chapter, we have only considered surface transport indirectly. For example, Langmuir's model assumes that if an incoming molecule happens to hit a bare site, the molecule will stick; if not, it will desorb. In effect, this is assuming that the surface transport of the incoming molecules is slow on the time scale of a trapping event. If surface transport were fast, the adsorbed layer could rearrange to create a site for the incoming molecule even if the incoming molecule hit an occupied site. Hence, Langmuir's model would not apply. Still, Langmuir's model does assume that surface transport is rapid on the time scale of the experiment. Section 5.10 showed that there are significant deviations from Langmuir's model when surface transport is so slow that the surface layer does not equilibrate over the time scale of the experiment.

In contrast, the mobile precursor model assumes that surface transport is so rapid that even if the incoming molecule hits a filled site, the incoming molecule can still find a site to adsorb. In Section 5.9.4 we discussed the case where the incoming molecule goes into a mobile precursor state before it adsorbs, and we implicitly assumed that the weakly adsorbed molecule moved across the surface until it found a bare site. Note, however, that we can also derive Equations 5.54–5.64 by assuming that extrinsic precursors are converted to intrinsic precursors whenever a strongly adsorbed molecule moves out of the way of the weakly held molecule. Hence, the mobile precursor model is essentially the rapid transport limit of the adsorption rate.

The immobile adsorbate model is the opposite case. In the immobile adsorption limit, we say that the adsorbate does not move around at all after it hits the surface. In that case, an incoming molecule is forced to stick where it lands and sometimes it cannot adsorb at all.

Unfortunately, the actual situation is more complex than would be inferred from any of these simplified models. When an incident molecule first collides with a surface, the incident molecule deposits a lot of energy into the surface, which results in local heating, and possibly some local rearrangements. The hot molecules often can diffuse around and find sites. However, the diffusivity is limited, so the adsorption rate can be changed in a substantial way due to surface transport.

At higher coverages, there are also some more subtle effects. In Chapter 3 we noted that the adsorbed layer often undergoes a series of phase transitions during the adsorption process. When a phase transition occurs, the adsorbate on the surface needs to move around to create space for additional adsorbate. The arrangements are often complex, which means that they take time to happen. Simulations have shown that under such circumstances, the rate of adsorption is limited by the rate at which the adsorbate rearranges to create sites. Hence, in principle, surface transport rates should be very important to adsorption.

Experimentally, one often observes characteristics which suggest that surface transport is important. For example, one normally has to go to fairly large exposures to saturate a surface; much larger than one would expect from Langmuir's model. Such behavior is characteristic of there being a transport limitation to adsorption. Unfortunately, so far, the role of surface transport in adsorption rates has not been quantified. Hence, while there are reasons to believe that surface transport is important to adsorption, many details are missing. As a result, there is still work to be done before we will understand sticking.

Our understanding of trapping is more complete than our understanding of sticking. There are several trapping models in Section 5.3. The Baule model explains many of the qualitative features of trapping data. The generalized Langevin model can be used to fit trapping data in a quantitative way. The Langevin model contains up to seven parameters, which means that one can usually fit data fairly well. The one limitation of the Langevin model is that with up to seven parameters, one can fit data even in situations where the model is inappropriate. For example, the Langevin model assumes that energy transfer occurs via vibrational excitation of the solid. In metals, thermal energy is normally carried by electronic excitations. As a result, one has to wonder whether the Langevin model is correct for trapping of gases on metal surfaces, even though the Langevin model has been used extensively for that purpose and fits the data. Of course, at present the attempts to do better have not gotten very far; in a sense, the main effect of electronic energy transport will be to modify the dissipative force in Equation 4.37, and people now fit the dissipative force to data anyway.

My own view is that at present we have a good qualitative view of the kinetics of trapping and sticking. There are models with lots of parameters that we can use to fit data. It is not clear whether the models are correct, and the role of surface transport needs to be more carefully considered. It also remains to be determined how all of the models fit together to provide a comprehensive picture of adsorption. At this point, however, we are somewhat limited by the data. People measure how much gas sticks, and the final state of the system after the surface layer has had time to equilibrate. However, reliable data on the dynamics of the process are just starting to appear. I do not think that we are going to develop better models for adsorption until we get better data, and I do not know how long that is going to take.

5.14 SOLVED EXAMPLES

Example 5.A   Using Adsorption/Desorption Measurements to Estimate Adsorption Isotherms

In Sections 5.9.2 and 5.12 we noted that one can use adsorption/desorption data to estimate adsorption isotherms. The objective of this example is to illustrate the method.

McCabe and Schmidt *Processings 7th International Vacuum Congress*, Vienna, 1201 [1977] examined the kinetics of hydrogen adsorption and desorption on a number of faces of platinum. At low temperature, the data were complex. However, near room temperature, the data were simple. For example, on Pt(111) McCabe and Schmidt found that at temperatures above 250 K, the sticking probability of hydrogen followed

$$S(\theta) = 0.1 \ (1 - \theta)^2 \tag{5.119}$$

while the rate of desorption followed

$$r_d = 1 \times 10^{11}/\text{sec} \ \theta^2 \exp\left(-\frac{17 \ \text{kcal/mole} - 3.5 \ \text{kcal/mole} \ \theta}{kT}\right) N_s \tag{5.120}$$

where r_d is the rate of desorption in molecules/cm² sec. Use these data to estimate an adsorption isotherm at 350 K assuming detailed balancing applies.

SOLUTION

Adsorption isotherms are calculated by setting $r_a = r_d$ and solving for θ as a function of pressure. From Equation 5.41

$$r_a = EXR(P)\, S(\theta) \qquad (5.121)$$

where $EXR(P)$ is the collision rate of molecules with the surface in molecules/cm^2 sec as a function of pressure, P. Note EXR is equal to the exposure per unit time. Setting r_a from Equation 5.121 equal to r_d from Equation 5.120 and solving for θ yields

$$\theta = \frac{kP^{1/2}}{1 + kP^{1/2}} \qquad (5.122)$$

where

$$kP^{1/2} = \left\{ \frac{EXR(P)}{10^{12}/\text{sec } N_s} \exp\left(\frac{17\ \text{kcal/mole} - 3.5\ \text{kcal/mole } \theta}{kT} \right) \right\}^{1/2} \qquad (5.123)$$

According to Equation 5.44, $EXR(P)$ can be calculated from

$$EXR(P) = \frac{\dfrac{2.97 \times 10^{16}\ \text{molecules}}{\text{cm}^2\ \text{K}^{1/2}\ \text{AMU}^{1/2}\ \text{Langmuir}} \left(\dfrac{\text{Langmuir}}{10^{-6}\ \text{torr sec}} \right)}{\sqrt{2}\ \text{AMU}\ 350\ \text{K}}\, P \qquad (5.124)$$

One can solve Equations 5.122 and 5.123 by successive substitution to calculate the adsorption isotherm.

We have written both a C and a FORTRAN program to do the calculations. The FORTRAN program comes first and the C program follows it. In these programs, p is pressure measured in torr. Theta is coverage, EX is exposure, EX_overNsK = Exposure/10^{12}/sec/N_s. The program picks a value of pressure, then uses the subroutine to calculate the coverage and exposure. The exposure is calculated by substituting into Equation 5.124. The coverage is calculated by guessing θ, then iteratively solving Equation 5.152 and 5.153. The following output was obtained:

Pressure (torr)	EXR (molecules/cm^2/sec)	Theta
1.00e−008,	1.12e+013	0.0154
5.00e−008,	5.61e+013	0.0324
1.00e−007,	1.12e+014	0.0440
5.00e−007,	5.61e+014	0.0850
1.00e−006,	1.12e+015	0.1099
5.00e−006,	5.61e+015	0.1858
1.00e−005,	1.12e+016	0.2260
5.00e−005,	5.61e+016	0.3328
1.00e−004,	1.12e+017	0.3834
5.00e−004,	5.61e+017	0.5050
1.00e−003,	1.12e+018	0.5588
5.00e−003,	5.61e+018	0.6779
1.00e−002,	1.12e+019	0.7257
5.00e−002,	5.61e+019	0.8228
1.00e−001,	1.12e+020	0.8575
5.00e−001,	5.61e+020	0.9201

FORTRAN Program for Example 5.A

```fortran
        program E4A
        real p,theta
        integer i
        p=1.e-9
        theta=0.
        write(6,100)
100     format(/////'        pressure    EX    theta')
        do 10 i=1,8
        p=p*10
        call cover(p,theta)
10      continue
        end

        subroutine cover(p,theta)
        real p,EX,EX_overNsK,KP,theta
        integer i
        EX=2.97e16*1.e6/sqrt(2*350.)
        EX_overNsK=Ex/1.5e27
        do 20 i=1,10
        KP = sqrt(EX_overNsK*
    +   exp((17-3.5*theta)/0.7))
        theta=KP/(1+KP)
20      continue
        return
        end
```

C Program for Example 5.A

```c
#include <conio.h>
#include <stdio.h>
#include <stdlib.h>
#include <math.h>

main(void);
cover(double);
main()
{ float p;int i;
  p=1.e-9;
  printf("\n\n\n\n%s"
      "        pressure    EXR        theta");
  for(i=0;i<8 ;i++)
  { p = p*10;cover(p);cover(5.*p);
}
printf("\n\n\n\n");
return 0;
}

float theta=0;
cover(p)
float p;
{   float EX,EX_over_NsK,KP;int i;
    EX=2.97e16*1.e6*p/sqrt(2*350.);
    /*calc exposure from equ 5.124*/
    printf("\n %6.2e, %7.2e ",p,EX);
    EX_over_NsK=EX/1.e12/1.5e15;
    /*note Ns = 1.5e15*/
    for(i=0;i<10;i++)
    {   KP=sqrt(EX_over_NsK*
            exp( (17-3.5*theta)/.7));
            theta=KP/(1+KP);
    }
    printf("%6.4f",theta);
return 0;
}
```

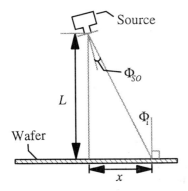

Figure 5.24 A diagram of the geometry used in physical vapor deposition.

Example 5.B Analysis of Physical Vapor Deposition

The equations in this chapter are quite useful in the semiconductor industry. The objective of this example is to give an example illustrating the utility of the methods.

We consider a process called *physical vapor deposition*. Physical vapor deposition is used to deposit metals onto integrated cricuits. A metal vapor is produced in a source crucible suspended in a machine called an *evaporator*. The metal vapor moves from the source crucible and is condensed onto a cold wafer that is positioned at some distance L from the source as indicated in Figure 5.24. The system is kept under high vacuum so that the metal atoms move in a straight line from the source to the wafer. The purpose of this problem is to develop some design rules for evaporators.

(a) Consider a point source of metal as shown in Figure 5.25. Assume that the source produces an equal flux of metal in all directions. Derive an equation for the flux of metal at some distance l from the source

SOLUTION

Let us define *Evap* as being the evaporation rate of metal out of the source in grams/hr. Now consider a mass balance on the control volume indicated by the dashed line in Figure 5.25, where the half-circle is used to denote a hemisphere.

If we assume that metal enters the control volume only from the source, and leaves only along the curved region of the control volume, then the mass flowrate in and out of the control volume would be equal to

$$\text{Mass Flowrate in} = Evap \tag{5.125}$$

$$\text{Mass Flowrate out} = (\text{area}) (\text{flux}) = 2\pi l^2 |\vec{v}_i \rho_g| \tag{5.126}$$

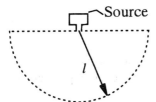

Figure 5.25 A diagram of the control volume for Example 5.B.

where $\vec{v}_i$ is the average velocity of the metal atoms in cm/sec, ρ_g is the density of the metal vapor in gm/cm³, and l is the distance from the source as defined in Figure 5.25. Setting Mass Flowrate in = Mass Flowrate out and rearranging yields

$$|\vec{v}_i\rho_g| = \frac{Evap}{2\pi l^2} \tag{5.127}$$

(b) Assume the condensation rate of metal onto the wafer (i.e., the rate at which metal sticks) follows Equation 5.42, and that the sticking probability of the metal is constant. Develop an expression for the thickness of the metal deposit as a function of the evaporation rate, the distance from the source to the wafer, L, and the position along the wafer x.

SOLUTION

Combining Equations 5.41 and 5.42 shows that r_a, the rate at which metal builds up on the surface in grams/cm² hr, is given by

$$r_a = |v_i\rho_g|S \cos(\phi_i) \tag{5.128}$$

where S is the sticking probability and ϕ_i is the incident angle.

$$\cos(\phi_i) = \frac{L}{\sqrt{L^2 + x^2}} \tag{5.129}$$

$$l = \sqrt{L^2 + x^2} \tag{5.130}$$

Combining Equations 5.127 through 5.130 yields

$$r_a = \frac{Evap\,(L)\,(S)}{(L^2 + x^2)^{3/2}} \tag{5.131}$$

The film growth rate, G_r, in cm/hr is given by

$$G_r = \frac{r_a}{\rho_S} \tag{5.132}$$

where ρ_S is the density of the solid in g/cm³. Substituting r_a from Equation 5.131 into Equation 5.132 yields

$$G_r = \frac{(Evap)\,(L)S}{\rho_S(x^2 + L^2)^{3/2}} \tag{5.133}$$

(c) Assume the wafer is 20 cm in diameter. How far does the wafer need to be from the source for the deposition rate at the edge of the wafer to be within 1% of the deposition rate at the center for the wafer?

SOLUTION

From Equation 5.133, the ratio of the growth at the center to the growth rate at the edge is given by

$$\frac{G_{edge}}{G_{center}} = \frac{L^3}{(L^2 + x^2)^{3/2}} = 0.99 \tag{5.134}$$

where x, the radius of the wafer, is 10 cm. Solving for L yields

$$L = \frac{x}{\left[\left(\frac{1}{0.99}\right)^{2/3} - 1\right]^{1/2}} = 122 \text{ cm} \tag{5.135}$$

Example 5.C   A Modified Soft Cube Calculation

As noted earlier, soft cube calculations were the earliest of the MD-type calculations. While cube calculations are no longer used to describe atom/surface collisions, the cube calculations have many features in common with more sophisticated calculations. Therefore we decided to include a simple soft cube calculation so that the reader could get a simple picture of how one uses the MD method to calculate a trapping or sticking probability.

Consider the simple system shown in Figure 5.26, where a spherical incident molecule impinges onto a flat group of surface atoms and the surface atoms are held to the bulk with a spring. The spring is assumed to have a vibrational frequency ω. In a standard cube calculation, one sets ω to zero before the collision. However, we will not do so.

The incident atom experiences two forces when it interacts with the surface, a force F_A due to a direct interaction with the exposed surface atoms and a force F_d due to a delocalized interaction with the electrons in the surface. For the purposes of this example we assume that

$$F_A = w \, range \, \exp(-range(z_{atom} - z_{surface} - collis\_diam)) \tag{5.136}$$

$$F_A = w_1 \, range_1[\exp(-2range_1(z_{atom} - z_{surface} - collis\_diam))$$
$$- \exp(-range_1(z_{atom} - z_{surface} - collis\_diam))] \tag{5.137}$$

where z_{atom} is the position of the incident atom, $z_{surface}$ is the position of the surface atom, $w \cdot range$ and $w_1 \cdot range$ are the strengths of the attractive wells, $range$ and $range_1$ are the range of the interactions, and $collis\_diam$ is the atomic distance at equilibrium. Our algorithm picks 1000 random initial conditions of the surface atoms and then integrates the equations

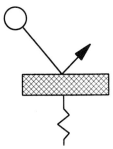

Figure 5.26 Collision of an incoming molecule with a soft cube.

of motion to see if atoms stick. The program uses the algorithm suggested by Tulley, Gilmer, and Shugard [1979]. The program is coded to make it as easy as possible to read, at the expense of program efficiency. A more efficient version is available from Professor Masel.

The one difficult part of the algorithm is deciding which molecules have been trapped. Figure 5.27 shows two typical trajectories computed with the model. Both of these trajectories were computed under conditions where most of the molecules should be trapped. However, most of the trajectories looked like the one on the left of Figure 5.27, where the incident molecule bounced off the surface a few times, but the molecule eventually desorbed. One does occasionally see a trajectory like that on the right of Figure 5.27, where the molecule stays on the surface for a long time. However, such trajectories are unusual.

The reason that one rarely gets a trajectory where the atoms stays on the surface for a long time with this model is an artifact of the model. In the trajectory on the right in Figure 5.27, the incident atom had initially transferred enough energy to the surface to be trapped. However, the model did not allow the surface to remove that energy from the point of impact. The energy was transferred back to the incident atom, and so the incident atom desorbed.

In more sophisticated calculations one does observe some trajectories where energy is transferred from the incident atom to the surface and back again. However, usually the incident atom transfers energy to a single atom at the point of impact and that energy spreads out in all directions. Simultaneously the incident atom moves down the surface, i.e., away from the point of impact. The model in this example does not allow the incident atom to move away from the point of impact. As a result, the energy stays localized near the incident atom where it can be easily transferred back to the atom. Hence, this model produces too many trajectories where the incident energy is transferred back to the incident atom and the incident atom desorbs.

In the algorithm that follows we avoid this difficulty by assuming that a molecule is trapped whenever it bounces about three times. However, that is at best an imprecise measure of when a molecule is trapped. Hence, there can be some errors. We include a C routine and a FORTRAN routine to do the calculations. The programs are written for clarity and not efficency. A more efficient version of the C program is available from Professor Masel.

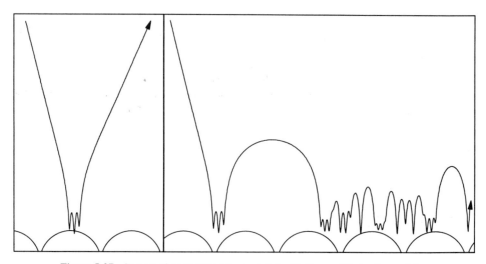

Figure 5.27   Some typical trajectories computed with the modified cube model.

C Program to Calculate Trapping Rates Via A Modified Via Cube Method
© 1993 R. I. Masel

```c
#include <stdio.h>
#include <conio.h>
#include <math.h>
#include <graph.h>
#include <signal.h>
#include <process.h>
#define NATOMS 2
float posit[NATOMS][3],veloc[NATOMS][3], accel[NATOMS][3];
float mass[NATOMS] ={(float)40,(float)195};
/* posit, veloc, and accel are the x,y, and z components of the position,
   velocity, and acceleration of the atoms, mass is the mass of each atom.
   The incident atom is atom 0, the other surface atoms are atoms 1 .. NATOMS-1
************************************************/
float old_accel[NATOMS][3], aver_velocity[NATOMS][3];
/* old_accel and aver_velocity are variables used in the numerical integration
scheme
**********************************/
float range=3; float range1=1; float collis_diam = 3.; float w_times_range = .4;
float w1_times_range1 = .13; float omega = 0.05;
/* range, range1, collision_diam, w_times_range, w1_times range, and omega are
   parameters in the interaction potential */
int trags,ntrapped;
/* trags = the total number of trajectories
   ntrapped = the number of trajectories which lead to trapping
********************/
float init_velocity=(float)0.03; float temp_surface=(float)100.;
float displace_init;float lattic_dimen = (float)3.;
int random[256]; int not_filled=1;
ranf()
/* Random number generator which mixes up the order of the random numbers
   to overcome a bug in the rand routine*/
{ int. i,ran;
  if(not_filled)
  { not_filled=0;
    for(i=0;i<256;i++) random[i]=rand();
  }
  i=rand()/128;i=random[i]/128; ran=random[i]; random[i]=rand();
  return ran;
}

main()
{ float timestep, tstepo6,zmin,zlast; char text[50];
  int steps,crossings; char text[50];
  srand(17)→nd(); read_parameters(); setup_graphics(); ntrapped=0;
  for(trags=0;trags<900;trags++)
  { pick_random_initial_conditions(); crossings=0;zmin=21; veloc[0][1]=(float)0;
    veloc[0][0]=int_velocity*.707;veloc[0][2]=-.707*init_velocity; start_plot();
    for(steps=0;steps<10000 && posit[0][2]<20.1 && crossings<4;steps++)
    { timestep= 0.3/sqrt(veloc[0][2]*veloc[0][2]+0.1);
      tstepo6=timestep/6.; zlast=posit[0][2];
      integrate_equations_of_motion( timestep,tstepo6);
      if(zmin>posit[0][2])zmin=posit[0][2];
      if(posit[0][2]>(zmin+0.5)&& zlast<(zmin+0.5))crossings++;
       if(posit[0][2]<(zmin+0.5)&&zlast>(zmin+0.5))crossings++;
      plot_trajectory();
    }
```

```
    if(posit[0][2]<16)ntrapped++;
    sprint(text,"\nOf %d total trajectories, %d atoms %s",
        trags+1,ntrapped,"were trapped");
    _settextposition( (short)24,(short)5);_outtext(text); while(kbhit())getch();
    }
    trags--; reset_scrn();
    return 123;
} /*end of main program*/

integrate_equations_of_motion(timestep,tstepo6)
/* integrates equations of motion for one time step
*/
float timestep, tstepo6;
{ int, i,j;
  for(i=0; i<NATOMS; i++)for(j=0; j<3; j++)
  {aver_velocity[i][j]=veloc[i][j]+tstepo6* (4*accel[i][j]-old_accel[i][j]);
   posit[i][j]= posit[i][j]+timestep*aver_velocity[i][j];
   old_accel[i][j]=accel[i][j];
  }
  set_accel_to_force_over_mass();
  for(i=0; i<NATOMS; i++)for(j=0; j<3; j++)
  veloc[i][j]= aver_velocity[i][j]+tstepo6*(2*accel[i][j]+old_accel[i][j]);
  return 0;
  /* note aver_velocity is the average velocity over the time step. veloc starts
     as the instantaneous velocity at the beginning of the time step. The algorithm
     changes veloc to the instantaneous velocity at the end of the time step
  */
} /* end of routine to integrate equations of motion
****************************************************/
set_accel_to_force_over_mass()
{ float force[NATOMS][3],forc,calc;int i, j ;
              force[0][0]=force[0][1]= (float)0;
              if(posit[0][2]<20)
          forc=w_times_range*exp(-range*(posit[0][2]-collis_diam-posit[1][2]));
              else forc=0;
              for(i=1;i<NATOMS;i++)for(j=0;j<3;j++)
               force[i][j]=-mass[i]*omega*omega*posit[i][j];
              force[1][2]= force[1][2]-force;
              calc=exp(-range1*(posit[0][2]-collis_diam));
          force[0][2]= forc+w1_times_range1*calc*(calc-1);
    for(i=0;i<NATOMS;i++)for(j=0;j<3;j++)
              accel[i][j]=force[i][j]/mass[i];
              return 0;
}/*end of acceleration calc*/
pick_random_initial_conditions()
{ int i,j; float phase;
  posit[0][0]=(float)0;posit[0][1]=(float)0; posit[0][2]=(float)100;
  velocity[0][2]=(float)0;
  for(i=1;i<NATOMS;i++)for(j=0;j<3;j++)
  { phase=6.2831853*(float)ranf()/32768.;posit[i][j]=displace_init*cos(phase);
    veloc[i][j]=omega*displace_init*sin(phase);accel[i][j]=-omega*omega*posit[i][j];
  }
  for(i=0;i<300;i++)integrate_equations_of_motion( 0.3, .05);
  posit[0][0]=((float)(trags%15)+(float)ranf()/32768.)*lattic_dimen/6.-20;
  posit[0][1]=((float)(trags%225)+ (float)ranf()/32768.)*lattic_dimen/6.;
  posit[0][2]=20.; return 0;
}
```

Fortran Program To Calculate Trapping Rates Via A Modified Cube Method
© 1992 R. I. Masel

```fortran
block data
real posit(2,3),veloc(2,3),init_velocity, accel(2,3)
real old_accel(2,3), aver_velocity(2,3)
real range , range1, collis_diam, w_times_range
real w1_times_range1, omega, mass(2)
real displace_init, lattic_dimen,temp_surface, randm(256)
integer not_filled
common/b1/posit,veloc,init_velocity
common/b2/old_accel,aver_velocity,accel
common/b3/range,range1,collis_diam,w_times_range
common/b4/w1_times_range1,temp_surface, mass
common/b5/displace_init,lattic_dimen,omega
common/r1/randm,not_filled
data not_filled/1/
data range,range1/3.,1./
data collis_diam,w_times_range/3.,0.4/
data init_velocity/0.3/
data temp_surface,lattic_dimen/100.,3./
data w1_times_range1,omega/0.13,0.1/
end


program cube
real posit(2,3),veloc(2,3), init_velocity
common/b1/posit,veloc,init_velocity
real timestep, tstepo6,zmin,zlast,trapp
integer steps,crossings,ntrapped,trags
real *4 ran
call seed(17)
call random(ran)
ntrapped=0
do 20 trags = 1,900
call pick_random_initial_conditions(trags)
crossings=0
zmin=21
veloc(1,2)=0
 veloc(1,1)=init_velocity*.707
  veloc(1,3)=-.707*init_velocity
steps=0
do while (steps.lt.10000.and.posit(1,3).GT.20.1.and. crossings.lt.4)
        steps=steps+1
    timestep=1
        tstepo6=timestep/6.
        zlast=posit(1,3)
        call integrate_equations_of_motion(timestep, tstepo6)
        if(zmin.gt.posit(1,3))zmin=posit(1,3)
 if(posit(1,3).gt.(zmin+0.5).and. zlast.lt.(zmin+0.5))crossings = crossings+1
    if(posit(1,3).lt.(zmin+0.5).and. zlast.lt.(zmin+0.5))crossings=crossings+1
  enddo
  if(posit(1,3).lt.16.)ntrapped =ntrapped+1
  write(6,100)trags,ntrapped
100 format(' Of ',i4,' total trajectories,',i4, 'atoms were trapped')
 20 continue
  trapp=float(ntrapped)/trags
  write(6,101)trapp
101 format(' The trapping probability is',f7.4)
  stop
  end

  subroutine integrate_equations_of_motion(timestep,tstepo6)
  real timestep, tstepo6, posit(2,3),veloc(2,3), init_velocity,accel(2,3)
```

Fortran Program To Calculate Trapping Rates Via A Modified Cube Method (Continued)
© 1992 R. I. Masel

```
      real old_accel(2,3), aver_velocity(2,3)
      common/b1/posit,veloc,init_velocity
      common/b2/old_accel,aver_velocity,accel
      integer i,j
      do 10 i=1,2
      do 10 j=1,3
      aver_velocity(i,j)=veloc(i,j)+tstepo6*(4*accel(i,j)-old_accel(i,j))
      posit(i,j)= posit(i,j)+timestep*aver_velocity(i,j)
      old_accel(i,j)=accel(i,j)
 10   continue
      call set_accel_to_force_over_mass
      do 20 i=1,2
      do 20 j=1,3
 20   veloc(i,j)= aver_velocity(i,j)+tstepo6*(2*accel(i,j)+old_accel(i,j))
      return
      end
       real function ranf()
       real randm(256)
       real*4 ran
       integer i,not_filled
       common/r1/randm,not_filled
       if( not_filled.ne.0)then
       not_filled=0
       do 10 i=1,256
       call random(ran)
 10    randm(i)=ran
       endif
       call random(ran)
       i=ran*256
       i=randm(i)*256
       ranf=randm(i)
       call random(ran)
       randm(i)=ran
       return
       end

      subroutine set_accel_to_force_over _mass
      real force(2,3),forc,calc, posit(2,3),veloc(2,3), init_velocity, acccel(2,3)
      real old_accel(2,3), aver_velocity(2,3),range, range1, collis_diam,omega,
      real w_times_range, w1_times_range1, mass(2),displace_init, lattic_dimen
      real temp_surface
      integer, i,j
      common/b1/posit,veloc,init_velocity
      common/b2/old_accel,aver_velocity,accel
      common/b3/range,range1,collis_diam,w_times_range
      common/b4/w1_times_range1,temp_surface,mass
      common/b5/displace_init,lattic_dimen,omega
            force(1,1)=0.
            force(1,2)=0.
            if(posit(1,3).lt.20) then
      forc=w_times_range_*exp(-range*(posit(1,3) -collis_diam-posit(2,3)))
            else
      forc=0
      endif
      do 10 i=2,2
         do 10 j=1,3
 10      force(i,j)=-mass(i)*omega*omega*posit(i,j)
      force(2,3)= force(2,3)-forc
      calc=exp(-range1*(posit(1,3)-collis_diam))
      force(1,3)= forc+w1_times_range1*calc*(calc-1)
```

```
      do 20 i=1,2
          do 20 j=1,3
 20   accel(i,j)=force(i,j)/mass(i)
      return
            end
    subroutine pick_random_initial_conditions(trags)
    integer i,j,trags
    real phase
    real posit(2,3),veloc(2,3),init_velocity,accel(2,3), old_accel(2,3)
    real aver_velocity(2,3), displace_init, lattic_dimen,omega
    common/b1/posit,veloc,init_velocity
    common/b2/old_accel,aver_velocity,accel
    common/b5/displace_init,lattic_dimen,omega
          posit(1,1)=0.
          posit(1,2)=0.
          posit(1,2)=100.
      do 10 i=2,2
          do 10 j=1,3
          phase=6.2831853*ranf()
      posit(i,j)=displace_init*cos(phase)
      veloc(i,j)=omega*displace_init*sin(phase)
      accel(i,j)=-omega*omega*posit(i,j)
 10 continue
          do 20 i=1,300
 20 call integrate_equations_of_motion(3.05)
          posit(1,1)=mod(trags,15)+ranf()*lattic_dimen/6.-20.
          posit(1,2)=mod(trags,225)+ranf()*lattic_dimen/6.
          posit(1,3)=20.
          return
          end
```

Example 5.D   An MD/Langevin Calculation

MD/Langevin calculations are very similar to the cube calculations in Example 5.C except that one keeps track of several layers of atoms, and one imposes stochastic boundary conditions to the atoms in the bulk. The algorithm in Example 5.C only needs two changes. We need to keep track of more atoms, and the force calculation is more complex. A full three-dimensional simulation requires a work station. As a result, we have only done a two-dimensional simulation. We use three layers of six surface atoms arranged on a line with periodic boundary conditions. We assume that the surface atoms are connected with a network of springs with periodic boundary conditions as indicated in Figure 5.28. There are three types of springs in the figure: those that link atoms in the X direction; those that link atoms in the Z direction; and those that move at a diagonal. (*Note:* Since we use periodic boundary conditions, we need diagonal springs to dissipate the X energy produced during the collision.

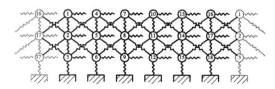

Figure 5.28 The atomic arrangement for Example 5.D.

We assume that the X and Z springs produce a force.)

$$F_x = m_s\omega_x^2 \, (X_j - X_i) \tag{5.138}$$

$$F_z = m_s\omega_z^2 \, (Z_j - Z_i) \tag{5.139}$$

where F_x and F_z are the forces along the X and Z directions; ω_x and ω_z are the frequency of the springs; and X_j, X_i, Z_j, and Z_i are the positions of the atoms at the ends of the spring.

To simplify the analysis we assume

$$\omega_x = \omega_z = \omega_1 \tag{5.140}$$

The diagonal springs are assumed to obey

$$F_z = F_x = 0.707 \times m_s\omega_z^2 \, \sqrt{(X_i - Z_j)^2 + (Z_i - Z_j)}$$

In order to keep the Z forces consistent with those in Example 5.C, we need to satisfy

$$m_s \, \omega_{4c}^2 = m_s \, \omega_1^2 + (2)(0.707)m_s \, \omega_2^2$$

where ω_{4c} is the frequency from Exmaple 5.C.

In the program, we also apply a random force and memory kernel to the Z motion of the third layer of atoms. Following Tulley et al. [1979], we approximate the random force as being random (white) noise, and we expand the memory kernel in a Taylor series,

$$R_D^n = -R_A^n Z_n - R_B^n \frac{\partial Z_n}{\partial t} + \cdots \tag{5.141}$$

and only keep the first two terms. Adleman and Doll [1976] show that one can approximate R_A^n and R_B^n by

$$R_A^n = \frac{m_s\omega_D^2}{3} \tag{5.142}$$

$$R_B^n = \frac{m_s\omega_D \, \pi}{6} \tag{5.143}$$

where ω_D is a constant called the Debye frequency of the solid. Values of ω_D are tabulated in Kittel [1986].

The random force is taken to be Gaussian noise satisfying

$$P(R_R) = \frac{\exp\left(-\dfrac{R_R^2}{2\sigma_R^2}\right)}{\sqrt{2\pi\sigma_R^2}} \tag{5.144}$$

where $P(R_R)$ is the probability that a given component of the random force has a value R_R. Adleman and Doll [1976] show that if

$$\sigma_R = 2kTR_B^n \tag{5.145}$$

the system stays in thermal equilibrium. Most of the resultant computer program is identical to the one in example 5.3. However, two subroutines need to be changed:

set_accel_to_force_over_mass and read_parameters. Corrected subroutines are given at the end of this example.

In the C program that follows, the incident atom is atom 0, while the surface atoms are numbered 1 through 18 as indicated in Figure 5.28. The indices posit[i][j] for j = 1, 2, 3 are x, y, and z components. The position of the ith atom, posit[(i+2)%18+1][j], is the position of the atom to the right of the ith atom, where the % 18 (i.e., modulus 18) is needed to keep track of the periodic boundary conditions. Similarly, the position of the atom to the left of atom i is posit[(i+14)%18+1][j]. The position of the atom below atom i is posit[(i+15)%18+1][j].

In the FORTRAN program, atom 1 is the incident atom, while atoms 2 through 19 are the surface atoms; posit(i,j) is the position of atom i; posit(mod(i+14,18)+2,j) is the position of the atom to the left of i; posit(mod(i+2,18)+2,j) is the position to the right of i, etc.

C Subroutine For The Force Calculations In The Langevin Program

---

```
set_accel_to_force_over_mass(timestep)
float timestep;
{    float force[NATOMS][3],forc,calc,x,z,dist;int i, j;
            force[0][1]=0;
            for(i=1; i,NATOMS; i++)/*x springs*/
            { force[i][1]= -ms_w1_squared*posit[i][1];
              if(i%3)force[i][0]=ms_wd_squared*(-posit[i][0]);
              else force[i][0]= -ms_w1_squared*(posit[i][0]);
            }
            for(i=1; i<NATOMS; i+=3)/*z springs*/
            { force[i][2]=ms_w1_squared*(posit[i+1][2]-posit[i][2]);
              force[i+2][2]=ms_w1_squared*(posit[i+1][2]-posit[i+2][2]);
              force[i+1][2]= -(force[i][2]+force[i+2][2]);
              force[i+2][2]=force[i+2][2]-ms_wd_squared*posit[i+2][2]-
                RB*( veloc[i+2][2]+0.5*timestep*(3*accel[i+2][2]
                -old_accel[i+2][2]))+sigma_R*gaussian_noise(sigma_R);
            }
            for(i=1; i<2*(NATOMS/3)+1; i++)
    /*diagonal springs*/
            { /* diagonal down right spring */
              x=posit[i][0]-posit[(i+3)%18+1][0]+6;
              z=posit[i][2]-posit[(i+3)%18+1][2]+6;
              forc=ms_w2_squared*(sqrt(x*x+z*z)-1.4142136*6);
              force[i][2]=force[i][2]-forc;
              force[i][0]=force[i][0]+forc;
              force[(i+3)%18+1][2]=force[(i+3)%18+1][2]+forc;
              force[(i+3)%18+1][0]=force[(i+3)%18+1][0]-forc;
    /*diagonal down left spring*/
              x=posit[i][0]-posit[(i+15)%18+1][0]+6;
              z=posit[i][2]-posit[(i+15)%18+1][2]+6;
              forc=ms_w2_squared*(sqrt(x*x+z*z)-1.4142136*6);
              force[i][2]=force[i][2]-forc;
              force[i][0]=force[i][0]-forc;
              force[(i+15)%18+1][2]=force[(i+15)%18+1][2]+forc;
              force[(i+15)%18+1][0]=force[(i+15)%18+1][0]+forc;
            }
    calc=exp(-range1*(posit[0][2]-collis_diam));
            force[0][2]=w1_times_range1*calc*(calc-1.);
            force[0][0]=force[0][1]=0;
            for(i=1; i<NATOMS; i+=3)
            { x=posit[0][0]-(i-1)/3*lattic_dimen;
              z=posit[0][2]-posit[i][2];
```

C Subroutine For The Force Calculations In The Langevin Program (Continued)

```
            while(x>lattic_dimen*3)x= x-lattic_dimen*6;
            while(x<-lattic_dimen*3)x= x+lattic_dimen*6;
            dist=sqrt(x*x+z*z);
            if(dist<9.)
            {forc=w_times_range*exp(-range*(dist-collis diam));
            force[i][2]= force[i][2]-forc;
            force[0][2]=force[0][2]+forc;
            }
        }
        for(i=0;i<NATOMS;i++)for(j=0;j,3;j++)
        accel[i][j]=force[i][j]/mass[i];
        return 0;
}
```

Abbreviated Version Of The Fortran Program

```
    subroutine set_accel_to_force_over_mass(timestep)
    external gaussian_noise
    real timestep,mass(19),RB,sigma_R,ms_w2_squared,ms_w1_squared
    real ms_wd_squared, lattic_dimen,force(19,3),forc,calc,x,z,dist
    real posit(19,3),accel(19,3),old_accel(19,3),init_velocity, veloc(19,3)
    integer i,j ,NATOMS
    real aver_velocity(19,3), range, range1, collis_diam, w_times_range
    real w1_times_range, displace_init, temp_surface
    common/b1/posit,veloc,init_velocity
    common/b2/old_accel,aver_velocity,accel
    common/b3/range,range1,collis_diam,w_times_range
    common/b4/w1_times_range1,temp_surface,mass
    common/b5/displace_init,lattic_dimen,omega
    common/b6/ ms_w2_squared,ms_w1_squared,ms_wd_squared
    NATOMS=19
    force(1,2)=0
    do 10 i=2,NATOMS
c X springs
    force(i,2)= -ms_w1_squared*posit(i,2)
    if( mod(i,3).eq.1)then force(i,1)=ms_wd_squared*(-posit(i,1))
    else force(i,1)= -ms_w1_squared*(posit(i,1))
    endif
 10 continue
    do 20 i=2,NATOMS
C Z springs
    force(1,3)=ms_w1_squared*(posit(i+1,3)-posit(i,3))
    force(i+2,3)=ms_w1_squared*(posit(i+1,3)-posit(i+2,3))
    force(i+1,3)= -(force(i,3)+force(i+2,3))
    forc(i+2,3)=force(i+2,3)-ms_wd_sqaured*posit(i+2,3)-RB*(velo(i+2,2)+0.5*
  + timestep*(3*accel(i+2,2)-old_accel(i+2,2)))+sigma_R*gaussian_noise(sigma_R)
 20 continue
    do 30 i=2,2*(NATOMS/3)+1
C diagonal down right spring
    x=posit(i,1)-posit(mod(i+3,18)+2,1)+6
    z=posit(i,3)-posit(mod(i+3,18)+2,3)+6
    forc=ms_w2_squared*(sqrt(x*x+z*z)-8.485416)
```

Abbreviated Version Of The Fortran Program (Continued)

```
   force(i,3)=force(i,3)-forc
   force(i,1)=force(i,1)+forc
   force(mod(i+3,18)+2,3)=force(mod(i+3,18)+2,3)+forc
   force(mod(i+3,18)+2,1)=force(mod(i+3,18)+2,1)-forc
C diagonal down left spring*/
   x=posit(i,1)-posit(mod(i+15,18)+2,1)+6
   z=posit(i,3)-posit(mod(i+15,18)+2,3)+6
   forc=ms_w2_squared*(sqrt(x*x+z*z)-8.485416)
   force(i,3)=force(i,3)-forc
   force(i,1)=force(i,1)-forc
   force(mod(i+5,18)+2,3)=force(mod(i+15,18)+2,3)+forc
   force(mod(i+15,18)+2,1)=force(mod(i+15,18)+2,0)+forc
30 continue
   calc=exp(-range1*(posit(1,3)-collis_diam))
   force(1,3)= w1_times_range1*calc*(calc-1.)
   force(1,1)=0
   force(1,2)=0
   do 40 i=2,NATOMS,3
   x=posit(1,1)-(i-2)/3*lattic_dimen
   z=posit(1,3)-posit(i,3)
   do while( x.gt.lattic_dimen*3 )
   x= x-lattic_dimen*6
   enddo
   do while(x.it.-lattic_dimen*3)
   x= x+lattic_dimen*6
   enddo
   dist=sqrt(x*x+z*z)
   if(dist.lt.9)then
   forc=w_times_range*exp(-range*(dist-collis_diam))
   force(i,3)= force(i,3)-forc
   force(1,3)=force(1,3)+forc
   endif
40 continue
   do 50 i=1,NATOMS
   do 50 j=1,3
50 accel(i,j)=force(i,j)/mass(i)
   return
   end
```

PROBLEMS

5.1 Define the following terms in three sentences or less:

(1) Sticking
(2) Trapping
(3) Scattering
(4) Elastic scattering
(5) Inelastic scattering
(6) Lattice model
(7) Hard sphere model
(8) Energy accommodation coefficient
(9) Head-on collision
(10) Trapping probability
(11) Impact parameter
(12) Sticking probability
(13) Initial sticking probability
(14) Reaction channel
(15) Langmuir's adsorption law
(16) Type A sticking behavior
(17) Type B sticking
(18) Mobile precursor

(19) Precursor moderate adsorption
(20) Immobile adsorption
(21) An exposure of one Langmuir
(22) Trapping-dominated adsorption

(23) Molecular dynamics
(24) Langevin method
(25) Random sequential adsorption

5.2   What are the key assumptions and advantages/disadvantages of

(a) Baule's model?

(b) Lattice model?

(c) Langevin model?

5.3   Assume that you have measured an uptake curve similar to Figure 5.13. Would you decide whether your data fit the Langmuir adsorption model, the precursor model, the immobile adsorption model, or some combination of the precursor and immobile adsorption model? What are the key differences between these models?

5.4   In the beginning of Section 5.8 we derived a number of equations for the impingement of gases on surfaces. The objective of this problem is to put the number of collisions in perspective.

(a) How many times will air molecules collide with your skin in the next second? How many collisions have there been so far in your lifetime?

(b) Assume for the moment that each air molecule hits your skin only once. How long would it take for a metric ton of air to collide with your skin? How many tons of air have collided with your skin so far in your lifetime? Why are you still alive?

(c) Repeat parts (a) and (b) for your eyes. Why can you still see?

(d) How long will it take for every air molecule in your bedroom to collide with your skin? Assume that every air molecule only collides once and there are no mass transfer limitations.

5.5   In Section 5.9.2 we showed that one can derive the Langmuir adsorption isotherm for a nondissociative adsorption process by equating the rate of adsorption and desorption. Consider a molecule that adsorbs dissociatively

$$A_2 + 2S \rightleftharpoons 2A_{ad}$$

Assume that the rate of desorption follows

$$r_d = k_d \, \theta_A^2 \tag{5.146}$$

Combine Equations 5.41, 5.48, 5.51, and 5.146 to derive the Langmuir adsorption isotherm for dissociative adsorption.

5.6   In the beginning of Section 5.4 we mentioned that years ago accommodation coefficients were determined by measuring the rate of heat loss from a hot filament. Consider the heat losses from a tungsten filament in a light bulb.

(a) Calculate the rate of heat loss from the filament due to collisions with gas molecules as a function of the filament temperature. Assume the filament has a surface area of 5 mm^2, the bulb is filled with 10^{-4} torr of argon, and the argon is at 300 K. Use Baule's model to estimate the accommodation coefficient.

(b) Compare your results in (a) to the heat losses due to radiation. How hot does the filament have to be before the radiant heat losses are much larger than the heat losses due to conduction?

(c) Langmuir was an employee of the General Electric Company, and the Langmuir adsorption isotherm was discovered as part of his work on the heat losses from light bulbs. One of Langmuir's more interesting observations was that when he exposed a hot tungsten filament to hydrogen, the hydrogen accommodated with the surface. At high temperatures, the hydrogen desorbed as atoms and not as molecules. How would conversion of hydrogen molecules into atoms affect the rate of heat loss? Calculate the fraction of hydrogen molecules that dissociate by assuming the reaction $H_2 \rightleftharpoons 2\,H$ reaches equilibrium.

5.7 In the beginning of Section 5.3 we noted that accommodation coefficients can be measured using a time-of-flight technique. Janda et al. *Surface Sci.* **93**, 270 [1980] directed a stream of hot nitrogen at a W(110) surface, and determined the velocity distribution of the scattered molecules by measuring the time it took for the molecules to fly 35.0 mm from the surface. The following data were obtained:

$T_g = 3400$ K $T_s = 1360$ K	Intensity	0	0	0	0.01	0.15	0.59	0.99	1.00
	Time of flight (μs)	101	139	156	177	202	224	239	251
	Intensity	0.95	0.78	0.53	0.30	0.19	0.10	0.05	0.01
	Time of flight (μs)	260	283	303	329	344	371	404	501
$T_g = 2330$ K $T_s = 1490$ K	Intensity	0	0.01	0.04	0.17	0.45	0.73	0.88	0.88
	Time of flight (μs)	186	202	219	241	263	281	302	317
	Intensity	0.75	0.61	0.45	0.32	0.21	0.12	0.09	0.08
	Time of flight (μs)	341	360	378	397	422	453	474	501
$T_g = 1020$ K $T_s = 1372$ K	Intensity	0	0.01	0.02	0.11	0.36	0.71	0.93	0.97
	Time of flight (μs)	250	303	325	351	378	400	425	446
	Intensity	0.91	0.80	0.68	0.50	0.35	0.26	0.22	0.14
	Time of flight (μs)	464	475	500	525	555	578	604	628

(a) Fit these data to the Maxwell–Boltzmann velocity distribution described in Appendix A to calculate the temperature of molecules leaving the surface

(b) Integrate the velocity distribution to calculate the average energy of the molecules.

(c) Calculate the thermal accommodation coefficient for the molecules at each set of conditions given here.

5.8 Lee and Masel (unpublished) examined the sticking of silane on a tungsten silicide. The following data were obtained:

Gas temperature (K)	300	400	450	500	550	600	650
Initial sticking probability on a 330 K surface	0.21	0.20	0.18	0.17	0.135	0.03	0.025

(a) What do you conclude from the drop in the sticking probability with increasing gas temperature?

(b) Can you fit these data with the Baule–Weinberg–Merrill model?

5.9   Lee and Masel also measured the sticking probability of silane on tungsten as a function of surface temperature. The following results were obtained:

Surface temperature (K)	330	400	500	600	700	800
Initial sticking probability of a 300 K gas	0.21	0.225	0.278	0.29	0.30	0.305

(a) What do you conclude from the increase in sticking probability with increasing surface temperature?

(b) How can you model these results in light of the findings in Problem 5.8.

(c) What additional data would you want to take to fit the data to the precursor model?

(d) Lee, MS thesis, University of Illinois [1987], measured uptake curves as a function of the gas and surface temperature. Are these data sufficient?

5.10   On page 383 we said that hydrogen adsorption on Pt(110) behaves qualitatively as we would expect for a trapping dominated process at $\Phi_i = 60°$.

(a) How well do the data actually fit the Baule–Weinberg–Merrill model?

(b) At $\Phi_i = 0°$, the data show evidence for direct adsorption. What do you conclude from the difference in behavior at $\Phi_i = 0°$ and $60°$? Can you say anything about the active site for adsorption?

5.11   Figure 5.13 shows an uptake curve for carbon monoxide adsorption on Pt(410). Integrate Equation 5.45 to fit that data to Langmuir's model ($n = 1$), Langmuir's model ($n = 2$), the precursor model, and the immobile adsorption model ($n = 2$). (Numerically integrate the latter two cases.) Can you distinguish between the various models from the data given in the figure? Which model fits best? How would a 5% uncertainty in the data in the figure affect your conclusions?

5.12   The heat of adsorption of hydrogen and CO is similar on many transition metals, but the sticking probability of hydrogen is usually less than that of CO. What factors cause the sticking probability of hydrogen to be less than that of CO?

5.13   In our discussions of Equations 5.17 and 5.26, we assumed that we knew the heat of adsorption of the incident molecules and were trying to calculate trapping probabilities. However, Weinberg and Merrill [1972] pointed out that with a very weakly bound species, it is sometimes easier to measure the trapping probability than the heat of adsorption. One can estimate the heat of adsorption from sticking probability data. As a result, it may be possible to use trapping measurements to measure heats of adsorption. A. G. Stoll, D. L. Smith, and R. P. Merrill *J. Chem. Phys.* **54**, 163 [1971] measured what they believed to be trapping probability Ne, Ar, Kr, and Xe on Pt(111). The following results were obtained:

Gas	Trapping Probability					
	$T_g = 300\ K$ $T_s = 373\ K$	$T_g = 300\ K$ $T_s = 573\ K$	$T_g = 300\ K$ $T_s = 773\ K$	$T_g = 973\ K$ $T_s = 373\ K$	$T_g = 973\ K$ $T_s = 573\ K$	$T_g = 973\ K$ $T_s = 773\ K$
Neon	0.15	—	—	0.11	0.03	0.09
Krypton	0.38	0.35	0.29	0.25	0.19	0.21
Argon	0.48	0.49	0.48	0.44	0.31	0.35
Xenon	0.67	0.63	0.60	0.50	0.44	0.42

(a) How well do these data fit Baule's model? Which experiment trends differ from the Baule model?

(b) Use these data to estimate the heat of adsorption of Ne, Ar, Kr, and Xe on Pt(111).

(c) How much confidence do you place in your numbers in light of Figure 5.5.

5.14 Seebauer, Kong, and Schmidt, *Surface Sci.* **176**, 134 [1986] measured the sticking probability of CO and NO on Pt(111). The following data were obtained:

Coverage (molecules/cm^2 × 10^{-15})	0.03	0.17	0.22	0.28	0.31	0.38	0.46	0.49
CO sticking probability on a 160 or 330 K surface	0.85	0.85	0.78	0.61	0.51	0.29	0.082	0.02

Coverage (molecules/cm^2 × 10^{-15})	0.03	0.06	0.12	0.25	0.28	0.34	0.38	0.40
NO sticking probability on a 160 K surface	0.80	0.80	0.72	0.52	0.45	0.24	0.087	0.02

Coverage (molecules/cm^2 × 10^{-15})	0.03	0.06	0.11	0.15	0.19	0.28	0.33	0.35
NO sticking probability on a 330 K surface	0.80	0.80	0.70	0.63	0.54	0.24	0.063	0.02

(a) Try fitting the data to the precursor model. How well does the precursor model fit?

(b) Notice that the sticking probability of NO varies with the surface temperature. However, the sticking probability of CO is independent of the surface temperature. What is the significance of the results?

5.15 Figure 5.23 shows how the sticking probability of hydrogen varies with coverage for hydrogen adsorption on Ni(111).

(a) Try to fit these data to Langmuir's model, with various values of n. From the data given, how many sites are taken up by each hydrogen atom.

(b) In fact, hydrogen adsorbs dissociatively, taking up two sites. How can you explain the discrepancy?

5.16 Bowker and King, *J. Chem. Soc. Faraday Trans. I* **75**, 2100 [1979] examined the adsorption of nitrogen on W(110). They found that the nitrogen adsorbs into a weakly bound molecular state and measured sticking probabilities as follows:

Coverage (×10^{14} molec/cm^2)	0.13	0.5	1.0	1.5	2.0	2.5	3.0	3.5	4.0
Sticking probability	0.24	0.30	0.35	0.38	0.37	0.32	0.24	0.16	0.09

(a) Try fitting these data to Equation 5.60. How well does the equation fit the data? What is the significance of your value of $K_1^{P_0}$?

(b) Bowker and King fit these data by assuming that K_1^P was unity and that attractive interactions between adjacent nitrogens reduced the barrier to diffusion of extrinsic precursors onto unoccupied sites so that K_{F2MT} increases with increasing coverage. What experiments would you do to see whether it is better to assume that K_1^P is larger than unity or that K_{F2MT} increases with increasing coverage? (*Hint:* What experiments would tell if the trapping probability were larger on an occupied site than on an unoccupied one?)

5.17 In Example 5.B we discussed physical vapor deposition. Consider the physical vapor deposition of an aluminum film.

(a) Assume that an 8-in.-diameter wafer is placed in an evaporator. How far does the wafer need to be from the source for the deposition rate at the edge of the wafer to be within 2% of the deposition rate at the center of the wafer?

(b) What evaporation rate do you need to get a growth rate of 10 μ/min? Assume that the sticking probability is unity.

(c) How would your results in part (b) change if there was a finite probability of the incident aluminum atoms accommodating with the growing aluminum surface?

(d) Use Baule's model to estimate a trapping probability for the case in part (c). Assume that the incident aluminum atoms impinge at normal incidence with gas temperature of 1400 K. Assume $T_s = 400$ K and $w = 10$ kcal/mole.

(d) How would your results in part (a) change if instead the flux out of the source followed

$$|\vec{v}_i \rho_g| = \frac{Evap \cos(\Phi_{SO})}{\pi L^2}$$

where the quantities are defined in Figure 5.29.

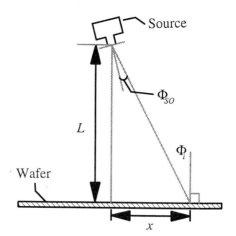

Figure 5.29 The geometry for Problem 5.17.

(e) If you wanted to include several wafers in the evaporator, how would you position the wafers so that the deposition rate is the same on all of the wafers (i.e., how should L vary with changing Φ_{SO})?

5.18 Gorte and Schmidt [1978] generalized the results in Section 5.9.4 to dissociative adsorption. Consider the kinetic scheme

$$A_{2,\,gas} \underset{2}{\overset{1}{\rightleftharpoons}} A_{2,\,precursor} \overset{3}{\to} 2A_{(ad)}$$

(a) Derive an equation for the sticking probability as a function of coverage assuming that the probability of finding two adjacent bare sites follows Equation 5.47. How do your results compare to Equation 5.60?

(b) Repeat part (a) assuming that the probability of finding two adjacent bare sites follows Equation 5.113.

(c) Assume you are measuring uptake data. How would you distinguish between the cases in parts (a) and (b)?

5.19 In Example 5.C we provided a program to do a cube calculation:

(a) Outline in words the steps the program executes to calculate a trapping probability.

(b) Show where the program

    (1) Picks initial conditions.

    (2) Integrates the equations of motion.

    (3) Calculates the forces on the atoms.

    (4) Calculates the accelerations of the molecule.

(c) Convince yourself that the program is integrating the equations of motion correctly.

    (1) Start with the case where the acceleration is zero. Show that the program is correctly computing the final position.

    (2) Now put in a constant acceleration. Show that the results are again correct.

    (3) Try instantaneous acceleration = accel + (t (old_accel-accel)/timestep).

(d) Run the program. Can you see any difference between the trajectories that lead to trapping and those which do not?

5.20 In Example 5.D, we provided a subroutine to do a Langevin calculation

(a) Outline in words the steps the program executes to calculate the forces on the atoms.

(b) Identify where the program

    (1) Calculates the forces due to the X springs.

    (2) Calculates the forces due to the Z springs.

    (3) Calculates the forces due to the diagonal springs.

    (4) Calculates the memory kernels.

    (5) Calculates the random forces.

(c) Show that the program correctly identifies the atom to the left and right of a given atom.

More Advanced Problems

5.21 In Section 5.3 we noted that Knudsen [1911] proposed that the energy accommodation coefficient α was equal to the momentum accommodation coefficient f_m. However, this is an approximation. Use a derivation similar to that in Section 5.4 to develop an expression for the fraction of an incident molecule's normal (i.e., z) and tangential (i.e., x) momentum that is transferred when a molecule of mass m_g collides with a surface atom with an effective mass of m_s. Assume the surface atom is isolated so the incident atom can hit anywhere on the surface of the surface atom. How does f_N compare to α?

5.22 The drag on a spacecraft is largely determined by momentum transfer to gas molecules as the spacecraft moves through outer space. Consider a 5 m^2 satellite orbiting at 100 miles above the earth. At that altitude, the atmosphere is almost all oxygen atoms. Calculate the force on the satellite due to collisions with the oxygen atoms. Assume the satellite is moving at 25,000 mi/hr and the oxygen pressure is 10^{-9} torr. (*Hint:* Use your results from Problem 5.21.)

5.23 In our deviation of Equation 5.26 we assumed that there was no conversion of tangential momentum (i.e., x momentum) into normal momentum (i.e., z momentum). However, in reality conversion occurs. The objective of this problem is to derive an approximation for the sticking probability, which includes some conversion of tangential momentum into normal momentum. Let our model be the one discussed in Section 5.4 except that we will keep track of both the normal and tangential momentum of the incident molecule.

(a) Consider a molecule with a velocity v_i that impinges onto the surface at some angle Φ_i. Calculate the new x and z components of the molecule's velocity when the incident molecule reaches the bottom of the attractive well. Assume that the entire heat of adsorption goes into z momentum.

(b) Now consider what happens as the molecule collides with the surface atom. Use a derivation similar to that in Section 5.4 or your results in Problem 5.21 to calculate the x and z components of the molecule's momentum after the collision. For the moment assume that the surface atom is isolated from all other surface atoms so that the molecule can collide anywhere on the surface atom.

(c) Use your results in part (b) to develop a criterion for the fraction of the molecules that have lost enough of their z momentum to stick.

(d) In reality, when an incident molecule impinges on a surface, the incident molecule can impinge on the tops of surface atoms. However, it is difficult for the incident atoms to get down to the sides of the surface atoms because other surface atoms get in the way. We call the blockage of one surface atom by another shadowing. Qualitatively, how will shadowing affect your result in part (c)?

(e) It is difficult to account for the shadowing analytically. However, it is easy to do so on a computer. Write a computer program to calculate the trajectory of a molecule as it impinges on a two-dimensional (111) hard sphere surface similar to that in Figure 5.3. Assume all of the surface atoms lie in a plane, and that the incident molecule has a diameter of one-half that of a surface atom. Start with a molecule with velocity v_i that is impinging on a random point on the surface. Use your results in part (a) to calculate the trajectory of the molecule as it interacts with the attractive well. Next calculate where the molecule hits the surface atom. Use your results in part (b) to calculate the trajectory of the incident atom as it bounces from the surface atom. Be sure to consider multiple collisions. Does the incident atom have enough z momentum to leave the surface?

(f) Repeat your results in part (b) for several thousand initial starting positions to calculate a sticking probability. How does the sticking probability vary with the incident angle? Assume that the incident molecules have a translational energy of 1 kcal/mole and the well depth is 2 kcal/mole.

(g) How would your results in part (f) change as the diameter of the incident atom changes?

(h) In reality the collision diameter of the surface atoms is larger than the spacing between surface atoms due to van der Waal's repulsions. How would that affect your results?

(i) What would change in your calculations if you used the Langevin method to calculate the trajectories of your incident molecules?

5.24 In Section 5.9.4 we assumed that there was a finite rate constant K_{MT} for direct conversion of intrinsic precursors (i.e., precursors over empty sites) into strongly bound species. Consider the adsorption of CO on Pt(111). The CO adsorbs molecularly. Why should K_{MT} be a finite number?

5.25 In Section 5.10.1 we derived an equation for adsorption of an immobile dimer on a line of sites. However, the derivation can be extended to other systems. Consider the adsorption of a trimer on a line of sites.

(a) What changes in the derivation?

(b) Is δ constant?

(c) How would you go about solving the equations? Do not solve!

5.26 In Sections 5.9 and 5.10 we derived equations for the adsorption in the mobile and immobile limits. The objective of this question is to use Monte Carlo calculations to see when the mobile and immobile limits apply.

(a) First, consider immobile adsorption of dimers on a 100×100 square lattice with periodic boundary conditions. Write a computer program to pick a surface site at random, and then choose a random nearest neighbor to the site. Assume that adsorption occurs only when both sites are vacant, and that if not, the incident molecule desorbs. Use your program to compute p_2 and the coverage as a function of exposure; compare your result to Equations 5.110 and 5.118. (*Hint:* You can keep track of the sites that are filled by defining a 100×100 occupancy matrix, and assuming that the occupancy is 0 if the site is empty and 1 if the site is filled.)

(b) Now consider the effects of mobility by allowing individual atoms to move along the surface once between each adsorbing Monte Carlo step. Use the following algorithm to examine the effects of diffusion: Choose a site at random, and a corresponding nearest neighbor site. Allow diffusion to occur if the original site is filled and the nearest neighbor site is empty. Diffusion corresponds to moving the atom from the original site to the nearest neighbor sites. How does diffusion affect the adsorption rate?

(c) In part (b), we are essentially assuming that the attempt rate for adsorption and diffusion are the same. In reality, however, the attempt rates can be quite different. Develop a computer program to explore this effect. First, choose a site and a nearest neighbor site. Then choose a random number between 0 and 1. If the number is less than p_d, the probability of diffusion, see if the adsorbate can diffuse. If the number is greater than p_d, attempt to adsorb a dimer onto the pair of sites. Try varying p_d. Over what range of coverages and values of p_d is Equation 5.48 valid? Over what range of values of p_d is Equation 5.118 valid?

5.27 In Section 5.10 we considered the adsorption of immobile adsorbates. However, the equations are applicable to other situations. Consider the adsorption of a molecule that covers two sites. Assume the molecule is mobile.

(a) Start with the low coverage limit. Is Langmuir's approximation accurate, given the results in Figure 5.19?

(b) Use your Monte Carlo program from the previous example to derive a correction to Langmuir's result.

(c) Compare your results to Equation 5.79. What do you conclude?

5.28 In Problem 5.21 you developed an expression for the momentum accommodation coefficient, f_m, and the thermal accommodation coefficient α. The objective of this problem is to consider the implications of a finite momentum accommodation coefficient on viscosity measurements. Consider a viscometer consisting of two concentric cylinders; assume that one of the cylinders is 2.0 cm in diameter, the other is 2.1 cm in diameter, both cylinders are 2 cm long, and the space between the cylinders is filled with 10^{-3} torr of nitrogen.

(a) Develop an expression for the force on the inner cylinder as a function of the rotation rate of the cylinder and the momentum accommodation coefficient of the gas. (*Hint:* At 10^{-3} torr there are no gas phase collisions.)

(b) Draw a diagram of the velocity profile for the case in part (a).

(c) Now consider raising the pressure so there are gas phase collisions. How will the velocity profile change?

(d) How high a pressure do you need in order for the no-slip condition to hold to within 1%?

5.29 DeAngelis, Glines, and Anton, *J. Chem. Phys.* **96**, 8582 [1992] examined the adsorption of CO on Ni(110). They found that when the CO first adsorbs, the layer shows c(8x2) periodicity. However, the structure of the layer changes to c(4x2) periodicity, then to c(2x1) periodicity as the coverage is increased. There are changes in the uptake curve at each stage. Anton et al. modeled the adsorption by assuming that the adsorption followed a modification of the precursor model, where intrinsic and extrinsic precursors were assumed to be equivalent, but there were three possible sites to hold CO, "C(8x2)" sites, "C(4x2)" sites, and "C(2x1)" sites. The objective of this problem is to evaluate this approximation

(a) Develop a modification of the precursor mechanism that considers the possibility of there being three sites to hold gas and allow there to be interconversion between adjacent sites.

(b) Integrate the resultant equations numerically to obtain an approximation for the coverage over each site as a function of time, using Anton's parameters in all cases.

(c) In Chapter 4 we noted that one can observe different adsorbate structures because there are interactions between adjacent molecules rather than there being multiple sites for adsorption. One can simulate the process using a modification of the algorithm in Problem 5.26. How would you extend your results in Problem 5.26 to this case?

(d) Write the computer program, and run it for one typical value of the parameters. Compare the results of the calculations to those in part (a). How would you explore which model is correct?

5.30 Pudney, Bowker, and Joyner, *Surface Sci.* **251**, 1106 [1991] found that the presence of oxygen substantially enhanced the sticking of ethanol on Cu(110). How would you go about modeling this effect? Write a computer program similar to the one in Example 5.C to model this situation. Use your program to reproduce the results in Figure 5 in Pudney et al.

5.31 Roberts [1938a and b] examined the effect of oxygen on the sticking of He on a (110)-oriented tungsten ribbon. The sticking probability varied from 0.02 on a "clean" ribbon on 0.6 on a ribbon that was covered by a monolayer of oxygen. Write a computer program similar to that in Example 5.C to model this effect. Use your program to see if you can explain Roberts' data.

Application of the Methods in Other Fields

5.32 In Section 5.2 we discussed trapping, and noted that the trapping rate is determined by the rate of energy transfer and not the rate that molecules cross a transition state. Look back at transition state theory in your P-chem book. How would a finite barrier to energy transfer affect the rate of reaction? Can you think of some examples where energy transfer may play an important role in a reaction rate in a gas phase reaction? A discussion of this topic is given in Chapter 7.

5.33 We derived Equations 5.113 and 5.118 for dissociative adsorption of a gas. However, these equations also apply to many other situations.

(a) Dawson and Peng [1972] applied these ideas to desorption of immobile molecules from a square lattice. The idea is to treat desorption as a bond-formation process between two adjacent sites, or equivalently adsorption of vacancy pairs onto a filled lattice. Derive an equation analogous to Equation 5.118 for this case. Fit your results to Dawson and Peng's data for nitrogen desorption from W(100).

(b) Several people have applied these models to multivalent binding of molecules to cells. References include C. DeLisi in C. Delisi and R. Blumenthal, eds., *Physical Chemical Aspects of Cell Surface Events in Cellular Regulation* Elsevier [1989] or C. A. Macken and A. S. Perelson, Lecture Notes in Biomathematics **50,** 1 [1980]. Discuss the application of Equations 5.109 and 5.113 to bioregulation.

(c) Flory [1939], Cohen and Reiss [1968], Gonzalez et al. [1974] studied the cross-linking of a polymer chain. Consider the disproportionation of polymethyl-vinyl-ketone and assume that the reaction follows Figure 5.30.

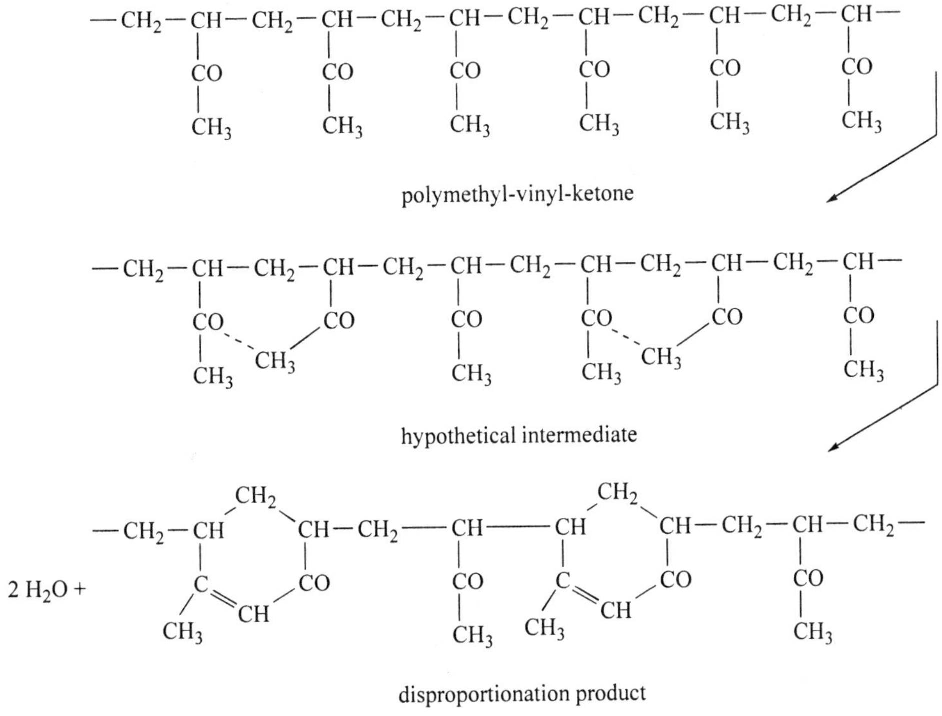

$$-CH_2-CH-CH_2-CH-CH_2-CH-CH_2-CH-CH_2-CH-CH_2-CH-$$

polymethyl-vinyl-ketone

hypothetical intermediate

$2 H_2O +$

disproportionation product

Figure 5.30

One can view the disproportionation process as a process where random links are added between two sites. Discuss the application of Equation 5.113 to this case.

Sample Midterm Problems

When Professor Masel gives this course, he gives a midterm exam after Chapter 5. In the exam he asks students to synthesize the ideas from Chapters 2 through 5. The problems below are some samples from these exams.

5.34 In the semiconductor literature, one often uses a hydrogen plasma to help deposit copper films on silicon substrates. The main reaction is $Cu(hfac)_2 + 2H \rightarrow Cu + 2Hhfac$.

The objective of this problem is to consider how different species in the plasma interact differently with different surfaces. First consider the hydrogen atoms.

(a) Will the hydrogen atoms bind more strongly to silicon or to copper? Assume that the Si is in a Si(111)(7x7) structure and that the copper is in a Cu(111) structure. Be sure to justify your answer.

(b) How will the bonding be different on silicon and on copper?

(c) Would you expect strong bonds (i.e., >25 kcal/mole) or weak bonds in each case?

(d) How will the trapping process be different in the two cases?

(e) Which surface will show a larger trapping probability? Be sure to justify your answer. (*Hint:* This is a trick question. Think about it again.)

(f) Would you expect the hydrogen atoms to form an ordered overlayer on both surfaces? Be sure to justify your answer.

Now consider the H_2 molecules.

(g) How will the interactions of H_2 be different on silicon and copper?

(h) Will the H_2 dissociate easier on Si or copper? Be sure to justify your answer.

5.35 The objective of this problem is to compare the reactivity of silicon (111) and gallium arsenide (111).

(a) Why are the arrangement of the atoms in Si(111) and GaAs(111)B different?

(b) Why are the dangling bond states different on the two surfaces?

(c) How do the differences in the dangling bond states affect the reactivity of the two surfaces? Which is more reactive?

(d) How would the trapping/sticking of Ga atoms be different on the two surfaces?

(e) How would the trapping/sticking of As_2 be different on the two surfaces?

(f) Would you expect gallium to form an ordered overlayer on the two surfaces? Be sure to justify your answer.

5.36 Figure 5.31 shows an overlayer structure for CO on Pt(111).

(a) Work out Wood's notation for the overlayer.

(b) What pairwise interactions would cause the structure to form?

(c) Explain how the interactions could arise.

(d) How should the precursor model be modified to account for these interactions?

(e) How will the interactions affect trapping of CO?

(f) How will the interactions affect sticking of CO?

Figure 5.31   The structure of a CO layer on Pt(111).

5.37 The objective of this problem is to compare the properties of Pt(110) and Cu(110).

(a) How are the arrangement of the atoms in these two surfaces different?

(b) How are the electronic structures of the two surfaces different?

(c) How are the electronegativities of the two surfaces different?

(d) How do the differences described in parts (a), (b), and (c) affect the ability of Cu(111) and Pt(110) to bind ethylidyne ($\equiv$CCH$_3$), ethan-1-yl2-ylidyne ($\equiv$C$-$CH$_2-$) and methoxy ($-$OCH$_3$)?

(e) How will the dissociation of hydrogen be different on the two surfaces?

(f) How will the binding of formaldehyde be different on the two surfaces?

(g) Which surface will be more effective in converting methoxy to formaldehyde?

(h) Which surface will allow the formaldehyde to desorb?

BIBLIOGRAPHY

Ashford, M. N. R., and C. Rettner, *Dynamics of Gas Surface Collisions*, Royal Society, London (1991).

Gasser, R. H. P., *An Introduction to Chemisorption and Catalysis by Metals*, Oxford University Press, Oxford (1985).

Goodman, F. O., and H. Y. Wachman, *Dynamics of Gas-Surface Scattering*, Academic Press, NY (1976).

Morris, M. A., M. Bowker, and D. A. King in C. H. Bamford, ed., *Comprehensive Chemical Kinetics*, Elsevier, Amsterdam Vol. 19, p. 163 (1984).

REFERENCES

Adleman, S. A., and J. D. Doll, *J. Chem. Phys.* **61**, 4242 (1974); **64**, 2375 (1976).

Amirau, A., and M. J. Cardillo, *Surface Sci.* **198**, 192 (1988).

Arumainayagam, C. R., and R. J. Madix, *Prog. Surface Sci.* **38**, 11 (1991).

Arumainayagam, C. R., R. J. Madix, M. C. McMaster, V. M. Suzawa, and J. C. Tulley, *Surface Sci.* **226**, 180 (1990).

Baule, B., *Annalen Der Physik* **44**, 145 (1914).

Brenig, W., in M. Grnze and H. J. Kreuzer, eds., *Kinetics of Interface Reactions*, p. 19, Springer-Verlag, NY (1987).

Brosilow, B. J., R. W. Ziff, and R. D. Vigil, *Physical Rev. A* **43**, 631 (1991).

Bruiuio, G. P., and T. B. Grinley, *Surf. Sci. Rep.* **17**, 1 (1993).

Cardillo, M. J., M. Balooch, and R. E. Stickney, *Surface Sci.* **50**, 263 (1975).

Cassuto, A., and D. A. King, *Surface Sci.* **102**, 388 (1981).

Ceyer, S. T., *Ann. Rev. Phys. Chem.* **39**, 479 (1988).

Cohen, E. R., and H. Reiss, *J. Chem. Physics* **38**, 680 (1968).

Dawson, P. T., and Y. K. Peng, *Surface Sci.* **33**, 565 (1972).

Dickman, R., J.-S. Wang, and I. Jensen, *J. Chem. Phys.* **94**, 8252 (1991).

Doll, J. D., and A. F. Voter, *Ann. Rev. Phys. Chem.* **38**, 413 (1987).

Ducros, R., and R. P. Merrill, *Surface Sci.* **55**, 227 (1976).

Ehrlich, G., *J. Physical Chem.* **59**, 473 (1955).

Ellett, A., and H. F. Olson, *Phys. Rev.* **31**, 643 (1928).

Flory, P. J., *J. Am. Chem. Soc.* **61**, 1518 (1939).

Gadzuk, J. W., and H. Metiu, *Proceedings 2nd International Conference on Vibrations on Surfaces*, 519 (1980).

Gonzalez, J. J., and P. C. Hemmer, J. S. Høye, *Chem. Physics* **3**, 228 (1974).

Goodman, F. O., *Prog. Surface Sci.* **5**, 261 (1974).

Goodman, F. O., *Surface Sci.* **7**, 391 (1967).

Goodman, F. O., *J. Physics Chem. Solids* **23**, 1269 (1962).

Gorte, R., and L. D. Schmidt, *Surface Sci.* **76**, 559 (1987).

Gunnarsson, O., and K. Schoenhamer, *Phys. Rev. B.* **25**, 2514 (1982).

Gunnarsson, O., and K. Schoenhamer, *J. Electron Spec.* **29**, 91 (1983).

Head-Gordon, M., J. C. Tulley, H. Schlichting, and D. Menzel, *J. Chem. Phys.* **95**, 9266 (1991).

Johnson, T. H., *J. Franklin Inst.* **206**, 308 (1928).

Joyce, B. A., and C. T. Foxton in C. H. Bamford, ed. *Comprehensive Chemical Kinetics* **19**, 181 (1984).

Kayser, H., *Wiederman's Annalen Der Physik und Chemie* **12**, 526 (1881).

Kisluik, P., *J. Physics Chem. Solids* **3**, 95 (1957).

Kittel, C., *Introduction to Solid State Physics*, Wiley, NY (1986).

Kneringer, G., and F. P. Netzer, *Surface Sci.* **49**, 125 (1975).

Knudsen, M., *Ann. Phys.* **34**, 593 (1911); **48**, 1113 (1915).

Kundt, A., and E. Warburg, *Poggendorf's Annalen Der Physik* **155**, 337 (1875).

Langmuir, I., *J. Am. Chem. Soc.* **38**, 221 (1916); **39**, 1848 (1917); **40**, 1361 (1918).

Langmuir, I., *Phys. Rev.* **8**, 149 (1916).

Langmuir, I., *Trans. Faraday Soc.* **17**, 111 (1921).

Lennard-Jones, J. E., *Trans. Faraday Soc.* **28**, 333 (1932).

Logan, R. M., and J. C. Keck, *J. Chem. Phys.* **49**, 860 (1968).

Logan, R. M., and R. E. Stickney, *J. Chem. Phys.* **44**, 195 (1966).

Mackenzie, J. K., *J. Chem. Phys.* **37**, 723 (1962).

Makuski, K., *J. Phys. C* **16**, 3617 (1983).

Maxwell, J. C., *Philos. Mag. J. Sci.*, ser. 4, **19**, 19 (1860).

Maltz, A., and E. E. Molta, *Surface Sci.* **115**, 599 (1982).

McCarroll, B., G. Ehrlich, *J. Chem. Phys.* **38**, 523 (1963).

McQuistan, R. B., and D. Lichtman, *J. Math. Phys.* **9**, 1680 (1968).

McQuistan, R. B., and D. Lichtman, *Surface Sci.* **20**, 401 (1970).

Nord, R. S., and J. W. Evans, *J. Chem. Phys.* **82**, 2795 (1985).

Page, E. S., *J. Roy. Stat. Soc.* **B21**, 364 (1959).

Pease, R. N., and R. J. Stewart, *J. Am. Chem. Soc.* **47**, 1235 (1925).

Pirig, G., G. Broden, and H. P. Bonzel, in *Proceedings, 7th International Vacuum Conference*, Vienna, Vol. 2, p. 906 (1977).

Rendulic, K. D., and A. Winkler, *Int. J. Mod. Phys. B* **3**, 941 (1989).

Roberts, J. K. *Nature* **135**, 1037 (1935).

Roberts, J. K., *Proc. Royal Soc. A* **152**, 445 (1937).

Roberts, J. K., *Proc. Cambridge Philos. Soc.* **34**, 399 (1938); **34**, 577 (1938).

Roberts, J. K., and A. R. Miller, *Proc. Cambridge Philos. Soc.* **35**, 293 (1939).

Rossington, D. R., and R. Borst, *Surface Sci.* **3**, 202 (1965); **12**, 501 (1968).

Rossington, D. R., and R. Borst, *J. Chem. Phys.* **3**, 202 (1965).

Schaaf, P., and J. Talbot, *J. Chem. Phys.* **91**, 4401 (1989).

Schaaf, P., J. Talbot, H. M. Rabeony, and H. Reiss, *J. Phys. Chem.* **92**, 4826 (1988).

Schonhamer, K., and O. Gunnarson, in D. Langreth and H. Suhl, eds. *Many Body Phenemenon at Surfaces*, p. 421, Academic Press, Orlando, FL (1984).

Schonhamer, K., and O. Gunnarson, in A. Yoshimari and M. Tsukada, eds., *Dynamic Processes at Surfaces*, Springer-Verlag, NY (1985).

Serri, J. A., M. J. Cardillo, and G. E. Becker, *J. Chem. Phys.* **77**, 2175 (1982).

Singh-Boparai, S. P., M. Bowker, and D. A. King, *Surface Sci.* **53,** 55 (1975).

Steele, W. A., *The Interaction of Gases with Solid Surfaces*, Pergamon, NY (1974).

Steinrück, H. P., K. D. Rendulic, and A. Winkler, *Surface Sci.* **152,** 323 (1985); **154,** 99 (1985).

Steinrück, H. P., K. D. Rendulic, and A. Winkler, *Phys. Rev. B* **32,** 5032 (1985).

Talbot, J., and P. Schaaf, *Phys. Rev. A* **40,** 422 (1989).

Talbot, J., and P. Schaaf, *Phys. Rev. Lett.* **62,** 175 (1989).

Talbot, J., G. Tarjus, and P. Schaaf, *Phys. Rev. A* **40,** 4808 (1989).

Taylor, H. S., *Proc. Royal. Soc A.* **108,** 105 (1925).

Taylor, H. S., *J. Phys. Chem.* **30,** 145 (1926).

Taylor, J., and I. Langmuir, *Phys. Rev.* **44,** 423 (1933).

Thomas, L. B., in C. L. Brundin, ed., *Rarefied Gas Dynamics*, p. 155, Academic Press, NY (1967).

Tulley, J. C., G. H. Gilmer, and M. Shugard, *J. Chem. Phys.* **71,** 1630 (1979).

Tulley, J. C., *Ann. Rev. Phys. Chem.* **31,** 319 (1980).

Vette, K. J., T. W. Orent, D. K. Hoffman, and R. S. Hansen, *J. Chem. Phys.* **60,** 4854 (1974).

Von Giesen, J., *Drud. Annalen Der Physik* **10,** 838 (1903).

Von Saussure, T., *Gilbert's Annalen Der Physik* **47,** 113 (1814).

Weinberg, W. H., and R. P. Merrill, *J. Vac. Sci. Tech.* **8,** 718 (1971).

Widom, B., *J. Chem. Phys.* **44,** 3888 (1966); **58,** 4043 (1973).

Zwanzig, R. W., *J. Chem. Phys.* **32,** 1173 (1960).

6 Introduction to Surface Reactions

PRÉCIS

So far in this book we have been discussing adsorption on solid surfaces. In this chapter, however, we leave our discussion of adsorption and begin our discussion of reactions on solid surfaces. The objective here is to provide a qualitative overview of reactions on surfaces and to introduce some of the key concepts that are developed later in this book. In particular we provide an overview of the mechanisms of surface reactions and briefly describe how the mechanisms vary with surface structure. We then briefly discuss rates of surface reactions and introduce some of the models we consider later. Most of the discussion is qualitative. A more quantitative treatment of this material is given in Chapters 7, 8, and 10.

6.1 INTRODUCTION

As noted in Chapter 1, reactions on heterogeneous catalysts have been studied since the pioneering work of Priestley [1790], van Marum [1796], Davy [1817], Henry [1824], Dobereiner [1823], Dulong and Theonard [1822, 1823], Faraday [1834], and Berzelius [1836]. At first, researchers did not know that heterogeneously catalyzed reactions occurred on the surface of the catalyst. Faraday [1834] and Berzelius [1836] both discuss the compression of molecules in the pores of a catalyst and suggest that the compression would lead to enhanced rates. Ostwald, in fact, showed that the compression of the reactants would often be sufficient to explain the enhancement of reaction rates by catalysts. However, in 1913 Sabatier expanded an earlier idea of Clément and Désarnes [1806] to propose that heterogeneously catalyzed reactions occur when molecules form an unstable complex with the surface of the catalyst. As a result, Sabatier proposed that catalytic reactions occurred on the surface of the catalyst.

This idea did not immediately meet with universal acceptance. According to Schwab [1981] and the *Dictionary of Scientific Biography*, two Nobel prize winners, Van't Hoff and Ostwald, originally thought that the idea that there could be an unseen intermediate during a reaction was preposterous. A third Nobel prize winner, Arrhenius [1899], proposed an alternative model where the catalyst would provide a continuous source of molecules with enough energy to climb over the activation barrier. However, in a series of important experiments, Langmuir [1912–1918] showed that heterogeneously catalyzed reactions occur on the surface of a catalyst. By 1925 the idea that catalytic reactions occur on surfaces was accepted by most workers.

Schwab [1931] noted that at first people concentrated on understanding how catalysts work. However, why they work was not understood.

It took many years and the advent of surface spectroscopy before people began to understand why catalysts work. However, today people recognize that *one* of the main roles of a catalyst is to stabilize intermediates of reactions. For example, the reaction $O_2 + 2H_2 \rightarrow 2H_2O$ can occur in the gas phase. According to Hinshelwood [1946], some of the key steps in the mechanism are

$$X + O_2 \rightarrow 2O + X$$
$$O + H_2 \rightarrow OH + H$$

$$H_2 + OH \rightarrow H_2O + H \tag{6.1}$$

$$H + O_2 \rightarrow OH + O$$

$$X + 2H \leftrightharpoons H_2 + X$$

Anton and Cadogan [1991] examined the mechanism of hydrogen oxidation on a Pt(111) surface. They found that the main reactions were

$$O_2 + 2S \rightarrow 2O_{(ad)}$$

$$O_{(ad)} + H_2 \rightarrow OH_{(ad)} + H_{(ad)}$$

$$2O_{(ad)} + H_2 \rightarrow 2OH_{(ad)}$$

$$O_{(ad)} + H_{ad} \rightarrow OH_{(ad)} \tag{6.2}$$

$$2OH_{(ad)} \rightarrow H_2O + O_{ad}$$

$$H_{ad} + OH_{ad} \rightarrow H_2O$$

$$H_2 + 2S \rightleftharpoons 2H_{ad}$$

Notice that there is a close similarity between the mechanism of hydrogen oxidation in the gas phase, Reaction 6.1, and the mechanism on the surface, Reaction 6.2. The elementary steps and intermediates are almost the same. There is an important difference between the mechanism in the gas phase and on a surface: during the surface reaction the reacting species are bound to the surface. They are stable species rather than radicals. However, species adsorbed on metal surfaces often follow reaction pathways that are quite similar to reaction pathways of radicals in the gas phase, although there are many exceptions.

People often picture surface reactions via a catalytic cycle, where surface intermediates are created and then destroyed. For example, Figure 6.1a and b show the catalytic cycles for

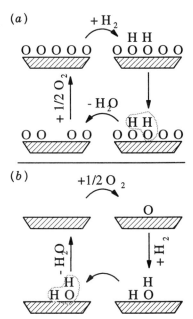

Figure 6.1 Catalytic cycles for the production of water (a) via disproportion of OH groups, (b) via the reaction $OH_{(ad)} + H_{(ad)} \rightarrow H_2O$.

the production of water via the mechanism in Equation 6.2. Water is produced in two reactions in Equation 6.2: the disproportion of $OH_{(ad)}$, and the direct reaction between H_{ad} and $OH_{(ad)}$. The first reaction dominates when there is excess oxygen on the surface, while the second dominates when the oxygen concentration is low. As a result, we can view the overall reaction as consisting of two reaction pathways. In the first pathway, the surface starts out covered by oxygen. Hydrogen adsorbs on the oxygen to yield OH groups. Two OHs then couple to yield H_2O and a bare site. Oxygen then adsorbs on the bare site to get back to the initial conditions. In the second pathway the surface starts out empty. Hydrogen and oxygen adsorb. The adsorbed species then react to yield water and regenerate the clean surface.

All steady-state catalytic reactions can be viewed as occurring via a catalytic cycle where surface species are formed and destroyed. In a film growth reaction one leaves atoms behind during the reaction and those atoms provide sites for further reaction.

Catalytic cycles occur in both the gas phase and on the surface. Reaction 6.1 can be considered a catalytic cycle where radicals are formed and then destroyed. However, there is an important difference between reactions in the gas phase and on a surface. In the gas phase, radicals are rather unstable species. Hence, one has to go to rather high temperatures before one gets a high enough concentration of radicals to get a reasonable rate. However, when a radical binds to a surface, the radical forms a chemisorbed complex that is stable. Hence, one can get a reasonable concentration of adsorbed radicals at low temperatures. For example, at 300 K, less than one hydrogen molecule in 10^{35} dissociates in the gas phase, while on a platinum surface, more than 99.9% of the hydrogen dissociates.

Of course, the hydrogen atoms on the surface are bound to the platinum and are no longer radicals. As a result, the hydrogen atoms on the surface are less reactive than free hydrogen atoms in the gas phase. However, the effect of the reduced reactivity is smaller than the effects of the enhanced concentration of hydrogen atoms. As a result, at moderate temperatures the hydrogen/oxygen reaction occurs much faster over a platinum catalyst than in the gas phase. At room temperature a hydrogen/oxygen mixture is stable for years. However, if one puts a platinum wire in the mixture, the mixture will explode.

In general, surfaces stabilize reactive intermediates. That is part of why a catalyst works. That is also why films grow at a surface in chemical vapor deposition (CVD) rather than the source gases (i.e., reactants) reacting to form powders in the gas phase.

The one limitation with the idea that surfaces stabilize intermediates is that sometimes the surfaces stabilize the intermediates so much that the intermediates become unreactive. For example, tungsten is not a good catalyst for hydrogen oxidation. The tungsten does dissociate hydrogen and oxygen. However, the resultant species are so strongly bound that they are fairly unreactive. The hydrogen/oxygen reaction occurs rapidly at 300 K on platinum. However, on tungsten one needs to go to temperatures over 1100 K to get a reasonable rate.

One can tell whether the bonding of the intermediates is appropriate by examining how free energy is dissipated during a reaction. For example, Figure 6.2 shows a plot of the free-energy change that occurs during hydrogen oxidation on platinum. The plot was calculated from data of Anton and Cadogan [1991]. During the reaction, oxygen adsorbs, OHs form, the species disproportionate, and water desorbs. Notice that each step goes downhill on the free-energy plot; the total free energy of the reaction is approximately equally partitioned between the various intermediates. The fact that the free energy is approximately equally partitioned makes platinum an excellent catalyst for hydrogen oxidation.

This is thought to be a general result. The best catalysts for simple reactions are thought to equally partition the free energy of a reaction over all of the individual steps in the reaction mechanism, although more data are needed to conclusively show that this is true. In Chapter 10 we show that in more complex reactions, one does not necessarily want to equally partition the free energy. With a complex reaction there will be special constraints to one of the steps in the reaction mechanism. One usually wants to maximize the thermodynamic driving force for that step. There are a few examples, e.g., ammonia synthesis, where the best catalysts do not follow these guidelines. Still, the observation that the best catalysts are materials that

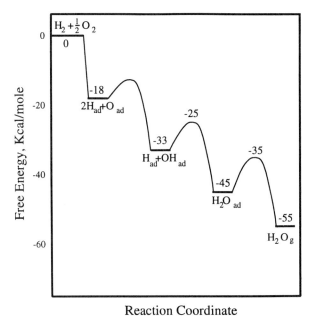

Figure 6.2 A free-energy diagram for the reaction in Figure 6.1 calculated from the data in Figure 6.3.

lower the free energy of the intermediates to somewhere between the free energy of reactants and that of the products allows one to identify interesting catalytic materials for study.

In the literature, one does not usually see free-energy plots. Rather, one instead sees a potential energy plot similar to that in Figure 6.3. The potential energy plot shows how the enthalpy (or internal energy) of the system changes as the reaction proceeds. Figure 6.3 shows a typical plot. This plot is typical of those in the literature.

There are some general rules that one can use to identify rapid reaction pathways from the potential energy plot. Generally, the adsorption steps need to be fairly exothermic for a surface reaction to occur at a reasonable rate. Example 4.C in the solved examples for Chapter 4 showed that if the heat of adsorption of a gas is less than about 20 kcal/mole, the equilibrium coverage of the gas will be negligible at room temperature in an ultrahigh vacuum (UHV). If the heat of adsorption is less than about 30 kcal/mole, the equilibrium coverage will be negligible at 300°C. That means that if a reaction is being run at room temperature, each reactant must have a heat of adsorption of at least 20 kcal/mole to get a significant amount of each reactant onto the surface. That goes up to 30 kcal/mole at 300°C. As a general rule of thumb, the most active catalysts have the equilibrium coverages of the reactants above 0.7, although there are some exceptions to this rule of thumb. The analysis in Example 4.C shows that the heats of adsorption must be at least 25 kcal/mole for a reaction at room temperature, and 37 kcal/mole at 300°C to get an equilibrium coverage of 0.7. Therefore, if one plots the potential energy of a reaction being run at room temperature in UHV, one would want the adsorption step to go downhill by about 25 kcal/mole for each mole of reactants at 300°C. See Example 4.C for further details.

Now consider the reaction $H_2 + \frac{1}{2}O_2 \rightarrow H_2O$. During the reaction a mole of H_2 reacts with a half a mole of oxygen. Therefore, if one wanted to run the reaction at room temperature in UHV, one would like the enthalpy of the system to go down by about $25 + \frac{1}{2}(25) = 37$ kcal/mole during the adsorption step. Rates of adsorption are higher at 1 atm. Hence, to get the reaction to go at 300 K and 1 atm, one would want the adsorption step to be only 11

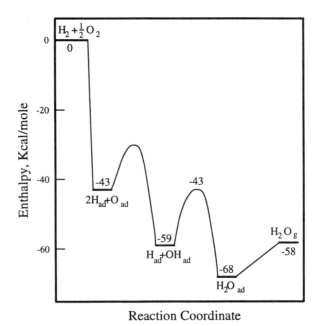

Figure 6.3 A potential energy diagram for the reaction $H_2 + \frac{1}{2}O_2 \rightarrow H_2O$ via the reaction in Figure 6.1b. (Adapted from Anton and Cadogan [1991].)

kcal/mole exothermic. Usually the heats of adsorption of both reactants need to be equivalent, or else one of the reactants will fill up all of the adsorption sites, and the other reactant will not be able to adsorb. Of course, one can adjust the surface concentrations by varing the concentrations of the reactants in the gas phase. However, typically one tries to balance the gas phase concentrations so that all of the key reactants have a significant surface concentration.

One also needs to design a catalyst so that each of the reaction steps taking place on the surface proceed at a high rate, but there are fewer guidelines to do so. It is often assumed that the reaction will be fastest if all of the reaction steps are exothermic. However, there are not enough data to tell if that is correct. Nor are there any guidelines to tell how exothermic the reaction steps should be. People generally assume that the fastest reactions occur when the intermediates equally partition the enthalpy of the reaction between the adsorbed reactants and adsorbed products. Once again, more data are needed to tell if that assumption is correct. In any case, it is clear that in the best catalysts, the enthalpy of the adsorbed intermediates is somewhere between that of the adsorbed reactants and that of the adsorbed products, although some of the details are unclear.

Experimentally, desorption steps are generally endothermic. Note that the results in Example 4.C show that the rate of desorption of a species will be negligible at room temperature if the heat of adsorption of the species is more than 30 kcal/mole. At 20 kcal/mole, rates are typically on the order of 1 molecule per exposed surface atom/sec at 300 K, which is equivalent to the rate of a typical catalytic reaction. Hence, one often finds that on a good catalyst, the products are bound by 5 to 20 kcal/mole, although there are exceptions. In general, the cases where the products are more strongly bound are slow reactions, although they may be very selective reactions.

There are a few surface reactions that do not obey these guidelines. Generally, one finds that if a surface reaction step in a given mechanism is endothermic, that step will be slow compared to the other steps in the mechanism. Similarly, if the desorption step is too endothermic, that step will be slow. Consequently, one can use the potential energy plot to identify possible bottlenecks in a reaction. As a result, potential energy plots are fairly useful for catalyst design.

Another key factor in determining the catalytic activity is the availability of d-electrons. Recall from Chapter 3 that the d-electrons only play a small role in binding molecules to surfaces. However, the d's play a key role in determining the barriers for dissociation. For example, the dissociative adsorption of hydrogen, i.e., $H_2 \rightarrow 2H_{ad}$ is quite exothermic on potassium, aluminum, germanium, or silicon. Yet, experiments have shown that the dissociation rate is negligible at 300 K. In contrast, platinum and nickel bind hydrogen much more weakly than potassium, aluminum, or silicon. Yet, hydrogen dissociates rapidly on platinum and nickel.

It is interesting to consider how the H_2 dissociation rate varies over the periodic table. Figure 6.4 is a plot showing which metals readily dissociate hydrogen at room temperature in UHV and which do not. Notice that the hydrogen readily dissociates on all of the metals near the center of the periodic table, but it does not dissociate on metals on either side of the periodic table. It happens that all of the metals that readily dissociate hydrogen have partially filled d-bands. In contrast, the metals with no d-electrons available for bonding do not really dissociate hydrogen. Instead, the dissociation process has a considerable activation barrier. To put these results in perspective, note that the heat of dissociative adsorption of hydrogen generally increases moving away from gold in the periodic table. Aluminum, potassium, sodium each bind atomic hydrogen strongly even though they do not dissociate H_2. Thus, there is no correlation between the dissociation rate and the heat of reaction for the dissociation process, i.e., $H_2 \rightarrow 2H_{(ad)}$. There is some general correlation between the ability of a surface to dissociate H_2 and the interstitial electron density. However, aluminum has a large interstitial electron density even though aluminum is inactive for hydrogen dissociation. The only property that is known to correlate with the data in Figure 6.4 is the availability of d-electrons. Therefore, it seems that d-electrons play a critical role in allowing dissociation processes to occur.

In the current literature there are two models of the role of the d-bands: a model discussed by Mango [1969], Masel [1986], and Nørskov [1990] that suggests that the main role of the d-bands is to stabilize antibonding orbitals, and a model discussed by Harris and Andersson [1985] that suggests that the main role of the d-bands is to lower Pauli repulsions.

Both of these models were briefly mentioned in Chapter 3. The idea of the orbital stabi-

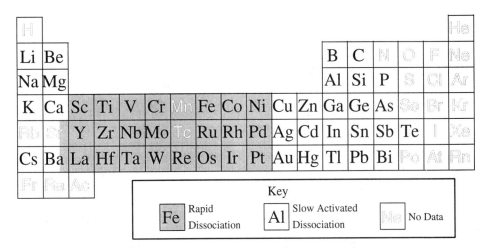

Figure 6.4 Part of the periodic table showing which elements dissociate hydrogen and which do not.

lization model is that when a reaction like H_2 dissociation occurs, the H_2 bond has to stretch and break. There is a large barrier to bond extension. As a result, anything one can do to make it easier to stretch the hydrogen–hydrogen bond will make it easier for the dissociation process to occur. One way to stretch a bond is to stabilize antibonding orbitals. Recall that when one puts electrons into the antibonding orbitals in a molecule, one weakens the bonds in the molecule. That makes it easier for bond scission to occur.

In Chapter 3 we noted that when molecules adsorb on surfaces, electrons flow to equalize the electronegativity of the system. In simple molecules, such as CO, NO, and H_2, the electron flow pulls the energy of antibonding orbitals down so the energy levels of the antibonding orbitals is close to the Fermi level (E_f) in the metal. Note that when the d-bands are partially filled, the d-bands also have energies close to E_f. As a result, the energies of the antibonding orbitals in simple molecules will be almost the same as the energy of the d-bands.

Now consider using the analysis in Section 3.10 to examine the interaction of the d-bands with the antibonding orbitals in H_2. Note that the antibonding orbitals will be stabilized by an amount given by Equation 3.88:

$$\Delta e_l = \frac{2(1 - f_d)}{E_{\text{d-band}} - \epsilon_l^0} \left| \int \Psi_l^*(\vec{r}) V_{sph}(\vec{r}) \Psi_k(\vec{r}) d\vec{r} \right|^2 \tag{3.88}$$

The largest stabilization occurs when ($E_{\text{d-band}} - \epsilon_l^0$) is small and the overlap integral is large. Notice that when the d-bands are partially filled, $E_{\text{d-band}} \approx \epsilon_l^0$. Therefore, according to Equation 3.88 there can be a strong interaction with the d-bands. In contrast, when the d-bands are filled ($E_{\text{d-band}} - \epsilon_l^0$) will be large, so one will have a much weaker interaction. Of course, the integral in Equation 3.88 will also affect the strength of the interaction. In Chapter 10 we will show that the integral produces some interesting quantum effects. However, the thing to remember for now is that partially filled d-bands have a strong interaction with the antibonding orbitals of molecules. That interaction weakens the bonds in the molecule, which facilitates bond scission processes.

In general, then, a surface has two main functions in promoting chemical reactions. The surface can lower the free energy of certain key intermediates. It can also interact with the antibonding orbitals in a molecule to facilitate the bond scission process. Both processes are important catalytically. As a result, one has to understand both processes to understand reactions on surfaces.

6.2   GENERAL MECHANISMS OF SURFACE REACTIONS

One way to learn about how surfaces facilitate reactions is to examine mechanisms of surface reactions. Much of the available data are summarized in the monographs of King and Woodruff [1982], Anderson and Boudart [1981], and Madix [1994]. In general, reactions on surfaces are not that much different from reactions in the gas phase or in solution. Later in this chapter, we show that the intermediates and reactions are similar. The one major difference is that the free energies of the intermediates are lower on a surface than in the gas phase. As a result, rates are often higher on the surface than in the gas phase or solution. One can also tailor a surface to selectively stabilize certain key reactive intermediates. As a result, one can design surfaces that promote certain key reactions without promoting other undesirable reactions.

There are three generic types of surface reactions: those that follow **Langmuir–Hinshelwood** mechanisms, those that follow **Rideal–Eley** mechanisms, and those that follow **precursor** mechanisms. Figure 6.5 shows a schematic of these three mechanisms for a hypothetical reaction A + B → AB. In the Langmuir–Hinshelwood mechanism A and B first

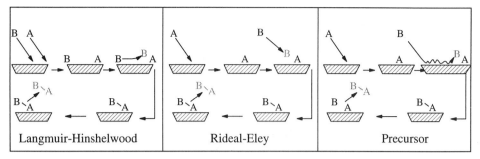

Figure 6.5 Schematic of (*a*) Langmuir–Hinshelwood; (*b*) Rideal–Eley; (*c*) precursor mechanism for the reaction A + B → A-B.

adsorb onto the surface of the catalyst. Next, the adsorbed A and B react to form an adsorbed A–B complex. Finally, the A–B complex desorbs. In what's now called Rideal–Eley mechanism* the reactant A chemisorbs. The A then reacts with an incoming B molecule to form an A–B complex. The A–B complex than desorbs. In the precursor mechanism A adsorbs. Next, B collides with the surface and enters a mobile precursor state. The precursor rebounds along the surface until it encounters an adsorbed A molecule. The precursor then reacts with the A to form an A–B complex, which desorbs.

Each of these reactions in Figure 6.5 can also occur in reverse, as shown in Figure 6.6. For example, one can run the Langmuir–Hinshelwood reaction in reverse by adsorbing an A–B molecule, heating to allow the A–B to decompose into adsorbed A and B, and then desorbing the A and B. Alternatively, if the adsorbed A–B molecule decomposes to yield an adsorbed A molecule and a gas phase species B, one would have the reverse of the Rideal–Eley mechanism. If a precursor forms on the way to the products, then one would have the precursor mechanism.

In the literature the precursor mechanism is sometimes referred to as the **trapping mediated** mechanism. Similarly, the Rideal–Eley mechanism is sometimes referred to as the **Eley–Rideal** mechanism.

All of the surface reactions that have been studied so far can be viewed as following a Langmuir–Hinshelwood mechanism, a Rideal–Eley mechanism, a precursor mechanism, or some combination of these three mechanisms.

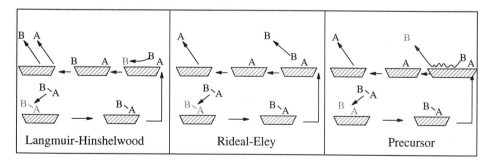

Figure 6.6 Schematic of (*a*) Langmuir–Hinshelwood; (*b*) Rideal–Eley; (*c*) precursor mechanism for the reaction A–B → A + B.

*In their original paper, Rideal and Eley did not distinguish between what is now called a Rideal–Eley mechanism and a precursor mechanism. However, more recent workers now make a distinction. See Weinberg [1992] for details.

Langmuir–Hinshelwood mechanisms are the easiest to examine experimentally. One technique used for this purpose is called temperature programmed desorption (TPD). In TPD one adsorbs the reactants onto a surface at low temperature, heats the surface, and watches the evolution of products. There have been thousands of papers that used this technique. For example, Yagasaki and Masel [1990] studied ethylene hydrogenation

$$C_2H_{4(ad)} + 2H_{(ad)} \rightarrow C_2H_{6(ad)} \rightarrow C_2H_6 \tag{6.3}$$

by adsorbing ethylene (C_2H_4) and hydrogen (H_2) onto a (2x1)Pt(110) surface at low temperature in UHV, heating the sample and looking for the ethane (C_2H_6) product. Yagasaki and Masel found that ethane was formed at high rates at 300 K. They proposed that the reaction occurred via a Langmuir–Hinshelwood mechanism, since they were starting with adsorbed reactants, the pressure was so low that gas phase collisions were negligible, and there was no evidence for a precursor.

The reviews in the series edited by King and Woodruff [1982] and Anderson and Boudart [1981] give many examples of Langmuir–Hinshelwood mechanisms. Most of the reactions of simple organic molecules on transition metal surfaces follow Langmuir–Hinshelwood mechanisms in UHV. As a result, there are more examples of Langmuir–Hinshelwood reactions than any other type of reaction. It is unclear, however, whether this is because Langmuir–Hinshelwood mechanisms really predominate under all conditions or whether instead Langmuir–Hinshelwood mechanisms simply predominate under UHV conditions. In UHV, the impingement rate of molecules on surfaces is small. In Chapter 7 we show that the Rideal–Eley and precursor mechanisms are attenuated when the pressure is low. Consequently, while Langmuir–Hinshelwood mechanisms dominate in UHV, it is unclear whether they will dominate under other conditions.

Precursor mechanisms are difficult to detect. We discussed precursor mechanisms in Section 5.9.4. Precursors are short lived species so they are difficult to observe spectroscopically. As a result, one often does not know for sure whether a reaction goes via a precursor mechanism.

One of the first examples of a precursor reaction was found by Ducros and Merrill [1976]. Ducros and Merrill were examining the adsorption of oxygen on Pt(100) and discovered that when they adsorbed a small amount of oxygen on (5x20)Pt(100), the chemisorbed oxygen would react with background CO (or hydrogen) with near unity probability. Ducros and Merrill found that under their conditions, chemisorbed CO was virtually unreactive. Nevertheless, a fast reaction was seen when the CO was in the gas phase. At first one might suppose that the reaction was occurring via a Rideal–Eley mechanism. However, Ducros and Merrill found that the reaction probability was near unity even when they went to low enough coverages that probability of a direct collision between the CO and oxygen was small. As a result, they proposed that the reaction occurred via a precursor mechanism. Under other conditions the reaction can also proceed via a Langmuir–Hinshelwood mechanism.

In Section 5.9.4 we show that there is an easy way to identify a precursor mechanism experimentally. One runs a steady-state reaction with a beam system so that one can vary the surface temperature and the gas temperature independently. One then measures the rate of reaction as a function of the gas temperature and the surface temperature. The rates of precursor-dominated mechanisms should go up as you heat the gas molecules, because there is a higher chance of surmounting the activation barrier. However, an increase in surface temperature produces a lesser enhancement in rate, or perhaps even a rate decrease, since the lifetime of the precursors goes down with increasing surface temperature.

Chapter 5 provides several examples of precursor-dominated dissociative adsorption processes. For example, in Problems 5.8 and 5.9, we ask the reader to show that the sticking of silane on tungsten silicide follows a precursor mechanism. Other examples that have been studied in great detail include the dissociative chemisorption of ethane on Pt(111) or

Ir(110)(1x2) (see Weinberg [1992] or Arumainayagain and Madix [1991]) or the dissociative chemisorption of methane (see Brass and Ehrlich [1987] and Ceyer [1988]). At present, we are not aware of any reaction of the type A + B → A–B that has definitely been shown to go via a precursor mechanism other than the CO-oxidation case mentioned two paragraphs ago. Even that case has been questioned (see Hall, Zoric, and Kasemo [1992]). That is not to say that precursor reactions do not exist. Precursor reactions are thought to be fairly common. However, one has to do a difficult experiment to clearly show that a reaction occurs via a precursor mechanism. So far the number of examples where people have done the careful experiments is small.

Nevertheless, there are a number of examples that were originally proposed to go by a Rideal–Eley mechanism, but probably really go via a precursor mechanism. Theoretically, reactions that go via a Rideal–Eley should be much rarer than reactions that go via either a Langmuir–Hinshelwood mechanism or a precursor mechanism. One can show that if a reaction has the possibility of going by both a Rideal–Eley mechanism and a precursor mechanism, the precursor mechanism will usually dominate. Typically, a gas/surface collision lasts for only about a picosecond, while the analysis in Section 5.9.4 shows that a molecule stays in the precursor state for perhaps 50 μsec. Hence, if all other factors were equal, an incident molecule would have about a factor of (50 μs/1 psec) = 5 × 10^7 times greater chance of reacting via a precursor than via a direct collision. One can overcome this factor of 5 × 10^7 by starting with excited molecules. However, under most conditions, the precursor mechanism will dominate over the Rideal–Eley mechanism.

At extreme conditions, however, Rideal–Eley reactions can be important. For example, in electronic materials production, one often uses a plasma process to deposit films on wafers. In a plasma process, one produces radicals or other highly excited species that react at the surface. The reactions of highly excited species are thought to often follow Rideal–Eley mechanisms.

For example, one grows CVD diamonds by heating methane to produce CH_3 radicals in the gas phase. The CH_3 groups then react with a cool surface to produce a diamond film. Harris [1990, 1993] examined the mechanism theoretically, and found that one first forms a hydrocarbon layer. Then CH_3 groups react with the hydrocarbon layer to form a hydrogen-terminated diamond surface. Finally, other CH_3 groups react with the layer to remove excess hydrogen (hydrogen also desorbs). Notice that each of these reactions follows the general reaction scheme:

$$CH_{3,g} + A_s \rightarrow products$$

where a gas phase species, $CH_{3,g}$, reacts with a surface intermediate, A_s, to form a product. Hence, classically, this is a Rideal–Eley mechanism. Of course, we do not yet know whether the CH_3 groups adsorbs into a mobile state before reacting. However, the available evidence is that diamond film growth follows a Rideal–Eley mechanism.

Another example of a Rideal–Eley mechanism arises in the decomposition of trimethyl-gallium (TMGa) $(CH_3)_3Ga$, on GaAs (4x2) during GaAs growth. Yu [1993] and Creighton et al. [1993] found that at low fluxes, the TMGa dissociatively adsorbed to eventually form an adsorbed mono-methyl gallium species, $(CH_3)Ga_{(ad)}$. The mono-methyl gallium then decomposes via what is believed to be a Rideal–Eley mechanism to yield adsorbed gallium and methyl radicals in the gas phase, i.e.,

$$Ga(CH_3)_{(ad)} \rightarrow Ga_{(ad)} + CH_{3(g)} \tag{6.4}$$

Note that some investigators do not consider Reaction 6.4 to be a Rideal–Eley reaction. However, it is a Rideal–Eley reaction in reverse (see Figure 6.6).

Film growth does not always go via a Rideal–Eley mechanism. For example, at higher fluxes, Lin and Masel [1991] found that one can also deposit gallium via a precursor mechanism

$$Ga(CH_3)_{(ad)} + Ga(CH_3)_{(p)} \rightarrow Ga_{(ad)} + Ga(CH_3)_{2_{(ad)}} \qquad (6.5)$$

$$Ga(CH_3)_{2_{(ad)}} \rightarrow Ga(CH_3)_{2_{(g)}}$$

where the subscript (ad) refers to adsorbed species, the subscript (p) refers to precursors, and the subscript (g) refers to species in the gas phase. However, at low pressures and high temperatures, the Rideal–Eley mechanism seems to dominate.

The only way to be certain that a reaction goes via a Rideal–Eley mechanism is to direct a pulsed beam of reactants at a surface, and measure the rate at which pulses of products are produced. If a reaction occurs via a Rideal–Eley mechanism, the product pulses will arrive at the detector almost instantaneously with the unreacted reactants. Alternatively, if the reaction goes via a precursor mechanism, the product pulses will be delayed by the time 100 μsec or so that the reactants spend in a precursor state.

As an example of this technique, Williams et al. [1992] examined the reactions of naphthalene ions with adsorbed CH_3 groups, in a pulsed beam. They found that the unreacted naphthalene ions and the methyl-naphthalene arrived at the detector at the same time. Their time resolution was such that they knew that the reactants had spent less than 160 nsec in the region of the surface before reacting. Hence, they concluded that the reaction followed a Rideal–Eley mechanism.

Experimentally, Rideal–Eley mechanisms seem to be much more common in the reactions important to semiconductor film growth than in the reactions important in catalysis. At present, it is not clear why this is so. However, experimentally semiconductor growth commonly goes via a Rideal–Eley mechanism, while catalytic reactions rarely go by that route.

6.3 SYSTEMATIC TRENDS IN THE RATES AND MECHANISMS OF SURFACE REACTIONS

Next we provide an overview of mechanisms of reactions on surfaces. That is actually a fairly large topic. At the time this book was being written there already were perhaps 100,000 papers that examine mechanisms of reactions on surfaces. The objective of the next section is to highlight the key themes that the reader should look for when reading the literature. We will discuss reactions on metals (e.g., Pt, Ni), reactions on insulators (e.g., Al_2O_3, SiO_2), and reactions on semiconductors (e.g., Si, NiO, ZnO, TiO_2) separately. Generally, one finds that mechanisms on metals all look very similar to each other, but quite different than reactions on semiconductors or oxides. Thus we can simplify the analysis if we treat metals separately from insulators or semiconductors. Because of page limitations, we cannot do more that a cursory overview of this material. More detailed presentations of parts of this material appear in the reviews listed in the references at the end of this chapter.

6.3.1 Reactions on Metals

Reactions on the surfaces of metals have been studied for almost 200 years. These reactions are important in heterogeneous catalysis, in the corrosion of metals, and the deposition of interconnects for electronic circuits. Consequently, reactions on metals have been studied in great detail.

Often reactions on metal surfaces are very similar to oxidation/reduction reactions in the gas phase. The hydrogen/oxygen example discussed earlier in this chapter is an example of that. In the gas phase the reaction follows the radical mechanism indicated in Equation 6.1, while on the surface the reaction follows the mechanism in Equation 6.2. Both are very similar. In general surface reactions involve a series of steps where first the reactants adsorb, then there are a series of bond scission processes where adsorbed intermediates are produced and destroyed, and finally products desorb. Generally reactions on metals follow chain propagation mechanisms where individual atoms are transferred one at a time from one adsorbed species to another. Then the resultant fragments recombine and desorb.

It is useful to consider a second example to illustrate the ideas. Barteau and Madix [1982] show that the predominate reactions in the oxidation of deuterated methanol to formaldehyde over an Ag(110) catalyst are

$$O_{2(g)} \rightleftharpoons O_{2(ad)} \tag{1}$$

$$CH_3OD_{(g)} \rightleftharpoons CH_3OD_{(ad)} \tag{2}$$

$$O_{2(ad)} \rightarrow 2O_{(ad)} \tag{3}$$

$$CH_3OD_{(ad)} + O_{(ad)} \rightarrow CH_3O_{(ad)} + OD_{ad} \tag{4}$$

$$CH_3OD_{(ad)} + OD_{(ad)} \rightarrow CH_3O_{(ad)} + D_2O_{(ad)} \tag{5}$$ $$(6.6)$$

$$CH_3O_{(ad)} \rightarrow CH_2O_{(ad)} + H_{(ad)} \tag{6}$$

$$CH_2O_{(ad)} \rightarrow CH_2O_{(g)} \tag{7}$$

$$D_2O_{(ad)} \rightarrow D_2O_{(g)} \tag{8}$$

$$2H_{ad} \rightarrow H_{2(g)} \tag{9}$$

The first two steps are the adsorption of the reactants. The third step is the production of two adsorbed radicals. The fourth and fifth steps are simple ligand extraction steps where an adsorbed radical extracts a ligand from a different adsorbed species. In the example shown, only a deuterium atom is transferred, although in other reactions more complex ligands (e.g., C_xH_y groups) can also be transferred. The sixth reaction is a radical decay. The seventh and eighth reactions are desorptions, while the ninth is a radical recombination reaction. Steps 3, 4, 5, 6, and 9 also occur during methanol oxidation in the gas phase, which shows the similarity of reaction mechanisms in the gas phase and on metal surfaces.

The only main difference between the reaction in the gas phase and on a surface is that in the gas phase one also observes the reactions

$$\cdot OH + CH_3OH \rightarrow \cdot CH_2OH + H_2O$$ $$(6.7)$$
$$\cdot O \cdot + CH_3OH \rightarrow \cdot CH_2OH + \cdot OH$$

where a hydrogen is removed from the CH_3 group in the methanol. However, no such reactions are seen in UHV. Interestingly, Zaleney et al. [1992] observed these reactions on a platinum electrode in solution. Physically, in the gas phase the $\cdot OH$ or $\cdot O\cdot$ can easily collide with the hydrogens in the CH_3, which makes the reaction facile. In contrast, the OH and CH_3OH are all bound with oxygen down on a clean metal surface. The hydrogens in the methyl group are not in close proximity to the OH. As a result, the reaction between the CH and the adsorbed OH or O is inhibited. The reactions in Equation 6.7 are not inhibited on a water-covered surface, because in such a situation the methanol can be oriented with the methyls pointing toward the surface (see Figure 6.7).

Still, the general result is that methanol oxidation on an Ag(110) surface is quite similar to the same reaction in the gas phase. However, it is not identical, because the surface orients molecules, while no preferred orientations are seen in the gas phase.

We generalize these results to a wide variety of reactions in Chapter 10. Generally, we

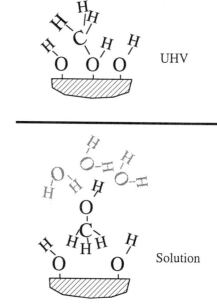

UHV

Solution

Figure 6.7 The geometry of the OH + CH₃OH reaction in the gas phase and in solutions.

find that oxidation and hydrogenation reactions on surfaces look very similar to oxidation and hydrogenation reactions in the gas phase. Simple decomposition reactions like

$$CH_3CH_3 \rightarrow Products$$

$$CH_3OH \rightarrow Products$$

look different in the gas phase and on a surface in that the initial step is C–C or C–O bond scission in the gas phase and CH or OH bond scission on the surface. However, the radicals that we produced in each reaction follow very similar reaction pathways in the gas phase and on a surface. Generally, the first step in a reaction on a metal surface is to produce atoms (i.e., adsorbed radicals) on the surface. The atoms can then extract atoms from molecularly adsorbed species, producing a new adsorbed radical. The adsorbed radicals then react in almost the same way that the radicals react in the gas phase. Of course, there are some differences between reactions in the gas phase and on surfaces. The initial step in the decomposition process is different in the two systems. Further, while the intermediates of reactions on surfaces look like radicals and react like radicals, they are not true radicals. Rather, they are species that are bound to the surface. As a result, the species can form at much lower temperatures than in the gas phase. Also species can have multiple adsorbate–surface bonds and the surface changes the orientation of the species, which inhibits certain key reactions. However, reactions as a general rule on *metal* surfaces are quite similar to reactions in the gas phase, and different than reactions in solution.

We can speculate why reactions on metal surfaces often look so much like radical reactions in the gas phase. In Section 3.4.1 we noted that while there are some exceptions, most ionic species are not stable on transition metal surfaces. Positive ions usually grab electrons when they come close to a metal surface, while negative ions usually lose electrons. Transition metal surfaces are generally weak Lewis acids. They lose electrons more easily than they gain electrons. As a result, species on metal surfaces are often slightly negatively charged. Consequently, usually only neutral or slightly negatively charged species participate in reactions on metal surfaces.

Those species are generally covalently bonded to the surface. In Chapter 3 we noted that there are two kinds of bonds between metals and adsorbed complexes; local bonds to the d-bands, and delocalized bonds to the s-band. Now consider what would happen if the delocalized bonding dominated. According to the effective medium model one can calculate the properties of a species held by a delocalized bond by embedding the species in jellium. Generally, within the context of this model most species look like radicals or slightly negatively charged species embedded in jellium.

Now consider the reactions of the adsorbed species. One can imagine three cases: one where the transition state has a $+1$ charge, one where the transition state has a -1 charge, and one where the transition state is nearly neutral. Note that the analysis in Section 3.4., i.e., electronegativity equalization, will apply to this case as well. Positively charged transition states will tend to grab electrons from a metal surface, while negatively charged species will tend to lose electrons to the surface. As a result, highly charged transition states are destabilized on a metal surface. Therefore, reactions on metals will tend to follow a reaction pathway where there are no large charge separations on the surface.

That is not to say that there cannot be some residual net charge in the transition state. For example, hydrogen is more electronegative than silver, but less electronegative than oxygen. Dai and Gellman [1993] examined the reaction.

$$CH_3CH_2O_{(ad)} \rightarrow CH_3CHO_{(ad)} + H_{(ad)} \tag{6.8}$$

on clean silver and found that the hydrogen was slightly negatively charged in the transition state. In contrast, Barteau and Madix [1982] found that in a number of oxidation reactions of the form

$$O_{ad} + H-R_{(ad)} \rightarrow OH_{ad} + R_{ad} \tag{6.9}$$

the rate scaled as the gas phase acidity of the hydrogen, which implies that the hydrogen in the transition state was slightly positively charged. Still, if one works out the numbers, one finds that the *metal* surfaces on the residual charges are small, i.e., ± 0.1 electron or less. Thus, one can view intermediates on metal surfaces as being nearly neutral although not exactly neutral.

It is useful to compare these results to results for the gas phase and in aqueous solution. In the gas phase, one often sees small charge transfers in the transition due to the difference in the electronegativities of various species. However, large charge separations are usually not observed because it costs too much energy to separate the charges. In contrast, in aqueous solution, the water molecules surround the ions and effectively screen them, so large charge separations are possible. According to the principle of electronegativity equalization, the charge separations in reactions on metal surfaces should look more like the charge separations in reactions in the gas phase than like the charge separations of reactions in solution. As a result, the mechanisms of reactions on *metal* surfaces are similar to reactions in the gas phase and quite different from reactions in aqueous solution.

Another way to look at a reaction on a metal surface is to note that at every point during the reaction, all of the species in the system are surrounded by a sea of electrons. The electrons, in effect act as a solvent for the reactants, products, and transition state. Just as in the solution, the presence of the sea of electrons can stabilize reactants, products, and transition states. Generally, the electron sea reacts more strongly with a soft species like an atom than a hard species like an ion. As a result, soft species and soft transition states dominate on metal surfaces. In contrast water is a hard base. As a result, hard species and hard transition states dominate in aqueous solutions. In the gas phase a soft species will follow a soft reaction pathway. As a result, the reactions of soft species on metal surfaces tend to follow a reaction pathway that is similar to the pathway in the gas phase and quite different from the reaction pathway in an aqueous solution.

The discussion in the last few pages should not be taken to imply that reactions on surfaces are identical to reactions in the gas phase. In the gas phase radicals are rather unstable species. One rarely sees an intermediate with more than one dangling bond. In contrast, it is quite possible for a species to have multiple dangling bonds that bind to the surface. As a result, intermediates on surfaces can be different from intermediates in the gas phase.

For example, in the gas phase Back and Martin [1979] found that ethylene decomposition follows a standard radical exchange mechanism:

$$2 \ H_2C{=}CH_2 \ \rightarrow \ H_2C{=}CH\bullet + H_3C{-}CH_2\bullet$$

$$H_3C{-}CH_2\bullet + H_2C{=}CH_2 \ \rightarrow \ C_2H_6 + H_2C{=}CH\bullet \qquad (6.10)$$

$$H_3C{-}CH_2\bullet + H_2C{=}CH_2 \ \rightarrow \ C_4H_9\bullet$$

$$C_4H_9\bullet + C_2H_4 \ \rightarrow \ \text{Higher hydrocarbons}$$

However, on a Pt(111) surface, Kesmodel, Dubois, and Somorjai [1979] and Ibach and Lehwald [1978] showed that the reaction follows the mechanism shown in Figure 6.8. First, the ethylene adsorbs molecularly in what is called a di-σ state ($-CH_2CH_2-$). Next, the ethylene loses a hydrogen and shifts another hydrogen to form a species called ethylidyne ($\equiv C{-}CH_3$). The ethylidyne then sequentially loses hydrogens, to eventually yield carbon and hydrogen.

In the gas phase ethylene does not decompose via the mechanism in Figure 6.8. As a result, it seems that the surface must be playing an important role in the chemistry.

It is easy to see why the reaction in Figure 6.8 does not occur in the gas phase. In the gas phase, an ethylidyne radical (CH_3C:) would have three dangling bonds. As a result ethylidyne would be quite unstable in the gas phase. However, on Pt(111) the ethylidyne is no longer a radical. Rather, the ethylidyne forms three bonds to the surface. As a result, ethylidyne is much more stable on Pt(111) than in the gas phase. Consequently, the mechanism in Figure 6.8 is unlikely to occur in the gas phase, even though the mechanism is followed on the Pt(111) surface.

It is harder to know why the reaction mechanism in Equation 6.10 occurs in the gas phase but not on the surface. All of the species in the mechanism in Equation 6.10 have been observed on platinum surfaces. Thus, it is not immediately obvious why the reaction follows the mechanism in Figure 6.8 rather than the mechanism in Equation 6.10. Note, however, that the intermediates in Equation 6.10 are seen only on stepped surfaces. Therefore, it seems that the chemistry is controlled by the surface geometry, i.e., the ability of the threefold sites on Pt(111) to form a threefold bond with the ethylidyne.

Experimentally, there is strong evidence that the surface geometry plays a key role in controlling the chemistry of ethylene decomposition. Yagasaki and Masel [1994] review the relevant data. Experimentally, ethylene decomposition follows the mechanism shown in Fig-

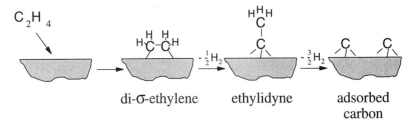

di-σ-ethylene   ethylidyne   adsorbed
                                            carbon

Figure 6.8   The mechanism of ethylene decomposition on Pt(111). (Proposed by Kesmodel, Dubois and Somorjai [1979] and confirmed by Ibach and Lehwald [1978].)

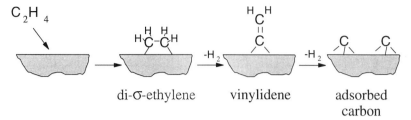

Figure 6.9 The mechanism of ethylene decomposition on (1x1)Pt(100). (Proposed by Hatzikos and Masel [1987] and confirmed by Sheppard [1988].)

ure 6.8 on the hexagonal faces of all the group VIII metals examined so far, except that the molecular ethylene sometimes is bonded in a π bound state and there is a second pathway on Ni(111). Nevertheless, the mechanism is often different on square surfaces. For example, Hatzikos and Masel [1987] found that on the square, (1x1), reconstruction of Pt(100) the reaction followed the mechanism shown in Figure 6.9 where the ethylene decomposed to form di-α-vinylidene at 300 K. Similar reactions are also observed on Pd(100) and Ni(100) (see Sheppard [1988] for details). Experimentally, ethylene decomposition follows the mechanism in Figure 5.6 on all of the hexagonal faces of group VIII metals. However, it follows a different mechanism on Ni(100), Pd(100), and (1x1)Pt(100). Further, Hatzikos and Masel [1987] found that if one heats the (1x1)Pt(100) surface to put it into the hexagonal, (5x20), reconstruction shown in Figure 2.4a, ethylene decomposes via the mechanism in Figure 6.8, i.e., the same mechanism as on Pt(111). Hatzikos and Masel [1987] found that one gets one reaction when the surface is in a square reconstruction and a different reaction when the same surface is in a hexagonal reconstruction. Hence, the surface geometry seems to play an important role in the chemistry.

Initially this would appear to be what is called a simple geometric effect. Notice that during ethylene decomposition on platinum, threefold symmetric intermediates are seen on threefold symmetric surfaces, while two- and fourfold symmetric intermediates are seen on square surfaces. There is a one-to-one correspondence between the geometry of the surface and the geometry of the reactive intermediates. Therefore, at first sight one might think that geometry alone is controlling the reaction mechanism.

However, that does not appear to be a general result. Sheppard [1988] shows that ethylene decomposition follows a mechanism similar to that in Figure 6.9 on Ni(100) and Pd(100). However, Slavin et al. [1988] found that on another square surface, Rh(100) ethylene, decomposition basically follows the mechanism in Figure 6.8, except that rather than forming an ethylidyne ($\equiv C-CH_3$) coordinated to three surface atoms, one forms a linearly bound ethylidyne where the carbon atom forms a triple bond to a single surface atom. The triple bond does have fourfold symmetry. However, the ethylidyne is not fourfold symmetric. Hence, geometry alone is insufficient to explain the variation in mechanism with crystal face.

In Chapter 10 we show that one has to consider both the geometry and the electronic structure of surfaces to predict decomposition mechanisms. However, the thing to remember for now is that surfaces are active participants in surface reactions. One can form multiply-bound intermediates on a surface, while species with multiple dangling bonds are very unusual in the gas phase. Of course, one can only form multiply-bound intermediates when the intermediates are properly configured to bond to surfaces. For example, one could never form a triply coordinated species if there were no triply coordinated sites on a surface. There are exceptions to this rule because the surface will sometimes reconstruct so that it is properly configured to hold the intermediates of a reaction. However, surface geometry often plays an important role in controlling mechanisms of reactions on surfaces.

Another way that geometry plays a role is in blocking reactions. For example, ethylene adsorbs molecularly onto a (1x1)Pt(110) surface, and there are threefold sites available for reaction. The sites are identical to those on Pt(111). Hence, initially one might think that ethylene decomposition on (1x1)Pt(110) would follow the mechanism as on Pt(111) where the ethylene decomposes via an ethylidyne intermediate, as shown in Figure 6.8. However, experimentally, no ethylidyne is seen. Instead Yagasaki and Masel [1990] found that the reaction follows the mechanism in Figure 6.10 where a mixture of carbon, CH_3, and $\equiv C-CH_2-$ intermediates are seen.

Such a result is easy to understand based on the geometry of the surface. Assume that ethylidyne tries to form on (1x1)Pt(110). Figure 6.10 shows the geometry of the surface when ethylidyne starts to form. Initially, there is no problem. However, ethylidyne would like to be held with three bonds to the platinum surface. The axis of the carbon–carbon bond should be perpendicular to the three bonding sites. Notice, however, that if one could put an ethylidyne into its desired configuration, the van der Waals radius of the ethylidyne would overlap the van der Waals radius of the platinum atom on the opposing step. As a result, there is insufficient space on the (1x1)Pt(100) surface for ethylidyne to fit. Experimentally, no ethylidyne is seen. Instead either a carbon–carbon bond stretches and breaks to yield a CH_3 species and a carbon atom, or a carbon–hydrogen bond breaks to yield ethane-1-*yl*-2-ylidyne ($\equiv C-CH_2-$) and hydrogen. Both pathways lead to species that fit on the (1x1)Pt(110) surface. Therefore, it seems that geometry is playing an important role in ethylene decomposition on (1x1)Pt(110).

Experimentally, surface atoms can move during reactions. However, their displacements are usually less than 0.1 Å, unless the entire surface reconstructs. As a result, one is usually constrained to only forming species that have the proper number of bonds to coordinate with the surface atoms, and fit into the spaces on the surface. Many of the details are still not understood. Nevertheless, the data show that surface geometry plays a key role in determining mechanisms of catalytic reactions.

In fact, presently, it seems that mechanisms of surface reactions vary more with surface structure than they do with changing metal. For example, Yagasaki and Masel [1994] note that the mechanism of ethylene decomposition is the same on the hexagonal faces of platinum, palladium, osmium, rhodium, palladium, cobalt, and iron. Slightly different chemistry is seen on nickel, but entirely different mechanisms are seen on the square faces of all of these metals except rhodium.

There have not been enough examples in the literature to know if this is a general phenomenon. However, another case that has been examined in some detail is methanol decomposition. Davis and Barteau [1989] proposed that the reaction followed the mechanism shown in Figure 6.11 on all of the faces of group VIII and Ib metals examined up to 1988 (mainly hexagonal faces) where the methanol sequentially loses hydrogens to eventually yield CO. However, Wang and Masel [1991] found that the reaction followed quite a different mechanism on (1x1)Pt(110), where the first step is C—O bond scission. The mechanism on (1x1)Pt(110) is shown in Figure 6.12. Hence, it appears that this is another case where surface structure controls the mechanism of a reaction.

In general, reactions on *metal* surfaces are very similar to reactions in the gas phase. Species adsorb molecularly and then dissociate to form adsorbed radicals. The adsorbed radicals react via chain propagation mechanisms, which are very similar to the chain propagation reactions seen in the gas phase. One major difference between reactions in the gas phase and on surfaces is that in the gas phase radicals are rather unstable species. However, on a metal surface the radicals are embedded in jellium. The radicals, in effect form a bond to the jellium so the radicals are stable. Multiply-bound radicals are also reasonably common on surfaces, while species with multiple dangling bonds are unstable in the gas phase. Also in Chapter 10 we find that the initial radical is different in the gas phase and on a surface. As a result, reactions on surfaces can be different from reactions in the gas phase, even though the nature of the species is similar in both cases.

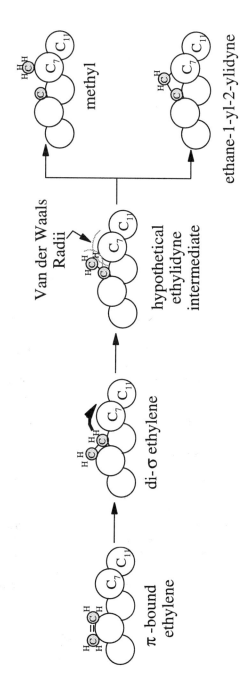

Figure 6.10 The mechanism of ethylene decomposition on (1x1)Pt(110). (Suggested by Yagasaki and Masel [1991].)

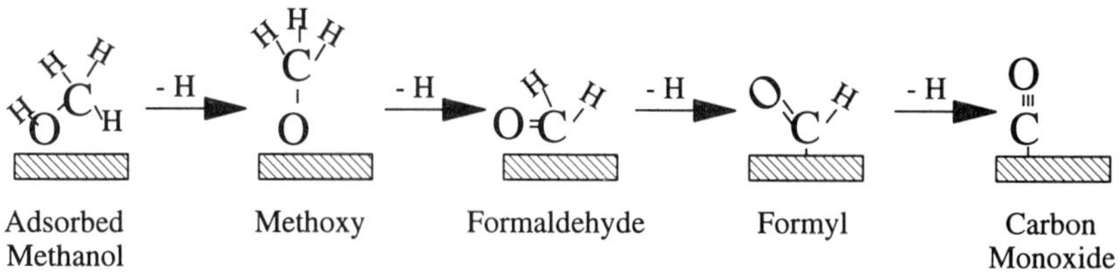

Adsorbed Methanol	Methoxy	Formaldehyde	Formyl	Carbon Monoxide

Figure 6.11 The mechanism of methanol decomposition on the hexagonal faces of group VII and 1b metals. (Proposed by Davis and Barteau [1989].)

Surface structure seems to play a key role in stabilizing multiply-bound intermediates. Generally, when multiply-bound intermediates form, different intermediates are seen on different faces of the same metal. Smaller but important differences are seen between metals. At present, there are no guidelines to tell when a reaction will go via a multiply-bound intermediate rather than via the mechanism seen in the gas phase. However, surface structure is quite important, as is described in detail in Chapter 10.

6.3.2 Structure-Sensitive Reactions

Surface structure has a second important effect as well, which is to change the rate of reactions. The idea that reaction rates were structure sensitive goes back to the 1925 measurements of Pease and Stewart, who found that if they adsorbed a small amount of mercury onto a copper catalyst, the rate of adsorption of hydrogen went down by a factor of 200 even though the equilibrium constant for the adsorption only changed by an order of magnitude. If the adsorption process were occurring uniformly over the surface of the copper, the decrease in rate would be the same as the decrease in the amount adsorbed. However, it was not. As a result, Pease and Stewart [1925] proposed that adsorption was not occurring uniformly over the solid surface. Rather, it was occurring on a series of distinct sites on the surface of the copper.

Taylor [1925, 1948] expanded these ideas substantially. Taylor's idea was that if one needed special sites for adsorption, one would also need special sites for reaction. Taylor showed experimentally that it was possible to poison a reaction without poisoning the adsorption process. As a result, Taylor suggested that most surface reactions occur only on special sites that he termed **active sites.**

The idea that there are active sites implies that reaction rates vary with surface structure. If the rate of a reaction were independent of structure, then by definition no site would be more able to catalyze a reaction than any other. As a result, when one says that there are active sites, one is also saying that the rate of reaction varies with surface structure.

In 1969 Boudart proposed that one could classify reactions into two general classes, **structure-sensitive reactions** and **structure-insensitive reactions.** In Boudart's definitions structure-sensitive reactions are those whose rate varies with surface structure, and structure-insensitive reactions are reactions whose rate does not depend on surface structure.

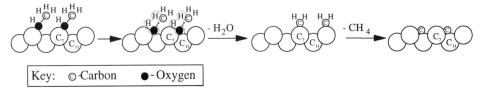

Key: ◎-Carbon   ●-Oxygen

Figure 6.12 The mechanism of methanol decomposition on (1x1)Pt(110) as proposed by Wang and Masel [1991].

I am not aware of any reaction that is structure insensitive under all conditions. Some reactions (e.g., ammonia synthesis from nitrogen and hydrogen) are structure sensitive under most conditions. Other reactions (e.g., ethylene hydrogenation) are structure sensitive only under special conditions. However, I am not aware of any truly structure-insensitive reaction. Hence, I do not believe it is that useful to classify reactions as being completely structure sensitive or structure insensitive.

A much better classification is to note that some reactions display structure sensitivity under a wider range of conditions than others and to call those reactions that show structure sensitivity under a wide range of conditions "structure sensitive."

One of the earliest ways to see if a reaction is structure sensitive was to run the reaction on a supported catalyst and see if the reaction rate varies with a particle size. Experimentally, the distribution of sites in a supported catalyst varies with particle size. Hence, if one observes a variation in rate with particle size, one knows that a reaction is structure sensitive.

For example, Figure 6.13 shows how the rate of the reaction

$$N_2 + 3H_2 \rightarrow 2NH_3 \qquad (6.11)$$

over an iron catalyst varies with particle size. Note that the specific rate (i.e., the rate per unit surface area) varies by four orders of magnitude with particle size. As a result, Boudart et al. [1975] concluded that Reaction 6.11 was structure sensitive.

There is a danger with this type of measurement, in that one cannot always say that because one does not observe a variation in rate with particle size, the reaction is structure insensitive. The distribution of sites in a supported catalyst varies with both the size of the catalyst particles and how the catalyst is processed. As a result, it is not obvious that if one simply varies the particle size, one samples all possible surface structures.

Another effect is that when one varies the particle size, one does not vary the distribution of all of the sites in the same way. For example, the concentration of kinks may change

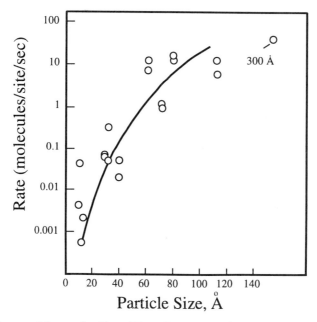

Figure 6.13 The rate of the reaction $N_2 + 3H_2 \rightarrow 2NH_3$ over an iron catalyst as a function of size of the iron particles in the catalyst. (Data of Boudart et al. [1975].)

significantly. However, the concentration of (111)-like C_9 sites may only change a small amount.

The effect of these uncertainties is that different investigators often come to different conclusions about whether a reaction is structure sensitive. For example, Bond [1956] and Sinfelt and Lucchesi [1963] found that the rate of ethylene hydrogenation varied with particle size, while Dorling, Eastlake, and Moss [1969] and Schlatter and Boudart [1972] found that it did not. For years it was thought that Bond's and Sinfelt and Lucchesi's measurements were incorrect. However, another possibility was that Bond's catalysts and Sinfelt and Lucchesi's catalysts had some surface structures that were different from those used by Dorling et al. Recently, Backman and Masel [1992] found that the rate of ethylene hydrogenation is about a factor of 8 higher on Pt(111) than on (5x20)Pt(100). Hence, it is quite possible that the reason different investigators came to different conclusions about the structure sensitivity of ethylene hydrogenation is that the catalysts used by these various investigators had a different distribution of sites.

Another way to look for structure sensitivity is to run reactions on a variety of faces of single crystals and look for a variation in rate with crystal face. For example, Figure 6.14 shows how the rate of nitric oxide decomposition varies as a function of crystal face for NO dissociation on the faces of platinum along the principle zone axes of the stereographic triangle. This reaction shows the largest variation in rate with crystal face observed so far. On Pt(410) Park, Banholzer, and Masel [1985] found that NO dissociatively adsorbed at temperatures above 130 K, and molecularly adsorbed below that. On Pt(310) Sugai et al. [1990] found that the NO dissociated at room temperature (their minimum temperature). In contrast, negligible dissociation is detected on Pt(100) below 400 K, while NO dissociation does not start on Pt(111) until the sample is heated above 1000 K. Analysis of these data indicates that the activation energy for N–O dissociation varies from 6 kcal/mole on Pt(410) to 45 kcal/mole on Pt(111). This is equivalent to a 10^{21} variation in the rate with crystal face at 450 K.

The NO example shows an especially large variation in rate with crystal face. However, there are several other examples that show 2 to 3 orders of magnitude variation in rate with crystal face. Key reactions in ammonia synthesis, ethane hydrogenolysis, C–O bond scission in methanol, and C–O bond scission in carbon monoxide have each been observed to show several orders of magnitude variation in rate with crystal face. Hence, the NO results are not an isolated case. Rather, structure-sensitive reactions are fairly common.

The one major problem with doing measurements like those in Figure 6.14 is that unless one is lucky or has a theoretical model, it is easy to miss some important chemistry. There are many examples that so far have only shown less than half an order of magnitude variation in rate with crystal face. However, it is unclear whether these examples really do show small variations in rate with crystal face or whether the measurements have not yet been run on the correct surface geometries. Note, for example, that NO decomposition shows much higher rates on Pt(310) and Pt(410) than on any other face. Pt(310) and Pt(410) are not special in any other way. Hence, it would have been easy to miss the unusual activity of Pt(310) and Pt(410).

Another effect is that it is unclear that one can access all possible surface geometries with single crystals. Recent work of Sachdev, Masel, and Adams [1992] has shown that small particles of metals have surface sites that are not present in bulk materials. Thomas [1976] has shown that irregular structures such as dislocations can have an important influence on the reactivity of a surface. Hence, it is unclear whether one can necessarily produce the particular sites that were especially active for reactions by cutting single crystals. Therefore, if one observes a large variation in rate with crystal face on a single crystal, one knows that the reaction is structure sensitive. However, one cannot say the opposite. A reaction that does not show large variation in rate with crystal face may or may not be structure insensitive.

I find it useful to classify reactions not by whether they are structure sensitive or insensitive, but rather to note that some reactions show more structure sensitivity than others. Table 6.1

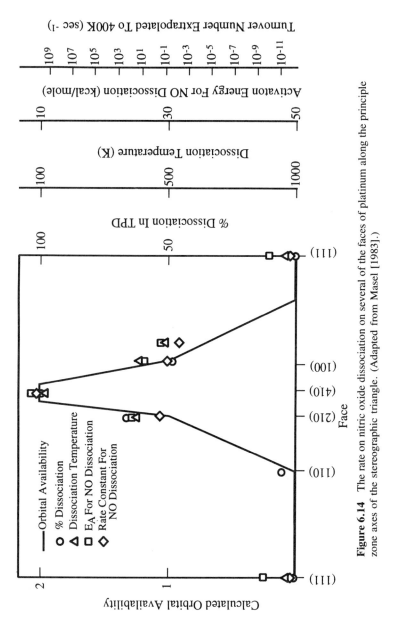

Figure 6.14 The rate on nitric oxide dissociation on several of the faces of platinum along the principle zone axes of the stereographic triangle. (Adapted from Masel [1983].)

459

TABLE 6.1   The Structure Sensitivity of a Series of Reactions

Reaction	Largest Variation in Rate with Geometry Observed Prior to 1996
$2CO + O_2 \rightarrow 2CO_2$	6
$C_2H_4 + H_2 \rightarrow C_2H_6$	8
$CH_3OH \rightarrow CH_{2_{(ad)}} + H_2O$	>100
$C_2H_6 + H_2 \rightarrow 2CH_4$	10^4
$N_2 + 3H_2 \rightarrow 2NH_3$	10^5
$2NO + 2H_2 \rightarrow N_2 + 2H_2O$	$\approx 10^{21}$

shows some typical results. Generally, reactions where carbon–hydrogen bond production or scission is rate determining (i.e., the slow step in the reaction) are less structure sensitive than reactions where the scission of carbon–carbon, carbon–oxygen, or carbon–nitrogen bonds is rate determining. The rate of scission of double or triple bonds is more structure sensitive than the scission of single bonds. Oxidation reactions show inconsistent trends. For example, Engel and Ertl [1979] found that at 400 K the rate of CO oxidation on palladium varies by a factor of 6 with crystal face. However, at 500 K the rate of the reaction was almost independent of crystal face. In later data, Ertl [1988] found that under the same conditions, the reaction rate oscillates on Pt(100), Pt(110), and Pt(210), but not on Pt(111).

Part of the reason for the inconsistent trends seen with oxidation reaction is that most oxidation reactions can go via either a precursor mechanism or a Langmuir–Hinshelwood mechanism. Generally, precursor mechanisms and Rideal–Eley mechanisms tend to be much less structure sensitive than Langmuir–Hinshelwood mechanisms. As a result, if one runs a reaction under conditions where the precursor mechanism dominates, one will see little structure sensitivity even though the same reaction shows structure sensitivity under conditions where the Langmuir–Hinshelwood reaction dominates.

There is another complication as well, which is that reactions may be structure sensitive for reasons that have nothing to do with the main reaction pathway. For example, a few paragraphs ago we discussed ethylene hydrogenation and noted that Backman and Masel [1988] found that ethylene hydrogenation showed a higher rate on Pt(111) than on (5x20)Pt(100). Note, however, that during ethylene hydrogenation a carbonaceous deposit builds up on the platinum surface. This carbonaceous deposit blocks the surface to further reaction. On Pt(111) the carbonaceous overlayer has a fairly open structure. There are many empty metal sites available to catalyze the reaction. However, on (5x20)Pt(100) the surface is more nearly covered by the carbonaceous fragments. As a result, one reason that (5x20)Pt(100) surface is less active than the Pt(111) surface is that under reaction conditions there are fewer metal sites exposed on the (5x20)Pt(100) surface than on the Pt(111) surface. Backman and Masel found that they could account for all of their measured structure sensitivity by keeping track of how many sites are exposed. Thus, even though Backman and Masel found that the overall rate of ethylene hydrogenation was structure sensitive, they did not find that the individual steps in the main reaction pathway were structure sensitive. Rather, they concluded that the reaction was structure sensitive because during the reaction the number of exposed metal sites was different on Pt(111) and (5x20)Pt(110).

This example illustrates an important point. Surface reactions are complicated. Sometimes one can find a reaction that would not be structure sensitive under idealized conditions. However, the reaction can still appear to be structure sensitive because of some secondary processes in the reaction mechanism leading to site blockage. Manogue and Katzer [1974] call this **secondary structure sensitivity.** Burwell [1967] and Butt [1987] have documented many examples of secondary structure sensitivity.

6.3.3 "Models" of Structure Sensitivity

In Chapter 10 we discuss some of the detailed models of structure-sensitive reactions. However, it is useful to provide an overview of the material now so the reader would knows where we are heading.

H. S. Taylor [1948] did the most important early work on structure-sensitive reactions. He discovered active sites, and with the use of selective poisons and radiotracers was able to show that the active sites only constituted a small fraction of the exposed metal surface. Taylor supposed that active sites were places on the catalyst that had an irregular geometry so that atoms were exposed with an unusual degree of coordination. However, Taylor never presented any data to support this supposition.

Thomas [1976], among others, did a considerable amount of work whose objective was to create defects in surfaces and examine the influence of these defects on rates. One of Thomas's accomplishments was to show that he could create defects on solids via cold working, and that produced an enhancement in the catalytic activity. Further, there was a correlation between the enhanced reactivity of the solids and the production of dislocations in the surface of the solid. Thomas also showed that lines of dislocations are preferentially etched on metal surfaces. Therefore, Thomas concluded that dislocations could be active sites for surface reactions.

The idea that active sites in catalysts are associated with dislocations, stacking faults, or other unusual structures was heavily discussed before 1970. However, more recently this model has fallen from favor. Part of the reason for that is that modern catalysts show activity far in excess of that which one can account for based on reactions on dislocations, stacking faults, etc. Still Thomas's work does provide clear evidence that dislocations and stacking faults can have unusual reactivity, and his work should not be forgotten when modeling catalytic reactions.

In 1929 Balandin proposed a model to understand structure sensitivity which did not require an active site to be associated with a defect. That model seems to have gained favor with modern investigators. Balandin's idea is that surface reactions involve a series of bond scission and bond production processes. At each stage in the reaction, the surface must be configured to hold each of the reactive fragments. As a result, only a special arrangement of atoms will be able to promote catalytic reaction.

Balandin defined what he called a **multiplet** for a reaction, where the multiplet is a group of surface atoms that are properly configured to hold the reactive fragments formed during the reaction. For example, on oxide catalysts, ethanol can react via two reaction pathways, a dehydrogenation pathway to form acetaldehyde

$$CH_3CH_2OH \rightarrow CH_3COH + H_2 \tag{6.12}$$

and a dehydration pathway to form ethylene

$$CH_3CH_2OH \rightarrow CH_2CH_2 + H_2O \tag{6.13}$$

Figure 6.15 shows a diagram of the multiplet for these two reactions from Balandin [1929]. Balandin [1929] proposed that Reaction 6.12 requires one to have two catalytic centers, one to couple the two hydrogens and one to hold the aldehyde fragments. Similarly, Reaction 6.13 requires two different reaction centers, one to couple the water and one to hold the olefin products. Balandin noted that one might need quite a different site to form water than to couple two hydrogens. Olefins would bind differently than aldehydes. Thus, one would need a different site to produce ethylene than to produce acetaldehyde.

Balandin expanded his model by assuming that all of the elementary steps in a catalytic reaction occur on the same site. As a result, he suggested that only a site that is configured to do all of the steps in a catalytic reaction would be catalytically active. Balandin noted that

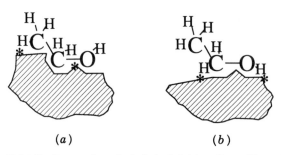

(a) (b)

Figure 6.15 The multiplet for the reaction of ethyl alcohol (a) to form either ethylene, (b) to form acetaldehyde and hydrogen, as discussed by Balandin [1929]. The asterisks in the figure represent catalytic centers.

in particular, bond lengths are critical. Thus, one might need a special configuration of atoms to get a given reaction to occur.

Since Balandin's initial work, it has become clear that while his ideas have some qualitative significance, quantitatively they often do not work so well. For example, it is now known the dehydrogenation of ethanol follows a mechanism similar to that in Figure 6.11, where the molecule sequentially loses hydrogens. Hence, the active site does not have to perform a number of tasks simultaneously. However, the idea that in order for a site to promote a given elementary reaction, the site must be properly configured to hold the reactants and products of the reaction seems to be consistent with the literature.

Over the years Balandin's multiplet model has been modified substantially. Dowden [1973] defined a **critical ensemble** for a reaction, where the critical ensemble is the minimum number of surface atoms necessary to promote a reaction. For example, consider the ethanol dehydrogenation case discussed earlier. When the reaction occurs, an adsorbed ethanol transfers two hydrogens to the surface. If all of the transfers would occur simultaneously, then one would need at least three bonding positions to get the reaction to happen: one to hold the ethanol, and two others to hold the hydrogens. Dowden supposed that one would need at least one surface atom per bond. Thus, the critical ensemble must contain at least three atoms.

One can probe the critical ensemble by isolating a group of surface atoms and seeing how they react (see Martin [1988], Ponec [1983], or Sachtler [1977] for details). Table 6.2 shows typical results. Generally, reactions that show little structure sensitivity show small critical ensemble sizes (i.e., one to three atoms), while reactions whose rate varies strongly with surface structure show larger critical ensemble sizes. The critical ensemble sizes are often considerably larger than one would expect based on the number of bonding positions one

TABLE 6.2   Measured Critical Ensemble Sizes for Some Important Catalytic Reactions

Reaction	Critical Ensemble Size
$C_6H_6 + 3H_2 \rightarrow C_6H_{12}$	3 ± 2
$CO + 3H_2 \rightarrow CH_4 + H_2O$	13 ± 2
$2CO + 5H_2 \rightarrow CH_4 + 2H_2O$	22 ± 2
$CH_4 + D_2 \rightarrow CH_3D + HD$	3 ± 2
$CH_4 + 4D_2 \rightarrow CD_4 + 4HD$	5 ± 2
$C_2H_6 + H_2 \rightarrow CH_4$	12 ± 2
$C_3H_8 + H_2 \rightarrow CH_4 + C_2H_6$	12 ± 2
$C_3H_8 + 2H_2 \rightarrow 3CH_4$	17 ± 2

As reported by Martin [1988].

needs to hold the reactants. Further, experimentally the arrangement of the atoms in the ensemble is at least as important as the ensemble size. Hence, it does not appear that one can understand active sites purely on the number of surface bonds that form during the reaction.

Over the years there have been several other models proposed to understand why this is so. There are three basic models, those that attribute structure sensitivity to geometric effects, those that attribute structure sensitivity to thermodynamic effects, and those that attribute structure sensitivity to electronic effects. The geometric models have the longest history. They were first discussed by Balandin [1929], and have been considered by many investigators since. The basic idea of these models is that in order to get a reaction to occur, the reactants must be held together in the right configuration to allow bonds to form. For example, Hoffmann et al. [1992] found that the reaction

$$3C_2H_2 \rightarrow C_6H_6 \tag{6.14}$$

occurs when two acetylenes polymerize to a C_4H_4 metallocycle, and then the metallocycle reacts with another acetylene to form benzene (C_6H_6). Gentile and Mutterties [1984] found that the reaction is rapid on Pd(111) where all of the acetylenes are oriented to form the various intermediates. However, there is little reaction on Pd(110) because there is no room for a metallocycle to react with an acetylene on Pd(110). Pd(100) is an intermediate case. The metallocycle can form and react with an acetylene. However, the metallocycle needs to reorient itself before it reacts and that slows the reaction.

In general, in order for a reaction to occur, one must put reactants into a configuration that allows them to react. As a result, surfaces that naturally put the reactants into the right configuration tend to be more active than those that do not.

Thermodynamic effects can also affect rates of reactions on surfaces. Tempkin [1957] noted that thermodynamic effects could have an important role in reactivity. These ideas were expanded by Tempkin [1965] and Boudart and Djego–Mirajossou [1984], Balandin [1969], and Shusturovich [1991] and are now well accepted by most investigators. The idea behind the thermodynamic models is that the heat of adsorption and desorption of the intermediates of a reaction can vary strongly with crystal face. Stepped surfaces are often presumed to bind intermediates more strongly than closed packed planes, although the data do not always show that. The variation in the binding energy of the intermediates can have an important effect on the reactivity of a catalyst.

For example, consider a reaction where the decomposition or desorption of a given intermediate is rate determining. Surfaces which bind the intermediate strongly will stabilize the intermediate and thereby prevent the intermediate from reacting. Consequently, those surfaces that bind the intermediate most strongly will tend to be less active than those surfaces that bind the intermediate less strongly.

As an example, consider the decomposition of methanol

$$CH_3OH \rightarrow CO + 2H_2 \tag{6.15}$$

over a platinum catalyst. The rate-determining step in methanol decomposition is the decomposition of a methoxy (CH_3O) intermediate

$$\tag{6.16}$$

The methoxy is held to the surface with a platinum oxygen bond. We do not know the heat of adsorption of methoxy. However, oxygen has a binding energy of 48 kcal/mole on Pt(111) and 63 kcal/mole on Pt(110). Hence, based on the thermodynamics of the reaction, one would expect methoxy decomposition to be faster on Pt(111) than on Pt(110). Experimentally, Wang and Masel [1991] found this to be true. Hence, thermodynamic effects can also influence the relative reactivity of various sites in a surface.

Electronic structure effects are important as well. The electronic structure models have been discussed by Boudart [1950], Tompkins [1976], Bond and Wells [1964], Kesmodel and Falikov [1975], Masel [1984], and Nørskov, Stoltze, and Nielsen [1991]. The idea behind these models is that the rate of a surface reaction is determined by the ability of the surface to form and break bonds. The ability of the surface to form and break bonds can be calculated from the Schrödinger equation, and therefore is related to the electronic properties of the surface. These ideas are discussed in detail in Chapters 9 and 10. However, basically, the ability of the surface to form and break bonds is determined by the overlap of key orbitals in the surface with those in the adsorbate, and the ability of the orbitals in the surface and the adsorbate to donate electrons to form bonds. Both the overlap integrals, and the ease with which electrons are donated change as the arrangement of the atoms in the surface change. This can lead to variations in rate.

In actual practice, geometric, thermodynamic, and electronic effects all contribute to the variation in rate with surface geometry. Sometimes one effect will dominate; other times another. In Chapter 10 we discuss some guidelines to indicate which effect dominates. However, at present these ideas are still being developed, so it is difficult to reach any definitive conclusions.

Experimentally, the situation is still unclear. There were many models proposed in the literature from 1930 to 1960. However, up until about 1970, one could not vary surface structure in a well-controlled systematic way, and so one could not test the models. People could vary particle sizes in catalysts, but that did not give a well-defined distribution of particle shapes. However, in the 1970s, with the advent of routine ultrahigh vacuum technology people began to be able to test these models experimentally. A typical experiment was to run a reaction on a series of single-crystal catalysts and try to correlate the measured reactivity with the surface structure of the crystal.

Blakely and Somorjai [1975] provided the first reliable data using the technique. Blakely and Somorjai examined the rate of cyclohexane dehydrogenation

$$C_6H_{12} \rightarrow C_6H_6 + 3 H_2 \tag{6.17}$$

and hydrogenolysis

$$C_6H_{12} + H_2 \rightarrow C_6H_{14} \tag{6.18}$$

on Pt(111), Pt(755), Pt(211), Pt(13 1 1), Pt(16 12 11), and Pt(976). Figure 6.16 shows a diagram of these faces. Pt(111) is a hexagonal surface: Pt(755), Pt(211), and Pt(13 1 1) are stepped surfaces along the [001] zone axes. Pt(755) and Pt(211) have (111) terraces and (100) steps, while Pt(13 1 1) has (100) terraces and (111) steps. Pt(16 12 11) and Pt(976) are kinked surfaces with (111) terraces, (100) steps, and (110) kinks.

Blakely and Somorjai plotted their data as a function of step and kink density. Figure 6.17 shows some of their results. Blakely and Somorjai found that initially they were not able to detect hexane formation. Instead, the cyclohexane completely dehydrogenated. However, after they ran the reaction for several minutes, a carbonaceous layer built up on the surface, and they were able to detect some hexane formation. Experimentally, the rate of hexane formation reached a maximum and then decayed. Figure 6.17 shows a plot of the maximum rate as a function of step density and kink density. The point for Pt(13 1 1) is approximate in the figure. Blakely and Somorjai took data for Pt(13 1 1) and included it in some of their plots. However, when they replotted the data as a function of step or kink density, they deleted the point for Pt(13 1 1) and rescaled everything. In plotting Figure 6.17 it was assumed that the scale factor for Pt(13 1 1) rate is the same as for Blakely and Somorjai's other data.

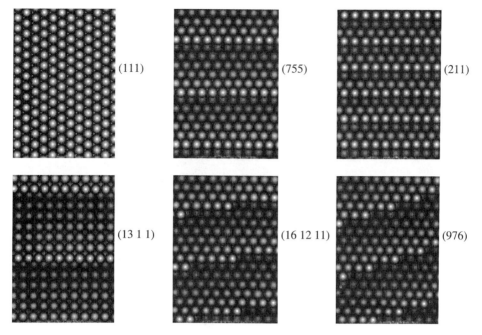

Figure 6.16 The ideal surface structure of the single-crystal faces examined by Blakely and Somorjai [1975].

Notice that there is a general trend in the data that as one increases the step density, the rate of hydrogenolysis goes up. Kinks have a smaller effect. The data are not fully consistent with the idea that steps and kinks lead to higher reactivity. Pt(13 1 1) is more active than Pt(755) even though Pt(13 1 1) has a considerably lower step density than Pt(755). However, the data in Figure 6.17 suggest that one might be able to correlate structure sensitivity with the concentration of steps and kinks, although the correlation is not perfect.

Note, however, that the data in Figure 6.17 are in some ways a special case in that the data probably represent an example of secondary structure sensitivity. Blakely and Somorjai are examining heavily carbon contaminated surfaces. Davis and Somorjai [1982] note that the Pt(111) surface is inactive because it is completely poisoned under Blakely and Somorjai's reaction conditions. The other surfaces examined by Blakely and Somorjai were poisoned too. However, the poisoning appears to occur more slowly on stepped surfaces than on Pt(111). Davis and Somorjai suggest that there are bare metal atoms at the steps in Blakely and Somorjai's samples, while few bare metal atoms are present on Pt(111). Thus, even though Blakely and Somorjai observed a correlation between step density and rate, their measurements are in some ways deceiving because they are examining a case where, under reaction conditions, the surface composition is different near their steps from elsewhere on their sample.

Since Blakely and Somorjai's work there have been a number of other examples where it was found that under reaction conditions, the surface composition was different near the steps than near the closed packed terraces. A particularly nice example is given in the work of Gellman Dunphy, and Salmeron [1992], who found that heterocyclization of acetylene to thiophene on a sulfided ''Pd(111)'' sample occurred mainly on steps that exposed bare palladium atoms. It is relatively easy to understand why steps might enhance rates of reactions under conditions where bare metal atoms are exposed only at the steps. However, such observations do not tell us whether steps will enhance rates when the surface is clean.

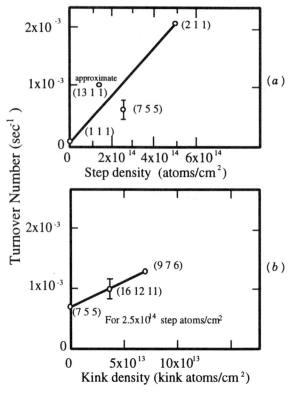

Figure 6.17   The rate of cyclohexane hydrogenolysis to hexane on Pt(111), Pt(755), Pt(211), Pt(13 1 1), Pt(16 12 11), and Pt(976). (Data of Blakely and Somorjai [1976].)

Still, there has been considerable speculation in the literature that steps and kinks should enhance rates of catalytic reactions, even under clean conditions. Experimentally, the situation is unclear. Herz et al. [1981], Kahn, Peterson, and Somorjai [1984], and Davis and Somorjai [1980] presented a number of examples where stepped surfaces had a different reactivity than closed packed planes. However, as of when this book was being written there was no data which clearly demonstrates a one to one correspondence between rate and step or kink density under conditions where the surface stays clean.

In fact, the data in some ways conflict with the notion that steps and kinks are especially active. For example, Herz et al. [1981] reexamined the hydrogenolysis of cyclohexane on Pt(111), Pt(557), Pt(10 8 7), and Pt(25 10 7) with excess hydrogen to avoid the effects of poisoning. Figure 6.18 shows a diagram of these faces. Pt(111) is a flat surface, Pt(557) is a stepped surface with (111) steps and (100) steps, while Pt(10 8 7) and Pt(25 10 7) are surfaces with (111) terraces, (100) steps, and (110) kinks. The kink density is 6% on Pt(10 8 7) and 9% on Pt(25 10 7).

Table 6.3 shows some of Herz et al.'s results. Herz et al. found that under their conditions the flat, Pt(111) face is the most active for hydrogenolysis. Steps and kinks *reduced* the reactivity for hydrogenolysis. By comparison Blakely and Somorjai's data in Figure 6.17 show precisely the opposite trends. Hence, the idea that steps and kinks are especially active for hydrogenolysis reactions is not supported by Herz and Somorjai's data.

Of course, the dehydrogenation rate measured by Herz and Somorjai does seem to be enhanced when there are steps and kinks on the surface. The maximum dehydrogenation rate is seen on Pt(25 10 7), which is also the most highly kinked surface that Herz and Somorjai

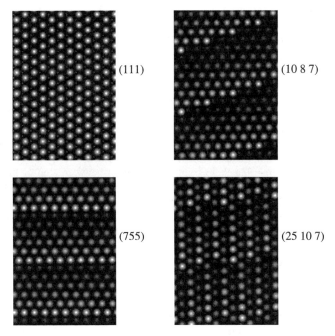

Figure 6.18 The surface structure of the surfaces examined by Herz, Gillespie, Peterson, and Somorjai [1981].

used. However, Davis and Somorjai [1980] found that when there is a trace of oxygen on the surface, Pt(654) is more active than any of the surfaces used by Herz et al., even though the Pt(654) surface only has 6% steps. Further, Blakely and Somorjai [1975] found that under other conditions the rate of dehydrogenation is unaffected by step or kink density. Thus at this point there is no evidence that steps or kinks have special properties for dehydrogenation reactions.

Experimentally, there have been a number of instances where the reactivity has reached a maximum either on a flat surface or on a surface with an intermediate step density. The NO data in Figure 6.14 show that if one starts with the Pt(100) surface and adds (111) steps, the rate goes down. Hence, the flat surface is more active than the stepped one. Of course, if one instead adds (110) kinks, the rate goes up. However, the rate reaches a maximum at an intermediate kink density. Pt(410) is much more active than Pt(210), even though both Pt(210) and Pt(410) have the same the step geometry and Pt(210) has a higher kink density. Siera et al. [1989] have found that Pt/Rh alloys also show similar behavior.

TABLE 6.3   The Initial Rate of Cyclohexane Conversion to Benzene, Cyclohexene, and *n*-Hexane on a Series of Platinum Surfaces[a]

Surface	Dehydrogenation to Benzene	Dehydrogenation to Cyclohexane	Hydrogenolysis to *n*-Hexane
(111)	1.3×10^{-6}	3.1×10^{-9}	2.2×10^{-10}
(755)	2.4×10^{-6}	2.1×10^{-9}	0.94×10^{-10}
(25 10 7)	4.7×10^{-6}	1.9×10^{-9}	0.98×10^{-10}
(10 8 7)	4.1×10^{-6}	2.9×10^{-9}	1.3×10^{-10}

[a]Measured by Herz et al. [1981]. Rates in moles/cm^2 min.

Herz et al. [1981] examined the dehydrocyclization and hydrogenolysis of n-heptane on the same surfaces explored by Kahn and Somorjai. Herz et al. found that the hydrogenolysis activity followed (25 10 7) > (111) > (10 8 7) > (7 5 5). Notice that the stepped (755) surface is less active than the flat (111) surface. One kinked surface, Pt(25 10 7) is slightly more active than the (111). However, another kinked surface, Pt(10 8 7) is less active than the (111).

Other examples include methanol decomposition, where Wang and Masel[1991] found that the flat Pt(111) surface is considerably more active than the stepped Pt(110) surface for methoxy decomposition, and cyclohexane dehydrogenation, where Blakely and Somorjai [1976] showed that the platinum (111) surface was more active than any of the stepped surfaces for the reactions that lead to poisoning of the surface.

Experimentally, it seems that there is little correlation between step or kink density and reactivity. Stepped surfaces can be more active than closed packed planes, but they can also be less active. Rates often reach a maximum at intermediate step densities. Hence, there is no experimental evidence that steps or kinks have anything to do with catalytic activities of surfaces, even though there has been considerable speculation to that effect in the literature.

6.3.4 The Effects of Varying Composition

In Chapter 10, we discuss models for structure sensitivity in more detail. However, for now we leave this topic and move on to review how surface composition affects reactivity. Studies of the influence of surface composition on reactivity have a long history. Priestley [1790] found that copper was a more effective catalyst than iron oxide for dehydration of ethanol. Davy [1817] found that platinum was uniquely able to catalyze the reactions of mine gases (e.g., $CO + H_2 + O_2$). Faraday [1834] discovered that small amounts of H_2S or PH_3 would effectively inhibit the H_2/O_2 reaction on platinum. Experimentally, it is relatively easy to examine the influence of surface composition on reactivity. One runs a reaction on a series of surfaces made of different materials (e.g., Pt, Ni) and sees how the rate changes. Alternatively, one can add an inactive component (e.g., Na^+, H_2S) and measure rate variations. Such experiments have been done for almost 200 years. As a result, there is considerable data on how surface composition affects reactivity.

Generally, surface composition affects reactivity in a similar way to the way surface structure affects reactivity. For example, Figure 6.19 shows Sinfelt's [1973] measurements of the rate of ethylene hydrogenolysis and ethylene hydrogenation on a variety of transition

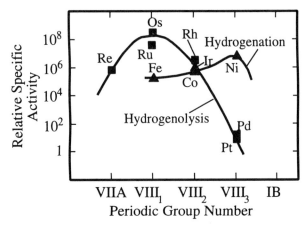

Figure 6.19 The relative rate of ethane hydrogenolysis and ethylene hydrogenation over several transition metal catalysts. (Data of Sinfelt [1963].)

TABLE 6.4 Catalysts for Some Common Reactions

Reaction	Catalyst	Reaction	Catalyst
Olefin hydrogenation	Pt, Pd, Ni, CrO_3	$CO + 3H_2 \rightleftharpoons CH_4 + H_2O$	Ni
Olefin polymerization	Ti^+, Cr^+, Zr^+	Aldehyde $\rightarrow$ Alcohol	Cu, Ni, Pt
Olefin metathesis	ReO, WO_2	Complete oxidation of hydrocarbons	Pd, Pt
Alcohol $\rightarrow$ Olefin	γ-Al_2O_3	$C_2H_4 + \frac{1}{2}O_2 \rightarrow C_2H_4O$	Ag
$CO + 3H_2 \rightarrow CH_3OH$	Cu/ZnO	Paraffin cracking, isomerization	Al_2O_3, SiO_2
$CO + xH_2 \rightarrow$ Hydrocarbons	Fe or Co	$\frac{1}{2}O_2 + SO_2 \rightarrow SO_3$	V_2O_5

metal catalysts. Notice that the rate of the hydrogenolysis reaction varies by five orders of magnitude with changing metal. The rate of ethylene hydrogenation varies to a lesser extent. The rate of ethane hydrogenolysis also seems to vary much more strongly with surface structure than the rate of ethylene hydrogenation.

Experimentally, if the rate of a reaction is structure sensitive, the rate of the reaction will also change significantly with surface composition. The converse is not true, however. There are several examples of reactions whose rate depends strongly on surface composition but only weakly on surface structure.

One of the consequences of the observation that rates of reactions often depend strongly on surface composition is that often only a single metal, or a few closely related metals are effective catalysts for a given reaction. For example, platinum is a good catalyst for complete oxidation of hydrocarbons. However, silver is a more effective partial oxidation catalyst. Table 6.4 lists some of the common catalysts used for a series of industrially important reactions. Gates [1992] notes that about half of the elements in the periodic table are used in industrial catalysts. Hence, without guidelines, it is difficult to know which metals will be the best catalysts for a given reaction.

In Chapter 10 we discuss some of the factors that cause some materials to be better catalysts for a given reaction than others. However, briefly, the key factors are the strength of the bond between the catalytic intermediates and the metal and the ability of the metal to form and break bonds with the adsorbed species. As noted in Section 6.1, catalytic reactions involve the sequential formation and destruction of adsorbate surface bonds. If the bonds are too weak, few reaction intermediates will form. If the bonds are too strong, the intermediates will be too stable to decompose at high rates. As a result, the best catalysts are those that form an intermediate strength bond with the catalytic intermediates. A discussion of why a given surface forms a strong bond or a weak bond to a given intermediate was given in Chapter 3.

6.3.5 The Effects of Alloy, Promoters, and Poisons

One of the ways one can try to tailor the strength of the bonding between a surface and an intermediate is by changing the composition of the surface. For example, one can combine two transition metals to form an alloy. That changes the electronic structure of the surface that leads to changes in reactivity. One can also add what is called a **poison** or a **promoter** to modify the reactivity. Generally poisons are things (e.g., sulfur) that bind to surfaces strongly and thereby inhibit undesirable reactions. Promoters are generally species such as alkalis that can donate electrons to the metal and thereby modify the metals properties. There are also things called structural promoters. Structural promoters are thought to modify the distribution of sites in a supported metal catalyst so that there is a higher proportion of active sites.

For example, iron catalysts are used for ammonia synthesis. Typically, the iron is promoted with potassium hydroxide and alumina, and often some ruthenium is added to modify the iron's properties. Alumina is thought to be a structural promoter. Strongin, Barey and Somorjai [1987] found that alumina will cause an iron catalyst to reconstruct in a way that exposes (111) facets. It happens that (111) facets are especially active for the formation of ammonia. Potassium is thought to be a simple electronic promoter. The potassium dissolves in the iron and gives up electrons, and thereby modifies the electronic properties of the iron. The ruthenium also dissolves in the iron and modifies the iron's properties, although the details are unclear.

One does not usually deliberately add poisons during ammonia synthesis. However, if there are as much as 200 PPM of H_2O or CO, or 100 PPM of H_2S in the feed, the catalyst is quickly poisoned because the surface of the catalysts is covered by oxygen, carbon or sulfur.

The influence of alloying agents on rates of catalytic reactions has been examined in great detail, as summarized in the reviews of Martin [1988], Ponec [1983], Sachtler and Van Sauten [1977], or Sinfelt, Carter, and Yates [1972]. For example, Figure 6.20 shows a plot of the reactivity of a copper–nickel alloy catalyst for ethane hydrogenolysis and cyclohexane dehydrogenation:

$$C_2H_6 + H_2 \rightarrow 2\ CH_4 \tag{6.19}$$

$$C_6H_{12} \rightarrow C_6H_6 + 3\ H_2 \tag{6.20}$$

where the reactivity is plotted as the rate per exposed nickel atom. Notice that the normalized rate of the dehydrogenation reaction is almost independent of the alloy composition, while the rate of the hydrogenolysis reaction depends on the alloy composition. When this work was done it was thought that copper was completely inert for both reactions, and as a result the main effect of the copper was to dilute the nickel sites. Now consider two different cases,

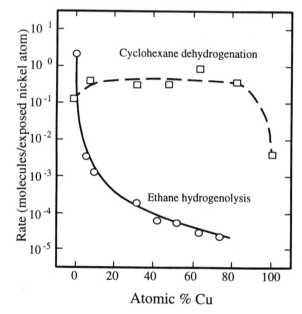

Figure 6.20 The rate of ethane hydrogenolysis and cyclohexane hydrogenation on a series of Cu–Ni alloy catalysts. (Adapted from Sinfelt et al. [1972].)

one where the critical ensemble size is 1 and a second where the critical ensemble size is 5 and assume that all 5 atoms must be nickel. In the former case, the rate of reaction is proportional to the number of nickel atoms on the surface. Therefore, the rate per exposed nickel atom will be constant. In contrast, in the later case, the rate will be proportional to the number of groups containing 5 adjacent nickel atoms. Notice that the number of 5-atom nickel ensembles will drop much more quickly with increasing copper concentration than the number of bare nickel atoms. Therefore, the implication of Figure 6.20 is that ethane hydrogenolysis requires a larger ensemble than cyclohexane dehydrogenation.

Recent data of Rodriguez, Campbell and Goodman [1991] call that interpretation into question. Rodriguez et al. find that when copper is deposited onto rhodium, the electronic structure of the copper changes so that the copper is active for dehydrogenation.

In general, alloying an active metal with an inactive one has two effects. It changes the ensemble size. However, it also modifies the electronic structure of each of the metals. Physically, each metal has its own electronegativity. When one forms an alloy, electrons flow to equalize the electronegativity of the system, as discussed in Section 3.4. The result is that the electronic structure of each of the atoms in the system is perturbed.

The perturbation has the most effect on the bond energy of surface species. Most of the ideas follow directly from ideas in Chapter 3. An electron withdrawing group lowers the interstitial electron density of the system, which tends to strengthen the interactions with electron-donating groups and weaken the strength of the interactions with electron-accepting groups. That changes the free-energy plot for the reaction (Figure 6.2). As a result, the reaction rate changes.

In order to work out the details of this model, one needs to understand the relationship between thermodynamics of reactions and rates of reactions. That material is discussed in Chapters 9 and 10. What is important to remember for now is that alloying changes the thermodynamics of reactions on surfaces. That, in turn, changes the reaction rate.

Promoters work similarly to alloys. When one adds a promoter to a catalyst, the promoter adsorbs on the surface of a catalyst. The adsorbed species produces a Friedel oscillation in the electronic structure of the catalyst similar to that shown in Figure 4.15. The Friedel oscillation produces a local perturbation in the electronic structure of the surface, and that modifies the properties of the catalyst.

In principle, promoters could enhance or inhibit reactions. However, in practice, one can often vary the properties of the promoters to get beneficial effects.

Poisons are the opposite of promoters. Poisons are generally species that bind very strongly to surfaces and thereby take up sites that would otherwise be available for reaction. Hence, they reduce reaction rates. For years it was thought that poisons were to be avoided, but now poisons are used to selectively act in inhibiting undesirable reactions without strongly inhibiting the main reaction pathway. Hence, poisons are often used to control surface reactions. For example, one often sulfides a reforming catalyst prior to using it. Sulfur is a strong catalyst poison. However, if one deposits sulfur in a controlled way, one can poison the undesirable coke-forming reactions without greatly affecting the main reactions in the reformer.

6.4   REACTIONS ON INSULATORS

In Section 6.3 we said we were discussing metals. However, a very similar discussion could have been made for insulators. Insulating metal oxides are used as partial oxidation catalysts, and catalysts for conversion of hydrocarbons. They are also important dielectric materials in materials processing. Metal sulfides are used as hydrodesulfurization catalysts. Thus, the catalytic properties of insulators are quite similar to those of metals.

There are some important differences, however. Insulators have more complex crystal structures than metals. There are two or more components, e.g., a metal and an oxygen, and often multiple stable phases. Insulators also can be formed into highly porous materials with unique cell structures. These complexities mean that there tends to be a wider distribution of sites available on insulators than on metals. Sometimes metal atoms are exposed; other times they are not. To complicate things further, reactions often occur not on the insulator itself; rather the reaction may occur on a Brønsted acid site (i.e., an adsorbed H^+) on the oxide surface. The effects of these complications is that reactions on insulators are not as well understood as reactions on metals.

Moffat [1990] and Henrich and Cox [1994] provide a good overview of reactions on insulators. There is insufficient room here to describe many of the details, so the reader should refer to those references if he/she is interested in reactions on oxides. However, so far it seems that one of the key differences between reactions on metals and reactions on insulators is that while reactions on metals tend to go via reaction pathways similar to those of radicals in the gas phase, reactions on insulators often go via pathways similar to those of ions in solution. For example, Figure 6.21 shows the mechanism of ethanol dehydration on an alumina catalyst. The reaction is standard carbonium ion chemistry of the type one would find in an elementary organic chemistry. The ethanol first reacts with an H^+ to yield a carbonium ion. The ion splits to yield water and an ethyl ion. The ethyl ion then undergoes a β-hydrogen elimination to yield ethylene. A surface proton reacts with the ethanol to form an adduct. The adduct then loses water. An identical reaction is seen in an acid solution, which suggests that reactions on insulators are very similar to ionic reactions in solution.

At present, there is still some controversy in the literature whether ionic intermediates can form during reactions on metal surfaces. There are a few examples where people have observed ionic intermediates at high coverages on metals. However, ionic intermediates are rarely seen on metal surfaces. In contrast, ionic intermediates are quite common on insulators. Hence, it appears that ionic species are much more stable on insulators than on metals.

The discussion in Section 3.4.1 can possibly explain why reactions on metal surfaces often follow pathways that are similar to radical reactions in the gas phase, while those on oxides often follow ionic pathways. Consider an arbitrary reaction A → B that can go via a radical

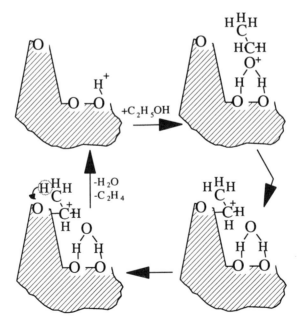

Figure 6.21 The mechanism of ethanol dehydration on an alumina catalyst. For details see Pines and Joust [1966].

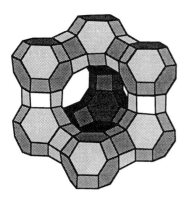

Figure 6.22 A diagram of the pore in a zeolite called faujasite.

transition state or an ionic transition state. The discussion in Section 3.4.1 shows that when a typical ion approaches a metal surface, the metal will tend to transfer an electron to the ion. The electron transfer destabilizes the ion. In contrast, a radical will simply form a strong bond to the metal surface. Thus, a clean metal surface will tend to destabilize an ionic transition state and stabilize a radical one.

In contrast, the discussion in Section 3.14 shows that insulators will have a much stronger interaction with ions than radicals. Thus, an insulator will tend to stabilize an ionic transition state.

Another way to look at this process is that an insulating surface acts like a solvent for a reaction in the same way that the electrons in a metal act like a solvent. However, insulators are hard acids and bases. Therefore, they tend to stabilize hard intermediates and hard transition states. Water stabilizes hard intermediates and hard transition states, too, while metal surfaces do not. As a result, reactions on insulators look much more like reactions in aqueous solution than do reactions on metal surfaces.

One does not want to carry these arguments too far because an adsorbate-covered metal surface can behave like an insulator, while a defect in an insulator can bond to a radical. Thus, there can be ionic pathways on metal surfaces or radical pathways on insulator surfaces. However, in general, insulators tend to promote ionic pathways while metals tend to promote radical pathways, which is a key difference in their behavior.

The other key difference between metals and insulators is that insulators can be formed into complex porous structures called zeolites. Figure 6.22 shows a diagram of the crystal structure of a zeolite called faujasite. There are a series of cavities connected by smaller pores. The result is a very complex structure. Reactions in zeolites are fundamentally different from reactions on extended surfaces. Mass transfer effects are quite important; if a reaction product cannot diffuse out of the zeolite, that product is trapped in the zeolite and eventually reacts to form something else. Hence, zeolites can be unusually selective for chemical reactions.

Another effect is a proximity effect. If a pore in a zeolite happens to have just the right shape to hold the transition state for a reaction, then entropy considerations described in Chapter 8 will catalyze the reaction. As a result, the shapes of the pores in a zeolite are more important than the specific interactions of the zeolite in determining the zeolite's reactivity. We will not be discussing zeolites in any detail in this book. A more complete discussion of the theory of reactions in zeolites is given in Moffat [1990].

6.5 REACTIONS ON SEMICONDUCTORS

Reactions on semiconductor surfaces have been studied to a much lesser extent than reactions on metals or insulators. However, so far it seems that reactions on semiconductors show properties somewhere between those of metals and those of insulators. Volkenshtein [1963,

1991] reviews much of the work up to 1960. Generally, two kinds of mechanisms are seen during reactions on semiconductors: radical mechanisms similar to those on metals and ionic mechanisms like those on insulators. For example, Volkenshtein proposes that the chlorination of ethane

$$C_2H_6 + Cl_2 \rightarrow C_2H_5Cl + HCl \tag{6.21}$$

follows a radical mechanism over a zinc chloride catalyst:

$$Cl_2 + 2S \rightarrow 2Cl_{(ad)}$$
$$C_2H_6 + 2S \rightarrow C_2H_{5(ad)}$$
$$C_2H_{5(ad)} + Cl_{(ad)} \rightarrow C_2H_5Cl + 2S \tag{6.22}$$
$$Cl_{(ad)} + H-(ad) \rightarrow HCl + 2S$$

Other examples of radical mechanisms are the hydrogenation of ethylene over ZnO or MoO_3 or NiO. On the other hand, the dehydration of ethanol (Reaction 6.13) seems to follow the ionic mechanism shown in Figure 6.21.

Semiconducting oxides differ from insulating oxides in that the semiconducting oxide is relatively easy to reduce to bare metal. That allows one to produce a catalytic cycle where one sequentially oxidizes and reduces the catalyst. For example, Schulz and Cox [1993] examined the oxidation of propylene to acrolein and water

$$H_2C=CH-CH_3 + O_2 \rightarrow H_2C=CH-CHO + H_2O \tag{6.23}$$

On a Cu_2O catalyst they found that the propylene reacts with the lattice oxygen to yield an allyloxy ($CH_2C=CHCH_2O$-) intermediate and a hydroxyl:

$$H_2C=CHCH_3 + 2O_{Latt} \rightarrow H_2C=CHCH_2O- + HO- \tag{6.24}$$

The $CH_2C=CCH_2-O$ and hydroxyl further react to form acrolein and water.

$$H_2C=CHCH_2O- + -OH \rightarrow H_2O + H_2C=CHCH_2O- + 2S \tag{6.25}$$

Then oxygen adsorbs to complete the cycle:

$$O_2 + 2S \rightarrow 2O_{Latt} \tag{6.26}$$

This type of chemistry where lattice oxygen is transferred to the products is thought to be very common, although the number of examples where one is sure that lattice oxygen is an active participant in the reaction is small.

Another interesting observation about semiconducting oxides is that as one changes the doping of the semiconductor, one often changes the electrical properties of the surface, and that produces a change in the rate of reaction. For example, Figure 6.24 shows how the rate and activation energy of CO oxidation varies as a function of the dopant concentration on a series of lithium–doped–nickel oxide catalysts. There is almost a three-order-of-magnitude change in the reaction rate, which does not show monotonic behavior with doping concentration. Interestingly, if one replots the data as a function of the work function, one gets a new linear correlation (see Figure 6.23). In Chapter 3 we noted that the workfunction change

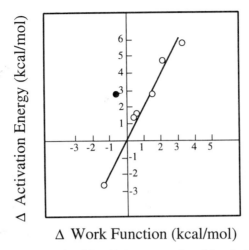

Figure 6.23 The activation energy for CO oxidation on a series of doped nickel oxide catalysts, as a function of the work function of the catalyst. (Adapted from Bielański and Derén [1969].)

is a measure of how the dopants affect the electronic properties of the catalyst. Therefore, the results in Figure 6.23 imply that changes in the surface electronic structure due to doping can have an important effect on rates of reactions on semiconductors.

There is some confusion about this point in the literature because, as we will see in Chapter 10, it is possible for a dopant to modify the bulk electronic properties of a semiconductor without strongly perturbing the surface electronic properties of the semiconductor. Further, most of the work on the effects of dopants on reactivity was done before 1950; Morrison [1990] and Kisklev [1990] note that impurities in the surface of the catalysts may have confused the doping effect. However, it still seems that the main difference between a reaction on a semiconductor and a reaction on a metal or insulator is that one can use dopants to

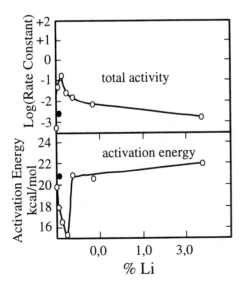

Figure 6.24 The rate and activation energy of CO oxidation over a series of lithium-doped–nickel oxide catalysts, as a function of the dopant of concentration. (Adapted from Bielański and Derén [1969].)

change the electrical properties of the semiconductor. That can have an important effect on the reactivity of the semiconductor.

6.6 SUMMARY AND PLANS FOR THE REMAINDER OF THIS BOOK

In summary, the results in this chapter show that reactions on surfaces are not all that much different from reactions in the gas phase or in solution. Mechanisms are quite similar. The one major difference is that surfaces can form bonds to intermediates, and thereby stabilize them. The stabilization of intermediates leads to enhanced rates. Of course, the surface is an active participant in a surface reaction. We usually only observe intermediates that are properly configured to bind to surfaces, and fit in the spaces in the surface. Details of the composition and electronic structure of surfaces play an important role in determining rates.

Most of the discussion in this chapter was qualitative. However, in the remainder of this book we quantify the ideas. In Chapter 7 we examine kinetic rate laws for reactions on surfaces. The chapter provides a framework to discuss rates of reactions on surfaces. It also discusses how adsorbate phase behavior relates to rates. Chapter 8 is a review of absolute rate theory. The main goal of the chapter is to review some of the methods that are used to estimate reaction rates and to review the application of those methods to reactions on surfaces. Chapter 9 is a more in-depth treatment of reaction rate theory, with an emphasis on why reaction rates change as the properties of the reactants change. This material is applied to predict rates and mechanisms of surface reactions in Chapter 10. A fair amount of review material is included in each of these chapters. However, it is assumed that the reader has already had an introduction to reaction rate theory at the level usually found in an introductory physical chemistry course.

PROBLEMS[*]

6.1   Define the following terms in three sentences or less

(1) Langmuir–Hinshelwood mechanism	(8) Structure-insensitive reaction
(2) Rideal–Eley mechanism	(9) Secondary structure sensitivity
(3) Eley–Rideal mechanism	(10) Multiplier
(4) Precursor mechanism	(11) Critical ensemble
(5) Trapping mediated mechanism	(12) Poison
(6) Structure-sensitive reactions	(13) Promoter
(7) Active site	

6.2   Figure 6.2 shows a free-energy diagram for the oxidation of hydrogen on Pt(111).

    (a) How would the diagram change if the heat of adsorption of oxygen were increased by 20 kcal/mole? Assume that the heat of adsorption of hydrogen and water does not change, but that the heat of adsorption of OH increases by 10 kcal/mole.

    (b) How would the rate change?

    (c) Now use the data in Tables 3.4 and 3.5 to compare hydrogen oxidation on nickel and platinum. Which would be a better catalyst?

    (d) Now generalize the results in part (c). How would you expect the rate of hydrogen oxidation to vary over the periodic table?

    (e) Copper and lead are not good hydrogen oxidation catalysts. Provide a mechanism to explain why they are inactive.

*Additional problems on this material can be found in Chapter 10.

6.3 Sachtler and Fahrenfort [1958] examined the decomposition of formic acid on a number of transition metal catalysts. They suggested that the reaction proceeded by the following mechanism:

$$HCOOH + 2S \rightarrow HCOO_{(ad)} + H_{ad}$$

$$HCOO_{(ad)} \rightarrow CO_{2(g)} + H_{ad}$$

$$2H_{ad} \rightarrow H_2$$

Based on the thermodynamic arguments in section 6.1, which of the following metals would be the most active for the reaction at 200°C in UHV?

Metal	Calculated Heat of Formation of Formate (kcal/mole)	Average Heat of Adsorption of Hydrogen (kcal/mole of H_2)
Copper	90	10
Silver	72	5
Rhodium	79	20
Platinum	78	18
Nickel	100	23
Iron	102	27
Tungsten	108	33

6.4 Describe in your own words how surface structure affects the mechanism and rates of surface reactions.

6.5 In Section 6.3.3 we noted that there is often little correlation between rates of reactions on surfaces and the step density of the surface.

(a) Replot the date in Table 6.3 as a function of step and kink density. What trends do you see?

(b) Compare the absolute rates to those in Figure 6.17. Why is there the large difference? (*Note*: 1 turnover number $\approx 10^{15}$ molecules/cm^2 sec).

6.6 Describe in your own words how surface composition affects rates of surface reactions.

6.7 Szabo, Henderson, and Yates, *J. Chem. Phys.* **96,** 6191 [1992] examined the active site for CO oxidation on a stepped surface, Pt(211).

(a) Summarize the key results in their paper.

(b) How well are their results in accordance with the general idea that steps enhance rates of surface reactions?

(c) Based on the results in Szabo et al.'s paper, would you conclude that CO oxidation is a structure-sensitive or structure-insensitive reaction?

6.8 The objective of this problem is to use the data in Figure 6.4 to estimate the critical ensemble size for ethane hydrogenolysis.

(a) Pick n adjacent sites on the surface. Use the material in Chapter 4 to show that if a random alloy forms (i.e., if $\beta h = 0$), the probability of n sites being covered by nickel atoms is θ_{Ni}^n where θ_{Ni} is the fractional coverage of nickel.

(b) Now consider a reaction that requires n adjacent nickel atoms to occur. Show that the rate of such a reaction per exposed nickel atom will be proportional to θ_{Ni}^{N-1}.

(c) In principle one could use the results in part (b) to estimate the critical ensemble size for the reactions. In Figure 6.4 the idea is to plot the ln rate vs. ln θ_{Ni} and use the slope to estimate n. The problem with doing that plot is that there is no guarantee that bulk concentrations of alloys are identical to surface concentrations.

Still, for the purpose of this problem, assume that θ_{Ni} equals the bulk nickel concentration to estimate the critical ensemble size for ethane hydrogenolysis and cyclohexane dehydrogenation.

More Advanced Problems

6.9 In Section 6.3 we said that reactions on metal surfaces often look quite like an oxidation/hydrogenation reaction.

(a) Look up the mechanism of ethane oxidation in the gas phase.

(b) Replace all of the OH and O radicals in part (a) with [s].

(c) Now look up the mechanism on a Ni(111) surface. What do you see?

6.10 Paul and Bent, *J. Catalysis* **14**, 264 [1994] examined alkyl coupling (e.g., $2CH_{3(ad)}$ → C_2H_6) on Cu(110). They found that there is a correlation between the bond energy of the adsorbed alkyl and the coupling rate.

(a) How would you expect the strength of the metal–alkyl bond to affect the coupling rate?

(b) How do your expectations continue to Paul and Bent's results?

6.11 Xu, Peck, and Koel, *J. Am. Chem. Soc.* **115**, p. 751 [1993] examined the reactions of acetylene on a platinum/tin alloy. They found that tin suppresses the decomposition of acetylene on a Pt(111) surface, thereby allowing the acetylene to trimerize to benzene. What do these results tell you about the critical ensemble for acetylene decomposition and benzene formation?

6.12 Martin [1988] describes the work that has been done to examine the critical ensemble for a number of reactions.

(a) Summarize his findings.

(b) How well do the ideas explain the data in Table 6.3?

6.13 Go to the library and do a computer search on the term Rideal–Eley.

(a) Are there any examples of Rideal–Eley reactions with stable molecules?

(b) What do you conclude from these findings?

REFERENCES

Anderson, J. R., and M. Boudart, *Catalysis Science and Technology*, Springer, NY (1981).

Anton, A. B., and D. C. Cadogan, *J. Vac. Sci. Tech. A* **9**, 1890 (1991).

Arrhenius, S. A., *Z. Physik. Chem.* **4**, 22 (1889); **28**, 317 (1899).

Arumainayagain, C. R., and R. J. Madix, *Prog. Surface Sci.* **38**, 1 (1991).

Back, M. H., and R. Martin, *J. Chem. Kinet.* **11**, 757 (1979).

Backman, A. L., and R. I. Masel, *J. Vac. Sci.Tech.* **6**, 1137 (1988).

Backman, A., and R. I. Masel, *J. Vac. Sci. Tech. A* **9**, 1789 (1992).

Balandin, A. A., *Z. Physik. Chem.* **132**, 289 (1929); 133, 167 (1929).

Balandin, A. A., *Adv. Catal.* **19**, 1 (1969).

Barteau, M. A., and R. B. Madix, in D. A. King and D. P. Woodruf, eds., *The Chemical Physics of Solid Surfaces and Heterogeneous Catalysis*, Vol. 4, p. 95, Elsevier, NY (1982).

Berzelius, J. T., *Edinb. New Philos. J.* **21**, 222 (1836).

Bielański, A., and J. Derén in F. F. Volkenshtein, ed., *Symposium on Electron Phenemena in Chemisorption and Catalysis on Semiconductors*, p. 49, DeGrapter, Berlin, (1969).

Bielański, A., and R. Dziembaj, *Bull. Pol. Acad. Sci. Ser. Sci. Chim.* **16**, 269 (1968).

Blakely, D. W., and G. A. Somorjai, *Nature* **258,** 580 (1975).

Bond, G. C., *Trans. Faraday Soc.* **52** 1235 (1956).

Bond, G. C., and P. B. Wells, *Adv. Catal.* **15,** 92 (1964).

Boudart, M. *J. Am. Chem. Soc.* **72,** 1040 (1950).

Boudart, M. *Adv. Catal.* **20,** 153 (1969).

Boudart, M. in *Proceedings 6th International Congress on Catalysis*, p. 1 (1975).

Boudart, M. (1975), *J. Catal.* A. Delboville, J. A. Dumesic, and S. Khammouma, H. Topsoe, *J. Catal.* **37,** 486 (1975).

Boudart, M., and G. Djega-Mirajossou, *Kinetics of Heterogeneous Catalytic Reactions*, Princeton University Press, New Brunswick, NJ (1984).

Brass, S. G., and G. Ehrlich, *J. Chem. Phys.* **87,** 4285 (1987).

Burwell, R. L., *Catal. Rev.* **57,** 1895 (1967).

Butt, J. B., and I. Onal, *Trans. Faraday Soc.* **78,** 1982 (1987).

Ceyer, S. T., *Ann. Rev. Phys. Chem.* **39,** 479 (1988).

Clément, and Désarnes, *Ann. Chim.* **59,** 329 (1806).

Creighton, J. R., B. A. Bansenauer, and T. Huett, and J. M. White, *J. Vac. Sci. Tech. A* **11,** 876 (1993).

Crowell, J. E., W. T. Tlsoe, and G. A. Somorjai, *J. Phys. Chem.* **89,** 1598 (1985).

Dai, Q., and A. J. Gellman, *J. Am. Chem. Soc.* **115,** 714 (1993).

Davis, J. L., and M. A. Barteau, *Surface Sci.* **197,** 123 (1988).

Davis, J. L., and M. A. Barteau, *J. Am. Chem. Soc.* **111,** 1782 (1989).

Davis, M. A., F. Zaera, and G. A. Somorjai, *J. Catal.* **85,** 206 (1984).

Davis, S. M., and G. A. Somorjai, *J. Catal.* **65,** 78 (1980).

Davis, S. M., and G. A. Somorjai, in D. A. King, and D. P. Woodruff, eds., *The Chemical Physics of Solid Surfaces and Heterogeneous Catalysis*, Vol. 4, p. 217, Elsevier, NY (1982).

Davy, H., *Philos. Trans.* **107,** 77 (1817).

Doberiener, M., *Annales de Chemie,* **24,** 93 (1823).

Donnelly, J. M., *J. Vac. Sci. Tech.* **9,** 2987 (1991).

Dorling, T. A., M. J. Eastlake, and R. L. Moss, *J. Catal.* **14,** 23 (1969).

Dowden, D. A., in *Proceedings, 5th International Congress on Catalysis*, p. 621 (1973).

Ducros, R., and R. P. Merrill, *Surface Sci.* **55,** 227 (1976).

Dulong, M. M., and L. J. Theonard, *Annales de Chemie,* **23,** 440 (1822); **24,** 330 (1823).

Elswirth, M., M. Möller, K. Wetzl, R. Imbihl, and G. Ertl, *J. Chem. Phys.* **90,** 510 (1989).

Engle, T., and G. Ertl, *J. Chem. Phys.* **69,** 1267 (1978).

Engle, T., and G. Ertl, *Adv. Catal.* **28,** 1 (1979).

Faraday, M., *Philos. Trans.* **124,** 55 (1834).

Gates, B. C., *Catalytic Chemistry*, Wiley, NY (1992).

Gellman, A. J., J. C. Dunphy, M. Salmeron, and I. Langmuir, **8** 534 (1992).

Gentile, T. M., and E. L. Mutterties, *J. Phys. Chem.* **87,** 2469 (1983).

Gow, T. R., and R. Lin, and R. I. Masel, *J. Cryst. Growth* **106,** 577 (1990).

Hall, J., I. Zoric, and B. Kasemo, *Surface Sci.* **270,** 460 (1992).

Harris, S. J., *Appl. Phys. Lett.* **56,** 2298 (1990).

Harris, S. J., and D. G. Goodwin, *J. Phys. Chem.* **97,** 23 (1993).

Harris, J., and S. Anderson, *Phys. Rev. Lett.* **55,** 1583 (1985).

Hatzikos, G., and R. I. Masel, *Surface Sci.* **185,** 479 (1987).

Henrich, V., and P. A. Cox, *Metal Oxides*, Cambridge University Press, NY (1994).

Henry, W. D., *Philos. Trans.* **114,** 266 (1824).

Herz, R. K., Ph.D. thesis, University of California, Berkeley, 1977.

Herz, R. K., W. D. Gillespis, E. E. Petersen, and G. A. Somorjai, *J. Catal.* **67,** 371 (1981).

Hinshelwood, C. N., *Proc. Roy. Soc.* **A188**, 1 (1946).

Hoffmann, H., F. Zaera, R. M. Olmerod, R. M. Lambert, J. M. Yao, D. K. Saldin, L. P. Wang, D. W. Bennett, and W. T. Tysoe, *Surface Sci.* **268**, 1 (1992).

Ibach, H., and S. Lehwald, *J. Vac. Sci. Tech.* **15**, 407 (1978); *Surface Sci.* **89**, 425 (1978).

Imbihl, R., and S. Ladas, and G. Ertl, *Surface Sci.* **215**, L307 (1989).

Kahn, D. R., E. E. Peterson, and G. A. Somorjai, *J. Catal.* **34**, 294 (1974).

Kesmodel, L. L., L. H. Dubois, and G. A. Somorjai, *J. Chem. Phys.* **70**, 2180 (1979).

Kesmodel L., and L. Falicov, *Solid State Communication* **16**, 1201 (1975).

King, D. A., and D. P. Woodruff, *The Chemical Physics of Solid Surfaces and Heterogeneous Catalysts*, Elsevier, NY (vol. 1, (1981); vol. 3b, (1991).

Kiselev, *Adsorption and Catalysis on Transition Metals and Their Oxides*, Springer Verlag, NY (1989).

Langmuir, I., *J. Am. Chem. Soc.* **34**, 1310 (1912); **35**, 105 (1913); **37**, 1139 (1915); **40**, 1361 (1918).

Lin, R., and R. I. Masel, *Surface Sci.* **258**, 225 (1991).

Madix, R. J., ed., *Surface Reactions*, Elsevier, NY (1994).

Madix, R. T., in D. A. King, and D. P. Woodruff, eds., *The Chemical Physics of Solid Surfaces and Heterogeneous Catalysts*, Vol. 4, p. 1, Elsevier, NY (1982).

Mango, F., *Adv. Catal.* **70**, 291 (1969).

Manogue, W. H., and J. R. Katzer, *J Catal.* **32**, 166 (1974).

Martin, G. A., *Catal. Rev.* **30**, 519 (1988).

Masel, R. I., *Catal. Rev.* **28**, 335 (1986).

McBaker, M., and G. I. Jenkins, *Adv. Catal.* **7**, 1 (1955).

Moffat, J., *Theoretical Aspects of Heterogeneous Catalysis*, Van Nostrand Reinhard, NY (1990).

Morrison, S. R., *The Chemical Physics of Surfaces*, Plenum, NY (1990).

Norskov, J K., *Rep. Prog. Phys.* **53**, 1253 (1990).

Norskov, J. K., and P. Stoltze, and U. Nielsen, *Catal. Lett.* **9**, 173 (1991).

Park, Y. O., and W. F. Banholzer, and R. I. Masel, *Surface Sci.* **155**, 341 (1985); *Adv. Surf. Sci.* **19**, 145 (1985).

Pease, R. N., and R. J. Stewart, *J. Am. Chem. Soc.* **47**, 1235 (1925).

Pines, H., M. Joust, *Adv. Catal.* **16**, 49 (1966).

Priestley, J., *Experiments on Different Kinds of Air*, 2nd ed., J. Johnson London, Vol. III (1790).

Ponec, V., *Adv. Catal.*, **32**, 149 (1977); (1983).

Ponec, V., and W. H. Sachtler, in *Proceedings, 5th International Conference on Catalysis*, p. 645, Elsevier, NY (1972).

Rodriguez, J. A., and D. W. Goodman, *Surface Sci. Rep.* **14**, 1 (1991).

Rodriguez, J. A., and R. T. Campbell, D. W. Goodman, *Surface Sci.* **244**, 211 (1991).

Sabatier, P., *Catalysis in Organic Chemistry*, Paris (1913), English translation, Van Nostrand, NY (1923).

Sachdev, A., R. I. Masel, and R. B. Adams, *J. Catal.* **136**, 320 (1992).

Sachtler, W. H. M., and J. Fahrenfort, *Proceedings, 5th International Congress on Catalysis*, p. 1 (1958).

Sachtler, W. M. H., and R. A. Van Santen, *Adv. Catal.* **26**, 69 (1977).

Schlatter, J. C., and M. Boudart, *J. Catal.* **17**, 482 (1972).

Schwab, G. M., in J. R. Anderson and M. Boudart, eds., *Catalysis Science and Technology*, Vol. 2, p. 1, Springer Verlag, NY (1981).

Schwab, G. M., and L. Z. Rudoff, *Z. Electrochem.* **37** 660 (1931).

Sheppard, N., *Ann. Rev. Phys. Chem.* **39**, 589 (1988).

Shultz, K. H., and D. F. Cox, *J. Catal.* **143**, 646 (1993).

Shusturovich, E., *Metal Surface Reaction Energetics*, VCH (1991).

Siera, J., B. E. Niewenhuys, H. Hirano, and K. I. Janaka, *Catal. Lett.* **3**, 179 (1989).

Sinfelt, J. H., *Bimetallic Catalysis: Discoveries, Concepts, Applications*, Wiley, NY (1983).

Sinfelt, J. H., H. L. Carter, and D. J. Yates, *J. Catal.* **24,** 283 (1972).

Sinfelt, J. H., and P. J. Lucchesi, *J. Am. Chem. Soc.* **85,** 3365 (1963).

Slavin, A. L., R. E. Bent, C. T. Kao, and G. A. Somorjai, *Surface Sci.* **202,** 388 (1988); **206,** 124 (1988).

Somorjai, G. A., *Adv. Catal.* **26,** 1 (1977).

Strongin, D. R., S. R. Bare, and G. A. Somorjai, *J. Catal.* **103,** 289 (1987).

Sugai, S., K. Shimuzu, H. Watanbe, H. Miki, and K. Kawasaki, *Surface Sci.* **287,** 455 (1993).

Sugai, S., K. Takeuchi, T. Ban, H. Miki, K. Kawasaki, and T. Kioka, *Surface Sci.* **282,** 67 (1993).

Taylor, H. S. *Proc. Royal Soc. London A* **108,** 105 (1925).

Taylor, H. S. *Adv. Catal.* **1,** 1 (1948).

Tempkin, M. I., *Zh. Fiz. Khim.* **31,** 3 (1957).

Tempkin, M. I., *Dokl. Akad. Nauk. SSSR* **161,** 160 (1965).

Thomas, J. A., *Adv. Catal.* **19,** 108 (1976).

Tompkins, F. C., in *Proceedings, 6th International Congress on Catalysis*, p. 32 (1976).

Van Marum, M., *Scheikd. Bibl. Roelofswaar* **3,** 209 (1796).

Van't Hoff, J. H., *Lectures on Theoretical Chemistry*, p. 215 (1898).

Volkenshtein, F. F., *The Electronic Theory of Catalysis on Semiconductors*, Oxford University Press, NY (1963).

Volkenshtein, F. F., *Electronic Processes on Semiconductor Surfaces during Chemisorption*, Consultants Bureau, NY (1991).

Wang, J., and R. I. Masel, *Surface Sci.* **243,** 199 (1991).

Weinberg, W. H., *J. Vac. Sci. Tech. A.* **10,** 2271 (1992).

Wolkenstein, T. H., *J. Adv. Catal.* **12,** 189 (1960).

Williams, E. R., G. C. Jones, L. Fang, R. N. Zare, B. J. Garrison, and D. W. Brenner, *J. Am. Chem. Soc.* **114,** 3207 (1992).

Yagasaki, E., and R. I. Masel, *Surface Sci.* **225,** 51 (1990).

Yagasaki, E., and R. I. Masel, Royal Society Catalysis series, 375 (1994).

Yu, M. L., *Thin Solid Films* **225,** 7 (1993a).

Yu, M. L., *J. Appl. Phys.* **73,** 716 (1993b).

Zaleney, P., A. Wieckowski, E. Herrero, K. Franszcuk, J. Wang, and R. I. Masel, *J. Phys. Chem.* **96,** 8509 (1992).

7 Rate Laws for Reactions on Surfaces I: Kinetic Models

PRÉCIS

In the previous chapter, we found that mechanisms of surface reactions can be put into three general categories: Rideal–Eley reactions, Langmuir–Hinshelwood reactions, and precursor reactions. The objective of this chapter is to examine the kinetic consequences of these three mechanisms and how interactions between the adsorbed reactants affect the rate. First we discuss rates of reactions on surfaces in general, and derive what are called Hougan and Watson kinetics for the system. Next, we derive corrections to these simple kinetics based on the material in Chapters 4 and 5. The result is a kinetic expression that can be compared to experiment. Much of the work builds on the material in Chapter 4, so the reader might reread that chapter before proceeding.

7.1 INTRODUCTION

As noted in the previous chapter, heterogeneously catalyzed reactions were first studied in the later part of the eighteenth century. At the time, however, people did not know that these reactions were surface reactions. Rather most investigators assumed that reactions were occurring uniformly in the bulk of the fluid in the pores in the catalyst.

The first real evidence that a reaction could happen on a surface came from the work of Naumann [1871]. Naumann was examining the decomposition of phosphine in a glass flask, and found that as he changed the shape of the flask, his rate data changed. Generally, Naumann found that if he had a long thin tube, the reaction rate was higher than on a round flask with identical volume. The highest rates were found when his reactor had a large surface area. Therefore, Naumann proposed that a reaction was occurring on the surface of his flask.

Van't Hoff [1896] suggested several experiments to check whether reactions were occurring on the surface of a reactor. The experiments included covering the surface of the reactor with an inert oil or soot, aging the vessel in reactants, and adding beads to change the surface area of the vessel. Thus, one can see if there is a correlation between the rate and the surface area. Naumann did the appropriate experiments with his reactor. He also showed that the rate of the decomposition process scaled as the surface area of his vessel. As a result, Naumann concluded that phosphine could decompose on the surface of quartz. This was the first time that someone had definitely showed that reactions could occur on surfaces.

Soon thereafter, Van't Hoff proposed that rates of surface reaction should be measured as a rate per unit area, r_A, in units of molecules converted/cm^2 sec. This proposal is still followed today.

Over the years, people have used a variety of different areas in the calculation of a rate per unit area. Naumann, for example, calculated a rate per unit surface area of his reactor where the surface area was calculated by measuring the physical dimensions (i.e., length, width, height of the reactor). Some years later, people realized that glass vessels and other materials were porous, so they began to report rates per total area of the material where the total area includes both the physical area of the material and the area of all of the pores. In the catalysis literature, it has now become common to measure a rate per unit *active surface area* where the active surface area is the portion of the surface that is catalytically active. For example, a supported metal catalyst consists of an active metal (e.g., platinum) on a relatively inactive support (e.g., Al_2O_3). In many cases only the platinum is catalytically active. In such a case, one might report the rate of the reaction as a rate per square meter of exposed metal. The metal area can be estimated by measuring an adsorption isotherm for a gas that only adsorbs on the metal, and then using the methods in Example 4.H to estimate a surface area. An example of this type of calculation is given in Example 7.A. Of course, there is no guarantee that the metal area is the active surface area. Some metal atoms may be inactive. There could also be reaction on the support. Still in heterogeneous catalysis it has become common to report a rate per unit metal surface area.

In film growth, one instead reports what is called a *growth rate*, g_A, in Å/min where the growth rate is related to the rate per unit area by

$$g_A = \frac{r_A}{\rho_f} \tag{7.1}$$

where ρ_f is the molar density of the growing film. The growth rate is proportional to the reaction rate in molecules/cm^2 but is generally measured in microns/min.

One also often sees a rate expressed as a **turnover number,** or turnover frequency, T_N, where the turnover number is defined as the rate of molecules converted per exposed surface atom per second. Note the turnover number, T_N, is related to the rate per unit area, r_A, by

$$T_N = \frac{r_A}{N_A} \tag{7.2}$$

where N_A is the number of exposed surface atoms/cm^2. In the literature it has become common to assume that there is one site per exposed surface atom. In that case, N_A is equal to the number of exposed surface atoms/cm^2. In such a case the turnover number will equal the reaction rate in molecules/surface atom/sec. In film growth, the turnover number is also equal to the growth rate in monolayers/sec.

Figure 7.1 shows a plot of some typical turnover numbers for a number of surface reactions. Turnover numbers for catalytic reactions of hydrocarbons (i.e., hydrogenation, dehydrogenation, cyclization) are typically in the order of 10^{-4} to 10^2/sec. Film growth reactions usually have turnover numbers in the order of 1 to 200/sec. One can occasionally observe reactions with turnover numbers greater than 200/sec. However, in practice most reactions become mass transfer limited at turnover numbers in the order of 100–300/sec.

7.2 KINETICS OF SURFACE REACTIONS

Over the years, there have been many attempts to determine how rates of surface reactions vary with the partial pressure of the reactants. At first, people measured rates of surface reactions empirically. One of the things that has been discovered is that rates of surface reactions do not bear any simple relationship to the stoichiometry. In the gas phase, rates of reactions are often proportional to the reactant concentrations to some simple powers. However, surface reactions follow much more complex rate equations. It is common for the rate of a catalytic reaction to be constant or even go down as the concentration of one of the reactants increases. This is quite different from gas phase reactions where rates generally increase with increasing reactant pressure.

In the discussion that follows we provide examples of this effect for two separate cases:

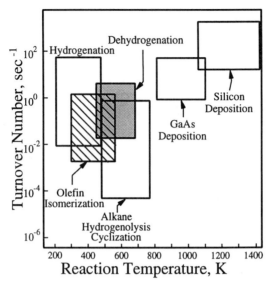

Figure 7.1 Approximate rates of hydrocarbon conversion and film growth reactions. (Adapted from Davis and Somorjai [1982].)

unimolecular surface reactions, where a single reactant rearranges or decomposes to yield products, and bimolecular surface reactions, where two or more reactants combine and rearrange to form products. Figure 7.2 shows some data for the rate of a simple unimolecular surface reaction: the decomposition of ammonia

$$NH_3 \rightarrow \tfrac{1}{2} N_2 + \tfrac{3}{2} H_2 \tag{7.3}$$

over a polycrystalline platinum wire. Notice that at moderate temperatures, the rate of reaction increases with increasing ammonia pressure, and then levels off. At higher temperatures the rate continues to increase over the pressure range shown. However, other data show that the rate will eventually level off at high enough pressures.

Figure 7.3 shows how the reaction rate varies with temperature at a series of fixed pressures. Notice that the rate increases with increasing temperature, and then begins to level off.

The results in Figure 7.2 and Figure 7.3 are typical of those for unimolecular surface reactions. Generally, one observes rates that increase with increasing temperature and pressure. However, the rate does not usually monotonically increase with pressure. Rather the rate increase slows with increasing pressure, and the rate eventually plateaus at high pressures. The rate also increases with temperature. However, the slope of the curve decreases with increasing temperature.

Note that gas phase reactions show much different behavior than that seen in Figure 7.2 and Figure 7.3. In the gas phase the rate of a simple unimolecular reaction increases continuously with increasing pressure and never levels off. The rate also increases continuously with increasing temperature according to Arrhenius's law:

$$\text{rate} = r_0 \exp\left(\frac{E_a}{kT}\right) \tag{7.4}$$

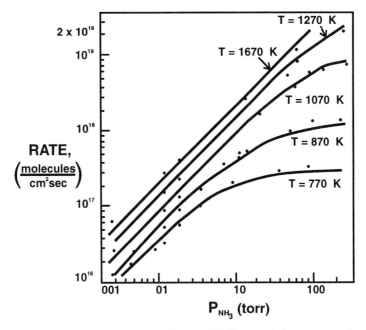

Figure 7.2   The rate of the reaction $NH_3 \rightarrow 1/2\ N_2 + 3/2\ H_2$ on a platinum wire catalyst at a variety of temperatures. (Data of Loffler and Schmidt [1976].)

According to Equation 7.4 the slope of the rate versus temperature curve should increase with increasing temperature. However, Figure 7.3 shows that the slope actually decreases. Therefore, it seems that the kinetics of unimolecular surface reactions are quite different from the kinetics of unimolecular reactions in the gas phase.

Even larger differences are seen in bimolecular reactions. For example, Figure 7.4 shows how the rate of the reaction

$$CO + \tfrac{1}{2}\ O_2 \rightarrow CO_2 \tag{7.5}$$

varies with temperature over a Rh(111) catalyst. Notice that at fixed reactant pressure the rate reaches a maximum with increasing temperature and then declines. Figure 7.5 shows the pressure dependence of reaction rate. The figure looks confusing, but if one starts with the 700 K curve on the left side of the figure, one finds that the rate first increases with increasing CO pressure, and then declines. At lower temperatures the rate continuously decreases with increasing CO pressure, while at higher temperatures the rate increases with increasing CO pressure.

The behaviors in Figure 7.4 and Figure 7.5 are typical of that in bimolecular surface reactions. The reaction rate generally shows complex behavior with temperature and concentration. The behavior is usually quite different than that of bimolecular gas phase reactions.

One additional complication is that surface reactions can show multiple steady states. For example, Figure 7.6 shows how the rate of the reaction $CO + \tfrac{1}{2}\ O_2 \rightarrow CO_2$ varies with temperature over a Rh(100) catalyst. Note that there are two different lines between 400 and 600 K. The reaction follows the lower of these lines if one starts the reaction at a low temperature, and slowly increases the temperature. In contrast, the reaction follows the higher

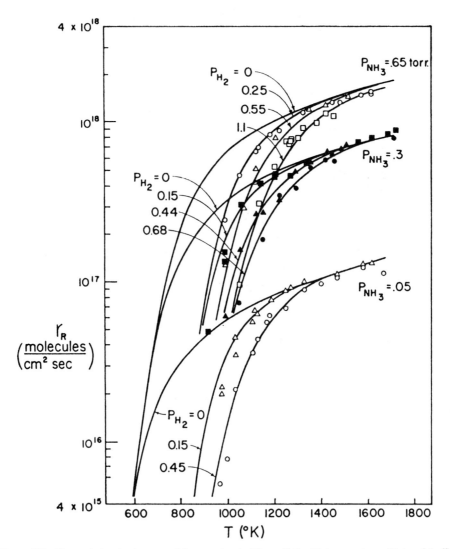

Figure 7.3 The variation in the rate of the reaction in Figure 7.2 with temperature. (Data of Loffler and Schmidt [1976].)

line, if one starts with a hot catalyst and cools. As a result, there are two possible steady-state rates at a fixed temperature and pressure.

Multiple steady states are not that common in surface reactions; they are usually seen over a limited range of temperature. Still, the presence of multiple steady states implies that the reaction rate is not uniquely determined by the temperature and pressure of the system and the concentration of all of the reactants, i.e., the reaction rate is not a thermodynamic state variable. Such complications are rarely seen in the gas phase.

Typically one fits data for the rate of a surface reaction with a complex rate equation. Table 7.1 shows a selection of the rate equations that have been fit to data on supported catalysts. The table is extracted from a longer compilation of Mezaki and Ioue [1991]. Note that the rate equations are often quite complex, and the rate can go up or down with increasing reactant pressure. The rate of the reaction $CO + 2H_2 \Rightarrow CH_3OH$ increases with increasing

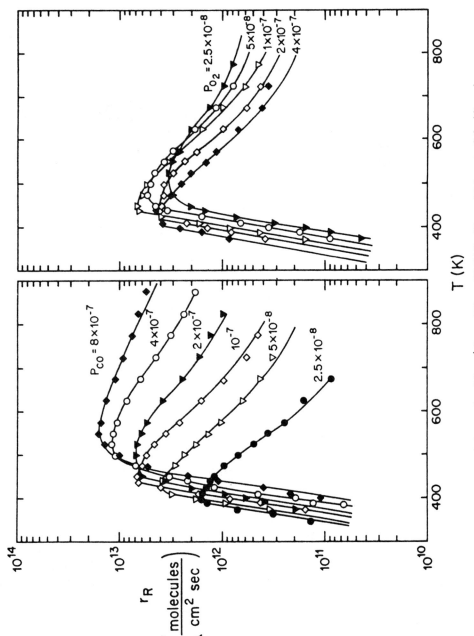

Figure 7.4 The rate of the reaction $CO + \frac{1}{2} O_2 \rightarrow CO_2$ on Rh(111). (Data of Schwartz, Schmidt, and Fisher [1986].)

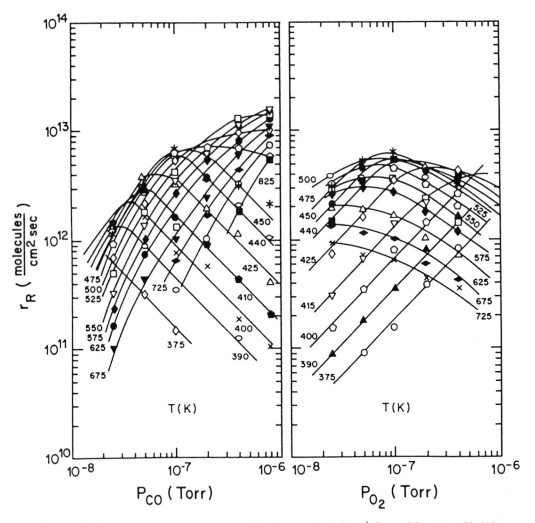

Figure 7.5 The reactant pressure dependence for the reaction $CO + \frac{1}{2} O_2 \rightarrow CO_2$ over a Rh(111) catalyst. (Data of Schwartz et al. [1986].)

CO pressure and then reaches a plateau. However, the rate continuously increases with continuous increase in H_2 pressure. In contrast, the rate of the reaction $SO_2 + \frac{1}{2} \Rightarrow SO_2$ plateaus as the partial pressure of either reactant increases, while the rate of the reaction $4NH_3 + 6NO \Rightarrow 5N_2 + 6H_2O$ reaches a maximum with increasing reactant pressure, and then declines. Generally, the kinetics of reactions on surfaces are quite different from the kinetics of reactions in the gas phase or in solution. Consequently, much different rate equations are generally used.

7.3 THE LANGMUIR RATE EQUATION

Langmuir [1912, 1918] proposed a simple model to explain why a rate of simple surface reactions shows such complex behavior. Langmuir assumed that the reaction between two

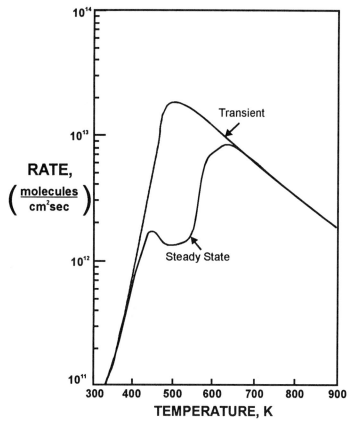

Figure 7.6 The rate of the reaction $CO + \frac{1}{2}O_2 \rightarrow CO_2$ over an Rh(100) catalyst. (Data of Schwartz et al. [1986].)

species A and B followed a Langmuir–Hinshelwood mechanism:

$$S + A \rightleftharpoons A_{ad} \quad (1)$$

$$S + B \rightleftharpoons B_{ad} \quad (2) \qquad\qquad (7.6)$$

$$A_{ad} + B_{ad} \longrightarrow Products + 2S \quad (3)$$

Langmuir also assumed that the last step in Reaction 7.6 was much slower than the rest, so the surface concentrations of A and B maintained a dynamic equilibrium. If Reaction 3 is elementary, then r_{AB}, the rate of the surface catalyzed reaction between two species A and B is equal to

$$r_{AB} = \kappa_{AB} N_A N_B \qquad\qquad (7.7)$$

where N_A and N_B are the concentrations of A and B on the surface of the catalyst in molecules/ cm^2, and κ_{AB} is a constant. If it is assumed that there are a fixed number of sites, N_S, to hold gas, then Equation 7.7 can be rewritten

$$r_{AB} = k_{AB} \theta_A \theta_B \qquad\qquad (7.8)$$

TABLE 7.1   A Selection of Some of the Rate Equations for Some Common Catalytic Reactions Extracted from the Compilation of Mezaki and Inoue [1991]

Reaction	Catalyst	Rate Equation
$SO_2 + \dfrac{1}{2} O_2 \rightarrow SO_3$	V_2O_5	$\dfrac{k_1 P_{SO_2} P_{O_2}^{1/2} - k_2 P_{SO_3}}{1 + K_3 P_{O_2}^{1/2} + k_4 P_{SO_2} + k_5 P_{SO_3}}$
$N_2 + \dfrac{3}{2} H_2 \rightarrow NH_3$	Fe/Al_2O_3	$k_1 P_{N_2} \left(\dfrac{P_{H_2}^3}{P_{NH_3}^2} \right)^a - k_2 \left(\dfrac{P_{NH_3}^2}{P_{H_2}^3} \right)^{1-a}$
$CO + 2H_2 \rightarrow CH_3OH$	$CuO/ZuO/Al_2O_3$	$\dfrac{k_1 P_{CO} P_{H_2}^2 - k_2 P_{CH_3OH}}{1 + k_3 P_{H_2} + k_4 P_{CO} + k_5 P_{CO} P_{H_2}^{3/2}}$
$C_2H_4 + H_2 \rightarrow C_2H_6$	Ni/Al_2O_3	$\dfrac{k_1 P_{H_2} P_{C_2H_4}}{(1 + k_2 P_{H_2} + k_3 P_{C_2H_4})^2}$
$C_2H_4 + \dfrac{1}{2} O_2 \rightarrow C_2H_4O$	Ag/Al_2O_3	$\dfrac{k_1 P_{O_2} P_{C_2H_4}}{1 + k_2 P_{O_2} P_{C_2H_4}}$
$CO + H_2O \rightarrow H_2 + CO_2$	Fe_2O_3/Cr_2O_3	$\dfrac{k_1 P_{CO} P_{H_2O} - K_2 P_{H_2} P_{CO_2}}{(1 + k_3 P_{CO} + k_4 P_{H_2O} + k_5 P_{CO_2} + k_6 P_{H_2})^2}$
$4NH_3 + 6NO \rightarrow 5N_2 + 6H_2O$	Pt	$\dfrac{k_1 P_{NO} P_{NH_3}^{1/2}}{(1 + k_2 P_{NO} + k_3 P_{NH_3}^{1/2})^2}$

where θ_A and θ_B are the fractional coverage of A and B, i.e.,

$$\theta_A = \frac{N_A}{N_S} \qquad \theta_B = \frac{N_A}{N_S} \tag{7.9}$$

and $\kappa_{AB} N_S^2 = k_{AB}$. Langmuir then assumed that the adsorption of A and B would follow a Langmuir adsorption isotherm given by Equations 4.24 and 4.25. Note that according to Equation 4.24 when P_A, the partial pressure of A in a reaction vessel increases, more A will adsorb. As a result, θ_A will increase. However, according to Equation 4.25, when P_A increases θ_B decreases because there are fewer vacant sites to hold B. Now consider the product of θ_A and θ_B in Equation 7.8. Note that since θ_A goes up and θ_B goes down the product of the two can increase or decrease with increasing pressure of A. If the fractional decrease in θ_B is smaller than the fractional increase in θ_A, the net rate will increase. However, if the fractional decrease in θ_B exceeds the fractional increase in θ_A, the net rate will decrease. Consequently, an increase in the partial pressure of A will not necessarily increase the rate of the reaction between A and B, and it may decrease it.

Langmuir quantified these ideas by substituting Equations 4.24 and 4.25 into Equation 7.7 to yield

$$r_{AB} = \frac{k_{AB} K_{equ}^A K_{equ}^B P_A P_B}{(1 + K_{equ}^A P_A + K_{equ}^B P_B)^2} \tag{7.10}$$

Figure 7.7 shows a plot of the r_{AB}/k_{AB} as a function of $K_{equ}^A P_A$ calculated from Equation 7.10, with $K_{equ}^B P_B = 1$. Notice that the rate goes up, reaches a maximum, and then declines

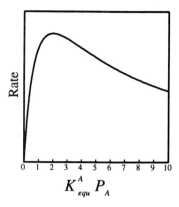

$$K^A_{equ} P_A$$

Figure 7.7 A plot of the rate calculated from Equation 7.10 with $K^B_{equ} P_B = 1$.

in qualitative agreement with the data in Figure 7.5. In fact, the fit is quantitative; the lines in Figure 7.5 are actual fits to the data via Equation 7.10. As a result, an increase in the pressure of A can produce either an increase or a decrease in the rate of reaction. Qualitatively, that is exactly what happens, which shows the utility of Langmuir's analysis.

Equation 7.10 is the key result that led to Langmuir's Nobel prize. Before proceeding, the reader should memorize Equation 7.10 and verify that it can reproduce all of the trends in Figures 7.2, 7.3, 7.4, and 7.5.

7.4 LANGMUIR, HOUGEN, AND WATSON RATE LAWS

Langmuir did all of his work between 1910 and 1930. However, it was not until Hougen and Watson's influential book in 1948 that Langmuir's model was used to routinely model catalytic reactions. In their book, Hougen and Watson [1943] used Langmuir's model to derive rate equations for a number of simple examples. Hougen and Watson assumed that all surface reactions proceeded via a catalytic cycle, where one of the steps called the **rate-determining step** was much slower than all of the other steps in the catalytic cycle. Hougen and Watson assumed that all of the steps before the rate-determining step were in equilibrium with the reactants, while all of the steps after the rate-determining step were in equilibrium with the product. They also assumed that all of the reactions were reversible.

As an example of the analysis, consider the reversible reaction sequence

$$S + A \underset{2}{\overset{1}{\rightleftharpoons}} A_{ad}$$

$$A_{ad} \underset{4}{\overset{3}{\rightleftharpoons}} B_{ad} \qquad (7.11)$$

$$B_{ad} \underset{6}{\overset{5}{\rightleftharpoons}} B + S$$

If the reaction $A_{ad} \rightleftharpoons B_{ad}$ is rate determining, then r, the overall rate of reaction, is given by

$$r = k_3[A_{ad}] - k_4[B_{ad}] \qquad (7.12)$$

where k_3 and k_4 and the rate constants for Reactions 3 and 4, and $[A_{ad}]$ and $[B_{ad}]$ are the concentrations of adsorbed A and B in molecules/cm^2 (or moles/cm^2). One can calculate $[A_{ad}]$

and $[B_{ad}]$ from the steady-state approximation

$$0 = r_{A_{ad}} = k_1 P_A[S] - k_2[A_{ad}] - k_3[A_{ad}] + k_4[B_{ad}] \qquad (7.13)$$

$$0 = r_{B_{ad}} = k_6 P_B[S] - k_5[B_{ad}] - k_4[B_{ad}] + k_3[A_{ad}] \qquad (7.14)$$

where $r_{A_{ad}}$ and $r_{B_{ad}}$ are the rates of formation of adsorbed A and B, P_A and P_B are the partial pressures of A and B over the reactive surface, and $[S]$ is the concentration of empty surface sites. If Reactions 3 and 4 are rate determining, k_3 will be much smaller than k_2 and k_6, while k_4 will be much smaller than k_5 and k_1. Therefore, the last two terms in Equations 7.13 and 7.14 will be negligible. Under such circumstances

$$[A_{\text{ad}}] = \left(\frac{k_1}{k_2}\right) P_A[S] \qquad (7.15)$$

$$[B_{ad}] = \left(\frac{k_6}{k_5}\right) P_B[S] \qquad (7.16)$$

One needs a site balance like that in section 4.3 to complete the analysis. Following the derivation in Section 4.3, let us define S_o as the total number of sites on the surface. Each site can be bare, or it can be covered by A or B. Therefore,

$$S_o = [S] + [A_{ad}] + [B_{ad}] \qquad (7.17)$$

Substituting Equations 7.15 and 7.16 into Equation 7.17, and then solving for $[S]$, yields

$$[S] = \frac{S_o}{1 + \dfrac{k_1}{k_2} P_A + \dfrac{k_6}{k_5} P_B} \qquad (7.18)$$

Substituting Equations 7.15 and 7.16 into Equation 7.12 and then substituting $[S]$ from Equation 7.18 yields

$$r = \frac{\left(\dfrac{k_1 k_3}{k_2}\right) P_A S_o - \left(\dfrac{k_6 k_4}{k_5}\right) P_B S_o}{1 + \dfrac{k_1}{k_2} P_A + \dfrac{k_6}{k_5} P_B} \qquad (7.19)$$

Table 7.2 shows many other examples. In the problem set we ask the reader to derive each of these examples.

There is a trick that makes the derivation easier. Notice that when k_3 is much smaller than k_2 while k_4 is much smaller than k_5 in Equations 7.13 and 7.14, the last term in Equations 7.13 and 7.14 will vanish. Consider Equation 7.13. If the last term in equation 7.13 is negligible,

$$k_1 P_a[S] \cong k_2[A_{ad}] \qquad (7.20)$$

Notice that the left side of Equation 7.20 is the rate of Reaction 1, while the right side of Equation 7.20 is the rate of Reaction 2. Therefore, the implication of Equation 7.20 is that when steps 3 and 4 are rate determining, the rate of Reaction 1 will almost equal the rate of

TABLE 7.2 Hougan, Watson–Langmuir Rate Equations for a Number of Reactions

Mechanism	Rate-Determining Step	Rate Equation	
		Langmuir–Hinshelwood Mechanisms	
$S + A \underset{2}{\overset{1}{\rightleftharpoons}} A_{ad}$ $A_{ad} \underset{4}{\overset{3}{\rightleftharpoons}} B_{ad}$ $B_{ad} \underset{6}{\overset{5}{\rightleftharpoons}} B + S$	$A \rightleftharpoons A_{ad}$	$\dfrac{k_1 S_o P_A - \left(\dfrac{k_6 k_4 k_2}{k_5 k_3}\right) P_B S_o}{1 + \left(\dfrac{k_6}{k_5}\right)\left(1 + \dfrac{k_4}{k_3}\right) P_B}$	(7.2.1)
	$A_{ad} \rightleftharpoons B_{ad}$	$\dfrac{\left(\dfrac{k_1 k_3}{k_2}\right) P_A S_o - \left(\dfrac{k_6 k_4}{k_5}\right) P_B S_o}{1 + P_A \left(\dfrac{k_1}{k_2}\right) + P_B \left(\dfrac{k_6}{k_5}\right)}$	(7.2.2)
	$B_{ad} \rightleftharpoons B$	$\dfrac{\dfrac{k_1 k_3 k_5}{k_2 k_4} P_A S_o - k_6 P_B S_o}{1 + \dfrac{k_1}{k_2}\left(1 + \dfrac{k_3}{k_4}\right) P_A}$	(7.2.3)
$S + A \underset{2}{\overset{1}{\rightleftharpoons}} A_{ad}$ $S + B \underset{4}{\overset{3}{\rightleftharpoons}} B_{ad}$ $A_{ad} + B_{ad} \underset{6}{\overset{5}{\rightleftharpoons}} C_{ad} + S$ $C_{ad} \underset{8}{\overset{7}{\rightleftharpoons}} C + S$	$A \rightleftharpoons A_{ad}$	$\dfrac{k_1 P_A S_o - \dfrac{k_2 k_3 k_6 k_7}{k_4 k_5 k_8} \dfrac{P_C}{P_B} S_o}{1 + \dfrac{k_4}{k_3} P_B + \dfrac{k_7}{k_8} P_c + \dfrac{k_3 k_6 k_7 P_c}{k_4 k_5 k_8 P_B}}$	(7.2.4)
	$A_{ad} + B_{ad} \rightleftharpoons C_{ad}$	$\dfrac{\dfrac{k_2 k_4 k_5 k_7}{k_1 k_3 k_6} P_A P_B S_o - \dfrac{k_5 k_7}{k_8} P_C S_o}{1 + \dfrac{k_2}{k_1} P_A + \dfrac{k_4}{k_3} P_B + \dfrac{k_7}{k_8} P_C}$	(7.2.5)
	$C_{ad} \rightleftharpoons C$	$\dfrac{\dfrac{k_2 k_4 k_5}{k_1 k_3} P_A P_B S_o - k_8 P_C S_o}{1 + \dfrac{k_2}{k_1} P_A + \dfrac{k_4}{k_3} P_B + \dfrac{k_2 k_4 k_5}{k_1 k_3 k_6} P_A P_B}$	(7.2.6)

TABLE 7.2 (*Continued*)

Langmuir–Hinshelwood Mechanisms (*Continued*)

Mechanism	Rate-Determining Step	Rate Equation
$A + S_1 \underset{2}{\overset{1}{\rightleftharpoons}} A_{ad^1}$ $B + S_2 \underset{4}{\overset{3}{\rightleftharpoons}} B_{ad^2}$ $A_{ad^1} + B_{ad^2} \underset{6}{\overset{5}{\rightleftharpoons}} C_{ad^1} + S_2$ $C_{ad^1} \underset{8}{\overset{7}{\rightleftharpoons}} C + S_1$ where S_1 and S_2 are different sites	$A \rightleftharpoons A_{ad}$	$\dfrac{k_1 P_A S_1^o - k_2 \dfrac{k_6 k_8}{k_7 k_5} P_C S_1^o \left(1 + \dfrac{k_3}{k_4 P_B}\right)}{1 + \dfrac{k_8}{k_7} P_C \left[1 + \dfrac{k_6 k_8}{k_7 k_5}\left(1 + \dfrac{k_3}{k_4 P_B}\right)\right]}$ (7.2.7)
	$B \rightleftharpoons B_{ad}$	$\dfrac{k_3 P_B S_2^o - \dfrac{k_2 k_4 k_6 k_8}{k_1 k_5 k_7}\dfrac{P_C}{P_A} S_2^o}{1 + \dfrac{k_6 k_8 k_2}{k_1 k_5 k_7}\dfrac{P_C}{P_A}}$ (7.2.8)
	$A_{ad} + B_{ad} \rightleftharpoons C_{ad}$	$\dfrac{\dfrac{k_1 k_3}{k_2 k_4} P_A P_B S_1^o - \dfrac{k_6 k_8}{k_7} P_C S_2^o}{\left(1 + \dfrac{k_3}{k_4} P_B\right)\left(1 + \dfrac{k_1}{k_2} P_A + \dfrac{k_8}{k_7} P_C\right)}$ (7.2.9)
	$C_{ad} \rightleftharpoons C$	$\dfrac{\dfrac{k_1 k_3 k_5 k_7}{k_2 k_4 k_6} P_A P_B S_1^o - k_8 P_C S_1^o}{1 + \dfrac{k_1}{k_2} P_A + \dfrac{k_1 k_3 k_5}{k_2 k_4 k_6} P_A P_B}$ (7.2.10)

Rideal–Eley Mechanisms

	$S + A \rightleftharpoons A_{ad}$	$\dfrac{S_o k_1 P_a - \dfrac{k_2 k_4 k_6}{k_3 k_5}\left(\dfrac{P_C}{P_A}\right) S_o}{1 + \dfrac{k_6}{k_5} P_C}$ (7.2.11)

494

Mechanism	Rate-determining step	Rate expression	
$S + A \underset{2}{\overset{1}{\rightleftharpoons}} A_{ad}$ $A_{ad} + B \underset{4}{\overset{3}{\rightleftharpoons}} C_{Ad}$ $C_{ad} \underset{6}{\overset{5}{\rightleftharpoons}} C + S$	$A_{ad} + B \rightleftharpoons C_{ad}$	$\dfrac{\dfrac{k_1 k_3}{k_2} P_A P_B S_o - \dfrac{k_4 k_6}{k_5} P_C S_o}{1 + \dfrac{k_1}{k_2} P_A + \dfrac{k_6}{k_5} P_C}$	(7.2.12)
	$C_{ad} \rightleftharpoons C + S$	$\dfrac{\dfrac{k_1 k_3 k_5}{k_2 k_4} P_A P_B S_o - k_6 P_C S_o}{1 + \dfrac{k_1}{k_2} P_A + \dfrac{k_1 k_3}{k_2 k_4} P_A P_B}$	(7.2.13)
$S + A \underset{2}{\overset{1}{\rightleftharpoons}} B_{Ad}$ $B_{ad} \underset{4}{\overset{3}{\rightleftharpoons}} B + S$	$S + A \rightleftharpoons B_{Ad}$	$\dfrac{k_1 S_o P_A - S_o \left(\dfrac{k_2 k_4}{k_3}\right) P_B}{1 + \dfrac{k_4}{k_3} P_B}$	(7.2.14)
	$B_{ad} \rightleftharpoons B + S$	$\dfrac{\dfrac{k_3 k_1}{k_2} P_A S_o - k_4 P_B S_o}{1 + \dfrac{k_1}{k_2} P_A}$	(7.2.15)
$A_2 + 2S \underset{2}{\overset{1}{\rightleftharpoons}} 2A_{ad}$ $A_{ad} + B \underset{4}{\overset{3}{\rightleftharpoons}} C + S$	$A_2 + 2S \rightleftharpoons 2A_{ad}$	$\dfrac{k_1 S_o^2 P_A - k_2 S_o^2 \dfrac{k_4^2}{k_3^2} \dfrac{P_C^2}{P_B^2}}{1 + \dfrac{k_4}{k_3} \dfrac{P_C^2}{P_B^2}}$	(7.2.16)
	$A_{ad} + B \rightleftharpoons C + S$	$\dfrac{k_3 \left(\dfrac{k_1}{k_2} P_A\right)^{1/2} S_o P_B - k_4 S_o P_c}{1 + \left(\dfrac{k_1}{k_2} P_A\right)^{1/2}}$	(7.2.17)

TABLE 7.2 *(Continued)*

Mechanism	Rate-Determining Step	Rate Equation
		Precursor Mechanisms
$A \underset{2}{\overset{1}{\rightleftharpoons}} A_p$ $S + A_p \underset{4}{\overset{3}{\rightleftharpoons}} A_{ad}$ $A_{ad} \underset{6}{\overset{5}{\rightleftharpoons}} B_{Ad}$ $B_{Ad} \underset{8}{\overset{7}{\rightleftharpoons}} B + S$	$A_p \rightleftharpoons A_{ad}$	$\dfrac{\dfrac{k_1 k_3}{k_2} P_A S_o - \dfrac{k_8 k_6 k_4}{k_7 k_5} P_B S_o}{1 + \dfrac{k_8}{k_7} P_B + \dfrac{k_8 k_6}{k_7 k_5} P_B}$ \quad (7.2.18)
	$A_{ad} \rightleftharpoons B_{Ad}$	$\dfrac{\dfrac{k_1 k_3 k_5}{k_2 k_4} P_A S_o - \dfrac{k_8 k_6}{k_7} P_B S_o}{1 + \dfrac{k_1 k_3}{k_2 k_4} P_A + \dfrac{k_8}{k_7} P_B}$ \quad (7.2.19)
	$B_{Ad} \rightleftharpoons B$	$\dfrac{\dfrac{k_1 k_3 k_5 k_7}{k_2 k_4 k_6} P_A S_o - k_8 P_B S_o}{1 + \dfrac{k_1 k_3}{k_2 k_4} P_A + \dfrac{k_1 k_3 k_5}{k_2 k_4 k_6} P_A}$ \quad (7.2.20)

reaction 2. A similar analysis shows that when Reactions 3 and 4 are rate determining, the rate of Reaction 5 must almost equal the rate of Reaction 6.

Note, however, that this is an approximation. The rate of Reaction 5 cannot be exactly equal to the rate of Reaction 6, because if that were the case, there would be no net production of products. Similarly, if the rate of Reaction 1 were exactly equal to the rate of Reaction 2, there would be no net consumption of reactants. However, the implication of the preceeding derivation is that one can calculate an accurate value of the surface concentration of each of the species by assuming that the rate of Reaction 1 approximately equals the rate of Reaction 2, and the rate of Reaction 5 approximately equals the rate of Rection 6, even though in reality the rates are not exactly equal.

One can extend these ideas to other situations. Consider, for the moment, how the arguments would change if Reactions 5 and 6 were rate determining in Reaction 7.11. If 5 and 6 are rate determining, then a steady-state approximation on B_{ad} gives

$$0 = r_{B_{ad}} = k_3[A_{ad}] - k_4[B_{ad}] - k_5[B_{ad}] \tag{7.21}$$

As before, the last term in Equation 7.21 is negligible.

$$k_3[A_{ad}] \cong k_4[B_{ad}] \tag{7.22}$$

The left side of Equation 7.22 is the rate of Reaction 3, while the right side is the rate of reaction 4. Therefore, the implication of Equation 7.22 is that one can calculate the concentrations of adsorbed A and B by assuming that the rate of Reaction 3 is equal to the rate of Reaction 4. Now consider Reactions 1 and 2. The steady-state approximation for A_{ad} implies

$$0 = r_{A_{ad}} = k_1 P_A[S] - k_2[A_{ad}] + k_4[B_{ad}] - k_3[A_{ad}] \tag{7.23}$$

Note, however, that according to Equation 7.20, the last two terms in Equation 7.21 are approximately equal when Reactions 5 and 6 are rate determining. Therefore, when Reactions 5 and 6 are rate determining,

$$k_1 P_A[S] \cong k_2[A_{ad}] \tag{7.24}$$

The left side of Equation 7.24 is the rate of Reaction 1 and the right side is the rate of Reaction 2. Therefore, the implication of Equation 7.24 is that when Reactions 5 and 6 are rate determining, the rate of Reaction 1 is approximately equal to the rate of Reaction 2. We note again that the rates are not exactly equal. However, the key result is that one can calculate accurate surface concentrations by assuming that the rates are almost equal even though the rates are not exactly equal.

One can generalize these results to say that if there is a single rate-determining step in a reaction, then one can calculate an accurate rate equation by assuming that all of the steps before the rate-determining step are in equilibrium with the reactants, while all of the steps after the rate-determining step are in equilibrium with the products. Note, however, that they are not truly in equilibrium. It is just that the rate equation comes out right when one assumes that the steps are in equilibrium.

Table 7.2 shows many other examples. The first case is a Langmuir–Hinshelwood reaction of the form A $\rightleftharpoons$ B. The second is a Langmuir–Hinshelwood reaction of the form A + B $\rightleftharpoons$ C. The third case is also a Langmuir–Hinshelwood reaction of the form A + B $\rightleftharpoons$ C. However, in the second case the reactants adsorb on a single type of site S, while in the third case the reactants adsorb on two types of sites, S_1 and S_2. The fourth case is a Rideal–Eley reaction for an overall reaction A + B $\rightleftharpoons$ C, while the fifth case is a Rideal–Eley reaction for an overall reaction A $\rightleftharpoons$ B. The last case is a precursor reaction for an overall reaction

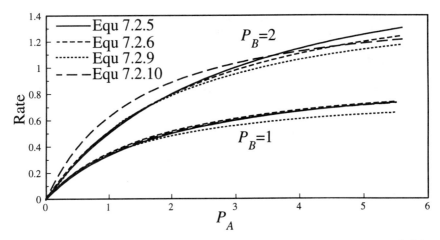

Figure 7.8 A comparison of the rate calculated via several of the equations in Table 7.2.

$A \rightleftharpoons B$. In most cases, we have assumed that there is only one site on the surface to hold gas. In Chapter 3 we noted that often adsorbate can be held on more than one binding site. Therefore, we included one case with two sites, and asked the reader to work out some other cases in the problem set.

Figure 7.8 shows plots of several of the rate equations from Table 7.2. Notice that the plots are quite similar, even though the models used to derive the rate equations are quite different. Experimentally, it is often hard to distinguish between the rate equations. As a result, it is difficult to use the rate equation to suggest a mechanism. In the problem set we ask the reader to verify that the rate equations in Table 7.2 are correct, and to indicate how the equations could be distinguished experimentally.

7.4.1 Comparison to Data

It is useful to compare the experimental equations listed in Table 7.1 to the theoretical rate equations in Table 7.2. Notice that in most of the cases, the form of the rate equation is just as predicted by the Langmuir model. This is a general result. For example, Mezaki and Inoue [1991] compile rate equations for about 1000 reactions on commercial heterogeneous catalysts, and all but 2 follow Langmuir–Hinshelwood or Rideal–Eley rate equations. Admittedly, in many cases people simply assume that their data follows a Langmuir–Hinshelwood or Rideal–Eley rate equation, and fit constants. Still, the key point from this section is the following:

Most rate data on heterogenous catalysts are accurately represented via a Langmuir–Hinshelwood rate equation.

The situation is less clear with single crystals. Schmidt and co-workers have examined a number of examples and showed they do follow Langmuir–Hinshelwood kinetics with possibly two parallel reaction paths (see Table 7.3). However, there are many other reactions that do not follow Langmuir–Hinshelwood rate equations on single crystals. For example, multiple steady states, such as those in Figure 7.6, are not possible with Langmuir–Hinshelwood kinetics. There are also other interesting deviations.

One anomaly is that Boudart [1956] has shown that even though the Langmuir–Hinshelwood equation is derived assuming that the adsorption of the reactants follows a Langmuir

TABLE 7.3   A Selection of Some of the Rate Equations for Some Reactions on Single Crystal Surfaces

Reaction	Catalyst	Rate Equation
$CO + 1/2\,O_2 \rightarrow CO_2$	Pt(111), Rh(111)	$\dfrac{k_1 P_{CO} P_{O_2}}{(1 + k_2 P_{CO} + k_3 P_{O_2})^2}$
$2NH_3 \rightarrow N_2 + 3H_2$	Pt(111), Rh(111), Fe(111)	$\dfrac{k_1 P_{NH_3}}{1 + k_2 P_{NH_3} + k_3 P_{H_2}^{1/2}}$
$NO + CO \rightarrow CO_2 + 1/2N_2$	Pt(100), Rh(111)	$\dfrac{k P_{CO}}{(1 + k_2 P_{CO})^2}$
$2NO \rightarrow N_2 + O_2$	Pt	$\dfrac{k_1 P_{NO}}{1 + k_2 P_{NO_3} + k_3 P_{O_2}}$
$NH_3 + CH_4 \rightarrow HCN + 3H_2$	Pt	$\dfrac{k_1 P_{NH_3}^{1/2}}{[1 + k_2 P_C H_4 / P_N^{1/2} H_3]^4}$
$1/2O_2 + H_2 \rightarrow H_2O$	Pt(111)	$\dfrac{k_1 P_{O_2} P_{H_2}}{1 + k_2 P_{H_2}}$
$CO + H_2O \rightarrow CO_2 + H_2$	Pt	$\dfrac{k_1 P_{CO} P_{H_2O}^{1/2}}{(1 + k_2 P_{CO})^2}$
$CH_4 + 3NO \rightarrow 3/2N_2 + CO + 2H_2O$	Pt	$\dfrac{k_1 P_{CH_4} P_{NO} + k_2 P_{CH_4}^{1/2} P_{NO}}{1 + k_3 P_{CH_4}}$
$NH_3 + 3/2NO \rightarrow 5/4N_2 + 3/2H_2O$	Pt	$\dfrac{k_1 P_{NO} P_{NH_3}^{1/2}}{(1 + k_2 P_{NO} + k_3 P_{NH_3}^{1/2})^2}$

Data extracted from papers by Schmidt and co-workers.

adsorption isotherm, Langmuir–Hinshelwood-type rate equations often work under conditions where the adsorption of the reactants does not follow a Langmuir adsorption isotherm. As a result, Langmuir–Hinshelwood-type rate equations are more general than they initially appear.

Boudart [1956] only considered reactions on heterogeneous catalysts. As noted in Chapter 4, typically heterogeneous catalysts contain a wide variety of sites. Some sites bind the reactants weakly, while others bind the reactants strongly. In Chapter 4 we found that there are important deviations from the Langmuir adsorption isotherm when there is a wide distribution of sites. Interestingly, however, the equations in Table 7.2 still usually work on heterogeneous catalysts, even though these equations were derived assuming that all sites are equivalent so that the adsorption process follows a Langmuir adsorption isotherm.

At this point, no one knows in detail why Langmuir–Hinshelwood-type rate equations work for reactions on heterogeneous catalysts under conditions where the Langmuir isotherm fails. Generally, it is thought that there is a cancellation of errors, although the details are not understood. One effect is that with a structure-sensitive reaction some sites on a heterogeneous catalyst are much more catalytically active than others. When an adsorption isotherm is measured, gas adsorbs on all of the sites on the catalyst. The adsorption process will follow a Langmuir adsorption isotherm only when all of the sites are equivalent and there are no important interactions between adjacent adsorbed molecules. However, when one measures a reaction rate, one selectively samples the most catalytically active sites. If the most catalytically active sites are equivalent, then the Langmuir adsorption isotherm will be appropriate for adsorption on these sites. If the Langmuir adsorption isotherm works on the active sites, then the derivation in Section 7.3 will still be valid for a reaction on the active sites, even though the Langmuir adsorption isotherm does not work when one considers adsorption on all of the sites on the catalyst. As a result, the kinetics of a reaction that occurs principally on a few especially active sites can follow Langmuir–Hinshelwood kinetics even though the overall adsorption process does not follow a Langmuir adsorption isotherm.

Still, the ideas in the last paragraph would only apply to a reaction that occurs on a few isolated active sites. There are many cases of reactions that occur on many different sites on the surface of a heterogeneous catalyst, and yet still follow Langmuir–Hinshelwood kinetics under conditions where the adsorption of the reactants does not follow a Langmuir–Hinshelwood rate equation. Doriswamy [1991] considers these cases in detail and suggests that there is a compensation effect on a commercial catalyst. Some sites bind the reactants strongly, while others bind the reactants weakly. At first one might think that the sites that bind the reactants strongly are the most catalytically active, because the concentration of the reactants is highest on those. However, in Chapters 8 and 10, we note that there is a compensation effect. When the reactants adsorb onto a site that binds the reactants strongly, a very stable complex forms. The stable complex is more difficult to decompose than the complex that forms when the reactants bind weakly. As a result, the rate constant for a reaction is lower on a site that binds the reactants strongly than on a site that binds the reactants weakly. Consequently, in a heterogeneous catalyst some sites have a high reactant concentration and a low rate constant, while other sites have a low reactant concentration and a high rate constant. According to Equation 7.8 the overall rate of a surface reaction is proportional to the product of the rate constant and the reactant concentrations. Doriswamy [1991] shows that the overall rate of reaction on a site with a low reactant concentration and a high rate constant will not be much different than the overall rate on a site with a high reactant concentration and a large rate constant. As a result, the site-to-site variation in the reactant concentration is partially compensated (i.e., canceled) by the site-to-site variation in the rate constant for the reaction. Doriswamy [1991] shows that if everything works out right, it is possible for the cancellation to be good enough that a reaction will follow Langmuir–Hinshelwood kinetics under conditions where the adsorption of the reactants does not follow a Langmuir adsorption isotherm.

Another interesting case is one where the adsorbate forms an ordered overlayer. In Chapter 4 we found that ordered adsorbate overlayers are quite common. The adsorption process does not follow a Langmuir adsorption isotherm. However, in Sections 7.18 and 7.19 we show that the system can still follow a Langmuir–Hinshelwood rate law. Thus, Langmuir–Hinshelwood rate equations are much more general than they would initially appear.

Experimentally, Langmuir–Hinshelwood kinetics work over a wide range of conditions even though the Langmuir adsorption isotherm fails. In some cases, one can explain the result because only a few sites are active; other times there is a compensation. There can also be an ordered overlayer. However, often one does not know why a Langmuir–Hinshelwood-type rate equation works as well as it does.

7.4.2 Mars–Van Klevan Kinetics

Still, there are a few systems where people say that Langmuir–Hinshelwood kinetics are insufficient to explain the experimental data. For example, in oxidation reactions over a metal oxide the surface of the metal oxide is reduced and reoxidized during the reaction. As a result, some people question whether Langmuir's assumption that the number of surface sites is fixed is valid under oxidation conditions.

Mars and Van Klevan [1954] have derived a modified model to account for the oxidation and reduction of the metal oxide. Consider the partial oxidation of a hydrocarbon such as propylene over an oxide like Bi_2O_3:

$$C_3H_6 + \tfrac{1}{2}O_2 \Rightarrow C_3H_6O \tag{7.25}$$

During the reaction oxygen from the Bi_2O_3 lattice reacts with propylene to form propylene oxide. Oxygen from the gas phase then reoxidizes the Bi_2O_3.

Mars and Van Klevan asserted that the rate of loss of lattice oxygen is proportional to the coverage of propylene θ_P times the partial pressure of propylene P_P to some power

$$r_{\substack{oxygen \\ consumption}} = bk_2\theta_P(P_P)^n \tag{7.26}$$

where b is the number of moles of oxygen lost when one mole of propylene is oxidized. Similarly, they asserted that the rate of reoxidation of the lattice is given by

$$r_{reoxidation} = k_1(P_{O_2})^m (1 - \theta_P) \tag{7.27}$$

where P_{O_2} is the partial pressure of O_2. At steady state,

$$r_{\substack{oxygen \\ consumption}} = r_{reoxidation} \tag{7.28}$$

Substituting Equations 7.26 and 7.27 into Equation 7.28, solving for θ_P, and then substituting back into Equation 7.26, yields

$$r_{\substack{oxygen \\ consumption}} = \frac{bk_1k_2P_{O_2}^m P_P^n}{k_1P_{O_2}^m + bk_2P_P^n} \tag{7.29}$$

which is very similar but not identical to a Langmuir–Hinshelwood form. Equation 7.29 is called the **Mars–Van Klevan rate equation.** In actual practice, one can fit data equally well via Equation 7.29 or by the corresponding Langmuir–Hinshelwood equation

$$r = \frac{bk_1k_2P_{O_2}^m P_P^n}{1 + k_1P_{O_2}^m + bk_2P_P^n} \tag{7.30}$$

However, workers in the oxidation literature tend to favor the Mars–Van Klevan rate equation.

7.4.3 Tempkin Kinetics

The other major example of a commercial catalytic reaction that may not follow Langmuir–Hinshelwood kinetics is the reaction of hydrogen and nitrogen over an iron catalyst to yield ammonia

$$N_2 + 3H_2 \xrightarrow{\text{Fe}} 2NH_3 \tag{7.31}$$

Reaction 7.31 was one of the first catalytic reactions that was used in large-scale chemical production. In the 1920s and 1930s many investigators tried to fit rate data for this reaction to Langmuir–Hinshelwood kinetics without success. Consequently, a concerted effort was made to see if the rate data could be fit by a non-Langmuir–Hinshelwood rate form. Eventually, Tempkin [1938] showed that the reaction could be fit by a power law expression

$$r_{net} = k_1^0 S_o P_{N_2} \left(\frac{P_{H_2}^3}{P_{NH_3}}\right)^m - \frac{k_1^0}{K_{equ}} S_o \left(\frac{P_{NH_3}^2}{P_{H_2}^3}\right)^{1-m} \tag{7.32}$$

Two years later, Tempkin and Pyzhev [1940] showed that one can derive Equation 7.32 by starting with a Langmuir–Hinshelwood rate equation, and then assuming that the equilibrium constant for the adsorption of nitrogen varies with coverage, due to interactions between species adsorbed on the iron surface. In the material that follows, we will review some of the analysis that was done to understand the kinetics of the ammonia synthesis reaction. Additional discussions can be found in Boudart and Djéga-Mariadassou [1984] or White [1990].

Tempkin and Pyzhev's [1940] derivation assumes that the formation of ammonia proceeds via a two-step mechanism

$$N_2 + S \underset{k_{-1}}{\overset{k_1}{\rightleftharpoons}} N_{2(ad)} \tag{7.33}$$

$$N_{2(ad)} + 3H_2 \underset{k_{-2}}{\overset{k_2}{\rightleftharpoons}} 2NH_3 + S \tag{7.34}$$

and that Reaction 7.33 is rate determining.

The derivation starts by noting that the net rate of reaction, r_{net}, for a reaction following the mechanism in Reactions 7.33 and 7.34 is

$$r_{net} = k_1[S]P_{N_2} - k_{-1}[N_{2(ad)}] \tag{7.35}$$

If Reaction 7.33 is rate determining,

$$[N_{2,ad}] = \frac{k_{-2}}{k_2} \frac{P_{NH_3}^2}{P_{H_2}^3} [S] \tag{7.36}$$

Combining Equations 7.35 and 7.36 yields

$$r_{net} = k_1[S]P_{N_2} - \frac{k_{-1}k_{-2}}{k_2} \frac{P_{NH_3}^2}{P_{H_2}^3} [S] \tag{7.37}$$

If all of the k's in Equation 7.37 were independent of coverage, one could use Equation 7.37 to derive a Langmuir–Hinshelwood rate form. However, Tempkin noted that if k_1 varies as

$$k_1 = K_1^o [N_{2,ad}]^{-m} \tag{7.38}$$

where K_1^o and m are constants, one can derive Equation 7.32. Tempkin then noted that one can justify Equation 7.38 if one assumes that there are interactions between various nitrogen atoms on the surface, as described in Chapter 4, so that the heat of adsorption of nitrogen decreased as more nitrogen adsorbed onto the surface. In order to simplify the algebra, Tempkin assumed that the heat of adsorption ΔH_{ad} varied logarithmically with the nitrogen coverage:

$$\Delta H_{ad} = \Delta H_{ad}^o + \alpha_T Ln([N_{ad}]) \tag{7.39}$$

where α_T and ΔH_{ad}^o are constants.

Tempkin then noted that as the heat of adsorption of nitrogen increases, it becomes easier for the nitrogen to adsorb. To some approximation the activation energy for adsorption, E_{ad}, will vary linearly with the heat of adsorption

$$E_{ad} = E_{ad}^o + \alpha_P(\Delta H_{ad}) \tag{7.40}$$

where α_P and E_{ad}^o are constants. See Chapter 9 for details. Substituting Equations 7.37, 7.39, and 7.40 into Arrhenius's law:

$$k_1 = \kappa_1 \exp(-E_{ad}/kT) \tag{7.41}$$

yields

$$k_1 = k_1^o \left(\frac{P_{H2}^3}{P_{NH_3}[S]} \right)^m \tag{7.42}$$

where

$$k_1^o = \kappa_1 \left(\frac{k_{-2}}{k_2} \right)^m \exp \left(-\frac{E_{ad}^o}{kT} + \alpha_P \Delta H_{ad}^o \right) \tag{7.43}$$

$$m = \alpha_P \alpha_T \tag{7.44}$$

Note that changes in the binding energy of the nitrogen intermediate cannot change the equilibrium constant, K_{equ} for the overall reaction. It is easy to show

$$K_{equ} = \frac{k_1 k_2}{k_{-1} k_{-2}} \tag{7.45}$$

Substituting Equation 7.42 into Equation 7.37 yields

$$r_{net} = k_1 [S] P_{N_2} - \frac{k_1}{K_{equ}} [S] \frac{P_{NH_3}^2}{P_{H_2}^3} \tag{7.46}$$

Substituting Equation 7.42 into Equation 7.46 yields

$$r_{net} = k_1^o P_{N_2} \left(\frac{P_{H_2}^3}{P_{NH_3}^2} \right)^m [S]^{1-m} - \frac{k_1^o}{K_{equ}} \left(\frac{P_{NH_3}^2}{P_{H_2}^3} \right)^{1-m} [S]^{1-m} \tag{7.47}$$

Tempkin then assumed

$$[S]^{1-m} = S_o \tag{7.48}$$

where S_o is a constant. Combining Equations 7.47 and 7.48 yields

$$r_{net} = k_1^o S_o P_{N_2} \left(\frac{P_{H_2}^3}{P_{NH_3}^2} \right)^m - \frac{k_1^o}{K_{equ}} S_o \left(\frac{P_{NH_3}^2}{P_{H_2}^3} \right)^{1-m} \tag{7.49}$$

Equation 7.49 is called the Tempkin rate equation for ammonia synthesis.

Experimentally, the Tempkin rate equation fits the available kinetic data on ammonia synthesis over most of the range used in the commercial reactors (e.g., 50–200 atm and 400°C), so it is commonly used for design. However, it does not fit data taken at one atmosphere.

The preceding derivation was done by assuming that the heat of adsorption varied logarithmically with the nitrogen coverage. That is usually not such a good assumption if there is only one type of site on the surface. However, Tempkin [1957] showed that one could instead get the same result by assuming that there are a distribution of sites to hold gas. At low nitrogen coverages only the sites that bind nitrogen most strongly are covered by gas, but most of the other sites are empty. As a result, the rate constant is one characteristic of the most tightly bound sites. However, as one raises the coverage, more sites contribute. As a result, the activation energy of desorption changes. If one assumes a logarithmic distribution of sites, then one still can derive Equations 7.42 and 7.49 (see Boudart and Djéga-Mariadassou

[1984] for details). As a result, Equation 7.49 holds whenever there is a logarithmic distribution of sites. It works less well when there is only one type of site on the surface.

Boudart and Djéga-Mariadassou [1984] and Doriswamy [1991] discuss much of the work that has been done to extend Tempkin's formalism to other catalyst systems. Generally, the approach that is taken is to use an adsorption isotherm to calculate the distribution of sites on a surface. One then averages the site distribution to calculate a rate. The interesting thing, though, is that in most cases only a small fraction of the sites on a surface are catalytically active at any given temperature. These sites are uniform so the Langmuir–Hinshelwood rate equation applies. As a result, a Langmuir–Hinshelwood-type rate equation often works quite well, even when there is a distribution of sites on a surface. The one complicating factor, however, is that often one group of sites has the largest contribution to the rate at some given temperature, while another group of sites has the largest contribution to the rate at a higher temperature. The net effect is that the apparent activation energy for the reaction is quite different from the one expected from measurements of the activation energy on any given site.

Another complicating factor is that even where Tempkin's formalism fits the data, there is no assurance that it is correct. For example, a few paragraphs ago we noted that Tempkin first proposed the Tempkin rate law to understand his data for ammonia synthesis. Some years ago, Ertl [1981] examined each of the steps in the ammonia synthesis reaction (or their inverse) in ultrahigh vacuum (UHV), and found that each step obeyed a Langmuir–Hinshelwood rate equation. Stoltze [1985, 1987] took Ertl's results and used them to derive a complex Langmuir–Hinshelwood-type rate equation with no adjustable parameters (i.e., all parameters were either taken from Ertl's measurements or calculated from first principles). Stoltze's rate equation fit data from a commercial reactor as well or better than the Tempkin equation. Further, Stoltz's equation works even at low-pressure conditions where Tempkin's equation fails. Hence, it does not appear that one needs to invoke non-Langmuir–Hinshelwood rate laws to understand the kinetics of the ammonia synthesis reaction. Still, Tempkin kinetics are discussed extensively in the literature, so it is important for the reader to be familiar with them.

At this point, this author is not aware of any example where kinetic data on a supported catalyst cannot be fit by a Langmuir–Hinshelwood-type rate equation. As noted previously, Mezaki and Inoue [1991] tabulated kinetic rate laws for over 1000 common catalytic reactions, and all of them except the ammonia synthesis reaction can be fit by Langmuir–Hinshelwood-type kinetics. Further, in the case of ammonia synthesis, Stoltze [1987] found that a Langmuir–Hinshelwood rate form fits the data over a wider range of conditions than the Tempkin equation. Hence, at this point it appears that much of the data for reactions on commercial-supported metal catalysts are easily explained by a Langmuir–Hinshelwood rate form.

That does not necessarily imply that Langmuir–Hinshelwood rate forms are correct, however. Boudart [1956] and Weller [1956] have pointed out that a typical Langmuir–Hinshelwood rate equation has perhaps 3–5 adjustable parameters, so one can often fit data quite well, even when the model is incorrect. Further power law rate equations (i.e. $r_A = kP_A^{\alpha_1}P_B^{\alpha_2}$ where k, α_1, and α_2 are parameters) often fit data on a supported catalyst as well as Langmuir–Hinshelwood rate forms. Still, Langmuir–Hinshelwood rate forms are well-accepted in the literature, even under conditions where the assumptions used to derive the rate forms (e.g., a Langmuir adsorption isotherm) are incorrect. As a result, Langmuir–Hinshelwood rate equations are quite useful.

7.5 RELATIONSHIP TO MECHANISMS OF REACTIONS

One does have to be cautious, however, and not overinterpret the fact that a given Langmuir–Hinshelwood rate equation fits an experimental data set. Years ago people thought that one could use kinetic data to infer mechanisms of surface reactions. The idea was to assume a mechanism, and then use a procedure like that in Section 7.4 to derive a rate equation for the mechanism. One then fits the rate equation to experimental data and if the data fit the rate equation, one assumes that the mechanism is correct. However, today people know that such a procedure is fraught with error. In Chapter 4 we noted that when gases adsorb on surfaces, adjacent molecules usually interact. As a result adsorption isotherms rarely follow the Langmuir adsorption isotherm. In Sections 7.20 and 7.21 we show that when there are pairwise interactions between the adsorbed molecules the apparent order of the reaction changes. The kinetics still follow a Langmuir–Hinshelwood rate law. However, it is a different Langmuir–Hinshelwood rate law than the one in Table 7.2, so there is not a good correspondence between the mechanism and the rate equation. Another complication is that it is not uncommon for an incorrect mechanism to also give a rate equation which fits data. As a result, one needs to be exceedingly cautious about using rate data to infer mechanisms of surface reactions.

For example, White and Hightower [1983] examined propylene oxidation over a Bi_2O_3 catalyst and found that they could fit their rate data with a Langmuir–Hinshelwood model provided they assume that in the rate-determining step, two adsorbed propylenes react with an adsorbed O_2 molecule to yield two propylene oxide molecules. However, the best available evidence is that in reality the reaction proceeds when a single propylene reacts with an adsorbed oxygen atom. In particular, the reaction proceeds when there is adsorbed atomic oxygen but no molecular oxygen in the system. Hence, there is an apparent contradiction where kinetically, it would appear that molecular oxygen is the reactive species, but all of the other evidence suggests that instead atomic oxygen is the reactive intermediate.

Many oxidation reactions show similar contradictions. In the literature there has been considerable speculation why this is so. Often the answer is that the real kinetics are rather complex. For example, Falconer and Madix [1975] and Shoofs and Benziger [1984] suggested that during the partial oxidation of formaldehyde, a dimer (i.e., formic anhydride) forms on the surface or in the gas phase. The decomposition of the dimer is rate determining. In the examples at the end of the chapter, we ask the reader to compare the kinetics of two different mechanisms.

$$O_2 + 2S \; \rightleftharpoons \; 2O_{ad}$$
$$HC_{(ad)} + O_{(ad)} \; \longrightarrow \; Products$$
(7.50)

and

$$O_2 + S \; \rightleftharpoons \; O_{2(ad)}$$
$$2HC_{(ad)} + O_{2(ad)} \; \longrightarrow \; Products$$
(7.51)

where HC is a hydrocarbon ligand. One can show that if the molecular oxygen takes up two sites, and there are attractive interactions between the adsorbed oxygens, then the kinetics will be virtually identical for these two mechanisms (see Sections 7.18 and 7.19 for a further discussion of this point). The net result is that one could easily confuse the mechanisms in Equations 7.50 and 7.51.

This is a general result. Often a given reaction will follow a Langmuir–Hinshelwood rate equation, which is different from the rate equation one derives from the analysis in Section 7.4.

Another difficulty with Langmuir–Hinshelwood reactions is that often many different mechanisms follow very similar rate equations. One cannot distinguish between the rate equations based on data taken over a limited range of conditions. As a result, it is difficult to distinguish between mechanisms based on kinetics.

Finally, there is the issue that typically the Langmuir–Hinshelwood rate equation might have four or more parameters. With that many parameters, one can often fit rate data quite well with a rate equation that has little mechanistic significance. As a result, it is easy to be fooled into thinking that one knows the mechanism of a catalytic reaction because the rate equation for that mechanism fits rate data. So far, it appears that if one knows the complete mechanism of a reaction, including all of the interactions between molecules, one can derive a suitable rate equation. However, one can often fit rate data with a rate equation that has little mechanistic significance. That leads to the key point from this section:

> The fact that a mechanistic model fits rate data does not imply that the mechanistic model is correct.

7.6   EVIDENCE FOR MORE COMPLEX KINETICS

So far in this chapter we have only been discussing Langmuir–Hinshelwood–Tempkin-type rate equations. While Langmuir–Hinshelwood–Tempkin rate equations seem to fit rate data taken on commercial-supported metal catalysts quite well, these rate equations do not seem to fit data on single crystals or so-called model supported catalysts as well. The reason one gets deviations from Langmuir–Hinshelwood kinetics on single crystals is associated with the phase behavior we discussed in Chapter 4. For example, consider a reaction between an adsorbed A molecule and an adsorbed B molecule. When the reaction occurs A and B must move together and react. Such a process is easy if the A and B molecules are adjacent to each other on the surface. However, the process is less likely if the A and B molecules are adsorbed in separate islands. Hence, the ordering of the adsorbate on the surface can play an important role in reactivity.

Experimentally, adsorbate phase transitions seem to be more important on single crystals and in so-called model catalysts than they are in commercial-supported metal catalysts. On single crystals, adsorbate phase transitions are normally associated with long-range ordering on the surface. One, for example, might observe a (2x2) adsorbate arrangement converting to a (2x1) arrangement as the phase transition proceeds. However, commercial catalysts consist of small metal particles on an inert support. If the metal particle sizes are small enough, there would not be room for a (2x2) or (2x1) phase. As a result, the finite size of the metal particles on a supported metal catalyst tends to attenuate the adsorbate phase behavior.

Another effect is that commercial catalysts often contain promoters, i.e., impurities that enhance catalytic activity. The adsorption of a promoter on a given site enhances the activity of an adjacent site as discussed in Chapter 6. Generally, there are special places on the catalyst that contain both the promoter and the right combination of surface atoms to promote a desired reaction. Those sites are especially catalytically active. However, the sites tend to be far apart, and fairly isolated from their neighbors. The phase transitions we discussed in Chapter 4 are a result of adsorbate/adsorbate interactions. Adsorbate/adsorbate interactions are negligible on isolated sites. As a result, the adsorbate phase behavior tends to be attenuated on the most active sites in commercial-supported metal catalysts. Consequently, the phase behavior is not that important in commercial catalysts. In contrast, adsorbate phase behavior is quite common on single crystals, and so-called model catalysts, i.e., catalysts with large particle sizes. This adsorbate phase behavior can have an important influence on the kinetics of surface processes in the latter systems even though the adsorbate phase behavior does not seem to be as important as commercial catalysts.

7.7 TEMPERATURE PROGRAMMED DESORPTION

A common way to explore the influence of adsorbate phase behavior on the kinetics of surface processes is to use a technique called **temperature programmed desorption** (TPD). The technique is also called **temperature programmed reaction** (TRP). TPD was described in Chapter 1. Basically, a gas or a mixture of gases is adsorbed onto a cold single crystal. The crystal is then heated in a way such that the crystal temperature varies linearly with time. One can often start at temperatures low enough to avoid significant reaction. In those cases, the reaction occurs only when the crystal heats up. Typically, some of the reaction products desorb from the surface. In a TPD experiment a mass spectrometer is used to monitor the desorption of products. The result is a so-called TPD spectrum that is a plot of the desorption rate of products, r_d, as a function of the sample temperature.

Generally, one observes a TPD spectrum like that in Figure 7.9 where there are a series of peaks corresponding to desorption of each of the reaction products. Each of the peaks in the TPD spectrum represents a different kinetic process. For example, the fact that there are four peaks in the hydrogen spectrum in Figure 7.9 implies that there are four kinetic processes that lead to desorption of hydrogen. Similarly, the fact that there is one peak in the ethane desorption spectrum implies that ethane desorption shows a single desorption pathway.

Generally, one finds that processes which have a low activation barrier have a low peak temperature in TPD, while processes which have a higher barrier show larger peak temperatures in TPD. Therefore, one can say that the process which produces hydrogen at 330 K in Figure 7.9 has a lower activation barrier than the process which produces hydrogen at 550 K. Physically, if a process has a high reaction barrier, one needs to go to a high temperature to get it to occur. These ideas are quantified in the next section.

Unfortunately, one cannot really tell much about what the kinetic processes are like from a TPD spectrum. For example, Van Spaendonk and Masel showed that the ethylene (C_2H_4) is sequentially losing hydrogens during the experiment in Figure 7.9 to yield C_2H_3, C_2H_2,

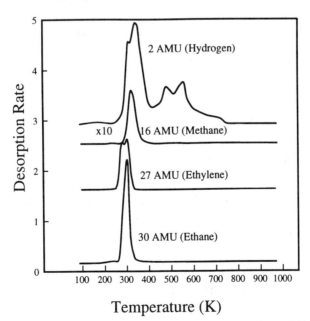

Figure 7.9 A TPD spectrum for ethylene decomposition on Pt(511). (Data of Van Spaendonk and Masel [1995].)

C_2H intermediates, and then eventually adsorbed carbon. If hydrogen would immediately desorb during each of those decomposition processes, then one might attribute each of the peaks in the hydrogen spectrum in Figure 7.9 to an individual reaction where a hydrogen is lost from a C_2H_x moiety. Note, however, that if hydrogen does not immediately desorb, and instead binds to the surface, the kinetic processes may be entirely different. Thus, one can often use TPD to distinguish between hydrogen that is being produced directly from the decomposition process and hydrogen that binds and then desorbs upon further heating. However, one cannot be sure what species are being formed on the surface without doing further spectroscopy.

Another complication is that the TPD spectrum from a single species can often show multiple peaks. For example, Figure 7.10 shows a TPD spectrum of hydrogen desorbing from Pt(100). A series of curves are shown in the figure. These curves are generated by dosing a clean Pt(100) sample with a variable amount of H_2 and then heating. The lowest curve in the figure corresponds to the lowest initial hydrogen dose, while the subsequent curves correspond to higher initial hydrogen doses. There are four peaks in the spectrum corresponding to four kinetic processes. This is a typical result. One usually observes multiple peaks in TPD even for simple desorption processes. We will show in the next few pages that the behavior is quite different than one would expect from Equation 7.8, i.e., Langmuir's model. Therefore, the presence of multiple peaks in TPD suggest that even simple surface processes have quite complex kinetics, with multiple kinetic processes contributing to an overall rate.

People discuss multiple peaks in TPD as though there are different "states" of the adsorbate on the surface accounting for each kinetic process seen in TPD. For example, if one had four different binding sites for the adsorbate, one might expect the desorption from each binding site to be different. As a result, there will be multiple desorption pathways and therefore multiple peaks in TPD.

Unfortunately, the real situation is more complicated than that. For example, consider adsorption of a gas from a square surface, and assume that the gas molecules interact as described in Chapter 4. The material in Chapter 4 shows that at any instant some of the adsorbed molecules will be isolated on the surface, while other adsorbed molecules will be surrounded by 1, 2, 3, or 4 nearest neighbors. If the adsorbate/adsorbate interactions are repulsive, then the adsorbed molecules that are surrounded by four nearest neighbors will be less strongly bound than the adsorbate molecules that are isolated. The kinetics of desorption will be different in the two cases. This produces two different kinetic processes and therefore

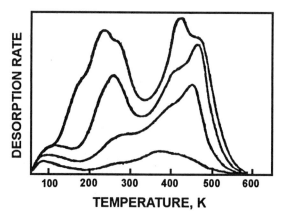

Figure 7.10   A TPD spectrum of hydrogen desorbing from Pt(100). (Data of McCabe and Schmidt [1977].)

two TPD peaks. Note, however, that in this example the binding sites for all the molecules are the same and the desorption is occurring from a single thermodynamic state. Thus, the idea that the presence of multiple TPD peaks implies that there are multiple "states" of the adsorbate on the surface is not correct using the thermodynamic definition of an individual state. In the literature people say that there are "multiple states" on the surface. However, what they mean is that there are multiple kinetic processes during desorption of the gas.

7.7.1 Redhead's Analysis of TPD

As noted earlier one of the main applications of TPD is in the measurement of the kinetics of surface processes. In a 1963 paper, Redhead [1963] showed how one could do this. Redhead's analysis assumes that the kinetics of the desorption process follow a simple power law:

$$\frac{r_d}{N_s} = -\frac{d\theta_A}{dt} = k_o \theta_A^n \exp(-E_A/kT) \qquad (7.52)$$

where r_d is the rate of desorption of some species A in molecules/cm^2; N_s is the concentration of surface sites in number/cm^2; θ_A is the coverage of the species; t is time; n is the order of the desorption reaction; k_o is the preexponential of the rate constant for desorption; E_A is the activation energy of desorption; k is Boltzmann's constant; and T is temperature.

Redhead's derivation assumes that one has designed the TPD experiment so the temperature, T, varies linearly with time, t,

$$T = T_0 + \beta_H t \qquad (7.53)$$

where T_0 is the initial temperature and β_H is the heating rate. Combining Equations 7.52 and 7.53 yields

$$\frac{r_d}{\beta_H N_s} = -\frac{d\theta_A}{dT} = \frac{k_o}{\beta_H} \theta_A^n \exp(-E_A/kT) \qquad (7.54)$$

Redhead also assumes that the design of the apparatus is such that one measures r_d directly during a TPD experiment. Both of these assumptions are reasonable in a modern TPD experiment.

Figure 7.11 shows a plot of a TPD spectra calculated by picking an initial coverage, and then numerically integrating Equation 7.54 to calculate the desorption rate as a function of temperature. There are three lines in the figure: the fractional coverage, θ, versus temperature; the desorption rate, r_d, versus temperature; and the rate constant, k_d, versus temperature, where k_d is defined as

$$k_d = k_o \exp(-E_A/kT) \qquad (7.55)$$

The figure shows that if the kinetics of the desorption process follows the rate law in Equation 7.54, one would expect to observe a single distinct peak in TPD. Physically, one observes a peak, rather than, for example, an exponential rise in the desorption rate, with temperature due to a balance between the various terms in equation 7.54. At low temperatures, the rate constant for desorption is small, and so little desorption is seen. As one raises the temperature, the rate constant rises, as shown in Figure 7.11, so initially, the desorption rate initial goes up. However, TPD is a non-steady state experiment. One starts with a surface which is covered by gas but as gas desorbs the coverage decreases as gas desorbs, as shown in Figure 7.11. The surface is eventually depleted of the desorbing species A. Once that happens, the

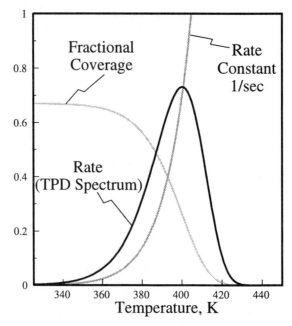

Figure 7.11   The coverage, rate constant for desorption, and desorption rate calculated by integrating Equation 7.48 for $k_o = 10^{13}$/sec, $E_A = 24$ kcal/mol, $\beta_H = 10$ K/sec, $n = 1$.

rate goes down. The maximum rate occurs somewhere in the middle where the rate constant is high, but there is still a significant amount of A on the surface.

Figure 7.12 shows what typical TPD peaks would be like if Equation 7.54 applied. There are several different lines in Figure 7.12. They were calculated by starting with some initial coverage θ_0, and then numerically integrating Equation 7.54 for $n = 1$ and $n = 2$. The figure shows that for a first-order desorption process (i.e., $n = 1$) the TPD peaks are asymmetric, but the peak position is independent of the initial coverage. In contrast, symmetric peaks should be seen with a second-order process. Further, the peak position varies with coverage.

Figure 7.13 and Figure 7.14 show how the peak temperature varies with the activation energy of desorption and the preexponential for desorption. Generally, the peak in the TPD curve shifts to higher temperatures as the activation energy for desorption increases. The peak also shifts to slightly higher temperatures as the heating rate increases, while the peak shifts to lower temperatures as the preexponential for desorption increases. Generally, the major effect is the change in the activation energy of desorption. Hence, one can also use TPD peak positions to infer activation energies of desorption.

As a general rule of thumb, the maximum in the TPD curve comes when the rate constant is on the order of 0.5/sec. i.e., at the maximum in a TPD curve, the desorption rate is usually on the order of 0.5 monolayers/sec times the initial coverage.

Redhead quantified these ideas. Redhead noted that the maximum in the desorption rate will occur when

$$\frac{dr_A}{dT} = 0 \tag{7.56}$$

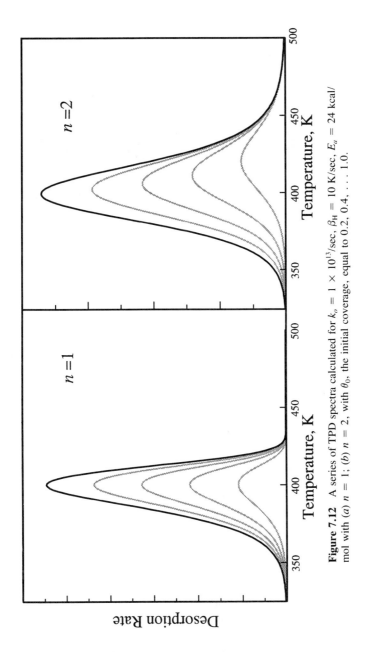

Figure 7.12 A series of TPD spectra calculated for $k_o = 1 \times 10^{13}$/sec, $\beta_H = 10$ K/sec, $E_a = 24$ kcal/mol with (a) $n = 1$; (b) $n = 2$, with θ_0, the initial coverage, equal to 0.2, 0.4, . . . 1.0.

511

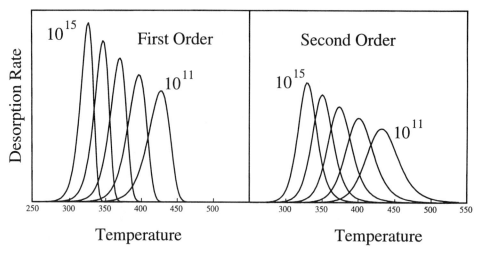

Figure 7.13 A series of TPD curves calculated for a first- and second-order desorption process with $E_a = 24$ kcal/mol, $\theta_0 = 0.67$ $k_o/\beta_H = 10^{11}, 10^{12}, \ldots, 10^{15}$.

Substituting Equation 7.54 into Equation 7.56 and differentiating assuming that n, k_o and E_A are independent of temperature yields

$$0 = \frac{dr_A}{dt} = k_o\, n\, \theta_p^{n-1} \left(\frac{d\theta}{dt}\right) \exp(-E_A/kT_0) + k_o\, \theta_p^n \left(\frac{E_A}{k}\right) \frac{1}{T_p^2} \exp(-E_A/kT_p) \left(\frac{dT}{dt}\right) \quad (7.57)$$

where T_p = the temperature of the maximum in the TPD curve and θ_p is the coverage at the peak maxima.

Substituting $\left(\dfrac{d\theta}{dt}\right)$ from Equation 7.52 into Equation 7.57, noting $\dfrac{dT}{dt} = \beta_H$, and canceling

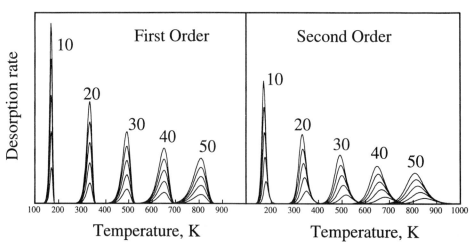

Figure 7.14 A series of TPD spectra calculated for a first- and second-order desorption process with $k_o/\beta = 10^{12}$ sec, and $E_a = 10, 20, \ldots, 50$ kcal/mol and $\theta_0 = 0.2, 0.4 \ldots 1.0$.

like terms yields

$$k_o\, n\theta_p^{n-1} \exp\left(\frac{-E_A}{kT_p}\right) = \frac{\beta_H E_A}{kT_p^2} \tag{7.58}$$

Rearranging Equation 7.58 for $n \neq 0$ yields

$$\frac{E_A}{kT_p} = \ln\left(\frac{k_o\, T_p n\theta_P^{n-1}}{\beta_H}\right) - \ln\left(\frac{E_A}{kT_P}\right) \tag{7.59}$$

For $n = 1$, the θ term in Equation 7.59 vanishes, while for $n = 2$, $\theta_P = \frac{1}{2}\theta_0$ where θ_0 is the initial coverage. Therefore, one can solve Equation 7.59 iteratively to calculate the activation energy corresponding to any TPD peak. It happens that the right side of Equation 7.59 is between 25 and 35 for most TPD data taken at near saturation coverages. As a result, the approximation

$$\frac{E_A}{kT_P} \cong 30 \text{ or } E_A \cong \left(.06\, \frac{Kcal}{Mol^\circ K}\right) T_P \tag{7.60}$$

allows one to estimate activation energies to $\pm 20\%$.

Figure 7.15 shows how T_P varies with E_A, for a first-order desorption process with k_o/β_H varying between 10^{10} and 10^{14}/K. Notice that the peak temperature increases as the activation energy for desorption increases. A TPD peak at 100 K corresponds to an activated process with an activation energy of 6 kcal/mole, while a peak temperature of 300 K corresponds to an activated process with an activation energy of 18 kcal/mole. Figure 7.15 is much more

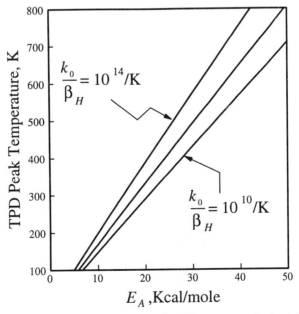

Figure 7.15 The variation in the peak temperature of a TPD spectrum calculated from Equation 7.44 for a first-order desorption process with $k_o/\beta_H = 10^{10}, 10^{12}, 10^{14}$/K.

general than it would first appear because the E_A on the left of Equation 7.59 is only a weak function of k_o and the order of the reaction. Figure 7.15 is exact for a second-order desorption process with $\theta_0 = 1$. Thus, Figure 7.15 is an approximate universal curve for any TPD process. As a result, one can use Figure 7.15 to get a reasonable approximation to the activation energy from any TPD spectrum.

Of course, Figure 7.15 is only an approximation in most situations. Sometimes one wants to calculate activation barriers. More accurately, Redhead showed that one could get an even better approximation to the activation energy of desorption by measuring how the peak temperatures vary with heating rate during a TPD experiment. The key equation is drived by starting with Equation 7.58, taking logs, and rearranging. The result is

$$-\left(\frac{E_A}{kT_p}\right) = \ln\left(\frac{\beta_H}{kT_p^2}\right) + \ln\left(\frac{E_A}{k_o\theta_p^{n-1}n}\right) \tag{7.61}$$

Du, Sarofin, and Longwell [1990] show that to a reasonable approximation one can rewrite Equation 7.61 as

$$-\left(\frac{E_A}{kT_p}\right) = \ln\left(\frac{\beta_H}{kT_p^2}\right) + \ln\left(\frac{E_A}{k_o\theta_0^{n-1}}\right) \tag{7.62}$$

where θ_0 is the coverage at the start of the TPD curve. Therefore a plot of $\ln\left(\frac{\beta_H}{T_p^2}\right)$ versus $\frac{1}{T_p}$ at constant θ_0^{n-1} should yield a line with slope $-E_A$.

Years ago Equation 7.61 was often used to measure the kinetics of the desorption process. Typically, one would do a series of experiments where one adsorbed a fixed amount of gas onto a surface, heat the surface at a variety of different heating rates, and then measure the variation in the peak temperature T_P with changing heating rate β_H. Typical data are shown in Figure 7.16. Experimentally, as one increases the heating rate, TPD peak temperatures go up. Redhead showed that for simple first- and second-order kinetics the last term in Equation 7.61 is constant during such an experiment. As a result, a plot of $\ln\left(\frac{\beta_H}{T_P^2}\right)$ versus $\frac{1}{T_P}$ from such an experiment should yield a line whose slope is the activation energy for desorption.

7.7.2 Comparison to Data

One of the difficulties with the analysis, however, is that TPD curves rarely look like Figure 7.12. Instead, there are multiple TPD peaks as described earlier (see Figure 7.10 and Figure 7.16). Generally, the multiple peaks arise because at any instant, the molecules on the surface have a series of different environments. Some molecules may be adsorbed on linear sites, while other molecules are adsorbed on bridgebound or triply coordinated sites. Some adsorbed molecules will be surrounded by four nearest neighbor molecules, while some other adsorbed molecules will have 3, 2, 1, or 0 nearest neighbor molecules. It works out that the rate constant for desorption of a molecule from a linear site is different from the rate constant for desorption from a bridged site. The activation energy for desorption from a site with four nearest neighbor molecules is different from the activation energy for desorption from a site with 3, 2, 1, or 0 nearest neighbor molecules. Later in this chapter we show that the variation in the activation energy from one site to another allows multiple peaks such as those in Figure 7.9 and Figure 7.16 to be seen in TPD.

The presence of inequivalent sites also has important implications to steady-state surface reactions. We discuss them later in this chapter.

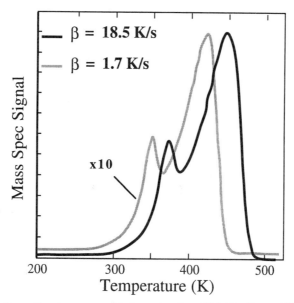

Figure 7.16 The effect of heating rate on CO desorption from Ni(110). (Data of Fiegerle, Desai, and Overbury [1990].)

7.8 IMPROVED ANALYSIS OF TPD DATA

Generally, one analyzes the effects of inequivalent sites and multiple surface environments by dividing up the molecules on the surface into a series of subpopulations, where all of the molecules in each subpopulation have an equivalent environment, i.e., the molecules are all adsorbed on equivalent sites and are all surrounded by the same number of nearest neighbors. One can then write an overall desorption rate as a sum of the rates of desorption from each subpopulation. For example, for a first-order desorption process one might have a series of subpopulations, j, each of which has a coverage ϑ_j. The net rate of desorption would be:

$$\frac{r_d}{N_s} = \sum_{j=0}^{4} k_j \vartheta_j \tag{7.63}$$

If each of the subpopulations were distinct, and there was no interconversion of molecules from one subpopulation to another, one could use Redhead's analysis on each of the terms in Equation 7.63 to calculate all of the kinetic parameters for the desorption process. One would get multiple peaks, because there are multiple terms in Equation 7.63. In reality there is conversion of one subpopulation to another. As a result, one needs a more complex analysis to model TPD spectra. However, the key feature is that one usually observes multiple peaks in TPD when the adsorbed molecules can be bound in a variety of different environments.

It is useful to consider an example to clarify these ideas. Figure 7.17 shows a TPD spectrum of CO desorbing from an Ni(110) surface. The reaction is a simple first-order reaction

$$CO_{(ad)} \longrightarrow CO_{(g)} \tag{7.64}$$

where $CO_{(ad)}$ is a CO molecule adsorbed on the surface and $CO_{(g)}$ is a CO molecule in the gas phase. Yet there are several peaks in the spectrum, and each of the peaks shifts with

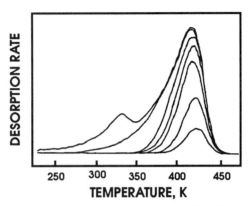

Figure 7.17 A series of TPD spectra for CO desorbing from Ni(110). (Data of Behm, Ertl, and Penka [1985].)

coverage. Note that according to Redhead's analysis, TPD peaks for first-order processes should not shift with coverage, so obviously Redhead's model does not fit the data. Physically, what is happening is that there are a series of phase transitions during the desorption process. The surface phase transitions lead to multiple peaks.

In the CO on Ni(110) case, at high coverages the CO adsorbs in a (2x1) phase. However, as CO is lost from the layer, the system changes to a c(4x2) phase, then a c(8x2) phase.

One can model this process by breaking the overall desorption process into a series of distinct steps, e.g.,

$$2CO_{(2x1)} \longrightarrow CO_{c(4x2)} + CO_{(g)} \tag{7.65}$$

$$2CO_{c(4x2)} \longrightarrow CO_{c(8x2)} + CO_{(g)} \tag{7.66}$$

$$CO_{c(8x2)} \longrightarrow CO_{(g)} \tag{7.67}$$

where $CO_{(2x1)}$, $CO_{c(4x2)}$, $CO_{c(8x2)}$ are CO molecules adsorbed in a (2x1), c(4x2) or a c(8x2) phase.

In the literature, it is common to model each phase transition as though it were a simple desorption process. For example, Figure 7.18 compares the adsorbate arrangement of the (2x1) and the c(4x2) overlayers of CO on Ni(110). Both structures are similar, but there are three extra COs in the 8x2 slice of the (2x1) structure shown in Figure 7.18. Therefore, to a first approximation one can model the conversion of a (2x1) structure to a c(4x2) structure as a simple first-order desorption process where two COs desorb from the (2x1) structure creating a c(4x2) structure. One can then identify two groups (i.e., subpopulations) of molecules on the surface, molecules that are present in the c(4x2) structure, and the extra molecules that desorb when the (2x1) structure is converted to the c(4x2) structure. These two populations are never really distinct on the surface (i.e., before the desorption process starts there is no way to tell whether a given molecule belongs in one population or the other). Nevertheless, if one treats the two subpopulations as distinct, one correctly predicts that there are multiple peaks in TPD. According to this analysis one can identify each of the peaks in the TPD spectrum in Figure 7.17 with an individual phase transition, e.g., the peak centered at 340 K is associated with the (2x1) → c(4x2) transition, the peak at 430 K is associated with desorption from the c(8x2) phase, while the shoulder at 370 K is associated with the c(4x2) to c(8x2) phase transition. In actual practice, TPD peaks often do not actually correlate with easily observed surface phase transitions. Still, there is the general result that multiple

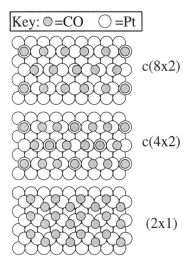

Key: ◎=CO ◯=Pt

c(8x2)

c(4x2)

(2x1)

Figure 7.18 A comparison of the (2x1) and c(4x2) structures of CO on Ni(110).

TPD peaks arise when there are multiple forms or multiple arrangements of the adsorbate on the surface.

Years ago people analyzed data like those in Figure 7.17 by measuring the peak temperature for each of the peaks in Figure 7.17 as a function of heating rate and then plotting the data according to Equation 7.61. Falconer and Madix [1975] showed that such a procedure yielded some useful information in a few simple cases. However, in most cases the shapes of the TPD curves changed with increasing heating rate; distinct peaks merge at high heating rates. As a result, Redhead's method [1963] often does not work except in the simplest of cases.

More recently King [1975] has noted that if one has TPD data taken at a variety of initial coverages, one can construct an Arrhenius plot at each coverage, and thereby calculate the activation energy of desorption as a function of coverage. This is called a **complete analysis** in the literature, and it is the most comprehensive way to analyze TPD data.

King's derivation starts by noting that the area under a TPD curve integrated from a temperature T to ∞ is proportional to $\theta(T)$, the coverage at that temperature i.e.,

$$\theta(T) = C_1 \int_T^\infty \frac{r_d(T)\, dT}{\beta_H} \tag{7.68}$$

where $r_d(T)$ is the desorption rate at temperature T, β_H is the heating rate, and C_1 is a constant. One can calculate C_1 by integrating the TPD curve for a saturated monolayer assuming $\theta = 1$ in that case. Thus, Equation 7.68 allows one to calculate a coverage for each point in a TPD curve.

King noted that in a TPD experiment, one measures the rate as a function of temperature with a variable coverage. Therefore, once one knows the coverages, one can create a table of the rate as a function of coverage and temperature. One can then rearrange those data to create a table of the rate versus temperature at a series of fixed values of the coverage. One can then plot those data on an Arrhenius plot (i.e., ln rate versus $1/T$ at fixed coverage) to calculate the activation energy for each of the fixed values of the coverage. Hence, one can obtain a plot of the activation energy as a function of coverage. Today people usually use a small modification of King's procedure, where they take data at a variety of heating rates as well as initial coverages. The use of multiple heating rates allows data to be taken over a wider range of temperatures. As a result, the accuracy of the method improves. Many in-

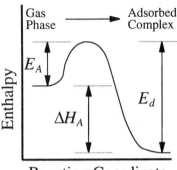

Figure 7.19   The enthalpy changes when gases adsorb.

vestigators still refer to this procedure as King's method. However, many investigators call it the **Taylor–Weinberg** method since Taylor and Weinberg [1978] were early proponents of the method. Another varient is to only examine the leading edge of the TPD curve, and assume that θ equals the initial coverage so one can make an Arrhenius plot directly from the data. This is called the initial slope method or the Habershaden–Küpers [1984] method. Both methods are illustrated in Example 7.B.

In the literature, it is common to assume that the absolute value of the heat of adsorption measured via the Taylor–Weinberg analysis equals the activation energy of desorption. Physically, if microscopic reversibility (see Chapter 5) works, then the equilibrium constant for adsorption should be the rate constant for adsorption divided by the rate constant for desorption (see Section 5.7 and Figure 7.19).

$$\frac{k_a}{k_d} = K_{equ} \tag{7.69}$$

If one assumes Arrhenius forms for each of the rate constants

$$k_a = k_a^o \exp(-E_A/kT) \tag{7.70}$$

$$k_d = k_d^o \exp(-E_d/kT) \tag{7.71}$$

$$K_{equ} = K_{equ}^o \exp(\Delta H_A/kT) \tag{7.72}$$

Solving Equations 7.69 to 7.72 simultaneously yields

$$\Delta H_A = E_A - E_d \tag{7.73}$$

If $E_A = 0$

$$\Delta H_A = -E_d \tag{7.74}$$

Figure 7.20 shows some typical results calculated using the Taylor–Weinberg procedure. Generally, one finds that the activation energy of desorption decreases with increasing coverage and that there are a series of breaks where the phase transitions occur. The breaks associated with phase transitions are found in most TPD spectra. Thus, they are quite important.

The difficulty with this analysis is that there is an inherent assumption that the phase behavior is only a function of coverage and not a function of temperature or some other parameters. This assumption often is questionable. Section 4.6 noted that the adsorbate ar-

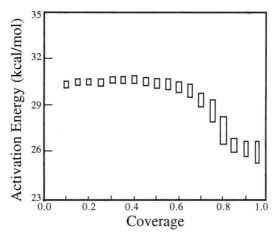

Figure 7.20 The activation energy for the desorption of CO from Ni(110) calculated using the method of Taylor and Weinberg. (Data of Fiegerle, Desai, and Overbuy [1990].)

rangement varies with temperature. For example, Figure 4.13 showed a phase diagram for oxygen and adsorption on W(110). Notice that the equilibrium arrangement of the adsorbed oxygen is temperature dependent. It happens that the phase transitions often occur right in the middle of a TPD spectrum. Thus, it is often incorrect to model a simultaneous phase transition and desorption as a single process. Physically, as one varies the heating rate, the phase behavior changes, which changes the TPD curve. The simple Taylor–Weinberg–Redhead analysis ignores the phase behavior, so it often makes errors.

7.9 BRAGG–WILLIAMS APPROXIMATION

An alternative analysis is based on the so-called Elovich equation:

$$r_A = k_o \, \theta^n \, \exp[-(E_A^E - \alpha_E\theta)/RT] \qquad (7.75)$$

where k_o, n, E_A^E, and α_E are parameters. The Elovich equation is derived by noting that interactions between adsorbate molecules, which cause the phase transitions to occur, also affect TPD spectra. If one can model the effects of the interactions on a TPD spectrum, one should be able to predict the effects of ordering on TPD spectra.

One can derive the Elovich equation from the Bragg–Williams approximation from Chapter 4. Recall that the Bragg–Williams approximation assumes that the presence of an adsorbate molecule on one site will not affect the probability that an adjacent site is filled. In this approximation the adsorbate fills the sites on the surface randomly; there are no ordered phases. However, there are interactions between molecules as though there are ordered phases.

According to Equation 4.127, the Bragg–Williams approximation, the free energy of adsorption on adjacent sites ΔG_A^i varies linearly with coverage:

$$\Delta G_A^i = \left(H_1 + \frac{\theta}{2} \sum_j h_{ij}\right) \qquad (7.76)$$

Note that if the free energy of adsorption increases, the molecules have to climb a large barrier before they can desorb (see Figure 7.21). Therefore, the activation energy for de-

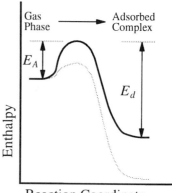

Figure 7.21 An illustration of how a change in the heat of adsorption will change the activation energy for desorption.

Reaction Coordinate

sorption will increase. Polanyi [1925] proposed that one can model these changes via what is now called the **Polanyi relationship**

$$E_A^i = E_d^o + \gamma_P(-\Delta G_A^i) \tag{7.77}$$

where γ_P is a constant called the transfer coefficient, $-\Delta G_A^i$ is the free energy of adsorption, and E_d^o is a constant. A derivation of Equation 7.77 is given in Chapter 9. Note that from macroscopic reversibility, the difference between the activation energy for adsorption and that for desorption must be the heat of adsorption. Therefore, if the adsorption process is unactivated, $E_d^o = 0$ and $\gamma_P = 1$.

Substituting Equation 7.76 into Equation 7.77 and canceling like terms yields

$$E_d^{BW} = E_d^o + \gamma_P \left(H_1 + \frac{\theta}{2} \sum_j h_{ij} \right) \tag{7.78}$$

where E_d^{BW} is the Bragg–Williams approximation to E_d. The implication of Equation 7.78 is is that if gas adsorbs on the surface, the apparant activation energy of desorption would vary linearly with coverage.

The Elovich equation, Equation 7.75, follows directly from Equations 7.76 and 7.78. Substituting Equation 7.78 into Equation 7.49 yields

$$r_d = k_o \, \theta \exp \left\{ -\left(E_d^o + \gamma_P H_1 + \frac{\gamma_P}{2} \theta \sum h_{ij} \right) \right\} \tag{7.79}$$

It is easy to show that Equation 7.79 is equivalent to the Elovich equation, Equation 7.75.

One interpretation of the Elovich equation is that there are interactions between adjacent molecules on the surface, and these interactions change as the coverage changes. If there are attractive interactions between the adsorbate molecules, the absolute value of the heat of adsorption will increase with increasing coverage. The activation energy of desorption will increase as well. Similarly, if there are repulsive interactions, the absolute value of the heat of adsorption will decrease with increasing coverage. The activation energy of desorption will also decrease.

Figure 7.22 shows a series of TPD spectra calculated by numerically integrating the Elovich equation for a system with strong repulsive interactions between molecules adsorbed on adjacent sites on the surface. Notice that the TPD peak shifts to lower temperatures with

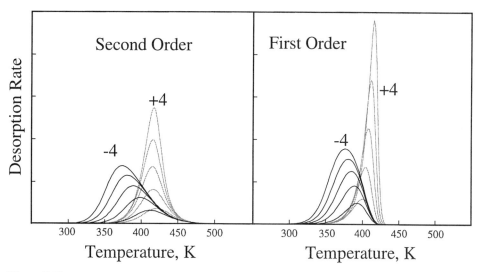

Figure 7.22 A series of TPD spectra calculated by numerically integrating the Elovich and initial coverages of 0.2, 0.4, 0.6, 0.8, and 1.0 for $E_d^o = 24$ kcal/mol, $\alpha_E = \pm 4$ kcal/mol.

increasing coverage but no new TPD peaks are seen. This is a general result. The Elovich equation predicts that the TPD peaks will shift to lower temperatures with increasing coverages when there are repulsive interactions between molecules adsorbed on adjacent sites. In contrast, the Elovich equation predicts that the peaks will shift to higher temperatures with increasing coverages when the interactions between adjacent molecules are attractive. More detailed simulations show that peak shifts are as expected from Figure 7.15, and no new TPD features arise if the adsorption process follows the Elovich equation.

The Elovich equation has been used extensively in the literature. Generally, it is a reasonable approximation when the adsorbate layer is quite disordered. However, it is now known that the Elovich equation is not a good approximation for an ordered overlayer. Note that in Chapter 4 we found that when there are strong interactions between adjacent molecules, the adsorbate layers will order. The Elovich equation ignores this ordering, which can lead to substantial errors.

Consider for the moment the case shown in Figure 7.23 where molecules are adsorbed on a square lattice with strong first nearest neighbor interactions. The material in Chapter 4 shows that when there are strong repulsive interactions between molecules on adjacent sites on the surface, the adsorbate will tend to form a c(2x2) overlayer at intermediate coverages. At coverages below 0.5 most of the molecules in the c(2x2) layer have zero nearest neighbors as shown in Figure 7.23. Thus, even though there are repulsive interactions, most of the molecules will not see the repulsions. In contrast, at coverages above 0.5, some molecules must reside in sites with three or more nearest neighbors. These additional molecules are bound more weakly than all the rest. In the next section we find that one should observe a two-peaked TPD spectrum in such a situation.

A typical TPD spectrum for this situation is shown on the right side of Figure 7.32. One can understand the qualitative features of Figure 7.32 with the aid of Figure 7.23 and Equation 7.54. At low coverage there are few nearest neighbors. As a result, the desorption process looks simple. According to Equation 7.54 if the adsorbate surface binding energy is about 24 kcal/mole, then one would expect a TPD peak at about 400 K, and indeed there is a peak at about 400 K at low coverages in Figure 7.32. The peak shifts with coverage because at moderate coverages there are, on average, a few nearest neighbors around each adsorbed molecule.

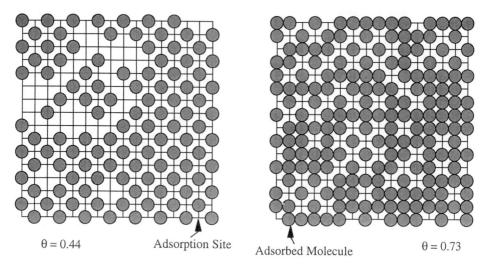

$\theta = 0.44$ Adsorption Site   Adsorbed Molecule $\theta = 0.73$

Figure 7.23 Some typical adsorbate configurations during the initial desorption from a layer with strong repulsive interactions.

The situation changes entirely at higher coverages. There most of the molecules that desorb are desorbing from a local environment where the molecule is surrounded by four nearest neighbors. In that case, the activation energy for desorption is about 24 kcal/mole $+ 4(-2$ kcal/mole) $= 16$ kcal/mole where -2 kcal/mole $= h$. According to Equation 7.54, desorption from a state with a binding energy of 16 kcal/mole should produce a TPD peak at about 270 K. Therefore, at high coverages one would expect two TPD peaks, one at about 270 K and one at about 400 K. That is exactly what we see in Figure 7.32.

Note that the Elovich equation ignores the formation of the c(2x2) phase, so it cannot reproduce the TPD spectrum in Figure 7.32. In general, the Elovich equation works well only when the assumption of a nearly random adsorption process works. The material in Chapter 4 indicates that such an assumption works only when $|\beta h| \leq 1.7627$. At larger values of $|\beta h|$ the surface layer orders. To put that in perspective, note that at 300 K a βh of 1.7627 corresponds to an h of 1.04 kcal/mole, i.e., a pairwise interaction of 1.04 kcal/mole is sufficient to get the surface layer to order. Experimentally, most adsorbate layers show ordered low-energy electron diffraction (LEED) patterns. The Elovich equation ignores the ordering. As a result, theoretically the Elovich equation should not adequately model desorption. A more detailed discussion of the limitations of the Elovich equation can be found in Section 7.12.2.

7.10 MONTE CARLO SIMULATION OF TPD

Monte Carlo calculations similar to those described in Sections 4.6 and 5.2 do a much better job of describing the effects of adsorbate/adsorbate interactions on TPD spectra. There are several different types of Monte Carlo calculations in the literature, but each of them starts with the basic premise that the rate of a simple desorption process

$$A_{ad} \rightarrow A \qquad (7.80)$$

can be written as a configuration average of the rate on each site, as described in Section

4.8. In this approximation,

$$r_d = \sum_n P_n \sum_i r_d^{i,n} \tag{7.81}$$

where r_d is the overall rate of desorption; $r_d^{i,n}$ is the rate of desorption on some site i when the system is in some configuration n; n is an index of all possible configurations of the system and P_n is the probability that the system is in that configuration. One then makes some assumptions about the form of $r_d^{i,n}$ to calculate a rate.

7.11 FIRST-ORDER DESORPTION

The simplest approximation applies to the case where the adsorbate, A, desorbs via a *first-order desorption process* without going through a precursor state or other weakly bound intermediate. In that case, one can calculate the rate of desorption on any site i via the rate equation

$$r_d^i = k_o \exp\left(-\frac{E_A^i}{kT}\right) \varepsilon_i \tag{7.82}$$

where k_o is the preexponential for the desorption process; E_A^i is the activation energy of desorption from site i; and ε_i is the occupancy number on site i, as defined in Section 4.7.1, $\varepsilon = 0$ for an empty site and $\varepsilon = 1$ for a filled site. E_A^i and k_o are usually dependent on the local arrangement of the molecules around site i, so Equation 7.81 is more complex than it initially seems.

People usually compute the sums in Equation 7.81 by assuming that the adsorbate arrangement stays at equilibrium throughout the desorption process. As noted earlier, the adsorbate may undergo a series of phase transitions during the desorption process. The implication of the equilibrium assumption is that as some of the adsorbate begins to desorb, the adsorbate remaining on the surface moves around very quickly so that the adsorbed layer can maintain a pseudo-equilibrium configuration (i.e., the configuration that would occur if there were gas over the surface to maintain the coverage at the given value).

If equilibrium is maintained, then one can use the Monte Carlo program in the supplemental material for Chapter 4 to calculate ε_i in all of the sums in Equation 7.81.

7.12 RESULTS FOR ISING UNIVERSALITY CLASS: FIRST NEAREST NEIGHBOR INTERACTIONS

In the next section we derive equations for the TPD from an equilibrium layer. We assume that the system follows Ising universality class with only first nearest neighbor interactions. We also assume a square lattice, although the results are easily generalized to hexagonal lattices.

For the purposes of derivation it is useful to rearrange Equation 7.81 as follows. If one has a square lattice, then each site on the lattice can have zero, one, two, three, or four nearest neighbor sites occupied. If one only has first nearest neighbor interactions, then all of the molecules with zero nearest neighbors are equivalent to all of the other molecules with zero nearest neighbors. Therefore, it is useful to look at the desorption process in terms of the number of adsorbed molecules with zero, one, two, three, or four nearest neighbors. Let's define N_0, N_1, N_2, N_3, and N_4 as the number of **occupied** sites with zero, one, two, three, or four of their nearest neighbor sites occupied. Equation 7.81 becomes

$$\frac{r_d}{N_s} = \sum_{j=0}^{4} k_j \vartheta_j \qquad (7.83)$$

where $j = 0$ corresponds to the occupied sites with 0 nearest neighbors, $j = 1$ corresponds to the occupied sites with 1 nearest neighbor, etc., k_j is the rate constant for desorption from a site with j nearest neighbors, and ϑ_j is defined as

$$\vartheta_j = \frac{\mathbb{N}_j}{N_s} \qquad (7.84)$$

where N_s is the total number of sites available to hold gas. Physically ϑ_0 is the probability that a given site is occupied (i.e., covered by adsorbate), but all of the nearest neighbor sites directly adjacent to the given site are empty. Similarly, ϑ_1 is the probability that a given site is occupied and exactly one of the sites directly adjacent to the given site is occupied. Notice that Equation 7.83 is identical to Equation 7.63, i.e., the rate for distinct populations on the surface. Therefore, the implication of Equation 7.83 is that at any instant there are a series of subpopulations on the surface: i.e., molecules with 0, 1, 2, 3, or 4 nearest neighbors and each of the subpopulations has a different rate constant for desorption. Note, however, that these subpopulations are only subpopulations in a kinetic sense. All of the molecules on the surface are thermodynamically equivalent and there is rapid exchange of molecules from one subpopulation to the next. Thus, even though one can identify different subpopulations on the surface, the desorption is occurring from a single thermodynamic state.

7.12.1 Qualitative Picture of the ϑ's

The principle of universality discussed in Chapter 4 allows one to greatly simplify the analysis. According to the principle of universality the ϑ_j's in Equation 7.83 are universal functions for adsorption on a square lattice with first nearest neighbor interactions. The ϑ_j depend on the coverage, βh, and the universality class of the system, but not other details. However, the ϑ_j do not change as the desorption rate changes (provided the layer stays in equilibrium). Therefore, the ϑ_j only need to be calculated once and for all and then they can be used to calculate a wide variety of TPD spectra.

One can use the methods discussed in Chapter 4 to calculate the ϑ_j. The simplest approximation is to assume that gas adsorbs randomly and that there are no interactions between adjacent molecules. In such a case one can get an analytical expression for the ϑ_j.

Consider ϑ_0. ϑ_0 is the probability that a given site is occupied and the four adjacent neighbor sites are empty. If all sites fill randomly, then ϑ_0 would be given by the product of θ, the probability that a given site is filled and $(1 - \theta)^4$, the probability that 4 adjacent sites are empty, i.e.,

$$\vartheta_0^M = \theta(1 - \theta)^4 \qquad (7.85)$$

where ϑ_0^M, is the value of ϑ_0 when sites fill randomly.

Similarly, ϑ_1 is the probability that a given site is occupied and one of the four nearest neighbor sites is also occupied. ϑ_1^M is given by the probability that a given site is occupied θ times 4θ, the probability that one of four nearest neighbor sites are occupied, times $(1 - \theta)^3$, the probability that the other three nearest neighbor sites are unoccupied, i.e.,

$$\vartheta_1^M = 4\theta^2(1 - \theta)^3 \qquad (7.86)$$

The factor of 4 in Equation 7.86 arises because any one of four sites can be occupied.

ϑ_2 is the probability that a given site is occupied and two of the four nearest neighbor sites are occupied. One can show

$$\vartheta_2^M = 6\theta^3(1 - \theta)^2 \tag{7.87}$$

The factor of 6 in Equation 7.87 arises because there are six ways to arrange two occupied and two unoccupied sites.

A similar analysis shows

$$\vartheta_3^M = 4\theta^4(1 - \theta) \tag{7.88}$$

$$\vartheta_4^M = \theta^5 \tag{7.89}$$

One can verify

$$\vartheta_0^M + \vartheta_1^M + \vartheta_2^M + \vartheta_3^M + \vartheta_4^M = \theta \tag{7.90}$$

Equation 7.90 implies that the probability that a given site is occupied with 0, 1, 2, 3, or 4 nearest neighbors equals the fractional coverage.

Figure 7.24 shows a plot of ϑ_0^M/θ, ϑ_1^M/θ, ϑ_2^M/θ, ϑ_3^M/θ, and ϑ_4^M/θ as a function of coverage. ϑ_0^M/θ, ϑ_1^M/θ, ϑ_2^M/θ, ϑ_3^M/θ, and ϑ_4^M/θ are the probability that a given adsorbed molecule is surrounded by 0, 1, 2, 3, or 4 nearest neighbors, respectively. ϑ_0^M/θ approaches unity at low coverages, which implies that if gas adsorbs randomly at low coverage most of the adsorbed molecules will have few nearest neighbors. In contrast, ϑ_4^M/θ approaches unity at $\theta = 1$, which implies that at saturation coverages most adsorbed molecules will have four nearest neighbors. There is also an interesting symmetry to the curves in Figure 7.24 in that many of the curves are mirror images of one another where the mirror plane is at $\theta = 0.5$. The

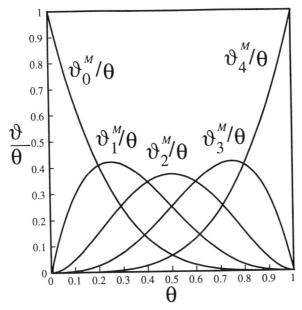

Figure 7.24 A plot of the mean field values of ϑ_0^M/θ, ϑ_1^M/θ, ϑ_2^M/θ, ϑ_3^M/θ, and ϑ_4^M/θ as a function of coverage.

ϑ_0^M curve is a mirror image of the ϑ_4^M curve, i.e., the high coverage behavior of ϑ_4^M is identical to the low coverage behavior of the ϑ_0^M curve and vice versa.

Of course, the results in Figure 7.24 are an approximation that assumed that gas adsorbed randomly. In reality the adsorbate layer orders. As a result, Figure 7.24 is only an approximation under most situations, and it can be a poor approximation when the adsorbate/adsorbate interactions are strong.

One can account for the ordering using the methods in Chapter 4. In particular, the algorithm in the supplemental material for Chapter 4 was used to calculate the adsorption isotherm, i.e., θ as a function of βH_1 for various values of βh. The algorithm actually calculates possible configurations of the adsorbate layer. If one knows the configuration of the adsorbate layer, one can directly calculate ϑ_0, ϑ_1, ϑ_2, ϑ_3, and ϑ_4, as a function of θ for various values of βh, where h is the strength of the pairwise interactions and $\beta = 1/kT$, where k is Boltzmann's constant and T is the adsorbate temperature.

We have done such calculations for adsorption on a 40x40 lattice. Figure 7.25 shows a plot of the results. One way to interpret this plot is that at a βh of 2.0, and a coverage of 0.5, approximately 72% of the adsorbed molecules have four nearest neighbors. In general, when $|\beta h| < 1.7627/2$, the results are similar to the mean field results in Figure 7.24. When βh is positive there is some enhancement in ϑ_3/θ and ϑ_4/θ (the mean probability is that an adsorbed molecule is surrounded by three and four nearest neighbors, respectively). In contrast, ϑ_0/θ and ϑ_1/θ are enhanced when βh is negative.

Figure 7.26 compares the exact and mean field results more carefully. Note that all of the ϑ's approach the mean field results at coverages greater than 0.9. However, there are significant deviations at lower coverages. Thus, one would expect the deviations from mean field kinetics at coverages below 0.9.

One can use the data in Figure 7.25 to calculate TPD spectra. The idea is to plug Equation 7.83 into Equation 7.54 to yield

$$\frac{r_d}{\beta_A N_s} = \frac{d\theta}{dT} = \frac{1}{\beta_H} \sum_{j=0}^{4} k_j \vartheta_j \tag{7.91}$$

Note that if one knows θ and T, one can calculate all of the ϑ_j by interpolating the data in Figure 7.25. Therefore, one can calculate the right side of Equation 7.91 as a function of θ and T. As a result, one can integrate Equation 7.91 to calculate a TPD spectrum.

The calculation are a little easier if one has an analytical approximation for the ϑ_j. One can derive an analytical approximation to the ϑ_j using the Bethe approximation described in Section 4.9.1. Recall that the Bethe approximation considers adsorption on a cluster of atoms, as shown in Figure 4.25. Adsorption within the cluster is treated exactly. Adsorption outside of the cluster is treated in an average sense. One can show that in the Bethe approximation if there are no islands on the surface, then ϑ_j, the probability that a given site is occupied and surrounded by j nearest neighbors is given by

$$\vartheta_j^{Bethe} = \frac{\binom{n_n}{j} \exp(\beta H_1)\,[\exp\{\beta(H_1 + 2g + h)\}]^j}{[1 + \exp\{\beta(H + 2g)\}]^{n_n} + \exp\{\beta H_1\}\,[1 + \exp\{\beta(H_1 + 2g + h)\}]^{n_n}} \tag{7.92}$$

where $\binom{n_n}{j}$ is the number of ways of arranging j molecules over n_n sites, and g is defined in Section 4.9.1.

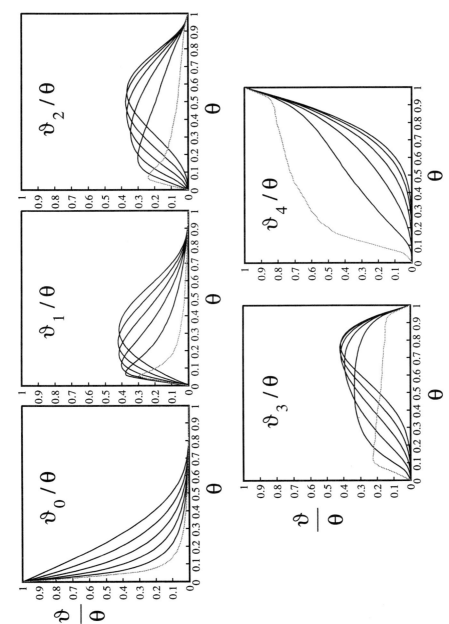

Figure 7.25 A plot of ϑ_j/θ, as a function of θ, for $\beta h = -0.5, 0, \ldots, 2.0$ for adsorption on a 40x40 lattice. The $\beta h = 2.0$ curve is dotted.

527

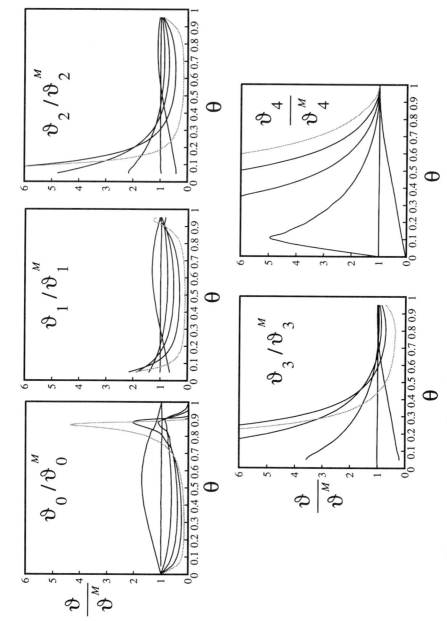

Figure 7.26 A plot of $\vartheta_0/\vartheta_0^M$, $\vartheta_1/\vartheta_1^M$, ... and $\vartheta_4/\vartheta_4^M$ as a function of θ, for $\beta h = -0.5, 0, \ldots, 2$. The $\beta h = -2.0$ curve is dotted.

Note

$$\binom{n_n}{j} = \frac{n_n!}{j!\,(n_n - j)!} \tag{7.93}$$

$$\binom{4}{0} = 1 \qquad \binom{4}{1} = 4 \qquad \binom{4}{2} = 6 \qquad \binom{4}{3} = 4 \qquad \binom{4}{4} = 1 \tag{7.94}$$

Following the derivation in Section 4.9 it is useful to define G_{Bethe} by

$$G_{Bethe} = \exp\{\beta(H_1 + 2g)\} \tag{7.95}$$

If one knows the total coverage, θ, and βh, then one can calculate G_{Bethe} and $\exp(\beta H_1)$ as described in Section 4.9.1. G_{Bethe} is the larger root of

$$G_{Bethe}^2 \{\exp(\beta h)\,(1 - \theta)\} + G_{Bethe}\,\{1 - 2\theta\} - \theta = 0 \tag{7.96}$$

while

$$\exp(\beta H_1) = \frac{G_{Bethe}\,[1 + G_{Bethe}]^{n_n - 1}}{[1 + \exp(\beta h)\,G_{Bethe}]^{n_n - 1}} \tag{7.97}$$

Combining Equation 7.92 through Equation 7.97 yields

$$\vartheta_j^{Bethe} = \frac{\binom{n_n}{j}\exp(\beta H_1)\,[\exp(\beta h)\,G_{Bethe}]^{\,j}}{[1 + G_{Bethe}]^{n_n} + \exp(\beta H_1)\,[1 + \exp(\beta h)\,G_{Bethe}]^{n_n}} \tag{7.98}$$

Figure 7.27 shows a plot of $\vartheta_0^{Bethe}/\theta$, $\vartheta_1^{Bethe}/\theta$, ..., $\vartheta_4^{Bethe}/\theta$, calculated using the Bethe approximation. These results look very similar to those calculated via the Monte Carlo method.

Figure 7.28 compares the systems more carefully. The figure shows how the ratio of $\vartheta_j/\vartheta_j^{Bethe}$ varies with coverage and βh. At first it might seem that Equation 7.98 shows significant errors, because the ratio is as large as 15 in some limits. It works out, however, that when $\beta h < 1.75$, the cases where $\vartheta_j/\vartheta_j^{Bethe}$ is greater than 1.1, the ϑ_j's are tiny, 0.02 or less. Equation 7.98 is within 10% of the exact results whenever ϑ_j is greater than 0.02. Therefore, the Bethe approximation is a fairly good approximation when $\beta h < 1.75$, although it usually underestimates the probability of unusual surface configurations.

There are significant deviations when $\beta h > 1.75$. Recall that islands form when $\beta h > 1.7627$. The Bethe approximation fails in the presence of islands.

In fact, there is some special behavior when $\beta h > 1.7627$ that is not represented by simple Monte Carlo calculations on a 40x40 lattice. In Example 4.G we found that $\beta h = 1.7627$ represents a critical point of the system. When $|\beta h| < 1.7627$ one cannot form any distinct ordered phases on the surface. However, when $\beta h > 1.7627$ the adsorbate phase separates into islands at moderate coverages. If one could actually reach equilibrium, all of the individual islands should agglomerate into a single large island of adsorbed gas. There should also be a few isolated adsorbed molecules outside of the islands, as shown in Figure 7.29. A molecule in an island will be completely surrounded by other molecules, while a molecule outside of the island will be isolated from all of the other molecules in the system. The result is that at equilibrium, ϑ_0 and ϑ_4 will be significant. However, ϑ_1, ϑ_2, and ϑ_3 will be negligible.

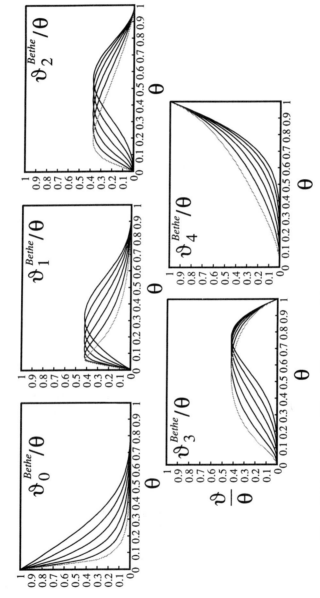

Figure 7.27 A plot of $\vartheta_0^{Bethe}/\theta$, $\vartheta_1^{Bethe}/\theta$, ..., $\vartheta_4^{Bethe}/\theta$ for $\beta h = -0.5$, 0, ..., 2.0. The $\beta h = 2.0$ curve is shaded.

530

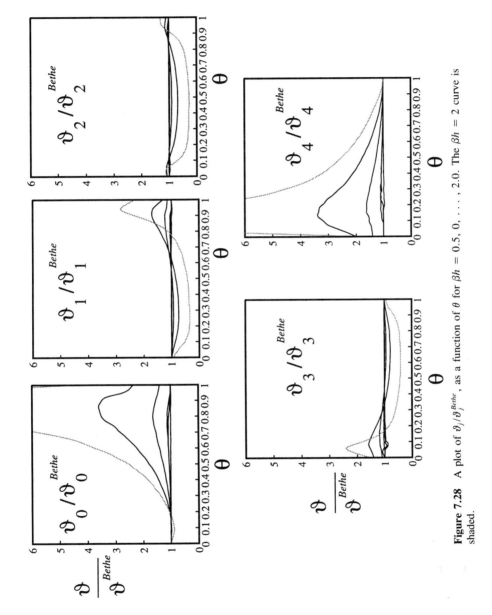

Figure 7.28 A plot of $\vartheta_j / \vartheta_j^{Bethe}$, as a function of θ for $\beta h = 0.5, 0, \ldots, 2.0$. The $\beta h = 2$ curve is shaded.

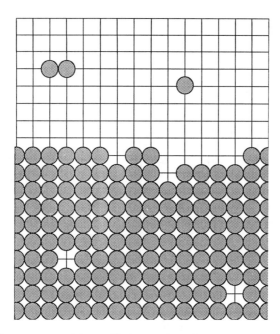

Figure 7.29 A diagram of the equilibrium configuration of the system for $\beta h = 3$.

There is some question whether in reality one gets large-sized islands in a TPD experiment, however. During an experiment it is easy to form a local island. However, it is much harder for islands to agglomerate into infinite condensed phases. As a result in a real system the islands look more like those in Figure 7.30 than those in Figure 7.29.

There is a very important difference between the islands in Figure 7.29 and Figure 7.30. The island in Figure 7.29 is quite large, so the number of molecules along the edge of the islands is small. In contrast, the island in Figure 7.30 is much smaller. With a smaller island, a much larger fraction of the adsorbed molecules lie along the edge of the island. In the next section we find that the rate of desorption along the edge of an island is much larger than the rate of desorption from the interior of the island. As a result, the TPD spectrum can be significantly different if one has small islands rather than large ones.

One can account for these effects by doing a modification of the Monte Carlo calculations in Chapter 4. In the calculations in Chapter 4, we varied βF and constant values of βh and calculated the phase behavior of the adsorbed layer. In that case, one finds that for $\beta h > 1.7627$, there is a first-order phase transition at $\beta F = 0$. When $\beta F < 0$ one only has a dilute adsorbate phase with no islands. In contrast, when $\beta F > 0$ the surface is nearly completely covered by adsorbate. The island size is infinite.

One can do the calculations a slightly different way, and get islands, however. The idea is to fix the coverage on a finite lattice, and then do a Monte Carlo calculation where individual molecules are allowed to move around on the surface, but no molecules are allowed to desorb. It is easy to show that one gets finite-sized islands like those in Figure 7.30 with such calculations. For example, consider adsorption on a lattice with 256 sites. At a coverage of 0.2 there are 51 molecules on the surface, so that an island on the lattice cannot contain more than 51 atoms. A spherical 51-atom cluster would be about 8 atoms wide and have perhaps 37 interior molecules and 24 edge molecules. Such an island would be quite different than the island in Figure 7.29. The island size always varies with coverage. However, one always finds finite-sized islands in such systems.

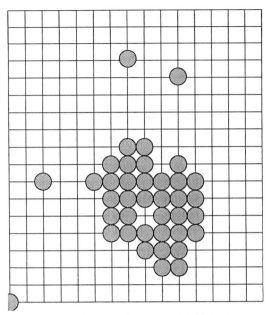

Figure 7.30 An island on a (16x16) lattice.

Figure 7.31 shows the results of calculations of ϑ_0, ϑ_2, . . . , ϑ_4 for desorption from a square lattice with finite islands. Notice that the results for a finite lattice are much closer to the mean field results than the results for the infinite lattice. Still there are significant derivations near $\theta = 0.5$ (i.e., near the top of the TPD peaks). Therefore, one would expect these deviations to have important effects on TPD spectra.

7.12.2 Effects of Ordering on TPD Spectra

One can quantify the effects by assuming a rate form for each of the k_j's in Equation 7.83. For the material which follows it will be convenient to assume that the activation energy for desorption from a given site varies only with the number of nearest neighbors around the site, and that the preexponential for desorption is constant, i.e.,

$$k_j = k_o \exp\left(-\frac{E_d^j}{kT}\right) \tag{7.99}$$

and E_d^j is the activation energy for desorption when an adsorbate is surrounded by j nearest neighbors. Combining Equations 7.77, 7.91, and 7.99 yields

$$\frac{r_d}{N_s} = \beta_H\left(\frac{d\theta}{dT}\right) = \sum_j k_o\, \vartheta_j \exp\left(-\frac{E_d^o}{kT} + \gamma_P \frac{jh}{kT}\right) \tag{7.100}$$

with $E_d = E_d^o + \gamma_P H_1$.

One can numerically integrate Equation 7.100 to calculate a TPD spectrum. A simple algorithm to do the integration is available from Professor Masel. The algorithm uses a slight

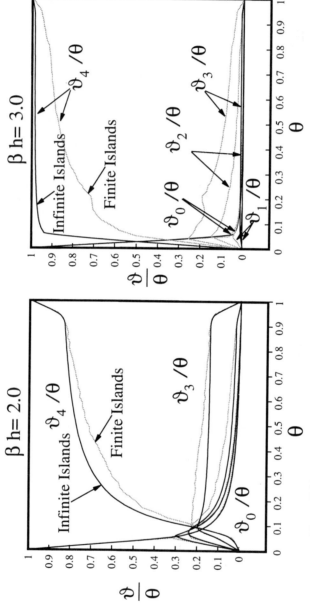

Figure 7.31 A comparison of ϑ_0/θ, ϑ_1/θ, ϑ_2/θ, ϑ_3/θ, and ϑ_4/θ calculated for an infinite and a finite (40x40) lattice with βh = 2.0 and 3.0.

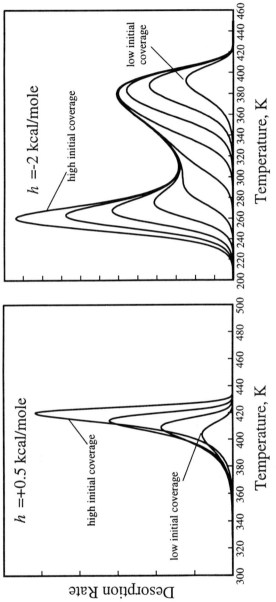

Figure 7.32 A series of TPD spectra calculated by numerically integrating Equation 7.100 for k_o/β_H $= 10^{12}$/sec, $E_d^o = 24$ kcal/mole, $\alpha_P = 1.0$.

535

modification of Equation 7.100.

$$\frac{r_d}{N_s} = \beta_H \left(\frac{d\theta}{dT}\right) = \sum_j k_o \, \mathbb{R}_j \, \vartheta_J^{Bethe} \exp\left(-\frac{E_d^o}{kT} + \gamma_P \frac{jh}{kT}\right) \qquad (7.101)$$

where $\mathbb{R}_j = \dfrac{\vartheta_j}{\vartheta_j^{Bethe}}$. It then performs a Runge–Kutta integration to calculate a TPD spectrum. Note that from Figure 7.28, $\mathbb{R}_j$ is close to unity for $\beta_h < 1.5$, and varies smoothly with coverage above that. Hence, Equation 7.101 can be easily numerically integrated.

Figure 7.32 shows a series of TPD spectra calculated in this way. The way to think of these spectra is to look at what happens with increasing initial coverage. Let's start with the spectrum on the left of Figure 7.32. The spectrum on the left of Figure 7.32 corresponds to a case where there are attractive interactions between the adsorbed molecules. All of the curves in the left spectrum show a single peak but each of the peaks is in a different position. One way to think about the peaks is to consider how the peaks shift with increasing coverage. The smallest peak (i.e., the one with the smallest area) in Figure 7.32 corresponds to the case with the lowest initial coverage, while the peak area increases as the coverage increases. Notice that the peaks in the left spectrum in Figure 7.32 shift to higher temperature with increasing coverage. Such peak shifts are expected when there are attractive interactions between the adsorbates. However, there is the unusual feature that the peaks seem to cross, which means that at certain temperatures, the desorption rate is higher when the coverage is low than when the coverage is high (i.e., the desorption rate shows apparent negative order kinetics).

The negative-order kinetics are a result of the attractive interactions between the molecules. Figure 7.33 shows the adsorbate configurations under conditions where inverted behavior is seen. When there are attractive interactions between adsorbed molecules, the molecules will tend to cluster together. At high coverages, all of the molecules will be surrounded by other molecules. The intermolecular interactions hold molecules on the surface, so initially the desorption rate is low. In contrast, at low coverages, a fair number of the molecules are isolated. The isolated molecules are bound less strongly, so they desorb. The net effect is

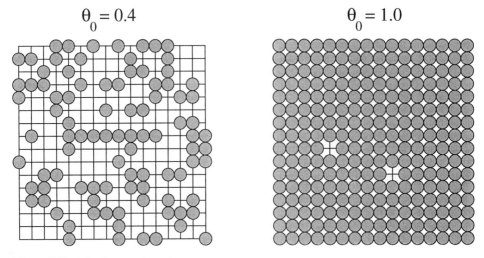

Figure 7.33 Adsorbate configurations in the left TPD spectrum in Figure 7.32. These configurations arise starting with an initial coverage of (a) 0.4, (b) 1.0, and then heating to 375 K while allowing desorption to occur.

that it is easier to desorb molecules at low coverages than at high coverage. As a result, the desorption rate is higher at low coverage than at high coverage.

The spectrum in the right in Figure 7.32 is calculated for a case where there are strong repulsive interactions. Note that when there are strong repulsive interactions the adsorbate will form a c(2x2) overlayer at $\theta = 0.5$. The overlayer will form an imperfect c(2x2) layer at $\theta < 0.5$, but the adsorbed molecules repel one another, so most of the adsorbed molecules will have zero nearest neighbors (see Figure 7.34 and Figure 7.23). Note, however, that the c(2x2) overlayer fills at $\theta = 0.5$. Therefore when θ is greater than 0.5, one must squeeze some molecules into sites with many nearest neighbors. These molecules are held more weakly than the rest so they desorb more easily. As a result, at high initial coverages one gets a two-peak TPD spectrum where one peak corresponds to desorption of molecules with no nearest neighbor molecules and the other peak corresponds to desorption of molecules with many nearest neighbor molecules.

At low initial coverages only one peak is seen because most molecules have no nearest neighbor throughout the desorption process.

One can also get a broad continuous desorption spectrum with repulsive interactions. The broad spectrum occurs because the c(2x2) phases are breaking up during the desorption process, which gives the possibility of many different activation energies for desorption.

Experimentally, one usually observes multiple peaks in a TPD spectrum, and the multiple peaks do correspond to the phase behavior discussed earlier. For example, Figure 7.35 shows a TPD spectrum for CO desorbing from Ni(100). It is interesting to compare this spectrum to calculated spectra on the right of Figure 7.32. Notice the close similarity between the spectra. There are two peaks in both spectra, and the peak heights roughly correspond. The experimental data show a smaller peak splitting than the Monte Carlo calculations. However, in other calculations, we have found that we can match the peak splittings exactly, by reducing α_p in the calculations. In contrast, the Bragg–Williams approximation predicts incorrect behavior (i.e., a single peak that shifts with coverage rather than a two-peak spectrum).

It happens that there is a slight difference in the phase behavior in the data in Figure 7.35 and the calculations. The calculations assumed that CO can adsorb on every site on the Ni(100) surface. However, experimentally, CO molecules actually cover two sites on the

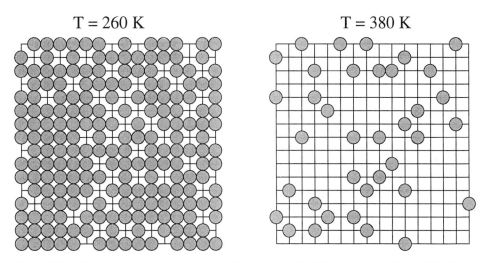

Figure 7.34 Adsorbate configurations calculated during the right TPD spectrum in Figure 7.32. These configurations arise by starting with a saturated surface flashing to (a) $T = 260$, (b) 380 K, and allowing desorption to occur.

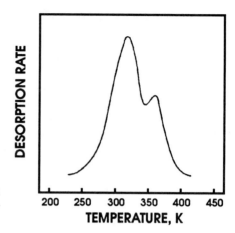

Figure 7.35 A TPD spectrum for CO desorption from the c(2x2)-CO phase on Ni(100). (Data of Johnson and Madix [1981].)

Ni(100) surface. However, the CO layer starts with all of the sites on a c(2x2) lattice filled, goes through an intermediate state where every other c(2x2) site is filled, and then to random adsorption. The calculations, on the other hand, assume complete filling of a (1x1) lattice at low coverages, and half-filling at intermediate coverages. Still, if we would have done the calculations assuming that adsorption occurs only on the c(2x2) sites, one would have predicted that first all of the c(2x2) sites are filled, then half are filled. Therefore, there is a close similarity between the calculated phase behavior and the experimental phase behavior.

Experimentally, one often observes close similarity between the Monte Carlo calculations and the data. The only cases where the correspondence is less good are cases where there are multiple binding sites. In those cases, one needs to do a more complex calculation to predict the adsorbate phase behavior and the TPD spectrum. The Bragg–Williams approximation, on the other hand, rarely gives correct results. Therefore, one needs to use it with caution.

There is one other important phenomenon that comes out of the Monte Carlo simulations: islanding on the surface can have a substantial effect on TPD curves. Figure 7.36 shows a TPD spectrum calculated under conditions where islands form. There are two sets of curves in the figure; the first set of curves is calculated in the limit, where all of the adsorbate has condensed into a single large island on the surface before the TPD spectrum begins. The second set of curves is for a case where the system starts out with a 400-molecule island that shrinks during the TPD experiment. The first curve shows an unusual looking TPD spectrum in that most of the desorption occurs over a narrow temperature range. These very narrow TPD curves are called surface explosions in the literature. The initial parts of all of the peaks lie on top of each other, which means that initially the rate of desorption is independent of coverage (i.e., zero order in the adsorbate coverage). Such effects are expected for infinite islands.

In reality one usually does not observe infinite islands in TPD spectrum. Instead, one observes peaks like those on the right of Figure 7.36. The peaks are still sharp. However, they also shift to higher temperatures with increasing coverage. At some temperatures the desorption rate decreases with increasing coverage. Such a shift is characteristic of a system with finite islands.

Physically, the surface explosions are analogous to the boiling of liquids. There is a critical temperature in the system, which I will call the boiling point above which the adsorbed phase is unstable. There is a finite boiling rate at temperatures below the boiling point. However, the boiling rate increases exponentially as one approaches the boiling point, so most of the desorption occurs over a small temperature range. With infinite islands the boiling curves are independent of coverage until you run out of adsorbate. Then boiling suddenly ceases.

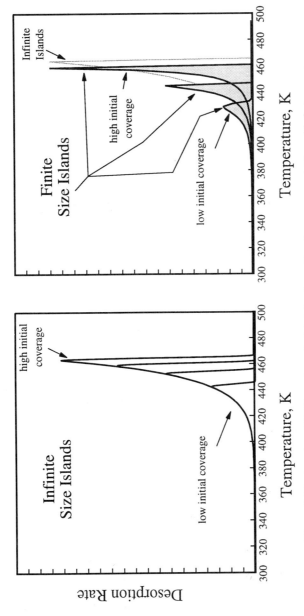

Figure 7.36 A series of TPD spectra calculated for desorption from a square lattice with first nearest neighbor interactions with $E_d^o = 24$ kcal/mole $k_o/\beta_H = 10^{12}$/K, $h = +2$ kcal/mole.

539

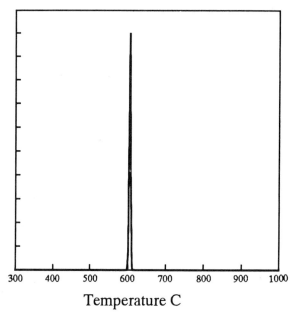

Temperature C

Figure 7.37 A TPD spectrum of oxygen from an oxidized Ge(100) sample. (Data courtesy of A. Rockett.)

The situation is different with finite islands. Small drops boil more easily than extended liquids due to surface tension. An atom at the edge of an island has fewer nearest neighbors than in the bulk, so it boils more easily than the bulk island. As a result, the TPD peaks shift to lower temperatures with decreasing coverage when finite-sized islands form.

Experimentally, surface explosions only occur rarely during TPD on metal surfaces. However, they are quite common during film growth (see Figure 7.37). The surface explosions generally occur when there are such strong attractive interactions between adsorbed species that the system has a finite boiling point. However, one can also get surface explosions when there is a first-order surface phase transition (i.e., a surface reconstruction) or when there is an autocatalytic reaction.

7.13 PRECURSOR-MODERATED DESORPTION

All of the models described in the last several sections assumed that the gas desorbed directly from the surface. However, in Section 5.9.4 we noted that gas often adsorbs via a precursor. If gas adsorbs via a precursor at a given temperature, then according to macroscopic reversibility (Section 5.12), the gas must also desorb via a precursor at that temperature. As a result, desorption via a precursor should be quite common.

It happens that the precursor does not change the qualitative features of the TPD results in the last two sections. However, it does change a few of the details. In the material that follows, we derive a series of equations for precursor-moderated desorption. The derivation will make use of the macroscopic reversibility arguments that were discussed in Section 5.12. The principle of macroscopic reversibility notes that at equilibrium, the rate of adsorption, r_a, is equal to the rate of desorption, r_a:

$$r_d = r_a \qquad (7.102)$$

Therefore, if one knows the rate of adsorption under certain conditions, one can calculate the rate of desorption at those conditions. In the derivation which follows, we assume that macroscopic reversibility applies under conditions far from equilibrium so we can use the equations that we have already derived for adsorption to model desorption. Combining Equations 5.41 and 5.43 shows

$$r_a = S(\theta) \frac{V_i}{4} \left(\frac{P_A^{Equ}}{kT} \right) \tag{7.103}$$

where $S(\theta)$ is the sticking probability, V_i is the molecular velocity of the adsorbate in the gas phase, k is Boltzmann's constant and T is temperature, and P_A^{Equ} is the equilibrium partial pressure of the gas and the coverage and temperature of the measurement.

Equation 5.60 gives an expression for $S(\theta)$. Substituting $S(\theta)$ from Equation 5.60 into Equation 7.103 and then substituting the resultant expression for r_a into Equation 7.102 yields

$$r_d = P_{trap}^{MT} \left(K_1^P \theta + \frac{\dfrac{(1-\theta)}{K_2^P}}{K_3^P + \dfrac{(1-\theta)}{K_2^P}} \right) \frac{V_i}{4} \frac{P_A^{Equ}}{kT} (1-\theta) \tag{7.104}$$

where P_{trap}^{MT}, K_1^P, K_2^P, and K_3^P are a series of constants defined in Section 5.9.1. The P_A^{Equ} term in Equation 7.104 needs further explanation. Physically, P_A^{Equ} is the partial pressure of gas one would need to maintain the coverage at a given value. Physically, P_A^{Equ} is the equivalent vapor pressure of the adsorbate over the surface. The vapor pressure is a function of the temperature and properties of the gas and surface. P_A^{Equ} can be calculated from the adsorption isotherm, e.g., Figure 4.21, 4.23, or 4.24.

When one actually does calculations, it is easier to work in terms of βH_1. From Equation 4.50

$$\exp(\beta H_1) = K_{equ}^{A, \theta = 0} P_A^{Equ} = k_0^{A, \theta = 0} P_A^{Equ} \exp(-\Delta H'_{Ad}/kT) \tag{7.105}$$

where ΔH_{Ad} is the heat of adsorption of A and $k_0^{A, \theta = 0}$ is the exponential of the entropy of adsorption of A.

Substituting Equation 7.105 into Equation 7.104 yields

$$r_d = P_{trap}^{MT} \left(K_1^P \theta + \frac{\dfrac{(1-\theta)}{K_2^P}}{K_3^P + \dfrac{(1-\theta)}{K_2^P}} \right) \frac{V_i}{4} \left(\frac{k_0^{A=0}}{kT} \right) \exp(\beta H_1) \exp(+\Delta H_{Ad}/kT) \tag{7.106}$$

It is easy to use Equation 7.106 to calculate a TPD spectrum. For example, at a given point in the TPD spectrum, the coverage is θ and temperature T are known. If one also knows values of βh and $K_{equ}^{A, \theta = 0}$ at that temperature, one can go to Figure 4.23 or 4.24 and read off a value of βH_1. One can then plug θ and βH_1 into Equation 7.104 to calculate the rate of desorption at that temperature, assuming one knows all of the parameters. One can then substitute that result into Equation 7.45 and integrate to calculate a TPD spectrum.

Figure 7.38 shows the results of the calculations using parameters similar to those in Figure 7.31*b*. Again one gets a two-peak TPD spectrum because of the phase transitions on the surface. Recall that with strong repulsive interactions one gets a c(2x2) overlayer at $\theta =$

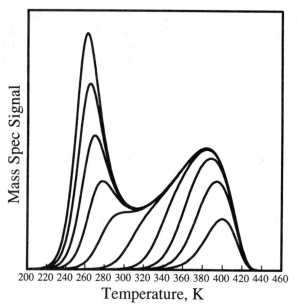

Figure 7.38 A TPD spectrum calculated from Equation 7.104 with $h = -2$ kcal/mole, and $K_1^P = K_2^P = K_3^P = 1$. See the text for details.

0.5. At $\theta > 0.5$ some of the molecules are surrounded by four nearest neighbors so they desorb easily. That increases P_A^{Equ}. In contrast, at $\theta < 0.5$ most of the molecules have zero nearest neighbors. That decreases P_A^{Equ}. The variations in P_A^{Equ} in Equation 7.104 cause the multiple peaks to occur. If only P_A^{Equ} was varying, the TPD spectra in Figure 7.38 would be identical to those in Figure 7.31b. In fact, the other terms also vary. As a result, the spectra in Figure 7.38 are slightly different than those in Figure 7.31b, but quite similar.

This brings up the key result from this section, which is that multiple TPD peaks arise because of the phase transitions described in Chapter 4, i.e., that phase transitions vary P_A^{Equ}, which causes the multiple peaks to occur. For example, the analysis in Example 4.E shows that under the conditions on the left spectrum in Figure 7.31, two TPD peaks are seen. One needs to be careful with the arguments, because when multiple phases are stable, the peaks corresponding to each phase are often difficult to distinguish due to peak overlaps. Still, experimentally there is a general correspondence between the number of surface phase transitions and the number of TPD peaks as predicted by the Monte Carlo and Bethe approximations.

7.14 COMPARISON OF THE METHODS

Surprisingly, multiple peaks are not seen in spectra calculated via the Elovich equation or via Redhead's method. For example, it is useful to compare the Monte Carlo results in Figure 7.32 to the results shown in Figure 7.12 which were calculated via Redhead's methods, and results in Figure 7.22, which were calculated from the Elovich equation. All three figures were calculated for a first-order desorption process. However, there is very little correlation between the results in the three figures. Redhead's analysis predicts that a first-order desorption process should produce a single TPD peak, with a peak position which is independent of coverage. In contrast, the Elovich equation predicts a peak that shifts with coverage, while the Monte Carlo results in Figure 7.32 show that in addition it is quite possible to observe multiple peaks in a simple first-order desorption process.

Note also that the peaks in Figure 7.32 do not shift nearly as much as predicted by the Elovich equation. In the literature, it has been quite common to use Redhead's analysis to analyze TPD data. One normally makes the assumption that when there are multiple peaks in the TPD spectrum, the peaks are associated with multiple desorption states. However, the results in Figure 7.32 call that assumption into question. As a result, one must be exceedingly careful when applying Redhead's analysis or the Elovich equation to TPD data.

What is not obvious is that the Elovich equation has difficulties even when there is only one peak in the TPD spectrum. In order to illustrate the difficulty, it is useful to define an apparent activation energy for desorption, E_d^{aP}, at any temperature and coverage by

$$r_d = r_a^o \exp\left(-\frac{E_d^{aP}}{kT}\right) \tag{7.107}$$

where r_d^o is a function of coverage, but not temperature. Solving Equation 7.107 for E_d^{aP},

$$E_d^{aP} = \frac{kT^2}{r_d}\left(\frac{\partial r_d}{\partial T}\right)_\theta \tag{7.108}$$

Substituting r_d from Equation 7.100 into Equation 7.108 yields

$$E_d^{aP} = \frac{\sum_j k_o\left[\vartheta_j\,(E_d^o + \alpha_P\Delta G_a^j) + \left(\frac{d\vartheta_j}{d\beta}\right)_\theta\right]\exp\left(\frac{-E_d^o + \alpha_P\Delta G_a^j}{kT}\right)}{\sum_j k_o\,\vartheta_j\exp\left(\frac{-E_d^o + \alpha_p\Delta G_a}{kT}\right)} \tag{7.109}$$

where β is Boltzmann's constant. Note that the denominator in Equation 7.109 is the rate of desorption. Therefore, the implication of Equation 7.109 is that the activation energy for desorption should be a weighted average of the activation energy on each site, where the weighting function is the rate of desorption at the site.

If one plugs numbers into Equation 7.109, one finds that the apparent activation energy for desorption varies with both temperature and coverage. Physically, as one changes the temperature or coverage, the arrangement of atoms on the surface changes. The changes in the arrangement of atoms change the apparent activation energy for desorption. As a result, the apparent activation energy for desorption varies with both temperature and coverage.

It is useful to discuss the coverage dependence first. Figure 7.39 shows a plot of the variation of the activation energy with coverage calculated from Equation 7.109 for TPD of a gas from a square surface. In the calculations, the temperature was calculated as a function of coverage by numerically integrating Equation 7.100. The temperature was then plugged into Equation 7.109 to get an apparent activation energy as a function of coverage. The calculations show that when there are repulsive interactions between adjacent molecules, the activation energy for desorption will decrease with increasing coverage, while if there are attractive interactions, the activation energy will increase. However, the effect can be very nonlinear because when there is a mixture of strongly and weakly bound molecules, the weakly bound molecules desorb preferentially. For example, Figure 7.33 shows two typical adsorbate arrangements for a case with strong attractive interactions between the molecules. At $\theta = 0.4$, most of the adsorbate molecules have many nearest neighbors. Still, there are a few isolated molecules. The isolated molecules desorb preferentially. The isolated molecules do not have any nearest neighbors, and so their activation energy of desorption is unaffected

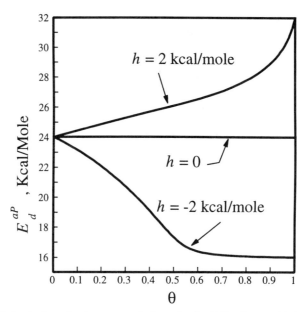

Figure 7.39 Variation in the activation energy of adsorption for TPD of a gas with first nearest neighbor interactions for a square lattice with $E_d^o = 24$ kcal/mole. (a) $h = +2$ kcal/mole; (b) $h = 0$; (c) $h = -2$ kcal/mole.

by the nearest neighbor attractions. As a result, at low coverage, the apparent activation energy of desorption is only slightly affected by the presence of adsorbate–adsorbate attractions. At high coverages, though, all of the adsorbate is surrounded by neighbors. The activation energy for desorption is increased. The result is a nonlinear curve, where the attractive forces have a much smaller effect on the activation energy of desorption at low coverage than at high coverage.

Figure 7.23 shows some typical adsorbate arrangements for the repulsive case. At coverages above 0.5, there are many molecules with four nearest neighbors. Those molecules desorb preferentially, and so at coverages above 0.5, most of the molecules which desorb are strongly affected by the repulsive interactions between the molecules. However, at coverages below 0.5, one starts to desorb molecules which have fewer nearest neighbors. Those molecules are less affected by the repulsions. The result is a nonlinear curve, where the repulsive forces have a smaller effect on the activation energy of desorption at low coverage than at high coverage.

Figure 7.40 compares the variation in the activation energy calculated from the Elovich equation to the activation energy calculated from Equation 7.83 for a case when there is only one TPD peak. The Elovich goes to the right limits at $\theta = 0$ and $\theta = 1$, but it fails in between. Hence, the Elovich equation is not a particularly good approximation to the exact result, except when θ is close to zero or one, because the Elovich equation ignores the preferential desorption of weakly bound molecules.

Experimentally, one usually observes many peaks in a TPD spectrum (see Figure 7.10). The peaks usually shift with coverage. People often discuss the multiple peaks as though one is desorbing gas from two or more different states. Sometimes that is reasonable. For example, if gas adsorbs on a square lattice, with strong first nearest neighbor repulsions, one might say that there are two states of the system: a c(2x2) state, where molecules are arranged as shown in the left of Figure 7.23, and a (1x1) state, where molecules are arranged as shown in the right of the figure. Such a description is quite reasonable when the strength of the nearest neighbor repulsions is much greater than kT so that the surface goes through a true c(2x2) phase. However, often the strength of the interactions is only a few times kT so that one gets

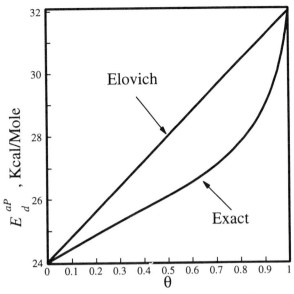

Figure 7.40 Variation in the activation energy calculated from the Elovich equation, Equation 7.109, and the exact Monte Carlo algorithm for desorption from a square lattice with first nearest neighbor interactions with $h = 2$ kcal/mole.

mixed phases. In that case, one still observes a TPD spectrum with two peaks, as shown in Figure 7.41. However, the peaks are not clearly distinct, and one cannot identify the peaks with individual surface phases.

Some investigators still try to fit the peak positions in such data with the Elovich equation. That usually works empirically, because when one focuses on a single TPD peak, one is only

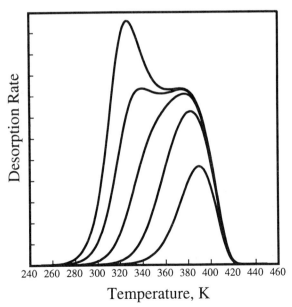

Figure 7.41 TPD spectra calculated for desorption from a square lattice with $h = -1$ kcal/mole. There is no distinct phase transition for this case, but one still observes multiple peaks.

looking at a limited range of coverage. One can always linearize the variation in the activation energy with coverage over a small range of coverage and thereby fit the data to the Elovich equation. The question, though, is whether the parameters one obtains in that way are physically meaningful. At this point the answer is unknown. Generally the Elovich results agree with the Monte Carlo results only when the interactions between the adsorbed molecules are weak (i.e., $|h| \leq kT/2$). With stronger interactions, the peak shifts are in the directions predicted by the Elovich equation, but the magnitudes of the shift can be much smaller or much larger than those predicted by the Elovich equation. Thus, one has to be very careful when applying the Elovich equation to the analysis of real data.

My viewpoint is that the Elovich equation was very useful years ago, but it is less useful today. The algorithm in the supplemental material allows one to do a Monte Carlo calculation for a TPD spectrum, and the program only takes a few seconds to run on a PC. Calculations using the Elovich equation are only marginally faster. As a result, there is no longer a significant advantage to using the Elovich equation, and the Elovich equation is far less accurate than the Monte Carlo results.

There is one other feature that is important in Equation 7.109, which is that the apparent activation energy for desorption is temperature dependent. Physically, as one varies the temperature, one changes the ordering on the surface, which in turn leads to a variation in the activation energy of desorption.

In a 1993 paper, Stuckless et al. estimated the magnitude of this effect. They measured the heat of adsorption of CO on Ni(110) calorimetrically and compared it to the adsorption energy of desorption heat measured via a Taylor–Weinberg analysis. Figure 7.42 shows some of Stuckless et al.'s results. Generally, the calorimetric heats of adsorption agree with the activation energies of desorption calculated by the Taylor–Weinberg analysis at high and low coverages. However, there are considerable differences in between. According to Equation 7.67 the two curves should be the same, but clearly they are not. Stuckless et al. attribute this difference to a change in the ordering of the adsorbed layer when the Ni(110) surface is heated from the temperature where the calorimetry was done to the temperature where the TPD work was done. These results show that the basic prediction of the Elovich equation, that the activation energy of desorption varies with coverage but not temperature, does not hold for CO adsorption on nickel. At this point, it is unclear whether the Elovich equation works in other cases, but it clearly does not work for the one case where it has been tested in detail.

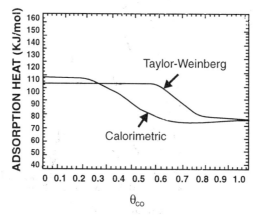

Figure 7.42 A comparison of the heat of adsorption of CO on Ni(110) measured calorimetrically to that calculated by the Taylor–Weinberg analysis. (After Stuckless et al. [1993].)

Before I close this section I do want to note that the Monte Carlo TPD algorithm in the supplemental material uses ϑ_j calculated for a square lattice with first nearest neighbor interactions. In Chapter 4 we found that the phase diagrams calculated for first nearest neighbor interactions were also applicable to other systems that obey Ising universality class (i.e., systems that show c(2x2), c(3x3) LEED patterns). However, there are deviations for systems that obey other universality classes. The changes are most significant for XY universality class (e.g., a system with a (2x1) or (3x1) structure due to strong second nearest neighbor repulsions and weaker first nearest neighbor attractions). The number of simulations that have been done for these systems is small.

However, in general TPD peaks show complex behavior. For example, if there are weak first nearest neighbor attractions and stronger second nearest neighbor repulsions, the TPD peaks shift to higher temperatures with increasing coverage at low coverages where the first nearest neighbor interactions dominate. However, the peaks shift to lower temperatures with increasing coverage at higher coverages because second nearest neighbor effects dominate. The result is complex behavior in the TPD spectrum.

Experimentally, one usually observes behavior that agrees with the behavior predicted by the program in the supplemental material. However, there are exceptions. There has been some work in using the quasichemical approximation to model those exceptions as reviewed by Zhdanov [1991]. However, the inaccuracies in the quasichemical method make such results questionable.

7.15 MONTE CARLO SIMULATION OF SURFACE REACTIONS

All of the analysis in the last two sections also applies to surface reactions. For example, consider a first-order reaction

$$A_{ad} \xrightarrow{\quad} B_{ad} \xrightarrow{\text{fast}} \text{Products} \qquad (7.110)$$

occurring on the surface of a catalyst. At any instant, there are molecules on the surface in many different environments. Some molecules will be isolated on the surface, while others will be surrounded by 1, 2, 3, or 4 nearest neighbors. Some molecules will be on linear sites, while other molecules will be adsorbed on bridgebound sites. The kinetics of any given surface reaction will be different on different sites. As a result, the presence of multiple environments should have important implications to steady-state surface reactions.

The analysis of this process is similar to the analysis in Section 7.10. It is useful to divide the molecules into subpopulations, such that the molecules in each subpopulation are equivalent (i.e., adsorbed on equivalent sites with equivalent numbers of nearest neighbors). One can then write the overall reaction rate as a sum of rates over all of the subpopulations. For a first-order reaction, the net rate of reaction becomes

$$\frac{r_{Net}}{N_s} = \sum_J k_J \, \vartheta_j \qquad (7.111)$$

Equation 7.111 is equivalent to Equations 7.63 and 7.83. One can calculate an effective activation energy for the reaction using Equation 7.92:

$$E_d^{aP} = \frac{kT^2}{r_d} \left(\frac{\partial r_d}{\partial T} \right)_\theta$$

If the ϑ_j are independent of temperature, then the activation energy works out to be a weighted average of the activation energies for each of the subpopulations on the surface (i.e., the j's).

In actual practice, the ϑ_j are temperature dependent. Variations in the ϑ_j change the apparent activation energy for reaction. As a result, the measured activation energy is not simply related to the k_j. The other major effect is that the apparent order of the reaction changes. One can get a reasonable picture of the effects from Tempkin's analysis in Section 7.4.3. As one changes the coverage, one changes the activation energy for reaction, as described in the previous paragraph. That changes the rate constant for the reaction. If the rate constant for a "first-order" reaction is coverage dependent, the rate will vary in some complex way with coverage. As a result, the reaction kinetics will not appear first order.

In actual practice, however, this is usually a minor effect for first-order reactions. First-order surface reactions usually follow first-order Langmuir–Hinshelwood kinetics. However, there are some deviations if one takes data over a wide coverage range.

7.16 DISSOCIATIVE ADSORPTION

The effects, however, can be important in other cases. The analysis in Sections 7.10 to 7.13 is for the case where a single molecule adsorbs molecularly on a surface, and then reacts or desorbs again via a first-order process. However, in many cases, real surface reactions involve two or more species that need to come together to react, i.e.,

$$A_{ad} + B_{ad} \longrightarrow \text{Products} \tag{7.112}$$

Most catalytic reactions are of the form in Equation 7.112, as are many film-growth reactions. Thus, reactions of the form in Equation 7.112 are quite important.

Reactions of the form in Equation 7.112 also arise in the desorption of many species. Recall that in Section 3.4 we noted that gases such as H_2, O_2, and N_2 adsorb dissociatively:

$$H_2 \rightleftharpoons 2H_{ad}$$

Now consider desorbing H_2 from a surface. In order for H_2 to desorb, two hydrogen atoms need to come together in a manner similar to that described by Equation 7.112. Thus, reactions of the form in Equation 7.99 are also important to TPD.

Reactions of the form in equation 7.112 are fundamentally different from the cases that were discussed in Sections 7.8 to 7.13 in that pairs of atoms or molecules need to come together to react. The result is a second-order desorption process, where the rate of desorption is proportional to the number of *pairs of atoms or molecules on the surface. Therefore, one needs to take account of pairs* of atoms or molecules on the surface to model the reaction.

In the next three sections we consider reactions where pairs of reactants need to come together to form products which desorb. Generally, we will assume that the rate equation follows

$$r_{Net} = \kappa N_{pairs} \tag{7.113}$$

where N_{pairs} is the number of appropriate pairs of atoms on the surface. We will then develop equations for the number of pairs of atoms on the surface to get a rate.

7.17 FORMATION OF PAIRS OF MOLECULES ON THE SURFACE

The simplest case is one where two identical adsorbed molecules react to form products, i.e.,

$$2A_{ad} \longrightarrow \text{Products} \tag{7.114}$$

Examples include desorption of molecules such as H_2, O_2, and As_2 where two adsorbed atoms recombine and desorb, or formation of ethane (C_2H_6) via recombination of two adsorbed CH_3 groups. We already considered this case in reverse in Section 5.10. Following the material in that section we note θ, the fractional coverage, is equal to the probability that any given site is filled. Further, we define δ_f as the probability that if the given site is filled, a given site adjacent to the first is also filled. In this notation, P_{filled} the probability that a given pair of sites is filled (i.e., occupied by adsorbate), is given by

$$P_{filled} = \delta_f \theta \qquad (7.115)$$

If we presume that the desorption rate r_d is proportional to the number of pairs of atoms on the surface, then

$$r_d = \kappa_d^1 P_{filled} N_{pairs}^{total} = k_d \delta_f \theta \qquad (7.116)$$

where N_{pairs}^{total} is the total number of pairs of sites on the surface and k_d is the rate constant for desorption. Note that k_d in Equation 7.116 depends on the number of neighbors adjacent to the given pair of sites in the same way that k_d for a first-order desorption process depends on the number of neighbors adjacent to a given site. Generally, k_d increases if there are repulsive interactions between adsorbed molecules that push the adsorbate off the surface, while k_d decreases when there are attractive interactions between adsorbed molecules that hold the adsorbate onto the surface. We described the effects of the adsorbate/adsorbate interactions on a first-order desorption process in the preceding two sections. The same effects also occur with a second-order desorption process. The TPD peaks shift with coverage and heating rate due to the interaction of molecules on the surface. One can also get multiple peaks in TPD due to the presence of ordered adsorbate phases. All of the analysis of these effects are identical in the second-order case as in the first-order case described in Sections 7.8 and 7.9. Therefore, the material will not be repeated here.

There is, however, one effect that is seen in a second-order desorption process which is not seen in a first-order desorption process: in a second-order desorption process, two adsorbed molecules must move to two adjacent sites on the surface before a reaction (desorption) can occur. Anything that affects how quickly the molecules come together affects the rate of the desorption process. For example, diffusion limitations can slow the rate that molecules come together as do repulsions between the molecules. That decreases the desorption rate. In contrast, attractions between the molecules can enhance the rate that molecules come together. That increases the desorption rate.

One can model these effects using Equation 7.116. In Section 7.19 we show that if there are no interactions between the adsorbed molecules or diffusion limitations

$$(\delta_f)_{field}^{mean} = \theta \qquad (7.117)$$

If there are interactions between the molecules or diffusion limitations, δ_f will change. That changes the rate. In the material that follows, we analyze these changes. To simplify the analysis we assume that k_d, the rate constant for desorption, is independent of the arrangement of the molecules on the surface. That can be a poor assumption. However, we have already discussed the effects of variations in k_d in Sections 7.10 to 7.13. Therefore, we do not repeat the discussion here.

7.18 THE IMMOBILE LIMIT

In the next several sections, we will describe how δ_f varies with coverage in two limits: the completely mobile limit, where diffusion of molecules is so rapid that the adsorbate maintains its equilibrium configuration during desorption, and the immobile limit where diffusion is slow, so that deviations from the equilibrium configuration are possible. Let's consider the immobile limit first. In the immobile limit, we start with some random distribution of atoms or molecules on the surface that then react and desorb. Reaction can occur only when there are adjacent pairs of atoms or molecules on the surface. That is not a problem at high coverage. However, when the surface coverage gets low, there is always some chance that a given site will be occupied, but all of the sites adjacent to the site will be empty. If the atom or molecule adsorbed on the given site is unable to diffuse, it will not be able to find another species to react with. As a result, the atom or molecule will be trapped on the surface, and not react. Therefore, in the immobile limit, the reaction rate will go to zero at a finite coverage.

For example, Dawson and Peng [1972] examined nitrogen desorption from W(100). They found that the desorption rate of nitrogen went to zero at a nitrogen coverage of about 0.1 because the adsorbed nitrogen atom diffused too slowly to find other atoms on the surface.

One can model this process using the analysis in Section 5.10. Consider starting with a filled lattice, and slowly desorbing gas. When the first two molecules desorb, a pair of adjacent vacant sites form on the surface. Additional adsorption produces additional vacancies. Notice, that one can formally treat this process as though one is adsorbing pairs of vacancies onto the surface: the desorption of the each pair of molecules produces a pair of vacancies. Therefore, the desorption of a pair of molecules is formally equivalent to adsorption of a pair of vacancies.

Now it is useful to look back to Figure 5.17. In Figure 5.17 we used a black oval to designate an adjacent pair of filled sites, but we could have just as well treated them as an adjacent pair of empty sites. We start off as shown in Figure 5.17a with all sites filled, slowly desorb pairs of sites at random, and eventually get to the situation in Figure 5.17d where there are a few isolated atoms on the surface. Notice that nothing in the derivations in Sections 5.10, 5.10.1, and 5.10.2 would change if one was adsorbing pairs of vacancies rather than adsorbing pairs of atoms. As a result, all of the analysis in these three sections would also apply. In particular, the desorption process will stop at a vacancy coverage of 0.90293 . . . (i.e., $\theta = (1 - 0.90293 \ldots) = 0.09707$). One can use the analysis in Section 5.10.1 to calculate an exact value of δ_f for adsorption on a one-dimensional lattice:

$$\delta_f = 1 + \tfrac{1}{2} \ln (\theta) \tag{7.118}$$

In two dimensions the value of δ_f varies with the local arrangement of the atoms on the surface. However, if one ignores that variation as described in Section 5.10.2, one finds that to a reasonable approximation

$$\delta_f \cong \tfrac{1}{2} [3 \sqrt{\theta} - 1] \tag{7.119}$$

where the approximation in Equation 7.119 breaks down for coverages below 0.15. Substituting Equation 7.119 into Equation 7.116, and noting that $r_d = (d\theta/dT)$, yields

$$\beta_H \frac{d\theta}{dT} = \frac{d\theta}{dT} = r_d = k_d \frac{\theta}{2} [3 \sqrt{\theta} - 1] \tag{7.120}$$

7.18.1   Implications for TPD

Figure 7.43 shows a plot calculated by numerically integrating Equation 7.119 for a case where

$$k_d = 10^{13}/\text{sec} \exp\left(-\frac{24 \; kcal/mole}{kT}\right)$$

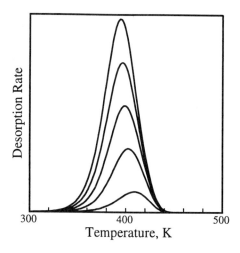

Desorption Rate

300   400   500
Temperature, K

Figure 7.43 A TPD spectrum calculated for the second-order desorption of an immobile adsorbate from a square surface.

independent of the neighbors on the surface. One finds asymmetric peaks that shift with coverage. Qualitatively, the results in Figure 7.43 are similar to the second-order results in Figure 7.12. However, the peak shift is smaller than expected for a second-order desorption process and the desorption process stops before the surface is completely depleted of adsorbate.

In reality, no real system is completely immobile. Instead, there is some mobility at higher temperatures. As a result, all of the adsorbate eventually desorbs at high temperatures. However, the key thing is that at low coverages, the desorption rate is unexpectedly small due to the difficulties in bringing two atoms together on the surface. This produces some significant changes to the shapes of the TPD curves.

7.19 THE COMPLETELY MOBILE LIMIT

The other limit discussed in detail in the literature is the completely mobile limit. In the completely mobile limit, the adsorbate is assumed to instantaneously rearrange to maintain the equilibrium configuration on the surface. The analysis of second-order desorption from a completely mobile lattice is essentially the same as the analysis of the first-order case discussed earlier in this chapter. One uses a Monte Carlo calculation to estimate the number of pairs of atoms on the surface. The one complication in the analysis is that an atom on site 0 in Figure 7.44 would like to react with an atom on site 1 in Figure 7.44; the atom on site 1 also has a finite probability of reacting with an atom on site 2 (i.e., a second nearest neighbor site). The rate constants are different for the 0–1 and 0–2 reactions. As a result, one needs to consider many different rate constants to model the system exactly.

In the literature, it is common to ignore the 0–2 reaction (i.e., to assume that a molecule needs to diffuse to a nearest neighbor site before any reaction occurs). In that approximation, δ_f in Equation 7.116 becomes the probability that site 1 in Figure 7.44 is filled, given that site 0 is filled. For the discussion that follows, it is useful to relate $\theta\delta_f$, the probability that the 0–1 pair of sites is occupied to ϑ_j, the probability that a given site is occupied and surrounded by j nearest neighbors. Note that ϑ_1 is the probability that a given site and one of four nearest neighbor sites are occupied, while ϑ_2 is the probability that a given site and two of four nearest neighbor sites are occupied, etc. Therefore, $\theta\delta_f$, the probability that a given site and one given nearest neighbor site are occupied is given by

$$\theta\delta_f = \tfrac{1}{4}\,(\vartheta_1 + 2\vartheta_2 + 3\vartheta_3 + 4\vartheta_4) \tag{7.121}$$

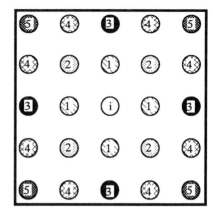

Figure 7.44 Some sites on a square lattice.

The reader should verify that if one substitutes Equations 7.77–7.81 (i.e., the mean field values of the ϑ_j) into Equation 7.121, one obtains

$$(\theta\delta_f)_{field}^{mean} = \theta^2 \tag{7.122}$$

One can rearrange Equation 7.122 to show that δ_f should be approximately proportional to θ. Figure 7.45 shows a plot of δ_f/θ versus θ calculated via a Monte Carlo method for a number of values of βh. Generally, δ_f/θ approaches unity at high coverage and varies between 0.71 and 6.0 at low coverages. δ_f/θ is greater than one when there are attractive interactions between the adsorbed molecules, and less than one when there are repulsive interactions between the adsorbed molecules.

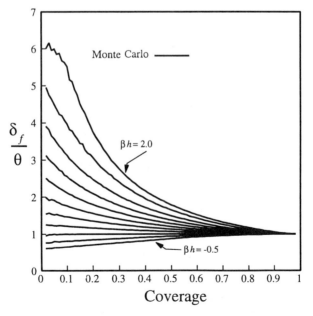

Figure 7.45 A plot of δ_f/θ versus θ calculated via a Monte Carlo method for adsorption on a (40x40) square lattice with $\beta h = -0.5, -0.25, \ldots, 2.0$.

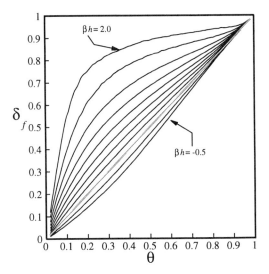

Figure 7.46   A plot of δ_f versus θ for $\beta h = 0, -0.5 \ldots 4$.

When $\beta h > 1.7627$, islands form on the surface. If islands form, δ_f depends on both the coverage and the sizes of the islands. Figure 7.45 shows two results for islands containing 1600θ adsorbed molecules. Similar curves are seen with larger islands.

Figure 7.46 shows a replot of the data from Figure 7.45 as a function of θ for various values of βh. At low coverages, δ_f is proportional to θ, with a proportionally constant that depends on βh. However, at higher coverage δ_f saturates. To put Figure 7.46 in perspective note that according to Equation 7.116 when δ_f is proportional to θ, the desorption rate will be second order in θ. In contrast, when δ_f is independent of θ, the desorption rate will be first order in θ, while if the lines in Figure 7.46 are parabolic the desorption process will switch from second to third order in θ with increasing coverage. Therefore, the implication of Figure 7.45 is that when there are attractive interactions between the adsorbed molecules, the desorption kinetics will switch from second order to first order as the coverage increases. In contrast, when there are repulsive interactions, the apparent order will switch from second to third order as the coverage increases.

7.19.1   Bethe Approximation

One can get a reasonable approximation to δ_f from the Bethe approximation. The key equation is derived by substituting Equation 7.91 into Equation 7.121. The result for the case where there is only one phase on the surface is

$$\delta_f^{Bethe} = \frac{G_{Bethe}\, \exp(\beta h)}{1 + G_{Bethe}\, \exp(\beta h)} \tag{7.123}$$

To keep Equation 7.123 in perspective note that when θ is small, $G_{Bethe} \approx \theta$. Therefore, the implication of Equation 7.123 is that δ_f should vary linearly with coverage at low coverages and then saturate at higher coverage in agreement with the exact results in Figure 7.46.

Of course, Equation 7.123 is an approximation that is not quite exact. Figure 7.47 compares Equation 7.123 to the Monte Carlo (i.e., exact) results for adsorption on a square lattice with first nearest neighbors interactions. One observes that Equation 7.123 is within a few percent of the exact results when $\beta h < 1.35$, but there are significant deviations at larger

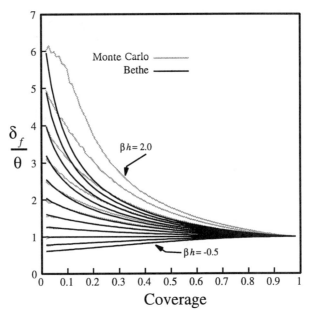

Figure 7.47 A comparison of Equation 7.123 to the exact Monte Carlo results for adsorption on a square lattice with $\beta h = -0.5, -0.25, \ldots, 2.0$.

values of βh, i.e., when the Bethe approximation predicts that there will be multiple phases. Still, Equation 7.123 is accurate enough for most purposes, even if it is not exact.

7.19.2 Implications for TPD

The results in the last section have some important implications for TPD. There are two key effects: a shift from second order to first order kinetics with increasing coverage due to clustering of molecules on the surface, and shifts in TPD peaks due to forces between molecules on the surface. Figure 7.48 shows a series of TPD spectra calculated by numerically integrating Equation 7.116, assuming that δ_f follows Figure 7.45 and that k_d is constant. Two sets of TPD curves are shown; one for $h = 0$ (i.e., $\delta_f = \theta$) and one for $h = 2$ kcal/mole. The TPD curves for $h = 0$ are identical to the $n = 2$ curves in Figure 7.12. The TPD curves for the $h = 2$ kcal/mole are qualitatively similar to those for $h = 0$ in that the TPD peaks shift to lower temperature with increasing coverage. However, the shifts are much smaller when $h = 2$ kcal/mole than when $h = 0$. The peak shifts are negligible at coverages between 0.2 and 1, which means that the desorption process looks first order over that coverage range.

One needs to be careful with Figure 7.48 because it was calculated assuming that k_d is independent of coverage. However, if one assumes that k_d varies with coverage, as described in Sections 7.10 to 7.12, one gets extra shifts due to the variations in k_d with coverage. Generally, there is an extra peak shift to higher temperatures with increasing coverages when there are attractive interactions between the adsorbed molecules, while the opposite occurs when there are repulsive interactions between the adsorbed molecules. Figure 7.49 illustrates these effects for $h = 2$ kcal/mole. At low coverages, the TPD peaks shift to lower temperatures with increasing coverages, since the desorption process is second order. However, at higher coverages, the TPD peaks shift to higher temperatures, since attractive interactions between the adsorbed molecules hold the adsorbed molecules on the surface. Therefore, the implication of the results in this section is that a second-order desorption process may show TPD peaks, which shift in a way that is quite different than expected from Redhead's analysis.

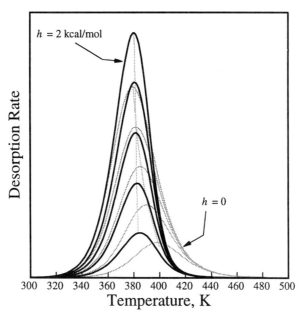

Figure 7.48 A series of TPD spectra calculated by numerically integrating Equation 7.116 assuming that k_d is constant.

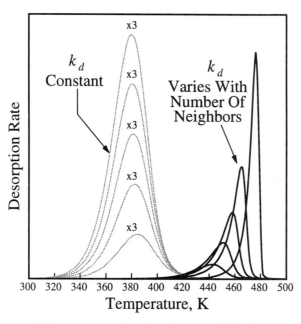

Figure 7.49 A series of TPD spectra calculated from Equation 7.116 with k_d varying as described in Sections 7.10–7.12 and $h = 2$ kcal/mol and $\alpha_P = 1$.

7.20   CHANGES IN THE ORDER OF STEADY-STATE REACTIONS

The processes described in the last section also have important implications to the kinetics of steady-state surface reactions. There are two important effects: (1) the apparent order of the reaction can be different from that given in Table 7.2, and (2) the rate constant changes to reflect the phase behavior on the surface.

The first of these, the change in the apparent order of the reaction, is the most important effect when there are attractive or repulsive interactions between molecules, the apparent order of the reaction is different from the order one would expect from Table 7.2.

Table 7.2 only considered one case where there was a dissociative adsorption process, the Rideal–Eley mechanism

$$A_2 + 2S \underset{2}{\overset{1}{\rightleftharpoons}} 2 A_{ad}$$

$$A_{ad} + B \underset{4}{\overset{3}{\rightleftharpoons}} C + S \tag{7.124}$$

The results in Table 7.2 show that when the dissociative adsorption of A, Reaction 1, is rate determining, the overall reaction rate will be first order in P_A, while if the surface reaction, Reaction 3, is rate determining, the overall reaction rate will be half-order in P_A. Note, however, that the derivation of Equation 7.217 assumed that the rate of Reaction 2, r_2, is given by

$$r_2 = k_2 \, \theta_A^2 \tag{7.125}$$

which is the mean field result. Now consider instead assuming that r_2 follows Equation 7.116, i.e.,

$$r_2 = k_2 \, \theta_A \, \delta_A \tag{7.126}$$

Note that since δ_A is not necessarily equal to θ, one will get deviations from Equation 7.125. For example, if δ_A is independent of θ_A, as will be the case when there are strong attractive interactions between the adsorbed A atoms, then the desorption of A (i.e., Reaction 2) will look first order, which means from macroscopic reversibility, that the adsorption of A will also look first order. After considerable algebra one can show that when Reaction 3 is rate determining, the net rate of reaction is

$$r_{net} = \frac{k_3 \dfrac{k_1}{k_2} S_o P_A P_B - k_4 S_o P_C}{1 + \dfrac{k_1}{k_2} P_A} \tag{7.127}$$

Notice that the reaction still follows a Langmuir–Hinshelwood rate law. However, the new rate of reaction is first order in P_A, while Table 7.2 says that it should be half-order in P_A. Thus, the apparent order of the reaction has changed due to adsorbate/adsorbate interactions.

At moderate to low coverage, the results change. The rate shows complex behavior with coverage. Interestingly, if one fits δ_A to a power law, one can show that the reaction should approximately follow a Tempkin rate equation over a moderate range of coverage similar to that described in Section 7.4.3.

The deviations from Langmuir–Hinshelwood behavior are the key point from the discussion in this section. Interactions between adsorbed molecules can change the apparent order of a reaction. The case here was one where the order increased, but in other cases it decreases. Such an effect is quite important to the ammonia synthesis case described earlier in this chapter in that it is responsible for the appearance of Tempkin kinetics. However, it also has important implications to the kinetics of most hydrogenation and oxidation reactions. For example, the kinetics of most oxidation reactions are first order in the oxygen pressure, as expected from Equation 7.127, even though the oxygen dissociatively adsorbs, and the rate-determining step is a surface reaction similar to Reaction 3 in Equation 7.124.

Adsorbate/adsorbate interactions change the apparent kinetics of surface reactions when the reactant concentrations are high. The reactions still follow Langmuir–Hinshelwood rate laws. However, the form of the rate equation is different from the one would expect from a Langmuir–Hougen and Watson analysis of the mechanism.

7.21   A+B REACTIONS

The analysis in the last section was for reactions of the form

$$2A_{ad} \longrightarrow \text{Products} \tag{7.128}$$

However, most catalytic reactions are of the form

$$A_{ad} + B_{ad} \longrightarrow \text{Products} \tag{7.129}$$

Again, one needs a pair of adsorbate molecules to come together on the surface in order for a reaction to occur. However, the analysis of the case in Equation 7.129 is fundamentally different from the analysis of the case in Equation 7.128 in that one needs to keep track of multiple surface species (i.e., A and B) in the case in Equation 7.129. The phase behavior of a two-component mixture is more complex than that of a single component, so the plots earlier in this chapter do not apply. Still, the qualitative features are the same in the case in Equation 7.128 as in 7.129. It is just the details (i.e., the phase behavior) that changes.

One can formulate the rate for $A_{ad} + B_{ad}$ reaction just as in Section 7.15. The rate can be written as a rate constant times the number of AB pairs on the surface:

$$r_{Net} = \kappa_{AB} N_{AB} \tag{7.130}$$

where N_{AB} is the number of A–B pairs on the surface and κ_{AB} is a constant. The number of AB pairs can be written as

$$N_{AB} = N_{\substack{total \\ pairs}} \, (\theta_A \, \delta_{AB} + \theta_B \, \delta_{BA})/2 \tag{7.131}$$

where θ_A and θ_B are the fractional coverage of A and B, which is equal to the probability that a given site is occupied by A or B. δ_{AB} is the probability that when a given site is occupied by A, a given nearest neighbor site is filled by B, while δ_{BA} is the probability that when a given site is occupied by B, a given nearest neighbor site is occupied by A. Note that if a given A molecule has a B molecule on a nearest neighbor site, then the B molecule must have an A on a nearest neighbor site. As a result one can show

$$\theta_A \, \delta_{AB} = \theta_B \, \delta_{BA} \tag{7.132}$$

Combining Equations 7.130, 7.131, and 7.132 yields

$$r_{Net} = k_{AB}\,\theta_A\,\delta_{AB} = k_{AB}\,\theta_B\,\delta_{BA} \tag{7.133}$$

where

$$k_{AB} = \kappa_{AB} N_{total \atop pairs} \tag{7.134}$$

As before, there are two key limits to Equation 7.133, a mobile and an immobile limit. The mobile limit is discussed in Section 7.21.1, while the immobile limit will be discussed in Section 7.21.2.

7.21.1   The Mobile Limit

If the adsorbate layer is mobile, then the adsorbed layer will maintain equilibrium during the reaction. In that case, δ_{AB} and δ_{BA} can be calculated as a function of θ_A and θ_B via Monte Carlo calculations similar to those in Chapter 4. In the case where there are only pairwise first nearest neighbor interactions, three additional parameters are needed to do the calculations, h_{AA}, h_{AB}, and h_{BB}, the strength of the pairwise AA, AB, and BB interactions. The calculations are straightforward, and a suitable program is available from Professor Masel. The only difficulty is that there are so many parameters in the calculations the results of the calculations are difficult to present in a simple way.

When this book was being written, universal plots of δ_{AB} and δ_{BA} had not yet appeared in the literature, and with five parameters in the calculations, we could not find a good way to present them.

One can, however, get most of the qualitative features of the solution from the Bethe approximation. The details of the derivation are not very important. However, the key result is that with two components, A and B, δ_{AB} and δ_{BA} are given by an expression that is analogous to that in Equation 7.123:

$$\delta_{AB} = \frac{\exp(\beta h_{AB})\,G_{Bethe,B}}{1 + \exp(\beta h_{AA})\,G_{Bethe,A} + \exp(\beta h_{AB})\,G_{Bethe,B}} \tag{7.135}$$

$$\delta_{BA} = \frac{\exp(\beta h_{AB})\,G_{Bethe,A}}{1 + \exp(\beta h_{AB})\,G_{Bethe,A} + \exp(\beta h_{AB})\,G_{Bethe,B}} \tag{7.136}$$

where $G_{Bethe,A}$ and $G_{Bethe,B}$ are solutions of

$$\theta_A = G_{Bethe,A}\,\frac{1 + \exp(\beta h_{AA})\,G_{Bethe,A} + \exp(\beta h_{AB})\,G_{Bethe,B}}{1 + 2\,(G_{Bethe,A} + G_{Bethe,B}) + 2\exp(\beta h_{AB})\,G_{Bethe,A}\,G_{Bethe,B} + \exp(\beta h_{AA})\,G^2_{Bethe,A} + \exp(\beta h_{BB})\,G^2_{Bethe,B}} \tag{7.137}$$

$$\theta_B = G_{Bethe,B}\,\frac{1 + \exp(\beta h_{AB})\,G_{Bethe,A} + \exp(\beta h_{BB})\,G_{Bethe,B}}{1 + 2\,(G_{Bethe,A} + G_{Bethe,B}) + 2\,\exp(\beta h_{AB})\,G_{Bethe,A}\,G_{Bethe,B} + \exp(\beta h_{AA})\,G^2_{Bethe,A} + \exp(\beta h_{BB})\,G^2_{Bethe,B}} \tag{7.138}$$

Notice that Equations 7.135 and 7.136 are similar to Equation 7.123. If h_{AB} is large so there are strong AB attractions, A and B will tend to cluster together. δ_{AB} and δ_{BA} tend to approach unity in such a case. There are additional effects when h_{AA} and h_{BB} are nonzero, however. If there are strong AA attractions, A molecules will tend to cluster together on the surface, which reduces the probability of producing an AB pair. Similarly, if there are strong BB attractions, the B molecules will tend to cluster together on the surface, which again reduces the probability of forming an AB pair. The situation is very complex at moderate coverages. If, for example, AA interactions are strongly repulsive, the A's will move apart, so the B's are forced to adsorb in the spaces between the A's. If the coverage is right, the B molecules will cluster around A, even when there are no direct interactions between A and B. As a result, δ_{AB} and δ_{BA} show complex behavior at moderate coverages.

Figure 7.50 illustrates this effect more carefully. The figure is calculated for a case where these are strong AA repulsions and moderate AB attractions, or repulsions. Notice that for all of the cases shown, δ_{BA} saturates at intermediate values of θ_A, while δ_{AB} reaches a maximum and then declines. Physically, the parameters in Figure 7.50 were chosen so that the A layer assumes a c(2x2) pattern at $\theta_A = 0.5$. The strong A–A repulsions force the B molecules to fit into the spaces in the c(2x2) pattern, so that at moderate coverages, all of the B molecules are completely surrounded by A's. Now, when one adds additional A molecules, the additional A molecules cannot be next to another B. Therefore, δ_{BA}, the probability that a given B adsorbed molecule is surrounded by A's is independent of θ_A, while δ_{AB}, the probability that a randomly chosen A molecule is adsorbed on a site that is next to a given site containing a B molecule goes down.

Now consider what happens to the rate of reaction. Figure 7.51 shows a series of plots of the rate of the reaction

$$A_{ad} + B_{ad} \xrightarrow{k_5} \text{Products} \tag{7.139}$$

calculated via Equation 7.131 for the case in Figure 7.50. Notice that the reaction rate is almost independent of θ_A over a wide range of θ_A, even though A is one of the reactants in Reaction 7.139. This example illustrates the key result in this section:

> When there are pairwise interactions between the adsorbed molecules, the kinetics of an *elementary* surface reaction may not bear a simple relationship to the stoichiometry of the reaction.

Figure 7.51 considered a case where A and B are involved in an elementary reaction. Yet at $\theta_A > 0.5$ the reaction rate is almost independent of θ_A.

It is useful to relate this result to the results in Table 7.2. In Table 7.2 we derived an expression for the rate of reaction following a standard Langmuir–Hinshelwood mechanism:

$$S + A \underset{2}{\overset{1}{\rightleftharpoons}} A_{ad} \tag{a}$$

$$S + B \underset{4}{\overset{3}{\rightleftharpoons}} B_{ad} \tag{b}$$

$$A_{ad} + B_{ad} \xrightarrow{5} \text{Products} \tag{c} \tag{7.140}$$

with no interactions between the adsorbed molecules. According to Table 7.2, when Reaction 7.140c is rate determining, the rate of the reaction should follow Equation 7.2.5. Now consider putting in strong pairwise repulsions between the A molecules. The results in Figure 7.51 show that in such a case, the rate of Reaction 7.140c will be essentially independent of θ_A, when $\theta_A > 0.5$. Equation 7.2.5 from Table 7.2 was derived assuming that the rate was proportional to θ_A. Therefore, there will be deviations from Equation 7.2.5 when there are significant interactions between the adsorbed molecules. Interestingly, however, after pages

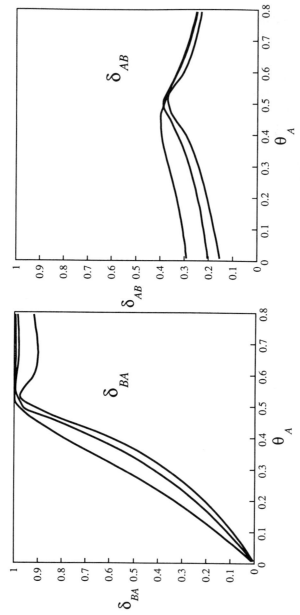

Figure 7.50 A plot of δ_{AB} and δ_{BA} calculated via MC calculations with $\theta_B = 0.2$, $\beta h_{AA} = -3$, $\beta h_{BB} = -0$, and $\beta h_{AB} = -0.5, 0, 0.5$.

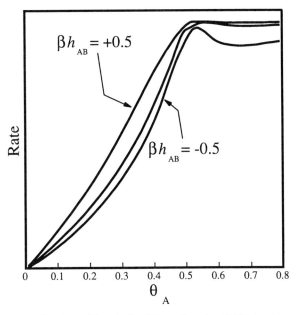

Figure 7.51 The rate as a function of θ_A calculated from Equation 7.131 for the case in Figure 7.50.

of algebra one can show that for $\theta_A > 0.5$ the rate is well approximated by a Langmuir rate expression

$$r_{Net} \cong k_5\theta_B \cong \frac{k_5\left(\dfrac{k_3}{k_4}\right)P_B}{2 + \dfrac{k_1}{k_2}P_A + \dfrac{k_3}{k_4}P_B} \tag{7.141}$$

where k_1, k_3, and k_4 are the rate constants for adsorption and desorption of A and B on a c(2x2) overlayer of A. Thus, the reaction still follows a Langmuir–Hinshelwood rate equation for $\theta_A > 0.5$. However, it is not the Langmuir–Hinshelwood rate equation for the mechanism in 7.140 with Reaction c rate determining.

Notice, however, that when k_6 and k_8 in Equation 7.2.3 are zero, Equation 7.2.3 in Table 7.2 reduces to Equation 7.141. Equation 7.2.2 is similar. Therefore, Equation 7.141 is the Langmuir rate equation for the mechanism in Equation 7.140 with Reaction 7.140b rate determining, even though Equation 7.139 was derived for the case when Reaction 7.140c was rate determining.

Physically, what is happening is that a c(2x2) A overlayer is forming on the surface. B molecules then adsorb on the open sites on the c(2x2) overlayer (see Figure 7.52). Each of the B molecules sees the same environment. As a result, the adsorption of B will follow a Langmuir adsorption isotherm even though h_{AB} are nonzero. Further, each adsorbed B molecule is completely surrounded by A molecules. Notice that as long as $\theta_A > 0.5$ so all the c(2x2) sites are filled with A, the number of AB pairs depends on θ_B but not θ_A. As a result, the rate of reaction is similar to the rate of Reaction 7.2.2 or 7.2.3, not 7.2.4.

This example illustrates an important point:

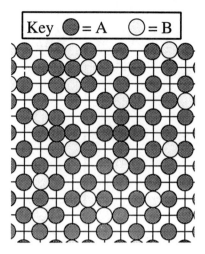

Figure 7.52 The arrangement of B atoms in a c(2x2) A overlayer.

> In the presence of pairwise forces, the rate equation for a simple surface reaction may be different from the rate equation one derives from the analysis in Section 7.4. Data can follow a simple Langmuir–Hinshelwood rate equation. Yet, the mechanism of the reaction or the rate-determining step can be quite different from the mechanism or rate-determining step one infers from Table 7.2. Therefore, one needs to be exceedingly cautious in using kinetic data to infer mechanisms of surface reactions.

The other key result from this section is that at moderate coverages $0.1 < \theta_A + \theta_B < 0.5$ Langmuir–Hinshelwood kinetics fail when there are strong interactions between the adsorbed molecules. Figures 7.50 and 7.51 show that at moderate coverage δ_{AB} and δ_{BA} and r_{AB} vary in a complex way with coverage. The results are not easily fit by a Langmuir–Hinshelwood rate form, even though the system follows a Langmuir–Hinshelwood rate form at higher coverages. Interestingly, one can fit the low coverage portion of the data to a power law (i.e., Tempkin kinetics), but that rate equation does not work at higher coverages.

This example illustrates the second key point from this section:

> In a typical surface reaction, the form of the rate equation often varies with coverage and temperature. Therefore, one needs to be exceedingly cautious about using a rate equation to extrapolate rates measured under one set of conditions to a different set of conditions.

7.21.2 The Immobile Limit

The discussion in the preceding section was for the mobile limit. However, strong deviations from Langmuir–Hinshelwood behavior also arise in the immobile limit. In the immobile limit, adsorbate sticks where it adsorbs, and does not diffuse around. Simulations have shown that islands form in such a case, even in the absence of strong adsorbate/adsorbate interactions (i.e., under conditions where no islands would be present at equilibrium). These islands affect the kinetics of reactions. The process is very analogous to the effects of islands described in Section 7.21.1.

For example, Figure 7.53 shows a snapshot of the configuration of CO and O_2 during the reaction $CO + \frac{1}{2}O_2 \Rightarrow CO_2$. The calculation was done assuming an immobile layer. Notice that some regions of the surface are covered by CO, while other regions are covered by O. Let's consider a reaction on the CO-covered region. If an oxygen adsorbs, the oxygen will be quickly reacted away. However, if a CO adsorbs, the CO will stick and not react. If the conditions are right, the CO- and O-covered regions of the surface remain CO or O covered for a long time. As a result, if a CO- or O-covered region forms by chance, that region will be stable, while any regions covered by a near equal mixture at CO and O_{ad} will quickly react away. The result is that in the immobile limits (i.e., no surface diffusion) islands will form on the surface. The islands are responsible for the multiple steady states similar to those in Figure 7.6.

Generally, these effects occur only at a very limited set of conditions, so they are not very

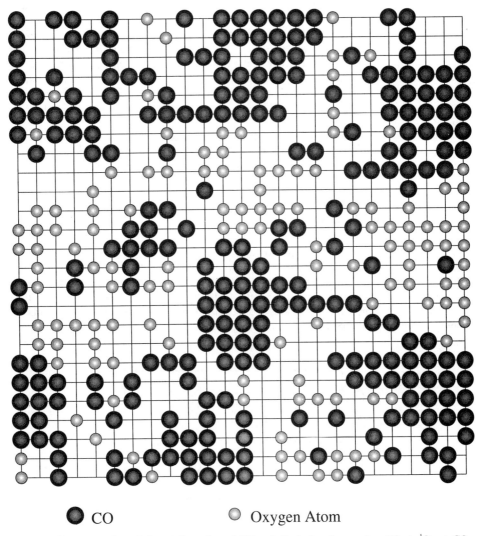

● CO   ○ Oxygen Atom

Figure 7.53 A snapshot of the configuration of CO and O_2 during the reaction $CO + \frac{1}{2}O_2 \rightarrow CO_2$. The calculation was done assuming an immobile layer.

important to steady-state catalytic reactions. The one exception is CO oxidation. It happens that during the startup of a catalytic converter in a car, one does get to a set of conditions where the transients are important. Thus, the CO oxidation case has been examined in great detail (see, for example, Wolf and Boudeville [1993] or Ertl [1993]).

At present people are still learning how to model reactions in the immobile limit. One complication is that experimentally one gets large-scale inhomogeneities on the surface during reaction. A reaction front moves back and forth across the surface. This behavior is difficult to model (see Luss [1994] or Ertl [1993] for details).

7.22 MULTIPLE STEADY STATES AND OSCILLATIONS

There is a related issue, which is that when there are different populations on the surface, a given surface reaction may not reach steady state. A classic case is carbon monoxide (CO) oxidation on Pt(100), where there are two surface phases, an inactive phase (i.e., one where the reaction rate is small), and an active phase (i.e., one where the reaction rate is large). At low coverage the CO adsorbs into the inactive phase. However, if the conditions are right, as the CO builds up on the surface, the surface undergoes a phase transition into the active phase. Then the reaction starts. The start of the reaction pushes the surface back toward the inactive phase. If the conditions are just right, the surface will oscillate between the active and inactive phases. As a result, the system never reaches steady state. Instead, the reaction rate oscillates as shown in Figure 7.54. One can also get multiple steady states like those in Figure 7.6 by such a mechanism.

In the recent literature there have been hundreds of papers written on such oscillations. However, they have little practical importance. Therefore, they will not be discussed in detail here. One should refer to the book by Gray and Scott [1990] for a detailed discussion of oscillations during surface reactions.

7.23 PRECURSOR REACTIONS

Another complication is that reactions often go via precursor mechanisms. One can use Equation 7.120 to examine the effects of precursors on steady-state reaction. Generally, one finds that if the adsorption or desorption steps in a catalytic reaction are rate determining, there are significant deviations from the results in Table 7.2 when adsorption or desorption occurs via a precursor. In those cases, one can use Equations 5.60 or 7.120 to directly calculate a rate. If on the other hand, the reactions between adsorbed species are rate deter-

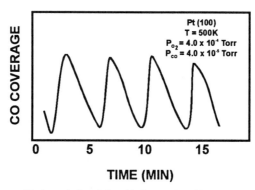

Figure 7.54 Rate oscillations during CO oxidation on Pt(100). (Data of Cox et al. [1983])

mining, the presence of the precursor does not significantly alter the rate (i.e., the rate looks just like a Rideal–Eley reaction), so one can use the analysis in the last few sections to calculate a rate. In my experience, many reactions go by precursor mechanisms. However, usually, the presence of the precursor only makes a small difference to the steady-state kinetics of the reaction. (It makes a much larger difference to the transient kinetics.) As a result, the presence of precursors, while important, do not affect steady-state kinetics very much.

7.24 DYNAMIC CORRECTIONS

In the last several sections, we considered the completely mobile and immobile limits. In the completely mobile limit, as gas molecules desorb or reaction occurs, the species remaining on the surface rearrange so the adsorbed layer is always in its equilibrium configuration. In the immobile limit no diffusion occurs, so the surface does not rearrange.

Now there is an intermediate region where diffusion occurs, but equilibrium is not maintained. This intermediate regime can be quite important.

For example, consider a TPD experiment. In a typical TPD experiment one adsorbs gas at low temperature, pumps away, and then heats the surface linearly and measures the rate of desorption of products with a mass spectrometer. At low temperatures, diffusion rates are small. Therefore, one does have to wonder whether one is reaching equilibrium in the initial stages of a TPD experiment.

However, later in the experiment, temperatures are higher and diffusion occurs rapidly. There is an intermediate temperature regime where the dynamics of the layer affect the rate.

One way to tell if this effect is important is to change the way one does the TPD experiment. In a typical TPD experiment one adsorbs gas at low temperatures and heats. However, one could, for example, dose at elevated temperatures and then cool to low temperatures and then heat, or do an annealing step prior to heating the surface linearly. If the equilibrium assumption were correct, then it should reach the same surface configuration no matter how the adsorbed layer is prepared. The equilibrium assumption implies that as long as the coverage and temperature are constant at the start of the temperature ramp, the initial state of the adsorbate will be the same, independent of how the layer was prepared. As a result, if the equilibrium assumption were valid, the TPD spectrum should be independent of how the initial layer is prepared.

Experimentally, however, one often finds that the way one prepared the initial layer has a substantial effect on the TPD spectrum. For example, Figure 7.55 compares a TPD spectrum

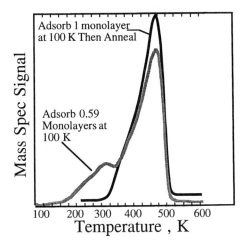

Figure 7.55 TPD spectra for CO from Ni(111) measured (a) by adsorbing 0.59 of a monolayer of CO at 100 K and (b) by adsorbing 1 monolayer of CO at 100 K, then flashing to reduce the coverage to 0.59. (Data of Vasquez, Muscat, and Madix [1994].)

taken by adsorbing half a monolayer of CO onto a 100 K Ni(111) sample and then heating at 14 K/sec to a spectrum taken by adsorbing a monolayer at 100 K, annealing to reduce the coverage to half a monolayer, cooling back to 100 K, and then heating at 14 K/sec. Note that if the equilibrium assumption were valid, both spectra should be identical, but they clearly are not. This example shows an unusually large difference between the two spectra. However, in the majority of cases where such an experiment has been reported, the TPD spectra showed some differences according to whether one simply doses at low temperature and then flashes (i.e., heats the sample rapidly), or when one doses at low temperatures and anneals to moderate temperatures and then flashes.

Experimentally, one often traps some gas in a weakly bound state when one doses at 100 K. Those molecules desorb very easily. As a result, when one saturates the surface with gas at 100 K, and then anneals to reduce the coverage, one often gets to a different situation than when one simply adsorbs an equivalent amount of gas at 100 K. Consequently, there is some question whether the equilibrium assumption is valid in a typical TPD experiment even though it is used routinely for analysis of TPD spectra.

At the time this book was being written, no one had reported a simulation that considered the effects of molecules trapped in weakly bound states in the analysis of TPD data. In a sense, the issue is as much an experimental problem as a theoretical one. One would like to take TPD spectra under conditions where the data are easily interpreted. Hence, if one could avoid the nonequilibrium processes, one should do so. The problem is that it is often unclear whether nonequilibrium processes are contributing to TPD data. As a result, the analysis of TPD data has some uncertainty.

The other case where the dynamics of the surface layer play a key role is in film growth, particularly molecular beam epitaxy (MBE). In MBE, one exposes a hot substrate to a continuous flux of atoms or molecules, e.g., Ga or As_2. The atoms or molecules react on the substrate to deposit a film. One can run an MBE system with a cold substrate. In that case, every atom that impinges on the substrate sticks and reacts. However, the result is a random distribution of atoms, and hence a rough film. One can get a much smoother film if one heats the substrate close to the desorption temperature of the impinging atoms. In that case atoms are trapped, as indicated in Figure 7.56. Some of those atoms desorb. However, if the atom

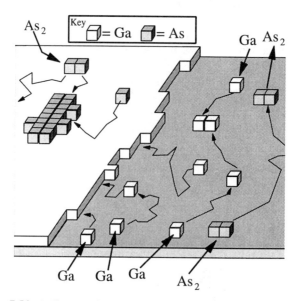

Figure 7.56 A diagram of the chemical processes occurring during MBE.

can diffuse to a step, it usually does not desorb (the binding energy of atoms is usually higher on the steps). Instead, the atom is incorporated into the growing film. Film growth occurs mainly at the step boundaries. This produces a much smoother, more uniform film.

One needs a complex calculation to adequately model the growth process. Generally, people use Monte Carlo calculations similar to those in Chapter 3, except that the surface is constantly changing. Further, one needs to consider adsorption, diffusion, and reaction explicitly so the effects are complicated. Madhokar and Ghaisas [1988] have a good description of the key findings. Generally, one finds that the nucleation process (i.e., nucleating a cluster of atoms on the surface, which can then grow) controls the growth rate, while growth of the cluster controls the film quality. There is a balance between growth rate and film quality. The best quality films are grown at conditions where most clusters evaporate, so the nucleation rate is small. Yet if the nucleation rate is zero, no film grows. The details are rather complex. The reader is referred to Madhokar and Ghaisas [1988] for further information.

7.25 SUMMARY

Let's now summarize where we are in our understanding of the kinetics of surface reactions. First, it is important to note that except in a few special cases (e.g., film growth) the kinetics of steady-state surface reactions are well understood. At coverages greater than 0.5, most steady-state surface reactions follow Langmuir–Hinshelwood rate laws. The results in Sections 7.21 and 7.21.1 indicate that often the reaction can follow a Langmuir–Hinshelwood rate equation, which is different from one derived using the simple analysis in Section 7.4 (e.g., Table 7.2). However, steady-state surface reactions usually follow Langmuir–Hinshelwood rate laws when the reactant coverage is greater than 0.5.

Deviations usually occur at lower coverages. The results in Figure 7.25 indicate that the ϑ's deviate significantly from the mean field results at low coverage. When there are strong attractive interactions between the adsorbed molecules, one can form clusters of atoms on the surface, even at low coverage. The clustering phenomenon allows one to nucleate a new layer in MBE, so it is quite important.

Repulsive interactions between the adsorbed molecules have a different effect. At very low coverages few molecules are close together, so the effects of the interactions are small. However, at moderate coverages (i.e., $0.1 < \theta < 0.5$), one finds that the kinetics are changing with coverage. These changes produce deviations from simple Langmuir–Hinshelwood behavior. A Polanyi-type rate equation can sometimes be used under such circumstances. However, often a Monte Carlo calculation is needed to represent the phase behavior.

Fortunately, one usually runs steady-state reactions in one of two regimes, a surface reaction limited regime, where the coverages are above 0.8, or a diffusion-controlled regime, where the coverages are below 0.1. In both cases, the rate equation reduces to a Langmuir–Hinshelwood rate form, although not necessarily the rate form in Table 7.2.

Now consider a transient experiment like TPD. In TPD one starts the experiment at high coverages where Langmuir kinetics work. However, one lowers the coverage during the experiment. Eventually, one gets to a situation where the deviations from Langmuir–Hinshelwood kinetics matter. The peaks in the TPD curves come just in the critical region. As a result, TPD peak positions are very sensitive to non-Langmuir–Hinshelwood behavior. They are much more sensitive than measurements of rates of steady-state reactions under real process conditions. As a result, TPD tends to over emphasize non-Langmuir–Hinshelwood behavior.

Now, there are a few industrial processes (e.g., automotive catalytic convertors) where the transient behavior matters. In those cases, one expects complex behavior. However, those systems are exceptions. Steady-state surface reactions normally follow Langmuir-type kinetics, except under unusual conditions. However, it is important to note again that it is often not the Langmuir–Hinshelwood rate equation expected from Table 7.2 (i.e., the Hougan and Watson analysis).

7.26 SOLVED EXAMPLES

Example 7.A Simulation of TPD Spectrum

TPD_SYM, a program to simulate TPD spectra, is available from Professor Masel. Because of length limitations we could not include a program listing (one is available from Professor Masel). Still, it is useful to include a discussion of how the program works. The program has three main parts, a top-level manager that controls the input/output (I/O) subroutine TPD_SIMULATE, which actually calculates the TPD spectrum, and a function RATE, which calculates the rate of desorption as a function of coverage, θ, temperature, T, and the strength of the interaction between adjacent adsorbed molecules, h.

The top-level manager gets the run data and starts the simulation. It is a simple loop that asks the reader to input data and then calls a subroutine TPD_SIMULATE to calculate the TPD spectrum.

The subroutine TPD_SIMULATE uses the function RATE to calculate the desorption rate as a function of temperature, coverage, and h. It then does a fourth-order Runge–Kutta integration of

$$\frac{r_d}{\beta_H N_s} = -\frac{d\theta}{dT} \qquad (7.142)$$

to calculate how the coverage varies with temperature during the TPD run. Once one knows how the coverage varies with temperature, one can plug into the rate equation to calculate the TPD spectrum (i.e., rate) as a function of temperature.

The hardest part of the program is the RATE function. The RATE function calculates the rate from Equation 7.101. RATE first calculates G_{Bethe} from Equation 7.96. RATE then plugs G_{Bethe} into Equations 7.97 and 7.98 to calculate the ϑ_j's. The program then uses a lookup table to calculate the $\mathbb{R}_j$ and plugs into Equation 7.101 to calculate the rate. Again copies of the program are available from Professor Masel.

Example 7.B Analysis of a TPD Spectrum

Earlier in this chapter we discussed several methods used to analyze TPD spectra. The objective of this example is to show how the methods work. Our approach is to start with the simulated TPD data in the right of Figure 7.32 and use the methods from this chapter to calculate the activation energy and preexponential for the reaction.

THE HABERSHADEN–KÜPERS METHOD

In the Habershaden–Küpers method one replots the TPD data on an Arrhenius plot, i.e., a plot of the log of the desorption rate (= TPD spectrum) versus $1/T$, and uses the initial slope of the line to infer the activation energy of desorption. Figure 7.57 shows a plot of the data in this way. Notice that the initial part of the curve is linear, which implies that the activation energy is constant. The simulations show that the coverage is nearly constant in the initial part of the curve as well. Therefore, one can calculate the activation energy of desorption as

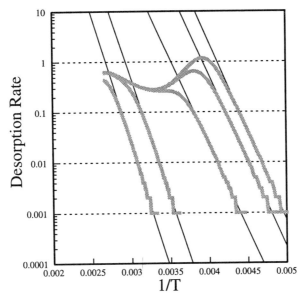

Figure 7.57 A plot of the data from the right side of Figure 7.32. (As suggested by Habershaden and Küpers [1984].)

a function of coverages from the slopes of the lines in Figure 7.57, i.e.,

$$E_a = k \; (slope) \tag{7.143}$$

where k is Boltzmann's constant. We have used a least squares procedure to calculate the slopes of the lines in Figure 7.57. The results were as given in Table 7.4. By Comparison, since the TPD "data" are simulations, one can determine the E_a exactly using Equation 7.102. The result is as given in Table 7.5.

Note that even though we were using simulated TPD data (with no noise or offset) we did not get the exact result with the Habershaden–Küpers method. Rather the calculated activation energies were as much as 1.6 kcal/mole off, and the preexponentials were off as much as a factor of 8. DeJong and Niemansuerdeit [1990] compared the results of the Habershaden–Küpers analysis to simulations in more detail and showed that the Habershaden–Küpers method

TABLE 7.4 The Activation Energy of Desorption as a Coverage Calculated with a Habershaden–Küpers Analysis of the TPD Curves in Figure 7.32

θ	0.2	0.4	0.6	0.8	1.0
E_a, kcal/mole	23.3	21.7	15.7	16.0	16.4
k_o, sec^{-1}	0.8×10^{13}	2.8×10^{13}	0.6×10^{13}	2.4×10^{13}	4.2×10^{13}

TABLE 7.5 The Exact Values of E_a and k_o for the Data in Figure 7.32

θ	0.2	0.4	0.6	0.8	1.0
E_a, kcal/mole	22.0	19.1	16.3	16.1	16
k_o, sec^{-1}	1×10^{13}	1×10^{13}	1×10^{13}	1×10^{13}	1×10^{13}

is very sensitive to small errors in the baseline. Therefore, one needs to be rather cautious in trusting the results of the Habershaden–Küpers analysis.

KING'S ORIGINAL METHOD

One could also try to analyze the data using King's original method, i.e., integrate the TPD curve to calculate the coverage as a function of temperature for each of the TPD curves. One picks out coverage points from each of the scans so that one can calculate the rate as a function of temperature for various fixed values of the coverage. For example, if one integrates the curves in Figure 7.32 one obtains the data in Table 7.6.

I was not able to make a useful Arrhenius plot of these data since the temperature range was so small.

TAYLOR–WEINBERG MODIFICATION OF KING'S METHOD

One can get a better Arrhenius plot by varying both the coverage and the heating rate as outlined by Taylor and Weinberg (1978). The integration of the rate data is given in Table 7.7.

Again, one can construct Arrhenius plots of these data to calculate the desorption rate as a function of coverage. The Arrhenius plot is shown in Figure 7.58. Again one gets an excellent fit to the Arrhenius plot. A least squares fit to the lines is given in Table 7.8.

Again the results are only modestly accurate, even though there was no noise or offset in the data. DeJong and Niemansuerdiet [1990] compared the results of the Taylor–Weinberg analysis to simulations in more detail and showed that the Taylor–Weinberg procedure is slightly less subject to error than the Habershaden–Küpers method. However, one has to measure data over a wide range of heating rate, which is difficult experimentally. In my

TABLE 7.6 The Results of King's Analysis of the Data in Figure 7.32

Integrated Coverage	Initial Coverage = 0.2		Initial Coverage = 0.4		Initial Coverage = 1.0	
	Temperature (K)	Rate	Temperature (K)	Rate	Temperature (K)	Rate
0.05	387	0.42	389	0.47	389	0.47
0.1	377	0.48	380.5	0.59	380.5	0.60
0.2	—	—	365	0.60	366	0.61

TABLE 7.7 The Results of the Taylor–Weinberg Analysis of the Data in Figure 7.32 Plus Additional Data Taken at a Variety of Heating Rates

Integrated Coverage	Heating Rate = 1/sec		Heating Rate = 10/sec		Heating Rate = 100/sec	
	Temperature (K)	Rate	Temperature (K)	Rate	Temperature (K)	Rate
0.2	343	0.068	363	0.61	392	5.5
0.4	308	0.037	328	0.35	351	3.4
0.9	231	0.099	247	0.91	264	7.8

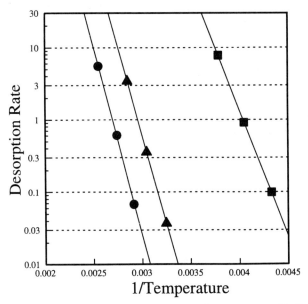

Figure 7.58   A plot of the data from the right side of Figure 7.32. (As suggested by Taylor and Weinberg [1978].)

TABLE 7.8   The Activation Energy of Desorption as a Coverage Calculated with a Taylor–Weinberg Analysis of the TPD Curves in Figure 7.32

θ	0.2	0.4	0.9
E_a, kcal/mole	24.5	22.7	16.2
k_o, sec^{-1}	2.8×10^{13}	16.8×10^{13}	14×10^{13}

experience, it is unclear that the extra work is justified, given the limited improvement in the results.

PROBLEMS

7.1   Define the following in three sentences or less.

(1) Langmuir rate equation
(2) Langmuir–Hougen and Watson rate law
(3) Rate-determining step
(4) Tempkin kinetics
(5) Elovich equation

(6) Surface explosion
(7) Precursor-moderated desorption
(8) Turnover number
(9) Turnover frequency
(10) Mars–Van Klevan kinetics

7.2   Explain in your own words:

(a) How and why are the kinetics of gas phase reactions different from the kinetics of surface and reactions?

(b) Why does the rate of reaction decline at high pressures in Figures 7.4 and 7.7?

(c) What are the key assumptions in deriving the Langmuir–Hougen and Watson rate equations?

(d) How would the Langmuir–Hougen and Watson rate equations change if there were multiple sites to hold gas.

(e) How do interactions between adsorbed molecules affect the kinetics of surface reactions?

(f) What are the pitfalls in using kinetics measurements to infer mechanisms of surface reactions?

(g) Why do TPD spectra show peaks as opposed to rises in rate with increasing temperature?

(h) Why do multiple peaks arise in TPD?

(i) Why is δ_{BA} constant at high coverage in Figure 7.50?

(j) How does adsorbate mobility affect TPD curves?

(k) Why does the order of steady-state reactions change when the adsorbate forms ordered phases?

(l) How do islands affect TPD curves?

7.3 (a) Compare the kinetics of Rideal–Eley and precursor kinetics. Under what conditions can one see a difference between the two?

(b) Compare the kinetics of Rideal–Eley and Langmuir–Hinshelwood. Under what conditions can one see a difference between the two?

7.4 Derive the various rate equations in Table 7.2. Assume that you had rate data for a reaction A + B → products. How would you find out which of the rate equations in Table 7.2 fit the data?

7.5 Figures 7.2 and 7.3 show data for the decomposition of ammonia over a platinum catalyst. Assume that the reaction obeys the following mechanism:

$$NH_3 + S \rightleftharpoons NH_{3\,(ad)}$$

$$NH_{3\,(ad)} \longrightarrow (1/2)\,N_2 + (3/2)\,H_2$$

$$H_2 + 2\,S \rightleftharpoons 2\,H_{(ad)}$$

(a) Derive a Langmuir–Hougen and Watson rate equation for the reaction.

(b) Notice that hydrogen does not directly participate in the reaction. Why does the hydrogen pressure in the reactor affect the rate of reaction?

(c) Fit your rate equation to the data in Figure 7.2. How well does the rate equation fit?

(d) Now try to fit your rate equation to the $P_{NH_3} = 6.5$ torr data in Figure 7.3. How well does your rate equation fit?

7.6 Figures 7.4 and 7.5 show data for the oxidation of CO over an Rh(111) catalyst. Assume that the reaction obeys the following mechanism:

$$CO + S \rightleftharpoons L\ CO_{(ad)}$$

$$O_2 + 2S \rightleftharpoons 2O_{(ad)}$$

$$CO_{(ad)} + O_{(ad)} \longrightarrow CO_2$$

(a) Derive a Langmuir–Hougen and Watson rate equation for the reaction.

(b) Fit your rate equation to the data in Figure 7.4. How well does the rate equation fit?

(c) Why does the rate decline at high CO pressures?

(d) Now try to fit your rate equation to the $P_{CO} = 6.5$ torr data in Figure 7.5. How well does your rate equation fit?

(e) Can you fit the data in figure 7.6 with your rate equation.

7.7 Creighton, *Thin Solid Films* **241**, 310 [1994] tried to model the chemical vapor deposition of tungsten. The main reaction is $WF_6 + 3H_2 \rightarrow W + 6HF$. Creighton modeled the reaction by assuming that the reaction obeyed the following mechanism:

$$WF_6 + 2S \longrightarrow WF_{5\,(ad)} + F_{(ad)}$$

$$H_2 + 2S \rightleftharpoons 2H_{(ad)}$$

$$H_{(ad)} + F_{(ad)} \longrightarrow HF$$

$$WF_5 + 5/2\,H_2 \longrightarrow W + 5HF \text{ (fast)}$$

(a) Use the steady-state approximation on $H_{(ad)}$ and $F_{(ad)}$ to derive a rate equation for the mechanism. Assume that the last reaction is fast, but do not assume that any of the reactions is rate determining.

(b) Show that the rate is zero below a critical value of the hydrogen pressure.

(c) Explain your result in part (b) physically.

(d) Sketch how you would expect the rate to vary with hydrogen pressure?

(e) Experimentally, the rate declines at high hydrogen pressures. What is the physical reason for the decline?

7.8 Malik, Gulari, Li, and Bhattacharya, *J. Appl. Phys.* **73**, 5193 [1993] proposed that the growth of germanium/silicon alloys obeys the following mechanism:

$$Si_2H_6 + 2S \rightleftharpoons 2SiH_{3\,(ad)}$$

$$GeH_4 + S \rightleftharpoons GeH_{4\,(ad)}$$

$$2SiH_{3\,(ad)} \longrightarrow 2Si + 3H_2 + 2S$$

$$GeH_{4\,(ad)} \longrightarrow Ge + 2H_2 + S$$

$$H_2 + 2S \rightleftharpoons 2H_{(ad)}$$

(a) Derive a Langmuir–Hougen and Watson rate equation for the reaction

(b) Malik et al.'s experimental results show that at low germane (GeH_4) flowrates the rate of film growth increases with increasing germane flow. However, at high germane flowrates the growth rate decreases again. Use your results in part (a) to show that one would expect the rate to decline at high germane concentrations.

(c) What is the physical reason for the decline?

7.9 H. Kuhne, *Semiconductor Sci. Tech.* **8**, 2018 [1993] examined the effect of the background hydrogen pressure on the deposition of silicon from silane (SiH_4). Assume that the reaction follows the following mechanism:

$$SiH_4 + S \rightleftharpoons SiH_{4\,(ad)}$$

$$SiH_{4\,(ad)} \longrightarrow Si + 2H_2$$

$$H_2 + 2S \rightleftharpoons 2H_{(ad)}$$

(a) Derive a Langmuir–Hougen and Watson rate equation for the reaction.

(b) Notice that hydrogen does not directly participate in the reaction. Why, physically, does the partial pressure of hydrogen affect the rate of the reaction?

(c) Kuhne suggested that in truth, hydrogen does follows a Freudlich adsorption isotherm rather than a Langmuir adsorption isotherm. Derive a modified rate equation for the reaction assuming that the fraction of the surface sites covered by hydrogen, θ_H obeys

$$\theta_H = K_1 (P_{H_2})^{\alpha/2} \theta_S$$

where α is a constant and θ_S is the coverage of bare sites.

(d) What experiments would you distinguish between the rate equations in parts (a) and (c)?

7.10 Haupfear and Schmidt, *J. Electrochem. Soc.* **140**, 1793 [1993] examined the kinetics of the reaction between $TiCl_4$ and propane to yield titanium carbide, i.e., $TiCl_4 + C_3H_8 \rightarrow TiC + 2\,CH_3Cl + 2\,HCl$. They found that the following rate equation fit their data:

$$r = \frac{k P_{C_3H_8} P_{TiCl_4}}{(1 + K_1 P_{C_3H_8} + K_2 P_{TiCl_4})^2}$$

(a) Find a mechanism to explain this rate equation.

(b) How much confidence would you place in your mechanism?

(c) What additional experiments would you do to see if the mechanism is correct?

7.11 Tables 7.1 and 7.3 list experimental rate equations for a number of reactions.

(a) Provide a mechanism that is consistent with each of those rate equations.

(b) What additional experiments would you do to see if the mechanism is correct?

7.12 Velasco et al., *Anales de Quimica* **88**, 466 [1992] examined the dehydrogenation of phenol to cyclohexanone. They found that they could fit their data equally well with either of three different Langmuir–Hinshelwood mechanisms or a Rideal–Eley mechanism. Each mechanism had five or so parameters. Discuss the pitfalls in fitting data to a rate equation with five or more parameters. If you were Velasco, how could you tell which of the mechanisms are correct? (*Hints:* (1) The answer is not to take more kinetic data. (2) The answer to this question does not require you to know the mechanisms proposed by Velasco et al.)

7.13 Davidster and Laszlo, *Tetrahedron Letters* **34**, 533 [1993] examined the alkylation of toluene in an alcohol solution with an acidic "clayzic" catalyst. They were hoping to get the reaction

$$R-Cl + C_7H_8 \longrightarrow RC_7H_7 + HCl$$

However, they found that instead a side reaction

$$R'-OH + C_7H_8 \longrightarrow R'C_7H_7 + H_2O$$

dominated. Interestingly, when they eliminated the alcohol from the system, the overall reaction rate increased, so evidently the first reaction can occur. It is just that the first reaction is poisoned in the presence of the alcohol.

(a) Derive a Langmuir–Hougen and Watson rate equation for the reaction, assuming that the rate-determining step in the mechanism is the reaction between the chloride or alcohol and the toluene. Assume that the alcohol and the chloride compete for the same sites, and that toluene adsorbs on an adjacent site.

(b) Under what conditions would the second reaction dominate? Assume that the rate constant for the first reaction is ten times that for the second reaction.

(c) What is the physical reason that the second reaction dominates?

(d) What kinds of experiments would you do to test the mechanism for the reaction?

7.14 In Problem 7.13 we noted that when two parallel reaction pathways are occurring, one surface reaction can dominate over another, even under conditions where the other reaction has a higher rate constant. That effect can lead to interesting reaction selectivities. Consider the hydrogenation of acetylene. When ethylene is made, it often contains an undesirable acetylene impurity. It is common practice to hydrogenate the acetylene to ethylene, i.e.,

$$C_2H_2 + H_2 \longrightarrow C_2H_4$$

However, one needs to avoid hydrogenating the ethylene

$$C_2H_4 + H_2 \longrightarrow C_2H_6$$

(a) Derive a Langmuir–Hougen and Watson rate equation for the hydrogenation of acetylene in the presence of ethylene. Assume that the reaction goes via the following mechanism. Assume that the third step is rate determining.

$$C_2H_2 + S \rightleftharpoons C_2H_{2\,(ad)}$$
$$H_2 + 2S \rightleftharpoons 2H_{(ad)}$$
$$C_2H_{2\,(ad)} + H_{(ad)} \longrightarrow C_2H_{3\,(ad)} + S$$
$$C_2H_{3\,(ad)} + H_{(ad)} \rightleftharpoons C_2H_{4\,(ad)} + S$$
$$C_2H_{4\,(ad)} \rightleftharpoons C_2H_4 + S$$

(b) Derive a Langmuir–Hougen and Watson rate equation for the hydrogenation of ethylene. Assume that the reaction goes via the following mechanism. Assume that the third step is rate determining.

$$C_2H_4 + S \rightleftharpoons C_2H_{4\,(ad)}$$
$$H_2 + 2S \rightleftharpoons 2H_{(ad)}$$
$$C_2H_{4\,(ad)} + H_{(ad)} \longrightarrow C_2H_{5\,(ad)} + S$$
$$C_2H_{3\,(ad)} + H_{(ad)} \rightleftharpoons C_2H_6 + S$$

(c) Under what conditions would the hydrogenation of acetylene dominate over the hydrogenation of ethylene? Assume that the rate constant for the third step in each mechanism is the same.

(d) Experimentally, the hydrogenation of the acetylene proceeds with a 99.9% selectivity over a palladium catalyst under conditions where the ethylene is in a 100 to 1 excess. What does that tell you about the interactions of acetylene and ethylene on palladium.

(e) Now think about the thermodynamics. If you fed acetylene and excess hydrogen into a reactor, would you expect to produce ethylene or ethane? Which product is thermodynamically favored? How is it possible to produce mainly ethylene?

(f) Generalize your results in part (e) to suggest how one can identify catalysts that are able to selectively produce a partially hydrogenated or partially oxidized species in a mixture of species. Be sure to consider that the rate constants for the reaction can change as you change the reactants.

7.15 The partial oxidation of ethylene-to-ethylene oxide is often run on a silver catalyst.

(a) Use your results from Problem 7.14 to suggest reasons why the reaction can be run selectively.

(b) What experiments would you do to tell why silver is selective, i.e., to tell which of the possibilities from part (a) is correct?

7.16 One of the purposes of a catalytic converter is to oxidize the CO in the engine exhaust to CO_2. The kinetics are actually complex. However, under limited sets of conditions one can fit them with a Langmuir–Hinshelwood rate equation.

$$r = \frac{k_1 \, K_{CO} \, K_{O_2} \, P_{CO} \, P_{O_2}}{(1 + K_{CO} \, P_{CO} + K_{O_2} \, P_{O_2})^2}$$

(a) What kind of a mechanism could lead to this rate equation?

(b) In reality oxygen dissociatively adsorbs. Provide a credible explanation for why the denominator varies linearly with the oxygen partial pressure rather than as the square root of the oxygen partial pressure.

(c) Experimentally, the reaction rate is very low below 100°C, but then raises rapidly above 100°C. TPD shows that CO also desorbs from the surface at 100°C. Provide a feasible mechanism for the rapid rise in the rate at 100°C.

7.17 Describe in your own words the kinds of information you get from a TPD experiment.

(a) How is TPD useful in identifying the mechanism of a surface reaction?

(b) How is TPD useful in determining the kinetics of a surface reaction?

7.18 (a) What is the key qualitative difference between the TPD spectrum for a first-order desorption process and a second-order desorption process? (*Hint:* See Figure 7.12.)

(b) How would your conclusions change if there were repulsive forces between the adsorbed molecules? (*Hint:* See Figure 7.22.)

7.19 Figure 7.16 shows a series of TPD spectra for CO desorption from Ni(110).

(a) Use Equation 7.60 to estimate E_A at low and high coverage from the $\beta = 18.5$ K/sec data in Figure 7.16.

(b) Use Equation 7.60 to estimate E_A at low and high coverage from the $\beta = 1.7$ K/sec data in Figure 7.16.

(c) Use Figure 7.15 to estimate E_A at low and high coverage from the $\beta = 18.5$ K/sec data in Figure 7.16.

(d) Use Figure 7.15 to estimate E_A at low and high coverage from the $\beta = 1.7$ K/sec data in Figure 7.16.

(e) Now try using the Taylor–Weinberg method to analyze the data. Assume that θ is the same at each of the lower temperature peaks in Figure 7.16. Use the figure to estimate the desorption rate at the low temperature peak in the $\beta = 18.5$ K/sec and $\beta = 1.7$ K/sec data. Plug into Arrhenius law to calculate E_A.

7.20 Figure 7.13 shows how the widths of a TPD peak varies with k_o.

(a) What key trends do you observe?

(b) What would be the implication of a very narrow TPD peak (i.e., one only 4 K wide)?

(c) Edwards, *Surface Sci* **54**, 1 [1976] has proposed using peak widths to infer desorption kinetics. How easy will peak widths be to measure?

(d) Apply Edwards' method to the spectrum for finite-sized islands and high initial coverage in Figure 7.35. How well does Edwards' method reproduce the correct behavior? (Note the peak width is 4 K.)

(e) Chan, Aris, and Weinberg, *Appl. Surface Sci.* **1**, 360 [1978] proposed a small variant of Edwards' procedure. Repeat part (d) for that method.

7.21 A data file TPD_HWRK.PRN is available from Professor Masel. The text file contains simulated TPD data taken at several heating rates and initial coverages. The format of the file is such that it can be imported into most spreadsheets.

(a) Use Equation 7.60 to estimate E_A as a function of coverage from the data in the file.

(b) Use Figure 7.15 to estimate E_A as a function of coverage from the data in the file.

(c) Make a plot of the data as suggested in Equation 7.61 to calculate E_A as a function of coverage.

(d) Make a plot of the data as suggested in Equation 7.62 to calculate E_A as a function of coverage.

(e) Use the Taylor–Weinberg method to calculate E_A as a function of coverage.

(f) Use the Habershaden–Küpers method to calculate E_A as a function of coverage.

(g) Use the data to estimate h, the pairwise interaction between two adjacent molecules.

(h) Verify your value of h by plugging back into the program TPD_SYM and seeing how well you reproduce the data.

7.22 Describe in your own words the advantages and disadvantages of the following methods as a way of analyzing TPD data.

(a) Plotting the data as in Figures 7.13 and 7.14 and seeing what fits.

(b) Equation 7.60.

(c) Figure 7.15.

(d) Plotting the data as suggested under Equation 7.62.

(e) Taylor and Weinberg's complete analysis.

(f) The Elovich equation.

(g) The Habershaden–Küpers method.

7.23 (a) What is the physical significance of the ϑ_j's in Figure 7.25?

(b) What is the physical significance of the ϑ_j/θ?

(c) Why does the $\beta h = 2$ curve look different from the other curves in Figure 7.25? Why are there the changes in slope? (*Hint:* Look at Figure 4.21.)

7.24 The program TPD_SYM is available from Professor Masel. Use the program to explore how the TPD spectrum varies with coverage. Use the set of parameters in Table 7.9 plus any others you want and report on your results.

TABLE 7.9 Parameters for Problem 7.24

Case	β, (K/sec)	E_A, (kcal/mole)	k_o (sec^{-1})	h (kcal/mole)	γ_P
1	1	24	10^{13}	-2	1
2	10	24	10^{13}	-2	1
3	100	24	10^{13}	-2	1
4	1	24	10^{12}	-2	1
5	100	24	10^{14}	-2	1
6	10	24	10^{13}	-1	1

TABLE 7.9   (*Continued*)

Case	β, (K/sec)	E_A, (kcal/mole)	k_o (sec^{-1})	h (kcal/mole)	γ_P
7	10	24	10^{13}	0	1
8	10	24	10^{13}	2	1
9	10	24	10^{13}	1	1
10	100	24	10^{13}	1	1
11	1	24	10^{13}	1	1
12	10	24	10^{13}	1	0.5
13	10	24	10^{13}	1	0
14	10	24	10^{13}	-4	0.5
15	10	24	10^{13}	-2	0.5
16	10	24	10^{13}	-2	0
17	10	24	10^{19}	-2	1
18	10	24	10^{16}	-2	1
19	10	26	10^{16}	-2	1
20	10	28	10^{13}	-2	1
21	10	20	10^{10}	-2	1
22	10	48	10^{13}	-2	1

7.25   In the discussion of TPD in this chapter, we considered cases where there were only first nearest neighbor interactions. However, there can also be second or higher nearest neighbors. Consider the adsorption of a gas on a square lattice with first and second nearest neighbor interactions with $h_{nn} = -1$, $h_{2nn} = -0.25$, $E_A^o = 26$ kcal/mole, $\gamma_P = 1$, $k_o = 10^{13}$/sec.

   (a) Use the analysis in Example 4.E in the solved examples for Chapter 4 to calculate the phase diagram for the system. What ordered phases would you expect to observe? How many phases are seen?

   (b) How many TPD peaks would you expect?

   (c) Estimate E_A for each of the peaks.

   (d) Use Equations 7.60 and 7.62 to estimate the peak temperature for each of the peaks.

7.26   Figure 7.9 shows a TPD spectrum for ethylene decomposition on Pt(511).

   (a) How many rate processes are present in the hydrogen spectrum?

   (b) Estimate the activation energy for each of the processes.

   (c) Repeat for the ethylene, ethane, and methane traces.

   (d) What does the observation of methane formation tell us about the surface reaction? What bonds break and form?

   (e) What does the observation of ethane formation tell us about the surface reaction? What bonds break and form?

7.27   Write a computer program to calculate a TPD spectrum for desorption of gas from an immobile layer assuming that the desorption rate is given by Equation 7.118. Compare your results to those calculated for the second-order desorption from a mobile layer, i.e., for when the desorption rate follows Equation 7.54.

   (a) Are the TPD spectra sufficiently different that one can use TPD to learn something about the mobility of molecules on surfaces?

   (b) How could you distinguish between the mobile and immobile cases experimentally?

7.28   What information would you need to use Figure 7.50 to predict rates of surface reactions?

More Advanced Problems

7.29 Write your own program to use the Bethe approximation to calculate a TPD spectrum.

(a) First, write a subroutine to calculate the ϑ_j's as a function of h, θ and T from Equation 7.98.

(b) Write a function for calculate r_d as a function of h, θ, and T, k_o and γ_P from Equation 7.100.

(c) Use a fourth-order Runge–Kutta algorithm to integrate the rate equation to calculate the desorption rate as a function of temperature. (*Hint:* (a) Be sure that θ decreases as gas desorbs. (b) The Runge–Kutta algorithm makes an error when θ is near to zero. I found it useful to set the rate equal to zero for $\theta < 0.0005$.)

(d) Plot the rate as a function of temperature to calculate the TPD spectrum. Your algorithm should have β, h, E_A^o, γ_P as input parameters and integrate from 100 to 800 K in 1K steps.

(e) As a test run, choose parameters identical to those in Figure 7.32 and report on your results.

7.30 This chapter reviewed many of the key methods used to analyze TPD data. However, there are a few other methods in the literature. Look up each of these methods and report on your findings.

(a) D. Edwards, *Surface Sci.* **138**, 279 [1976].

(b) C. M. Chan, R. Aris, and W. H. Weinberg, *Appl. Surface Sci.* **1**, 360 [1978].

7.31 DeJong and Niemansuerdiet, *Surface Sci.* **233**, 355 [1990] compare a series of methods for the analysis of complex TPD data. Read the paper and report on their findings.

7.32 Khankar and Argarwal, *J. Chem. Phys.* **99**, 9237 [1993] model the role of surface diffusion on the kinetics of a simple surface reaction. Read the paper and report on their findings.

REFERENCES

Behm, R. J., G. Ertl, and V. Penka, *Surface Sci.* **160**, 387 (1985).

Boudart, M., *AIChE J.* **2**, 62 (1956).

Boudart, M., and Djégo-Mariadassou, *Kinetics of Heterogeneous Catalytic Reactions*, Princeton University Press, Princeton, NJ (1984).

Boudeville, Y., and E. E. Wolf, *Surface Sci.* **297**, L127 (1993).

Bozso, F., G. Ertl, M. Gronze, and M. Weis, *J. Catal.* **49**, 18 (1977); **50**, 519 (1977).

Cox, M. P., G. Ertl, R. Imbihl, and J. Rüstig, *Surface Sci.* **134**, L517 (1983).

Davis, M. A., and G. A. Somorjai in D. A. King and D. P. Woodruff eds., *The Chemical Physics of Solid Surfaces and Heterogeneous Catalysis*, Vol. 4, p. 217, Elsevier, NY (1982).

Dawson, P. T., and Y. K. Peng, *Surface Sci.* **3**, 565 (1972).

deJong, A. M., and J. W. Niemansuerdeit, *Surface Sci.* **233**, 355 (1990).

Doraiswami, L. K., *Prog. Surface Sci.* **37**, 1 (1991).

Du, Z., A. F. Sarofin, and S. P. Longwell, *Energy and Fuels* **4**, 296 (1990).

Ertl, G., in *Proceedings, 7th International Congress on Catalysis*, Tokyo 21 (1981).

Ertl, G., *Adv. Catal.* **37**, 213 (1990).

Ertl, G., in J. R. Jennings, ed., *Catalytic Ammonia Synthesis*, Plenum, NY (1991).

Ertl, G., *Surface Sci.* **2**, 287, 1 (1993).

Falconer, J. L., and R. J. Madix, *Surface Sci.* **46**, 473 (1974); **48**, 393 (1975); **45**, 393 (1976).

Feigerle, C. S., S. R. Desai, and S. H. Overbury, *J. Chem. Phys.* **93**, 787 (1990).

Gray, P., and S. K. Scott, *Chemical Oscillations and Instabilities*, Oxford University Press, NY (1990).

Habershaden, E., and J. Küpers, *Surface Sci.* **138**, L147 (1984).

Hahnman, J., and M. A. Passler, *Surface Sci.* **203**, 449 (1988).

Horiuti, J. R., *Inst. Catalysis Hokkaido Univ.* **5**, 1 (1957).

Horiuti, J. R., and T. Nakamura, *Adv. Catal.* **17**, 1 (1967).

Hougen, A. O., and K. M. Watson, *Ind. Eng. Chem.* **35**, 529 (1943).

Hougen, A. O., and K. M. Watson, *Chemical Process Principles*, Wiley, NY (1943).

Huang, Z. Q., Z. Hussain, W. T. Huff, E. J. Moler, and D. A. Shirley, *Phys. Rev. B* **48**, 1709 (1993).

Imbihl, R., *Prog. Surface Sci.* **44**, 185 (1993).

Johnson, S., and R. J. Madix, *Surface Sci.* **108**, 77 (1981).

King, D. A., *Surface Sci.* **47**, 384 (1975).

Langmuir, I., *J. Am. Chem. Soc.* **34**, 1310 (1912); **40**, 1361 (1918).

Lauterback, I., G. Haas, H. H. Rotermund, and G. Ertl, *Surface Sci.* **294**, 116 (1993).

Loffler, D. G., and L. D. Schmidt, *J. Catal.* **41**, 440 (1976); *Surface Sci.* **59**, 195 (1976).

Lombardo, S. J., and A. T. Bell, *Surface Sci. Rep.* **13**, 1 (1991).

Love, C. A., S. L. Schultz, and C. S. Feigere, *Surface Sci.* **244**, 143 (1991).

Luss, D., U. Middya, and M. Sheintuch, *J. Phys. Chem.* **101**, 4688 (1994).

Madden, H. H., J. Kuppers, and G. Ertl, *J. Chem. Phys.* **58**, 3401 (1973).

Madhukar, A., and S. V. Ghaisas, *CRC Crit. Rev. Mat. Sci.* **14**, 1 (1988).

Mars, P., and D. W. Van Klevan, *Chem. Eng. Sci.* **3**(suppl.), 41 (1954).

McCabe, R. W., and L. D. Schmidt, *Surface Sci.* **66**, 101 (1977).

McCabe, R. W., and L. D. Schmidt in *Proceedings, 7th International Vacuum Congress*, Vol. 2, p. 1201 (1977).

Mezaki, R., and H. Inoue, *Rate Equations of Solid-Catalyzed Reactions*, University of Tokyo Press, Tokyo (1991).

Naumann, A., *Deuts. Chem. Gesel.* **3**, 702 (1870).

Redhead, P. A., *Vacuum* **12**, 203 (1963).

Schwartz, S. B., L. D. Schmidt, and G. B. Fisher, *J. Phys. Chem.* **90**, 6194 (1986).

Shoofs, G., and J. B. Benziger, *J. Phys. Chem.* **88**, 4439 (1984); *Surface Sci.* **187**, 359 (1984).

Stoltz, P., *Phys. Scripta* **36**, 824 (1987).

Stoltz, P., and J. K. Nørskov, *Phys. Rev. Lett.* **55**, 2502 (1985).

Stuckless, J. T., N. Alsarraf, C. Wartnaby, and D. A. King, *J. Chem. Phys.* **99**, 2202 (1993).

Taylor, J. L., and W. H. Weinberg, *Surface Sci.* **188**, 70 (1978).

Taylor, J. L., D. E. Ibbotson, and W. H. Weinberg, *J. Chem. Phys.* **69**, 4299 (1978).

Tempkin, M. I., *Zh. Fiz. Khim. SSSR* **14**, 1241 (1938).

Tempkin, M. I., and U. Pyzhev, *Acta Physico Chim.* **12**, 327 (1940).

Tempkin, M. I., *Zh. Fiz. Khim.* **31**, 1 (1957).

Tempkin, M. I., *Adv. Catal.* **28**, 173 (1979).

Van Spaendonk, V., and R. I. Masel, unpublished results (1995).

Van't Hoff, J. H., *Etudes de Dynamique Chemie*, Paris (1884). (Reprinted in English, Williams and Northgate Press, London, 1896.)

Vasquez, N., A. Muscat, and R. J. Madix, *Surface Sci.* **301**, 83 (1994).

Watson, K. M., *Principles of Reactor Design*, CRC Press, Cleveland (1946).

Weller, S., *AIChE J.* **2**, 61 (1956).

Weller, S., *Catal. Rev.* **34**, 227 (1992).

White, M. G., *Heterogeneous Catalysis*, Prentice Hall, Englewood Cliffs, NJ (1990).

White, M. G., and J. W. Hightower, *J. Catal.* **82**, 185 (1983).

Wolf, E. E., and Y. Boudeville, *Surface Sci.* **297**, L127 (1993).

Zhdanov, V. P., *Surface Sci.* **111**, 63 (1981).

Zhdanov, V. P., *Elementary Physico Chemical Processes on Solid Surfaces*, Plenum, NY (1991).

Zhdanov, V. P., and B. Kasemo, *Chem. Phys.* **177**, 519 (1993).

Zhdanov, V. P., and B. Kasemo, *Surface Sci. Rep.* **20**, 111 (1994).

8 A Review of Reaction-Rate Theory

PRÉCIS

In Chapter 7 we discussed rate laws for surface reactions. It was assumed that the mechanism of a surface reaction and the rate constants for each of elementary steps in the mechanism were known and then predictions were made about the rate of the reaction. One of the reasons that surfaces are interesting is that the rates of many reactions are much higher on a surface than in the gas phase. That is why surfaces are useful catalysts. The rate enhancement also leads to better quality deposited films. The objective of the remainder of this book is to understand that rate enhancement and its implications for catalysis and film growth.

In this chapter we provide an overview of reaction-rate theory. Reaction-rate theory is a big subject and there are already many excellent books on the subject. There is no room here to review the bulk of the material. However, our objective in this chapter is to provide a bare bones overview of the theory of reactions so that the reader will have enough background to understand the material in the later chapters. We will briefly review collision theory, transition-state theory, and unimolecular-rate theory. We show how the ideas can be applied to gas phase reactions. We then indicate how the ideas have been applied to surface reactions. Our discussion is necessarily brief. A more detailed discussion can be found in Moore and Pearson [1986], Laidler [1987], Levine and Bernstein [1987], Steinfeld, Fransisco, and Hase [1989], Bamford [1969–1993], or Connors [1990].

8.1 HISTORICAL INTRODUCTION

Kinetics became a well-identified part of chemistry in the later part of the nineteenth century. The first systematic measurements of rates of reactions as a function of temperature and concentration were done by Wilhelmy in 1850. In 1884, van't Hoff published a very influential book, *Etudes de Dynamique Chemique*, where he laid out the field of chemical kinetics. In that book, van't Hoff showed that reactions follow simple rate laws and that one can get some information about the order of a reaction from the molecularity of the reaction. Van't Hoff also proposed that reactions would obey what is now called Arrhenius's law:

$$k_1 = k_0 e^{-E_a/kT} \tag{8.1}$$

where k_1 is the rate constant for a reaction, k_0 is the preexponential for the reaction, E_a is the activation energy for the reaction, k is Boltzmann's constant, and T is temperature.

Arrhenius did many early studies that tried to understand why reactions obey Equation 8.1. In an 1889 paper, Arrhenius proposed that one could divide the molecules in a system into two classes, the molecules that could react and the molecules that could not. Arrhenius proposed that equilibrium was maintained between the two groups of molecules and that the activation energy in Equation 8.1 corresponded to the free-energy difference between the molecules that could react and those that could not. In a later paper Arrhenius [1899] proposed that the difference between the molecules that could react and those that could not was the kinetic energy of the molecules. Those molecules with a large kinetic energy could react, while those molecules with a low kinetic energy could not.

8.2   COLLISION THEORY

A few years later Trautz [1916, 1918] and Lewis [1916, 1918] independently proposed the collision theory of reactions. Their idea was that during a reaction, molecules in the gas phase collided with one another. If molecules came together with enough energy, reaction could occur. If not, the molecules would simply scatter. In this approximation, the rate of the reaction between a molecule A and another molecule BC is given by the number of collisions between the molecules times the probability that the molecules react.

One can quantify these ideas by assuming that the reaction probability is a function of $v_{A \to BC}$ the velocity at which the A molecule approaches the B—C molecule, and E_{BC}, the internal state (i.e., energy) of the BC molecule before collision occurs. From statistical mechanics one can write the reaction rate $r_{A \to BC}$ as

$$r_{A \to BC} = \int \int P_{reaction}(v_{A \to BC}, E_{BC}) \, r_{collision}(v_{A \to BC}, E_{BC}) \, D(v_{A \to BC}, E_{BC}) \, dv_{A \to BC} \, dE_{BC}$$

$$(8.2)$$

where $D(v_{A \to BC}, E_{BC})$ is the probability that a given pair of molecules will have a velocity $v_{A \to BC}$ and an internal energy E_{BC}, $r_{collision}(v_{A \to BC}, E_{BC})$ is the collision rate of those molecules; and $P_{reaction}(v_{A \to BC}, E_{BC})$ is the probability the molecules react when collision occurs. From kinetic theory, the collision rate, $r_{collision}(v_{A \to BC}, E_{BC})$, is given by

$$r_{collisions}(v_{A \to BC}, E_{BC}) = v_{A \to BC} \, \sigma^c_{A \to BC} \, C_A C_{BC} \qquad (8.3)$$

where C_a and C_{BC} are the concentrations of A and BC and $\sigma^c_{A \to BC}$ is a constant called the collisional cross section. The reaction rate becomes

$$r_{A \to BC} = k_{A \to BC} \, C_A C_{BC} \qquad (8.4)$$

where $k_{A \to BC}$ is the rate constant for the reaction. It is easy to show that $k_{A \to BC}$ is given by

$$k_{A \to BC} = \int \int v_{A \to BC} \, \sigma^c_{A \to BC} \, P_{reaction}(v_{A \to BC}, E_{BC}) \, D(v_{A \to BC}, E_{BC}) \, dv_{A \to BC} \, dE_{BC} \qquad (8.5)$$

Trautz [1916, 1918] and Lewis [1916, 1918] assumed that the reaction rate was zero if the internal energy of the incident molecules was less than E_a, and unity if the internal energy of the molecules was above E_a. Trautz and Lewis also assumed that D was given by a Boltzmann distribution. After considerable algebra, they were able to show

$$k_{A \to BC} = \bar{v}_{A \to BC} \, \sigma^c_{A \to BC} \, e^{-E_a/kT} \qquad (8.6)$$

where $\bar{v}_{A \to BC}$ is the average velocity of A moving toward BC.

The collision theory of reactions has had considerable success. It explained why the rate of gas phase reactions changed as the concentration of the reactants changed. It also explained why reactions had finite rates. Thus, it was quite useful.

The one difficulty with the model was that some of the model assumptions could not be justified based on what was known in 1916. Note that the preceding derivation assumes that the reaction rate is zero when the molecules come together with an energy less than E_a, so in effect the model assumes that there is a finite barrier to reaction. In the early 1900s there were several papers that tried to explain why the barrier to reaction arose. Trautz [1916, 1918] and Lewis [1916, 1918] proposed that molecules needed a certain critical energy to get a reaction to occur. When Trautz and Lewis did their work, it was thought that the energy could be obtained via radiation (i.e., infrared) from the walls of the reaction vessel. However, earlier Arrhenius [1899] had shown that critical energy could also come through gas phase collisions. Still, none of this work provided an explanation of why one needed the incident molecules to have more than a critical amount of energy before a reaction would occur.

8.3 REACTIONS AS MOTION ON POTENTIAL ENERGY SURFACES

The first real insights into why molecules needed a critical energy to react came in a series of papers by Marcelin [1914–1920]. Marcelin suggested that one could consider a reaction to be a trajectory on a potential energy surface. For example, Figure 8.2 shows a potential energy surface for a reaction where an atom A reacts with a molecule BC to produce a new molecule AB and an atom C, i.e.,

$$A + BC \rightarrow AB + C$$

Marcelin noted that in order for a reaction to occur A, B and C must come close together. While AB and BC are stable molecules the ABC complex must not be stable because if it were stable some ABC would form. Therefore the energy of the system must go up when A, B, and C come close together. Marcelin proposed that this additional energy was responsible for the activation barrier for the reaction.

Marcelin also suggested that one could understand the interaction of A, B, and C by examining $V(\vec{R}_A, \vec{R}_B, \vec{R}_C)$ the potential energy for the interaction of A, B, C, as a function of $\vec{R}_A$, $\vec{R}_B$, and $\vec{R}_C$, the positions of atoms A, B, and C. To simplify the analysis Marcelin only considered a linear reaction, i.e., he confined A, B, and C to lie on a line, as shown in Figure 8.1. Under these circumstances $V(\vec{R}_A, \vec{R}_B, \vec{R}_C)$, the potential for the interaction between A, B, and C, is a function of two variables: R_{AB}, the distance from A to B, and R_{BC}, the distance from B to C. For future reference, R_{AB} and R_{BC} are related to $\vec{R}_A$, $\vec{R}_B$, and $\vec{R}_C$ by

$$R_{AB} = |\vec{R}_B - \vec{R}_C|$$
$$R_{BC} = |\vec{R}_C - \vec{R}_B|$$
(8.7)

Figure 8.2 shows a typical potential energy surface for the reaction between A and BC. The lines in the figure are contours of constant energy. The potential is zero in the upper right corner. The potential goes down, moving along the right side of the figure, reaching a minimum at point A. The potential then goes up again, moving to the lower right corner of the figure. Initially, the potential is almost constant, moving left from the lower right corner of the figure, but when one gets half-way across the figure, the potential starts to rise again. The potential eventually reaches a maximum at the lower left corner of the figure. Similarly, the potential starts out at zero at the upper right corner of the figure. The potential goes down,

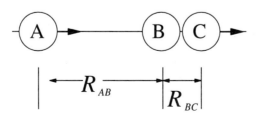

Figure 8.1 The initial configuration for A, B, and C confined to a line.

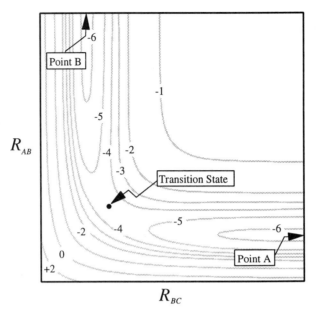

Figure 8.2 A potential energy surface for the reaction A + BC → AB + C when the reactants are confined to a line.

moving left from there, reaching a minimum at point B. Then the potential rises again. Physically, in the top left portion of the figure, R_{AB} is large, while R_{BC} is small. Therefore the top left portion of the figure corresponds to the initial state of the reaction, where an A atom is just beginning to approach a BC molecule. When A is far away from BC there will be little interaction between A and BC. As a result, the potential energy surface will look just like that in a BC molecule. The potential will be attractive at long B–C distances and repulsive at short B–C distances. The potential will not depend on R_{AB}. In a similar way, in the bottom right portion of the figure, R_{AB} is small, while R_{BC} is large. Therefore the bottom right portion of the figure corresponds to the final state of the reaction where A has reacted with BC to form a stable AB molecule and an isolated C atom. In that case there will be little interaction between AB and C. As a result, the potential energy surface will look just like that in an AB molecule. The potential will be attractive at long A–B distances and repulsive at short A–B distances. The potential will not depend on R_{BC}.

The lower left portion of Figure 8.2 is complex. The lower left portion of the curve corresponds to the case where A, B, and C are all close together. Obviously, if A, B, and C get too close, the potential will be repulsive. However, at intermediate distances there will be a metastable AB bond and a metastable BC bond. As a result, the potential energy of the system will be lower than in the case when A, B, and C are all far apart. It is not immediately obvious whether the minimum in the potential will occur when A, B, and C are all close together so that there is a stable ABC molecule, or when A and B are close together and C is far away or when B and C are close together and A is far away. However, it is easy to show that provided no ABC molecules form (i.e., ABC molecules are unstable) the potential will be higher when A, B, and C are all close together than when A is far from BC or C is far from AB. See Chapter 9 for a further discussion of this point. For future reference we call the saddle point in the potential in Figure 8.2 the **transition state** for the reaction. Later in this chapter we will note that an understanding of the properties of transition states is critical to the understanding of rates of chemical reactions.

Now consider what happens when A collides with BC. In that case, the system will evolve

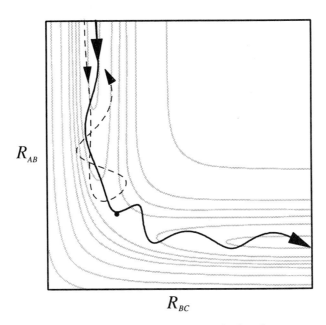

Figure 8.3 Trajectories for the reaction A + BC → AB + C when the reactants are confined to a line.

in time; R_{AB} and R_{BC} will change. One can plot that change as a trajectory (i.e., line) on the potential energy surface as shown in Figure 8.3. The solid dark line in Figure 8.3 starts at the reactants (i.e., the upper left) and ends at the products (i.e., the lower right). We call that line a reactive trajectory, or a reactive collision since if the system follows that trajectory, reaction will occur. Similarly, the heavy dashed line starts out at the reactants (i.e., the upper left) and loops back to the upper left, i.e., back to the reactants. We call such a trajectory, a nonreactive collision, since if the system follows that line, A will scatter from BC and no reaction will occur.

One can calculate a reaction rate from Figure 8.3 by starting with a distribution of molecules, computing a series of trajectories like those in Figure 8.3, and adding up how quickly the molecules react. Note that in the case shown in Figure 8.3 the potential goes up when A, B, and C come close enough to react. Therefore, it will take a certain amount of energy to get A, B, and C close enough to react. This example illustrates a key point:

It usually costs energy to bring the reactants close enough that they can react. As a result, there is a barrier to most chemical reactions.

In Chapter 9, we will show that additional barriers arise because one needs to stretch or break bonds during chemical reactions.

8.4 MOLECULAR DYNAMICS SIMULATION OF REACTIVE COLLISIONS: LINEAR A + BC → AB + C

The discussion in the last section was qualitative, but one can also expand the ideas in a quantitative way. The idea is to use molecular dynamics to calculate $P_{reaction}(v_{A \to BC}, E_{BC})$ as a function of $v_{A \to BC}$ and E_{BC}. One then plugs $P_{reaction}(v_{A \to BC}, E_{BC})$ and an estimate of $\sigma_{A \to BC}$ into Equation 8.5 to calculate a reaction rate.

The hard part in the computation is the calculation of $P_{reaction}(v_{A \to BC}, E_{BC})$. Consider the interaction between three atoms A, B, and C. If A, B, and C are confined to a line, then one can replace $\vec{R}_A$, $\vec{R}_B$, and $\vec{R}_C$, the position vectors for A, B, and C, by scalar quantities, R_A, R_B, and R_C. The classical equations of motion for A, B, and C are

$$m_A \frac{d^2 R_A}{dt^2} = F_A = -\frac{\partial V(R_A, R_B, R_C)}{\partial R_A} \tag{8.8}$$

$$m_B \frac{d^2 R_B}{dt^2} = F_B = -\frac{\partial V(R_A, R_B, R_C)}{\partial R_B} \tag{8.9}$$

$$m_C \frac{d^2 R_C}{dt^2} = F_C = -\frac{\partial V(R_A, R_B, R_C)}{\partial R_C} \tag{8.10}$$

where R_A, R_B, and R_C are the positions of atoms A, B, and C; m_A, m_B, and m_C are the masses of A, B, and C, t is time; and F_A, F_B, and F_C are the net forces on atoms A, B, and C.

If one knows the potential $V(R_A, R_B, R_C)$, one can numerically integrate Equations 8.8, 8.9, and 8.10 to calculate $P_{reaction}(v_{A \to BC}, E_{BC})$ as a function of the initial velocity and internal energy of the molecules. A program to integrate the trajectories, called REACT_MD, is available from Professor Masel. Figure 8.4 shows what typical trajectories are like. The trajectories on the left were calculated for a case where the energy was -5, while on the transition state energy was -4.5. In that case, A moves toward BC, but A does not have enough energy to get over the barrier in the potential energy surface. In that case, A simply scatters from BC. No reaction occurs.

This is a general result. If a molecule does not have enough energy to cross the activation barrier, no reaction occurs in the absence of quantum effects discussed in Section 8.11.

At higher energies, one does have some probability of reaction. However, surprisingly, when the incoming molecules have barely enough energy to cross the barrier, the reaction probability is relatively small. As one raises the energy of the incident particles, the reaction probability goes up. However, the rate of production of AB goes down again at very high energy because A is moving so fast that the A never sticks to B.

A more careful analysis shows that concerted motions of the atoms are needed for the A + BC → AB + C reaction to occur. At the start of the reaction, the BC molecule is vibrating. C moves away from B, and then back toward B. Simulations show that if A approaches B just as C moves away, there is a high probability of reaction. If not, the reaction probability is low.

One can generalize these effects of reactions where the individual species A, B, and C are molecules rather than atoms. Generally, a reaction involves concerted destruction and formation of bonds. Both processes must occur simultaneously. When a molecule A collides with a molecule BC, and A only stays in close proximity to BC for a short time: usually less than 10^{-11} sec. No reaction occurs unless the C can move away from B before A leaves. As a result, it helps to have C moving away from B when A moves in toward B.

Another issue is that in order for reaction to occur, the B–C bond must somehow acquire enough energy to break before A leaves. Therefore the rate that energy accumulates in the B–C bond has an important influence on the reaction rate. Generally, one finds that when the system has barely enough energy to cross the activation barrier reaction rarely occurs, because

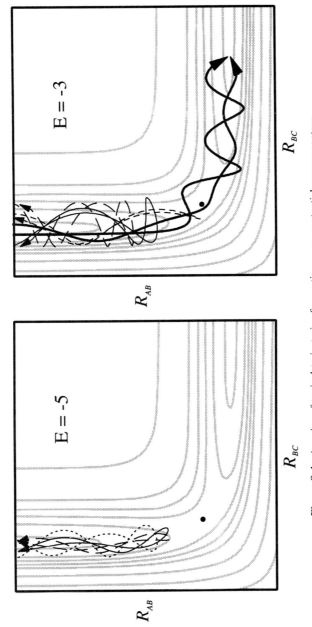

Figure 8.4 A series of typical trajectories for motion over a potential energy contour.

587

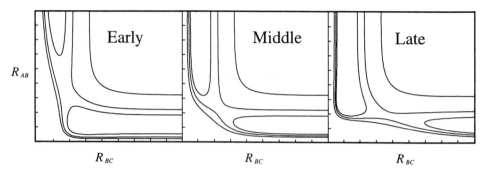

Figure 8.5   Potential energy surfaces with early, middle, and late transition states.

the collision energy usually does not get localized in the B–C bond. The reaction rate increases with increasing energy. Note, however, that if A comes in too fast, the A will leave before there is time for reaction to occur. In that case, the reaction rate will go down again.

Another interesting effect is that the trajectories that lead to reaction change as one changes the position of the transition state. In Chapter 9, we will note that one can characterize reactions according to whether they have an early, middle, or late transition state. A reaction with an early transition state will reach the transition state before the reactants are significantly deformed. The B–C bond will not be stretched significantly. However, the A–B bond will be much longer than in a stable AB molecule (see Figure 8.5). In contrast a reaction with a late transition state will reach the saddle point near the end of the reaction. The B–C bond will be almost broken, while the A–B bond will be about the same length as in a stable AB molecule. A reaction with a middle transition state is somewhere between the two, where the bonds in both the reactants and products are stretched by similar amounts in the transition state.

Simulations show that $P_{reaction}(v_{A \rightarrow BC}, E_{BC})$ changes according to whether the reaction has an early, middle, or late transition state. If the reaction has an early transition state, $v_{A \rightarrow BC}$ has a much bigger effect on $P_{reaction}(v_{A \rightarrow BC}, E_{BC})$ than E_{BC}. In contrast, if the reaction has a late transition state, E_{BC} has a somewhat bigger effect on $P_{reaction}(v_{A \rightarrow BC}, E_{BC})$ than $v_{A \rightarrow BC}$. Physically, when there is an early barrier to the reaction, most of the barrier is associated with getting A and BC in close proximity. In that case, the faster A moves toward BC the more momentum A will carry into the collision. A's momentum allows it to get in close proximity to B–C. That increases the rate of reaction. In contrast, if the reaction has a late transition state, most of the barrier is associated with breaking the B–C bond. In that case, it is better to have energy in the B–C bond before the reaction, rather than having to transfer energy during the collision.

All of these trends can be explored with the REACT_MD program. If the reader has access to this program he/she is advised to run the program and vary the various parameters before proceeding with this chapter.

8.5   THE NONLINEAR CASE

The discussion in the previous three sections considered the reaction A + BC → AB + C where A, B, and C are confined to a line. With a linear collision, all of the A atoms that hit the BC molecule react with a constant probability. However, in reality, A, B, and C are not confined to a line. With a nonlinear system the reaction probability varies with how close of a collision occurs. Generally, the reaction probability is high when there is a head-on (i.e., linear) collision between A and B–C, and lower when A approaches B–C with a glancing

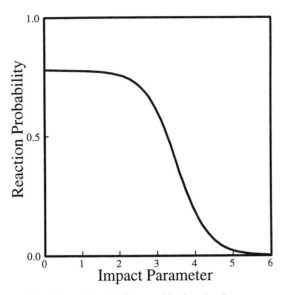

Figure 8.6 The variation in $P_{A \to BC}$ with changing impact parameter.

trajectory. More precisely, it is useful to define a quantity b called the impact parameter. The impact parameter is the point that atom A would intersect the plane going through the center of the BC bond if there were no interaction between A and BC (see Figure 8.7). Figure 8.6 shows how the reaction probability varies with the impact parameter. Generally, the reaction probability is highest with an impact parameter of 0. The reaction probability is relatively insensitive to the impact parameter when the impact parameter is small. However, there is a sharp cutoff at a larger impact parameter. As a result, the reaction rate is negligible unless atoms hit with a small impact parameter.

One can understand this behavior qualitatively with the aid of Figure 8.7. When the molecules collide with an impact parameter of zero, all of the momentum of atom A is directed toward BC. In contrast, when the impact parameter is nonzero, one can divide A's momentum into two components as indicated in Figure 8.7. One component $v_{\parallel}$ carries A toward BC, while the second component $v_{\perp}$ carries A away from BC. As the impact parameter

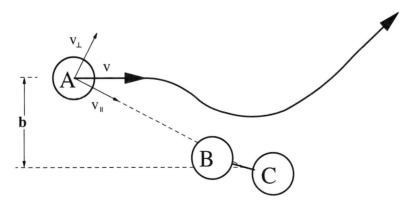

Figure 8.7 A typical trajectory for the collision of an A atom with a BC molecule.

increases, $v_\perp$ increases. As a result, the probability that atom A will move away from BC increases as the impact parameter increases.

In order to fully model this process for the reaction $A + BC \rightarrow AB + C$, one has to solve the equations of motion (i.e., Equations 8.7, 8.8, and 8.9). In general, that cannot be done analytically. However, one can get the structure of the solution by solving the equations of motion for a simpler reaction, $A + B \rightarrow AB$. A detailed derivation of the key equations is given in the next section. However, the key result is that when A and B collide with a nonzero impact parameter, there will be an extra barrier to the reaction, V_{extra}, given by

$$V_{extra} = \frac{b^2 E_{kE}}{R_{AB}^2} \tag{8.11}$$

where E_{kE} is the kinetic energy of A moving toward B, b is the impact parameter, and R_{AB} is the distance from A to B. Notice that the extra barrier grows as the impact parameter increases. As a result, the barrier prevents the reaction from occurring at larger impact parameters.

8.5.1   Derivation of the Angular Momentum Barrier to Reaction

Equation 8.11 arises because whenever $v_\perp$ is nonzero, the AB bond has angular momentum. One needs to overcome that angular momentum for A and B to react. The purpose of the next several pages is to derive Equation 8.11. We include it because many of our students need to see a derivation before they accept the result. However, the reader could skip to the next section without loss of continuity.

The derivation starts by looking at the classical equations of motion of A and B:

$$m_A \frac{d^2 \vec{R}_A}{dt^2} = \vec{F}_A = -\nabla_A V(\vec{R}_A, \vec{R}_B) \tag{8.12}$$

$$m_B \frac{d^2 \vec{R}_B}{dt^2} = \vec{F}_B = -\nabla_B V(\vec{R}_A, \vec{R}_B) \tag{8.13}$$

From Newton's second law, the force that A exerts on B must be equal and opposite to the force that B exerts on A, i.e.,

$$\vec{F}_B = -\vec{F}_A \tag{8.14}$$

It is useful to change coordinates to what is called **center-of-mass coordinates.** Let's define $\vec{X}$ via

$$\vec{X} = \frac{m_A \vec{R}_A + m_B \vec{R}_B}{m_A + m_B} \tag{8.15}$$

$\vec{X}$ is the position of the center of mass of A and B. Taking the second derivative of Equation 8.15,

$$\frac{d^2 \vec{X}}{dt^2} = \left(\frac{1}{m_A + m_B}\right)\left(m_a \frac{d^2 \vec{R}_A}{dt^2} + m_B \frac{d^2 \vec{R}_B}{dt^2}\right) \tag{8.16}$$

Substituting Equations 8.12, 8.13, and 8.14 into Equation 8.16 yields

$$\frac{d^2\vec{X}}{dt^2} = \frac{1}{m_A + m_B}(\vec{F}_A + \vec{F}_B) = 0 \tag{8.17}$$

Therefore, the center of mass of the system moves with a constant velocity throughout the collision. Now let's develop an expression for $\vec{R}_{AB}$ the distance from A to B. The definition of $\vec{R}_{AB}$ is

$$\vec{R}_{AB} = \vec{R}_B - \vec{R}_A \tag{8.18}$$

Taking the second derivative of Equation 8.18 yields

$$\frac{d^2\vec{R}_{AB}}{dt^2} = \frac{d^2\vec{R}_B}{dt^2} - \frac{d^2\vec{R}_A}{dt^2} \tag{8.19}$$

Substituting Equations 8.12 and 8.13 into Equation 8.19 yields

$$\frac{d^2\vec{R}_{AB}}{dt^2} = \frac{1}{m_B}\vec{F}_B - \frac{1}{m_A}\vec{F}_A \tag{8.20}$$

but $\vec{F}_A = -\vec{F}_B$, therefore

$$\frac{d^2\vec{R}_{AB}}{dt^2} = \left(\frac{1}{m_A} + \frac{1}{m_B}\right)\vec{F}_B = \frac{1}{\mu_{AB}}\vec{F}_B \tag{8.21}$$

where μ_{AB} is defined by

$$\frac{1}{\mu_{AB}} = \frac{1}{m_B} + \frac{1}{m_A} = \frac{m_A + m_B}{m_A m_B} \tag{8.22}$$

μ_{AB} is called the **reduced mass** of the system. Rearranging Equation 8.21 yields

$$\mu_{AB}\frac{d^2\vec{R}_{AB}}{dt^2} = \vec{F}_B \tag{8.23}$$

Note:

$$\vec{F}_B = -\nabla_B V = -\left(\frac{d}{dX_B}, \frac{d}{dY_B}, \frac{d}{dZ_B}\right)V \tag{8.24}$$

but

$$\vec{R}_{AB} = \vec{R}_B - \vec{R}_A$$
$$\vec{F}_B = -\left(\frac{d}{dX_{AB}}, \frac{d}{dY_{AB}}, \frac{d}{dZ_{AB}}\right)V \tag{8.25}$$

where X_{AB}, Y_{AB}, and Z_{AB} are the x, y, and z components of $\vec{R}_{AB}$.

It is useful to convert Equations 8.23 to spherical coordinates; R_{AB}, θ_{AB}, and ϕ_{AB}, where R_{AB}, θ_{AB}, and ϕ_{AB} are related to X_{AB}, Y_{AB}, and Z_{AB}, the X, Y, and Z components of R_{AB}, by

$$X_{AB} = R_{AB} \cos \theta_{AB}$$

$$Y_{AB} = R_{AB} \sin \theta_{AB} \cos \phi_{AB} \qquad (8.26)$$

$$Z_{AB} = R_{AB} \sin \theta_{AB} \sin \phi_{AB}$$

For the moment we will only consider $\phi_{AB} = 0$.

Let's define L_{AB} as the angular momentum of the AB pair by

$$L_{AB} = \mu_{AB} \, v_{A \to B} R_{AB} = \mu_{AB} \left(\frac{d\theta_{AB}}{dt} \right) R_{AB}^2 \qquad (8.27)$$

Where $v_{A \to B}$ is the instantaneous velocity of A toward B. Next let's demonstrate that angular momentum is conserved during the collision. When $\phi_{AB} = 0$

$$\theta_{AB} = \arctan \left(\frac{Y_{AB}}{X_{AB}} \right) \qquad (8.28)$$

$$\frac{d\theta_{AB}}{dt} = \left(\frac{1}{1 + \left(\frac{Y_{AB}^2}{X_{AB}^2} \right)} \right) \frac{d}{dt} \left(\frac{Y_{AB}}{X_{AB}} \right) = \left(\frac{1}{1 + Y_{AB}^2} \right) \left(\frac{1}{X_{AB}} \frac{\partial Y_{AB}}{dt} - \frac{Y_{AB}}{X_{AB}^2} \frac{\partial X_{AB}}{dt} \right) \qquad (8.29)$$

Multiplying the top and bottom of Equation 8.29 by X_{AB}^2 yields

$$\frac{d\theta}{dt} = \frac{X_{AB} \dfrac{dY_{AB}}{dt} - Y \dfrac{dX_{AB}}{dt}}{X_{AB}^2 + Y_{AB}^2} = \frac{X_{AB}}{R_{AB}^2} \frac{dY_{Ab}}{dt} - \frac{Y_{AB}}{R_{AB}^2} \frac{dX_{AB}}{dt} \qquad (8.30)$$

Now consider

$$\frac{dL_{AB}}{dt} = \frac{d}{dt} \left(\mu_{AB} R_{AB}^2 \frac{\partial \theta_{AB}}{dt} \right) \qquad (8.31)$$

Substituting Equation 8.30 into Equation 8.31 yields

$$\frac{dL_{AB}}{dt} = \frac{d}{dt} \left(\mu_{AB} \left(X_{AB} \frac{dY_{AB}}{dt} - Y_{AB} d \frac{X_{AB}}{dt} \right) \right) \qquad (8.32)$$

Performing the algebra

$$\frac{dL_{AB}}{dt} = \mu_{AB} \frac{dX_{AB}}{dt} \frac{dY_{AB}}{dt}$$

$$+ \mu_{AB} X_{AB} \frac{d^2 Y_{AB}}{dt^2} - \mu_{AB} \frac{dX_{AB}}{dt} \frac{dY_{AB}}{dt} - \mu_{AB} Y_{AB} \frac{d^2 X_{AB}}{dt^2} \qquad (8.33)$$

The first and third terms cancel. Therefore

$$\frac{dL_{AB}}{dt} = \mu_{AB} X_{AB} \frac{d^2 Y_{AB}}{dt^2} - \mu_{AB} Y_{AB} \frac{d^2 X_{AB}}{dt^2} \tag{8.34}$$

Equations 8.23 and 8.25 imply

$$\mu_{AB} \frac{d^2 X_{AB}}{dt^2} = \cos\theta_{AB} \frac{\partial V(R_{AB})}{\partial R_{AB}} = -\frac{X_{AB}}{R_{AB}} \frac{\partial V(R_{AB})}{\partial R_{AB}} \tag{8.35}$$

$$\mu_{AB} \frac{d^2 Y_{AB}}{dt^2} = \sin\theta_{AB} \frac{\partial V(R_{AB})}{\partial R_{AB}} = -\frac{Y_{AB}}{R_{AB}} \frac{\partial V(R_{AB})}{\partial R_{AB}} \tag{8.36}$$

Substituting Equations 8.35 and 8.36 into Equation 8.34 yields

$$\frac{dL_{AB}}{dt} = X_{AB} \left(-\frac{Y_{AB}}{R_{AB}} \frac{\partial V(R_{AB})}{\partial R_{AB}} \right) - Y_{AB} \left(-\frac{X_{AB}}{R_{AB}} \frac{\partial V(R_{AB})}{\partial R_{AB}} \right) \tag{8.37}$$

Note that all of the terms on the right of Equation 8.37 cancel. Therefore

$$\frac{dL_{AB}}{dt} = 0 \tag{8.38}$$

Equation 8.38 is a key result because it shows that angular momentum is a conserved quantity. A similar derivation shows that the energy of the system, E_{AB}, is a conserved quantity where the energy is given by

$$E_{AB} = \tfrac{1}{2}\mu_{AB}\, v_{A \to B}^2 + V(R_{AB}) \tag{8.39}$$

where $v_{A \to B}$ is the instantaneous velocity of A relative to B and $V(R_{AB})$ is the A–B potential. Equation 8.39 can be rewritten as

$$E_{AB} = \tfrac{1}{2}\,\mu_{AB}\, (v_\parallel^2 + v_\perp^2) + V(R_{AB}) \tag{8.40}$$

where $v_\parallel$ and $v_\perp$ are the two components of the particles' velocity as defined in Figure 8.7. Next, it is useful to derive an expression for $v_\perp$ during the A–B collision. $v_\perp$ provides angular momentum to the A–B pair, so let's start by calculating the angular momentum of the system. Consider the situation long before a collision occurs. Atom A is moving toward atom B with a velocity v_o. Combining Equations 8.26 and 8.29 shows

$$L_{AB}^o = \mu_{AB} R_{AB}^2 \left(\frac{d\theta}{dt} \right) = X_{AB} \left(\frac{dY_{AB}}{dt} \right) - Y_{AB} \left(\frac{dX_{AB}}{dt} \right) \tag{8.41}$$

where L_{AB}^o is the initial value of L_{AB}. Initially,

$$Y_{AB} = b, \qquad \frac{dY_{AB}}{dt} = 0, \qquad \frac{dX_{AB}}{dt} = v_o \tag{8.42}$$

where v_o is the velocity of A toward B at the beginning of the reaction (i.e., when R_{AB} is large so $V(R_{AB}) = 0$). Substituting Equation 8.39 into Equation 8.40 shows

$$L_{AB}^o = -b \; \mu_{AB} v_o \tag{8.43}$$

Note that angular momentum is a conserved quantity. Therefore, L_{AB}, the value of the angular momentum at any point during the A–B collision, must equal the initial value, i.e.,

$$L_{AB} = L_{Ab}^o = -b \; \mu_{AB} \; v_o \tag{8.44}$$

For future reference, it is useful to note that when R_{AB} is large $V(R_{AB})$ is zero. Equation 8.40 becomes

$$E_{AB} = \tfrac{1}{2} \; \mu_{AB} \; v_o^2 \tag{8.45}$$

Next, it is useful to derive an expression for $v_\perp$ in terms of the angular momentum for the collision:

$$v_\perp = R_{AB} \frac{d\theta_{AB}}{dt} \tag{8.46}$$

Substituting Equations 8.27 into Equation 8.46 yields

$$v_\perp = \frac{L_{AB}}{R_{AB}\mu} \tag{8.47}$$

Substituting Equation 8.44 and 8.47 into Equation 8.40 shows

$$E = \frac{1}{2} \; \mu_{AB} \left(\frac{dR_{AB}}{dt} \right)^2 + \frac{L_{AB}^2}{2\mu_{AB}R_{AB}^2} + V(R_{AB}) \tag{8.48}$$

Substituting Equations 8.44 and 8.45 into Equation 8.48 yields

$$E = \frac{1}{2} \; \mu_{AB} \left(\frac{dR_{AB}}{dt} \right)^2 + \left[\frac{E_{AB}b^2}{R_{AB}^2} + V(R_{AB}) \right] \tag{8.49}$$

It is useful to define an effective potential V_{eff} by

$$V_{eff}(R_{AB}) = \frac{E_{AB}b^2}{R_{AB}^2} + V(R_{AB}) \tag{8.50}$$

Substituting Equation 8.50 into Equation 8.49 and rearranging yields

$$\frac{dR_{AB}}{dt} = \pm \sqrt{\frac{2}{\mu_{AB}} (E - V_{eff}(R_{AB}))} \tag{8.51}$$

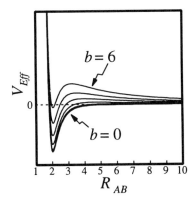

Figure 8.8 A plot of the effective potential as a function of R_{AB} or a modified Lennard-Jones potential with $b = 0, 1, 2, 3, 4, 5, 6$.

8.6 THE ANGULAR MOMENTUM BARRIERS TO REACTION

Equation 8.51 can be integrated numerically, to calculate R_{AB} as a function of time. Therefore, one can use Equation 8.51 to calculate whether a reaction occurs. Figure 8.8 shows a plot of the effective potential as a function of R_{AB}. We chose a case where $V(R_{AB})$ was an offsetted–Lennard-Jones potential, so when $b = 0$, there is no barrier to prevent A from approaching B. Note, however, that as b increases a barrier arises. This barrier occurs because when b is nonzero, the angular momentum, L_{AB}, in Equation 8.48 is nonzero. Therefore, the barrier is called the angular-momentum barrier to reaction.

If one considers a more complex reaction, one finds that the arguments are more subtle. For example, with the reaction A + BC → AB + C, one can transfer angular momentum from A to C, which slightly reduces the angular-momentum barrier to reaction. Still, angular-momentum barriers arise in all gas phase reactions:

> The angular-momentum barrier prevents reactions from occurring when molecules approach with large impact parameters. As a result, the barrier is quite important to reactions.

8.6.1 Influence on the Overall Rate

Now, it is useful to go back and see how the angular-momentum barrier changes the analysis in Section 8.2 (i.e., collision theory). Note that the analysis in Section 8.2 assumed that the reaction rate was equal to the collision rate times the probability that molecules which collided reacted. The probability is assumed to be constant for all molecules that hit each other. However, the results in the previous section show that the probability is not constant. Thus, some modification of the results in Section 8.2 is needed.

One can account for the variations in the reactive probability by defining a reaction cross section $\sigma^r_{A \to BC}$ by integrating the reaction probability over the impact parameter, b, and ϕ_{AB} yields

$$\phi^r_{A \to BC} = \int \int P_{A \to BC}(\phi_{AB}, b) b \, db \, d\phi_{AB} \qquad (8.52)$$

For a spherically symmetric case

$$\phi^r_{A \to BC} = 2\pi \int P_{A \to BC}(b)b \; db \tag{8.53}$$

The reaction probability is already in Equation 8.53. Therefore, Equation 8.5 for the rate constant becomes

$$k_{A \to BC} = \int \int v_{A \to BC} \sigma^r_{A \to BC} D(v_{A \to BC}, E_{BC}) \; dv_{A \to BC} \; dE_{BC} \tag{8.54}$$

Equations 8.53 and 8.54 can be used to calculate the rate constant for the reaction if $P_{reaction}(b)$ is known. Hence, they are quite useful.

8.7   ENERGY TRANSFER BARRIER TO UNIMOLECULAR REACTIONS

The analysis in Section 8.6 discussed the angular momentum to reaction. However, there are also energy transfer barriers to reactions. The energy transfer barriers to reactions were mentioned in Secion 8.4. In order for a reaction like A + BC → AB + C to occur, energy must be localized within the B–C bond. There often is an energy transfer limitation to the localization, as discussed in Section 8.4. This type of energy transfer barrier arises in all reactions, and must be considered in a complete analysis as discussed in Section 8.4. However, there is another type of energy transfer barrier that arises only in reactions where two reactants A and B combine to form a single product AB, or a single product AB reacts to form one or more species. Typical examples are

$$Na + Cl \to NaCl$$

$$HCl \to H + Cl$$

$$Cyclopropane \to Propylene$$

Consider the generalized reaction A + B → AB. At the start of the reaction A approaches B with some energy E_{AB}. For the purposes of discussion assume that the A–B potential $V(R_{AB})$ is a Morse potential and that b is small so the effective potential looks like that in Figure 8.9. As A approaches B, A is first accelerated since $V(R_{AB})$ is attractive at long distances. However, A begins to slow down again as A and B get to be very close. One can calculate A's velocity from Equation 8.51 It is useful to also calculate the net force on A from Equation 8.8. Note that in order for the reaction A + B → AB to occur, A must fall into B's attractive well. A's velocity must be small (i.e., like a vibration). Further, the net force on A must be small or else A will be pushed away from B. It works out, however, that when one actually calculates the velocity from Equation 8.51 and the net force from Equation 8.8, one finds that one can never get to the situation where the force and velocity are small at the same time.

Consider the case where an A atom collides with a B atom. Figure 8.9 shows a plot of A's velocity and the net force on A during the collision. A starts moving out slowly. However, it is accelerated by B's attractive well. The net result is that A's velocity is quite large when A is sitting over the minimum in the A–B potential. As a result, A flies past the potential well. A will eventually stop when the A–B bond is highly compressed. However, the net force on A is tremendous there and so A is pushed away. One can never get to the position where the net force on A is small and A is moving slowly enough so that A stays bound to B.

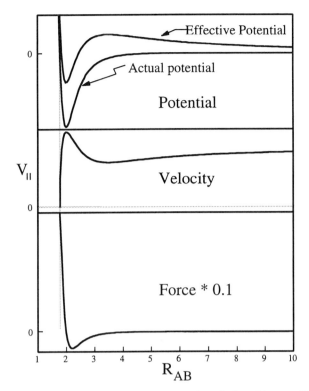

Figure 8.9   A typical V_{eff} for the reaction A + B → AB. The velocity of A toward B is given, as well as the net force on A.

As a result, A cannot react with B. The preceding analysis was for a Morse potential. However, detailed simulations with the program REACT_MC show that the reaction A + B → AB never occurs. A always scatters from B independent of the form of A–B potential.

> As a result, elementary reactions of the form
>
> $$A + B \rightarrow AB$$
>
> are impossible.

Physically, what is happening is that A is approaching B with some energy, E_{AB}. A has to lose that energy in order to fall into B's attractive well. However, there is no place for the energy to go. As a result, A simply flies out again.

Of course, experimentally one often observes reactions of the form A + B → AB. Note, however, that such reactions are never elementary reactions. Instead one finds that the reaction actually follows a mechanism A + B + X → AB + X, where X is something called a **collision partner.** X can be another molecule in the system, or the walls (i.e., surfaces) of the reaction vessel. The function of the collision partner is to remove energy and momentum from the reactive complex so that A can fall into B's attractive well. Experimentally, the rate of a reaction of the form A + B → AB goes to zero at low pressures, because there are no

TABLE 8.1 **Arrhenius Parameters for the Reaction O + NO + X →**
NO$_2$ + X

Collision Partner	Pre-exponential ($\mathring{A}^3$/(mole^2-sec))	Activation Energy (kcal/mole)
None	0	?
Ar	1.7×10^{14}	1.88
O$_2$	1.7×10^{15}	1.88
CO$_2$	3.84×10^{15}	1.88
H$_2$O	1.1×10^{16}	1.88

Data from Westley [1980]

collision partners. However, one can observe reactions of the form A + B → AB at moderate pressures. At moderate pressures, collisions with other molecules in the system occur. Those collisions carry away energy and momentum from the AB complex. As a result, the reaction has a finite rate. In the problem set we ask the reader to show that the reaction looks elementary at high pressures. However, the reaction is not really elementary. In particular the rate constants for the reaction vary with the environment since the collision partner plays an important role in the reaction. For example, Table 8.1 shows data for the rate constant for the reaction NO + O + X → NO$_2$ + X. Notice that the rate constant is about a factor of 60 lower when the reaction is run in argon than when the reaction is run in water vapor. Clearly, this is a significant effect. Therefore, it is incorrect to view reactions of the form A + B → AB as elementary. Rather, it is better to view the reactions as needing a collision partner.

A similar argument also applies to unimolecular reactions, i.e., reactions where a single reactant reacts to form one or more products, e.g., reactions of the form A → C or AB → A + B. In each case the reactants start off in a stable configuration, and so the reactants need to accumulate energy before any reaction can occur. That energy comes from collision partners, i.e., other gases in the system or photons. Again, reactions of the form A → C or AB → A + B are not elementary reactions. Instead, the reactions follow A + X → C + X or AB + X → A + B + X. As with the A + B → AB case, the rate constant varies according to the nature of the collision partner. For example, Table 8.2 shows data for the reaction O$_2$ + X → 2O + X. Again one finds that the rate constant varies by almost a factor of 80 according to whether argon or oxygen is used as the collision partner. This leads to the key point in this section:

All elementary chemical reactions must have two or more reactants and two or more products. Reactions of the form A + B → AB, AB → A + B, A → C are not elementary; collision partners are needed before such reactions can occur.

TABLE 8.2 **Arrhenius Parameters for the Reaction O$_2$ + X →**
2O + X

Collision Partner	Preexponential ($\mathring{A}^3$/(mole-sec) at 4000 K)	Activation Energy (kcal/mole)
None	0	?
Ar	7.5×10^{14}	118.8
O$_2$	5.8×10^{16}	118.8

Data from Westley [1980].

8.8 MD/MC CALCULATIONS OF REACTION-RATE CONSTANTS

One can in principle combine Monte Carlo (MC) and molecular dynamics (MD) calculations to estimate rate constants for each of the types of reactions discussed in the previous two sections. The general procedure is to start with an ensemble of reactant molecules, and assume that the internal states of the reactant molecules are in equilibrium before reaction occurs. One then uses a modified Monte Carlo routine to choose initial conditions for the molecules, while a molecular dynamics routine is used to calculate the reactive cross section. For example, if one was examining the reaction A + BC → AB + C one would use the Monte Carlo routine to pick initial values of $v_{A \to BC}$ and E_{BC} where $v_{A \to BC}$ and E_{BC} are defined at the beginning of Section 8.4. The calculations are easier if one modifies the Monte Carlo routine to only pick values of $v_{A \to BC}$ and E_{BC} that meet the constraint that the total energy of the molecules is sufficient to traverse the activation barrier for reaction. One then picks several values of the impact parameter b and integrates Equations 8.8, 8.9, and 8.10 to calculate $P_{A \to BC}(b)$. One then plugs into Equation 8.53 to calculate a reactive cross section. The process is repeated for thousands of values of $v_{A \to BC}$ and E_{BC}. One then takes an average using Equation 8.54 to calculate a rate. Such calculations are easy to do provided one knows an accurate potential energy surface for the reaction. Generally the results of the calculations agree quite well with experiment, provided the potential energy surface is accurate and something called tunneling is negligible. Tunneling is described in Section 8.11. As a result, such calculations are quite common.

8.9 TRANSITION-STATE THEORY

The problem with doing calculations like those in the previous section is that they require a nonnegligible amount of computer time. If one knows the potential energy surface for the reaction, one can do the calculations accurately. However, if one does not know the potential energy surface for a reaction precisely, the calculations of rates are not particularly accurate. In those cases it is difficult to justify using a nonnegligible amount of computer time to calculate a rate. As a result, people often use approximations to calculate rate equations for reactions where one only knows the potential energy surface approximately.

One of the simplest approximations was proposed by Eyring [1935] and Evans and Polayni [1935]. Prior to Eyring and Evans and Polayni's work, Pelzer and Wigner [1932] had noted that the transition state (i.e., saddle point) in the potential energy surface in Figure 8.2 plays an important role in reactions. Consider the reaction A + BC → AB + C. A series of trajectories for this reaction are seen in Figure 8.3 and Figure 8.4. Notice that in order for a trajectory to go from reactants to products, the trajectory must cross the saddle point in the potential energy surface. It is possible for a trajectory to cross the saddle point, and then go back again. As a result, not all trajectories that cross the saddle point lead to reaction. However, most trajectories that cross the saddle point lead to reaction, while none of the other trajectories lead to reaction. Thus, the saddle point plays a key role in determining which trajectories lead to reaction.

Eyring [1935] proposed that one can estimate a rate by assuming that the reactive complex will maintain thermal equilibrium with its surroundings when the collision between reactants occurs. In effect this assumes that heat transfer between molecules is so rapid that the internal modes of the reactive complex can thermally equilibrate with their surroundings during the 10^{-12} sec where a reactive collision occurs. If this approximation works, then one can write the rate of a typical reaction A + BC → AB + C as a velocity times an equilibrium constant for formation of reactive complexes. A derivation in Eyring, Walter, and Kimble [1944], Steinfeld, Fransisco and Hase [1989], or Laidler [1987] shows that in the absence of a process called tunneling

$$k_{A \to BC} = \left(\frac{kT}{h_p}\right) K^{\ddagger}_{equ} \tag{8.55}$$

where $k_{A \to BC}$ is the rate constant for the reaction A + BC → AB + C, k is Boltzmann's constant, T is the absolute temperature, h_p is Planck's constant and $K^{\ddagger}_{equ}$ is the equilibrium constant for production of molecules in the transition state. One can estimate $K^{\ddagger}_{equ}$ from statistical mechanics. A derivation in Appendix A shows

$$K^{\ddagger}_{equ} = \frac{q^{\ddagger}}{q_A q_{BC}} e^{-\Delta U^{\ddagger}/kT} \tag{8.56}$$

where q_A and q_{BC} are the molecular partition functions for the reactants A and BC, $q^{\ddagger}$ is the molecular partition function for the transition state, and ΔU is the difference between the zero-point energy of the reactants and the transition state.

Combining Equations 8.55 and 8.56 yields

$$k_{A \to BC} = \left(\frac{kT}{h_p}\right)\left[\frac{q^{\ddagger}}{q_A q_{BC}}\right] e^{-\Delta U^{\ddagger}/kT} \tag{8.57}$$

Equation 8.57 is the key result from transition-state theory. According to this equation, the activation energy for a given reaction is $U^{\ddagger}$, the energy of the transition state. Therefore, the implication of the equation is that activation barriers to reactions are associated with giving molecules enough energy to traverse the transition state. Such a prediction is quite useful in that it allows one to manipulate the activation energy for reactions.

Anything one does to lower the energy of the transition state for a reaction lowers the activation energy for the reaction.

One does not want to overstate this conclusion because in MD simulations, the activation energy for a reaction is generally greater than the energy of the transition state. Physically, one needs to localize energy in a specific bond in the reactive complex to get the reaction to go. Typically there is a barrier to intramolecular energy transfer, so unless the reactive complex has some excess energy, it is unlikely for reaction to occur. The net result is that molecules that cross the saddle point and go on to form products are on the average hotter than one would expect from equilibrium. As a result, activation barriers for reactions are generally somewhat higher than one would expect from Equation 8.57.

The other major prediction of Equation 8.57 is that $k^o_{A \to BC}$, the preexponential for the reaction A + BC → AB + C, is given by

$$k^o_{A \to BC} = \left(\frac{kT}{h_p}\right)\left[\frac{q^{\ddagger}}{q_A q_{BC}}\right] \tag{8.58}$$

In the solved examples for this chapter we will calculate $k^o_{A \to BC}$. It happens that the term in square brackets in Equation 8.58 is usually within an order of magnitude or two of unity in units of molecules and angstroms, while kT/h_p is approximately equal to 10^{13}/sec. As a result, one of the key predictions of transition-state theory is that preexponentials for reactions are always within an order of magnitude or two of 10^{13} in units of molecules, angstroms, and seconds, i.e., first-order reactions have preexponentials in the order of 10^{13}/sec, second-order reactions have preexponentials in the order of 10^{13}Å^3/(molecule-sec), etc. The numerical value of preexponentials changes when one changes units. However, it is easiest to remember that the theory predicts a preexponential of 10^{13} in units of molecules, angstroms, and seconds, independent of the order of the reaction.

Experimentally, preexponentials actually vary from 10^{10} to 10^{16}. Years ago people explained this variation in terms of transition-state theory. In the discussion in the previous paragraph we said that the term in brackets in Equation 8.58 was within an order of magnitude or two of unity. But still the term in brackets varies from 10^{-2} to 10^{+2}. One can show that the term in brackets is given by

$$\left[\frac{q^{\ddagger}}{q_A q_B} \right] = e^{\Delta S^{\ddagger}/k} \tag{8.59}$$

where $\Delta S^{\ddagger}$ is the entropy change in approaching the transition state. Therefore, the term in square brackets represents the effects of the transition-state entropy on the reaction. The term is small when there is a very narrow transition state (i.e., a sharp valley) and large when the transition state is wide. More precisely, the term in brackets in Equation 8.58 is greater than one when the system gains degrees of freedom approaching the transition state. In contrast, the term is less than one when the system loses degrees of freedom when approaching the transition state. Years ago people thought that such variations explained why preexponentials for reaction were different than 10^{13}.

Unfortunately, such ideas have almost no correlation with experiment. For example, Table 8.3 and Figure 8.10 show data for the quantity $h_p k^o_{A \rightarrow BC}/kT$ for a number of reactions. According to Equation 8.58

$$\frac{h_p k^o_{A \rightarrow BC}}{kT} = \frac{q^{\ddagger}}{q_A q_{BC}} \tag{8.60}$$

However, Table 8.3 and Figure 8.10 show that in fact there is no correlation between $h_p k^o_{A \rightarrow BC}/kT$ and $q^{\ddagger}/q_A q_{BC}$. If Equation 8.60 had worked, then all of the data in Figure 8.10

TABLE 8.3   A Comparison of Experimental Values of $(h_p k^o_{A \rightarrow BC})/kT$ to Those Calculated from Transition State Theory

Reaction	Experiment $\dfrac{h_p k^o_{A \rightarrow BC}}{kT}$	Transition-State Theory $\dfrac{q^{\ddagger}}{q_A q_{BC}}$
$NO + O_3 \rightarrow NO_2 + O_2$	0.15	0.07
$NO + O_3 \rightarrow NO_3 + O$	1.05	0.02
$NO_2 + F_2 \rightarrow NO_2F + F$	0.26	0.02
$NO_2 + CO \rightarrow NO + CO_2$	0.32	1.05
$2NO_2 \rightarrow 2NO + O_2$	0.33	0.83
$NO + NO_2Cl \rightarrow NOCl + NO_2$	0.13	0.13
$2NOCl \rightarrow 2NO + Cl_2$	1.66	0.07
$NO + Cl_2 \rightarrow NOCl + Cl$	0.66	0.21
$H + H_2 \rightarrow H_2 + H$	2.34	8.33
$H + CH_4 \rightarrow H_2 + CH_3$	10.47	3.31
$CH_3 + C_2H_6 \rightarrow CH_4 + C_2H_5$	0.093	0.167

The transition-state calculations are from Laidler [1987]. The experimental result are from Westley [1980] when available, and if not, from Laidler [1987]

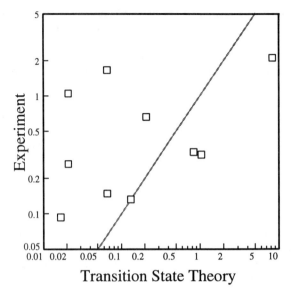

Figure 8.10 A plot of the results in Table 8.3. If the simple formation of transition state theory with constant transmission coefficients had worked, the results should have scattered around the line in the figure.

should have scattered around the line in the figure. However, clearly the data do not scatter around the line. In fact, for the examples shown there is almost a negative correlation between the transition-state theory predictions and the experiment.

To put the data in Table 8.3 in perspective, we chose these data because Laidler [1987] uses them to show that transition-state theory works so much better than collision theory. These are cases where transition-state theory works unusually well. As a result, the fact that the experimental values of $h_p \, k^o_{A \to BC}/kT$ show almost no correlation with those predicted by transition-state theory suggests that the corrections associated with changes in $q^{\ddagger}/q_A \, q_{BC}$ do not have a correlation with variations in rates of reactions.

8.10 THE ROLE OF THE TRANSMISSION COEFFICIENT

The reason that Equation 8.60 failed is that it implicitly assumed that the molecules in the "transition state" are in thermal equilibrium with their surroundings, and all of those molecules react to form products. In reality, one needs to get energy into a specific bond in order for a reaction to occur. Energy transfer is inherently a nonequilibrium process. Molecules with enough energy to cross the activation barrier do not necessarily react. As a result, preexponentials and activation energies for reactions can be different than those one predicts from an equilibrium analysis.

Steinfeld et al. [1989] explained why transition-state theory fails. Consider the reaction $A + BC \rightarrow AB + C$. To some approximation one can view this reaction as one where A and BC collide to form a "reactive complex." Then the "reactive complex" decays either onto reactants or back to products, i.e.,

$$A + BC \rightarrow (ABC)^{\ddagger} \underset{\longrightarrow}{\overset{\longrightarrow}{}} \begin{array}{l} AB + C \\ A + BC \end{array} \qquad (8.61)$$

If energy transfer within the molecule is rapid so that all of the reactive complexes decay to products, then Steinfeld et al. [1989] show that transition state theory will be a good approximation in the absence of a process called tunneling described in Section 8.11. The problem is that many of the reactive complexes decay back to reactants rather than into products. As a result, transition-state theory can make sizable errors.

People usually account for the errors by defining a **transmission coefficient** κ_T, where the transmission coefficient is defined as the fraction of the reactive complexes that decay to products. In this formulation the rate becomes

$$k_{A \to BC} = \left(\frac{kT}{h}\right) \kappa_T \frac{q^{\ddagger}}{q_A q_{BC}} e^{-\Delta U^{\ddagger}/kT} \qquad (8.62)$$

The transmission coefficient is always less than unity, but it is temperature dependent. Generally, molecules with some energy in excess of that needed to cross the activation barrier have a higher probability of reacting to form products than molecules with barely enough energy. One gets more hot molecules with increasing temperature, so κ_T generally increases with increasing temperature. If one takes data over a wide range of temperatures, one finds that κ_T shows complex behavior with temperature. Still, one can fit κ_T to an Arrhenius form over a limited range of temperature, i.e.,

$$\kappa_T = \kappa_T^o \, e^{-E_\kappa/kT} \qquad (8.63)$$

Combining Equations 8.62 and 8.63 shows that the apparent preexponential for reaction $k_{A \to BC}^o$ is given by

$$k_{A \to BC}^o = \left(\frac{kT}{h_p}\right) \kappa_T^o \left[\frac{q^{\ddagger}}{q_A q_{BC}}\right] \qquad (8.64)$$

While κ_T is always less than one, κ_T^o can be greater than one. Generally, κ_T^o varies from 10^{-3} to 10^3. A κ_T^o greater than one corresponds to data above the line in Figure 8.10, so clearly values of κ_T^o greater than one are quite common. Further, Figure 8.10 shows that the variations in κ_T^o are comparable to the variations in the terms in square brackets in Equation 8.64. Physically, one changes κ_T by changing the shape of the potential energy. The highest values of κ_T come when the potential energy surface is shaped like a racetrack with gently banked turns and no sharp corners. There is some subtlety to the arguments, because changes to the potential energy surface generally change both κ_T^o and the term in brackets in Equation 8.64. Often the variations in κ_T^o are larger than the variations in the term in square brackets in Equation 8.64. As a result, one does not necessarily speed up a reaction when one increases the terms in square brackets in Equation 8.64 (i.e., when one gains degrees of freedom when approaching the transition state), nor does one necessarily slow down a reaction when one reduces the terms in square brackets in Equation 8.64.

There are more complex formulations of transition-state theory that allow one to calculate all of the terms in Equation 8.64 explicitly. Miller [1976] has a particularly convenient formulation. However, such methods are just starting to be applied to surface reactions (see, for example, Salfrank and Miller [1993] or Wonchuba and Truhlar [1993]). As a result, these methods will not be reviewed here. The reader is referred to Truhlar, Hase, and Hynes [1983], Pechukas [1981], Marcus [1987], or Miller [1993] for details.

8.11 TUNNELING

There is one special effect that is quite important to reactions on surfaces, which is tunneling. Tunneling is a quantum effect where molecules that do not quite have enough energy to go

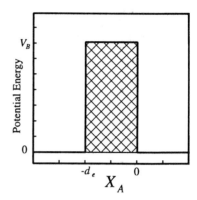

Figure 8.11 A plot of the square-well barrier.

over an activation barrier tunnel through the barrier. The net effect is that molecules that do not have quite enough energy to cross the barrier to reaction can still sometimes react.

It is interesting to solve a simple example to illustrate this effect. Consider scattering of an atom through the square wave potential barrier shown in Figure 8.11. For the purposes of derivation, it will be assumed that the barrier is given by

$$V(X_A) = \begin{cases} 0 & \text{for } X_A > 0 \\ V_B & \text{for } 0 > X_A > -d_e \\ 0 & \text{for } X_A < -d_e \end{cases} \tag{8.65}$$

Consider the collision of a beam of molecules onto the barrier. Assume that molecules start out at $X_A = \infty$ moving toward the barrier. The molecules will hit the barrier and then either reflect from the barrier or be transmitted through it. It is easy to show that the wavefunction for the incident molecules is

$$\psi_i = e^{-iK_iX_A} \qquad \text{for } X_A > 0 \tag{8.66}$$

where $i = \sqrt{-1}$ and K_i is related to the incident energy E_i and mass m_i of the molecules by

$$K_i = \sqrt{\frac{2m_iE_i}{h^2}} \tag{8.67}$$

For the moment, assume that the barrier is infinitely thick (i.e., d_e is infinite). In that case, a derivation in the supplement or material shows that the solution of the Schrödinger equation is

$$\psi_T = e^{-iK_iX_A} + \left(\frac{K_i - p_i}{K_i + p_i}\right) e^{-iK_iX_A} \qquad \text{for } X_A \geq 0 \tag{8.68}$$

$$\psi_T(X_A) = \left(\frac{2K_i}{K_i + p_i}\right) e^{-ip_iX_A} \qquad \text{for } X_A \leq 0 \tag{8.69}$$

where

$$p_i = \sqrt{\frac{2m_i(E_i - V_B)}{h^2}} \tag{8.70}$$

When E_i is greater than V_B, the solution looks quite ordinary. Incident molecules hit the potential barrier. Some of the incident molecules scatter, producing scattered wavefunction.

$$\psi_r = \left(\frac{\kappa_i - p_i}{\kappa_i + p_i} \right) e^{+i\kappa_i x_A} \tag{8.71}$$

The remainder of the incident molecules traverse the barrier. The wavefunction for the molecules that traverse the barrier is given in Equation 8.69. The surprising thing, however, is that even when E_i is less than V_B, the wavefunction is nonzero for $X_A < 0$ (i.e., within the barrier). As a result, quantum mechanically, molecules penetrate into the barrier, even though classically no molecules should penetrate into the barrier (i.e., to $X_A < 0$). Figure 8.12 shows a plot of the wavefunctions when $E_i < V_B$. Notice that the wavefunction shows an exponential decay into the barrier. Thus, the probability that a molecule penetrates through the barrier decays as the barrier gets thicker. More precisely, $P(d_e)$, the probability of finding a molecule some distance $-d_e$ in the barrier is

$$P(-d_e) = \psi^*(-d_e)\, \psi(-d_e) \tag{8.72}$$

Substituting Equation 8.69 into 8.72 yields

$$P(-d_e) = \left(\frac{2\kappa_i}{\kappa_i + p_i} \right)^2 \exp\left(-2d_e \sqrt{\frac{2m_i(V_B - E_i)}{h^2}} \right) \tag{8.73}$$

If the step in the potential ends at d_e, then there will be some probability of a molecule moving through the barrier. If one assumes a barrier of the form in Equation 8.65, there is an artifact associated with the quantum reflection at the sharp drop in potential at $X_A = -d_e$. However, the key result is that the tunneling rate can be approximated by Equation 8.73. Tunneling rates are highest if the barriers are thin, (i.e., d_e is small) and the species that

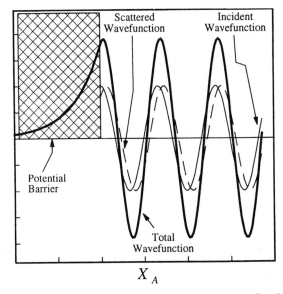

Figure 8.12 The real part of the incident (ψ_i) scattered (ψ_r) and total wavefunction (ψ_T) for the square-well barrier.

tunnels has a low mass. Generally, tunneling is only important in hydrogen transfer reactions, i.e.,

$$D + H_2 \rightarrow HD + H$$

$$O + CH_4 \rightarrow OH + CH_3$$

Tunneling rates are much higher with hydrogen than deuterium, so one can easily detect them by looking for a large isotope effect. Generally, one can only detect tunneling at low temperatures. At higher temperatures, the molecules traversing the barrier swamp those that tunnel.

8.12 UNIMOLECULAR REACTIONS

Another special topic that has important applications in surface reactions is the theory of unimolecular reactions. A unimolecular reaction is a reaction that starts with a single molecule AB, which either isomerizes to form a new species C or splits apart to form two or more products. Examples of unimolecular reactions include the isomerization of cyclopropane

$$\underset{H_2C-CH_2}{\overset{CH_2}{\diagup \diagdown}} \Rightarrow CH_2{=}CH{-}CH_3$$

and the scission of H_2 into atoms

$$H_2 \rightarrow 2H$$

Most desorption reactions can be considered to be a unimolecular decomposition of a surface complex. As a result, the theory of unimolecular reactions should be quite important to reactions on surfaces.

Most kinetics texts provide a treatment of the rates of unimolecular reactions in the gas phase. The gas phase process is usually fairly simple. Generally the reactants collide with a collision partner to form an excited intermediate. The excited intermediate can either isomerize to produce products, or lose energy via collisions with other collision partners to go back to the ground state. The key point, however, is that the excited complex lasts for perhaps 10^{-8} seconds at 1 atm pressure. By comparison, the transition state for a bimolecular reaction (e.g., $A + BC \rightarrow AB + C$) typically only lasts for 10^{-11} seconds or less. As a result, the excited complex formed during a unimolecular reaction has many more chances to react than the transition state for a bimolecular reaction. The net effect is that in the high pressure limit (i.e., many collisions) the rates of unimolecular reactions are larger than one would expect from transition-state theory. Typically, preexponentials for unimolecular gas phase reactions vary between 10^{14} and 10^{17}/sec in the high-pressure lmit. By comparison transition-state theory predicts preexponentials of about 10^{13}/sec.

8.13 APPLICATION TO SURFACE REACTIONS

So far we have only been discussing gas phase reactions in this chapter. However, all of the ideas can (and have) been applied to surface reactions. One of the key predictions of the models is that preexponentials for bimolecular surface reactions should be within an order of magnitude or two of 10^{13} Å^2/molecule sec. Table 8.4 shows a selection of the data. Notice that the preexponentials are within two orders of magnitude of 10^{13}.

Table 8.5 shows a selection of the data for unimolecular surface reactions. Unimolecular

TABLE 8.4 A Selection of Preexponentials for Bimolecular Surface Reactions

Reaction	Surface	Preexponential (sec^{-1})
$2H_{ad} \rightarrow H_{2(g)}$	Pd(111)	10^{12}
$2H_{ad} \rightarrow H_{2(g)}$	Ni(100)	$10^{13.5}$
$2H_{ad} \rightarrow H_{2(g)}$	Ru(001)	$10^{14.5}$
$2H_{ad} \rightarrow H_{2(g)}$	Pd(110)	$10^{13.5}$
$2H_{ad} \rightarrow H_{2(g)}$	Mo(110)	10^{13}
$2H_{ad} \rightarrow H_{2(g)}$	Pt(111)	10^{12}
$2N_{ad} \rightarrow N_{2(g)}$	Ni(100)	10^{14}

Data from the compendiums of Bowker and King [1983], Seebauer, Kang, and Schmidt [1987], Zhandov [1992], and Cristman [1988].

surface reactions often show preexponentials greater than 10^{13}/s. The preexponentials are often similar to those for unimolecular reactions in the gas phase. At present, there is some question in the literature why preexponentials for unimolecular surface reactions are so large. In the gas phase, preexponentials are unusually large because one produces an excited intermediate, which is stable for perhaps 10^{-8} see due to energy transfer limitations. The extra long lifetime in the excited state gives the excited intermediate extra chances to react. As a result, the preexponential for the reaction is unusually large. Energy transfer rates are higher on a surface than in the gas phase because adsorbed molecules are always in close proximity to the surface. Thus, excited complexes are less stable on a surface than in the gas phase. Still, in Chapter 5 we found that one type of unimolecular surface reaction, desorption, often goes via a mobile precursor. A mobile precursor has a long lifetime compared to a collision, so in some ways it is analogous to the excited molecules seen during unimolecular reactions in the gas phase. It is unclear whether other metastable species form during unimolecular surface reactions.

In the literature, people have also discussed the influence of the term in square brackets in Equation 8.64 on rates of unimolecular surface reactions. It is thought that generally rates should increase when the transition-state entropy (i.e., the term in square brackets in Equation 8.64) increases. The effect is most important if a mobile species reacts to form an immobile one or vice versa. In that case, the transition-state entropy is unusually large and may be able to swamp the variations in the transmission coefficient.

I used "may" in the last sentence, because there is some controversy about whether the effects really occur. Recently, there have been many calculations of rates on embedded

TABLE 8.5 A Selection of Preexponentials for Unimolecular Surface Reactions

Reaction	Surface	Preexponential (sec^{-1})
$CO_{ad} \rightarrow CO_g$	Cu(001)	10^{14}
$CO_{ad} \rightarrow CO_g$	Ru(001)	$10^{19.5}$
$CO_{ad} \rightarrow CO_g$	Pt(111)	10^{14}
$Au_{ad} \rightarrow Au_g$	W(110)	10^{11}
$Cu_{ad} \rightarrow Cu_g$	W(110)	$10^{14.5}$
$Cl_{ad} \rightarrow Cl_g$	Ag(110)	10^{20}
$NO_{ad} \rightarrow NO_g$	Pt(111)	10^{16}
$CO_{ad} \rightarrow CO_g$	Ni(111)	10^{15}

Data from the compendiums of Bowker and King [1983], Seebauer, Kang and Schmidt [1987], and Zhandov [1992].

clusters. In the calculations, one does not find that variations in the mobility of the transition state make that much difference to the rate because variations in energy transfer limitations dominate over variations in the transition-state mobility. The effect is very similar to that in the gas phase. Of course, a cluster approximation in effect assumes that the surface behaves like the gas phase. Experimentally, people say that there appears to be a correlation between the mobility of the reactants and the preexponential for the reaction, although data are also not very convincing. At this point it is unclear whether the transition-state entropy makes much difference to rates, but people discuss it nevertheless.

The other key prediction from this chapter is that anything one does to stabilize the transition state for a reaction will enhance the rate of the reaction. While this prediction may not be absolutely correct in all cases, generally people in the surface science literature just accept it and use it as a basis for catalyst design. In the remainder of this book we develop this idea.

8.14 SOLVED EXAMPLES

8.A The Evaluation of $\dfrac{q^{\ddagger}}{q_A q_B}$

Earlier in this chapter, we said that typically $\dfrac{q^{\ddagger}}{q_A q_B}$ is within an order of magnitude or two of unity. The object of this problem is to evaluate $\dfrac{q^{\ddagger}}{q_A q_B}$ using the methods from Example 4.C.

Consider a second-order reaction A + BC → AB + C

$$q_A = q_T^3 \tag{8.74}$$

$$q_{BC} = q_T^3 q_v q_r^2 \tag{8.75}$$

If we assume a linear transition state

$$q_A^{\ddagger} = q_T^3 q_v^4 q_r^2 \tag{8.76}$$

$$\frac{q^{\ddagger}}{q_A q_{BC}} = \frac{q_v^3}{q_T^3} \tag{8.77}$$

Using the numbers in Table 4.14

$$\frac{q^{\ddagger}}{q_A q_{BC}} = \left(\frac{4}{10/\text{Å}}\right)^3 = 0.06 \ \text{Å}^3 \tag{8.78}$$

If instead we assume a nonlinear transition state,

$$\frac{q^{\ddagger}}{q_A q_{BC}} = \frac{q_v^2 q_r}{q_r^2} \frac{4^2 50}{(10/\text{Å})^3} = 0.8 \ \text{Å}^3 \tag{8.79}$$

8.15 SUPPLEMENTAL MATERIAL

8.15.1 Tunneling

In Section 8.11, we noted that tunneling is an important process for hydrogen transfer reactions. The purpose of this supplement is to derive an expression for tunneling. First consider the square-wave barrier in Equation 8.65.

It is easy to show that the general solutions of the Schrödinger equation is

$$\psi_i = \begin{cases} e^{-iK_iX_A} + A_Te^{+iK_iX_A} & \text{for } X_A > 0 \\ B_Te^{-ip_iX_A} + C_Te^{+ip_iX_A} & \text{for } 0 > X_A > -d_e \\ D_Te^{-iK_iX_A} + E_Te^{+iK_iX_A} & \text{for } X_A \le -d_e \end{cases} \tag{8.80}$$

where A_T, B_T, C_T, D_T, and E_T are constants. The boundary condition that particles come from the $X_A = \infty$ but not $X_A -\infty$ implies

$$E_T = 0 \tag{8.81}$$

The other constants come from the condition that the wavefunction and its derivatives must be continuous at $X_A = 0$ and $X_A = -d_e$.

At $X_A = 0$

$$\Psi_i = e^{-iK_iX_A} + A_Te^{+iK_iX_A} = B_Te^{-ip_iX_A} + C_Te^{+ip_iX_A} \tag{8.82}$$

$$\frac{d\Psi_i}{dX_A} = -iK_ie^{-iK_iX_A} + iK_iA_Te^{iK_iX_A}$$

$$= -ip_iB_Te^{-ip_iX_A} + ip_iC_Te^{+ip_iX_A} \tag{8.83}$$

At $X_A = -d_e$

$$\Psi = B_Te^{+ip_id_e} + C_Te^{-ip_id_e} = D_Te^{+iK_id_e} \tag{8.84}$$

$$\frac{\partial \Psi_i}{\partial X_i} = -ip_iB_Te^{+ip_id_e} + ip_iC_Te^{-ip_id_e} = iK_iD_Te^{iK_id_e} \tag{8.85}$$

Solving Equations 8.82, 8.83, 8.84, and 8.85 simultaneously yields

$$A_T = \frac{(K_i + p_i)(K_i - p_i)(1 - e^{2ip_id_e})}{(K_i + p_i)^2 - e^{2ip_id_e}(K_i - p_i)^2} \tag{8.86}$$

$$B_T = \frac{2K_i(K_i + p_i)}{(K_i + p_i)^2 - e^{2ip_id_e}(K_i - p_i)^2} \tag{8.87}$$

$$C_T = \frac{2K_i(p_i - K_i)e^{2ip_id_e}}{(K_i + p_i)^2 - e^{2ip_id_e}(K_i - p_i)^2} \tag{8.88}$$

$$D_T = \frac{4K_ip_i e^{i(p_i - K_i)d_e}}{(K_i + p_i)^2 - e^{2ip_id_e}(K_i - p_i)^2} \tag{8.89}$$

People do not usually use a square-wave potential to model tunneling. Instead they assume a Ekardt potential.

$$V(X_A) = \frac{V_B}{\cosh(\omega_B X_A)} \tag{8.90}$$

The solution of the Schrödinger equation for this potential is given in Landau and Lifshitz [1965, pp. 72 and 79].

PROBLEMS

8.1 Define the following terms in three sentences or less

(1) Activation barrier   (6) Tunneling
(2) Collision cross section   (7) Collision partner
(3) Reactive cross section   (8) Unimolecular reaction
(4) Impact parameter   (9) Activated complex
(5) Transition state   (10) Transmission coefficient

8.2 Consider two reactions:

$$Na + Cl \rightarrow NaCl$$

$$Na + LiCl \rightarrow NaCl + Li$$

(a) Which is usually more reactive, an individual chlorine atom or a lithium chloride molecule?

(b) In fact, if you run both reactions in a vacuum system, with no collision partners present, which reaction will show a higher rate?

(c) Explain your answer in part (b) physically.

8.3 Why is there an energy transfer barrier to reaction? Why does it arise? Where is it important? How does the energy transfer barrier affect the reaction $H + O_2 \rightarrow OH + O$? How about the reaction $Na + Cl \rightarrow NaCl$?

8.4 Why is there an angular-momentum barrier to reaction? Why does it arise? Where is it important? How does the angular-momentum barrier affect the reaction $H + O_2 \rightarrow OH + O$? How about the reaction $Na + Cl \rightarrow NaCl$?

8.5 In 1934, Rice and Herzfeld proposed that the pyrolysis of ethane

$$C_2H_6 \rightarrow C_2H_4 + 2CH_4$$

$$C_2H_6 \rightarrow C_2H_4 + H_2$$

follows the following mechanism:

$$C_2H_6 \rightarrow 2CH_3 \tag{1}$$

$$CH_3 + C_2H_6 \rightarrow CH_4 + C_2H_5 \tag{2}$$

$$C_2H_5 \rightarrow C_2H_4 + H \tag{3}$$

$$H + C_2H_6 \rightarrow C_2H_5 + H_2 \tag{4}$$

$$H + C_2H_5 \rightarrow C_2H_6 \tag{5}$$

(a) Which steps need collision partners? Be sure to explain why the steps need collision partners. Also if you decide that a step does not require a collision partner, explain why it does not need a collision partner.

(b) How do the energy transfer barriers to reaction affect each of the reactions.

(c) How do angular-momentum barriers affect each of the reactions above.

(d) Use transition-state theory to estimate the preexponentials for each of the reactions.

(e) Use the steady-state approximation to derive an equation for the overall rate of the reaction.

(f) How would your answer in part (e) change if no collision partners were present?

(g) How does the character of the collision partners affect the overall reaction? (*Hint:* Look at Table 8.1 and Table 8.2.)

8.6 Repeat Problem 8.5 for the reaction $H_2 + Br_2 \rightarrow 2HBr$. Assume that the reaction follows the following mechanism:

$$Br_2 \rightarrow 2\ Br \tag{1}$$

$$Br + H_2 \rightarrow HBr + H \tag{2}$$

$$H + Br_2 \rightarrow HBr + Br \tag{3}$$

$$2Br \rightarrow Br_2 \tag{4}$$

8.7 Use transition-state theory and the methods in Example 8.A to estimate the preexponentials for the following reactions at 300 K. Assume that each of the reactions is elementary

(a) $H + CH_4 \rightarrow CH_3 + H_2$

(b) $H + H_2 \rightarrow H_2 + H$

(c) $CH_3 + CD_3H \rightarrow CH_4 + CD_3$

(d) $CH_3 + CN \rightarrow CH_2 + HCN$

(e) $H_2 \rightarrow 2H$

(f) $O_2 + 2NO \rightarrow 2NO_2$

(g) $NO + O \rightarrow NO_2$

Is there reason to suspect that some of the reactions would not be elementary? Be sure to justify your answer.

8.8 Discuss the application of transition-state theory to the prediction of reaction rates. What are the key predictions of the model? According to transition-state theory, how do changes in the properties of the transition state affect rates of reactions? How well do the predictions of transition state work? What are the advantages of transition-state theory compared to MD? When would you have to use MC/MD rather than transition-state theory?

8.9 Explain, in your own words

(a) Why preexponentials for unimolecular reactions are generally larger than those predicted by transition state theory.

(b) Why unimolecular reactions need collision partners.

8.10 P. G. Schultz and R. A. Lerner, *Accts Chem. Res.* **26**, 391 [1993], L. C. Hsein, S. Yonkovich, L. Kochersperger, and P. G. Schultz, *Science* **260**, 337 [1993], R. A. Lerner, S. J. Benovich, and P. G. Schultz, *Science* **252**, 659 [1991] examined the production of "catalytic antibodies." The idea of a catalytic antibody is to build a molecule (antibody) that binds strongly to the transition state of a reaction, and less strongly to the reactants and products.

(a) Do you think that the idea will work?

(b) What do Schultz and Lerner find?

(c) How could the antibody bind more strongly to the transition state, and less strongly to the reactants and products?

(d) What would happen if the antibody bound the reactants strongly?

(e) What would happen if the antibody bound the products strongly?

8.11   In our discussion in Section 8.10, we said that the term in brackets in Equation 8.62 changed as the transition state got wider or narrower.

    (a) Specifically, how does the width of the transition state affect the term in brackets in Equation 8.62? (*Hint:* How does making the transition state wider or narrower affect the size of the rotational and vibrational partition functions for the transition state?)

    (b) Explain, in your own words, how these effects change rates of reactions.

8.12   Describe the role of tunneling on rates of chemical reactions. How do you detect tunneling? Why does it arise? Verify that tunneling actually occurs by showing that Equations 8.86 to 8.89 are correct.

8.13   Laidler [1987] (Section 7.7.4) discusses how transition-state theory applies to reactions on surfaces. He looks at the role of adsorbate mobility on the rate of reaction.

    (a) Where does adsorbate mobility come into Equation 8.62?

    (b) According to Equation 8.62, how should a change in the adsorbate mobility affect the rate of a reaction? Consider the case where a mobile species is reacting to form an immobile transition state.

    (c) How do you think changes in the adsorbate mobility will affect the transmission coefficient?

    (d) Look carefully at the data in Table 8.4. Is there any evidence for mobility effects?

8.14   In Section 8.20 we noted that variations in the transmission coefficient are often as large as variations in the terms in square brackets in Equation 8.64.

    (a) How does one change κ_T^o?

    (b) What is the physical picture of the terms in square brackets in Equation 8.64? How do changes in the potential energy surface affect the term in square brackets in Equation 8.64?

    (c) Use the data in Table 8.3 to compute κ_T^o. Make a plot of κ_T^o vs. the term in square brackets in Equation 8.64. Do you see any interesting trends?

    (d) How does the order of magnitude of κ_T^o compare to that of the term in square brackets in Equation 8.64?

8.15   The objective of this problem is to use the program REACT_MD available from Professor Masel to determine which trajectories lead to reaction.

    (a) Start with the case of linear $A + BC \rightarrow AB + C$ plot trajectories on a potential energy surface. Using the default values of the parameters look carefully at the simulations. What is different about the trajectories that lead to reaction and those that do not? Look particularly at the concerted motions of A, B, and C. Is C moving in or out when a reactive collision occurs?

    (b) Now vary the incident energy of the molecules. The transition-state energy is 4.5 in dimensionless units. Try an energy of 4.7. Vary the fraction of the initial energy that goes into translation and vibration. Can you find any trajectories where reaction occurs? Try raising the incident energy until you can find trajectories that lead to reaction.

    (c) Interpret your result in part (b) in terms of the barriers to energy flow within the ABC complex.

    (d) Given your results in part (b), what do you think about the transition-state theory prediction that the activation energy for the reaction is equal to the energy needed to bring the molecules to the transition state?

    (e) Now speculate about how changes in the shape of the potential energy surface will change the rate of reaction. How would you shape the potential energy surface to direct molecules from reactants to products? Draw an analogy to designing a high-way or a racetrack. In a racetrack you want to bank hairpin turns. Use that idea to design a potential energy surface for a reaction.

(f) Are the effects described in part (e) considered in transition state theory? If so, where? What term in Equation 8.62 accounts for this difference?

8.16 Earlier in this chapter, we discussed reactions as trajectories on potential energy surfaces. If you actually look at the trajectories, they do not quite look how you might expect. The difficulty is that the classical equations of motion, Equations 8.8, 8.9, and 8.10, do not separate in R_{AB} and R_{BC} coordinates. The objective of this example is to explore this effect.

Consider the scattering of an atom A from a molecule BC. Assume that A does not interact with BC.

(a) If A starts out at some initial point $(-3 \text{ Å}, 0 \text{ Å})$, with a velocity of 1000 cm/sec toward BC, use the classic equations of motion to show that the position of a follows

$$X_A(t) = 10 \text{ Å/psec t} - 3 \text{ Å}$$

(b) B and C are in a vibrational well. Show that if B and C have identical masses, an equilibrium distance of 2 Å, and a vibrational period of 0.4 Å, one can express the position of B and C by

$$X_B(t) = -0.2 \text{ Å} \sin(\omega t) - 1 \text{ Å}$$

$$X_C(t) = +0.2 \text{ Å} \sin(\omega t) + 1 \text{ Å}$$

(c) Now calculate R_{AB} and R_{BC} as a function of time assuming that $\omega = 10^{12}/\text{sec}$ and that A, B, and C are colinear. Use a spreadsheet to plot your result as a function of time for times between 0 and 2×10^{-10} sec. Why are there oscillations in R_{AB}?

(d) How do the results change for a noncolinear case. (i.e., assume that the center of the B–C bond is at $(0 \text{ Å}, 2 \text{ Å})$).

(e) Replot your results in parts (c) and (d) on a potential energy surface, i.e., R_{AB} vs. R_{BC}. What does your trajectory look like? Does A ever appear to be moving away from BC? Why does that occur?

(f) In what way do your results in part (e) help you interpret trajectories on potential energy surfaces?

8.17 Earlier in this book we discussed how the position of the transition state affects the rate of a chemical reaction.

(a) Use the program REACT_MD to explore the difference in the reactivity of molecules with early and late transition states and report on your findings.

(b) How would transition-state theory need to be modified to account for the presence of early and late transition state?

8.18 A. B. Anton, *J. Phys. Chem.* **97**, 1942 [1993] describes how changes in the adsorbate arrangement can lead to unusual preexponentials for desorption.

(a) Carefully read the paper and report on the findings.

(b) How will changes in the adsorbate configuration affect the partition function for the transition state?

(c) How could you account for these effects in the context of transition-state theory?

8.19 Mills and Jónsson, *Phys. Rev. Lett.* **72,** 1124 [1994] examined the role of tunneling on hydrogen dissociation on copper. Their calculations indicate that the activation barrier for hydrogen adsorption should drop by 30% at temperatures below 600 K.

(a) How could you look for that change experimentally?

(b) How else could you show that tunneling was important?

(c) Why is the activation energy temperature dependent when tunneling occurs?

More Advanced Problems

8.20 Consider the reaction between a sodium atom and chlorine atom. Assume that the atoms interact with a potential

$$V(R_{AB}) = \frac{1}{R_{AB}^{10}} - \frac{1}{R_{AB}} \tag{8.91}$$

Derive an equation for the trajectory of the atoms (i.e., R_{AB} and θ_{AB} vs. time) and show that no reaction occurs.

8.21 The objective of this problem is to show that energy is conserved during the reaction $A + B \rightarrow AB$, i.e.,

$$\frac{dE_{AB}}{dt} = 0 \tag{8.92}$$

(a) Plug Equation 8.40 into Equation 8.92 to obtain an expression for $\left(\dfrac{dE_{AB}}{dt}\right)$.

(b) Substitute Equations 8.35 and 8.36 into your expression to show that $\dfrac{dE_{AB}}{dt} = 0$.

8.22 Darling and Holloway, *J. Electron Spectroscopy Relat. Phenom.* **64,** 571 [1993] examined the role of the curvature of the potential energy surface on the rate of hydrogen dissociation on a metal surface.

(a) According to the discussion in Section 8.9, how should the curvature of the transition state affect the activation barrier for the reaction.

(b) Carefully examine Darling and Holloway's findings. How do their calculations agree with your expectations?

(c) Why does the dissociation of hydrogen on Pt(111) have a late transition state? (*Hint:* Compare the hydrogen–hydrogen bond length to the spacing between adjacent sites on the Pt(111) surface.)

8.23 The objective of this question is to guess whether reactions on surfaces will have early, middle, or late transition states. Recall that reactions with an early transition state have only small bond distortions in the forward direction and larger bond distortions in the backward direction. Consider the following reactions. Which reactions would you expect to have early, middle, or late transition states?

(a) A diffusion process where a hydrogen atom jumps from a threefold hollow on Pt(111) to an adjacent threefold hollow on the surface.

(b) Transfer of a hydrogen from an adsorbed methoxy (CH_3O-) to the Cu(111) surface yielding adsorbed hydrogen and formaldehyde. Assume that the methoxy stands with the C–O axis perpendicular to the surface and the methoxy binds with the C–O axis parallel to the surface.

(c) Transfer of a hydrogen from an adsorbed methoxy (CH_3O-) to an oxygen atom adsorbed on the Cu(111) surface, yielding an adsorbed OH and an adsorbed formaldehyde. (*Hint:* Work out the distances in all cases.)

8.24 J. Ellis and J. P. Toennies, *Phys. Rev. Letters* **70**, 2118 [1993] measured the rate of sodium diffusion on Cu(100). They argued that energy transfer during the reaction determine the preexponential for the reaction. Carefully review Ellis and Tonnies's paper and report on the findings.

8.25 A. C. Luntz and J. Harris, *J. Chem. Phys.* **96**, 7054 [1992] examine the role of tunneling on alkane dissociation on transition metals. They suggest that tunneling is quite important.

(a) Carefully review this paper and report on the findings.

(b) How do the theoretical results compare to the experimental findings of M. C. McMaster and R. J. Madix, *Surface Sci.* **294**, 420 [1993]?

8.26 S. E. Wonchoba and D. G. Truhlar, *J. Chem. Phys.* **99**, 9637 [1993] examined the role of surface vibrations in promoting chemical reactions.

(a) Carefully read their paper and report on the findings

(b) How could the surface act as a heat sink to speed chemical reactions?

(c) How would surface vibrations affect the potential energy surface for (1) dissociation of hydrogen on Cu(100) and (2) diffusion of hydrogen on Cu(100)?

(d) If the vibrational modes of the surface changed according to the adsorbate positions, would the rate change?

8.27 K. D. Dobbs and D. J. Doren, *J. Chem. Phys.* **99**, 100041 [1993] examined the role of energy coupling on CO diffusion on Ni(111).

(a) Carefully read the paper and report on the findings.

(b) Dobbs and Doren report that the coupling of the adsorbate bending mode with the translation mode played an important role in the dynamics of the reaction. Why would bending of the CO–surface bond enhance surface diffusion?

(c) Could you account for the effects using transition-state theory?

8.28 Kaukonen, Landman, and Cleveland, *J. Chem. Phys.* **95**, 4997 [1991] used molecular dynamics simulations to show that an argon cluster could catalyze the reaction Na + Cl $\rightarrow$ NaCl. Read the paper carefully, then answer the following questions: (*Note:* This problem is from the final for our graduate chemical kinetics course.)

In the second paragraph of the first page of the article, the authors discuss the idea that a cluster could act like a heat bath.

(a) Why is a heat bath needed for the reaction $A^+ + B^- \rightarrow AB$?

(b) Why could the argon in the cluster act as a heat bath?

(c) How does the argon in the cluster compare to a collision partner as a way of dissipating the heat of reaction?

The athors actually do calculations for the reaction of a Cl^- ion with an $Na_xCl_{x-1}^+$ cluster.

(d) Consider a linear collision of a bare cluster with no argon. What would the potential contour look like for this reaction? Would there be a barrier to the reaction? Be sure to justify your answer.

(e) How would the potential barrier change as you went to a nonlinear collision? Sketch the potential contour for a nonlinear collision as a function of the impact parameter.

(f)   How would the presence of argon affect your results in part (d)? Does the argon have a large or a small effect on the activation barrier for the reaction?

Now go to the top of page 4009 in Kaukonen et al., where the authors describe their results.

(g)   The authors suggest a way to choose impact parameters. Why did they choose an impact parameter that is proportional to the square root of s? (*Hint:* Look at the expression for the reaction cross section as a function of the reaction probability for each impact parameter.) Show that if the authors choose impact parameters in the way they have, the average over all of their trajectories (i.e., the ensemble average) is equal to the correct average over the impact parameters.

Now focus on Figure 1 of Kaukonen et al., where the authors plot the total energy of the cluster as a function of the vibrational temperature of the cluster.

(h)   How should the vibrational energy vary with temperature? Use your result to show that the authors have calculated the vibrational energy correctly.

(i)   Notice that there are some nonlinearities in Figure 1 in the two-phase region. Do you think that those nonlinearities are real, or are they associated with computational error? Be sure to justify your answer.

The authors discuss the cooling of their clusters on the fourth page of their article.

(j)   Qualitatively, why does the presence of a large argon cluster rather than a single argon atom lead to additional cooling of the cluster?

(k)   How would you expect the number of argon atoms to affect the branching ratio between production of the cubic (i.e., lowest energy) structure vs. production of the ring (i.e., higher energy) structure.

Now focus on Table Ia.

(l)   Why does the reaction probability decrease as the collisional energy increases?

(m)   Why does P_C/P_R (i.e., the probability of producing molecules in the low energy cube structure) decrease as the collisional energy increases?

Now skip to page 5003 of Kaukonen et al.

(n)   The authors point out that an increase in the vibrational energy of the cluster at constant total energy leads to an increase in the production of ring compounds. What does that tell you about the transition state for forming ring compounds, i.e., is it an early or late transition state?

Now skip to page 5007 where the authors are discussing the reaction of chlorine with a giant cluster. They discuss the reaction of a chlorine atom with a bare cluster and claim that the reaction probability is 84%.

(o)   Is that possible? Be sure to justify your answer.

(p)   The calculations show that one does form a cluster that is stable for the time scale of the calculations (i.e., 25 picoseconds). Where did the energy of the collision go?

(q)   What would happen if one ran a longer simulation? Would the final cluster stay together or would it eventually fall apart?

REFERENCES

Arrhenius, S. A., *Z. Phy. Cheme.* **4**, 226 (1889); **28**, 317 (1899).

Bamford, C., ed. *Comprehensive Chemical Kinetics* (many volumes; 1969–1993).

Berthelot, M., *Poggendorf's Ann.* **66**, 110 (1862). (Now *Annalen Der Phys.*).

Bertholl, A., *J. de Chem. Phys.* **10**, 573 (1912).

Christman, K., *Surf. Sci. Rep.* **9**, 1 (1988).

Connors, C. A., *Chemical Kinetics*, VCH (1990).

Evans, M. G., and M. Polayni, *Trans. Faraday Soc.* **31**, 875 (1935).

Eyring, H., *J. Chem. Phys.* **3**, 107 (1935).

Eyring, H., and S. M. Lin, *Chemical Kinetics*, Wiley, NY (1980).

Eyring, H., J. L. Walter, and G. E. Kimball, *Quantum Chemistry*, p. 307, Wiley, NY (1944).

Hernandez, R., and W. H. Miller, *Chem. Phys. Lett.* **214**, 129 (1993).

Laidler, K., *Chemical Kinetics*, Harper & Row, NY (1987).

Landau, L. D., and E. M. Lifshitz, *Quantum Mechanics*, Pergamon Press, NY (1965).

Levin, R. D., and R. B. Bernstein *Molecular Reaction Dynamics and Chemical Reactivity*, Oxford, NY (1987).

Lewis, W. C. M., *J. Chem. Soc.* **109**, 796 (1916); **113**, 471 (1918).

Marcelin, R., *C.R.* **158**, 116, (1914), **158**, 407 (1914); R. Marcelin, *Comptes Render*, **161**, 1052 (1920).

Marcelin, R., *J. de Chem. Phys.* **12**, 451 (1914).

Miller, W. H., *Acc. Chem. Res.* **9**, 306 (1976); **26**, 174 (1993).

Moore, J. W., and R. J. Pearson, *Kinetics and Mechanism*, Wiley, NY (1986).

Pechukas, P., *Ann. Rev. Phys. Chem.* **32**, 159 (1981).

Pechukas, P., *Ber. Bundesgeb. Phys. Chem.* **86**, 372 (1982).

Pelzer, H., and E. Wigner, *Z. Physik Chem.* **B15**, 445 (1932).

Salfrank, P., and W. H. Miller, *J. Chem. Phys.* **98**, 9040 (1993).

Steinfeld, J. I., J. S. Fransisco, and W. L. Hase, *Chemical Kinetics and Dynamics*, Prentice-Hall, Englewood Cliffs (1989).

Trautz, M., *Z. Anorg. Chem.* **96**, 1 (1916); **102**, 81 (1918).

van't Hoff, J. H., *Etudes de Dynamique Chemique*, Muller, Amsterdam (1884).

Wilhelmy, R. L., Poggendorf's Ann. **81**, 413 (1850). (Now *Annalen Der Phys.*)

Wonchuba, S. E., and D. G. Truhlar, *J. Chem. Phys.* **99**, 9637 (1993).

Wynne-Jones, W. K., and H. Eyring, *J. Chem. Phys.* **3**, 492 (1935).

9 Models of Potential Energy Surfaces: Reactions as Curve Crossings and Electron Transfer Processes

PRÉCIS

In Chapter 8, we provided an introduction to reaction-rate theory. We showed that we can understand a reaction by examining the motion of a particle over an activation barrier. If one knows the activation barrier, one can use transition-state theory to estimate the rate of the reaction. In this chapter we consider some of the properties of activation energies. In particular, we consider how the activation energy of a reaction changes as we make systematic changes to the reactants and their environment. Much of the work will build on the work of Evans and Polayni [1936], Marcus [1955], Fukui [1975], and Woodward and Hoffmann [1970], who showed that one can model activation barriers for reactions by considering the bonds that break and form during the reaction and the exchange of electrons that occurs as the reaction proceeds. In this chapter we provide an overview of the material. One should refer to Klopman [1974], Fukui [1975], or Pearson [1976] for further information.

9.1 INTRODUCTION

An understanding of potential energy surfaces is critical to an understanding of reaction rates. In the last few chapters we have assumed that the potential energy surface is known. In fact, the potential energy surface for a surface reaction is rarely known accurately. In principle, potential energy surfaces can be calculated from first principles by solving the Schrödinger equation for the system at every possible configuration of the reactants. However, in practice, the number of accurate potential energy surfaces that have been determined from first principles is very limited. Generally, only reactions of relatively simple molecules have been considered. There is some limited work on the calculation of potential energy surfaces for surface reactions. However, their accuracy is questionable. So far, no one has determined how a change in the reactants or in the surface affects the activation energy for the reaction. As a result, while in principle *ab-initio* calculations should give useful insights into surface reactions, so far their utility has been limited.

A similar situation exists in organic chemistry. Potential energy surfaces can be calculated for simple organic reactions. However, most real cases are too difficult to solve exactly.

In order to fill this gap, workers in the physical-organic chemistry community have developed a series of empirical or semiempirical methods to try to learn something about potential energy surfaces for reactions. These ideas have begun to be adapted by workers in surface chemistry. The objective of this chapter is to review some of these empirical and semiempirical methods that have been proposed by the workers in the physical-organic chemistry community to correlate rate data. We then try to understand why the correlations work so that we can relate the findings from physical-organic chemistry to reactions on surfaces.

9.2 EMPIRICAL CORRELATIONS FOR REACTION RATES: THE POLAYNI RELATIONSHIP AND BRØNSTED CATALYSIS LAW

Van't Hoff first defined the activation energy in 1883. However, it was not until Arrhenius [1888] provided a model to explain Van't Hoff's results that the general relationship

$$k = k_o \exp\left(-\frac{E_a}{kT}\right) \tag{9.1}$$

was accepted by most scientists. At first activation energies were simply tabulated. However, people soon began to wonder if there were any correlations between activation energies and other properties that they could measure. The earliest successful attempt to correlate rate data that I know about is the work of Taylor [1914], who proposed that there was a correlation between the rate of a series of acid-catalyzed reactions and the strength of the acid. Taylor's original correlation did not prove to have wide applicability. However, Brønsted and Pederson [1924] expanded on Taylor's work and were able to arrive at some correlations that have wide applicability.

Brønsted and Pederson [1924] were examining the acid-catalyzed decomposition of nitramide:

$$NH_2NO_2 \Rightarrow N_2O + H_2O$$

They found that when they plotted the log of the rate constant for the decomposition process, k_{na}, against the log of the equilibrium constant for the dissociation of the acid catalyst, K_{ac}, the data fell onto a straight line, as shown in Figure 9.1. Brønsted [1928] quantified these

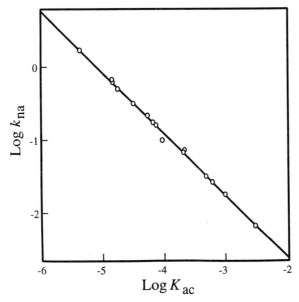

Figure 9.1 The rate constant for the acid-catalyzed dissociation of nitramide, k_{na}, as a function of the dissociation constant for the acid K_{ac}. (Data from Brønsted and Pederson [1924].)

data and proposed that, in general, the rate constant of an acid-catalyzed reaction, k_{ac}, was related to the equilibrium constant for the dissociation of the acid, K_{ac}, by

$$\ln[k_{ac}] = \gamma_{Br\phi n} \ln(K_{ac}) + \ln(\beta_{Br\phi n}) \tag{9.2}$$

where $\gamma_{Br\phi n}$ and $\beta_{Br\phi n}$ are constants. Similar correlations have since been proposed by many other investigators. They are now called Brønsted relationships in the literature to acknowledge the pioneering contributions of Brønsted.

Over the years a number of other reactions have been examined (see Brønsted [1928], Pederson [1934], Bell [1973] for details). It has been found that there often is a general correlation between the log of the rate constant of an acid-catalyzed reaction and the acid strength of the catalyst as shown in Figure 9.2 and Figure 9.3. Most Brønsted plots are only linear over a limited range of pH. Bell and Higginson [1949] have found one case, the dehydration of $CH_3CH(OH)_2$ that shows a linear Brønsted plot over a wide range of pH. This case has been widely reproduced in the literature and so we include it in Figure 9.2. However, most of the examples that have been examined in the literature show behavior like that shown in Figure 9.3, where the Brønsted plot is nonlinear but can be approximated by a linear relationship, provided that the data are taken over a limited range of pH. There are a few examples, such as those in Figures 9.25 and 9.26, where the Brønsted relationship does not work at all. However, the Brønsted relationship has proved to be quite useful (see Bell [1973] for details).

Houriti and Polayni [1935] proposed an extension of the Brønsted relationship. They noted that the acid-catalyzed reactions studied by Brønsted all followed the same general scheme where one extracted a proton from the acid and the proton catalyzed the reaction. From thermodynamics, K_{ac} is given by

$$\ln[K_{ac}] = -\frac{\Delta G_{ac}}{kT} \tag{9.3}$$

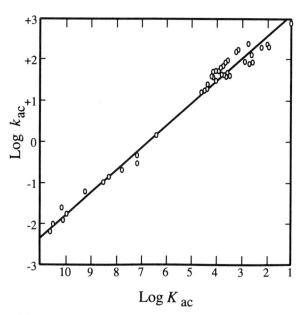

Figure 9.2 A plot of the rate constant of the acid-catalyzed dehydration of $CH_3CH(OH)_2$ in acetone. (Replot of data of Bell and Higginson [1949].)

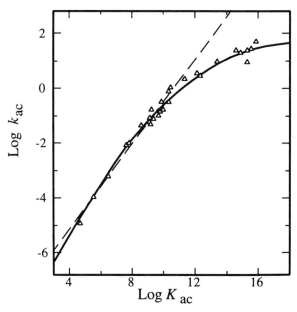

Figure 9.3 A Brønsted plot for the base-catalyzed enolization of $NO_2(C_6H_4)O(CH_2)_2COCH_3$. The solid line is a fit of the data via Equation 9.22. (Data from Hupke and Wu [1977].)

where ΔG_{ac} is the free-energy change that occurs when the acid dissociates, k is Boltzmann's constant, and T is the absolute temperature. Substituting k_{ac} from Equation 9.1 and K_{ac} from Equation 9.3 into Equation 9.2 and taking the log of both sides yields

$$E_a = \gamma_{Br\phi n}(\Delta G_{ac}) + kT \ln\left[\frac{k_0^{ac}}{\beta_{Br\phi n}}\right] \tag{9.4}$$

Therefore the implication of the Brønsted relationship is that there is a linear relationship between the activation energy for the reaction and the free energy it takes to remove a proton from the acid. Houriti and Polayni [1935] generalized this idea to propose that for any homologous series of reactions, the activation energy for the reaction, E_a, is related to the free-energy change during the reaction $(-\Delta G_r)$ by

$$E_a = \gamma_P(\Delta G_r) + E_a^0 \tag{9.5}$$

where γ_P and E_a^0 are constants. Various references call γ_P the reaction coefficient or the transfer coefficient. E_a^0 is called the intrinsic barrier to the reaction. Equation 9.5 is called the **Polayni relationship** or the linear free-energy relationship. Linear free-energy relationships are also sometimes written as

$$kT \ln\left[\frac{k_{ac}}{\frac{kT}{h}}\right] = \Delta G_r^{\ddagger} = \gamma_P(\Delta G_r) + E_a^0 \tag{9.6}$$

where $\Delta G_r^{\ddagger}$ is the free energy of activation and h is Planck's constant. Equation 9.6 implies that if you make a change that changes the free energy of the reaction, the free energy of activation also undergoes a corresponding change. Physically, Equation 9.6 arises because the transition state has properties of both the reactants and products. As a result, when you make a change in either the reactants or products, the change will be reflected in the properties of the transition state.

9.3 REACTIONS AS BOND EXTENSIONS PLUS CURVE CROSSINGS

In 1936 Evans and Polayni proposed a very simple model to explain Equation 9.6. This model has since been extended to explain a wide variety of linear free-energy relationships. Evans and Polayni's idea was that in the rate-determining step of an acid-catalyzed reaction, a proton is transferred from the acid to the reactant. One can separate this process into two parts: the scission of the bond between the proton and its conjugate base (i.e., the negative ion produced when the proton dissociates), and the formation of a bond between the proton and the reactant.

It is useful to use Figure 9.4 to model this idea. Consider the transfer of a proton from an acid BH to a reactant R. We can define two quantities: r_{BH}, the length of the B–H bond, and r_{HR}, the length of the R–H bond as indicated in Figure 9.4. Following Evans and Polayni we assume that during the reaction the distance between B and R does not change. We justify this assumption in Example 9.A.

During the reaction the proton-conjugate base bond breaks; r_{BH} increases while r_{RH} decreases. Figure 9.4 shows how the energy of the B–H and R–H bonds changes as the reaction proceeds. The energy of a typical B–H bond looks like the solid line in Figure 9.4, where the energy of the proton increases as the proton moves from its equilibrium position. We have drawn a Morse potential, but the same arguments could be made for a Lennard-Jones potential. The energy of the R–H bond also looks like a Morse or Lennard-Jones potential. However, when we plot the data onto Figure 9.4, we need to be careful, because the x-axis in Figure 9.4 is the length of the B–H bond. When r_{BN} increases, r_{RH} decreases. As a result, when we plot the potential energy function of the R–H bond as a function of r_{BH}, the curve will be backwards, as indicated by the dotted line in Figure 9.4.

Now consider what happens during the reaction. Initially the B–H bond breaks so the

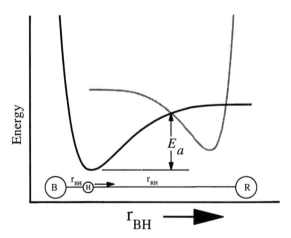

Figure 9.4 The energy changes occurring when a proton H is transferred between a conjugate base B and a reactant R. The solid line is the energy of the B–H bond, while the dotted line is the energy of the H–R bond.

potential will ride up the solid contour in Figure 9.4. However, once the R–H bond begins to form, the potential can go back down the dotted line in Figure 9.4. During the reaction, the scission of the B–H bond occurs simultaneously with the formation of the R–H bond. Thus, in principle the total energy should be the sum of the two curves. However, Evans and Polayni suggested that one can obtain a reasonable approximation if one assumed that the potential energy of the system went up the B–H curve as the reaction begun, and then went down the R–H curve as the reaction went to completion. Evans and Polayni also identified E_a in Figure 9.4 as the activation energy for the reaction.

Evans and Polayni's model, although not quantitative, is quite important. It shows that one can consider a reaction as a sum of two processes, one involving bond scission, and the other involving bond formation. One can then define the potential energy surface for each of the bond scission processes, and by considering the curve crossing between the two potential energy surfaces, one can get some information about rates. Inherent in this model is the idea that the transition state for a reaction has properties that are somewhere between those of the reactants and the products. This idea has been quite important in physical-organic chemistry and it will be used widely throughout the remainder of this book.

As an example of the utility of Evans and Polayni's model, consider how Figure 9.4 changes when one changes the strength of the acid by, for example, replacing one of the ligands on the acid with a different ligand. For the moment assume that the acid gets stronger. By definition, when the acid strength increases, the B–H bond becomes easier to break in solution. This corresponds to an increase in the free energy of the B–H bond. The free energy could increase if either the position or the shape of the B–H curve in Figure 9.4 changes. However, in Example 9.B we will show that changes in the shape of the B–H curve will usually not make enough of a difference in the free energy to make a significant change in the acid strength. Evans and Polayni noted that significant changes in the acid strength could only occur when the B–H curve in Figure 9.4 shifted up or down. Therefore they proposed that one can get a qualitative picture of how a change in acid strength affected the reaction by simply displacing the solid curve in Figure 9.4 up or down. Figure 9.5 shows the result of an upward displacement. Notice that as the solid curve is displaced up the activation energy for proton transfer goes down. For small displacements, the activation energy varies linearly with the change in the energy of the reaction. Evans and Polayni showed that if one linearizes all of the equations, one can derive Equation 9.5.

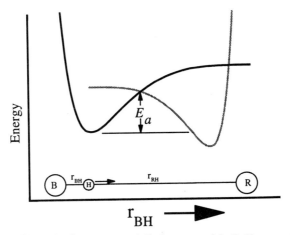

Figure 9.5 A diagram illustrating how an upward displacement of the B–H curve affects the activation energy.

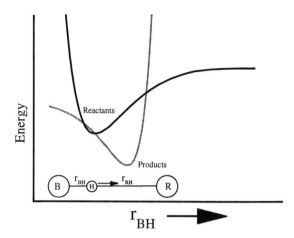

Figure 9.6   A diagram illustrating a case where the activation energy is zero.

On a more fundamental level, Evans and Polayni's model explains why activation energies arise in reactions. The idea is that when a reaction occurs, one bond breaks and another forms. One can distinguish between two cases: the case in Figure 9.5 where the process is activated, and the case in Figure 9.6 where the curves cross at the minimum in the BH potential so the activation energy is zero. Fundamentally, the two cases are similar. However, in the first case, when the reaction starts, the increase in the energy due to scission of the B–H bond is larger than the decrease in energy due to formation of the R–H bond. As a result, the total energy of the system goes up as the reaction proceeds. That causes the reaction to be activated. In contrast, in the case in Figure 9.6 the lowering of the total energy of the system due to bond formation is larger than the energy increase due to bond scission. As a result, there is no activation barrier to reaction.

It happens that most real cases look closer to the case in Figure 9.5 than the case in Figure 9.6. Consider reaction $A + BC \rightarrow AB + C$. In the initial part of the reaction, A is moving in toward B–C. At long range the A–BC potential is attractive, but as A gets closer, the A–BC potential is usually repulsive: The electron cloud of A overlaps with the electron cloud of B–C. That produces an electron–electron repulsion (i.e., a Pauli repulsion) of the type discussed in Appendix B. The Pauli repulsions tend to keep the reactants from getting closer than their van der Waals radii. Now as the reaction proceeds, the B–C bond begins to break, and the A–B bond begins to form. However, the van der Waals radii of the molecules are usually larger than the covalent radii of the atoms. As a result, when atom B begins to be transfered to A, atom A is still pretty far way from B as illustrated in Figure 9.7. The B–C bond loses energy as the B–C bond is stretched. However, the A–B bond is still very long, so the net bonding between A and B is relatively small. This produces a situation that is analogous to the situation in Figure 9.5 where the B–C bond starts to break, but the A–B bond barely starts to form. Figure 9.5 shows that in such a case, there is a finite barrier to reaction. Figure 9.5 and Figure 9.7 apply to most reactions. As a result, most reactions are activated.

A more careful analysis indicates that activation energies can also arise because of bond distortions. However, the basic implication of the Evans and Polayni model is that activation barriers arise because of the Pauli repulsions, the bond scissions, and the bond distortions that occur during reaction. The activation is zero only when no bonds break or distort during reaction, or when the reaction is so exothermic that bond formation dominates over bond scission. Both cases are rare. As a result, most reactions are activated.

Beginning of the reaction:

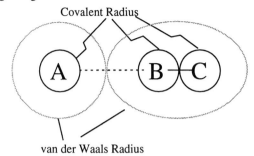

Transition state:

End of the reaction:

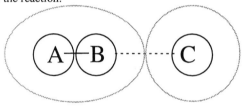

Figure 9.7 An illustration of the geometry for the reaction A + BC → AB + C: (*a*) A approaching BC; (*b*) the transition state; (*c*) near the end of the reaction.

9.4 THE POLAYNI RELATIONSHIP

The analysis in the previous section was all qualitative. However, Polayni and Evans showed that one can use the model to understand Equation 9.5 in a quantitative way. In the material that follows we reproduce Polayni and Evans's derivation to see how the ideas arise. Consider the reaction A + BC → AB + C. One can model the reaction as shown in Figure 9.8 and Figure 9.9, where the reactants come together at a constant value of r_{BC}, then atom *B* is transferred, and then the reactants move away along a line of constant r_{AB}. First it is useful to fit lines to the potential energy contour for the reaction as shown in Figure 9.8, where the lines are chosen to fit the tangent of the AB and BC potential at the transition state. For the purposes of the derivation it will be useful to assume that the energies of lines 1 and 2, E_1, and E_2 are given by

$$E_1 = Sl_1 (r_{AB} - r_1) \tag{9.7}$$

$$E_2 = Sl_2 (r_2 - r_{AB}) + \Delta G_r \tag{9.8}$$

where Sl_1 and Sl_2 are the slopes of the two potential energy curves at the transition state and the energies are measured from the horizontal line at the bottom of the A–B potential curve. Note that it is not immediately obvious whether the last term in Equation 9.8 should be ΔG_r

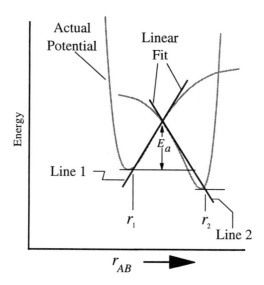

Figure 9.8   A linear approximation to the Polayni diagram used to derive Equation 9.9.

or ΔU. However, Evans and Polayni used ΔG_r because their results fit data better when they did so.

Notice from Figure 9.8 that E_a is equal to the energy at which line 1 intersects line 2. Solving Equations 9.7 and 9.8 simultaneously for $E_a = E_1 = E_2$ yields

$$E_a = \left(\frac{Sl_1 \, Sl_2}{Sl_1 + Sl_2} \right) (r_2 - r_1) + \frac{Sl_1}{Sl_1 + Sl_2} \Delta G_r \qquad (9.9)$$

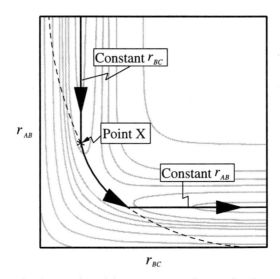

Figure 9.9   An approximation to the minimum energy trajectory for the reaction A + BC → AB + C.

Equation 9.9 is a linear free-energy relationship with

$$\gamma_P = \frac{Sl_1}{Sl_1 + Sl_2} \qquad (9.10)$$

Equation 9.9 is called the Polayni relationship. Equations 9.9 and 9.10 are Evans and Polayni's key results. The implication of Equation 9.9 is that if one had some way to change either the enthalpy or the free energy of a reaction, the activation energy for the reaction would also change. Physically, one can change the enthalpy of a reaction by adding a substituent group to one of the reactants. For example, consider the reaction between acetic acid and ethanol to yield ethyl acetate:

$$CH_3COOH + CH_3OH \rightarrow CH_3COO\ CH_3 + H_2O$$

One can vary the acid strength of the acetic acid by fluorinating the methyl group. That makes the reaction more exothermic, which in turn leads to a lowering of the activation barrier for the reaction.

In the same way, people often use the Polayni relationship to model reactions on surfaces. In Chapters 3 and 4 we found that one can change the thermodynamics of a surface reaction by changing the structure or composition of the surface, or by changing the coverage and thereby changing the pairwise adsorbate/adsorbate interactions. The Polayni model predicts that each of these changes will lead to a change in the activation energy for reaction.

9.5 THE MARCUS EQUATION

In actual practice Equation 9.9 does not fit qualitatively. However, Marcus [1955, 1968] derived a better approximation while Marcus was a professor at the University of Illinois. Marcus considered a general reaction

$$A + BC \rightarrow AB + C$$

and derived an equation for the rate using a modification of Polayni's methods.

In the remainder of this section we derive a formula for the free energy of activation using Marcus's work as a guide. Our derivation assumes that the reaction follows the solid trajectory in Figure 9.9, where the reactants come together, a reaction occurs, and then the products fly apart. We divide the trajectory into three parts, a part where the reactants come together without the reactants being significantly distorted, a part where atom B is transferred, and a part where the reactants fly away. In this approximation, the free energy of activation is w_r^1, the work it takes to bring the reactants to point X in Figure 9.9, plus E_a^1, the free energy it takes to transfer atom B, i.e.,

$$\Delta G_r^{\ddagger} = E_a^1 + w_r^1 \qquad (9.11)$$

Marcus postulated that w_r^1 would be nearly constant for a group of closely related reactions, and that E_a^1 could be calculated using a modification of Polayni's derivation described earlier in this section. For the purposes of derivation, we define a quantity r_X as the distance along the dashed line in Figure 9.9 with $r_X = 0$ at point X. We will then assume that the energy of the A-B and B-C bonds show a Lennard–Jones dependence on r_x as indicated in the dotted lines in Figure 9.10. To simplify the analysis we will fit each of the potential contours in Figure 9.5 with a parabolic function near the transition state, as indicated in Figure 9.10.

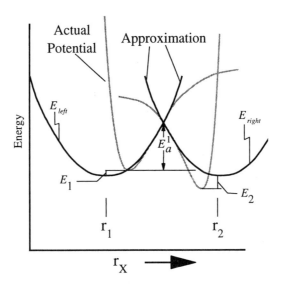

Figure 9.10 Marcus's approximation to the change in the potential energy surface, which occurs when ΔG_r changes.

In the next few pages we review Marcus's derivation. For the purpose of derivation, we assume that $E_{left}(r_X)$ and $E_{right}(r_X)$, the energies of the left and right parabolas in Figure 9.10, are given by

$$E_{left}(r_X) = SS_1(r_X - r_1)^2 + E_1 \tag{9.12}$$

$$E_{right}(r_X) = SS_2(r_X - r_2)^2 + \Delta G_r + E_2 \tag{9.13}$$

where SS_1, SS_2, r_1, and r_2 are fitting parameters. In Equations 9.12 and 9.13 we have noted that E_{left} and E_{right} are a function of r_X. We note in the material to follow that SS_1 and SS_2 are related to the vibrational frequency of the atom B as it is being transfered. Note that since we are measuring energies from the horizontal line at the bottom of the A–B potential curve in Figure 9.10, E_a^1 is equal to the energy where E_{left} equals E_{right}.

For the purpose of derivation, it is useful to define $r^\ddagger$ as the value of r_x where $E_{left} = E_{right}$.

Substituting $r^\ddagger$ for r_x in Equations 9.12 and 9.13 and equating E_{left} and E_{right} yields

$$SS_1(r^\ddagger - r_1)^2 + E_1 = SS_2(r^\ddagger - r_2)^2 + E_2 + \Delta G_r \tag{9.14}$$

Marcus simplified Equation 9.14 by assuming

$$SS_1 = SS_2 \tag{9.15}$$

$$E_2 = E_1 \tag{9.16}$$

Substituting SS_2 from Equation 9.15 into Equation 9.14, solving for $r^\ddagger$, and then substituting E_2 from Equation 9.16, yields

$$r^\ddagger = \frac{(r_1 + r_2)}{2} + \frac{E_2 - E_1 + \Delta G_r}{2SS_1(r_2 - r_1)} = \frac{r_1 + r_2}{2} + \frac{\Delta G_r}{2SS_1(r_2 - r_1)} \tag{9.17}$$

Substituting $r^{\ddagger}$ into Equation 9.12 yields

$$E_A^1 = E_{left}(r^{\ddagger}) = \left(\frac{r_2 + r_1}{2} - r_1 + \frac{\Delta G_r}{2SS_1(r_2 - r_1)} \right)^2 SS_1 + E_1 \qquad (9.18)$$

Rearranging Equation 9.18

$$E_A^1 = \left(1 + \frac{\Delta G_r}{SS_1(r_2 - r_1)^2} \right)^2 \frac{(r_2 - r_1)^2}{4} SS_1 + E_1 \qquad (9.19)$$

This equation can be put into a standard form by defining a quantity $\Delta G_0^{\ddagger}$ by

$$\Delta G_0^{\ddagger} = \frac{(r_2 - r_1)^2}{4} SS_1 \qquad (9.20)$$

Substituting Equation 9.20 into Equation 9.19 yields

$$E_A^1 = \left(1 + \frac{\Delta G_r}{4\Delta G_0^{\ddagger}} \right)^2 \Delta G_0^{\ddagger} + E_1 \qquad (9.21)$$

Substituting Equation 9.21 into Equation 9.20 yields a key result called the **Marcus equation:**

$$\Delta G_r^{\ddagger} = \left(1 + \frac{\Delta G_r}{4\Delta G_0^{\ddagger}} \right)^2 \Delta G_0^{\ddagger} + w_r \qquad (9.22)$$

with

$$w_r = w_r^1 + E_1 \qquad (9.23)$$

Marcus was awarded the 1993 Nobel Prize in chemistry for his work on the Marcus equation.

9.5.1 An Alternate Derivation: Bond Energy–Bond Order Relationship

In the derivation in the last section, we used some knowledge of how bond distances vary with bond length to derive Equations 9.9 and 9.22. However, we can derive all of the same results if we instead consider how the energy of a bond varies with something that is called the **bond order** of the A–B bond.

The bond order was proposed by Pauling [1960] as a way of quantifying the fractional bonding in the A–B bond. For example, if A and B are held by a σ-bond, then one can define the bond order as 0.5 times the number of electrons in the σ-orbital minus the number of electrons in the σ^* orbital. With this definition, single bonds have a bond order of 1.0, double bonds have a bond order of 2.0, and triple bonds have a bond order of 3.0.

Johnston [1963] proposed using the changes in bond order to understand what happens during a reaction. The idea is that when a generalized reaction such as

$$A + BC \rightarrow AB + C$$

occurs, the B–C bond breaks and the A–B bond forms. One can view that process as one where the bond order of the B–C bond decreases, while the bond order of the A–B increases. Johnston [1963] showed that by keeping track of each of the parts of the reaction, one can derive an equation for the activation energy as a function of the properties of the reactants using a derivation similar to the derivation of Marcus.

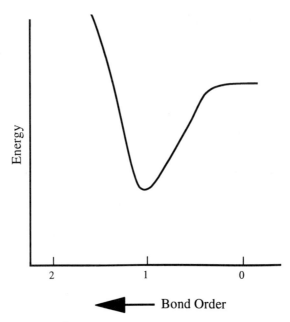

Figure 9.11   A schematic of the energy of the B–C bond as a function of the bond order.

To see how the method works, it is useful to consider the formation of the A–B bond and the scission of the B–C bond separately. First consider the B–C bond. If atoms B and C would like to form a σ-bond, then as we stretch the bond, which reduces the bond order, the energy of the B–C complex will increase. We do not know, a priori, what the shape of the curve will be like. However, it is not unreasonable to assume that the energy will increase either parabolically or as a Morse potential with bond order.

Now consider what happens if we try to form a bond with a bond order of more than unity. In this case, we need to rehybridize B and C, so again the energy will go up. Therefore, qualitatively, if we plot the energy of the B–C bond as a function of bond order, it will look like the curve given in Figure 9.11.

Notice that the curve in Figure 9.11 is very similar to the curves in Figure 9.4. Therefore, we could have derived the Marcus equation and the Polayni relationship based on a consideration of the changes in bond order that occur during the reaction. Hence, the Marcus equation and the Polayni relationship also apply to systems that conserve bond order.

The advantage of bond order methods is that Pauling [1960] has provided some ways to estimate the bond energy–bond order (BEBO) curve (i.e., Figure 9.11) for reactions where a hydrogen atom is transferred. As a result, one can use the BEBO to quantitatively estimate activation energies, although admittedly the methods are not terribly accurate.

9.5.2   Qualitative Features of the Marcus Equation

It is useful to consider the qualitative features of the Marcus equation. Figure 9.12 shows a plot of the variation in $\Delta G_r^{\ddagger}$ with varying ΔG_r calculated from Equation 9.22 for a typical set of parameters. $\Delta G_r^{\ddagger}$ actually varies parabolically with ΔG_r. However, if we examine a limited range of ΔG_r, $\Delta G_r^{\ddagger}$ varies approximately linearly with ΔG_r, and it is only when we work over an extended range of ΔG_r that nonlinearities are seen. Therefore, according to the Marcus equation one would expect reactions to obey linear free-energy relationships over a reasonable range of free energy. However, one would expect there to be some deviations from linearity if data are taken over a wide range of free energy.

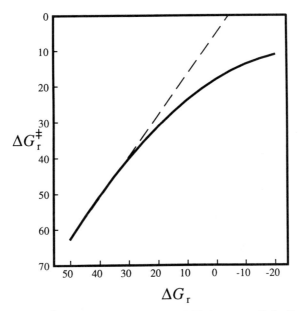

Figure 9.12 A plot of $\Delta G_r^{\ddagger}$ calculated from Equation 9.22 for $w_r = 10$ kcal/mol and $\Delta G_0^{\ddagger} = 8$ kcal/mol.

Experimentally, one often finds that $\Delta G_r^{\ddagger}$ varies linearly with ΔG_r in data taken over a limited range of ΔG_r. However, one usually observes some curvature if one takes data over a wide range of free energy. For example, consider the data in Figure 9.3. Notice that the data follow a curved line that is similar to the curved lines shown in Figure 9.12. Hupke and Wu [1977] fit the data in Figure 9.3 to Equation 9.22, and the results are given in Figure 9.3. Notice that the fit is quite good. Hence, we can explain the curvature in Figure 9.3 using the Marcus equation.

Marcus has shown that one can predict the fitting parameters in his equation from first principles, and fit real data. Further, it has been found that all of the examples in the literature showing curved free-energy plots can be explained via the Marcus formalism except those showing a change in mechanism with changing ΔG_r. Hence, the Marcus equation is quite useful in explaining data.

The Marcus equation is also a useful predictive tool. For example, Marcus [1964] and Albery [1975] showed that they could use the Marcus equation to predict the rate of a number of electron transfer reactions of the form

$$Fe^{3+} + I^{X+} \rightarrow Fe^{4+} + I^{(X-1)+}$$
$$Fe^{4+} + I^{(X-1)+} \rightarrow Fe^{3+} + I^{X+} \qquad (9.24)$$

where Fe^{3+} is an iron ion and I^+ is some metal ion, using only a series of standard cell potentials and rate data for one electron transfer reaction

$$Fe^{4+} + Fe*^{3+} \rightarrow Fe^{3+} + Fe*^{4+} \qquad (9.25)$$

where $Fe*$ is an isotopically labeled iron atom. Interestingly, they could even predict how the rate of reactions changed when it occurred on an electrode rather than in solution using only data on the rate of Reaction 9.25 in solution. These results were generalized to a series of other electron transfer reactions. Figure 9.13 shows how well the data fit. Notice that for

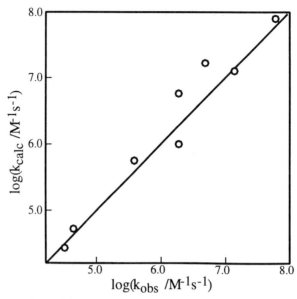

Figure 9.13 A comparison of the rate of a series of electron transfer processes to the predictions of the Marcus equation. (Replot of data given in Albery and Kreevoy [1978].)

this example the Marcus equation fits the rate data to almost the accuracy at which they are measured. There have been a number of other examples that fit this well. Hence, the Marcus equation has been adopted by many investigators.

There is one unusual prediction of the Marcus equation. Notice that according to Equation 9.22 the rate of reaction should reach a maximum (i.e., $\Delta G_r^{\ddagger}$ should reach a minimum) when $\Delta G_r = \Delta G_r^{max}$ where ΔG_r^{max} satisfies

$$1 + \frac{\Delta G_r^{max}}{8\Delta G_r^{\ddagger}} = 0 \qquad (9.26)$$

The rate should then decrease with increasing $-\Delta G_r$ when $-\Delta G_r \geq -\Delta G_r^{max}$.

Figure 9.14 illustrates why the maximum occurs. Most reactions show Polayni diagrams like those in Figure 9.14a where the reaction needs to go up a potential energy contour and back down again to occur. However, if the reaction is sufficiently exothermic, one can get to a situation where the energy curves for the reactants and products cross at the equilibrium point for the reactants, as illustrated in Figure 9.14b. Notice that at this particular potential, the activation barrier is zero. Now consider what happens when the reaction becomes more exothermic, as illustrated in Figure 9.14c. Note that as we continue to increase the potential, the intersection of the energy curves for the reactants and products moves to the left, which causes the activation energy for the reaction to increase.

We call the region where the rate decreases with increasing driving force the **Marcus inverted region.** Up until 1984, no one had observed Marcus inverted behavior. However, a number of electron transfer reactions that show inverted behavior have been discovered since 1984 (see Miller [1991] for details). There is a complication in electron transfer reactions in that the rate of the reaction is moderated by something called a Frank–Condon factor, which can decrease with increasing $-\Delta G_r$. We discuss this effect in Appendix B. At present, the best available data is that Frank–Condon factors are quite important in the Marcus inverted

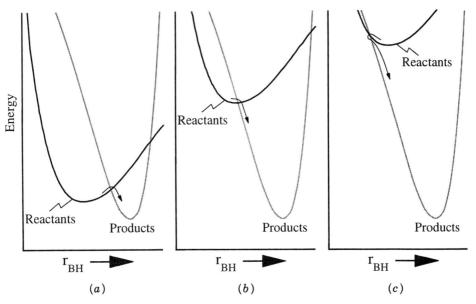

Figure 9.14 A Polayni diagram illustrating how changes in the driving force for reaction change the behavior of the reaction. (*a*) Normal case. (*b*) Saturation. (*c*) Marcus inverted region.

region. As a result, one cannot explain the data with the Marcus equation alone. Still, the Marcus equation gives the correct qualitative behavior even in the inverted region, and so it is quite powerful.

9.5.3 The Hammond Postulate

One of the issues that has been discussed at great length in the literature is the physical significance of the transfer coefficient. Bokris [1970] points out that the transfer coefficient is a measure of the asymmetry of the transition state. Referring back to Figure 9.8 and Equation 9.10, note that for a symmetric transition state, $Sl_1 = Sl_2$ so $\gamma_P = 0.5$. Therefore, $\gamma_P = 0.5$ for a symmetric transition state and γ_P is only significantly different from 0.5 when the transition state is highly asymmetric.

Now, let's derive an expression for γ_P from Equation 9.22. Our approach is to expand Equation 9.22 in a Taylor series

$$\Delta G^{\ddagger} = \Delta G_0^{\ddagger} + (\Delta G_r) \frac{\partial \Delta G_r^{\ddagger}}{\partial \Delta G_r} + \cdots \tag{9.27}$$

and only keep the first term. A comparison of Equations 9.6 and 9.27 shows

$$\gamma_P = \frac{\partial \Delta G_r^{\ddagger}}{\partial \Delta G_r} \tag{9.28}$$

Substituting $\Delta G_r^{\ddagger}$ from Equation 9.22 into Equation 9.28 yields

$$\gamma_P = \frac{1}{2} + \frac{\Delta G_r}{8\Delta G_0^{\ddagger}} \tag{9.29}$$

Notice that according to Equation 9.29 when ΔG_r is small, γ_P will be approximately 0.5; γ_P will only significantly differ from 0.5 when ΔG_r is large. Therefore, γ_P should be close to 0.5 for moderately endothermic or exothermic reactions and the transfer coefficient should differ significantly from 0.5 only in very exothermic or endothermic reactions.

Figure 9.14b shows a Polayni diagram for a very exothermic reaction. Notice, that the slopes are significantly different in this case, so γ_P is much less than 0.5. The transition state also lies close to the reactants. Similarly, when the reaction is very endothermic, the transfer coefficient will be larger than 0.5. The transition state also lies close to the products.

The observation that transfer coefficients are small when the transition state is close to the reactants and near unity when the transition state lies close to the products led Hammond [1955] to propose that the transfer coefficient was a measure of where the transition state lies relative to the reactants and products. If γ_P is small (i.e., less than 0.3), the transition state will likely be close to the reactants; if γ_P is large (greater than 0.7), the transition state will likely be close to the products. Most experiments yield transfer coefficients between 0.4 and 0.6, which suggests that most transition states lie somewhere between the reactants and products. Of course, no one has actually proved the Hammond postulate; if we allow SS_1 to be significantly different from SS_2 in Equation 9.14, γ_P will be significantly different from 0.5, even when the transition state lies halfway between the reactants and products. Hence, it is possible to get asymmetric transition states, even when the geometry of the transition state differs significantly from both the reactants and products. However, when we transfer atom B from AB to C, SS_1 will be related to the frequency of the A–B bond, while SS_2 will be related to the frequency of the B–C bond. Thus, SS_1 can only differ significantly from SS_2 when there is a large change in the vibrational frequency of the atom being transferred. In most cases, there is not enough of a change in the vibrational frequency to have a significant effect on γ_H. As a result, the Hammond postulate has proved to be a useful tool to learn about the properties of transition states even though it is not exact.

9.5.4 Application of the Marcus Equation: Tafel Kinetics

The earliest applications of the Polayni relation and the Marcus equation came in the electrochemical literature. In an electrochemical cell, one can vary the free energy of the reactions in the cell by varying the potential on the electrodes. As a result one can see if the Polayni relationship works by varying the potential in the cell and seeing how the reaction rates change.

We can derive an equation for the change as follows: Consider a simple electron transfer reaction, where a reactant A, receives n_e electrons to produce a species B:

$$A + n_e\, e^- \rightarrow B^{n_e^-} \tag{9.30}$$

According to Faraday's law, the free energy of the reaction in the cell ΔG_r varies with the applied potential, η_V, according to

$$\Delta G_r = \Delta G_r^0 + n_e\, \mathfrak{F}\eta_V \tag{9.31}$$

where $\mathfrak{F}$ is Faraday's constant and ΔG_r^0 is the free-energy change at zero applied potential. Combining Equations 9.6 and 9.31 yields

$$\text{Ln}\left[\frac{k_r}{\left(\frac{kT}{h}\right)}\right] = \left(\frac{E_a^0 + \gamma_P\, \Delta G_r^0}{kT}\right) + \left(\frac{n_e\, \mathfrak{F}\gamma_P}{kT}\right)\eta_V \tag{9.32}$$

Hence, if Reaction 9.30 followed a linear free energy law, then a plot of the log of the rate constant for the reaction versus the applied potential should be linear with slope, Sl, given by

$$Sl = \frac{n_e \, \mathfrak{F} \gamma_P}{kT} \tag{9.33}$$

Note $(kT/\mathfrak{F}) = 26$ meV at 300 K. Hence, according to Equation 9.32, a 120-meV change in potential will produce about an order of magnitude change in rate when $n_e = 1$, and $\gamma_P = 0.5$.

In an electrochemical experiment, one usually does not measure the rate of a reaction directly. Rather, one builds an electrochemical cell with an anode and cathode, and measures the current produced by the cell as a function of the applied voltage. We show below that if we measure potentials relative to the equilibrium point of the cell, and work at potentials that satisfy

$$|\eta_V| > 5 \, \frac{kT}{\mathfrak{F} \gamma_P} \tag{9.34}$$

The current will be proportional to the rate of the reaction. Hence, if the electrochemical reaction follows a linear free-energy relationship, a semilog plot of the current versus voltage should be a straight line for potentials that satisfy Equation 9.34.

Experimentally, Tafel [1905] showed that the current was linear for a number of oxidation reactions. For example, Figure 9.15 shows how the rate of hydrolysis of water varies with potential under a variety of different conditions. Notice that the data are linear over an extended region of potentials. There are deviations at high and low potential and the exact magnitude of the current varies with conditions. However, the linear free-energy relationship works over 11 orders of magnitude in rate. Tafel [1905], Houriti and Polayni [1930], Conway [1952], and Vetter [1967] review results for hundreds of electrochemical reactions, and all of them show the same general trends shown in Figure 9.15. The current is nonlinear at low potentials, but shows near-linear behavior over an extended range of potential. Further, the transfer coefficient, γ_H, is between 0.4 and 0.6 for all but one of the examples cited in these reviews. Hence, it appears that in electrochemical systems linear free-energy relationships work over an extended range of potential, and that the transfer coefficient is nearly constant for all of the examples. Electrochemical systems that follow linear free-energy relationships are said to obey **Tafel kinetics.** Most electrochemical systems obey Tafel kinetics. The main exceptions so far are systems that show a change in mechanism with changing potential.

The one unexpected feature is that transfer coefficients are slightly temperature dependent. This occurs because the solvent molecules in the neighborhood of the electrode order in the presence of an electric field, and the amount of ordering is temperature dependent (see Conway [1985] for details).

9.5.5 Nonlinear Behavior

Of course, the current does not vary linearly with voltage at low and high potentials. Physically, the deviations at low potential occur because at low potential the measured current is not proportional to the rate of reaction on the anode. In contrast, the deviations at high potential are caused by changes in the curvature of the potential surface. In this section we briefly discuss why one observes nonlinear behavior in these two limits. One is referred to Bokris [1970] or Dogandze [1971] for further details.

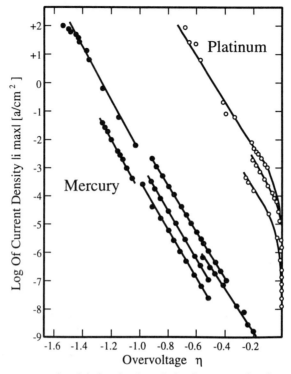

Figure 9.15 The current produced during the electrolysis of water as a function of the applied potential over mercury and platinum electrodes. The various lines are data taken at different conditions. (Data from Vetter [1967].)

First, let's consider the deviations at low potential. The deviations at low potential occur because of an artifact in the electrochemical system. An electrochemical cell contains an anode and a cathode, and reactions occur on both the anode and the cathode. The measured current in the cell, i_{net}, is equal to the sum of the currents on the anode and cathode, i_a and i_c, respectively,

$$i_{net} = i_a + i_c \tag{9.35}$$

where by convention i_c is negative. In the work to follow we fix the zero of our potential scale so that at $\eta_V = 0$, the anodic reaction is exactly balanced by the cathodic reaction, so the net current is zero.

One can derive an expression for the net current by considering the anodic and cathodic reactions. The current for the anodic reaction, i_a, is given by

$$i_a = n_e r_a \tag{9.36}$$

where r_a is the rate of the anodic reaction.

For the derivation to follow we assume

$$r_a = k_a[A] = \left(\frac{kT}{h}\right) e^{-\Delta G_r^{\ddagger}/kT} [A] \tag{9.37}$$

where $[A]$ is the concentration of A in molecules/cm^2. Substituting r_a from Equation 9.37 into Equation 9.36 yields

$$i_a = n_e \, [A] \left(\frac{kT}{h}\right) e^{-\Delta G_r^\ddagger/kT} \tag{9.38}$$

Substituting ΔG_r from Equation 9.31 into Equation 9.6 shows that the activation energy for the reaction $\Delta G_r^\ddagger$ is given by

$$\Delta G_r^\ddagger = E_A^0 + \gamma_a \Delta G_r^0 + n_e \, \mathfrak{F} \eta_V \gamma_a \tag{9.39}$$

where γ_a is the transfer coefficient for the reaction on the anode.
Substituting $\Delta G_r^\ddagger$ from Equation 9.39 into Equation 9.36 yields

$$i_a = i_0 \, e^{|1-(n_e \mathfrak{F} \gamma_a/kT)\eta_V|} \tag{9.40}$$

where

$$i_0 = n_e[A] \, \frac{kT}{h} \, e^{-(E_A^0 + \gamma_a \Delta G_r^0)/kT)} \tag{9.41}$$

Physically i_0 is the current when there is no applied potential.
It is useful to scale the potential so that when $\eta_V = 0$, i_0 for the anodic and cathodic reactions are the same. If so, i_{net} becomes

$$i_{net} = i_0 \exp\left[-\frac{n_e \, \mathfrak{F} \gamma_a}{kT} \, \eta_V\right] - i_0 \exp\left[-\frac{n_e \, \mathfrak{F} \gamma_c}{kT} \, \eta_V\right] \tag{9.42}$$

where γ_a and γ_c are the transfer coefficients for the anodic and cathodic reactions.
Equation 9.42 is called the Butler–Volmer equation, since it was first derived by Butler [1932] based on work by Volmer [1930]. Figure 9.16 is a plot of the current predicted by the Butler–Volmer equation as a function of the applied potential. The figure shows that if one made a semilog plot of the current versus potential, one would expect to observe nonlinearities at low potential, due to the reaction on the cathode. The shapes are similar to the experimental curves in Figure 9.15. Hence, one can use Equation 9.42 to qualitatively explain

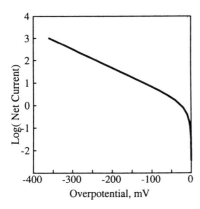

Figure 9.16 A plot of the current predicted by the Butler–Volmer equation as a function of the applied potential.

the deviations from linearity seen in Figure 9.15. Vetter [1967] shows that in fact the fit is quantitative. As a result, he concludes that the data in Figure 9.15 follow a linear free-energy relationship even at low potentials, even though the plot of current versus potential is nonlinear.

The situation is different at high potentials, however. Notice that the mercury data in Figure 9.15 begin to turn over at high potentials. This saturation effect is often attributed to a mass transfer limitation. However, calculations described below indicate that the rate will saturate even in the absence of mass transfer limitations. It is now thought that there are real deviations from linear free-energy behavior in the high potential regime.

Physically these deviations occur because the Polayni diagrams show nonlinearities when data are taken over a wide region of potential. Referring back to Figure 9.14, note that the Marcus equation predicts that at high potential the rate will reach a maximum and then decline. Hence, if Figure 9.14 applied to the electrochemical situation, one would expect the rate to reach a maximum at some high potential and then decline.

In reality, Figure 9.14 is an oversimplification of what actually is happening in an electrochemical reaction. During an electrochemical reaction an electron is transferred from the electrode into the solution. The electrode is usually a metal. In Chapter 3 we noted that the electron states in the electrode form a band. Electrons from everywhere in the band contribute to the rate of reaction. As a result the Polayni diagram for the reaction must consider a band of levels as indicated in Figure 9.17. It is easy to show that the driving force (i.e., free-energy change) for the reaction is largest if the electron comes from the top of the band, and so under normal circumstances the rate of electron transfer will be higher if the electron comes from the top of the band than if it comes from elsewhere in the band. However, if we choose a potential like that shown in Figure 9.17, the activation energy for the reaction will be less if the electron comes from the middle of the band than if the electron comes from the top of the band.

Now consider what happens as we change the potential in the example depicted in Figure 9.17. When we change the potential the energy curves for the electrode will be shifted up or down. As a result the activation energy for the removal of electrons from each of the states in the band will change. However, if the potential is as shown in Figure 9.17, some of the activation energies go up while some go down. As a result, the overall activation energy for

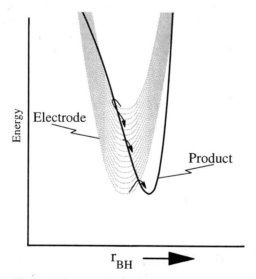

Figure 9.17 Polayni diagram for a band of levels.

the reaction only changes slightly. Dogandze [1971] analyzed the high potential case in some detail and showed that in fact, at high potentials, the decrease in activation energy from states below the top of the band almost exactly compensate the increases in activation energy at the top of the band. As a result, the rate is almost independent of the applied potential. Hence, according to Dogandze, the rate of an electrochemical reaction should eventually saturate even in the absence of a mass transfer limitation.

Experimentally, the situation is unclear. There is some evidence for a saturation phenomenon at high potentials. However, it is unclear whether the saturation is caused by the influence of the band as described earlier or by a Frank–Condon barrier to electron transfer, which will be described in Appendix B. Usually, the data show linear behavior over an extended range of potential. Hence, linear free-energy relationships usually work in electrochemical systems except at extreme conditions.

9.6 APPLICATIONS OF LINEAR FREE-ENERGY RELATIONSHIPS AND THE MARCUS EQUATION IN ORGANIC CHEMISTRY

Linear free-energy relationships and the Marcus equation have also found wide applicability in the physical-organic chemistry literature. In 1924, Hammett speculated that the free-energy relationships that were already well established in the electrochemical literature might also be useful in organic chemistry. At the time, no one had suitable thermodynamic data for decomposition of organic molecules. However, Hammett realized that one can get a measure of a molecule's free energy by looking at the molecule's acidity. Hence, he supposed that one might be able to create Brønsted plots for organic reactions using acidity as a free-energy scale.

It took several years to develop this idea. However, in 1933 Hammett and Pfluger examined the reaction of trimethylamine with a series of substituted methylbenzoates:

$$X-\text{C}_6\text{H}_4-COOCH_3 + N(CH_3)_3 \xrightarrow{H^+} X-\text{C}_6\text{H}_4-COO^- + N(CH_3)_4^+$$

and found that there was a general linear correlation between the log of k_X, the rate constant for the hydrolysis of X-methylbenzoate, and the log of K_X, the dissociation constant of the corresponding benzoic acid. Hammett expanded this work to include a wide variety of substituted benzoic acids. He found that in general there was a simple linear relationship between the rate constant for reactions of meta- and para-substituted benzoic acids and the dissociation constant of the acid. For example, Figure 9.18 shows a plot of the log of the rate constant for the reaction

$$X-\text{C}_6\text{H}_4-COOCH_2CH_3 + H_2O \xrightarrow{OH^-} X-\text{C}_6\text{H}_4-COOH + CH_3CH_2OH$$

$$X-\text{C}_6\text{H}_4-COOH + H_2O \longrightarrow X-\text{C}_6\text{H}_4-COO^- + H_3O^+$$

and the log of the equilibrium constant for the dissociation of the corresponding substituted benzoic acid.

The data in Figure 9.18 have been normalized to the values for X = hydrogen, i.e., the rate constant for the hydrolysis of ethylbenzoate, k_H, and the dissociation constant for benzoic

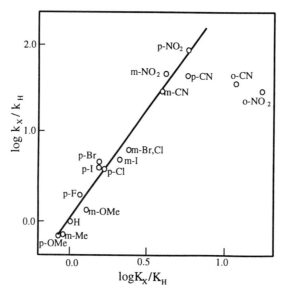

Figure 9.18 A plot of the rate constant for the hydrolysis of a series of substituted ethylbenzoates as a function of the dissociation constant of the corresponding benzoic acid. (Data of Ingold and Nathan [1936] and Evans, Gordon, and Watson [1937].)

acid, K_H. Notice that all of the points for the meta and para- groups lie close to the same line. There are significant deviations from the line for ortho- substitutions. We show below that the fact that such linear relationships exist implies that one can understand the effect of meta- and para- substitutions by looking at how the substituent affects the free energy of the reaction. The fact that there are deviations in the ortho- position implies that there is an additional effect of an ortho-substituted group. In the example in Figure 9.18 there is a steric hindrance; a group bound to the ortho-position gets in the way of the reaction.

Most linear free-energy plots look like Figure 9.18. One observes a near linear relationship between the rate of the reaction and the free energy of dissociation of the corresponding benzoic acid for meta- and para-substituted species, but not ortho-substituted species.

Hammett fit these data to a line of the form

$$\ln \left(\frac{k_X}{k_H} \right) = \gamma_H \ln \left(\frac{K_X}{K_H} \right) = \gamma_H \sigma_1^o \tag{9.43}$$

where γ_H is a constant, and

$$\sigma_1^o = \ln \left(\frac{K_X}{K_H} \right) \tag{9.44}$$

σ_1^o is now called the Hammett constant.

For the work to follow, it is interesting to relate Equation 9.43 to Equation 9.5. Let's first consider the left side of the equation. If one substitutes expressions for k_X and k_H from Equation 9.1 into $\ln(k_X/k_H)$, one obtains

$$kT \ln \left(\frac{k_X}{k_H} \right) = E_a^H - E_A^X + kT \ln \left(\frac{k_X^o}{k_H^o} \right) \tag{9.45}$$

where E_a^H and E_a^X are the activation energies for the hydrolysis of ethylbenzoate and X-substituted ethylbenzoate and k_H^o and k_X^o are the preexponentials for the same reactions.

Similarly, substituting expressions for K_X and K_H from Equation 9.3 into $\ln(K_X/K_H)$ shows

$$kT \ln \left(\frac{K_X}{K_H} \right) = \Delta G_{ac}^H - \Delta G_{ac}^X \tag{9.46}$$

where ΔG_{ac}^H and ΔG_{ac}^X are the free energies of dissociation of benzoic acid and X-substituted benzoic acid, respectively.

Substituting Equations 9.45 and 9.46 into Equation 9.43 shows

$$E_a^X = \gamma_H(\Delta G_{ac}^X - \Delta G_{ac}^H) + E_a^H + kT \ln \left(\frac{k_H^o}{k_X^o} \right) \tag{9.47}$$

Hence, Equation 9.43 is almost a linear free-energy relationship with

$$E_a^0 = E_a^H + \gamma_H \Delta G_{ac}^H + kT \ln \left(\frac{k_H^o}{k_X^o} \right) \tag{9.48}$$

Of course, in principle the last term in Equation 9.48 can vary with the substituent, X. Hence, Equation 9.43 is not quite a linear free-energy relationship. It is easy to show, however, that if one substitutes

$$\sigma^0 = \ln \left(\frac{K_X}{K_H} \right) + \frac{1}{\gamma_H} \ln \left(\frac{k_H^o}{k_X^o} \right) \tag{9.49}$$

for σ_1^o, the last term in Equation 9.48 cancels, and so one does get an exact linear free-energy relationship. The implication of Equation 9.49 is that if the Polayni relationship works, one should adjust σ_1^o slightly to account for variations in preexponential factors.

People in the physical-organic chemistry community have adjusted the values of σ_1^o to account for these effects. Many of the adjustments were done empirically. Reviews of this work include Chapman and Shorter [1972, 1978] and Wells [1968]. The corrected values are labeled σ^0 in Table 9.2. Figure 9.19 shows a replot of the data from Figure 9.18 to illustrate the kind of fit that is obtained. Generally, one can fit a wide range of data provided that one adjusts the constants appropriately.

Of course, there is a very important difference between Equation 9.47 and Equation 9.5; ΔG_r in Equation 9.5 is the overall free-energy change for the reaction, while ΔG_{ac}^X is the free energy of dissociation of benzoic acid. Hence, Equation 9.47 is only a true linear free-energy relationship when ΔG_r is linearly dependent on ΔG_{ac}.

There is some important physics in this assumption. The hydrolysis of ethylbenzoate occurs via electrophylic attack of a hydroxyl group onto the α carbon

$$\text{X} - \bigcirc \hspace{-1.5em} \underset{\text{OH}^-}{\overset{\displaystyle \overset{\text{O}}{\parallel}}{\text{C} - \text{OCH}_2\text{CH}_3}} \tag{9.50}$$

The transition state is a charged complex

$$\text{X} - \bigcirc \hspace{-1.5em} \underset{\underset{\text{H}}{\overset{|}{\text{O}}}}{\overset{\displaystyle \overset{\text{O}}{\parallel}}{\text{C} = \text{OCH}_2\text{CH}_3}} \tag{9.51}$$

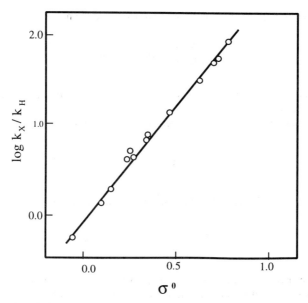

Figure 9.19 A replot of the meta- and para- data from Figure 9.18 using the modified values of σ^0 from the tabulation of Exner [1978].

Notice that if the substitutent X withdraws electrons, it will stabilize the transition state and thereby enhance the rate. If the species donates electrons, it will destabilize the transition state and thereby decrease the rate.

Now consider the effect of an electron-donating or -withdrawing group on the dissociation of benzoic acid:

$$X - \langle \bigcirc \rangle COOH + H_2O \longrightarrow X - \langle \bigcirc \rangle COO^- + H_3O^+$$

Notice that the dissociation process produces a charged species. Hence, an electron-withdrawing substituent will enhance the dissociation of benzoic acid, while an electron-donating substituent will reduce the dissociation of benzoic acid. As a result, the influence of a given substituent group in stabilizing the transition state shown in Equation 9.51 is similar to the influence of the substitutent group in enhancing the dissociation of benzoic acid. Hence, one would expect there to be a correlation between the effect of a given substituent on the acidity of benzoic acid, and the rate of the dissociation of ethylbenzoate, as is seen in Figure 9.18.

One can imagine many generalizations of this idea. For example, if a given reaction has a negatively charged transition state similar to the one in Reaction 9.51, then a given substituent may have a similar effect on the rate of the given reaction as on the rate of the dissociation of benzoic acid. In such a case, one would expect there to be a simple relationship between the rate constant for the reaction and the dissociation constant of benzoic acid.

As an example of this, consider the reaction between hydrogen peroxide and a series of substituted benzenesulfinic acids:

$$X - \langle \bigcirc \rangle SO_2^- + H_2O_2 \longrightarrow X - \langle \bigcirc \rangle SO_3^- + H_2O$$

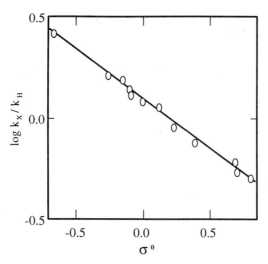

Figure 9.20 A plot of the log of the rate constant for the reaction of a series of substituted benzene-sulfinic acids with hydrogen peroxide against σ^0 values determined for substituted benzoic acids. (Data of Lindberg [1966].)

This reaction also has a negatively charged transition state, so one might expect it to follow a correlation similar to those described earlier in this section. Figure 9.20 shows a plot of Lindberg's [1966] measurements of the rate constant for the reaction against σ^0 values determined for substituted benzoic acids. Notice that one still gets a linear fit even though there is no benzoic acid in the reaction. The slope of the plot is negative because the benzene ring is less charged in the transition state than at the beginning of the reaction. As a result, an electron-withdrawing group destabilizes the transition state relative to the reactants.

Experiments have shown that there are many reactions that follow the Hammett correlation with σ^0 for substituted benzoic acids, even though there is no benzoic acid in the system. For example, Table 9.1 lists a series of acids whose dissociation constants correlate with σ^0 for benzoic acid. The constant γ_{K_a} is the slope of a log-log plot of the dissociation constant of the acid versus the dissociation constant of benzoic acid. Notice that many of the acids are quite different from benzoic acid. However, a given substituent changes the dissociation constant of the acid in the same way as the substituent changes the dissociation constant of benzoic acid. The implication of Table 9.1 is that if we wanted to construct a linear free-energy plot for a reaction of any of the acids listed in table, we could simply correlate the data to σ^0 values calculated for benzoic acid.

In fact, however, one should not push this idea too far because in a standard Hammett plot, one is assuming that a substituent affects the transition state in the same way if affects benzoic acid. If molecules are enough different from benzoic acid, the substituents can have quite different effects than they have on benzoic acid. Hence corrections would be needed.

For example, Figure 9.21 shows a Hammett plot for the reaction

$$(XAr)_3C-Cl \rightleftharpoons (XAr)_3C^+ + Cl^- \tag{9.52}$$

Notice that the measured rate is often 1 to 2 orders of magnitude different from the rate predicted by the linear free-energy curve. Hence, one could not use the line in Figure 9.21 to make useful predictions.

The deviations from the line in Figure 9.8 occur because the transition state for Reaction 9.52 is a positively charged species that has little in common with the negatively charged transition state seen during benzoic acid dissociation.

TABLE 9.1   A Series of Acids Whose Dissociation Constants Correlate with the Dissociation Constants of Substituted Benzoic Acids at 25°C

Acid	Solvent	γ_{K_a}
$XC_6H_4COOH \rightleftharpoons XC_6H_4COO^- + H^+$	H_2O	1.00
	50%aq. C_2H_5OH	1.60
	C_2H_5OH	1.96
$XC_6H_4CH_2COOH \rightleftharpoons XC_6H_4CH_2COO^- + H^+$	H_2O	0.49
$XC_6H_4CH_2CH_2COOH \rightleftharpoons XC_6H_4CH_2CH_2COO^- + H^+$	H_2O	0.21
$XC_6H_4CH{=}CHCOOH \rightleftharpoons XC_6H_4CH{=}CHOO^- + H^+$	H_2O	0.47
$XC_6H_4(NH_3)^+ \rightleftharpoons XC_6H_4NH_2 + H^+$	H_2O	2.77
	30%aq. C_2H_5OH	3.44
$XC_6H_4OH \rightleftharpoons XC_6H_4O^- + H^+$	H_2O	2.11
$XC_6H_4PO(OH)_2 \rightleftharpoons XC_6H_4PO \cdot OH \cdot O^-$	H_2O	0.76
	50% aq. C_2H_5OH	0.99

Source: Johnson [1973].

In order to account for these differences, workers in the physical-organic chemistry community have determined different values of σ for different classes of reactions as indicated in Table 9.2. σ_m^0 and σ_p^0 are the adjusted Hammett constants for meta- and para-substituted benzoic acid. σ_m^+ and σ_p^+ are used to correlate data for nucleophilic reactions (i.e., reactions with negative ions). σ_m^+ and σ_p^+ are values of the Hammett constants that have been adjusted for aryl chlorides. They are usually used to correlate data for reactions with electrophiles, or reactions with a positively charged transition state. σ_I and σ_R are constants used to correlate data for alkanes. People have found that reactions of para-substituted amines with nucleophiles do not follow any of these plots due to the presence of the lone pair on the amine, and so they have defined yet another Hammett constant, σ_p^-, for this situation. Note that we only list values of σ_p^- where the values deviate significantly from σ_p^0.

Figure 9.22 shows how an example of well-corrected values of σ fit actual data. A comparison of Figure 9.21 and Figure 9.22 shows that indeed one does fit data better if one uses

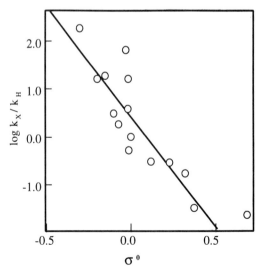

Figure 9.21   Hammett plot for the ionization of $(XAr)_3C$–Cl in liquid SO_2. (Replotted from data in Isaacs [1987].)

TABLE 9.2   A Selection of the Hammett Constants[a]

Substituent	σ_m^0	σ_p^0	σ_p^+	σ_p^-	σ_m^+	σ_I	σ_R
CH$_3$	−0.06	−0.14	−0.31		−0.07	−0.04	−0.11
−C$_2$H$_5$	−0.08	−0.13	−0.30			−0.02	−0.10
−(CH$_2$)$_3$ CH$_3$	−0.08	−0.16				+0.04	−0.12
−C(CH$_3$)$_3$	−0.09	−0.15	−0.26		−0.06	−0.03	−0.12
−OCH$_3$	+0.10	−0.12	−0.76		+0.05	+0.29	−0.45
−OCH$_2$ CH$_3$	+0.34	−0.14	−0.82			+0.26	−0.44
−F	+0.37	+0.15	−0.07		+0.35	+0.50	−0.34
−Cl	+0.37	+0.24	+0.11		+0.40	+0.47	−0.22
−Br	+0.37	+0.26	+0.15		+0.41	+0.47	−0.20
−I	+0.34	+0.28	+0.13		+0.36	+0.41	−0.16
−NO$_2$	+0.71	+0.81		+1.23	+0.67	+0.72	−0.15
−CN	+0.62	+0.70		+0.99	+0.56	+0.57	−0.13
−N(CH$_3$)$_2$	−0.10	−0.63	−1.67			+0.07	−0.33
−N(CH$_3$)$_3^+$Cl$^-$	+0.71	+0.57	+0.41		+0.36	+0.73	+0.15
−NH$_2$	−0.09	−0.30	−1.31		−0.16	+0.12	−0.48
−CO CH$_3$	+0.36	+0.47	−0.78	+0.82		+0.28	+0.16
−COOR	+0.35	+0.44	−0.83	+0.84	+0.37	+0.32	+0.14
−CF$_3$	+0.46	+0.53	−0.65	+0.65		+0.42	+0.08
−OH	+0.02	−0.22	−0.92	+0.64		+0.25	−0.40

[a]Recommended by Exner [1978], Jonson [1973], and Hansch, Leo, and Taft [1991].

modified values of σ. Hence, there is some advantage in using a modified σ scale. The disadvantage of this approach is that if we need to define a new σ scale for each set of reactions, we have in effect limited the generality of the methods. Therefore, people generally limit themselves to about four σ scales. There is one big advantage of having several σ scales, however. One can use the fact that there are multiple σ scales to learn something about transition states for reactions. Reactions that follow the σ^0 scale have transition states similar to those in the dissociation of benzoic acid. The carbon center can have a small residual positive or negative charge, but generally the charges are small. In contrast reactions that

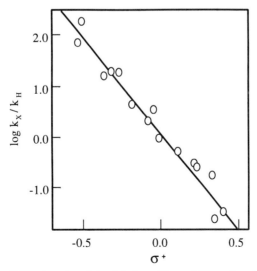

Figure 9.22   A replot of the data from Figure 9.21 on the σ^+ scale.

correlate with the σ^- scale have a partial lone pair (i.e., a large negative charge) in the transition state, while reactions that correlate to the σ^+ have empty orbital (i.e., a large positive charge) in the transition state. Hence, if we know that a reaction follows a specific σ scale, we can learn some important information about the transition state for the reaction. This is the key result in this section. Applications of this idea are illustrated in Section 9.15.

There are two other key σ scales in Table 9.2, and σ_R scale and the σ_I scale. The σ_I scale is called the inductive scale. It measures the ability of substituents to withdraw electrons to a reaction center. A species that has a strong interaction with the negative charge on the carbon center in the transition state in Reaction 9.51 would have a large value of σ_I. The σ_R scale is called the resonant scale. It measures the ability of a substituent to distribute the charge in a pi bond or ring. A species that has a large value of σ_R would increase the rate of in Reaction 9.50 when the species adsorbs in the para- position, but much smaller interactions will be seen in the meta- position (see Isaacs [1987] for a good description of the resonant interaction).

9.7 GROUP CONTRIBUTION METHODS

The idea that a given substituent can exert a similar influence on many different reactions deserves further comment. Some years ago Pauling [1960] examined all of the thermochemical data in the literature and noted that there are some general trends. Atoms that make strong bonds to one material tend to make strong bonds to another material. For example, Figure 9.23 shows a plot of the heat of formation of a series of metal nitrides, carbides, and formates as a function of the heat of formation of the corresponding oxide. Notice that materials that form a strong oxide also form a strong nitride, carbide, formate, etc. Hence, one can get a reasonable estimate of the heat of formation of, for example, a tungsten–carbon bond by first

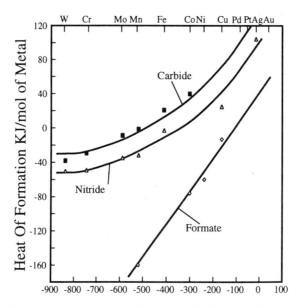

Figure 9.23 The heat of formation of a series of metal nitrides and carbides and the heat of reaction for the production of the corresponding metal formates (from formic acid) as a function of the heat of formation of the corresponding oxide.

looking up the strength of a tungsten oxygen bond, and then correcting the data for the difference between a metal–carbon bond and a metal–oxygen bond. These ideas lead to the Tamaru–Tanaka plots discussed in Chapter 3.

Benson [1958, 1978] expanded Pauling's idea and showed that one can estimate a wide variety of thermodynamic properties of molecules by adding up the contributions of all of the bonds in the molecule, then correcting for the influence of various side chains. These methods are illustrated in Example 9.C. Benson also showed that one can use group contribution methods to estimate entropies of activation.

The implication of Pauling and Benson's results is that if one wants to create a linear free-energy plot, one does not necessarily need to know the free energy of the reactants and products. Rather, one can use an analogy where one sees how the free energy of a closely related species varies, and use the data for a closely related species to calculate the free-energy plot. We see how to use these ideas for reactions on surfaces in Chapter 10.

9.8 RELATIONSHIP BETWEEN THE VARIOUS TRANSFER COEFFICIENTS

The weakness with using data other than the actual free energy of reaction to compute the linear free-energy plot is that when one uses a quantity other than the real free energy of reaction, the transfer coefficient, γ_H, is difficult to interpret. As noted in Section 9.5.3, according to the Marcus equation

$$\gamma_P = \frac{1}{2} + \frac{\Delta G_r}{8\Delta G_0^{\pm}} \tag{9.53}$$

Therefore, we can interpret γ_P, the transfer coefficient derived from Equation 9.5, very easily. However, a comparison of Equations 9.5 and 9.47 shows that

$$\gamma_H = \frac{\gamma_P(\Delta G_r^X - \Delta G_r^H)}{\Delta G_{ac}^X - \Delta G_{ac}^H} \tag{9.54}$$

where ΔG_r^X and ΔG_r^H are the actual free energy of reaction when the reactants are substituted with a given substituent, X, and when $X = H$. If we assume that γ_P is about 0.5, then we can use Equation 9.54 to calculate typical values of γ_H. When one does that one finds that γ_H can have essentially any value, which implies that it is difficult to extract any physics from a numerical value of γ_H.

Still, if one knows all of the free energies, one can use Equation 9.54 to back calculate γ_P and then get some useful information. An alternative approach is to measure rates and activation energies for a number of reactions, calculate an average value of E_a^0, then use Equation 9.5 to back calculate the transfer coefficient.

Albery and Kreevoy [1978] did that for a number of methyl transfer reactions of the form

$$Nu + CH_3Le \rightarrow CH_3Nu + Le \tag{9.55}$$

where Nu is a nucleophile and Le is a leaving group. Table 9.3 gives a selection of the results they obtained. Notice that as with the electrochemical examples, all of the values of γ_H except one are between 0.35 and 0.6, implying that all of the reactions go via fairly symmetric transition states. Of course, the transfer coefficients for the displacement of leaving groups with CN^- and OH^- are smaller than the rest. The latter cases are very exothermic, which leads to an asymmetric, early transition state.

This is a general trend. Experimentally, Polayni transfer coefficients are normally close to

TABLE 9.3   Transfer Coefficients, γ_P, for a Number of Methyl Transfer Reactions

Nucleophile	Leaving Group					
	$C_6H_5SO_3^-$	NO_3^-	F^-	Cl^-	Br^-	I^-
NO_3^-	0.43	0.50	0.50	0.50	0.49	0.50
Cl^-	0.43	0.50	0.50	0.50	0.49	0.50
Br^-	0.43	0.50	0.50	0.51	0.50	0.51
I^-	0.42	0.50	0.49	0.50	0.49	0.50
H_2O	0.43	0.50	0.49	0.49	0.49	0.50
CN^-	0.34	—	0.38	0.38	0.37	0.38
OH^-	0.37	0.42	0.42	0.41	0.41	0.41

Source: Albery and Kreevoy [1978].

0.5 except in the cases of very exothermic or very endothermic reactions, in agreement with the predictions of the Marcus equation.

There are a few exceptions to this general trend, however. For example, Bordwell and Boyle (1975) examined the reactions of the form

$$B^- + X-ArCH_2NO_2 \rightarrow BH + X-ArCH=NO_2^- \tag{9.56}$$

and found that the Polayni transfer coefficient, γ_P, was 1.5. A transfer coefficient of 1.5 implies that a given substituent stabilizes the transition state more than it stabilizes the reactants or products of the reaction. Note that according to Equation 9.10, it is impossible to get a transfer coefficient greater than 1. Hence one cannot explain the magnitude of the transfer coefficient for Reaction 9.56 based on the Polayni model. The Marcus model does allow transfer coefficients to be greater than 1 (see Equation 9.29). However, γ_P is greater than 1.0 only in very endothermic reactions. Further, the free-energy plot is curved in such cases. However, Reaction 9.56 shows a high transfer coefficient even though it is not unusually endothermic. Further, the free-energy plot is not curved. Hence, one cannot explain the data for Reaction 9.56 based on the Marcus equation.

Pross [1985] provides one other example that shows anomalous behavior. To keep this in perspective, so far these are the only two examples in the thousands of reactions that have been studied that show such anomalies. However, non-Marcus behavior does exist.

9.9   APPLICATIONS OF THE MARCUS EQUATION IN INORGANIC CHEMISTRY

There also have been important deviations from the simple Marcus–Polayni relationship in inorganic chemistry. Bolton, Magatagar, and McLendon [1991], Jordon [1991], Sutin [1968, 1983], and Cannon [1980] discuss applications of the Marcus equation to rates of electron transfer reactions in solution. Generally, one uses a model to estimate w_r in Equation 9.22. One then takes measured values for the activation energy for a self-exchange reaction such as Equation 9.25, and plugs into Equation 9.22 to calculate $\Delta G_0^{\ddagger}$. One can then calculate the activation energy for other reactions such as those in Reaction 9.24. Generally, one finds that activation energies calculated in this way are accurate to within a kcal/mole or so. Absolute rates are not as accurately predicted, however, because the transmission coefficients can be substantially less than unity (see Sutin [1983] or Jordan [1991] for details).

There have also been some attempts to apply linear free-energy relationships to ligand exchange reactions. A typical case is the reaction of a metal ion, M^{+z}, with a ligand, L^-, to

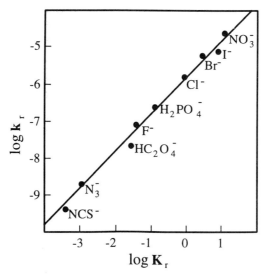

Figure 9.24 A plot of the rate constant for the hydrolysis of a series of cobalt(III) $[(NH_3)_5CoL_2]^{2+}$ complexes against the equilibrium constant for the reaction. (Data of Haim [1970].)

form a complex

$$(M^{+z})(H_2O)_n + L^- \rightarrow (ML)^{+z-1}(H_2O)_{n-1} + H_2O \qquad (9.57)$$

where the notation $(H_2O)_n$ is used to keep track of the solvent cage for the metal atom in solution.

There have been a few examples where people have measured both the rate constant and the equilibrium constant for a reaction. For example, Figure 9.24 shows a plot of the rate of the hydrolysis of a series of cobalt(III) salts

$$[(NH_3)_5CoL_2]^{2+} + H_2O \rightleftharpoons [(H_2O)(NH_3)_4CoL_2]^{2+} + NH_3 \qquad (9.58)$$

against the equilibrium constant for the reaction. Note that the data do follow a straight line as expected from the Polayni relationship (Equation 9.5).

Plots like those in Figure 9.24 are unusual, however. For example, Figure 9.25 shows a Brønsted plot for the equilibrium constant of the reactions of Cu^{+2} with a series of substituted hydroxyquinilines. Note that while some of the points lie on the line, there are substantial deviations. This is typical. As a result, it has been difficult to apply simple one-parameter Brønsted/Hammett-type relationships to inorganic reactions.

Pearson [1968] and Drago and Wayland [1965], Drago, Vogel, and Needham [1971], and Drago [1990] have done considerable work to understand when the simple Brønsted/Hammett-type relationships work in inorganic chemistry, and when the relationships fail. The basic conclusion from their work is that ligands are held to inorganic molecules by a mixture of "ionic" and "covalent" forces, and in general one needs to consider both ionic and covalent effects to adequately model the system. Usually, a two-parameter model is needed. However, one can get useful one-parameter correlations by limiting the range of ligands one considers so that only the ionic or covalent terms change.

Pearson [1963, 1968] developed a scheme to choose groups of ligands with the right properties. Pearson viewed ligand exchange reactions as processes where a Lewis acid reacted

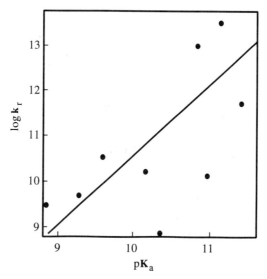

Figure 9.25 A plot of the equilibrium constant for the reaction between a Cu^{2+} ion and a series of substituted hydroxyquinolones vs. the pK_a of the hydroxyquinolone. (Adapted from Irving and Rissotti [1956].)

with a Lewis base. He then proposed dividing metals and ligands into three classes: **soft** acids and bases, which interact mainly by covalent forces; **hard** acids and bases, which interact mainly by ionic forces; and **borderline** acids and bases, which interact by a mixture of ionic and covalent forces, as discussed in Section 3.7.

Pearson [1968] showed that generally, hard acids interact strongly with hard bases, and soft acids interact strongly with soft bases. However, the interaction between a hard acid and a soft base or a soft acid and a hard base is relatively weak.

Table 3.8 shows a selection of the acids and bases considered by Pearson. Note that fairly ionic species are hard bases, while less ionic species are soft bases. Pearson's [1963, 1968] ideas, while important, did not prove to be quantitative. Drago and Wayland [1965], Drago et al. [1971], and Drago [1970, 1990] showed that all Lewis acids and bases have a mixture of ionic and covalent character. As a result, one cannot explain the behavior of acids and bases by focusing on only the ionic or only on the covalent part of the interaction.

Drago and Wayland [1965] and Drago et al. [1975] proposed an empirical way to account for the interaction. The method is called Drago's E and C method. Drago defined two parameters for a ligand E_{Drago} and C_{Drago} where E_{Drago} is a measure of the ability of a ligand to form ionic bonds, while C_{Drago} is a measure of the ability of a ligand to form covalent bonds. He then showed that one can estimate ΔH_{AB}, the strength of the interaction between a ligand A and B from

$$\Delta H_{AB} = E^A_{Drago}E^B_{Drago} + C^A_{Drago}C^B_{Drago} \tag{9.59}$$

where E^A_{Drago}, C^A_{Drago}, E^B_{Drago} and C^B_{Drago} are the Drago parameters for A and B.

Unfortunately, while Drago's parameters have proved to be useful ways to estimate heats of reaction, they have not proved useful for kinetics. Edwards and Pearson [1961] tried modified parameters. The modified parameters work for reactions involving only covalent interactions. However, they do not work for ionic reactions or mixed ionic/covalent reactions. Hence, we need to look elsewhere to quantitatively model the deviations from Brønsted/Hammett behavior.

9.10 REACTIONS AS ELECTRON TRANSFER PROCESSES

Molecular orbital (MO) theory is the place to start. As noted in Section 8.1, one can, in principle, calculate the entire potential energy surface exactly from MO theory. As a result, one should be able to understand all of the characteristics of the potential energy surface from MO theory.

All of the models we discuss here will view reactions as electron transfer processes. The idea is that when a reaction occurs, some bonds break and others form. One can model that process as an exchange of electrons between one orbital and another or equivalently as a sharing of electrons between one orbital with another and make useful predictions.

The idea of treating reactions as electron transfer processes was developed mainly by Fukui et al. [1952], Woodward and Hoffman [1965, 1970], Pearson [1969, 1976], and Klopman [1974]. In the sections below we describe some of the key results of the analysis. One is referred to Klopman [1974], Woodward and Hoffmann [1970], Fukui [1975], Pearson [1976], Pross [1985], or Parr and Yang [1988] for further information.

9.10.1 Perturbation Theory

The earliest useful results came from perturbation theory. In 1952, Kukui et al. noted that reactions can be often viewed as an electron transfer process where electrons are exchanged between a Lewis acid to a Lewis base. Fukui et al. then noted that it is easiest to exchange charge if charge is taken out of the highest occupied molecular orbital (HOMO) on the electron donor (i.e., Lewis base) and put it into the lowest unoccupied molecular orbital (LUMO) in the electron acceptor (i.e., Lewis acid). Fukui et al. labeled the HOMO and LUMO the **frontier orbitals** for the reaction, and proposed that one can learn how substituents affect rates of reactions by understanding how the energy of the frontier orbitals change when substituents are added to the molecule.

To see how the ideas work, consider the reaction in Equation 9.50 where a hydroxyl group interacts with a ethylbenzoate molecule to form the charged transition state shown in Equation 9.51. In this reaction, the hydroxyl group is the electron donor, while the ethylbenzoate is the electron acceptor. During the reaction, there is a partial charge transfer from the HOMO on the hydroxyl group and the LUMO on the benzoic acid.

Now consider what happens when we add a substituent that lowers the energy of the LUMO of ethylbenzoate. Note that such a substituent will make it easier for the ring to accept charge, so it will enhance the rate of reaction. Similarly, a substituent that raises the energy of the LUMO will make it harder for the ring to accept charge, so it will decrease the rate of reaction. Fukui et al. [1957] showed that electron-withdrawing groups decrease the energy of the LUMO, while electron-donating groups increase the energy of the LUMO. Further, one can correlate the Hammett constants σ_p^0 and σ_m^0 to the energy of the LUMO, and there is a one-to-one correspondence between the rate of Reaction 9.50 and the changes in the energy level of the LUMO. Hence, Fukui et al. concluded that one can qualitatively understand Hammett plots by understanding how the energies of the LUMOs and HOMOs change during a reaction.

Similar ideas can also be used to understand why different reactions need to use different σ scales. For example, during the reaction in Equation 9.52, the final state is an sp_2 hybrid with a vacant p orbital, which means that a vacant p orbital must form during the reaction. Interactions that lower the LUMOs on the aromatic ring can also stabilize the vacant p orbital. However, the energy shift of the p-orbital is different than the energy shift of the LUMO in Reaction 9.50. As a result, Reaction 9.52 follows a different Hammett plot than the reaction in Equation 9.50.

Brønsted/Hammett plots only work for organic reactions. When one has an inorganic reaction, several orbitals contribute to the rate. Further, one needs to examine the shapes and polarizabilities of the orbitals as well as their energies to predict reaction rates. Those complexities lead to deviations from the simple Brønsted/Hammett-type relationships.

Klopman [1968, 1974] analyzed these deviations. His analysis used a simplified model which assumed that the main interaction between a Lewis acid and a Lewis base was an interaction between the unoccupied orbitals localized on an atom in the acid and the occupied orbitals localized on a different atom in the base. Details of the analysis were given in Section 3.4. Klopman showed that one could classify reactions into two categories: **charge-controlled reactions,** where ionic forces dominate, and **frontier-orbital-controlled reactions,** where covalent forces dominate. Generally, hard acids and bases are more reactive for charge-controlled reactions, while soft acids and bases are more reactive for frontier-controlled reactions. Hence, the two classifications have proved to be useful.

Unfortunately, it is often difficult to make an a priori distinction between a charge-controlled reaction and a frontier-controlled reaction. Klopman tried to distinguish between the two by examining the properties of the orbitals available for reaction. However, he had to adjust parameters in the calculations to fit data, so it is unclear how general his results are. Nevertheless, Klopman concluded that reactions with soft centers such as peroxide oxygen or saturated carbon atoms will be a frontier-orbital-controlled reaction, while acylation of a carbonyl carbon will be a charge-controlled reaction, which does agree with experiment. Still, in the absence of data, it is difficult to distinguish between charge-controlled reactions and frontier-controlled reactions.

Fortunately, it is easy to distinguish between the two types of reactions experimentally. Generally, log of the rate of charge-controlled reactions varies linearly with pH. However, the rate of frontier-controlled reactions does not correlate well with pH. In contrast, the log of the rate constant for a frontier-controlled reaction varies linearly with the "Edwards parameter," where the Edwards parameter is a measure of the ability of nucleophiles to bind covalently. In contrast, the rate of charge-controlled reactions does not vary monotonically with the Edwards parameter. As a result, one can distinguish between charge-controlled reactions and frontier-controlled reactions by making Brønsted and Edwards plots and seeing which is linear.

As an example of this analysis consider the three groups of reactions of nucleophiles: reactions with peroxide centers, reactions with saturated carbon atoms, and reactions with carbonyl carbon atoms. Figure 9.26 shows Brønsted and Edwards plots for these reactions. Notice that the reaction of the nucleophiles with the carbonyl carbon follows a linear Brønsted plot where the rate increases monotonically with increasing pK_a. In contrast, the data for the other two reactions do not show a consistent trend with pK_a. On the other hand, if one plots the data against the Edwards parameter, the data for the reactions of the peroxide and alkyl carbon show consistent trends, while the data for the acylation of the carbonyl carbon scatter. These results show that acylation of the carbonyl carbon is a charge-controlled reaction, while the reactions of the peroxide oxygen and saturated carbon are frontier-controlled reactions.

There is some important physics in these observations. Generally, the pH or pK_a is a good measure of the propensity of a ligand to react with a hard center. However, when one reacts with a soft center, the data are better explained via some other energy scale, related more to the ionicity (i.e., reaction to produce H^-) rather than the acidity. Such results will be very important for surface reactions. Hence, Klopman's classification are quite important.

The problem with Klopman's results, however, is that they are not very quantitative. One cannot tell a priori which reactions will be charge controlled and which will be frontier controlled. Hence, a more quantitative model is needed.

9.10.2   Some Results from Density Functional Theory

Parr and Yang [1989] rederived Klopman's results from a more fundamental basis in an attempt to make them more quantitative. These efforts were discussed in Section 3.4.1 and

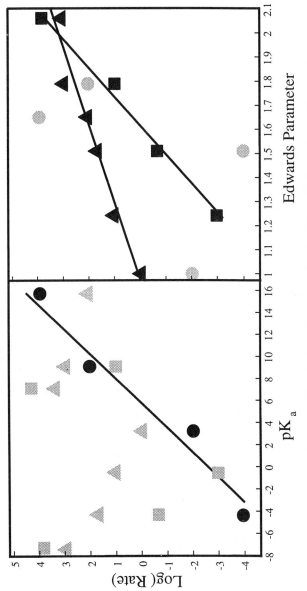

Figure 9.26 Brønsted and Edwards plots of reactions of nucleophiles with (squares) peroxide oxygen, (triangles) attack on a saturated carbon atom, and (circles) attack on a carbonyl carbon.

will not be repeated here in detail. However, the key result is that the reactivity of an inorganic molecule with a ligand can be calculated by examining three key parameters: the chemical potential of the electrons in the system

$$\mu_M = \left(\frac{\delta E_M}{\delta N_e}\right)_{\Omega(r)} \tag{9.60}$$

the hardness of the electrons

$$\frac{1}{2}\left(\frac{\partial \mu_M}{\partial N_e}\right)_{V(r)} = \eta_M \tag{9.61}$$

and the Fukui functions

$$\left(\frac{\partial \mu_M}{\partial \Omega(r)}\right)_{N_e} = Fu(r) \tag{9.62}$$

Parr also defines a softness, Sf_M, by

$$Sf_M = \frac{1}{2\eta_M} = \left(\frac{\partial N_e}{\partial \mu}\right)_{\Omega(r)} \tag{9.63}$$

Physically, the softness is a measure of the ease at which the molecule gives up and accepts charge due to changes in chemical potential. Politzer [1987] has shown that the softness is proportional to the polarization of the molecules. If Sf_M is small, the molecule will not accept or give up charge easily. Therefore, the "ionic" interactions will dominate. On the other hand, if the softness is large, there will be easy exchange of electrons so covalent bonds will form easily. Hence, covalent interactions will dominate.

The hardness η_M is equal to 0.5 over the softness. Therefore, the analysis in the last paragraph indicates that ionic contributions will dominate when the hardness is high, while covalent interactions will dominate when the hardness is low.

Pearson [1988] computed μ_M and η_M for a number of molecules, and showed that there was a correlation between the calculated values of μ_M and η_M and rate data. He also showed that there is a qualitative agreement between his model and his earlier divisions between hard and soft acids and bases and the calculated values of η_M. Hence, he proposed that the hardness and chemical potential of the reactants should be able to be used to predict behavior. At this point it is unclear to what extent these ideas are quantitative. However, they do explain many observations, including important observations in the field of surface reactions.

The reason Pearson's model might not fit quantitatively is that by only considering the hardness we are only considering the effects of the first term in Equation 3.46 and ignoring the effects of the second term. Fukui et al. [1952] and Fukui [1975] showed that the second term can also be quite important to reaction rates.

The second term in Equation 3.46 measures how the chemical potential of a molecule changes as we bring up another molecule and thereby change the atomic core potential around the original molecule. Physically, when we move an atom, which we will label A, close to a molecule M, the atomic core of A will attract some of M's electrons, while the outer electrons of A will repel M's electrons. The net extent of attraction and repulsion will depend on the electronegativities and hardnesses of the A and M. However, it will also depend on the detailed placement of the atomic cores as measured by $\Omega(r)$. The last term in Equation 3.46 accounts for these variations. Therefore, the Fukui functions are a measure of the effects of the geometry on the reaction rate.

In fact, Fukui [1975] discussed Fukui functions before density functional theory was invented. Parr and Yang [1989] present a derivation of the key results and we will not reproduce the derivation here. However, there are three key Fukui functions for electron transfer reactions: Fu^+, Fu^0, and Fu^-. Fu^+ is the Fukui function for adding an electron (i.e., the reaction for a nucleophile), Fu^0 is for reaction of a neutral species (i.e., a radical), and Fu^- is for loss of an electron (i.e., reaction with an electrophile). Parr shows that

$$Fu^+(r) \approx \rho_{LUMO}(r)$$

$$Fu^-(r) \approx \rho_{HUMO}(r)$$

$$Fu^0(r) \approx \tfrac{1}{2}[\rho_{LUMO}(r) + \rho_{HUMO}] \tag{9.64}$$

where $\rho_{LUMO}(r)$ and $\rho_{HUMO}(r)$ are the change in electron density when one adds an electron to the LUMO and HUMO, respectively.

Figure 9.27 shows a plot of Fu^+, Fu^0, and Fu^- for formaldehyde. It is not completely obvious from the plots, but Fu^+ is larger near the carbon atom than near the oxygen. Fu^0 and Fu^- are much larger near the oxygen atom than near the carbon. Therefore, one would expect nucleophiles to preferentially react with the carbon atom on the formaldehyde. In contrast, electrophiles and radicals should preferentially react with the oxygen. This prediction agrees with experiment.

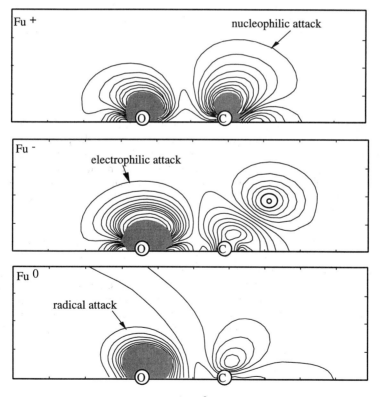

Figure 9.27 A plot of the Fukui functions Fu^+, Fu^0, and Fu^- for formaldehyde (CH_2O). (Adapted from Parr and Yang [1989].)

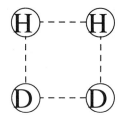

Figure 9.28 A stick diagram of a possible transition state for the reaction $H_2 + D_2 \rightarrow 2HD$.

Experimentally, one can explain many features of simple reactions by considering the hardness and electronegativity, plus the nature of the Fukui functions for the reaction. One can predict relative rates of reactions, selectivities, and stereospecificities. Further, the analysis is often quantitative. Hence, this analysis has proved to be quite useful.

9.11 LIMITATIONS OF THE SIMPLE MODEL

Still the analysis breaks down occasionally. The analysis so far in this chapter works for a reaction that follows Figure 9.10 where one bond breaks and another forms *simultaneously* during the reaction. However, there are special cases where bond scission and bond production cannot occur simultaneously. For example, one can imagine that during the reaction $H_2 + D_2 \rightarrow 2$ HD, the H–H bonds and D–D bonds are breaking and H–D bonds are forming simultaneously, as shown in Figure 9.28. However, quantum-mechanical arguments on page 674 show that both processes cannot occur simultaneously. Instead, one must completely break the H–H bond before the H–D bond can form. Consequently, there is an extra barrier associated with the bond-scission process which is not considered in the Polayni/Marcus model.

9.12 THE CONFIGURATION MIXING MODEL

Many investigators have developed rules to determine when it is possible for bonds to break and form simultaneously during the course of a reaction. Pross [1985] summarized the key ideas from several groups in what he calls the **configuration mixing model.** The idea is to treat a reaction as a process where molecular orbitals in the reactants are transformed into molecular orbitals in the products. One then analyzes the changes in electronic structure of the molecules which occur during reaction to see if bonds can break and form simultaneously as the reaction proceeds.

In the material that follows, we use the configuration mixing model to derive an equation analogous to the Polayni/Marcus equation. The derivation shows that there are some inherent assumptions in the Polayni/Marcus equation that sometimes work and sometimes fail. The failures lead to deviations from Polayni/Marcus behavior.

We start by reviewing the configuration mixing model. The analysis builds on concepts from MO theory. Recall that in MO theory one can write the wavefunction, ψ, for any MO of a system as an antisymmetrized product of the wavefunctions, ϕ_a, of all of the atoms in the system. For example, ψ_σ, the wavefunction for the σ bond in a diatomic molecule AB is given by

$$\psi_\sigma = \frac{A}{\sqrt{2}} (\phi_A + \phi_B)(\bar{\phi}_B + \bar{\phi}_A) \tag{9.65}$$

where ϕ_A is the wavefunction for an electron in the bonding orbital of A with the spin pointing up; $\bar{\phi}_A$ is the wavefunction for an electron in the bonding orbital of A with the spin pointing down; ϕ_B is the wavefunction for an electron in the bonding orbital of B with the spin pointing up; $\overline{\phi}_B$ is the wavefunction for an electron in the bonding orbital of B with the spin pointing down; and A is the antisymmetrizer. There are three important MOs for excited states of this system: $\psi_{\sigma*}$ the wavefunction for the $\sigma*$ orbital of the A–B molecule

$$\psi_{\sigma*} = \frac{A}{\sqrt{2}} (\phi_A - \phi_B)(\overline{\phi}_A - \overline{\phi}_B) \tag{9.66}$$

$\psi_{A:^-B^+}$, the wavefunctions for both electrons being on A

$$\psi_{A:^-B^+} = A\phi_A\overline{\phi}_A \tag{9.67}$$

and $\psi_{A:^+B:^-}$, the wavefunctions for both electrons being on B

$$\psi_{A:^+B^-} = A\phi_B\overline{\phi}_B \tag{9.68}$$

According to the configuration interaction (CI) model described in Section 3.5.1, we can write the wavefunction for any state of the system as a sum of the Slater determinates for these MOs, plus the Slater determinates for additional excited states. For example, if A and B form a bond with a significant dipole moment, then the wavefunction for the bonding state of the system, Ψ_{A-B} can be written as

$$\Psi_{A-B} = c_\sigma\psi_\sigma + c_{A:^-B^+}\,\psi_{A:^-B^+} + c_{A^+B:^-}\,\psi_{A^+B:^-} + c_{\sigma*}\psi_{\sigma*} \tag{9.69}$$

where the c's are constants. We used Ψ rather than ψ in Equation 9.69 to indicate that Ψ_{A-B} is a wavefunction that has been computed with the CI model.

Similarly, $\Psi_{A\cdot B\cdot}$, the wavefunction for one electron being on A and one being on B with no interaction between the two is

$$\Psi_{A\cdot B\cdot} = \frac{1}{\sqrt{2}} (\psi_\sigma + \psi_{\sigma*}) \tag{9.70}$$

For the discussion to follow we designate the wavefunction in Equation 9.70 as being the [A·B·] state of the system, and the wavefunction in Equation 9.69 as being the [A–B] state of the system. In some cases it will also be useful to discuss the [A$^+$B$^-$] state of the system. The [A$^+$B$^-$] state of the system is just the [A–B] state in the limit $|c_{A^+B:^-}| \gg |c_\sigma|$, $|c_{A:^-B^+}|$, and $|c_{\sigma*}|$.

In the material that follows, we consider how the electronic configuration of the A–B molecule changes as a reaction proceeds and use that information to predict the activation barrier for the reaction between A and B.

The derivation will start with a simple example where a sodium atom approaches a chlorine atom and reacts to form sodium chloride:

$$Na\cdot + Cl\cdot \rightarrow Na^+Cl^- \tag{9.71}$$

When the two atoms are far apart, the [Na·Cl·] configuration is the ground state of the system, while [Na$^+$Cl$^-$] configuration is the first excited state. In contrast, in sodium chloride the [Na$^+$Cl$^-$] state is the ground state and the [Na·Cl·] state is the first excited state.

Now consider doing a thought experiment where a sodium atom approaches a chlorine atom but the sodium atom and chlorine atom are prevented from exchanging electrons. According to calculations when the sodium atom approaches the chlorine atom the energy of both the [Na·Cl·] and [Na$^+$Cl$^-$] configurations go down as indicated in Figure 9.29. How-

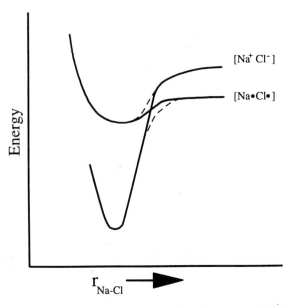

Figure 9.29 A schematic showing how the energy of the $[Na \cdot Cl \cdot]$ and the $[Na^+Cl^-]$ configuration of NaCl change as a sodium atom approaches a chlorine atom.

ever, the $[Na^+Cl^-]$ configuration is stabilized much more than the $[Na \cdot Cl \cdot]$ configuration. That is why the equilibrium structure is $[Na^+Cl^-]$.

Now let us do the same process but allow the sodium atom and chlorine atom to exchange electrons. When the sodium atom and chlorine atom initially approach one another they start out in the $[Na \cdot Cl \cdot]$ configuration. However, during reaction there is an exchange of electrons between the sodium and the chlorine to yield a $[Na^+Cl^-]$ species. The exchange of electrons occurs gradually during the course of the reaction, and so at any point along the reaction coordinate, one can represent the electronic structure of the system as a mixture of the $[Na \cdot Cl \cdot]$ and the $[Na^+Cl^-]$ configuration. For future reference we will indicate the reaction pathway as the lower dashed line in Figure 9.29. The reactants start out in the $[Na \cdot Cl \cdot]$ configuration. However, as the reaction proceeds there is a continuous change in the configuration of the system from the $[Na \cdot Cl \cdot]$ state into the $[Na^+Cl^-]$ state.

Note that if there was no mixing of the configurations, the sodium and chlorine would stay in the $[Na \cdot Cl \cdot]$ configuration. The atoms could not get into the $[Na^+Cl^-]$ configuration. Hence, one would not get any reaction. As a result it is the mixing of configurations that allows the sodium and chlorine react to form (metastable) sodium chloride.*

This example illustrates the idea that reactions can be represented as a conversion of the electronic structure of the reactants from one configuration to another. In the material that follows we model the changes in electronic structure and use that information to make predictions about reactions.

We start by noting that the conversion of one state into another can be modeled as something called an **avoided crossing**. Notice that if there were no interactions between the $[Na \cdot Cl \cdot]$ and $[Na^+Cl^-]$ configurations of sodium chloride, the curves for the energies of the two states in Figure 9.29 would cross. In reality, however, there are strong interactions between the two configurations. In Section 9.13.1 we show that when two states interact,

---

*As noted in Chapter 7, when an isolated sodium atom reacts with an isolated chlorine atom, a considerable amount of energy is released. The energy needs to be dissipated before a stable NaCl crystal can form. As a result, initially Reaction 9.71 produces a hot metastable complex, not crystalline sodium chloride.

they mix to form an upper state, whose energy is greater than that of either of the two individual states, and a lower state, whose energy is lower than that of either of the two individual states. The energy of the two states is given as the dashed line in Figure 9.29. Notice that because of the interactions between the [Na·Cl·] and [Na$^+$Cl$^-$] configurations, the energies of the eigenstates of the system never cross, hence the name "avoided crossing." One does not always get avoided crossings when energy curves for reactions cross. In Section 9.13, we provide several examples where the curves actually cross. However, the significance of the avoided crossing is that it provides a mechanism to allow the system to change from one configuration to another. When avoided crossings happen, one can continuously transfer an electron from one state of the system to another. That allows bonds to rearrange so a reaction can take place. If there is no avoided crossing, there is no change in the configuration of the reactants, and hence no reaction. As a result, avoided crossings are very important to the theory of reactions.

Woodward and Hoffmann [1970] show that one can model most reactions as a change in the electronic configuration of the reactants and an avoided crossing. (The main exception is in photochemical reactions, where the change in configuration can be driven by a photon rather than an avoided crossing.) Hence, the idea of describing a reaction as a change in the electronic configuration of the reactants with an avoided crossing is quite useful.

In the material to follow it is useful to draw what is called a **configuration mixing diagram** for a reaction, where the configuring mixing diagram is a plot showing the energy of each of the configurations of the system change as function of the reaction coordinate (e.g., bond order). The diagram also shows how the states mix.

Figure 9.30a shows a configuration mixing diagram for the reaction of a sodium atom with a chlorine atom. The figure contains the same information as was given in Figure 9.29. There are two solid lines in the coordination mixing diagram, one starting at the [Na·Cl·] state of the reactants and going to the [Na·Cl·] state in the products, the other starting at the [Na$^+$Cl$^-$] state in the reactants and going to the [Na$^+$Cl$^-$] state in the products. There are also dashed lines indicating the avoided crossing. The implication of this diagram is that during the reaction the [Na·Cl·] and [Na$^+$Cl$^-$] states mix continuously to convert the reactants to products.

Figure 9.30 also shows the configuration mixing diagram for the conversion of *cis*-2-

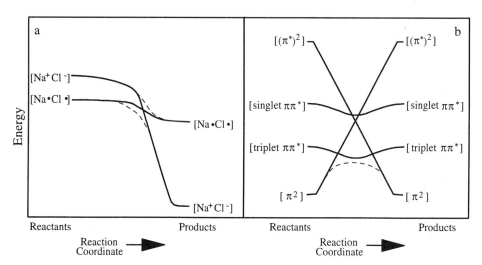

Figure 9.30 Configuration mixing diagram for (*a*) Na· + Cl· → NaCl and (*b*) the conversion of *cis*-2-butene to *trans*-2-butene.

butene into *trans*-2-butene:

$$
\begin{array}{ccc}
\underset{H}{\overset{H_3C}{\diagdown}}C{=}C\underset{H}{\overset{CH_3}{\diagup}} & \longrightarrow & \underset{H}{\overset{H_3C}{\diagdown}}C{=}C\underset{CH_3}{\overset{H}{\diagup}}
\end{array}
\qquad (9.72)
$$

The system starts out with two p-electrons in a π^2 orbital. For future reference, we note that there are several excited states of the system, a triplet $\pi\pi^*$ state, a singlet $\pi\pi^*$ state, and a singlet $(\pi^*)^2$ state. Now consider what happens when one rotates the :CHCH$_3$ group in the right half of *cis*-2-butene around the carbon–carbon bond, leaving the CH$_3$HC: group on the left fixed. Note that the rotation reverses the sign of all of the p orbitals on the right :CHCH$_3$ group. We start out with an electron in a π orbital with a wavefunction, ψ_π, given by

$$
\psi_\pi = \frac{1}{\sqrt{2}} (\phi_{p\,left} + \phi_{p\,right})
\qquad (9.73)
$$

where $\phi_{p\,left}$ and $\phi_{p\,right}$ are the wavefunctions for the p orbitals on the left and right :CHCH$_3$ group. Notice that when we rotate the right :CH$_2$CH$_3$ group, $\phi_{p\,right}$ is converted into $-\phi_{p\,right}$. As a result, and π orbital is converted into a π^* orbital with

$$
\psi_{\pi^*} = \frac{1}{\sqrt{2}} (\phi_{p\,left} - \phi_{p\,right})
\qquad (9.74)
$$

Similarly, the π^* orbitals are converted to bonding π orbitals. As a result, the original $(\pi^*)^2$ state is converted into the ground state of the system, while the original π^2 state is converted into a $(\pi^*)^2$ state.

Now consider what happens as the reaction proceeds. We start within the π^2 state. However, as we begin to twist the molecule, we begin to mix in some of the excited states. First the triplet $\pi\pi^*$ state begins to interact with the π^2 state. Later the original $(\pi^*)^2$ state plays a role. This diagram illustrates two important conclusions from the configuration mixing model: (1) the mixing of excited states with the ground state plays an important role in determining the potential energy surface for a reaction, and (2) often many different excited states play a role in the reaction.

The idea that the mixing of the ground state with the excited states of the system determines potential energy surfaces is quite important. For example, one can use the idea to explain why reactions are activated. When people first started to do reaction-rate theory, it was not obvious why activation barriers arose during reactions. The reaction of a hydrogen atom with a deuterium tritium molecule illustrates the difficulty.

$$
\text{H} + \text{DT} \rightarrow \text{HD} + \text{T}
\qquad (9.75)
$$

Note that as this reaction proceeds, a deuterium–tritium bond breaks and a hydrogen–deuterium bond forms. Initially, there is a single D–T bond. Halfway through the reaction there is half a H–D bond and half a D–T bond, while at the end of the reaction there is a single H–D bond. Hence, the total bond order is conserved during reaction.

Notice that there is no net bond scission at any point in the reaction. Hence, when people first began to study reactions, it was very difficult to understand why Reaction 9.75 would be activated. Yet, experimentally, the activation barrier is 8 kcal/mol.

The configuration mixing model explains in part why activation barriers arise during reactions. The general idea is that as a reaction proceeds, excited states are mixed into the ground-state wavefunction. The mixing of excited states raises the energy of the system, and hence produces an activation barrier. Physically, when excited states get mixed into the ground-state wavefunction, bonds are distorted. Hence, one might want to think about a barrier as being associated with distortion of bonds. However, it is better to think about the process in the converse way: When the molecules are distorted, the excited states of the molecule are mixed into the ground state. The mixing produces the barrier to molecular distortion, which in turn produces a barrier for a reaction. However, if a molecule could be distorted without mixing in any excited states, no barrier would arise. There are examples where there is no barrier to molecular distortion (e.g., a rotation around a bond) because the system stays in the ground state in all cases. Quantum mechanically, it is really the mixing of excited states that causes activation barriers to occur. Hence, the mixing of excited states is quite important to reactions.

It is useful to consider how the Marcus/Polayni relationship arises from the configuration mixing model. In the next few pages we show that the Marcus equation arises naturally from the configuration mixing model. Our approach will be to start with a two-state model where one of the states represents the reactant configuration, while the other state represents the product configuration. We then consider how a substituent affects the energy of each of the states. An analysis similar to that in Section 9.4 will be used to derive an equation analogous to the Marcus equation.

Consider the reaction

$$A\cdot\ +\ B\cdot\ \rightarrow\ A^+\ +\ B^- \tag{9.76}$$

A configuration mixing diagram for the reaction is shown as the solid line in Figure 9.31. The $[A\cdot B\cdot]$ configuration is the ground state of the reactants and an excited state of the products, while the $[A^+B^-]$ configuration is an excited state of the reactants and the ground state of the products. During the reaction the configuration of the system moves up the $[A\cdot B\cdot]$ curve, then down the $[A^+B^-]$ curve.

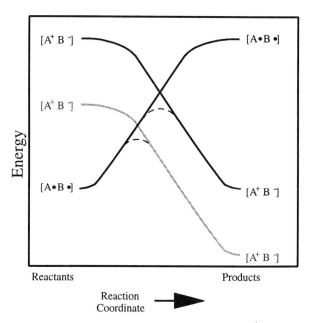

Figure 9.31 Configuration mixing model for the reaction $A\cdot\ +\ B\cdot\ \rightarrow\ A^+\ +\ B^-$, showing how changes in the energy of the $[A^+B^-]$ configuration affects the reaction.

Now consider making a change that displaces the curve for the $[A^+B^-]$ state up or down. Notice that such a displacement will cause the activation barrier to change. One can model the change by examining how the intersection between the curves shifts up or down as the reaction proceeds.

Notice that there is a close correspondence between the configuration mixing diagram in Figure 9.31 and the Polayni diagram in Figure 9.6. In both cases, the activation barrier is given as a curve crossing on a reaction coordinate diagram. Hence, the analysis in Section 9.4 also applies to the configuration mixing diagram Figure 9.31. In particular, if we approximate the curves in Figure 9.31 as lines, we can derive the Polayni relationship, while if we approximate the curves as parabolas, we will derive the Marcus equation. Hence, the Polayni relationship and Marcus equation apply equally well to the situation depicted in Figure 9.31 as to the situation depicted in Figure 9.6.

Of course, there will be a small correction to the analysis due to the lowering of the energy barrier at the avoided crossing. The derivation of the Marcus equation in Section 9.4 did not consider the avoided crossing. However, one can easily modify the derivation to consider the reduction of the barrier at the avoided crossing. For example, if reduction is a constant, the only effect of the reduction is to change the value of w_r. If the reduction changes with substituent, one will get extra terms in the derivation, which lead to deviations from the Marcus equation. However, normally, this is not a large effect, and so one can often get data to fit the Marcus equation by simply adjusting w_r.

Still, there are occasional deviations. The most important deviations come in the Marcus inverted region. In the Marcus inverted region, the activation barrier is very thin, which means that tunneling through the barrier becomes significant. In such systems small changes in the shape of the barrier lead to changes in the effective barrier. This can have a significant effect on the rate.

Of course, in the derivation of the Marcus/Polayni relationship in the last few paragraphs, we have assumed that the substituent displaces the energy of the excited state of the reactants by the same amount that it displaces the energy of the products. That is not an obvious assumption, but it is inherent in the Marcus formulation. Note that if the energy of the $[A^+B^-]$ state in the reactants moved relative to the $[A^+B^-]$ state of the products, the change in the activation energy would be different than if the $[A^+B^-]$ state in the reactants moved in concert the $[A^+B^-]$ state of the products. As a result, one would predict deviations from Marcus behavior.

A similar prediction can be made based on the original derivation of the Marcus/Polayni relationship, which we gave in Section 9.4. To see that, it is useful to consider how a change in the splitting between the ground state and the excited state at the A–B bond would be reflected in Figure 9.6. Note that at the minimum in the potential contour, the excited state does not contribute significantly to the energy. However, when the A–B bond stretches, there is a significant contribution from the excited state. Therefore, a change in the splitting between excited state and the ground state will cause a much larger change in the energy of A–B when the A–B bond is stretched than when A–B is in its equilibrium configuration. As a result, a substituent that changes the energy splitting between the ground state and the excited state of A–B will distort the potential contours in Figure 9.6. The Polayni/Marcus derivation did not allow the potential contours to change. As a result, an inherent assumption in the Marcus/Polayni formulation is that the excited states move in concert with the initial and final state of the system.

Of course, physically, we are talking about moving two states that are very similar. Note that in Figure 9.31 we are displacing the $[A^+B^-]$ configuration. The $[A^+B^-]$ configuration is the ground state of the products and an excited state of the reactants. However, it is still a single configuration of the system. A substituent that stabilizes the $[A^+B^-]$ state in the products would tend to also stabilize the $[A^+B^-]$ state in the reactants. As a result, the $[A^+B^-]$ state in the reactants tends to move in concert with the $[A^+B^-]$ state in the products.

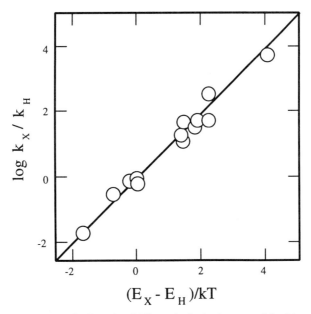

Figure 9.32   The rate constant for Reaction 9.77 vs. the ionization potential of the excited state of the X–C_6H_5 reactant. (Adapted from Fukuzumi and Kochi [1981].)

However, the two states do not have to move exactly in concert. That leads to deviations from Marcus behavior.

There have been few attempts to correlate data directly to the energy of the excited state and thereby avoid the errors associated with assuming that the excited states move in concert with the reactants. For example, Figure 9.32 shows a plot of Fukuzumi and Kochi's [1981] data for the rate constant for the reaction between mercury trifluoroacetate and a series of substituted benzenes

$$X-\!\!\langle\bigcirc\rangle\!\!-H + Hg(OOCF_3)_2 \longrightarrow X-\!\!\langle\bigcirc\rangle\!\!-Hg(OOCF_3) + HOOCF_3$$

$$(9.77)$$

versus the energy of the excited state of the substituted benzene. The energy of the excited state was measured spectroscopically. Their data included some points for ortho-substituted species that did not fit the correlations very well because of steric effects. However, when Fukuzumi and Kochi excluded the ortho points they found an excellent fit, as shown in Figure 9.32. The fit to the data is better than one gets with a standard Hammett plot. Hence, the idea that the changes in the energies of the excited states of the reactants is important in determining activation barriers to reactions has some validity, at least in this example.

The key assumption in the Marcus formulation is that the energy of the excited states moves in concert with the energy of the products. That assumption often works. However, it does not always work and, unfortunately, it is difficult to tell, a priori, when the assumption will fail. Hence, one does not always know, a priori, when there will be deviations from Marcus behavior.

Nevertheless, there are some situations where the assumption that the excited states move in concert with the ground state is particularly tenuous. Consider the example in Reaction

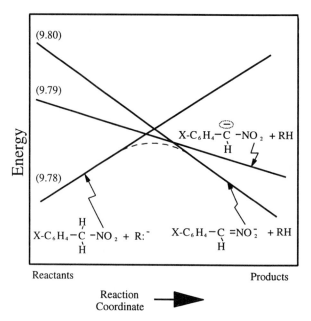

Figure 9.33 A configuration mixing diagram for Reaction 9.56. (Adapted from Pross [1985].)

9.56. Figure 9.33 shows a configuration mixing diagram for the reaction. The reaction starts when a nucleophile, R^- reacts with an arylnitrate:

$$\text{X} - \phi - \underset{\underset{\text{H}}{|}}{\overset{\overset{\text{H}}{|}}{\text{C}}} - \text{NO}_2 \qquad \overset{\text{R:}^-}{} \tag{9.78}$$

Pross [1985] shows that the ground state of the reactants has the electronic structure shown in Figure 9.34a with a negative charge on the nucleophile and neutral bonds. Pross also shows that the first excited state of the reactants has a

$$\text{X} - \phi - \underset{\underset{\text{H}}{|}}{\overset{(\dot{\ominus})}{\text{C}}} - \text{NO}_2 \qquad \text{R:H} \tag{9.79}$$

configuration and an electronic structure as shown in Figure 9.34b, while the second excited state has a

$$\text{X} - \phi - \underset{\underset{\text{H}}{}}{\text{C}} = \text{NO}_2^{\overline{}} \qquad \text{R:H} \tag{9.80}$$

configuration and an electronic structure as shown in Figure 9.34c. The products show the opposite pattern: The structure in Formula 9.80 is the ground state of the products, the structure in Formula 9.79 is the first excited state, while the structure in Formula 9.78 is the

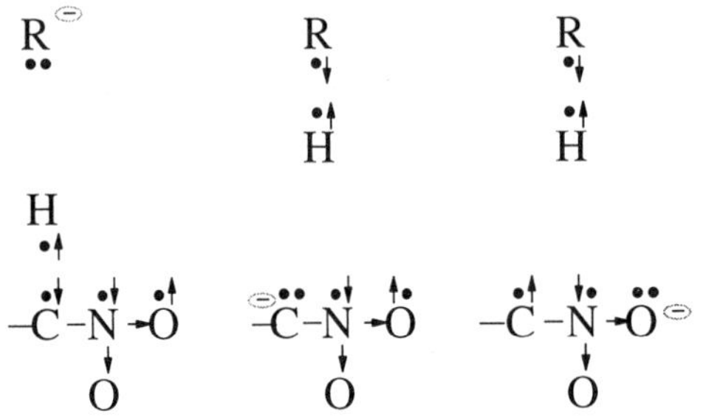

Figure 9.34 Some of the key configurations for Reaction 9.56 as proposed by Pross [1985].

second excited state. As a result, Pross concluded that the configuration mixing diagram will look like Figure 9.33, where the initial configuration will have an electronic structure like Figure 9.34a, and the final state will have an electronic structure like Figure 9.34c. However, the transition state will have a significant contribution from the intermediate energy structure 9.79.

Now consider what happens when an electron-withdrawing X-group is added to Species 9.78, 9.79, and 9.80. Note that an electron-withdrawing group will have little effect on the stability of the 9.78 configuration. The group will stabilize the 9.80 configuration of the molecule, but the stabilization will be slight because the charge is far from the ring. However, there will be a significant stabilization of the 9.79 configuration.

Note that the 9.79 configuration has a larger influence on the transition state than on the reactants or products. As a result, Pross concludes the electron-withdrawing substituents change the activation barrier for Reaction 9.56 much more than the substituents change the overall free-energy change of the reactant. That explains why the measured transfer coefficient for Equation 9.56 is larger than one expects from the Marcus equation.

Admittedly, it is not clear that Pross's analysis of Reaction 9.56 is completely correct. Bordwell and Boyle [1975], for example, suggests that instead a stable intermediate with a structure similar to that in 9.79 forms. However, the key point from Pross's analysis is that the transition state for a reaction may have contributions from configurations that do not contribute significantly to the reactants or products. As a result, the properties of the transition state can be different from the properties of the reactants or products. The basic assumption in the Polayni/Marcus relationship is that the transition state has properties somewhere between those of the reactants and products. Hence, the Polayni/Marcus formulation will fail when the transition state has properties that are not seen in the reactants or products.

Interestingly, Hammett-type relationships sometimes work under conditions where the Polayni relationship and Marcus equation fail. We noted in Section 9.5 that people have measured several different sets of Hammett constants. If one can find a Hammett constant that adequately describes the transition state, then one can correlate data to that Hammett constant, even though the Hammett constant is unrelated to ΔG_r for the reaction. Hence, the Hammett relationship is more general than the Polayni/Marcus relationship even though the Hammett relationship has less theoretical basis than the Polayni relation or Marcus equation.

9.13 SYMMETRY FORBIDDEN REACTIONS

There also can be some deviations from the Marcus/Polayni relationship during something called a **symmetry forbidden reaction.** Throughout the last section we assumed that when the energy contours for two states of the system cross, the states will mix to provide a pathway for a reaction. However, in 1965 Woodward and Hoffmann noted that sometimes states cannot mix. When states do not mix, one does not have a convenient pathway for a reaction. Hence, the reaction rate is negligible. Woodward and Hoffmann showed that the states will not mix when the symmetries of the states are wrong. They called reactions that show negligible rates because the reactant and product configurations do not mix **symmetry forbidden reactions.**

Woodward and Hoffmann [1970] wrote a famous book, *The Conservation of Orbital Symmetry*, which describes, in detail, the role of symmetry in determining rates of reactions. In the material that follows, we summarize the key ideas from Woodward and Hoffmann's analysis. We also review some other interpretations of the ideas due to Pearson [1976] and Fukui [1975]. One should refer to Woodward and Hoffmann [1970], Pearson [1976], or Fukui [1975] for further details.

9.13.1   Forbidden Crossings

In the next several sections we discuss symmetry forbidden reactions, i.e., reactions that cannot occur because symmetry prevents the key states from mixing. To start off, therefore, it is useful to consider when states can mix. Consider a reaction where molecules A and B come together and react. When the two molecules are far apart, we can describe the electronic structure of the system by a Hamiltonian, $\mathcal{H}_0^0$, with a ground-state wavefunction, Ψ_0, and a first excited-state wavefunction Ψ_1^0, where Ψ_0^0 and Ψ_1^0 satisfy

$$\mathcal{H}_0^0 \Psi_0^0 = E_0^0 \Psi_0^0 \tag{9.81}$$

$$\mathcal{H}_0^0 \Psi_1^0 = E_1^0 \Psi_1^0 \tag{9.82}$$

with $E_0 < E_1$.

Now consider doing a thought experiment where we move the two molecules together but do something to prevent the two molecules from reacting. Let's define a Hamiltonian for the nonreactive case as $\mathcal{H}_0(\Gamma)$, where we have noted that $\mathcal{H}_0$ is a function of Γ, where Γ is a generalized reaction coordinate. We assume that when $\Gamma = 0$, $\mathcal{H}_0(\Gamma)$ will equal the reactant Hamiltonian, and when $\Gamma = 1$, $\mathcal{H}_0(\Gamma)$ will equal the product Hamiltonian. We will also define $\Psi_0(\Gamma)$ and $\Psi_1(\Gamma)$ to be the eigenstates of the Hamiltonian that need to mix for the reaction to occur, and define $E_0(\Gamma)$ and $E_1(\Gamma)$ to be the corresponding eigenvalues of $\mathcal{H}_0(\Gamma)$.

Figure 9.35 shows a configuration mixing diagram for the reaction between A and B. The solid lines in the figure are plots of $E_0(\Gamma)$ and $E_1(\Gamma)$ as a function of the reaction coordinate. We have set up the system so that during the reaction Ψ_1 changes from the ground state to the first excited state, while Ψ_0 changes from the first excited state to the ground state.

Now consider moving the molecules together and trying to get the reaction to occur. When the reaction occurs the Hamiltonian of the system will be different from $\mathcal{H}_0$, so let's call the new Hamiltonian $\mathcal{H}_1(\Gamma)$. For the derivation that follows it is useful to define a constant $H(\Gamma)$ by

$$\mathcal{H}_1(\Gamma) = \mathcal{H}_0(\Gamma) + H(\Gamma) \tag{9.83}$$

The question that we want to address is whether the two states can mix when the atoms move together, so let us assume that the new ground state of the system has a wavefunction $\Psi'(\Gamma)$ that is a mixture of the wavefunctions for the two original configurations.

$$\Psi'(\Gamma) = c_0(\Gamma)\Psi_0(\Gamma) + c_1(\Gamma)\Psi_1(\Gamma) \tag{9.84}$$

where c_0 and c_1 are constants. We want the new wavefunction to satisfy

$$\mathcal{H}_1(\Gamma)\Psi'(\Gamma) = E'(\Gamma)\Psi'(\Gamma) \tag{9.85}$$

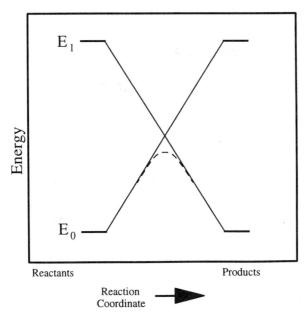

Figure 9.35 A configuration mixing diagram for a hypothetical reaction between A and B.

Substituting Ψ' from Equation 9.84 into Equation 9.85 yields

$$\mathcal{K}_1(\Gamma)\,(c_0(\Gamma)\Psi_0(\Gamma) + c_1(\Gamma)\Psi_1(\Gamma)) = E'(\Gamma)\,(c_0(\Gamma)\Psi_0(\Gamma) + c_1(\Gamma)\Psi_1(\Gamma)) \qquad (9.86)$$

Multiplying Equation 9.86 by $\Psi_0^*(\Gamma)$ and integrating over all space and noting that

$$\int \Psi_0^*(\Gamma)\,\Psi_0(\Gamma)\,d\vec{r} = 1 \qquad (9.87)$$

$$\int \Psi_0^*(\Gamma)\Psi_1(\Gamma)\,d\vec{r} = 0 \qquad (9.88)$$

yields

$$(E_0'(\Gamma) - E'(\Gamma))\,c_0(\Gamma) + \beta_{01}(\Gamma)\,c_1(\Gamma) = 0 \qquad (9.89)$$

with

$$E_0'(\Gamma) = \int \Psi_0^*(\Gamma)\,\mathcal{K}_1(\Gamma)\,\Psi_0(\Gamma)\,d\vec{r} \qquad (9.90)$$

$$\beta_{01}(\Gamma) = \int \Psi_0^*(\Gamma)\,\mathcal{K}_1(\Gamma)\,\Psi_1(\Gamma)\,d\vec{r} \qquad (9.91)$$

Similarly, multiplying Equation 9.86 by $\Psi_1^*(\Gamma)$ and integrating over all space, yields

$$(E_1'(\Gamma) - E'(\Gamma))\,c_1(\Gamma) + \beta_{10}(\Gamma)c_0(\Gamma) = 0 \qquad (9.92)$$

with

$$E_1'(\Gamma) = \int \Psi_1^*(\Gamma) \, \mathfrak{K}_1(\Gamma) \Psi_1(\Gamma) \, d\vec{r} \tag{9.93}$$

$$\beta_{10}(\Gamma) = \int \Psi_1^*(\Gamma) \, \mathfrak{K}_1(\Gamma) \Psi_0(\Gamma) \, d\vec{r} \tag{9.94}$$

Before we proceed it is useful to simplify Equations 9.91 and 9.94. Substituting $\mathfrak{K}_1(\Gamma)$ from Equation 9.83 into Equation 9.91 yields

$$\beta_{01}(\Gamma) = \int (\Psi_0^*(\Gamma) \, \mathfrak{K}_0(\Gamma) \Psi_1(\Gamma) + \Psi_0^*(\Gamma) \, H(\Gamma) \, \Psi_1(\Gamma)) \, d\vec{r} \tag{9.95}$$

Substituting $\mathfrak{K}_0(\Gamma)\Psi_1(\Gamma)$ from Equation 9.82 into Equation 9.95, yields

$$\beta_{01}(\Gamma) = E_0(\Gamma) \int \Psi_0^*(\Gamma) \, \Psi_1(\Gamma) \, d\vec{r} + \int \Psi_0^*(\Gamma) \, H(\Gamma) \, \Psi_1(\Gamma) \, d\vec{r} \tag{9.96}$$

Note that according to Equation 9.88 the first integral in Equation 9.96 is zero. Therefore

$$\beta_{01}(\Gamma) = \int \Psi_0^*(\Gamma) \, H(\Gamma) \, \Psi_1(\Gamma) \, d\vec{r} \tag{9.97}$$

A similar relationship also holds for $\beta_{10}(\Gamma)$:

$$\beta_{10}(\Gamma) = \int \Psi_1^*(\Gamma) \, H(\Gamma) \, \Psi_0(\Gamma) \, d\vec{r} \tag{9.98}$$

Note that since $H(\Gamma)$ is real,

$$\beta_{10}(\Gamma) = \beta_{01}^*(\Gamma) \tag{9.99}$$

Next we solve Equations 9.89 and 9.92 simultaneously to calculate c_0 and c_1. Putting Equations 9.89 and 9.92 into matrix form, yields

$$\begin{bmatrix} (E_0'(\Gamma) - E'(\Gamma)) & \beta_{01}(\Gamma) \\ \beta_{10}(\Gamma) & (E_1'(\Gamma) - E'(\Gamma)) \end{bmatrix} \begin{pmatrix} c_0(\Gamma) \\ c_1(\Gamma) \end{pmatrix} = \begin{pmatrix} 0 \\ 0 \end{pmatrix} \tag{9.100}$$

Equation 9.100 has two eigenvalues:

$$E_-'(\Gamma) = \frac{E_0'(\Gamma) + E_1'(\Gamma)}{2} - \frac{\sqrt{(E_0'(\Gamma) - E_1'(\Gamma))^2 + 4\beta_{01}(\Gamma)\beta_{10}(\Gamma)}}{2} \tag{9.101}$$

$$E_+'(\Gamma) = \frac{E_0'(\Gamma) + E_1'(\Gamma)}{2} + \frac{\sqrt{(E_0'(\Gamma) - E_1'(\Gamma))^2 + 4\beta_{01}(\Gamma)\beta_{10}(\Gamma)}}{2} \tag{9.102}$$

The dashed line in Figure 9.35 is a plot of E as a function of Γ for a typical value of $\beta_{10}(\Gamma)$. Note that, when $\beta_{10}(\Gamma)$ is nonzero, the energy of the lower eigenstate is transformed continuously from E_0 to E_1.

The eigenvectors of Equation 9.100 are

$$\Psi_-(\Gamma) = \begin{cases} \Psi_a(\Gamma) & \text{when } E_0(\Gamma) \leq E_1(\Gamma) \\ \Psi_b(\Gamma) & \text{otherwise} \end{cases} \tag{9.103}$$

$$\Psi_-(\Gamma) = \begin{cases} \Psi_a(\Gamma) & \text{when } E_0(\Gamma) \leq E_1(\Gamma) \\ \Psi_b(\Gamma) & \text{otherwise} \end{cases} \tag{9.104}$$

with

$$\Psi_a(\Gamma) = \frac{\Psi_0(\Gamma) + \left(\dfrac{\beta_{10}(\Gamma)}{E'_-(\Gamma) - E'_1(\Gamma)} \right) \Psi_1(\Gamma)}{\sqrt{1 + \left(\dfrac{\beta_{10}(\Gamma)}{E'_-(\Gamma) - E'_1(\Gamma)} \right)^2}} \tag{9.105}$$

$$\Psi_b(\Gamma) = \frac{\Psi_1(\Gamma) + \left(\dfrac{\beta_{01}(\Gamma)}{E'_+(\Gamma) - E'_0(\Gamma)} \right) \Psi_0(\Gamma)}{\sqrt{1 + \left(\dfrac{\beta_{01}(\Gamma)}{E'_+(\Gamma) - E'_0(\Gamma)} \right)^2}} \tag{9.106}$$

Now let's consider what happens when reaction occurs. At the start of the reaction $\beta_{10}(\Gamma) = 0$, $\Psi_-(\Gamma) = \Psi_0(\Gamma)$. Now let's assume that when the molecules come together, $\beta_{10}(\Gamma)$ is nonzero. Note that according to Equation 9.105, when $\beta_{10}(\Gamma)$ is nonzero, we get mixing of states. Further, at the end of the reaction $E_- \approx E'_1$. If $\beta_{10}(\Gamma)$ is nonzero, the final wavefunction will approach $\Psi_1(\Gamma)$. Therefore, a reaction can occur whenever $\beta_{10}(\Gamma)$ is nonzero when the molecules collide.

Note, however, that there are some cases where $\beta_{10}(\Gamma)$ is zero when two molecules collide. When $\beta_{10}(\Gamma)$ is zero throughout the collision process, $\Psi_-(\Gamma)$ is equal to Ψ_0 as the reactants come together. Hence there would be no coupling of the states. Therefore we conclude that the motion of the nuclei will convert one configuration of a system into another *only* when $\beta_{10}(\Gamma)$ is nonzero during the collision.

For future reference, we will call $\beta_{10}(\Gamma)$ the **coupling constant**, and note that reactions occur only when the coupling constant is nonzero.

9.13.2 Reactions with Negligible Coupling

In Nobel Prize-winning work, Woodward and Hoffmann [1970] provided several examples where $\beta_{10}(\Gamma)$ was zero during a reaction. In this section, we consider a simple example that shows how $\beta_{10}(\Gamma)$ can be zero. One should refer to Woodward and Hoffmann [1970] or Pearson [1976] for many other examples.

Consider a simple reaction:

$$\text{H}_2 + \text{D}_2 \rightarrow 2 \text{ HD} \tag{9.107}$$

One can imagine that the reaction can go by either the four-centered reaction shown in Figure 9.36, or by the chain-propagation mechanism shown in Figure 9.37. Wright [1976] has also proposed a concerted six-centered reaction (e.g., 2 H$_2$ + D$_2$ → H$_2$ + 2 HD).

At first sight one might think that the four-centered reaction would dominate. The six-centered reaction requires three molecules to come together simultaneously in the gas phase.

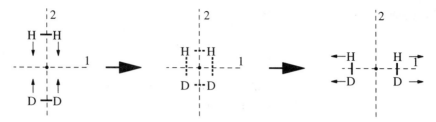

Figure 9.36 A hypothetical four-centered mechanism for H_2/D_2 exchange. The dotted lines in the figure denote symmetry planes that are preserved during the reaction (see text). This reaction is symmetry forbidden.

That is highly improbable. The first step in the chain-propagation mechanism (Figure 9.37) is hydrogen–hydrogen bond scission. Hydrogen–hydrogen bond scission is highly activated. In contrast, superficially, during the four-centered reaction one is simply exchanging bonds within the molecule. As we have drawn the picture, there is no net change in bond order anywhere in the reaction. Hence, from a superficial analysis, one might conclude that the four-centered reaction is the most probable.

In fact, however, the four-centered reaction has never been observed. At high temperatures, Reaction 9.107 goes via the chain-propagation mechanism. There has been some discussion about the mechanism at low temperatures where the six-centered reaction may dominate. However, there is no evidence that the reaction ever occurs at a measurable rate via a four-centered reaction. Calculations of Conroy and Malli [1969] indicate that the four-centered reaction has an activation energy of 515 kJ/mol! By comparison, the H–H bond strength is only 435 kJ/mol.

In the next few paragraphs we show that the reason the reaction does not occur via the four-centered reaction is that the β_{10}'s for the transformation of some of the reactant configurations in Reaction 9.107 into product configurations are zero. As a result, the reaction in Figure 9.36 cannot occur via a simple gas phase collision.

In order to derive the key results, it will be useful to view the reaction by sitting on a point halfway between the hydrogen and deuterium, as indicated by the dot in Figure 9.36 and Figure 9.38. One can imagine a transition state where there are both H–H and H–D bonds. If we model this transition state as one big molecule, then we can use molecular orbital (MO) theory to tell us how the configuration of the molecule changes as the reaction proceeds. There are four key MOs in the system. They are depicted in Figure 9.38. For future reference we label the MOs in terms of the bonding of one of the hydrogen atoms. The $\sigma\sigma$ MO will be an MO where the reference hydrogen is bound to the other hydrogen via a sigma bond and bound to a deuterium with a σ bond. The $\sigma\sigma^*$ state will have a σ bond between the reference hydrogen and the other hydrogen and a σ^* bond between the hydrogens and the deuteriums. Figure 9.38 shows a schematic of the wavefunctions for the various states. All of the orbitals in the wavefunction for the $\sigma\sigma$ state, $\psi_{\sigma\sigma}$, have the same sign. In $\psi_{\sigma\sigma^*}$ the orbitals on both hydrogens have the same sign, but the orbitals on the two deuteriums have the opposite sign. In $\psi_{\sigma^*\sigma}$ the orbitals on each H–D pair have the same sign, but the

$$H_2 + X \rightarrow 2\,H + X$$
$$H + D_2 \rightarrow HD + D$$
$$D + H_2 \rightarrow HD + H$$
$$2H + X \rightarrow H_2 + X$$

Figure 9.37 The chain-propagation mechanism for H_2/D_2 exchange.

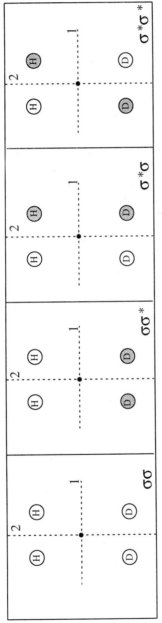

Figure 9.38 A schematic of the key molecular orbitals for the transition state of Reaction 9.107. Positive atomic orbitals are depicted as open circles; negative orbitals are depicted as shaded circles.

orbitals on the H–H's and D–D have the opposite sign. In $\psi_{\sigma^*\sigma^*}$ the wavefunction for the orbitals has alternating signs.

Now consider what happens when we move the H_2 and D_2 together. We start out with there being no net interaction between the H_2 and D_2 so the wavefunction for the system $\Psi_{reactants}$ can be approximated by

$$\Psi_{reactants} = \psi_{\sigma\sigma}\psi_{\sigma\sigma^*} \qquad (9.108)$$

At the end of the reaction there are H–D bonds, but there is no net interaction between the two HDs. Therefore the wavefunction for the products $\Psi_{products}$ can be approximated by

$$\Psi_{products} = \psi_{\sigma\sigma}\psi_{\sigma^*\sigma} \qquad (9.109)$$

As a result, during the reaction electrons are moved from the $\sigma\sigma^*$ to the $\sigma^*\sigma$ MO.

Woodward and Hoffmann [1970] show that it is convenient to represent the transfer of electrons by a **correlation diagram,** where the correlation diagram is very similar to a configuration mixing diagram except that in the correlation diagram, we keep track of what happens to each of the individual MOs (i.e., the ψ's) while in the configuration mixing diagram we keep track of what happens to the total wavefunction for the system (i.e., the Ψ's).

Figure 9.39 shows a correlation diagram for the four-centered reaction depicted in Figure 9.36. The reaction starts out with $\sigma\sigma$ and $\sigma\sigma^*$ MOs occupied and the $\sigma^*\sigma$ and $\sigma^*\sigma^*$ MOs empty. However, during the reaction, electrons are moved from the $\sigma\sigma^*$ to the $\sigma^*\sigma$ MO.

Now, the question is 'Can the reaction occur as depicted in Figure 9.38?' According to the results in Section 9.13.1, the motion of the H_2 and D_2 toward one another in the configuration shown in Figure 9.36 will allow the electrons in the $\sigma\sigma^*$ MO to be moved into the $\sigma^*\sigma$ MO whenever β is nonzero. Therefore, we can tell if reaction can occur by computing β. For future discussion we will note that there are two symmetry planes in Figure 9.36, one

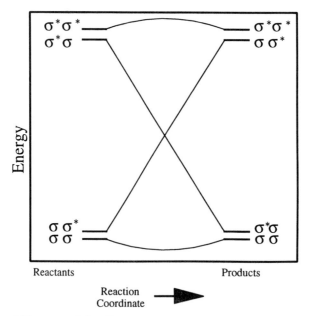

Figure 9.39 A correlation diagram for the reaction depicted in Figure 9.36.

that lies halfway between the H_2 and D_2 and one that goes between the midpoints of the H–H and D–D bonds. We designate those two symmetry planes as planes 1 and 2.

According to Equation 9.97 the coupling constant for conversion of the $\sigma\sigma^*$ state into the $\sigma^*\sigma$ state $\beta_{(\sigma\sigma^*)(\sigma^*\sigma)}$, is given by

$$\beta_{(\sigma\sigma^*)(\sigma^*\sigma)} = \int \psi_{\sigma\sigma^*}^* H \psi_{\sigma^*\sigma} \, d\vec{r} \tag{9.110}$$

Note that H is completely symmetric, assume that the intersection of the two symmetry planes is the origin for Figure 9.36 and that if we move up, we move in the y direction, while if we move to the right, we move in the positive x direction. Note that since hydrogen and deuterium are chemically equivalent,

$$H(x, y) = H(-x, y) = H(x, -y) = H(-x, -y) \tag{9.111}$$

It is useful to rewrite Equation 9.110 as

$$\beta_{(\sigma\sigma^*)(\sigma^*\sigma)} = \int_0^\infty \int_0^\infty \psi_{\sigma\sigma^*}^* H \psi_{\sigma^*\sigma} \, dx \, dy + \int_0^\infty \int_{-\infty}^0 \psi_{\sigma\sigma^*}^* H \psi_{\sigma^*\sigma} \, dx \, dy$$
$$+ \int_{-\infty}^0 \int_0^\infty \psi_{\sigma\sigma^*}^* H \psi_{\sigma^*\sigma} \, dx \, dy + \int_{-\infty}^0 \int_{-\infty}^0 \psi_{\sigma\sigma^*}^* H \psi_{\sigma\sigma^*} H \Omega_{\sigma^*\sigma} \, dx \, dy \tag{9.112}$$

For the purposes of derivation, it is useful to define a quantity *INT* by

$$INT = \int_0^\infty \int_0^\infty \psi_{\sigma\sigma^*}^* H \psi_{\sigma^*\sigma} \, dx \, dy \tag{9.113}$$

Now consider a quantity *INT'* given by

$$INT' = \int_0^\infty \int_{-\infty}^0 \psi_{\sigma\sigma^*}^* H \psi_{\sigma^*\sigma} \, dx \, dy \tag{9.114}$$

INT' is related to *INT* by a transformation that switches y with $-y$. Notice that when we switch y to $-y$, H and $\psi_{\sigma^*\sigma}$ do not change while $\psi_{\sigma\sigma^*}$ is converted into $-\psi_{\sigma\sigma^*}$. Therefore,

$$INT' = -INT \tag{9.115}$$

A similar argument shows

$$\int_{-\infty}^0 \int_0^\infty \psi_{\sigma\sigma^*}^* H \psi_{\sigma^*\sigma} \, dx \, dy = -INT \tag{9.116}$$

$$\int_{-\infty}^0 \int_{-\infty}^0 \psi_{\sigma\sigma^*}^* H \psi_{\sigma^*\sigma} \, dx \, dy = INT \tag{9.117}$$

Substituting Equations 9.113, 9.114, 9.115, 9.116, and 9.117 into Equation 9.112, yields

$$\beta_{(\sigma\sigma^*)(\sigma^*\sigma)} = 0 \tag{9.118}$$

Therefore, if we move an H_2 molecule toward a D_2 molecule in the configuration shown in Figure 9.36, there will be no coupling between the $\sigma\sigma^*$ and $\sigma^*\sigma$ MOs. As a result, according to Equation 9.105 no reaction will occur. One can get some small amount of coupling if one distorts the geometry (e.g., twist the H_2 relative to the D_2). However, $\beta_{(\sigma\sigma^*)(\sigma^*\sigma)}$ is always small. As a result, one does not get a significant reaction rate.

Now, it is useful to go back and consider why the picture of the transition state in Figure 9.36 was wrong. In Figure 9.36 we assumed that it will be possible to simultaneously break an H–H bond and form an H–D bond. However, if one multiplies out all of the wavefunctions, one will find that during the reaction we need to transform wavefunctions that look like ϕ_1 $\overline{\phi_2}$ ϕ_3 $\overline{\phi_4}$ into wavefunctions that look like ϕ_1 $\overline{\phi_2}$ ϕ_3 ϕ_4, where ϕ_1, ϕ_2, ϕ_3, and ϕ_4 are the atomic orbitals on atoms 1, 2, 3, and 4 (e.g., the two hydrogen atoms and the two deuterium atoms, respectively). Physically, if we try to share an electron between the two states, as one would do if there was a partial H–H and a partial H–D bond, we would find that we need a single electron to be spin up and spin down at the same time. The electron cannot be in both spin up and spin down simultaneously. Hence, it is impossible to simultaneously break the H–H bond and form an H–D bond. The only way for the reaction to occur will be to first break one of the H–H bonds and then form an H–D bond. That explains why Conroy and Malli found that the four-centered reaction has an activation energy of 515 kJ/mole.

Another way to get to the same result is to go back to the stick model of the four-centered reaction in Figure 9.28. In the stick model one is saying that in the transition state for the reaction, there are four σ bonds in the system with four total electrons. According to MO theory, the only way for the four-centered bonding to be stable would be for the four-centered bond to be one of the eigenstates of the system. However, the eigenstates of the system are the four states shown in Figure 9.38. Of those, only the $\sigma\sigma$ state is bonding in both directions. One can put two electrons in the $\sigma\sigma$ state, but the other two electrons must be put into other molecular orbitals. These other MOs are antibonding orbitals. As a result the four-centered bonding in Figure 9.28 is not an eigenstate of the system. Therefore, quantum mechanically, the four-centered bonding cannot occur.

A more subtle analysis shows that the reaction

$$H_2^+ + D_2^+ \longrightarrow \begin{bmatrix} H-H \\ \diagdown \diagup \\ D-D \end{bmatrix}^{2+} + \longrightarrow 2HD^+ \tag{9.119}$$

where H_2^+ and D_2^+ are positively charged molecules, is allowed. In the reaction in Equation 9.119 there are only two electrons in the transition state. One can put both electrons in the SS state so the bonding is possible.

This example illustrates an important feature of Woodward and Hoffmann's analysis: there are many simple reactions that look plausible on paper which cannot occur because they involve bonding schemes that are not possible. In a more general way, we can draw a stick diagram like that in Figure 9.28 for many proposed transition states. However, there is no guarantee that the bonding shown in the stick diagram is possible. If it is not, the proposed reaction will have a very high activation barrier. Hence, one has to do some analysis before one assumes that one can get a proposed reaction to occur. A generalization of the analysis in this section shows that four-centered reactions of species with σ bonds are usually symmetry forbidden. (There are exceptions where a nonbonding orbital can change the symmetry of the system.) Therefore, one would generally not expect four-centered reactions to be seen experimentally.

9.13.3 Conservation of Orbital Symmetry

One can imagine repeating the analysis in the last section for a series of reactions. One could simply work out all of the integrals. However, Woodward and Hoffmann found a trick that simplifies the analysis. The trick comes from group theory. Recall from Chapter 2 that we can classify surface structures based on the symmetry elements in the surface. In the same way, we can classify reaction paths in terms of their symmetry elements. For example, if we assume that hydrogen and deuterium are equivalent, the reaction in Figure 9.36 has three mirror planes, the two lines indicated in the figure and the plane of the paper. There also are three twofold axes, two at the plane of the paper at the mirror planes and one perpendicular to the plane of the paper at the dot. The dot is also an inversion center. In a similar way, one can classify the symmetry of all of the MOs of the system. Woodward and Hoffmann [1970] show that the symmetry of the MOs determine whether a reaction can occur. If orbital symmetry is conserved, the coupling constants will be nonzero so the reaction can occur. However, if the orbital symmetry is not conserved, no reaction can occur.

9.13.4 Some Results from Group Theory

We need to review some group theory to see how this idea arises. Recall, that in group theory one classifies the symmetry of molecules and the reaction pathways in terms of their point group. In Chapter 2 we noted that there are only 13 three-dimensional Bravais lattices and 32 space groups. In the same way, there are only 47 important point groups. There are more point groups than space groups because one can have 5, 7, 8, . . . fold axes in molecules, but not in a repeated three-dimensional structure. Cotton [1971] provides an excellent overview of the application of group theory to molecules and reactions, but there is not enough room to discuss all of the important ideas here. However, the key feature of the analysis is that if one knows the symmetry elements in a molecule or reaction path, one can use the tables in Cotton to find the point group.

The most important tables in Cotton's book are the character tables near the end of the book. The character tables include a listing of all of the important point groups and their symmetry elements. For example, Table 9.4 is a character table for the D_{2h} point group. The D_{2h} point group has three perpendicular twofold axes, three mirror planes, an inversion center, and an identify element (i.e., do nothing). Note that the transition state in Figure 9.36 has all of these symmetry elements and no extras. Therefore, we say that the reaction in Figure 9.36 follows the D_{2h} point group.

The character table also lists something called the **irreducible representations** of the point group. The idea of a group representation is more subtle than there is room to discuss here. However, if a molecule (or transition state) is a member of a given point group, then all of the solutions of the Schrödinger equations for the molecule must be members of one of the irreducible representations of the point group, i.e., the irreducible representations are just a way to classify the solution of the Schrödinger equation in terms of their symmetry properties.

TABLE 9.4 A Character Table for the D_{2h} Point Group

Irreducible Representation	Identity	z Twofold Axis	y Twofold Axis	x Twofold Axis	Inversion Center	xy Mirror Plane	xz Mirror Plane	yz Mirror Plane	Rotations and Orbitals
					Symmetry Element				
A_g	1	1	1	1	1	1	1	1	x^2, y^2, z^2
B_{1g}	1	1	-1	-1	1	1	-1	-1	r_z, xy
B_{2g}	1	-1	1	-1	1	-1	1	-1	r_y, xz
B_{3g}	1	-1	-1	1	1	-1	-1	1	r_x, yz
A_u	1	1	1	1	-1	-1	-1	-1	
B_{1u}	1	1	-1	-1	-1	-1	1	1	z
B_{2u}	1	-1	1	-1	-1	1	-1	1	y
B_{3u}	1	-1	-1	1	-1	1	1	-1	x

Adapted from Cotton [1971].

Each independent MO of a system will belong to a single irreducible representation. Hence, one can classify the symmetry of MOs in terms of their irreducible representation.

The character table makes it easy to find the irreducible representation for each MO. Notice all of the ones and minus ones in the character table (Table 9.4). The ones and minus ones tell what happens to the sign of the wavefunction when each of the symmetry operations of the point group act on a given MO. For example, the xy mirror plane converts y to $-y$. If we look down the xy mirror plane column in Table 9.4, we find that B_{2u} irreducible representation has a character of -1. Therefore $\Psi_{B_{2u}}$, the wavefunction of an MO of B_{2u} symmetry in the D_{2h} point group will be converted to $-\Psi_{B_{2u}}$ when y is converted to $-y$. One can therefore determine the irreproducible representation of a given MO by examining how the symmetry elements in the system affect the given MO.

For example, consider $\psi_{\sigma^*\sigma}$ in Figure 9.38: $\psi_{\sigma^*\sigma}$ does not change when it is reflected across the xy or yz mirror planes (i.e., the plane of the paper and plane 2 in Figure 9.38, respectively). However, it switches sign when it is reflected across the xz mirror plane (i.e., plane 1 in Figure 9.38). Therefore, the irreducible representation of $\psi_{\sigma^*\sigma}$ should have a 1 in the column for the xy and yz mirror planes and -1 in the column for the xz mirror plane. Table 9.4 shows that only the B_{2u} irreducible representation has the correct symmetry. Therefore, we conclude that $\psi_{\sigma^*\sigma}$ has B_{2u} symmetry. The reader may want to show that $\psi_{\sigma\sigma}$ has A_g symmetry, $\psi_{\sigma\sigma^*}$ has B_{3u} symmetry, and $\psi_{\sigma^*\sigma^*}$ has B_{1g} symmetry.

Now, it is useful to go back to our analysis of Equation 9.112. Note that we used the symmetry of the various wavefunctions to show that $\beta_{(\sigma\sigma^*)(\sigma^*\sigma)}$ is zero. Note that one could have done the analysis directly from the character table. For example, if we start with INT from Equation 9.113 and let y go to $-y$ (i.e., reflect across the xz mirror plane), $\psi_{\sigma^*\sigma}$ does not change while $\psi_{\sigma\sigma^*}$ is converted into $-\psi_{\sigma\sigma^*}$. H is completely symmetric (i.e., a member of the A_g irreducible representation), so it does not change either. Therefore,

$$INT' = (-1)(1)(1)INT \qquad (9.120)$$

As a result one could have arrived at the results in Section 9.13.2 directly from the character table without doing detailed analysis.

One can generalize this result to show that the coupling coefficients are always zero unless orbital symmetry is conserved. Consider transferring an electron from $\psi_{\sigma\sigma^*}$ to some other wavefunction ψ_x. The transfer coefficient for the transfer β_x is given by:

$$\beta_x = \int \psi_{\sigma\sigma^*}^* H \psi_x \, d\vec{r} \qquad (9.121)$$

If we start with a wavefunction of B_{2u} symmetry such as $\psi_{\sigma\sigma^*}$ and multiply by H, we still end up with a function of B_{2u} symmetry, since H is totally symmetric. Now if we multiply by ψ_x, we get a new function that we call Ж. One can show from something called the **Great Orthogonality theorem** that if Ж changes sign under any of the group operations, when we integrate over all space, half the integral will be positive and half will be negative. As a result, β_x will be zero. Notice, that we can tell if Ж changes sign directly under all of the group operations directly from the character table. If the irreducible representation of ψ_x has ones in any of the places where the irreducible representation of $\psi_{\sigma\sigma^*}$ has minus ones, or if the irreducible representation of ψ_x has minus ones in any of the places where the irreducible representation of $\psi_{\sigma\sigma^*}$ has ones, Ж will switch signs under at least one of the symmetry operations of the D_{2h} space group. As a result, β_x will be zero. The only instance where β_x will not be zero is when the irreducible representations of ψ_x and $\Psi_{\sigma\sigma^*}$ have ones and minus ones in just the same places in the character table. That happens only when ψ_x and $\psi_{\sigma\sigma^*}$ have the same orbital symmetry (i.e., are members of the same irreducible representation).

The conclusion from the analysis in the last paragraph is that during a reaction, we can convert one molecular orbital into another only when both orbitals are of the same symmetry. (i.e., members of the same irreducible representation). As a result, orbital symmetry is conserved during elementary (concerted) chemical reactions.

9.13.5 Examples of Symmetry Allowed and Symmetry Forbidden Reactions

It is useful to use these ideas to make some predictions about chemical reactions. Clearly, the first example is the four-centered reaction in Figure 9.36. Notice that we start with electrons in A_g and B_{2u} molecular orbitals and end up with electrons in the A_g and B_{3u} molecular orbitals. The orbital symmetry changes during the reaction. Hence the reaction is forbidden.

Now consider the reaction

$$H + D_2 \rightarrow HD + D \qquad (9.122)$$

where a hydrogen atom reacts with D_2. For the purpose of analysis, let's assume that the reaction occurs with the hydrogen atom and the two deuterium atoms lying in a line. Again there are four key molecular orbitals. They are depicted in Figure 9.40. The reaction starts out with the $\sigma\sigma$ and the $\sigma^*\sigma$ states filled and the $\sigma\sigma^*$ and $\sigma^*\sigma^*$ states empty. During the reaction there is an exchange of electrons between the $\sigma^*\sigma$ and $\sigma\sigma^*$ states. Notice that the states are the same in Reaction 9.122 as in the reaction in Figure 9.36. Therefore, at first sight one might think that Reaction 9.122 is also symmetry forbidden. There is an important difference between Figure 9.36 and Figure 9.40. There are many symmetry elements in Figure 9.36. However, the only symmetry elements in Figure 9.40 are the line through the HDD axis, and the plane of the paper. We call this point group $C_{\infty v}$. Notice that each of the molecular orbitals in Figure 9.40 does not change when the system is rotated around the HDD axis or when the system is reflected about the plane of the paper. Therefore, all of the molecular orbitals in Figure 9.40 are completely symmetric under all of the group operations. As a result, they have the same irreducible representation in the $C_{\infty v}$ point group. Thus orbital symmetry is conserved in Reaction 9.122.

Calculations indicate that the reaction in Figure 9.40 has an activation energy of 34 kJ/mol, while the reaction in Figure 9.36 has an activation energy of 515 kJ/mol.

This example illustrates an important point: we need to look carefully at a reaction pathway before we decide whether a reaction is allowed or forbidden. Many reactions that look feasible on paper do not occur because of symmetry constraints. In particular, four-centered reactions are usually forbidden, so they are rarely seen.

Woodward and Hoffmann [1970] and Pearson [1976] provide many other examples of symmetry allowed and symmetry forbidden reactions. One should refer to their work for further information on this topic.

9.14 PREDICTION OF MECHANISMS OF GAS PHASE REACTIONS

The most important implications of the methods in this chapter come in trying to use the methods to predict mechanisms of reactions in the gas phase. It is useful to consider a simple example to illustrate the ideas. Consider the gas phase reaction $H_2 + I_2 \rightarrow 2HI$. This reaction has been studied in great detail in the literature. Such reactions generally go by a chain-propagation mechanism. Therefore, one can guess a mechanism by assuming that the H_2 and I_2 break into atoms, and then looking at the reactions of the hydrogen and iodine atoms with

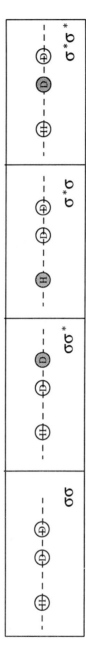

Figure 9.40 A schematic of the key molecular orbitals for the reaction $H + D_2 \rightarrow HD + D$. Positive orbitals are indicated by open circles, while negative orbitals are denoted by shaded circles. According to molecular orbital theory, only three of these orbitals are independent. Consequently, there is no orbital symmetry barrier to reaction.

all of the molecules in their system. The result is the following group of elementary reactions:

$$H_2 + I_2 \longrightarrow \begin{matrix} I-I \\ \backslash \ / \\ H-H \end{matrix} \longrightarrow 2HI \qquad (1)$$

$$M + H_2 \longrightarrow 2H + M \qquad (2)$$

$$M + I_2 \longrightarrow 2I + M \qquad (3)$$

$$H + I_2 \longrightarrow HI + I \qquad (4)$$

$$I + H_2 \longrightarrow HI + H \qquad (5) \qquad (9.123)$$

$$M + I + H_2 \longrightarrow H_2I + M \qquad (6)$$

$$I + H_2I \longrightarrow 2HI \qquad (7)$$

$$M + 2H \longrightarrow H_2 + M \qquad (8)$$

$$M + 2I \longrightarrow I_2 + M \qquad (9)$$

where M is a collision partner. Some of the reactions also can go in reverse. Over the years all of these reactions have been postulated to occur. However, the analysis in this section allows us to exclude some of these reactions, and therefore find a list of feasible elementary reactions for this system.

The basis of the analysis is to use the Marcus equation and the Woodward–Hoffmann rules to estimate activation barriers for each of the reactions. One then looks for reactions that have particularly high barriers and eliminates them. All remaining reactions are included in the mechanism.

Proceeding with the analysis, notice that the first reaction can be eliminated simply. In Section 9.12 we found that four-centered reactions are symmetry forbidden. Therefore, it is unlikely that the reaction would proceed via Reaction 1 except at very high temperatures. This prediction agrees with the data of Benson and Srinivansan [1955] and Sullivan [1963] and the calculations of Jaffe, Henry, and Anderson [1976]. All of the remaining reactions are symmetry allowed. Therefore they might occur.

One can rank the activation energies for each of the reactions using the material earlier in this chapter. Table 9.5 lists the free energy of formation of several of the intermediates in the mechanism given earlier. Note that Reactions 8 and 9 are highly exothermic. No bonds break during the reactions, so one would expect both reactions to have little or no activation barrier. This prediction agrees with experiment. Reaction 7 is also highly exothermic. However, bonds break during the reaction. Therefore, there could be a small activation barrier for the reaction. In actual practice, however, Sullivan [1963] found the barrier to be negligible. Reaction 4 is moderately exothermic. Bonds break and new bonds form, so the reaction could be activated. However, the activation barriers would be expected to be relatively small. Sullivan [1959] measured 0.5 kcal/mole. Reaction 6 is also moderately exothermic. No bonds break during the reaction. There are some bond distortions, so one might expect a small barrier. However, Sullivan [1963] found that the reaction occurs spontaneously at room temperature with little or no barrier. Reaction 5 is 19 kcal/mole endothermic. Therefore, one would expect some barrier. Sullivan measured 28 kcal/mole. In contrast, Reactions 2 and 3

TABLE 9.5 Free Energies of Formation for Some Key Species in the Reaction $H_2 + I_2 \rightarrow 2$ HI

Species	H_2	I_2	HI	H	I
ΔG_f	0	0	+6.2	+48.58	+36.4

are highly endothermic. Both would be expected to have sizable barriers. In particular, Reaction 2 must have an activation barrier of at least 104 kcal/mole (i.e., the heat of reaction). Therefore, one would expect Reaction 2 to be rather slow. Reaction 3 also has a large barrier. However, the barrier is only 70 kcal (i.e., the heat of reaction, since the reverse reaction is unactivated). Therefore, one would expect Reaction 3 to dominate over Reaction 2 under most conditions. In fact, Benson [1955] and Sullivan [1963] found that Reaction 3 does dominate over Reaction 2 as expected from the material in this chapter. In summary, then, one would guess that Reactions 3 to 9 are important to the reaction. This supposition agrees with experiment.

This is a general result. One can start with a simple mechanism and write down all of the possible reactions. One can then use the analysis in this chapter to exclude some of the reactions either because they are symmetry forbidden or because the reactions will have too high a barrier. The interesting thing is that if one is clever with the procedure and includes enough reactions, one usually finds that in the gas phase the predicted mechanisms agree with those in the experiments.

One should, in principle, be able to do the same thing on surfaces, i.e., start with a proposed mechanism with a large number of possibilities, and then use the analysis in this chapter to eliminate some of the reactions. So far, no one has done that. However, I expect that in the future this will be a fairly common occurrence, as it is in the gas phase.

The one thing that complicates surface reactions is that there are extra geometry constraints which are not seen in the gas phase. Molecules often have specific orientations on surfaces, and unless the bonds that break can come in close proximity to each other, the reaction is inhibited. Earlier in this chapter, we mentioned the methanol case. Methanol can be oxidized via two pathways:

$$2CH_3OH + O_{(ad)} \rightarrow 2CH_3O_{(ad)} + H_2O$$

$$2CH_3OH + O_{(ad)} \rightarrow 2CH_2OH_{(ad)} + H_2O$$

The first pathway dominates at low temperatures in UHV, while the second pathway dominates in an electrochemical cell. Physically, in UHV the CH_3OH adsorbs with the oxygen bonded to the surface. The hydroxyl hydrogen is much closer to the surface than the methyl hydrogen. As a result, the hydroxyl hydrogen is easier to transfer than the methyl hydrogen. The data are less clear in solution. However, Zelenay et al. [1992] postulated that the OH would hydrogen bond to the water solution, and so the CH_3 group would point toward the surface. One finds that the methyl hydrogen is much easier to transfer in such a situation.

9.15 APPLICATIONS: LEARNING ABOUT TRANSITION STATES FOR REACTIONS

The other major application of the material in this chapter is to use the Hammett equation to learn about the transition state for reactions. Kraus [1967] shows how this idea can be applied in catalysis. Kraus' general idea is that reactions which have similar transition states will all follow the same line on the Hammett plot, while reactions with different transition states will follow a different line on the Hammett plot. Figure 9.41 shows an example cited by Kraus. In this example methyl groups are being hydrogenated off of a series of aromatic hydrocarbons. Kraus noted that one gets a good fit to the data with the Taft parameter σ_R for each of the substituents X, suggesting that resonant interactions control the transition state. Note that if the reaction occurred via a direct interaction with the methyl group, one would expect such a correlation to work. On the other hand, if the reaction occurred via the hydrogenation of a carbon center on the X group, then the reaction would follow a different Hammett/Taft plot where the reaction rate would correlate with the Taft parameters for the ligands on the carbon

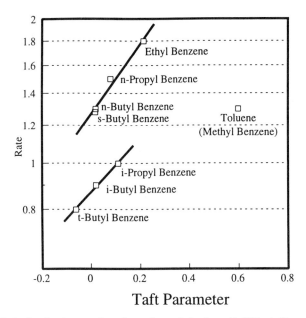

Figure 9.41 A Taft plot for the a series of reactions of the form X–CH₃ + H₂ → CH₄ + XH on a nickel/alumina catalyst. (Data of Beranek and Kraus [1966].)

center on the X group. Therefore, the fact that the data fit the line in Figure 9.42 implies that the reaction occurs via a direct interaction with the methyl group, i.e., via formation of a CH_4^+.

Note, however, that there are two lines in the figure, and one point that does not lie on either line. In the case of ethyl, n-propyl, n-butyl, and s-butyl benzene, one is removing a primary methyl group (i.e., one bound to a CH_2), while in the case of i-propyl, i-butyl, and t-butyl benzene, one is removing a secondary or tertiary methyl group. In toluene (methyl-

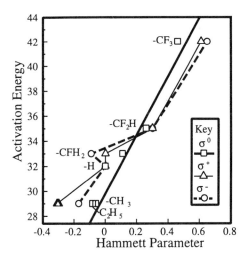

Figure 9.42 A Hammett plot of the Dai and Gellman's [1993] data for the decomposition of a series of alkoxides on Cu(111).

benzene) one is removing a methyl group that is directly bound to a benzene. The implication of these results is that the nature of the transition state is different in the three cases, perhaps because the charge is distributed differently on the three cases.

One can extend this idea to get direct information about the nature of transition states in catalytic reactions. For example, Dai [1992] and Dai and Gellman [1993] measured the activation energy for the decomposition for a series of reactions of the form

$$\begin{array}{ccc} \underset{\displaystyle O}{\overset{\displaystyle R\ H}{\underset{\displaystyle |}{\overset{\displaystyle \backslash /}{\underset{\displaystyle C}{\overset{\displaystyle C-H}{}}}}}} & \longrightarrow & \underset{\displaystyle O\ \ H}{\overset{\displaystyle R\ H}{\underset{\displaystyle \|}{\overset{\displaystyle \backslash /}{C}}}} \end{array} \tag{9.124}$$

with R = H, CH_3, CH_2F, CHF_2, CF_3. Figure 9.42 shows a Hammett plot for their data. Notice that the data fit the σ^0 scale much better than the σ^+ or σ^- scales. Therefore, one can conclude that in the transition state, the complex is almost neutral. The carbon in the alkoxide might be slightly positively or negatively charged. However, the charge should be small. There should not be any dangling bonds or unpaired orbitals.

In order to determine the charge of the transition state, one has to create a Taft plot (i.e., a plot of the activation energy versus σ_R and σ_I) for the reaction. Figure 9.43 shows such a plot. The data are not completely clear, but it appears that the data are better fit by an inductive rather than a resonant effect. This is consistent with the discussion in Isaacs [1987], which indicates that CF_3 interacts mainly by inductive effects.

To keep these results in perspective, consider the transition state shown in Figure 9.44a where the transition state has a δ^+ charge on the carbon and a δ^- charge on the hydrogen. In that case, an electron-withdrawing group such as a CF_3 will make it harder for the positive charge in the carbon to be stabilized by sharing electrons with the rest of the alkoxide. As a result, the effect of the CF_3 group will be to increase the activation barrier for a reaction with a transition state like that in Figure 9.44a.

The interaction with the hydrogen is similar. In this case, the positive charge on the carbon

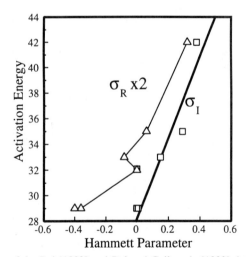

Figure 9.43 A Taft plot of the Dai [1992] and Dai and Gellman's [1993] data for the decomposition of a series of alkoxides on Cu(111).

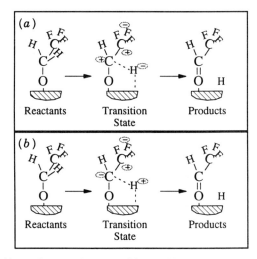

Figure 9.44 A diagram of two possible transition states for Reaction 9.124.

in the CF_3 group attracts the H^-, so it makes it harder for the H^- to leave the molecule. As a result, the CF_3 will increase the barrier for the reaction.

One can make a similar analysis for the transition state shown in Figure 9.44b. In this case, the effects discussed in the last two paragraphs lower the activation energy for the reaction. Experimentally, the addition of the CF_3 group raises the activation barrier, which suggests that the reaction has a transition state like that in Figure 9.44a. The differential charge δ must be small, since the reaction follows the σ^0 plot. This example illustrates how Hammett plots can be used to learn about transition states for reactions.

9.16 SUMMARY

In summary, then, in this chapter we reviewed why activation barriers arise during chemical reactions. Generally, it was found that the activation barriers arise when bonds break or when bonds are distorted during chemical reactions. The Polayni/Marcus model views this process as a simple bond extension where one bond stretches and breaks while another bond forms. In contrast, the configuration mixing model views the process as one where excited states are mixed into the ground state as the reaction proceeds. Both formulations are equivalent under most circumstances. However, there is the special case of a symmetry forbidden reaction where bonds break but no new bonds can form. Symmetry forbidden reactions do not occur except at very high temperatures.

In this chapter we also reviewed how substituents change the activation energy for a reaction. We saw that usually anything we do to make a reaction more exothermic will speed up the reaction. However, there are exceptions for very exothermic reactions.

9.17 SUPPLEMENTAL EXAMPLES

Example 9.A Justification of the Constant R–B Assumption in Section 9.3

In Section 9.3 we derived the Polayni relationship for the reaction

$$BH + R \rightarrow B + RH$$

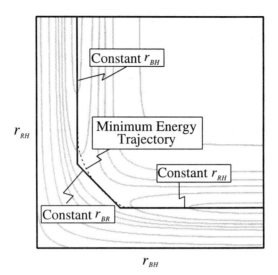

Figure 9.45   A comparison of a constant r_{BH} trajectory to the minimum energy trajectory for the reaction BH + R → B + RH.

assuming that reaction followed a trajectory where first the two reactants came together, then the hydrogen was transferred under conditions where r_{BR} was constant, and then the reactants flew apart. The objective of this example is to show that the activation barrier is almost the same with a constant r_{BH} trajectory as with the minimum energy trajectory.

Our approach will be to start with a potential energy surface for a simple reaction H + D_2 → HD + D and plot out the various trajectories to see how large an error is created by this approximation.

Figure 9.45 shows the potential energy surface for the reaction H + D_2 → HD + D and the two trajectories. The solid line in Figure 9.45 is a diagram of the trajectory used to derive the results in Section 9.3, while the dashed line is the minimum energy trajectory. If the reactants would follow the solid line, the reactants would come together with a fixed B–H distance, and then the reaction would occur with a fixed R–B distance, and finally the reactants would fly apart. Alternately, the minimum energy trajectory does not conserve the R–B distance. Nevertheless, both trajectories cross the saddle point in the potential energy surface in the same way. If one assumes that the reaction follows the solid line in Figure 9.45, one does make a small error in the work it takes to bring the reactants together. However, this error is exactly compensated for by a change in E_A^1, so the energy it takes to cross the barrier does not change.

Notice that we have not assumed anything special about the potential energy contour in deriving this result. In all cases there is an exact compensation so that the energy of the system at the saddle point in the potential energy surface is the same when we calculate it via the dashed line or the solid line in Figure 9.45. The only complications occur if the energy at the intersection of the constant r_{BH} and constant r_{BR} lines occurs at a point where the energy is comparable to the energy of the saddle point. However, that would be unusual. As a result, in most cases the assumption that the hydrogen transfer occurs at a fixed B–R distance does not lead to any error in the calculated value of the activation barrier for reaction.

Example 9.B   Contributions of Potential Shape Changes to Acidity

In Section 9.3 we asserted that changes in shape of the Morse potential would have a much smaller effect on the strength of a carboxylic acid than changes in the internal energy of the

TABLE 9.6 Partial Bond and Group Contributions for the Estimation of C_p, S^0, ΔH_f of a Gas Phase Species at 25°C and 1 Atom Extracted from Benson [1978]

Bond	ΔH_f (kcal/mol)	S^0 (cal/mol °K)	C_p (cal/mol °K)	Bond	ΔH_f (kcal/mol)	S^0 (cal/mol °K)	C_p (cal/mol °K)
C–H	−3.83	12.9	1.74	(C=C)–H	3.2	13.8	2.6
C–D	−4.73	13.6	2.06	(C=C)–C	6.7	−14.3	2.6
C–C	2.73	−16.4	1.98	(C=C)–Cl	−27	35.2	7.2
C–F	−52.5	−16.9	3.34	(CO)–H	−13.9	26.8	4.2
C–O	−12.0	−4.0	2.7	(CO)–C	−14.4	−0.6	3.7
O–H	−27.0	24.0	2.7	(CO)–O	−50.5	9.8	2.2
C–(C)(H$_3$)	−10.20	30.41	6.19	C–(C)$_4$	0.50	−35.10	4.37
C–(C)$_2$(H$_2$)	−4.93	9.42	5.50	C–(C=O)(C)(H)$_2$	−4.76	9.80	5.12
C–(C)$_3$(H)	−1.90	−12.07	4.54	C(C=C)$_2$(H)$_2$	−4.29	10.2	4.7

acid. The data in Figure 9.1 show that the pK_a values of carboxylic acids can change from 3 to 6 by changing the substituents on the acid. The question is how much do changes in the vibrational frequency of the O–H affect that. Note that the pK_a is equal to the log of the equilibrium constant for the dissociation process. If one substitutes the expression for the equilibrium constant from Appendix A into the definition of pK_a one obtains

$$\Delta pK_a = \frac{\Delta U_{Sub}}{kT} + \ln\left(\frac{q_v^{Sub}}{q_v^0}\right) \tag{9.125}$$

where ΔpK_a is the change in pK_a, ΔU_{Sub} is the change in internal energy when a substituent is added, and q_v^0 and q_v^{Sub} are the vibrational partition functions before and after addition of the substituent. The O–H vibrational frequency in carboxylic acids varies from 2500 to 3000 cm^{-1}. Plugging numbers into the partition function shows that the last term is no greater than 0.1.

Example 9.C The Use of Bond and Group Contribution Methods to Estimate Properties

We noted earlier in this chapter that Benson [1978] proposed that one an estimate properties of simple molecules by adding up the contributions of the bonds in the molecules, and correcting for the contributions for various groups. The purpose of this problem is to illustrate Benson's methods to estimate the heat of formation and specific heat of isobutane, $(CH_3)_3$–CH, at 300 K.

SOLUTION

Benson [1978] provides tables of the partial bond and group to ΔH_f, C_p, and S^0. Table 9.6 shows a selection of the values.

BOND CONTRIBUTION SOLUTION

Benson's first method is to add up contributions for all of the bonds in the molecule. Isobutane contains 10 C–H bonds, and 3 C–C bonds. Therefore, in this approximation

$$\Delta H_f = 10\,(\Delta H_f\,C–H) + 3(\Delta H_f\,C–C)$$

$$= 10 \cdot (−3.83\ \text{kcal/mol}) + 3(2.73\ \text{kcal/mol}) = −30.11\ \text{kcal/mole}$$

$$C_p = 10(C_p\ C\text{–}H) + 3(C_p\ C\text{–}C)$$

$$= 10(1.74) + 3(1.98) = 23.34\ \text{cal/mol K}$$

GROUP CONTRIBUTION SOLUTION

An alternate approach described by Benson is to add up the contributions from the multivalent atoms in the molecule. Isobutane contains four multivalent ions, a carbon bonded to three other carbons and a hydrogen, and three carbons bonded to three hydrogens and a carbon. In this approximation

$$\Delta H_f = 3[C\text{–}(C)(H_3)] + [C\text{–}(C)_3(H)]$$

where the notation $[C\text{–}(C)(H_3)]$ designates a carbon atom bound to a carbon and three hydrogens.

$$\Delta H_f = 3[-10.20] + [-1.90] = -32.50\ \text{kcal/mol}$$

$$C_p = 3[6.19] + [4.54] = 23.11\ \text{cal/mol K}$$

Experimental data $\Delta H_f = -32.0$ kcal/mol, $C_p = 23.1$ cal/mol K.

PROBLEMS

9.1 Define the following terms in three sentences or less:

    (1) Polayni relationship
    (2) Brønsted catalysis law
    (3) Marcus inverted region
    (4) Hammett postulate
    (5) Tafel kinetics
    (6) Charge-controlled reaction
    (7) Frontier-orbital-controlled reactions
    (8) Configuration mixing model
    (9) Configuration mixing diagram
    (10) Transfer coefficient
    (11) Hard acid
    (12) Soft acid
    (13) HUMO
    (14) LUMO
    (15) Frontier orbital
    (16) Fukui function
    (17) Avoided crossing
    (18) Symmetry forbidden reaction
    (19) Resonance scale
    (20) Inductive scale

9.2 Why are there orbital symmetry barriers to reaction? Where do they arise and where are they important? How does the orbital symmetry barrier affect the reaction $H_2 + D_2 \rightarrow 2\ HD$?

9.3 Why do activation barriers arise during chemical reactions? What are the key factors in determining whether a reaction is activated? Why are activation energies reduced for fairly exothermic reactions? What happens with very exothermic reactions? Why?

9.4 Describe the Marcus equation. What is it? How does it arise? What are the key predictions of the Marcus equation? Why is it useful?

9.5 Figure 9.3 shows a Brønsted plot for the base-catalyzed enolization of $NO_2(C_6H_4)O(CH_2)COCH_3$. Why is there curvature in the plot?

9.6 Figure 9.7 shows that van der Waals radii are larger than covalent radii. Review the discussion in Section 3.5.

(a) Physically, why are van der Waals radii larger than covalent radii?

(b) How would a change in the van der Waals radii of a molecule B–C affect the reactivity of the molecule?

9.7 How do Pauli repulsions affect barriers to reactions? Would there be a barrier in the absence of Pauli repulsions? Explain.

9.8 (a) Show mathematically that the four-centered reaction $A_2 + B_2 \rightarrow 2AB$ is symmetry forbidden.

(b) What is the physical significance of this result? (*Hint:* Show that in the four-centered configuration, it is impossible to simultaneously have an A–A, B–B, and two A–B bonds.)

9.9 Consider the concerted reaction

$$C_2H_4 + C_2H_4 \longrightarrow \begin{array}{c} H \ H \\ | \ \ | \\ H-C-C-H \\ | \ \ | \\ H-C-C-H \\ | \ \ | \\ H \ H \end{array}$$

(a) Draw a correlation diagram for the reaction.

(b) Draw a configuration mixing diagram for the reaction.

(c) Show that the reaction is symmetry forbidden.

(d) What is the physical significance of this result? Why does the symmetry constraint arise?

9.10 Discuss the various σ scales listed in Table 9.2. What types of reactions follow each σ scale? What are the transition states like in each case?

9.11 In 1934 Rice and Herzfeld proposed that the pyrolysis of ethane

$$C_2H_6 \rightarrow C_2H_4 + 2CH_4$$
$$C_2H_6 \rightarrow C_2H_4 + H_2$$

follows the following mechanism:

$$C_2H_6 \rightarrow 2\ CH_3 \qquad (1)$$

$$CH_3 + C_2H_6 \rightarrow CH_4 + C_2H_5 \qquad (2)$$

$$C_2H_5 \rightarrow C_2H_4 + H \qquad (3)$$

$$H + C_2H_6 \rightarrow C_2H_5 + H_2 \qquad (4)$$

$$H + C_2H_5 \rightarrow C_2H_6 \qquad (5)$$

(a) Use the analysis in Chapter 8 to indicate which steps need collision partners. Be sure to explain why the steps need collision partners. Also if you decide that a step does not require a collision partner, explain why it does not need a collision partner.

(b) Which of the steps have activation barriers? Be sure to explain why the activation barrier arises.

(c) Which of the steps do not have activation barriers? Be sure to explain why there is no activation barrier.

(d) Rank each of the steps in regard to the sizes of the barriers for each reaction, i.e., which step has the largest barrier, the second largest barrier, etc. Be sure to justify your answer.

(e) How do the energy transfer barriers to reactions described in Chapter 8 affect each of the reactions?

(f) How do angular momentum barriers to reactions described in Chapter 8 affect each of the reactions just given?

(g) Use transition-state theory to estimate the preexponentials for each of the reactions.

(h) Why does the reaction follow this mechanism rather than the reverse four-centered reaction $M + C_2H_6 \rightarrow H_2-C_2H_4^{\ddagger} \rightarrow C_2H_4 + H_2 + M$ (*Hint:* Prove that the four-centered reaction has a large activation barrier.)

(i) If you changed from ethane to 1,2-dichloroethane, how would the rates of each of the reactions change? Assume that the products are ethylene, chlorine, and chloromethane.

9.12 The reaction $H_2 + Br_2 \rightarrow 2\,HBr$ follows the following mechanism:

$$Br_2 \rightarrow 2\,Br \qquad\qquad (1)$$

$$Br + H_2 \rightarrow HBr + H \qquad\qquad (2)$$

$$H + Br_2 \rightarrow HBr + Br \qquad\qquad (3)$$

$$2\,Br \rightarrow Br_2 \qquad\qquad (4)$$

(a) Which steps need collision partners? Be sure to explain why the steps need collision partners. Also, if you decide that a step does not require a collision partner, explain why it does not need a collision partner.

(b) Which of the steps have activation barriers? Be sure to explain why the activation barrier arises.

(c) Which of the steps do not have activation barriers? Be sure to explain why the reaction does not have an activation barrier.

(d) Rank each of the steps in regard to the sizes of the barriers for each reaction, i.e., which step has the largest barrier, the second largest barrier, etc. Be sure to justify your answer.

(e) How do the energy transfer barriers to reactions described in Chapter 8 affect each of the reactions?

(f) How do angular momentum barriers affect each of the reactions just given?

(g) Use transition-state theory to estimate the preexponentials for each of the reactions.

(h) Why does the reaction follow this mechanism rather than the simple four center reaction? $H_2 + Br_2 \rightarrow H_2Br_2^{\ddagger} \rightarrow 2\,HBr$ (*Hint:* Prove that the four center reaction has a large activation barrier.)

(i) If you changed from bromine to chlorine, how would the rates of each of the reactions change?

9.13 The acid–base theory described in Sections 3.4.1 and 9.10.2 has many important implications for surface reactions. CO and H_2S are soft bases while H_2O is a hard base and HF is a hard acid–base.

(a) Consider the adsorption of these molecules on a platinum surface (i.e., a soft acid). Which of these intermediates should bind the most strongly?

(b) Consider the adsorption of these molecules on Al_2O_3 (i.e., a hard acid–base). Which should bind the most strongly?

(c) Consider the adsorption of these molecules on a silicon surface (i.e., a surface with both hard and soft properties, but more hard than soft contribution). Which should bind the most strongly?

9.14 The acid–base theory described in Sections 3.4.1 and 9.10.2 is useful for reactions on surfaces. Consider a simple hydrogen elimination reaction

$$R-H_{(ad)} + Bare\ Site \rightarrow R_{(ad)} + H_{(ad)}$$

(a) Assume you have data for the rate of reaction on a silver surface (i.e., a soft base). Would you expect the reaction to correlate better to the acidity (i.e., hard base properties) or ionicity (i.e., soft base properties)?

(b) Would the hydrogen be positively or negatively charged? How would the charge change on an oxygen-covered surface?

(c) Repeat for an acid site on the MgO surface (i.e., a hard acid).

9.15 Figure 9.24 is a plot of the log of the rate constant of a series of ligand exchange reactions as a function of the log of the equilibrium constant for the reaction.

(a) Show that if the plot was exactly linear, the data would follow the Polayni relationship, Equation 9.6.

(b) Use the data to estimate the transfer coefficient.

(c) What is the physical significance of the value of the transfer coefficient in part (b)? That is, what does the numerical value tell you about the transition state for the reaction?

9.16 Figure 9.20 is a Hammett plot for the reaction between hydrogen peroxide and a series of substituted benzene sulfinic acids. What does the slope of the plot tell us about the transition state for the reaction?

9.17 Moreau et al. *ACS Preprints* **33**, 298 [1987] examined the hydrogenolysis ($C_6H_5X + H_2 \rightarrow C_6H_6 + HX$) and hydrogenation ($C_6H_5X + 4H_2 \rightarrow C_6H_{12} + HX$) of a number of substituted benzenes over a $NiMo/Al_2O_3$ catalyst and the following data were obtained:

Compound	k, Hydrogenolysis	k, Hydrogenation	σ_R	σ_I
Thiophenol	400	Negligible	−0.17	0.27
Diphenyl sulfide	300	Negligible	−0.13	0.31
Phenol	2	33	−0.42	0.24
Diphenylether	3	3	−0.34	0.40
Aniline	1	10	−0.47	0.17
Diphenylamine	1	9	−0.50	0.30
Chlorobenzene	75	~1	−0.21	0.47
Bromobenzene	100	~1	−0.19	0.47
Fluorobenzene	8	2	−0.33	0.54
Benzene	—	2	0	0

(a) Do the hydrogenolysis data fit the σ_I or σ_R scale better?

(b) What do you conclude from part (a) about the transition state for the reaction?

(c) Do the hydrogenation data fit the σ_I or σ_R scale better?

(d) Neither the σ_I nor the σ_R fit very well. What does that suggest?

(e) Moreau et al. suggest that there is a better correlation to σ_R than σ_I. However, the correlation is nonlinear. What is the significance of the nonlinearity?

9.18 Benkeser, DeBoer, Robinson, and Sauve, *J. Am. Chem. Soc.* **78**, 682 [1956] examined the rate of a series of Metchuchan reactions

$$XC_6H_4N(CH_3)_2 + CH_3I \rightarrow XC_6H_4N^+(CH_3)_2 + I^-$$

in acetone. The following data were obtained:

Substituent	k, lit/mole-hr	Substituent	k, lit/mole-hr
p-CH$_3$O	5.40	p-Cl	0.23
p-CH$_3$	2.04	p-Br	0.22
m-CH$_3$	1.11	m-Cl	0.10
H	0.71	m-Br	0.085
m-CH$_3$O	0.60		

(a) Construct a Hammett plot for the reaction. Which σ scale fits best?

(b) What do you conclude from this?

(c) Why is the transfer coefficient so large?

(d) What is the significance of a negative transfer coefficient?

9.19 H. Yoshitake and Y. Iwasawa, *J. Phys. Chem.* **95**, 7368 [1991] examined the hydrogenation of a series of substituted ethylenes ($X-CH=CH_2$) over a Pt/SiO$_2$ catalyst. They measured both the rate of the reaction and the partial charge transfer to the platinum catalyst. The following data were obtained:

Substituent	CH$_3$	H	COCH$_3$	CN	CF$_3$
Log rate	2.2	1.8	−2	−3	2.3
Partial charge transfer	0.018	0.075	0.132	0.154	0.06

(a) Plot the data for the extent of charge transfer against the σ_P^0, σ_R and σ_I parameters for each of the substituents. Which parameter is the best indicator of the charge transfer? What do you infer from this result?

(b) How would the partial charge transfer affect the strength of the binding of a substituted ethylene to the surface?

(c) Consider the reaction $CH_2CHX_{(ad)} + 2 H_{(ad)} \rightarrow CH_3CH_2X$. How should an increase in the binding substituted ethylene affect the rate?

(d) How well do your results in part (c) explain Iwasaya's rate data.

(e) What additional experiments would you do to see if your hypothesis in part (d) is correct?

9.20 Smith and Pennekamp, *JACS* **67**, 276 [1945] examined the hydrogenation of a number of substituted benzenes $X-C_6H_5$ over a platinum catalyst. The following data were obtained:

X	$k \times 10^4$ (min)$^{-1}$	X	$k \times 10^4$ (min)$^{-1}$
H	650	i-C$_3$H$_7$	216
CH$_3$	350	n-C$_4$H$_9$	244
C$_2$H$_5$	294	s-C$_4$H$_9$	189
n-C$_3$H$_7$	265	t-C$_4$H$_9$	166

(a) How well do these data fit the Hammett correlation?

(b) Which σ scale works best?

(c) What do you conclude from your answer in part (b)?

9.21 M. A. Barteau, *Catalysis Letters* **8,** 175 [1991] summarized the data for formate decomposition on transition metals. Generally the formate decomposes via the reaction

$$HCOO_{(ad)} \rightarrow CO_2 + H_{ad} \qquad (9.126)$$

(a) Use the heats of adsorption of hydrogen and oxygen from Chapter 3 to estimate the ΔH for the reaction on each of the surfaces. Assume that the heat of adsorption of the formate is related to the heat of adsorption of the oxide via Figure 9.23.

(b) Use the material in Example 4.C to estimate ΔG for the reaction at 400 K and 10^{-9} torr.

(c) Experimentally the decomposition temperature is given by the following:

Surface	Decomposition Temperature (K)	Surface	Decomposition Temperature (K)
Pt(111)	260	Ni(111)	375
Pd(111)	270	Ag(110)	425
Rh(111)	290	Cu(111)	475
Au(110)	340	Cu(110)	475
Ru(001)	355	Fe(100)	490

Assume that the decomposition temperature, T_{Dec}, related to the activation energy for decomposition via

$$E_A = \left(0.07 \, \frac{Kcal}{Mole \, °K}\right)(T_{Dec})$$

Do the data fit a linear free-energy relationship?

(d) What is the significance of there being two lines in the plot? (*Hint:* Refer back to the discussion of Figure 6.4. The discussion starts on page 443.)

(e) What is the significance of the transfer coefficient you calculated in part (c)?

9.22 Tanaka and Ozaki, *J. Catalysis* **8,** 1 [1967] examined the relationship between the catalytic activity of a number of sulfates for hydration of propylene (i.e., CH$_3$CH=CH$_2$ + H$_2$O → C$_3$H$_9$OH) and the electronegativity of the metal center. The reaction is acid catalyzed, where water reacts with the metal center to form an acid site, then the acid reacts with the propylene

The following data were obtained for the temperature for 1% conversion:

Acid	Na	Mg	Ni	Zn	Co	Al	K	Mn	Cd	Co	Cr	Fe
Temp, °C	230	75	155	145	138	128	220	160	150	140	133	115

(a) Plot these data against the electronegativity of the ion. What do you see?

(b) Use the analyses in Sections 3.4.1, and 9.2, and 9.10.2 to show that the electro-negativity of the ion is proportional to the acidity of the ion.

(c) What is the physical significance of the curvature in the plot in part (a)?

9.23 Jacobs, Mortier, and Uylterhoeven, *J. Inorg. Chem.* **40,** 1919 [1978] examined the catalytic activity of a number of zeolites for dehydration of isopropanol. The following data were obtained:

Zeolite	HX	HY	HL	HClin	HZ
Acid strength	0.0194	0.110	0.126	0.140	0.145
Turnover rate $\times 10^4$/sec	4	6	8	9.5	10

Show that these data fit the Brønsted relationship.

More Advanced Problems

9.24 In the derivation of the Marcus equation we assumed that we could approximate the A–B interaction as a Lennard-Jones function of r_x. Is that reasonable? Plot the potential on the potential contour shown in Figure 9.7. Is it a Lennard-Jones potential? How could we modify the r_x coordinate so the curves follow a Lennard-Jones potential?

9.25 Figure 9.11c shows a one-dimensional diagram of the potential surface in the Marcus inverted region. Is this a reasonable potential diagram? What would the two-dimensional potential energy surface have to be like for this diagram to apply?

9.26 Barteau and Madix, *Surface Sci.* **120,** 262 [1980] examined the displacement of a number of Brønsted acids from silver surfaces. They found that the equilibrium for the displacement process scaled as the gas phase acidity.

(a) Relate this observation to a linear free-energy relationship like that in Table 9.1.

(b) What is the significance of the observation that the displacement rate varies as the gas phase acidity, and not the gas phase ionicity?

9.27 In Section 9.15 we showed that while Dai and Gellman's data, *JACS* **115,** 714 [1993] could be explained by an inductive effect, there may also be a resonant effect. What additional experiments would you do to see whether the resonant effect was important? What substituents would you use? What measurements would you make?

BIBLIOGRAPHY

Chapman, N. B., and J. Shorter, *Advances in Linear Free Energy Relationships*, Plenum, NY (1972).

Chapman, N. B., and J. Shorter, *Correlation Analysis in Chemistry*, Plenum, NY (1978).

Lowry, T. H., and K. S. Richardson, *Mechanism And Theory In Organic Chemistry*, 3rd ed, Harper & Row, NY (1987).

Parr, R. G., and W. Yang, *Density-Functional Theory of Atoms and Molecules*, Oxford University Press, NY (1989).

REFERENCES

Albery, W. J., *Electrode Kinetics*, Clarendon Press, Oxford (1975).

Albery, W. J., and H. M. Kreevoy, *Adv. Phys. Org. Chem.* **16,** 87 (1978).

Bell, R. P., *Acid Base Catalysis*, Oxford University Press, Oxford (1941).

Bell, R. P., *The Proton in Chemistry*, Chapman Hall, London (1973).

Bell, R. P., and W. C. E. Higginson, *Proc. Royal Soc.* **A197,** 141 (1949).

Benson, S. W., and J. H. Buss, *J. Chem. Phys.* **29,** 546 (1958).

Benson, S. W., *Thermochemical Kinetics*, Wiley, NY (1978).

Benson, S. W., and R. Srinivansan, *J. Chem. Phys.* **23,** 200 (1955).

Bokris, J. O., *Modern Electrochemistry*, Vol. 21, Plenum, NY (1970).

Bolton, J. R., N. Magataga, and G. McLendon, *Electron Transfer in Inorganic, Organic, and Biological Systems*, Advances in Chemistry, Vol. 228, American Chemical Society (1991).

Bordwell, F. W., and W. J. Boyle, *J. Am. Chem. Soc.* **97,** 3447 (1975).

Brønsted, J. N., and K. Pederson, *Z. Physik Chem.* **108,** 185 (1924).

Brønsted, J. N., *Chem. Rev.* **5,** 231 (1928).

Butler, T. A., *Trans. Faraday Soc.* **28,** 379 (1932).

Cannon, R. D., *Electron Transfer Reactions*, Butterworths, London (1980).

Chapman, N. B., and J. Shorter, *Advances in Linear Free Energy Relationships*, Plenum, NY (1972).

Chapman, N. B., and J. Shorter, *Correlation Analysis in Chemistry*, Plenum, NY (1978).

Cohen, A. O., and R. A. Marcus, *J. Phys. Chem.* **72,** 4249 (1968).

Conroy, H., and G. Malli, *J. Phys. Chem.* **50,** 5049 (1969).

Conway, B. E., *Electrochemical Data*, Elsevier, London (1952).

Conway, B. E., *Mod. Asp. Electrochem.* **16,** 103 (1985).

Cotton, F. A., *Chemical Applications of Group Theory*, Wiley, NY (1971).

Dai, Q., PhD Dissertation, University of Illinois (1992).

Dai, Q., and A. Gellman, *J. Am. Chem. Soc.* **115,** 714 (1993).

Dogandze, R. R., in N. S. Hush, ed., *Reactions of Molecules at Electrodes*, p. 135, Wiley, NY (1971).

Drago, R. S., *Coord. Chem. Rev.* **33,** 251 (1980).

Drago, R. S., *Inorg. Chem.* **29,** 1379 (1990).

Drago, R. S., and B. B. Wayland, *J. Am. Chem. Soc.* **87,** 3571 (1965).

Drago, R. S., G. C. Vogel, and T. E. Needham, *J. Am. Chem. Soc.* **93,** 6014 (1971).

Edwards, J. O., and R. G. Pearson, *J. Am. Chem. Soc.* **83,** 1743 (1961).

Evans, D. P., J. J. Gordon, and H. B. Watson, *J. Chem. Soc.* 1430 (1937).

Evans, M. G., and M. Polayni, *Trans. Faraday Soc.* **32,** 1333 (1936).

Exner, O., in N. B. Chapman and J. Shorter, eds., *Correlation Analysis in Chemistry*, p. 439, Plenum, NY (1978).

Fukui, K., *The Theory of Orientation and Stereospecificity*, Springer-Verlag, Berlin (1975).

Kukui, K., T. Yonezawa, and H. Shingo, *J. Phys. Chem.* **20,** 722 (1952).

Fukui, K., T. Yonezawa, and H. Shingo, *J. Chem. Phys.* **26,** 831 (1957).

Haim, A., *Inorg. Chem.* **9,** 426 (1970).

Hammett, L. P., *J. Am. Chem. Soc.* **59,** 96 (1937).

Hammett, L. P., and H. L. Pfluger *J. Am. Chem. Soc.* **55,** 4079 (1933).

Hammond, G. S., *J. Am. Chem. Soc.* **77,** 334 (1955).

Hansch, C., A. Leo, and R. W. Taft, *Chem. Rev.* **91,** 165 (1991).

Houriti, J., and M. Polayni, *Acta Physicochim. URSS* **2,** 505 (1935).

Hupke, D. J., and D. Wu, *J. Am. Chem. Soc.* **99,** 7653 (1977).

Ingold, C. K., and W. S. Nathan, *J. Chem. Soc.* **58,** 222 (1936).

Irving, H., and H. Rossohi, *Acta Chem. Scand.* **10,** 72 (1956).

Isaacs, N. S., *Physical Organic Chemistry*, Longman, Essex (1987).

Jaffe, R. L., J. M. Henry, and J. B. Anderson, *J. Am. Chem. Soc.* **98,** 1140 (1976).

Johnson, C. D., *The Hammett Equation*, Cambridge University Press, Cambridge (1973).

Johnston, H. S., and C. Parr, *J. Am. Chem. Soc.* **85,** 2544 (1963).

Jordan, R. B., *Reaction Mechanisms of Inorganic and Organometallic Systems*, Oxford University Press, Oxford (1991).

Kraus, M., *Adv. Catal.* **17**, 75 (1967).

Klopman, G., *J. Am. Chem. Soc.* **90**, 233 (1968).

Klopman, G., *Chemical Reactivity and Reaction Paths*, Wiley, NY (1974).

Kohn, W., and L. J. Sham, *Phys. Rev.* **140**, 1133 (1965).

Lindberg, B. J., *Acta Chem. Scand.* **20**, 1843 (1966).

Marcus, R. A., *J. Chem. Phys.* **24**, 966 (1955).

Marcus, R. A., *Ann. Rev. Phys. Chem.* **15**, 155 (1964).

Marcus, R. A., *J. Phys. Chem.* **72**, 891 (1968). Marcus, R. A., *J. Am. Chem. Soc.* **91**, 7224 (1969).

Miller, W. H., in J. R. Bolton, N. Magataga, and G. McLendon, eds., *Electron Transfer in Organic, Inorganic and Biological Systems*, Advances in Chemistry, Vol. 228, American Chemical Society, Washington, D.C. (1991).

Mulliken, R. S., *J. Chem. Phys.* **2**, 782 (1934).

Pauling, L., *The Nature of the Chemical Bond*, 3rd ed., Cornell University Press, Ithaca, NY (1960).

Pearson, R. G., *J. Am. Chem. Soc.* **85**, 3533 (1963).

Pearson, R. G., *J. Chem. Educ.* **45**, 643 (1968).

Pearson, R. G., *J. Am. Chem. Soc.* **91**, 4947 (1969).

Pearson, R. G., *Symmetry Rules for Chemical Reactions*, Wiley, NY (1976).

Pearson, R. G., *Inorg. Chem.* **27**, 734 (1988).

Pederson, K. J., *J. Phys. Chem.* **38**, 581 (1934).

Politzer, P., *J. Chem. Phys.* **86**, 1072 (1987).

Pross, A., *Adv. Inorg. Chem.* **21**, 99 (1985).

Sullivan, J. H., *J. Chem. Phys.* **30**, 1292 (1959); **39**, 3001 (1963).

Sutin, N., *Acc. Chem. Res.* **1**, 57 (1968).

Sutin, N., *Prog. Inorg. Chem.* **30**, 441 (1983).

Tafel, J., *Z. Phys. Chem.* **50**, 641 (1905).

Taylor, H. S., *Z. Elektrochem.* **20**, 201 (1914).

Van't Hoff, J. H., *Studies in Chemical Dynamics*, Edward Arnold, London (1883, 1896).

Vetter, K. J., *Electrochemical Kinetics*, Academic Press, NY (1967).

Volmer, M., and T. Erdey-Gruz, *Z. Physik Chem.* **150**, A.203 (1930).

Wells, R. P., *Linear Free Energy Relationships*, Academic Press, London (1968).

Woodwad, R. B., and R. Hoffmann, *J. Am. Chem. Soc.* **87**, 395, (1965a); **87**, 2046, (1965b); **87**, 2511 (1965c).

Woodward, R. B., and R. Hoffmann, *The Conservation of Orbital Symmetry*, Verlag Chemie, Deerfield Beach, FL (1970).

Wright, J. S., *Can. J. Chem.* **53**, 549 (1975).

Zelenay, P., C. K. Rhee, M. Rubel, A. Wieckowski, J. Wang, and R. I. Masel, *J. Phys. Chem.* **96**, 8509 (1992).

10 Rates and Mechanisms of Surface Reactions

Précis

So far in this book we have been reviewing adsorption and reaction on solid surfaces. In Chapter 6 we noted that surfaces speed up reactions by stabilizing intermediates and by lowering barriers to reaction. Chapter 7 showed that there are also important differences between the kinetics of surface reactions and the kinetics of gas phase reactions. Chapters 8 and 9 reviewed reaction rate theory. Now we are in a position to discuss how surfaces promote chemical reactions in more detail.

The purpose of this chapter is to examine some general principles that allow one to understand rates and mechanisms of reactions on surfaces. In saying that, one has to be cautious because there are few, if any, well-established principles of surface reactions in the literature. What there are instead are a series of qualitative models. None of the models works all the time. However, the models have been discussed in the literature and can be generalized to yield interesting insights into surface reactions. In some cases the ideas are old enough that most investigators consider them useful. In other cases the ideas are newer and not as well established. In this chapter we review many of the ideas. Still, since the ideas are evolving, I cannot say that the findings are well established.

10.1 HISTORICAL INTRODUCTION

As noted in Chapters 1, 6, and 7, surface reactions were discovered by accident by Priestley [1790] and Van Marum [1796]. There were many studies of surface reactions throughout the nineteenth century. However, it was not until the work of Naumann [1878] that people knew that reactions were occurring on surfaces, and not just in the pores of a highly porous material. In the latter part of the nineteenth century many theories were advanced that attempted to predict how the presence of surfaces changed rates of reactions. However, most of the ideas were quickly discarded. Still, in 1919, Rideal and Taylor wrote a very influential book, *Catalysis in Theory and Practice*, where they examined the key ideas of the day. Rideal and Taylor pointed out that surfaces could change rates of reactions by three basic mechanisms: the surfaces could act as an efficient medium for energy transfer; the surface could form a bond with various species in the system, thereby stabilizing what we now call reactive intermediates; the surface can push molecules together, thereby increasing the probability that the molecules react.

The details of these models have undergone some revisions and improvements since Rideal and Taylor wrote their book. However, the amazing thing is that our basic understanding of surface reactions has not changed that much since 1920. People still say that surfaces modify reactions by acting as efficient collision partners to initiate reactions, by stabilizing reactive intermediates, and by holding the reactants in a configuration that is conducive to reaction. The one really new idea which has come out since 1920 is that transition metal surfaces also influence rates by stabilizing antibonding orbitals, thereby lowering the intrinsic barriers to bond scission. This idea was first discussed by Mango [1969] in the context of transition metal clusters and it was applied to metal surfaces by Masel [1986] and Nørskov [1990]. Still, most of the key ideas about the reactivity of surfaces date back to the work of Langmuir, Sabatier, Ostwald, and Van't Hoff before 1920. It is no wonder then that all four of these people won Nobel prizes.

In the remainder of this chapter we summarize the key ideas that are used to understand reactions on surfaces. This topic is still evolving. Therefore, many of our conclusions will have to be somewhat tentative. However, an attempt is made to provide a current snapshot of what we understand, and what still needs to be explained. Our approach is to start by discussing some general principles of reactions on surfaces, and then move on to some specific cases. Just to orient the reader, Sections 10.2 to 10.17 are a discussion of general principles, while sections 10.18 to 10.22 are a discussion of some specific cases.

10.2 RADICAL REACTIONS IN THE GAS PHASE

We are going to start the discussion in a roundabout way. Recall from Chapter 3 that one can often view molecules adsorbed on metal surfaces as radicals solubilized in jellium. In Chapter 6 we noted that the mechanisms of reactions on metal surfaces often follow mechanisms that are similar to those for radical reactions in the gas phase. The intermediates and reaction pathways are similar. Therefore, a good place to start our discussion is to consider gas phase reactions. We then try to understand how the reactions change in the presence of the surface.

The mechanisms of gas phase reactions have been studied for almost 90 years. In 1956 Hinshelwood won the Nobel Prize in chemistry for showing that many gas phase reactions go via a radical propagation mechanism. First, there is an initiation step where a stable molecule R_2 breaks into radicals

$$X + R-R \longrightarrow 2R\cdot + X \qquad (10.1)$$

where X is a collision partner. Then there are a series of propagation steps where the radicals exchange ligands with a second stable molecule M_2:

$$R\cdot + M-M \longrightarrow RM + M\cdot \qquad (10.2)$$

$$M\cdot + R_2 \longrightarrow RM + R\cdot \qquad (10.3)$$

Finally, there are termination steps where radicals react to form stable species:

$$X + 2R\cdot \longrightarrow X + R_2 \qquad (10.4)$$

Benson [1976] notes that generally, one finds that the weakest bonds in the reacting molecules break preferentially during the initiation steps. However, during the propagation steps, one does not necessarily break the weakest bonds. Instead, one usually transfers atoms one atom at a time.

It is useful to consider why gas phase reactions follow the mechanism they do. Let's examine the initiation step first. In the initiation step the weakest bond in the R_2 molecule breaks. It is easy to understand why that occurs. If we look at the initiation step in reverse (i.e., the termination step), two radicals will come together to produce a stable species. The hybridization of the radicals usually does not change significantly during the reaction, which means that according to Chapter 9 one does not expect any orbital symmetry barriers to the reaction. Further, the reverse of the reaction is highly exothermic. There are no thermodynamic barriers to the reaction. The net result is that the termination reaction would be expected to be essentially unactivated.

Now recall that from the principle of microscopic reversibility (Chapter 5), at equilibrium the activation energy for a forward reaction, $E_a^{forward}$, is related to the activation energy for the reverse reaction, $E_a^{reverse}$ by

$$E_a^{forward} - E_a^{reverse} = \Delta H_a \tag{10.5}$$

where ΔH_a is the heat of reaction. If the reverse reaction is unactivated, then the activation energy for the forward reaction is equal to the heat of reaction.

The implication of that result is that the activation barrier for the initiation reaction will equal the energy of bond scission. The activation barrier will be lower if a weak bond breaks than if a strong bond breaks. Therefore, in the initiation step, the weakest bond in a molecule usually breaks.

Interestingly, Benson [1976] also notes that in the propagation steps one does not generally find that the weakest bonds in the molecule break more easily than the strongest bonds. Instead, one finds that reactions where a single atom is moved from one radical to the next dominate over reactions where a molecular ligand moves. For example, consider the interaction of a deuterium atom with an ethane molecule. Two reaction pathways are possible:

$$D + C_2H_6 \longrightarrow CH_3D + CH_3 \tag{10.6}$$

$$D + C_2H_6 \longrightarrow HD + C_2H_5 \tag{10.7}$$

ΔH for Reaction 10.6 is -15.7 kcal/mole; ΔH for Reaction 10.7 is -6 kcal/mole. Therefore, Reaction 10.6 is thermodynamically favored. According to Westley [1980] Reaction 10.7 is quite facile; it has an activation barrier of only 9.7 kcal/mole. However, Reaction 10.6 is not observed experimentally. Nicholas, Bayrakceken, and Fink [1972] and Oldershaw and Gould [1985] examined the reaction of hot hydrogen and deuterium atoms with ethane. Reaction 10.7 can be observed at collision energies above 9.8 kcal/mole. However, Reaction 10.6 was not observed at any collision energy (the experiments had an upper limit of 49.5 kcal/mole). Therefore, experimentally Reaction 10.6 has a much higher activation barrier than Reaction 10.7 even though Reaction 10.6 is more exothermic than Reaction 10.7. This is just the opposite of what one would expect from the Polayni relationship.

One can understand this result based on the principle of conservation of orbital symmetry described in Chapter 9. Reaction 10.7 is simple. A hydrogen atom comes in and reacts with a hydrogen in the ethane. The $H-C_2H_5$ stretches and simultaneously a hydrogen–deuterium bond forms. The situation is analogous to the situation in Figure 9.39. There is a small barrier to the reaction because a bond breaks. However, orbital symmetry is conserved during the reaction so the reaction is easy.

Reaction 10.6 is more difficult. Figure 10.1 shows two possible geometries for the reaction. In the case on the left of Figure 10.1, the deuterium approaches the ethane so the deuterium collides with one of the CH_3 groups. In that case, it is easy to extract a hydrogen. However, if one instead wants to form deuterated methane, the incoming deuterium has to move the hydrogens in the ethylene out of the way before the deuterium can form a D–C bond. There is a strong Pauli repulsion between the deuterium and the hydrogens. As a result, the reaction on the left of Figure 10.1 has a large activation barrier.

The reaction on the right of Figure 10.1 is different. The deuterium approaches the ethane perpendicularly to the C–C axis. In that case, the Pauli repulsions between the hydrogens are less than the Pauli repulsions in the case on the left of Figure 10.1. However, one still expects a large barrier to bond scission. Notice that the bond on the left carbon atom starts out pointing to the right carbon atom, but the bond has to end up pointing toward the deuterium.

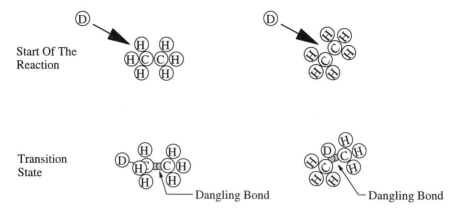

Figure 10.1 Two possible geometries for Reaction 10.6.

One can show that quantum mechanically a change in the orientation of a bond is equivalent to a change in the orbital symmetry of the bond. Therefore, the reaction on the right of Figure 10.1 does not conserve orbital symmetry. In Chapter 9 we found that reactions that do not conserve orbital symmetry generally have large activation barriers, because the new bond does not form until the original bonds break. As a result, one would expect Reaction 10.6 to have a large activation barrier, as is observed.

There is another way to arrive at the same result. Note that one can draw a stick diagram, where a C–C bond breaks at the same time that a C–D bond forms. However, in reality, the deuterium atom will need to overlap the dangling bonds on the carbon before the C–D bond can form. If the carbon–carbon bond in ethane is at its equilibrium position, the deuterium cannot squeeze in to overlap the dangling bond. As a result, the carbon–carbon bond needs to lengthen and twist before significant overlap occurs. That costs considerable energy. In fact, it works out that the carbon–carbon bond in the ethane needs almost to break before the dangling bond in the carbon can significantly overlap the deuterium. As a result, the reaction shown on the right of Figure 10.1 has a large activation barrier.

Another way to look at that effect is that in the transition state for the reaction in the right of Figure 10.1, one has to put three electrons into bonding orbitals: one electron from each carbon atom and one from the hydrogen. One cannot put all three electrons in a single bonding orbital, and there is no way to form two bonds because the orbitals are pointing in the wrong directions. As a result, quantum mechanically Reaction 10.6 should have a large activation barrier. Experimentally, Reaction 10.6 has never been observed.

So far we have been discussing Reactions 10.6 and 10.7, but the results are general. Benson [1976] shows that when a radical reacts with a stable molecule in the gas phase, it is usually much easier for the radical to extract a single atom (e.g., a hydrogen) than to extract a molecular ligand (e.g., a CH_3 group). The extraction of a single atom conserves orbital symmetry, while the extraction of the molecular ligand often does not. As a result, in the gas phase, it is very difficult for radicals to extract molecular ligands. Typically, the activation energy for extraction of molecular ligands is in the order of the energy of the bond between the molecular ligand and the rest of the molecule. As a result, in the gas phase, atom transfer reactions dominate over ligand transfer reactions.

In a more general way, I would like to propose what I believe to be a general principle for gas phase and surface reactions, which I call the **principle of maximum bonding:** The preferred pathways for gas phase and surface reactions are the ones that maximize the bonding in the transition state. Consider two parallel reaction pathways such as those in Equations 10.6 and 10.7. When either of those reaction pathways occur, the activation energy for the reaction pathway will be equal to the difference between the bond energy in the reactants and the bond energy in the transition state. Therefore, as one increases the net bonding in the transition state, one lowers the barrier to the reaction. Consequently, the way to get low barriers to reaction is to maximize the bonding in the transition state. If there are two parallel pathways for reaction, the step that maximizes the net bonding in the transition state will usually have a lower activation barrier than a step that has less net bonding. One does have to consider the orbital symmetry and orbital orientation to understand which steps maximize the net bonding. Stick diagrams like those in Figure 9.28 often give incorrect results. Still, one can distinguish between parallel reaction pathways by looking at the net bonding in the transition state. If all else is equal, the pathway that has the most net bonding in the transition state will dominate over the pathways that have less net bonding.

Later in this chapter, we note that this idea also allows one to understand why some elementary reactions have lower barriers on one surface than another. The idea is that surfaces that stabilize the transition state for a reaction will tend to speed up the reaction. Therefore, rates of elementary surface reactions tend to be highest on surfaces that maximize the bonding in the transition state for a reaction.

Another way to modify the barriers for reactions is to put charges on molecules. In Section 9.13.2 we showed that the four centered reaction

$$\text{H}_2 + \text{D}_2 \longrightarrow \begin{array}{c} \text{H}-\text{H} \\ \backslash \ / \\ \text{D}-\text{D} \end{array} \longrightarrow 2\text{HD} \qquad (10.8)$$

is symmetry forbidden. However, the same argument shows that the reaction

$$\text{H}_2^+ + \text{D}_2^+ \longrightarrow \left[\begin{array}{c} \text{H}-\text{H} \\ \backslash \ / \\ \text{D}-\text{D} \end{array} \right]^{2+} \longrightarrow 2\text{HD}^+ \qquad (10.9)$$

is symmetry allowed. Reaction 10.9 is different than Reaction 10.8 in that there are only two electrons in the transition state. One can put those two electrons in a single symmetric molecular orbital. Therefore, unlike the case in Reaction 10.8, there are no orbital symmetry constraints to Reaction 10.9.

In a more general way, when one has positive ions, one has less net bonding in the reactants, and so one does not need as much bonding in the transition state. As a result, reactions that are forbidden can occur when charged species are used.

Now let's apply that idea to the reaction in Equations 10.6 and 10.7. Consider the reaction

$$\text{H}^+ + \text{C}_2\text{H}_6 \longrightarrow \text{Products}$$

An H^+ ion has no orbital symmetry, and so there are no barriers to the interaction of an H^+ ion with an ethane molecule to form an asymmetric C_2H_7^+ ion with a geometry on the left of Figure 10.1. Hiraoka and Kebarle [1976] and Yeh, Prize, and Lee [1989] examined the reaction of H^+ with ethane. They found that the reaction of H^+ with ethane to form a stable C_2H_7^+ ion:

$$\text{C}_2\text{H}_6 + \text{H}^+ \longrightarrow \text{C}_2\text{H}_7^+ \qquad (10.10)$$

is unactivated when the H^+ approaches the complex as in the left of Figure 10.1. Interestingly, the ion rearranges to form a bridged complex with a geometry like that on the bottom right of Figure 10.1. The C_2H_7^+ ion is stable. However, if it is heated, it will decay via one of two pathways:

$$\text{C}_2\text{H}_7^+ \longrightarrow \text{C}_2\text{H}_5^+ + \text{H}_2 \qquad (10.11)$$

$$\text{C}_2\text{H}_7^+ \longrightarrow \text{CH}_4 + \text{CH}_3^+ \qquad (10.12)$$

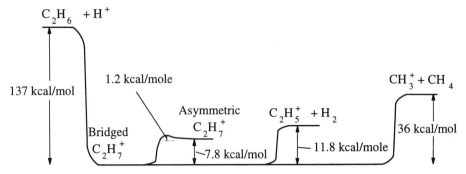

Figure 10.2   Potential for the decomposition of $C_2H_7^+$ as suggested by Yeh et al. [1989].

Yeh et al. report that Reaction 10.11 is 11.8 kcal/mole endothermic, while Reaction 10.12 is 36 kcal/mole endothermic. Both reactions proceed with moderate activation barriers as indicated in Figure 10.2.

In a more general way, the orbital symmetry effects and Pauli repulsions are much smaller with positively charged ions than with neutrals. As a result, the barriers to ligand exchange are lowered when positively charged ions are used.

In a similar way, one would expect the barriers to be enhanced when negatively charged ions are used, although suitable data to demonstrate that expectation has been difficult to locate.

10.3   INTRODUCTION TO THE ROLE OF THE SURFACE: STABILIZATION OF INTERMEDIATES

Now it is useful to consider how the presence of a surface would modify the conclusions in the last section. Later in this chapter we find that a surface has three effects: the surface stabilizes intermediates, the surface acts as a source of radicals or ions, and the surface interacts with antibonding orbitals to help overcome orbital symmetry limitations. All three of these effects have been discussed at length in the literature. However, the first effect, the role of the surface in stabilizing reactive intermediates has been discussed in the most detail. In Chapter 6 we noted that the stabilization of intermediates allows surfaces to catalyze reactions. The stabilization of intermediates also allows low defect films to grow. Therefore, the stabilization of intermediates is quite important. Many investigators assume that the stabilization of intermediates is the key reason that surfaces promote reactions. Later in this chapter we find that there are other effects which are as important as the stabilization of intermediates. Before we can understand those other effects, however, we need to quantitatively understand how the stabilization of intermediates increases rates. Therefore we discuss the stabilization of intermediates next.

In the next several sections, we review how the stabilization of intermediates leads to enhanced rates. We then quantify the effects, and try to show when factors other than the stabilization of intermediates are important to surface reactions.

First, let's show that if one can stabilize the intermediates produced during a reaction, one can enhance the rate of the reaction. Consider a simple gas phase reaction where a molecule A reacts to form a reactive intermediate $A^{\ddagger}$, and then $A^{\ddagger}$ decays to a product B, i.e.,

$$A \xrightleftharpoons[-1]{1} A^{\ddagger} \ (1)$$

$$A^{\ddagger} \xrightarrow{2} B \ (2) \tag{10.13}$$

If one uses the steady-state approximation, one can show that in the gas phase the overall rate of production of B is given by r_B:

$$r_B = \frac{k_1 k_2 P_A}{k_{-1} + k_2} \tag{10.14}$$

where k_1, k_2, and k_{-1} are the rate constants for steps 1, 2, and -1 in Reaction 10.13, and P_A is the partial pressure of A.

For the purpose of derivation, it is useful to express k_1 and k_2 via Arrhenius's law,

$$k_1 = k_1^o \exp\,(-E_{A1}/kT) \tag{10.15}$$

$$k_2 = k_2^o \exp\,(-E_{A2}/kT) \tag{10.16}$$

where k_1^o and k_2^o are the preexponentials for Reactions 1 and 2, and E_{A1} and E_{A2} are their corresponding activation energies, k is Boltzmann's constant, and T is the absolute temperature.

Substituting the Polayni relationship, Equation 9.5 into Equations 10.15 and 10.16 yields

$$k_1 = k_1^a \exp\,(-\gamma_p \Delta G_1/kT) \tag{10.17}$$

$$k_2 = k_2^a \exp\,(-\gamma_p \Delta G_2/kT) \tag{10.18}$$

where ΔG_1 and ΔG_1 are the standard free energies of Reactions 1 and 2, γ_p is the transfer coefficient, and k_1^a and k_2^a are given by

$$k_1^a = k_1^o \exp\,(-E_{A,1}^o/kT) \tag{10.19}$$

$$k_2^a = k_2^o \exp\,(-E_{A,2}^o/kT) \tag{10.20}$$

where $E_{A,1}^o$ and $E_{A,2}^o$ are the intrinsic activation barriers to Reactions 1 and 2.

Similarly, at equilibrium,

$$\frac{k_1}{k_{-1}} = K_{equ}^1 = \exp\,(\Delta G_1/kT) \tag{10.21}$$

where K_{equ}^1 is the equilibrium constant for Reaction 10.13. Substituting Equations 10.19, 10.20, and 10.21 into Equation 10.14 yields

$$r_B = \frac{k_1^a k_2^a \exp\,[-\gamma_p(\Delta G_1 + \Delta G_2)/kT]P_A}{k_1^a \exp\,[-(\gamma_p + 1)\Delta G_1/kT] + k_2^a \exp\,(-\gamma_p \Delta G_2/kT)} \tag{10.22}$$

Note since the free energy is a state property

$$\Delta G_1 + \Delta G_2 = \Delta G_{overall} \tag{10.23}$$

where $\Delta G_{overall}$ is the overall free-energy change in the reaction A $\rightarrow$ B. Combining reactions 10.22 and 10.23 yields

$$r_B = \frac{k_1^a k_2^a \exp\,[-\gamma_p \Delta G_{overall}/kT]P_A}{k_1^a \exp\,[-(1 + \gamma_p)\Delta G_1/kT] + k_2^a \exp\,[\gamma_p(-\Delta G_{overall} - \Delta G_1)/kT]} \tag{10.24}$$

Equation 10.24 allows one to calculate how stabilization of the intermediates of the reaction affects the rate of reaction. In a typical gas phase reaction ΔG_1, the free-energy change to produce the reactive intermediate is positive. ΔG_1 could be as large as $+20$ to $+50$ kcal/mole. If one plugs data into Equation 10.24, one finds that the rate is low under such situations. However, the presence of a surface can stabilize the reactive intermediates. The stabilization of the intermediates leads to a decrease in ΔG_1. Figure 10.3 shows a plot of the rate as a function of ΔG_1 calculated from Equation 10.24. Notice that initially, as one stabilizes the intermediates of a reaction (i.e., decreases ΔG_1) the rate of the reaction increases. The rate eventually reaches a maximum and then declines.*

These results show that as one stabilizes the intermediates of a reaction (i.e., decreases ΔG_1), one generally increases the rate of the reaction. As a result, if a surface stabilizes the intermediates of a reaction, the surface will catalyze the reaction. However, one does not want to overdo it, because at low enough ΔG_1, the rate declines again.

The idea that a surface acts to stabilize reactive intermediates was first discussed in a series of papers by Sabatier between 1890 and 1910. Much of the work is summarized in Sabatier's book, *Catalysis in Organic Chemistry* [1913]. Sabatier's idea was that a catalyst reaction involves sequential formation and destruction of bonds between the adsorbate and the surface. If the adsorbate/surface bonds are too weak, no reactive complexes form, so the overall reaction is slow. On the other hand, if the bonds are too strong, the adsorbate–surface bonds are difficult to break, so again the reaction is slow. Therefore, Sabatier suggested that the best catalysts are substances that have an intermediate bond strength between the adsorbate and the surface. The idea that the best catalysts have intermediate adsorbate–surface bond strengths is now called the **Principle of Sabatier.**

Balandin [1958, 1969] quantified this idea for reactions on surfaces. He first started with a derivation like the one earlier in this section. However, he found that it did not reproduce the observed pressure dependence of the rate, and it not fit data for reactions on surfaces when ΔG_1 was negative. Balandin pointed out that surface reactions are fundamentally different from gas phase reactions. In the gas phase, the concentration of reactive intermediates continuously increases with increasing pressure or decreasing ΔG_1. However, on a surface,

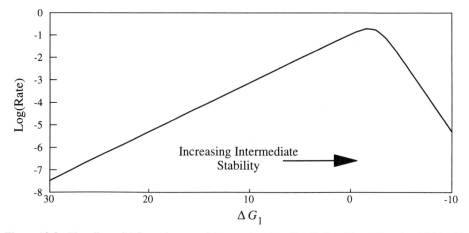

Figure 10.3 The effect of ΔG_1 on the rate of the reaction A $\rightarrow$ B calculated from Equation 10.24 with $\gamma_p = 0.5$ and $\Delta G_{overall} = -10$.

*The decline occurs because when ΔG_1 is negative the intermediate is too stable to react on to products.

the concentration of intermediates saturates at high pressures or when ΔG_1 is very negative. As a result Equation 10.24 does not apply to surface reactions. Consequently Balandin derived a new equation that explicitly accounted for the fact that the concentration of various species saturates at high pressures.

The derivation starts with a simple reaction

$$S + A \underset{-1}{\overset{1}{\rightleftharpoons}} A_{ad} \qquad (1)$$

$$A_{ad} \overset{2}{\longrightarrow} B + S \qquad (2)$$

$$(10.25)$$

where A is the reactant, B is the product, and S is a surface site, and notes that if the reaction follows Langmuir, Hougan, and Watson kinetics, and Reaction 2 is rate determining, the net rate of reaction is

$$r_{net} = \frac{k_2 K_1 P_A}{1 + K_1 P_A} \qquad (10.26)$$

where K_1 is the equilibrium constant for adsorption, k_2 is the rate constant for Reaction 2 in Equation 10.25, and P_A is the partial pressure of A. Similarly, the coverage is given by

$$\theta = \frac{k_1 P_A}{1 + k_1 P_A} \qquad (10.27)$$

Next, Balandin used the Polayni relationship, Equation 9.5, to see how the rate of Reaction 10.25 varied with ΔG_A, the free energy of adsorption of A. Recall from thermodynamics that the equilibrium constant for adsorption, K_1, is given by

$$K_1 = e^{-\Delta G_A/kT} \qquad (10.28)$$

Now consider Reaction 2 in Equation 10.25. The free-energy change during Reaction 2, ΔG_2^o, is given by

$$\Delta G_2 = \Delta G_{overall} - \Delta G_A \qquad (10.29)$$

where $\Delta G_{overall}$ is the free-energy change for the overall reaction $A \rightarrow B$. Substituting Equation 10.29 into Equation 9.5, substituting that into Arrhenius's law, Equation 10.16, yields

$$k_2 = k_2^b e^{+\gamma_p \Delta G_A/kT} \qquad (10.30)$$

where γ_p is the transfer coefficient and

$$k_2^b = k_2^o \exp\left(-(E_{A,2}^o + \gamma_p \Delta G_{overall}^o)/kT\right) \qquad (10.31)$$

Combining Equations 10.26, 10.28, and 10.30 yields

$$r_{net} = \frac{k_2^b \exp\left(-\Delta G_A^o (1 - \gamma_p)/kT\right) P_A}{1 + \exp\left(-\Delta G_A^o/kT\right) P_A} = \frac{k_2^b K_1^{(1-\gamma)} P_A}{1 + K_1 P_A} \qquad (10.32)$$

Figure 10.4 shows a log-log plot of the rate calculated from Equation 10.32 as a function of K_1. Note that the net rate of reaction goes up linearly, reaches a maximum, and then declines. Physically, K_1 is the equilibrium constant for adsorption. As K_1 increases, the surface is

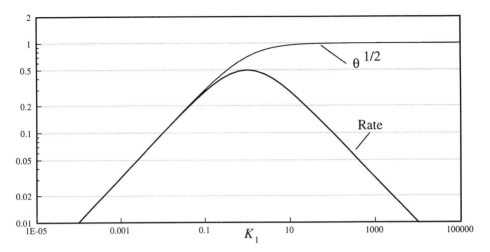

Figure 10.4 A log-log plot of the rate and coverage as a function of K_1 calculated from Equations 10.27 and 10.32 with $\gamma_p = 0.5$, $P_A = 1$.

stabilizing the intermediates of Reaction 10.25. The reaction rate is small when K_1 is small, since little gas adsorbs. The rate rises on the left side of the curve because as K_1 increases more gas adsorbs, so in effect, one is increasing the concentration of intermediates. However, the rate rises more slowly than θ (i.e., as $\theta^{1/2}$) because k_2, the rate constant for Reaction 2 in Equation 10.25, is decreasing at the same time that θ is increasing. The coverage saturates at $K_1 = 20$. At higher values of K_1 the coverage is almost constant. However, the stability of the reactive intermediates are changing. As K_1 increases, the adsorbed complex is becoming increasingly stable (i.e., harder to decompose), which causes the rate to decline. Eventually, at very large K_1, the adsorbed complex is completely stable, so no reaction is seen. The net effect is that the rate will increase with increasing K_1. However, then it will saturate and decline again as shown in Figure 10.4. Figure 10.4 is a key result because it says that if surfaces can stabilize the intermediates in a reaction, they can speed up the reaction. However, one does not want to overdo it, or the rate will decline. Consequently, the best catalysts for steady-state reactions are things that bind the reactants strongly, but not too strongly. This idea generally corresponds to experiment.

Now Equation 10.32 was derived for the reaction in Equation 10.25 where A adsorbs and reacts to form B, and B immediately desorbs. Further, the reaction was presumed to be completely irreversible. A more typical reaction is

$$S + A \overset{1}{\rightleftharpoons} A_{ad}$$

$$A_{ad} \overset{2}{\rightleftharpoons} B_{ad} \qquad (10.33)$$

$$B_{ad} \overset{3}{\rightleftharpoons} B + S$$

The case in Equation 10.33 is different from Equation 10.25 in that the reactions are all reversible. The analysis of Equation 10.33 is very similar to that for Equation 10.25. However, the details are different due to the presence of reversible reactions. There are three key free energies in the system, the free energy driving force for Reactions 1, 2, and 3, ΔG_1, ΔG_2, and ΔG_3

$$\Delta G_1 = \Delta G_A^o - kT \ln P_A \qquad (10.34)$$

$$\Delta G_3 = kT \ln P_B - \Delta G_B^o \tag{10.35}$$

$$\Delta G_2 = \Delta G_{overall} - \Delta G_1 - \Delta G_3 \tag{10.36}$$

where ΔG_A^o and ΔG_B^o are the standard free energies of adsorbed A and B, P_A and P_B are the partial pressures of A and B, and $\Delta G_{overall}$ is the standard free energy for the overall reaction A → B.

One can calculate the activation energy for the forward steps in each of these reactions via Equation 9.5. One can show that if all of the intrinsic activation energies (i.e., E_a^o's) and transfer coefficients, i.e., γ_p are equivalent, then the highest rate will come when ΔG_1, ΔG_2, and ΔG_3, are equal. If one of the ΔG's is smaller in absolute value than the rest, then that step will have a larger activation barrier than the other steps in the mechanism. It works out that the net rate of reaction is reduced. If everything is equal, one wants ΔG_1, ΔG_2, and ΔG_3 to be equal, which means that one wants to equally partition the free energy of the system over all the steps in the mechanism, as we stated previously in Chapter 6.

Now for complex reactions, the intrinsic activation energies are not all equal, as described later in this chapter. Therefore, one would not necessarily want to equally partition the free energy of the reaction. However, for simple reactions equal partitioning works. Therefore, it is a useful concept for catalyst design.

This brings up the second key point from this chapter:

One way surfaces enhance rates of reaction is to stabilize the intermediates of the reaction. One does not want to bind the intermediates too strongly because then rates decline. However, surfaces with moderate bond strengths often make excellent catalysts.

10.4 THERMODYNAMIC EFFECTS

Just to see how that all works, one needs to compare the data to see to what extent variations in rate with changes in surface composition or structure correlates with the stabilization of intermediates.

Many investigators have used Equation 10.32 to quantitatively examine the influence of the thermodynamics of adsorption on the rate of surface reactions. The general idea is that as one changes the structure or composition of a surface, one also changes the free energy of adsorption in Equation 10.29. This changes K_1 in Equation 10.29, which in turn changes the rate of Reaction 10.32.

Such variations are called **thermodynamic effects** in the literature, and they are quite important to surface reactions. For example, the diagram on the left of Figure 10.5 shows how the rate of the decomposition of formic acid

$$HCOOH \Rightarrow CO_2 + H_2 \tag{10.37}$$

varies with surface composition on a series of supported metal catalysts. The figure is a plot of the temperature at which one obtains 50% conversion in a plug flow reactor versus the heat of formation of the formate. One can show that if the reaction would follow a simple Arrhenius's law, the temperature at which one obtains 50% conversion is proportional to the activation energy for reaction. Therefore, the vertical axis in Figure 10.5 is approximately proportional to the activation energy for the reaction. In a similar way, if formic acid dissociative adsorbs, the heat of formation of the formate is proportional to $\ln (K_1)$. Therefore, the horizontal axis in Figure 10.5 is, in effect, $\ln (K_1)$. Consequently, Figure 10.5 is, in effect, a plot of the data according to Equation 10.32. Note that the rate reaches a maximum

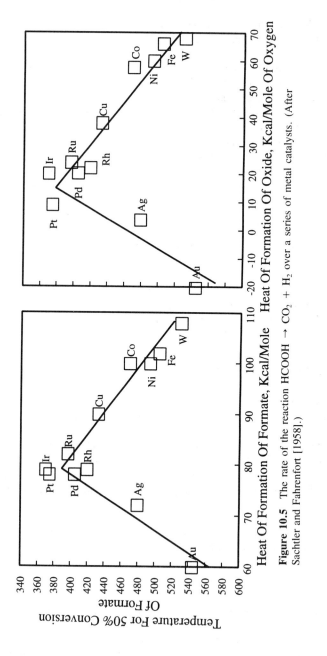

Figure 10.5 The rate of the reaction HCOOH → CO_2 + H_2 over a series of metal catalysts. (After Sachtler and Fahrenfort [1958].)

and then declines in qualitative agreement with Equation 10.32. This is general behavior. When one compares the reactivity of various group VIII metals, one usually finds that the rate reaches a maximum at intermediate bond strengths, in qualitative agreement with Equation 10.32. Balandin called curves like those in Figure 10.5 **volcano curves,** and the name stuck in the literature. Volcano curve behavior is quite common in many reactions. However, it is generally limited to group VII and VIII metals.

There is a little subtlety in Figure 10.5 that even though the axis of the figure is labeled "heat of adsorption of the formate," Sachtler and Fahrenfort only had data for the heat of adsorption of formic acid on nickel. All of the other "heat of adsorption" values in the figure were estimated from a plot like Figure 3.23, i.e., a plot of heat of adsorption of formates versus heat of formation of the most stable oxide. In other work, Sachtler and Fahrenfort decided that it was not really necessary to label their plots heat of formation of some reactive intermediate. Instead, they noted that the heats of adsorption were approximately proportional to heats of formation of bulk oxides (see Figure 3.23). Therefore, they plotted rates as a function of the heats of formation of bulk oxides, where they calculated heats of formation per mole of oxygen for the strongest oxide in the system. These plots are called **Sachtler–Fahrenfort** plots in the literature.

The right side of Figure 10.5 shows a Sachtler–Fahrenfort plot of the data from the left side of the figure. The qualitative trends are very similar in the plot of the right of the figure as in the plot on the left of the figure. However, the advantage of the plot on the right is that one knows the heat of formation of oxide values much more accurately than the heat of adsorption of formate values. As a result, the diagram on the right of Figure 10.5 is much more reliable.

In the literature one often sees an alternate plot called a **Tanaka–Tamaru plot.** Tanaka–Tamaru plots are very similar to Sachtler–Fahrenfort plots. However, the x-axis in the Tanaka and Tamaru [1963] plot is the heat of formation for the highest oxide (i.e., the one with the most oxygens) *per mole of metal*, while the x-axis in the Sachtler–Fahrenfort plot is the heat of formation of the oxide *per mole of oxygen*.

Figure 10.6 shows Sachtler–Fahrenfort and Tanaka–Tamaru plots for a simple reaction, ethane hydrogenolysis (i.e., $C_2H_6 + H_2 \rightarrow 2CH_4$). The Sachtler–Fahrenfort plot shows simple behavior with a sharp maximum in qualitative agreement with the predictions of Equation 10.31. The Tanaka–Tamaru plot shows less clear trends. This is typical behavior. The Sachtler–Fahrenfort plots often fit the data better than the Tanaka–Tamaru plots. However, no one knows why. One does need to be careful with the plots because the exact rates vary significantly with catalyst formulation. For example, Somorjai [1994] summarizes all of the data for ethane and propane hydrogenolysis reported to date. Under similar conditions, the rate of propane hydrogenolysis varies from 4×10^9 molecules/cm^2 sec on a nickel film to 10^{15} molecules/cm^2 sec on a nickel on alumina catalyst. Therefore, one does have to worry whether one is comparing similar catalysts when constructing Sachtler–Fahrenfort plots. Still, Sachtler–Fahrenfort plots work very well for reactions on group VIII metal.

A more serious difficulty with the Sachtler–Fahrenfort plots is that trends in the plots cannot be extrapolated to non-group VIII metals. For example, if one believed the trends in Figure 10.5, one might conclude that copper ($\Delta H_f = 39.8$ kcal/mole of oxygen) or mercury ($\Delta H_f = 21.7$ kcal/mole of oxygen) should be reasonable catalysts for ethane hydrogenolysis. In fact, Sinfelt [1973] found that copper is completely inactive for the reaction. Emmett [1954] reports that mercury poisons the reaction (i.e., decreases the activity of an active catalyst). Similarly, Figure 10.7 shows a Sachtler–Fahrenfort and Tanaka–Tamaru plots for ethylene hydrogenolysis. Based on the plots, one might conclude that silver ($\Delta H_f = 7$ kcal/mole of oxygen) or mercury ($\Delta H_f = 21.7$ kcal/mole of oxygen) should be active catalysts. In fact, Emmett [1958] reports that silver and mercury are completely inactive. Sachtler–Fahrenfort plots do correlate a wide range of literature on group VII and VIII metals. There-

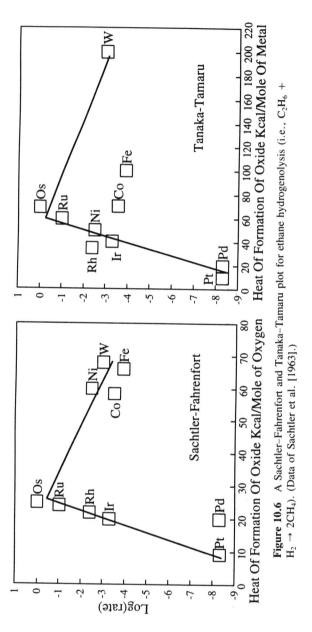

Figure 10.6 A Sachtler–Fahrenfort and Tanaka–Tamaru plot for ethane hydrogenolysis (i.e., C_2H_6 + $H_2 \rightarrow 2CH_4$). (Data of Sachtler et al. [1963].)

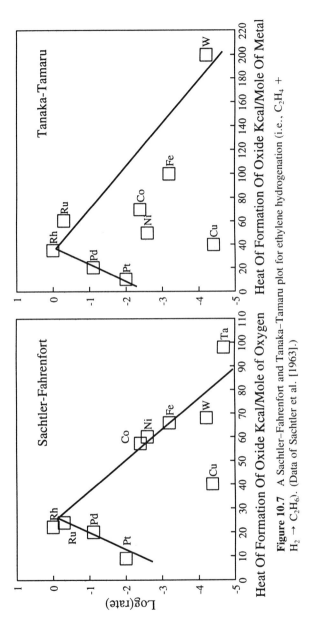

Figure 10.7 A Sachtler–Fahrenfort and Tanaka–Tamaru plot for ethylene hydrogenation (i.e., $C_2H_4 + H_2 \rightarrow C_2H_6$). (Data of Sachtler et al. [1963].)

fore, the plots are quite useful. However, the plots have serious limitations for non-group VII and VIII metals. As a result, the predictive value of the plots is limited.

10.5 FUNDAMENTAL LIMITATIONS OF THE ANALYSIS IN SECTION 10.3

The results in Section 10.4 highlight both the strengths and weaknesses of using the idea that catalysts stabilize reactive intermediates to find catalysts that are very active. On one hand, the best catalysts do form intermediate strength bonds with the adsorbate. Metals that form very weak or very strong bonds are inactive, as predicted by the analysis in Section 10.3. However, there are metals (e.g., mercury) that form intermediate strength bonds, and yet are still lousy catalysts. That brings us to the second key point in this chapter:

> A substance will not be a good catalyst unless it binds the intermediates of a reaction strongly, but not too strongly. However, a substance that binds the intermediates of a reaction strongly but not too strongly will not necessarily be a good catalyst for the reaction.

A more detailed analysis shows that the basic derivation in Section 10.3 is correct for all metals. It is just that E_2^o, the intrinsic barrier to Reaction 2, varies from one surface to the next. For example, hydrocarbon reactions have high intrinsic barriers on lead surfaces. As a result, lead is generally not a good catalyst for the reactions of hydrocarbons, even though, in inorganic complexes, lead binds hydrocarbon intermediates about as strongly as platinum.

In a more fundamental way, the intrinsic activation barrier E_2^o or E_4^o is the key unknown parameter in Equations 10.31 and 10.32. The derivation in Section 10.3 treated the intrinsic activation barrier as a constant that is known. However, typically one does not know the intrinsic barrier to a given reaction a priori. Instead, one has to measure the intrinsic barriers. Later in this chapter we show that the intrinsic barrier can vary with the surface geometry and composition. Thus there is a fundamental limitation to the analysis in Section 10.3 in that there is an unknown parameter that varies from one surface to the next.

In my view, an understanding of the intrinsic barriers to reactions is critically important to the modeling of reactions on surfaces. Later in this chapter we will find that variations in the intrinsic barriers usually control the mechanisms and selectivities of reactions on surfaces. For example, earlier in this chapter we noted that Reaction 10.6 is thermodynamically favored over Reaction 10.7. However, only Reaction 10.7 is observed. Physically, Reaction 10.6 has a large intrinsic activation barrier which prevents Reaction 10.6 from occurring in the gas phase. Reaction 10.7 has a much smaller intrinsic barrier. As a result Reaction 10.7 has a reasonably high rate. This example shows that variations in the intrinsic barriers are quite important to reactions. Therefore, it is quite important to understand how the intrinsic barriers arise, and how the height of the intrinsic barrier varies with conditions.

There are several physical processes that contribute to the intrinsic activation barrier, including the energy it takes to bring the reactants together, the orbital symmetry barriers, and various steric effects. For example, in Reaction 10.7 the main barrier is the energy needed to bring the reactants together. The analysis in Section 9.4 and in Appendix B, Section B.4, shows that if one can get the hydrogen in close enough to the ethane, there would be no barrier to Reaction 10.7. In reality, there is a Pauli (i.e., electron–electron) repulsion between the incoming hydrogen atom and the ethane. That produces a barrier to reaction. There is no barrier to Reaction 10.10 because there is no Pauli repulsion in Reaction 10.10.

Reaction 10.6 is entirely different. In Reaction 10.6 there is a contribution to the intrinsic barrier associated with the Pauli repulsions when the reactants come together. However, a much larger contribution comes from the orbital distortions (i.e., the orbital symmetry effects).

If the reaction occurs as in the left of Figure 10.1, the hydrogen–carbon bonds need to twist. That costs energy. If instead the reaction occurs as on the right of Figure 10.1, the carbon–carbon bond needs to distort. Again, that costs energy. Orbital distortions raise the intrinsic barriers to reaction, which is why Reaction 10.6 has a much higher barrier than Reaction 10.7.

Later in this book we will show that the intrinsic barriers to Reaction 10.6 vary from one metal to the next. The intrinsic barriers also vary from one crystallographic orientation of a given metal to the next. As a result intrinsic barriers often control which reactions occur on a given surface.

If there are high intrinsic barriers to a reaction on a given surface, the surface will be inactive, even when the thermodynamics of the reaction are favorable. Therefore, an understanding of the intrinsic barriers to reaction is essential to understanding rates of reaction.

In the next several sections we review the things that cause intrinsic barriers to arise and show how they can vary from one metal to the next. In that way we can start to predict something about catalytic activity and the preferred pathway for reactions on surfaces.

10.6 BOC-MP MODELING

The simplest kinds of reactions we can discuss are reactions like Reaction 10.7 where one transfers an atom from one species to the next and there are no orbital symmetry barriers to reaction. Marcus [1968] analyzed these types of reactions in some detail. A synopsis of the work appears in Sections 9.3, 9.4, and 9.5. Basically Marcus found that intrinsic activation barriers arise in simple atom transfer reactions because as the reaction proceeds, the bonds in the reactants need to stretch before new bonds can form. The bond-stretching process produces an intrinsic barrier to reaction.

Marcus's original derivation was for a reaction in the gas phase or in solution and not for a reaction on a surface. In the gas phase or solution, molecules do not have any fixed binding sites. However, in Chapters 3, 4, and 7 we noted that during a surface reaction, the reactant molecules are often held on a fixed site. As a result, the bond-stretching processes are different on a surface than in the gas phase or solution. Therefore, the derivation in Sections 9.3, 9.4, and 9.5 does not strictly apply to a surface reaction.

Fortunately, Shustorovich [1986] has proposed an extension of the Marcus equation that he calls the **BOC-MP model,** which accounts explicitly for the fact that the reactants in a surface reaction are bound at fixed sites. The model is quite like the Marcus model described in Section 9.5. However, the key difference is that Shustorovich's model assumes that the reactants adsorb at fixed sites on the surface, and uses experimental data rather than calculations to determine the geometry of the adsorbed reactants and products, the bond energies in the adsorbate, and the heats of adsorption of all the species. The formalism has never been derived from first principles, and it has a significant empirical component. However, for certain simple reactions, particularly reactions where a hydrogen atom is transferred from one ligand to another, the formalism seems to work quite well.

Shustorovich's derivation of the BOC-MP model was very similar to the derivation of the Marcus model in Section 9.5. One treats the reaction as a process where an atom or ligand is transferred from one adsorbed species to another and keeps track of all of the energy changes as some bonds break and others form.

Consider a simple reaction where an atom, A, is transferred from an adsorbed B–A molecule to a surface S:

$$B-A_{ad} + S \longrightarrow B_{ad} + A_{ad} \tag{10.38}$$

where S is a bare surface site.

Figure 9.4 shows a potential for the reaction. As noted previously, when Reaction 10.38 occurs, the B–A bond breaks and the S–A bond forms. If one knows all of the potentials, one can calculate the rate using the formalism in Sections 9.4 and 9.5.

Shustorovich makes the assumption that all of the atomic interactions can be modeled by Morse potentials. So for example, the B–A potential, V_{BA} is assumed to follow:

$$V_{BA}(r_{BA}) = D_{BA} \left[(\exp\left[-\alpha_{BA}(r_{BA} - r_{BA}^o)\right] - 1)^2 - 1 \right]$$

$$= D_{BA} \left(\exp\left[-2\alpha_{BA}(r_{BA} - r_{BA}^o)\right] - 2 \exp\left[\alpha_{BA}(r_{BA} - r_{BA}^o)\right] \right) \tag{10.39}$$

where D_{BA} is the bond dissociation energy of the BA bond, r_{BA} is the length of the B–A bond, r_{BA}^o is the equilibrium bond length, and α_{BA} is related to the force constant for the B–A bond.

Similarly, the B-surface and A-surface bonds are assumed to follow Morse potentials:

$$V_{BS}(r_{BS}) = Q_B \left[(\exp\left[-\alpha_{BS}(r_{BS} - r_{BS}^o)\right] - 1)^2 - 1 \right] \tag{10.40}$$

$$V_{AS}(r_{AS}) = Q_A \left[(\exp\left[-\alpha_{AS}(r_{AS} - r_{AS}^o)\right] - 1)^2 - 1 \right] \tag{10.41}$$

where V_{BS} and V_{AS} are the energies of the B–S and A–S bonds, r_{BS} and r_{AS} are the corresponding bond lengths, r_{BS}^o and r_{AS}^o are the equilibrium bond lengths, and Q_B and Q_A are the strengths of the B-surface and A-surface bonds.

Shustorovich also makes the assumption that "bond-order" of atom A in Reaction 10.38 is conserved, where the bond order of atom A is defined as

$$\text{Bond order} = \exp\left[-\alpha_{AB}(r_{AB} - r_{AB}^o)\right] + \exp\left[-\alpha_{AS}(r_{AS} - r_{AS}^o)\right] \tag{10.42}$$

This assumption is described in greater detail in the next several pages.

Next, Shustorovich uses a derivation very similar to that in Section 9.5 (i.e., calculating the energy as a function of r_{BA} and r_{AS}, and seeing where the curves cross) to calculate an expression for $E_A^{BA_{ad} \rightarrow B_{ad} + A_{ad}}$.

After doing considerable algebra, Shustorovich [1986, 1991] obtains

$$E_A^{BA_{ad} \rightarrow B_{ad} + A_{ad}} = \frac{1}{2}\left(D_{BA} + \frac{Q_A Q_B}{Q_A + Q_B} + Q_{BA} - Q_A - Q_B \right) \tag{10.43}$$

where Q_{BA} is the heat of adsorption of BA molecules, and all of the other terms in Equation 10.43 are defined on the previous two pages.

Note that ΔH_r which is the heat of reaction for Reaction 10.38, is equal to

$$\Delta H_r = D_{BA} + Q_{BA} - Q_A - Q_B \tag{10.44}$$

Therefore, Equation 10.43 is equivalent to a Polanyi relationship

$$E_A^{BA_{ad} \rightarrow B_{ad} + A_{ad}} = E_A^o + \gamma_P(\Delta H_r) \tag{10.45}$$

with

$$E_A^o = \frac{1}{2}\frac{Q_A Q_B}{Q_A + Q_B} \tag{10.46}$$

and $\gamma_P = 0.5$.

TABLE 10.1 Comparison of Activation Barriers Calculated via Equation 10.30 to Experimental Values after Bell [1991] with Additions and Corrections

Reaction	Surface	E_a (kcal/mol)	
		Calculation	Experiment
$H_{2(g)} \longrightarrow 2H_{(ad)}$	Cu(111)	7	14
$H_2O_{(ad)} \longrightarrow H_{(ad)} + OH_{(ad)}$	Cu(111)	26	27
$O_{(ad)} + H_{(ad)} \longrightarrow OH_{(ad)}$	Cu(111)	21	22
$CO_{(ad)} + O_{(ad)} \longrightarrow CO_{2(g)}$	Pd(111)	24	25
$CH_3O_{(ad)} \longrightarrow H_2CO_{(g)} + H_{(ad)}$	Cu(111)	25	24
$OH_{(ad)} + CO_{(ad)} \longrightarrow HCOO_{(ad)}$	Rh(100)	10	8
$NO_{(g)} \longrightarrow N_{(ad)} + O_{(ad)}$	Pt(111)	2	52
$C_2H_{4(ad)} \rightarrow CH_{3(ad)} + CH_{(ad)}$	Pt(110)	27	6

Shustorovich and Bell have done extensive tests of Equation 10.43 for a wide range of surface reactions. Table 10.1 shows some of their results. Generally, they find that the methods work quite well when one is examining the scission of O–H or C–H bonds, and less well for scission of C–C bonds or N≡O bonds. So far all of the tests have been limited to the group VIII metals. However, in the homework set we ask the reader to show that Equation 10.43 predicts a negative activation barrier for H_2 dissociation on K(110). By comparison, experimentally Sprunger and Plummer [1994] find that H_2 does not dissociate on K(110). Earlier work by Langmuir shows that the activation barrier to adsorption is more than 50 kcal/mole. Thus there is a large discrepancy between the BOC-MP predictions and the data. As a result, one would not want to use Equation 10.43 on a nontransition metal surface. Still, Equation 10.43 has proved to be moderately useful for C–H and O–H bond scission/bond formation reactions on group VIII metals.

10.6.1 Derivation of the BOC-MP Method

The easiest way to learn about the strengths and weaknesses of Equations 10.39 to 10.43 would be to actually derive the equations from first principles. At present, a derivation of Equations 10.39 through 10.42 has not yet appeared in the literature. However, one can at least justify the equations based on the analysis in Sections B.3 and B.4 of Appendix B.

Let's start by deriving Equation 10.39. In order to derive Equation 10.39, one needs to assume that the A–B and A–S bonds are formed from s-orbitals of the form

$$\psi_B(r) = \text{const } e^{-0.3\alpha_{BA}(r - r_B)} \tag{10.47}$$

where ψ_A and ψ_B are the wavefunctions for the s-orbitals, r_A and r_B are the positions of A and B, α_{BA} is the orbital exponent and const is a normalization constant. If one plugs Equation 10.47 into the density functional expression for the interaction between two atoms, Equations B.68 and B.94 of Appendix B, one finds that the atomic interaction potential is given as a sum of exponentials. The resultant expression is complex when $r_{BA} < \frac{1}{2} r_{BA}^o$. However, after pages of algebra, one can show that when $r_{BA} > r_{BA}^o$, the potential reduces to equation 10.39. Figure 9.4 shows that in most cases the activation barrier is only a function of the long-range behavior of the potential, so provided Equation 10.47 is valid, Equation 10.39 should work too. Therefore, one can derive Equation 10.39 by assuming that the adsorbate is bonded in an s-orbital.

Of course, there are limitations to the use of s-orbitals. s-Orbitals are completely spherical. The wavefunction does not change with angle, so the potential does not vary with angle (i.e., the orientation of bonds) either. As a result, by using s-orbitals one is ignoring the energy to bend a bond. Now if one considers a hydrogen transfer reaction like that in Reaction 10.7, the hydrogen wavefunctions are spherical, so the use of Equation 10.47 is a reasonable assumption. However, Reaction 10.6 is different in that bonds to the carbon atoms are not well represented by spherical orbitals. For example, in the transition state on the right of Figure 10.1, the sp³ hybrids on the carbon need to be reoriented before reaction can proceed. Equation 10.39 ignores that reorientation energy, so it underpredicts the barrier to Reaction 10.7. In fact, in the problem set, we ask the reader to show that based on Equation 10.43, one would expect Reaction 10.6 to be essentially unactivated in the gas phase, while experimentally Oldershaw and Gould [1985] found that the activation energy is more than 49.5 kcal/mole (i.e., the upper limit of Oldershaw and Gould's experiments). In Section 10.15.6 we will show that this difference is attenuated on a group VIII metal. However, the example in Reaction 10.6 shows that there is some difficulty in using Equation 10.43 when nonspherical orbitals (e.g., carbon–carbon bonds) matter to the reaction.

A second key assumption in Shustorovich's model is that when the B–A bond breaks, one can immediately form an A–S bond, i.e., there are no orbital symmetry limitations to bond formation. Later in this chapter we find that to be a good assumption for hydrogen transfer reactions on a transition metal (e.g., Pt, Ni) but not on a simple metal like potassium or aluminum. Therefore, one would expect the methods to work quite well for hydrogen transfer reactions on group VIII metals, but less well on simple metals.

Finally, Equation 10.42 has never been derived in the literature. The closest I have been able to come to a derivation is to start with the effective medium model in Section 3.10, to derive an expression for the embedding energy as a function of homogeneous electron gas density, i.e., ρ_e in Figure 3.43. It works out that for orbitals of the form in Equation 10.47 the equivalent homogeneous electron gas density, ρ_e, is given by

$$\rho_e = \exp\left[-0.6\ \alpha_{AB}(r_{AB} - r^o_{AB})\right] + \exp\left[-0.6\ \alpha_{AS}(r_{AS} - r^o_{AS})\right] \qquad (10.48)$$

which is very similar to the expression for the bond order in Equation 10.42. Therefore, to some approximation, conservation of bond order is equivalent to assuming that ρ_e is constant. Recall, from the discussion in Section 3.10 that each atom (e.g., hydrogen) has an optimal value of ρ_e (see Figure 3.43). If the hydrogen can maintain that optimal value during reaction, the barriers will generally be lower than in other cases. Thus, the optimal reaction pathway will often be the one that conserves ρ_e.

Now the problem, however, is that if the electron density is low, it may not be possible to find a pathway from reactants to products, which conserves ρ_e. For example, consider transferring a hydrogen from a C–H bond 30 Å above the surface. In that case, as the C–H bond breaks, the electron density on the hydrogen atom goes down, and the ρ_e does not recover until the hydrogen atom gets close to the surface. Of course, from a practical standpoint reactions where atoms jump by 30 Å never occur, so the fact that the assumption that ρ_e in Equation 10.48 is constant fails for these reactions is not particularly important. Figure 3.43 shows that even a small variation in ρ_e will produce a large change in the embedding energy, which in turn will lead to a large increase in barrier to the reaction. Thus, from a practical standpoint, most real reactions will conserve ρ_e, even though one cannot show that ρ_e must be constant during reaction.

Therefore, based on the derivation in this section, one would expect the BOC-MP method to work well for reactions involving transfer of atoms such as hydrogen or oxygen on transition metal surfaces and less well in other systems. That would seem to limit the methods. However, in fact, the biggest use of transition metal catalysts is for hydrogen and oxygen transfer reactions. Therefore, one would expect the BOC-MP method to work in many of the most important cases.

10.7   FUNDAMENTAL LIMITATIONS OF THE BOC-MP AND POLAYNI METHODS

It is useful to go back and compare to data to quantify that idea. It is easy to show that the BOC-MP method predicts the existence of Sachtler–Fahrenfort and Tanaka–Tamaru plots.

As noted in Section 10.4 the Sachtler–Fahrenfort plots do correlate with many of the key trends in rates of reactions on surfaces. For example, earlier in this chapter we noted that a catalyst will be inactive unless it is able to form intermediate strength bonds to the adsorbate. The BOC-MP model allows one to quantify that idea. Note that in principle one can calculate all of the intrinsic barriers. Therefore, under favorable circumstances one can use the BOC-MP method to identify potential catalysis. The analysis in Section 10.3 suggests that if everything else is equal, the best catalysts will be surfaces that equally partition the free energy of the reaction over all of the steps in the mechanism. That prediction seems to agree with experiment. Further, Figures 10.5, 10.6, and 10.7 show that the Polayni relationship and modifications can be used to correlate the variations in the rate of many simple reactions with changing metal. The BOC-MP method predicts these trends and therefore provides a useful framework to think about reactions on surfaces.

Still, there are limitations to the methods. For example, lead binds hydrocarbon ligands about as strongly as platinum. As a result, BOC-MP predicts that lead and platinum should have similar catalytic activities. Experimentally, however, platinum is a much better catalyst than lead for hydrogenation or hydrogenolysis of hydrocarbons. Thus BOC-MP fails. Later in this chapter we show that accessible d-electrons in platinum facilitate reactions. The facilitation by the d-electrons is not considered in the Polayni or BOC-MP model. As a result, the Polayni and BOC-MP models fail under some circumstances. In practice, the Polayni model and the BOC-MP method work well if one wants to compare the reactivity of a series of surfaces that all have similar electronic properties, e.g., all with accessible d-electrons. However, the methods work less well if one extends the results to surfaces of dissimilar materials (see below) or to different faces of the same material. Further, in the discussion below we will show that the methods have some important limitations when one tries to distinguish between competing reaction pathways. Therefore, while the Polayni model and BOC-MP method are useful ways to think about the trends in many surface reactions, the methods often fail in detail.

The failure of the methods to distinguish between behavior of dissimilar metals has not been discussed at length in the literature. However, the failures are well-documented in a few key examples. Barteau [1991] tried to see if Tanaka–Tamaru plots would provide a useful framework to discuss the behavior of methanol in temperature-programmed desorption (TPD). For the discussion that follows, it is important to note that Equation 10.32 only applies to a steady-state reaction. If one runs the experiment in a TPD mode, where one adsorbs gas at low temperature, and then heats, the adsorption process is not in equilibrium, so Equation 10.32 would not apply. Instead, one would expect the rate to follow Equation 10.30, where the rate varies exponentially with the free energy. Barteau tested this idea for a number of reactions and Figure 10.8 shows some of his results. The surprising thing in the figure is that there are two lines in the figure: one line for the group VIII metals and a second line for the group Ib metals. However, Equation 10.30 predicts that all of the data should fall on a single line. Therefore, the implication of Figure 10.8 is that there are deviations from the Polayni relationship when one compares data on groups VIII and Ib metals.

In a similar way, H_2 dissociation is unactivated on group VIII metals, while the dissociation process shows a large activation barrier on potassium, aluminum, or silicon. Yet Equation 10.30 predicts a lower activation barrier on potassium, aluminum, or silicon than on group VIII metals. These results show that the Polayni relationship cannot be used to compare reactions on the group VIII metals to reactions on other metals or semiconductors.

In a more fundamental way, the catalytic activity of metals with unpaired d-electrons (i.e., partially filled d-bands) is fundamentally different than catalytic activity of materials with no unpaired d-electrons. Later in this chapter we show that the d-bands stabilize the transition state for many bond scission processes. The stabilization of the transition state lowers the intrinsic barriers to reaction (i.e., the E_a^o in Equation 9.5). The interactions with the d-bands are much stronger when there are unpaired d-electrons than when the d-bands are completely filled. As a result, metals with partially filled d-bands tend to be much more catalytically

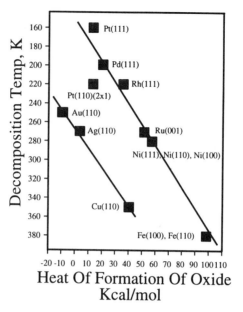

Figure 10.8   Tanaka–Tamaru plot for the reaction $CH_3OH_{(ad)} \rightarrow CH_3O_{(ad)} + H_{(ad)}$. (After Barteau [1991] with corrections due to Yagasaki and Masel [1994].)

active for bond scission processes than metals with completely filled d-bands. In particular, E_a^o in Equation 10.31 is lower on d-band metals than on other surfaces. The variations in E_a^o produce the two lines in Figure 10.8. However, most of the variations in E_a^o are ignored in the Polayni model and the BOC-MP method. The net effect is that while the Polayni model and the BOC-MP method are quite useful in comparing surfaces with equally accessible d-bands, the Polayni model and the BOC-MP method are not useful in comparing metals with partially filled d-bands to those without partially filled d-bands.

Another thing is that in general, the Polayni relationship and the BOC-MP method are often not useful in distinguishing between various reaction paths. For example, consider methanol oxidation:

$$CH_3OH_{(g)} + O_{(ad)} \longrightarrow H_2CO_{(g)} + H_2O_{(g)} \tag{10.49}$$

Methanol oxidation has been examined on a variety of different surfaces. Generally, the methanol adsorbs molecularly, and then reacts via a number of steps as shown in Figure 6.11, p. 456, and Equation 6.6, p. 449. It is useful to consider the fourth step in the methanol oxidation pathway in Equation 6.6. During the fourth step, the adsorbed methanol reacts with adsorbed oxygen to yield methoxy and an OH. Wachs and Madix [1978] studied the oxidation process in detail and found that on Ag(110) and Cu(110) the main reaction was

$$CH_3\,^{16}OD_{(ad)} + {}^{18}O_{(ad)} \longrightarrow CH_3\,^{16}O_{(ad)} + {}^{18}OD_{(ad)} \tag{10.50}$$

where a deuterium atom is transferred from the methanol to an adsorbed oxygen. In contrast, the reaction

$$CH_3\,^{16}OD_{(ad)} + {}^{18}O_{(ad)} \longrightarrow CH_3\,^{18}O_{(ad)} + {}^{16}OD_{(ad)} \tag{10.51}$$

was not observed. It is interesting to consider why Reaction 10.50 occurs rather than reaction 10.51. In Reaction 10.51 a methyl group is transferred from an adsorbed methanol to an

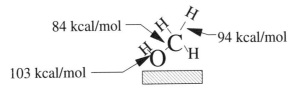

Figure 10.9 Bond energies in methanol. (Data from the *CRC Handbook.*)

adsorbed oxygen. Notice that Reactions 10.50 and 10.51 are very similar. The reactants and products are chemically the same. The only difference is that in Reaction 10.50 the O–D bond breaks, while in Reaction 10.51 the C–O bond breaks. In the gas phase the C–O bond is 19 kcal/mole weaker than the O–D bond, i.e., D_{AB} in Equation 10.43 is 19 kcal/mole less for Reaction 10.50 than for Reaction 10.51 (see Figure 10.9). All of the other terms in Equation 10.43 are identical for the two reactions. Therefore, according to Equation 10.43, Reaction 10.51 should have a lower activation barrier than Equation 10.50. However, Wachs and Madix's [1978] data show just the opposite.

In the literature, no one has considered why Reaction 10.51 is slower than Reaction 10.50. However, notice that the situation in Reaction 10.51 is very similar to the situation in Reaction 10.6. The bond on the $-CH_3$ group in the methanol has a specific direction. If $C-^{16}O$ bond scission would occur, the bond on the CH_3 group would need to be reoriented toward the $^{18}O_{ad}$. As a result, the $C-^{16}O$ bond needs to break (i.e., the orbital symmetry needs to change), before the $^{18}O-CH_3$ bond forms. In Section 10.2 we noted that this is a symmetry forbidden process. In contrast, hydrogen atoms have no preferred bonding directions. The $^{16}O-H$ and $H-^{18}O$ bonds can break and form simultaneously. As a result, the intrinsic barrier to reaction is lower in Reaction 10.50 than in Reaction 10.51. Thermodynamics still favor Reaction 10.51. However, the differences in the intrinsic barriers evidently are larger than the differences in the thermodynamics because only reaction 10.50 is seen.

Note that the Polayni relationship and the BOC-MP method ignore the role of orbital orientation on the intrinsic barriers to reaction. As a result, the BOC-MP method predicts that Reaction 10.51 should be favored over Reaction 10.50, and that Reaction 10.6 should be favored over Reaction 10.7. These predictions contradict the experiments.

This example shows that the Polayni relationship and the BOC-MP method have serious limitations in trying to predict preferred reaction pathways or reaction selectivities. The fundamental reason that the Polayni relationship and the BOC-MP methods fail in these more complex reactions is that they view bonds in a stick model, and ignore that there are quantum mechanical limitations to bond formation. Bonds have specific directions but that directionality is ignored in the BOC-MP and Polayni models. One also gets orbital symmetry limitations because one cannot put more than two electrons into a single molecular orbital. Both of these effects make significant perturbations to the intrinsic barriers to reaction. Therefore, they are very important to rates of surface reactions. Reactions can have high intrinsic barriers under conditions where according to Reaction 10.46, the intrinsic barriers should be low. In a fundamental way, the variations in intrinsic barriers cause the rates of surface reactions to be different than one expects from the Polayni relationship and the BOC-MP method.

10.8 SURFACE REACTIONS AS CHAIN-PROPAGATION REACTIONS

One way to start to think about surface reactions in a more careful way is to assume that a surface reaction consists of a series of chain-propagation reactions. When a radical reaction like that in Equations 10.1 through 10.4 occurs in the gas phase, there is always an initiation step where a neutral species is converted into a radical. However, a surface reaction is different in that a reactive metal or semiconductor surface has unpaired electrons. Therefore, the surface is, in effect, a radical. On an insulating surface the active site is instead an ion. As a result, there is no need to create radicals or ions during the reaction (i.e., you do not need an initiation step). The surface can act as a source of radicals or ions.

Now that is important because it means that all of the steps in reactions on surfaces are, in effect, propagation steps. A stable molecule reacts with an active site (radical or ion) to produce a new species. The new species reacts via a catalytic cycle (e.g., Figure 6.1, p. 439) to eventually regenerate the active site.

Another important difference between a gas phase reaction and a surface reaction is that with a reaction on a surface one does not need collision partners. Rather, the surface atoms can take up all of the energy and momentum during reactions of adsorbed species. Therefore, surfaces are in effect very efficient collision partners.

The implication of the analysis in the last paragraph is that reactions on surfaces will behave like propagation steps in the gas phase. In each step in the process, one bond breaks and another forms. If the process occurs in a way that maximizes the bonding in the transition state, then one would expect the process to have a low intrinsic barrier. If not, the intrinsic barrier will be high. As with the gas phase on a metal surface one would normally expect steps that transfer a single sigma-bonded atom from one species to the next to have a lower intrinsic barrier than steps that transfer a molecular ligand. However, there are exceptions to that rule due to residual charges on the species and interactions with the d-bands. These effects are discussed in greater detail later in this chapter. The key point is that most of the steps in a catalytic cycle involve formation of one bond and destruction of another. In such a case, one does not necessarily expect the weakest bonds in a molecule to have the lowest intrinsic barriers. Rather, the steps that conserve bonding (i.e., orbital symmetry) during the reaction process would be expected to have lower intrinsic barriers than steps where bonds need to be broken or reoriented during the reaction. It is this difference that explains why Reaction 10.50 dominates over Reaction 10.51.

Experimentally, the intrinsic barriers to surface reaction vary with the type of reaction and the composition and structure of the surface. As a result, the situation is quite complex. In the next several sections, we review how the intrinsic barriers vary with the properties of the reactants and of the surface.

10.9 VARIATION IN THE INTRINSIC BARRIERS TO REACTION WITH CHANGING REACTION TYPES

The first question we need to address is how do the intrinsic barriers to reaction vary with reaction type. According to the analysis in Section 10.2, one can answer that question by examining how the orientation and direction of orbitals will change during reaction. For example, in Section 10.4 we found that during a propagation reaction in the gas phase the intrinsic barrier to transfer a molecular fragment is higher than the intrinsic barrier to transfer an atom because bonds need to be reoriented when molecular fragments are transferred. One would expect a similar effect during reactions on a metal surface. Bonds need to be reoriented when molecular ligands are transferred. As a result, one would expect the intrinsic barriers to transfer of a molecular ligand to be higher than the intrinsic barriers to the transfer of an atomic ligand. For example, Yagasaki and Masel [1994] review all of the existing data for methanol, ethanol, and ethylene decomposition on the transition metals. The data in Yagasaki and Masel show that methanol and ethylene generally dehydrogenate on all of the surfaces of the transition metals examined so far, except (1x1)Pt(110), even though transfer of an alkyl ligand is thermodynamically favored by 30–60 kcal/mole. These results show that on a transition metal surface the intrinsic barrier for transfer of a molecular ligand is at least 30 kcal/mole higher than the intrinsic barrier for the transfer of an atomic ligand. By comparison if one plugs the data in Benson [1976] into the Polayni relationship, one finds that in the gas phase the intrinsic barriers to transfer of molecular ligands are 50–100 kcal/mole higher than the intrinsic barriers to the transfer of atomic ligands. Still, atomic transfers usually have much lower intrinsic barriers than molecular ligand transfers on a transition metal surface.

The situation is entirely different on an insulating oxide surface. In Section 6.4 we noted that most species on insulators are charged. For example, when hydrocarbons adsorb on an insulating oxide the resultant species generally have a positive charge. One common species is called a carbenium ion. In Section 10.21 we will note that carbenium ions have empty orbitals. Quantum mechanically, the intrinsic barriers to transferring a ligand to an empty orbital are small because one can form a hybrid that bridges between the empty and filled orbital. As a result, it is much easier to transfer molecular ligands in carbenium ions, than to transfer molecular ligands in neutral molecules. Consequently, the intrinsic barriers to transfer of a molecular ligand are much lower on an oxide surface than on a metal.

10.9.1 An Example of the Prediction of Reaction Mechanisms

It is interesting to consider a simple example to illustrate these ideas in greater detail. The decomposition of ethanol (CH_3CH_2OH) has been examined on many surfaces. On Pt(111) and Pd(111) Davis and Barteau [1987] and Sexton, Rendulic, and Hughes [1982] found that the main reaction pathway starts out with a series of simple atomic transfers:

$$S + CH_3CH_2OH_{(ad)} \longrightarrow CH_3CH_2O\backslash_{(ad)} + H_{(ad)} \tag{10.52}$$

$$S + CH_3CH_2O\backslash_{(ad)} \longrightarrow CH_3\overset{\displaystyle O}{\overset{\|}{C}}H_{(ad)} + H_{(ad)} \tag{10.53}$$

$$S + CH_3\overset{\displaystyle O}{\overset{\|}{C}}H_{(ad)} \longrightarrow CH_3\overset{\displaystyle O}{\overset{\|}{C}}\backslash_{(ad)} + H_{(ad)} \tag{10.54}$$

However, the next step is the transfer of a molecular ligand:

$$S + CH_3\overset{\displaystyle O}{\overset{\|}{C}}\backslash_{(ad)} \longrightarrow {}_{/}CH_{3\,(ad)} + CO_{(ad)} \tag{10.55}$$

In contrast, Pines [1966] reports that on the Brønsted acid site on an acidic oxide surface the main reaction is to transfer the ethyl group, to eventually yield ethylene:

$$H^+_{(ad)} + CH_3CH_2OH \longrightarrow CH_3CH_2^+{}_{(ad)} + H_2O$$
$$CH_3CH_2^+{}_{(ad)} \longrightarrow C_2H_4 + H^+_{(ad)} \tag{10.56}$$

It is interesting to use the ideas from Section 10.8 to understand these results. Reactions 10.52, 10.53, and 10.54 are all unexpected thermodynamically. Note that in Reactions 10.52 and 10.53 the C–O bond is the weakest bond in the reactants, so according to Equation 10.37 the C–O bond should break first. However, instead the adsorbate dehydrogenates. Reaction 10.54 is different in that the C–O bond is strong; the C–C bond is the weakest bond in the molecule. Nevertheless, a C–H bond breaks.

Of course, Reactions 10.52, 10.53, and 10.54 involve transfer of an atomic ligand. In contrast, if the C–O or C–C bond would break, one would transfer a molecular ligand.

According to the discussion in Section 10.8, the intrinsic barriers to the transfer of a molecular ligand are much higher than the intrinsic barriers for the transfer of an atom. Evidently, the high intrinsic barriers to transfer molecular ligands prevent C–O or C–C bond scission from occurring, since only C–H or O–H bond scission is seen.

Reaction 10.56 also occurs because of low intrinsic barriers. Note that Reaction 10.56 is analogous to Reaction 10.12, and so according to the discussion in Section 10.2 one would expect Reaction 10.56 to be rapid. Pines [1960] reports that it goes almost instantaneously on an acidic silica/alumina at 400 K.

Reaction 10.55, however, is an exception to the general idea that atomic transfers dominate over molecular transfers on metals. It is interesting to see why Reaction 10.55 occurs rather than

$$S + CH_3\overset{\overset{\displaystyle O}{\|}}{C}_{(ad)} \longrightarrow H_{(ad)} + H_2C{=}C{=}O_{(ad)} \qquad (10.57)$$

while in the gas phase Laidler and Lin [1967] find that a reaction analogous to Reaction 10.57 dominates (with a CH_3 radical replacing the surface). Well, thermodynamically, Reaction 10.55 is much more favorable than Reaction 10.57. According to data in the *CRC Handbook*, the C–C bond in acetaldehyde has a bond strength of only 70 kcal/mole, while the CH bond is 96 kcal/mole. Therefore, in the gas phase reaction 10.55 is favored over Reaction 10.57 by 26 kcal/mole. The situation is even more favorable on a surface. The products of Reaction 10.55 all bind strongly to Pt(111). However, one would expect the ketene ($H_2C{=}C{=}O$) product in Reaction 10.57 to only bind weakly on Pt(111) (CO_2, which is isoelectronic to ketene only binds weakly, and isoelectronic compounds often have similar bond energies). Therefore, Reaction 10.55 is strongly thermodynamically favored over Reaction 10.57. Just to plug in numbers, if one estimates the heat of adsorption of CH_3 groups as described in Example 3.A, assumes that the heat of adsorption of ketene is the same as that for CO_2, and uses data for everything else, one finds that on Pt(111) Reaction 10.55 is thermodynamically favored by 45 kcal/mole over Reaction 10.57. Evidently, on Pt(111) the 45-kcal/mole driving force is able to overcome the higher intrinsic barriers to Reaction 10.55, so Reaction 10.55 dominates over Reaction 10.57. In contrast, in the gas phase the intrinsic barriers do not allow Reaction 10.55 to occur.

These results demonstrate that on a metal surface the intrinsic barriers to the transfer of a molecular ligand are much larger than the intrinsic barriers to the transfer of an atomic ligand, in qualitative agreement with Benson's results for gas phase reactions. Nevertheless, if the thermodynamics of molecular ligand transfer are sufficiently favorable, the transfer of a molecular ligand can still occur.

A more detailed analysis shows that the intrinsic barriers to transfer of a methyl ligand are considerably lower on a Pt(111) surface than in the gas phase. Thus, on a metal surface it is much easier to transfer a molecular ligand than in the gas phase.

In a series of papers summarized in Dumesic [1993], Dumesic has estimated the intrinsic barriers for a number of reactions. There is not enough information yet to draw any general conclusions. However, Dumesic has suggested that the intrinsic barrier for the transfer of a methyl group is about 30 kcal/mole on a nickel surface. By comparison, one can plug the data in Westley [1980] in Equation 10.45 to show that the intrinsic barriers for transfer of a methyl group are between 80 and 90 kcal/mole in the gas phase. These results indicate that the presence of a surface can substantially modify the intrinsic barriers to reaction from their gas phase values.

In a more general way, even though on a metal surface the intrinsic barriers for transfer of a molecular (i.e., alkyl) ligand are generally higher than those for transfer of an atomic ligand, the differences are much smaller than in the gas phase. As a result, one can overcome those differences in intrinsic barriers, if the thermodynamic driving force for reaction is sufficiently favorable. As a general rule of thumb, on a transition metal surface the intrinsic barriers for transfer of a molecular ligand are perhaps 30 kcal/mole higher than those for transfer of a sigma-bonded atomic ligand. Therefore, atomic transfers dominate unless the thermodynamics favor molecular transfers by more than 30 kcal/mole. The situation is entirely different on a metal oxide (or carbide). Later in this chapter we note that transfers of molecular ligands have low intrinsic barriers on metal oxides and carbides.

10.10 A MODIFIED BOC-MP METHOD

One can understand all of these trends via a modification of the BOC-MP method described in Section 10.6. Note that the derivation of Equation 10.43 considered the intrinsic activation barriers associated with stretching one bond and forming another. The assumption that ρ_e in Equation 10.48 is constant also, in effect, includes the influence of Pauli repulsions on the intrinsic barriers. However, Equation 10.43 ignores the effects of changes in orbital symmetry or orbital orientation on the intrinsic barriers to reaction. It also ignores that species could acquire a charge during reaction. Thus the BOC-MP method is inaccurate for reactions where a molecular ligand is transferred.

One way to overcome that difficulty is to add an extra term into Equation 10.43, i.e.,

$$E_A^{BA_{ad} \to B_{ad} + A_{ad}} = \frac{1}{2}\left(D_{BA} + \frac{Q_A Q_B}{Q_A + Q_B} + Q_{BA} - Q_A - Q_B\right) + E_A^{orb} \qquad (10.58)$$

where E_A^{orb} is the contribution to the intrinsic barriers associated with the changes in orbital symmetry or orbital orientation or changes in the charge on species during reaction. If E_A^{orb} is constant, one can use the Polayni relationship and the BOC-MP method to compare a series of reactions. However, variations in E_A^{orb} change the intrinsic barriers to reaction, which in turn changes the reaction rate.

Generally, one would expect reactions on metal surfaces to follow trends similar to those propagation reactions in the gas phase. For example, Benson [1976] shows that if one compares a series of similar reactions in the gas phase, where a single σ-bonded atom is transferred

$$D\cdot + CH_3R \longrightarrow DH + \cdot CH_2R$$
$$HO\cdot + CH_3R \longrightarrow H_2O + \cdot CH_2R$$
$$D\cdot + ICH_2R \longrightarrow DI + \cdot CH_2R \qquad (10.59)$$
$$D\cdot + HOR \longrightarrow DH + \cdot OR$$

one finds that the intrinsic barriers are low. It works out that in the gas phase E_A^{orb} is negligible. Such reactions would be expected to have low intrinsic barriers on transition metal surfaces. In contrast, one can use the results in Benson [1976] to show that in the gas phase, reactions such as

$$D\cdot + CH_3R \longrightarrow CH_3D + R\cdot$$
$$HO\cdot + CH_3R \longrightarrow CH_3OH + R\cdot \qquad (10.60)$$
$$D\cdot + HOR \longrightarrow DOH + R\cdot$$

where a single σ-bonded molecular ligand is transferred have a large (~ 90 kcal/mole) value of E_A^{orb}. Such reactions usually do not occur in the gas phase. They can occur on a clean metal surface, but the intrinsic barriers are 30 kcal/mole or more, so reactions are inhibited. In contrast, such reactions can occur easily on a metal oxide or carbide.

Even higher intrinsic barriers are observed when π-bonded atoms are transferred in the gas phase. The gas phase results show that the intrinsic barriers to bond scission vary significantly with the type of bond being broken. In general, E_A^{orb} increases as the bonding in the reactants becomes more directional so the intrinsic barriers increase too. Clearly one would like to quantify these effects for surface reactions, but unfortunately one cannot do so at present. However, one can say that the intrinsic barriers for the transfer of a molecular ligand are generally higher than the intrinsic barriers for the transfer of a σ-bonded atomic ligand. Therefore, one can often make qualitative, but perhaps not quantitative predictions about the relative barriers to various reaction pathways.

10.11 THE PROXIMITY EFFECT

One of the reasons that it is difficult to make any quantitative predictions is that while the intrinsic barrier to bond scission varies with the type of bond being broken, the intrinsic activation barrier also depends on the orientation of the bond relative to the surface. The main issue is what I call a **proximity effect.** Bonds break only when they are in close proximity to the surfaces. As a result, the intrinsic barriers to reaction depend on how a molecule is bound to a surface as well as the type of bond being broken. Generally, the intrinsic barriers to bond scission are larger when a bond is far away from the surface than when it is near to the surface.

It is useful to consider an example to show the kinds of things that are observed. Earlier in this chapter we noted that methanol and ethanol readily dehydrogenate on the group VIII metals. Now it happens that this is a special chemistry which is not always seen. For example, data summarized in Ceyer [1988] shows that ethane and propane do not react at ordinary ultrahigh vacuum (UHV) conditions; a reaction occurs only when the ethane or propane molecule is accelerated toward the surface in a molecular beam. The question then is: How is ethane decomposition different from methanol or ethanol decomposition? Consider two different reactions

$$CH_3OH_{(ad)} \longrightarrow CH_3O_{(ad)} + H_{(ad)} \tag{10.61}$$

$$CH_3CH_3 \longrightarrow CH_3CH_{2(ad)} + H_{(ad)} \tag{10.62}$$

Reactions 10.61 and 10.62 are very similar in that one is transferring a hydrogen in both cases and the thermodynamics of both reactions are similar. However, in Reaction 10.62 the surface needs to form a bond to a carbon atom, while in Reaction 10.61 the surface binds to an oxygen atom. Note that the carbon atom in Reaction 10.62 starts out surrounded by hydrogens. The hydrogen atoms keep the carbon atom far away from the surface. Consequently, it is hard to form the carbon–surface bond, i.e., the intrinsic activation barrier for bond scission is high. In contrast the oxygen in Reaction 10.61 only has one hydrogen. The oxygen can get close to the surface so an oxygen–metal bond can form easily. The net result is that Reaction 10.61 has a much lower intrinsic activation barrier than Reaction 10.62 on most transition metal surfaces.

At first, one might think, "So what?" An O–H bond is breaking in Reaction 10.61, while a C–H bond is breaking in Reaction 10.62. Perhaps the intrinsic barriers to C–H bond scission are higher than the intrinsic barriers to O–H bond scission. Note, however, that Zaera [1992] has examined the decomposition of ethyl groups on the same surfaces discussed by Ceyer. Zaera finds that the ethyl groups undergo rapid C–H bond scission (i.e., dehydrogenation) under conditions where little bond scission is seen in ethane. Therefore, Reaction 10.62 is inhibited not because there is a general barrier to C–H bond scission. Rather, there is an extra intrinsic barrier to C–H bond scission in ethane that is not present during C–H bond scission in ethyl groups due to the proximity effect.

Ceyer [1988] and co-workers quantified this idea for the dissociative adsorption of methane on nickel surfaces. They noted that in order for methane to dissociate, the carbon atom in the methane needs to get close to the surface. In order for that to happen, the hydrogens must move out of the way. Therefore, Ceyer et al. proposed that the intrinsic activation energy for adsorption is equal to the energy it takes to move the hydrogens out of the way (i.e., to distort the methane as shown in Figure 10.10). Ceyer calls this the splat model (i.e., the methane needs to go splat to stick). Ceyer [1988] shows that this model seems to have a reasonable correlation to data for methane adsorption on nickel surfaces. At the point this book was being written, it was not clear how to extend the model to other systems. However, an extension would be clearly quite useful.

One can generalize the ideas in the previous two paragraphs to many other reactions. Consider the reaction

$$CH_3OH_{(ad)} \longrightarrow CH_2OH_{(ad)} + H_{(ad)} \tag{10.63}$$

Yagasaki and Masel [1994] noted that Reaction 10.63 is thermodynamically favored over Reaction 10.61 on most conditions. Yet Reaction 10.63 has never been observed in UHV. Instead, Reaction 10.61 is seen. Interestingly, Zeleney et al. [1992] showed that in solution, Reaction 10.63 dominates over Reaction 10.61.

One can understand this effect with the aid of Figure 6.7, p. 450. In UHV methanol is held onto the platinum surface with a dipole-induced-dipole bond. The dipole is strongest near the oxygen. As a result, the interaction with the surface is maximized if the oxygen in the methanol or ethanol points down toward the surface, as shown in Figure 6.7. In such a configuration it is much easier to form a metal–oxygen bond than to form a metal–carbon bond. As a result, O–H bond scission dominates over C–H bond scission.

The opposite effect occurs in solution. In the presence of water the OHs would rather hydrogen bond with the water than point toward the surface (see Figure 6.7). In the preferred configuration the carbon is closer to the surface than the oxygen. As a result, it is easier to form a metal–carbon bond than a metal–oxygen bond. Consequently, in solution a C–H bond in methanol breaks before the O–H bond breaks. The reaction has a significant barrier in solution since even in the best geometry, the carbon atom is still fairly far away from the surface.

All of these results show that the proximity effect plays a major role in determining the intrinsic barrier to reaction.

At present, no one has *quantified* the influence of the proximity effect on the intrinsic barriers to reaction. Generally, one would expect that the farther a bond is from a surface,

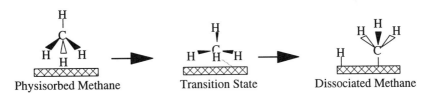

Physisorbed Methane   Transition State   Dissociated Methane

Figure 10.10 The splat model for methane adsorption.

the harder it would be for the bond to break (i.e., the intrinsic barrier would be expected to increase the farther a bond has to stretch). The proximity effect would be expected to be smaller on surfaces where the electron cloud spills a long way out into the vacuum (see Figure 3.32 and 3.33) than in cases where the electrons are localized near the surface.

Unfortunately, the proximity effect is difficult to quantify in a simple way. The main difficulty is that molecules need to be reoriented as reactions occur, and the reorientation energy increases the intrinsic barrier to reaction. For example, during Reaction 10.63 the methanol starts with its filled dangling bonds on its oxygen pointing down toward the surface. However, when Reaction 10.63 occurs, one creates a dangling bond on the carbon. That dangling bond needs to bind to the surface. As a result, when Reaction 10.63 occurs the methanol needs to be reoriented. The reorientation energy increases the intrinsic barrier to reaction. When this book was being written we did not have any good guidelines to predict barriers to reorientation of molecules on surfaces. As a result, it has been difficult to quantify the proximity effect.

Experimentally, the proximity effect seems to be especially important in structure-sensitive reactions. The general idea is that the presence of steps changes the configuration of molecules, and those changes in configuration lead to changes in reactivity. For example, a methoxy (CH_3O-_{ad}) can bind either on the top of the steps or in the valleys on a step surface (see Figure 10.11). If the methoxy binds at the top of the steps, its reactivity is decreased, while if the methoxy binds on the valleys, its reactivity increases. Such an effect can lead to very unusual coverage/structure sensitivities where small changes in step density or coverage force some methoxies down into the valleys, and thereby enhance the methoxy's reactivity.

Another aspect to the proximity effect is that it can produce a maximum in reactivity at intermediate step densities. Figure 10.12 illustrates a case where an atom X in a reactant reacts at an active site at the bottom of a step. If the step density is low, the molecule can bind to the terrace and position X over the active site. However, if the step density gets too high, the reactant will no longer fit on the terraces. As a result, the reactants need to bridge over the steps, as illustrated on the left molecule on the right diagram shown in Figure 10.12. In the configuration in the right diagram in Figure 10.12, the adsorbed molecule needs to break a metal–carbon bond before the reaction can proceed. As a result, the reaction rate is lower on the surface, on the right of Figure 10.12, i.e., on the surface with a higher step density. Therefore, at high step densities, an increase in step density can produce a decrease in reactivity.

Masel [1986] quantified this effect for one case: NO decomposition on platinum. Figure 6.14 shows how the rate of NO decomposition varies over the stereographic triangle. Notice

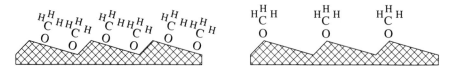

Figure 10.11   Some possible configurations for methoxy adsorption on stepped surfaces.

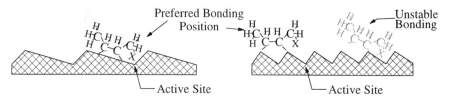

Figure 10.12   An illustration showing how an increase in step density can produce a decrease in reactivity.

that the rate of NO decomposition is highest on Pt(410). The rate is much lower on Pt(210). Yet Pt(410) and Pt(210) have very similar geometries (see Figure 2.38, p. 44). The only difference is that Pt(210) has twice the step density as Pt(410). The steps are so short that the NO can no longer fit. As a result, Pt(210) is much less reactive than Pt(410) for NO decomposition.

In Chapter 6 we noted that experimentally there is often little correlation between step density and reactivity. The maximum reactivity often comes at intermediate step density. No one has a good model to explain that effect. However, one key factor seems to be the proximity effect. At very high step densities molecules cannot squeeze down into the valleys on the surface. If the active site is down at the valleys, the rate will be lower on highly stepped surfaces than on surfaces with few steps and longer terraces. Unfortunately, at the time this book was being written, the influence of the proximity effect on the structure sensitivity of reactions had not been characterized in a quantitative way except in the case of NO decomposition on platinum. As a result, it is not known whether the experimental observation of a maximum in reactivity at intermediate step density discussed in Section 6.3.3 is caused by the proximity effect or some other factor.

Still, we now know that the proximity effect can have a substantial influence on the intrinsic barriers to reaction. Experimentally, the size of the intrinsic barriers vary with the type of reaction, the configuration of the adsorbed reactants, and the geometry of the surface. No one has a general quantitative model for the proximity effect. However, experimentally, it seems to be quite important for reactions.

10.12 EFFECTS OF SURFACE GEOMETRY ON THE INTRINSIC BARRIERS TO REACTION

The proximity effect is one example of a wider class of effects called **geometric effects.** The idea of geometric effects is that the geometry of the surface plays a key role in determining the intrinsic barriers to reactions.

Experimentally, there are often important steric effects when molecules react on surfaces. For example, ethylene ($H_2C{=}CH_2$) reacts to form ethylidyne ($\equiv C{-}CH_3$) over the threefold hollow sites on Pt(111) and Pd(111). There is no room to form ethylidyne over the threefold sites on (1x1) Pt(110) or (1x1) Pd(110), so Yagasaki and Masel [1989] and Yoshinobou et al. [1990] found that instead either ethan-1-*yl*-2-ylidyne ($\equiv C{-}CH_2{-}$) forms or the C–C bond in the ethylene breaks (see Figure 6.10 and the discussion on pp. 454 to 456). At present, it is not clear how general the steric effects are. However, one would think that these steric effects would be quite important for reactions on stepped surfaces.

Steric effects can substantially raise the intrinsic barriers to reaction. Therefore, they should be quite important. Unfortunately, no one knows how to quantify the steric effects, so our discussion of them is limited.

Years ago, people also considered how the spacings between the surface atoms would affect the rate of reactions on surfaces. The idea is that as one changes the spacing between the surface atoms, one changes the distances that bonds need to stretch during reaction that produces a change in the intrinsic barrier to reaction, and hence the rate. For example, the dehydrogenation of ethoxy

$$S + CH_3CH_2O_{\diagdown(ad)} \longrightarrow CH_3\overset{\overset{\displaystyle O}{\|}}{C}H_{(ad)} + H_{(ad)} \tag{10.64}$$

is a key step in the conversion of ethanol to acetylaldehyde. Figure 10.13 shows a possible transition state for the reaction. During the reaction a hydrogen is transferred from the ethoxy

Figure 10.13   The transition state for the dehydrogenation of ethoxy on a stepped and flat surface. (After Balandin and Robinstein [1934].)

to the surface. If everything else is equal, the barrier for reaction will be higher if the hydrogen has to stretch a long distance than if the C–H bond does not have to stretch significantly. Balandin [1929] noted that this effect can produce significant structure sensitivity. If the length of the surface bonds are such that the hydrogen naturally sits over the hydrogen adsorption site, the reaction should occur with a low activation barrier, while if not, the reaction should have a higher barrier. Balandin assumed that the lengths between adjacent sites are related to the distances between the atoms in the catalysts. Therefore, he proposed that there should be a correlation between metal–metal bond lengths and rates.

At present, there has been little experimental verification of this idea. Figure 10.14 shows a plot of the methoxy decomposition temperature over a series of metal surfaces as a function of the bond length of the metal. The figure is basically a scatterplot. There is no obvious correlation between bond length and reaction. Balandin [1969] has suggested that good correlations could be obtained for ethoxy decomposition on transition metal oxides. However, convincing experimental data are lacking.

The reason that changes in bond lengths do not have a good correlation with changes in rates is that the only way to change the bond lengths significantly is to change the composition of the surface, i.e., making an alloy or changing the metal. When one changes the composition of the surface, one also changes the thermodynamics of each of the surface reactions, and the local electronic structure around the surface sites. In Section 10.4 we noted that the thermodynamic effects are quite large with changing metal. The electronic effects are, too. Experimentally, the electronic and thermodynamic effects tend to be larger than the effects of the varying bond length. As a result, while in principle, changes in bond length could affect the intrinsic barriers to reaction, in actual practice, other factors play a much larger role.

10.13   HOLDING THE REACTANTS IN A POSITION CONDUCIVE TO REACTION

There is a different geometric effect that has not been extensively discussed in the current literature, but still can have an important influence on the intrinsic barriers to reaction. Recall from Section 10.1 that one of the roles of a surface is to hold the reactants in a position conducive to reaction. As one changes the surface geometry, one changes the relative orientation of the reactants. In principle, changes in the relative orientation of the reactants can have an important influence on the intrinsic barriers to reaction.

For example, consider the trimerization of acetylene to benzene C_6H_6.

$$3 \ HC{\equiv}CH \ \longrightarrow \ C_6H_6 \tag{10.65}$$

During this reaction three acetylenes are thought to come together to form the trycyclic intermediate shown in Figure 10.15. The tricyclic intermediate then desorbs as benzene. The tricyclic intermediate forms easily on a hexagonal surface since the hexagonal geometry matches that of the intermediate. The tricyclic intermediate can also form on a square surface

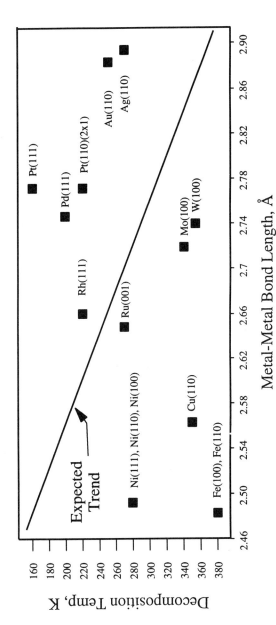

Figure 10.14 A plot of the temperature for methoxy decomposition as a function of the bond length in the metal. References to the data can be found in Yagasaki and Masel [1994].

Hexagonal Surface   ## Square Surface

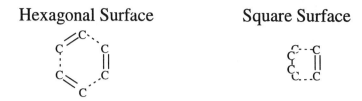

Figure 10.15 The tricyclic intermediate during the trimerization of acetylene.

by first dimerizing two acetylenes to form a complex that lifts off the surface, and then adding a third acetylene. However, that is a much less favorable process. Experimentally, Gentle and Muelleties [1983] found that Reaction 10.65 occurs rapidly on Pd(111) and at a much lower rate on Pd(100). Lambert and Ormerod [1994] suggest that the main reaction pathway on Pd(100) involves hexagonal defects.

To put this finding into the nomenclature used earlier in this chapter, one might say that the intrinsic barriers to forming a tricyclic intermediate are much lower on a hexagonal surface than on a square surface. Physically, the intrinsic barriers are low because the reactants are being held in a geometry conducive to reaction.

Experimentally, the relative orientation of the intermediates seems to have an important influence on the intrinsic barriers to Reaction 10.65.

No one knows if this is a general phenomenon. However, recently Yagasaki and Masel [1994] pointed out that when ethylene decomposes on a hexagonal or pseudohexagonal surface, the intermediates are all bound to the surface via sp^3 hybrids. In contrast, sp^2 hybrids are seen on square surfaces (see Figure 10.16 and Figure 10.17). Thus, there is the interesting observation that on a threefold symmetric surface, one forms threefold symmetric bonds, while on a twofold symmetric surface one forms twofold symmetric bonds. Thus, there seems to be a general correlation between the nature of the intermediates and the symmetry of the

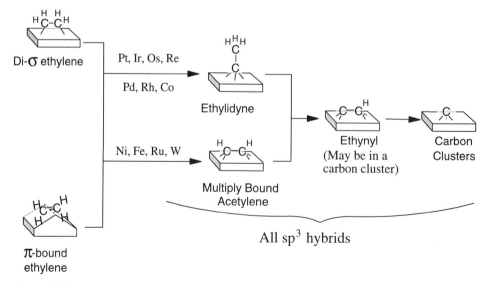

Figure 10.16 The mechanism of ethylene decomposition on the close packed faces of transition metals. (After Yagasaki and Masel [1994].)

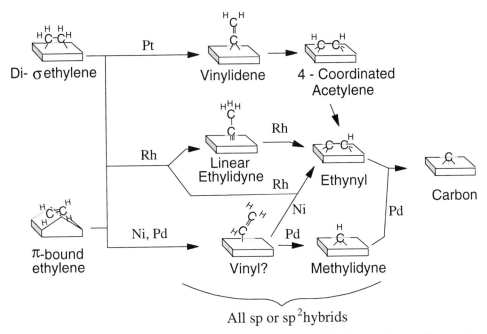

Figure 10.17 The mechanism of ethylene decomposition on the (100) faces of the transition metals. (After Yagasaki and Masel [1994].)

surface. Unfortunately, when this book was being written, it was unclear why such a correlation exists.

We discuss Figure 10.16 and Figure 10.17 is more detail later in this chapter. However, the thing to remember for now is that the reaction pathways for ethylene decomposition seem to correlate with the geometry of the surface. No one understands why. However, apparently surface geometry influences the intrinsic barriers to reaction in ways that we do not understand.

10.14 THE ROLE OF SURFACE ELECTRONIC STRUCTURE ON THE INTRINSIC BARRIERS TO REACTION

One of the difficulties with understanding the geometric effects is that whenever one changes the geometry of the surface, one also changes the electronic structure of the surface. According to quantum theory, the electronic structure of the surface should play a key role in determining the reactivity of the surface. Therefore, electronic structure effects should be quite important to the intrinsic barriers to reaction.

In the next four sections, we discuss the influence of surface electronic structure on intrinsic barriers to reaction in detail. We start by first considering the overall trends and then move on to specifics.

Let's start with the general ideas. In Chapter 3 we noted that adsorbates are held to surfaces by localized bonds on semiconductors and a mixture of localized and delocalized bonds on metals. Now consider a simple reaction

$$H_3C-CH_{3\,(ad)} \longrightarrow 2\,CH_{3\,(ad)} \qquad (10.66)$$

Figure 10.18 shows a diagram of the atomic motions during Reaction 10.66. If Reaction

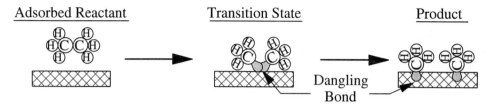

Figure 10.18   A diagram of the atomic motions during Reaction 10.66.

10.66 would occur in the gas phase, one would need to create two dangling bonds. That is a highly activated process (E_A = 90 kcal/mole). In contrast, it has been said that on a metal surface the dangling bonds are stabilized by interactions with the jellium. As a result, the overall barriers (but not the intrinsic barriers) to reaction are much lower on a metal surface than in the gas phase. In actual practice Reaction 10.66 is rarely seen because the ethane would rather transfer a hydrogen atom to the surface than a methyl ligand, as described in Section 10.8. However, reactions like

$$H_3C-CH_{3\,(ad)} \longrightarrow H_3C-CH_{2\,(ad)} + H_{(ad)}$$

$$H_3C-CH_{2\,(ad)} \longrightarrow CH_{3\,(ad)} + CH_{2\,(ad)} \tag{10.67}$$

are thought to occur. These reactions are thought to be facilitated by jellium.

The situation is more complex on a semiconductor. Reaction 10.67 is not seen on silicon until one gets to very high temperatures, because there are never two adjacent dangling bonds. However, the reaction is observed on partially reduced nickel oxide, presumably because there are dangling bonds in close proximity.

10.14.1   The Influence of Interstitial Electron Density on the Intrinsic Barriers to Reaction

Note that the interactions with the jellium are thought to be stabilizing the transition state in Reaction 10.66. Consequently, one might expect the interstitial charge density (i.e., the density of the jellium) to influence the intrinsic barrier for the bond scission process. For example, one might suppose that increases in the interstitial charge density will tend to increase the stability of the transition state for carbon–carbon bond scission, which in turn will lead to a decrease in the intrinsic barriers to reaction. In such a case, one would expect the rate to eventually plateau at high interstitial electron densities because at high enough electron densities, all of the dangling bonds will be saturated. However, one might expect a general increase in intrinsic barriers to reaction at low to medium interstitial electron density, and perhaps some leveling off at high interstitial electron density.

Figure 3.45 shows how the interstitial charge density varies over the left half of the periodic table. Generally, the interstitial charge density increases in moving from gold, silver, and copper to tungsten, molybdenum, and cobalt. The interstitial charge density then continuously decreases, reaching a minimum at the alkali metals.

Now, it is useful to compare the changes in interstitial electron density to changes in reactivity. Figure 3.7 shows data for the dissociation of CO, NO, N_2, and O_2 on metals in the middle part of the periodic table. To put the figure in perspective, note that the thermo-dynamic driving force for CO, N_2, O_2, and NO decomposition generally increases, moving to the left in the periodic table. As a result, even if the intrinsic barriers to bond scission were independent of metal, one would expect the bond scission rates to increase moving to the left in the periodic table. However, changes in the intrinsic barriers to bond scission could amplify that effect. Thus, if the intrinsic barriers change, one could enhance the rate of bond scission. Notice that the rate of dissociation is low on gold, silver, and copper, where the interstitial electron density is low. In contrast, the rate of dissociation is high on cobalt, molybdenum, and tungsten, where the interstitial electron density is high. Thus, there seems to be a qualitative correlation between the changes in interstitial electron density and changes in reaction.

Still, one does not know whether the changes in reactivity are caused by changes in interstitial electron density, or by some other factor. As noted earlier the data in Table 3.5 show that the thermodynamics of bond scission also get increasingly favorable in moving from gold to tungsten. Therefore, the variations in bond scission rates could also be caused by variations in the thermodynamics of bond scission.

Another factor is the interaction with the d-bands. We see below that interactions with the d-bands also get more favorable in moving from gold to tungsten. Therefore, by just focusing on data from gold to tungsten one cannot tell whether the variation in rate with surface composition is caused by a variation in the interstitial electron density or by some other factor.

Unfortunately, there are few experimental data on metals to the left of tungsten in the periodic table. The only evidence that the rate may track the interstitial electron density comes from work with evaporated potassium and cesium films. Langmuir found that hydrogen, ethylene, and carbon monoxide will not adsorb on potassium or cesium films on tungsten, even though potassium and cesium form stable hydrides, carbides, and oxides so the dissociative adsorption of hydrogen, ethylene, and carbon monoxide is quite thermodynamically favorable. Potassium and cesium have a low interstitial electron density which could cause a low reactivity. Thus, there is correlation between the rate of ethylene, hydrogen, and carbon monoxide dissociation and the interstitial electron density. There is no evidence for a cause-and-effect relationship, however.

In fact, I do not believe that changes in interstitial electron density are making potassium and cesium so unreactive. As noted earlier, the d-bands can play a key role in promoting chemical reactions. There are no d-electrons available for bonding on potassium or cesium. As a result, one would expect potassium and cesium to be relatively unreactive for H_2 dissociation, even if they did not have a low interstitial electron density. Thus, it is by no means clear that a metal with a low interstitial electron density will be unreactive, or that a metal with a high interstitial electron density will be especially reactive. Aluminum, for example, has a moderate interstitial electron density. Yet we note in the material that follows that aluminum surfaces are fairly unreactive in hydrocarbon environments. Experimentally, the occupancy in the d-bands seems to be more important than the role of the interstitial electron density in promoting reactions.

10.15 THE ROLE OF d-BANDS IN MODIFYING THE INTRINSIC BARRIERS TO REACTION

It is useful to consider an example that illustrates the role of the d-bands in modifying the intrinsic barriers to reactions. Consider the dissociative adsorption of H_2

$$H_2 \longrightarrow 2\ H_{(ad)} \qquad (10.68)$$

Figure 6.4, p. 443, shows a portion of the periodic table with sections shaded to indicate the relative reactivity of the surfaces for H_2 dissociation. Hydrogen dissociates readily at 100 K on all of the metals in the part of the periodic table that is shaded, while the H_2 dissociation process has an activation barrier of 24 kcal/mole or more on all of the other metals except copper. On copper Campbell and Campbell [1991] and Rettner, Michaelson, and Auerbach [1993] find an activation barrier of about 14 kcal/mole.

It is useful to consider several cases to put the results in Figure 6.4 in perspective, Sprunger and Plummer [1994] examined the adsorption of hydrogen atoms and H_2 molecules on Mg(0001), Mg(11$\bar{2}$0), Li(110), and K(110). In all cases, data on the adsorption of hydrogen atoms suggests that the H_2 dissociation process is more than 50 kcal/mole exothermic. Yet no dissociation is observed. Berger and Rendulic [1992] examined H_2 dissociation on Al(111) with molecular beams. The dissociation process had an activation barrier greater than 25 kcal/mole, even though the dissociation process is more than 50 kcal/mole exothermic. Many investigators have examined H_2 dissociation on Si(111). No H_2 dissociation has been observed at moderate temperatures even though the dissociation process is 54 kcal/mole exothermic. In contrast, H_2 dissociation is essentially inactivated on Pt(110) and Pd(110) (there may be a small barrier in Pt(111)). Yet, McCabe and Schmidt [1977] report that the dissociation process is only 13 kcal/mole exothermic on Pt(110).

To put these results into the context of our earlier discussions, if one assumes that $\gamma_P = 1$ in Equation 10.45, then one calculates that the intrinsic barrier to H_2 dissociation is over 80 kcal/mole on silicon and on the order of 90 kcal/mole on sodium and aluminum. By comparison, the intrinsic barrier is less than 13 kcal/mole on Pt(110). Thus, the intrinsic barriers to H_2 dissociation vary significantly from one material to the next.

A more detailed comparison shows that the intrinsic barriers to H_2 dissociation are much lower on the metals with a partially filled d-band than on all of the other metals in the periodic table. H_2 dissociates readily on all of the metals with partially filled d-bands. The activation barrier is 2 kcal/mole or less. In contrast, the dissociation process has a barrier of 25 kcal/mole or more on all of the other metals in the periodic table except copper. Campbell and Campbell [1991] find that on copper the activation barrier is 14 kcal/mole. If one plugs their data into Equation 10.45, one finds that the intrinsic barriers to H_2 bond scission are small on metals with unfilled d-bands and large (3 eV or more) on all of the other metals in the periodic table except copper and silver. Therefore, it seems that the d-bands in some way are able to lower the intrinsic barriers for the dissociation of H_2.

In Section 3.10.1 we noted that there are two basic models that attempt to explain the role of the d-bands in the dissociation of H_2: a model by Harris and Anderson [1985], which suggests that in nickel and platinum the d-bands act as a sink for electrons and thereby lower the Pauli (i.e., electron–electron) repulsions when H_2 (or He) approach the surface and a model by Masel [1986] and Nørskov [1990], which suggests that the role of the d-bands is to stabilize antibonding orbitals in the H_2. In Section 3.10.1, we noted that Rettner et al. [1992, 1993] did helium scattering experiments that show that the lowering of the Pauli repulsions is not significantly different on platinum and copper. However, experimentally, the barrier for H_2 dissociation is 14 kcal/mole lower on platinum than on copper. Therefore, the Harris and Anderson model does not seem to correlate with the data. We see below that the models of Masel [1986] and Nørskov [1990] do correlate with the data. Therefore, the best available evidence is that the d-bands promote H_2 bond scission in H_2 by interacting with antibonding orbitals in the H_2.

The easiest way to understand why the antibonding orbitals are important is to note that the dissociative adsorption of H_2

$$H_2 + 2S \longrightarrow \begin{array}{c} H-H \\ / \quad \backslash \\ S-S \end{array} \longrightarrow \begin{array}{c} H \quad H \\ | \quad | \\ S-S \end{array} \qquad (10.69)$$

is analogous to the four center reactions discussed in Section 9.13.2, p. 669. In Section 9.13.2 we noted that when there are only s orbitals or sp^3 hybrids available for bonding, a four center reaction is symmetry forbidden. One has to break bonds in the reactant, or create antibonding electrons before the reaction can proceed. The same argument can be directly applied to H_2 dissociation on surfaces. If only s orbitals and sp hybrids are available for bonding, the reaction will be symmetry forbidden. The intrinsic barrier to bond scission will be on the order of the bond energies. Therefore, the reaction will be inhibited. For example, on silicon only sp^3 hybrids are available for bonding. Therefore, the argument in Chapter 9 implies that H_2 adsorption on silicon is symmetry forbidden, i.e., one needs to break silicon–silicon bonds or create antibonding (i.e., conduction band) electrons before reaction can proceed. Experimentally, no H_2 dissociation is detected on silicon at temperatures up to 800 K, even though the dissociation process is over 50 kcal/mole exothermic. This result suggests that H_2 dissociation is symmetry forbidden on silicon, as expected from the discussion in Section 9.10.2.

The arguments are more subtle on metals because the electrons in metals are held in bands. However, in the next two sections we show that H_2 dissociation is symmetry forbidden on those metals that interact weakly with the antibonding orbitals in H_2 and symmetry allowed on metals that interact strongly with the antibonding orbitals in H_2. The results in Section 9.10 imply that a symmetry allowed reaction will have a much lower intrinsic barrier than a symmetry forbidden reaction. Thus, the interactions with the antibonding orbitals play a key role in determining the intrinsic activation barriers for a reaction.

In the next several sections we review how the d-bands affect H_2 dissociation. Most of what we discuss will require some knowledge of quantum mechanics. Note that if one ignores quantum effects, s-electrons and d-electrons are exactly equivalent. Therefore, in the absence of quantum effects, one would not expect d-electrons to have unique capabilities. Experimentally, metals with unfilled d-electrons have unique properties, and to understand those unique properties one needs to understand quantum effects.

In the discussion that follows we review those effects. First, we show that the key step in H_2 bond scission is the transfer of electrons into the antibonding orbitals of H_2. Next, we show that the antibonding orbitals interact much more strongly with d-bands than with s-bands. Finally, we show that these effects lead to substantially enhanced rates. In all cases, I have derived the key results. However, if the reader is unfamiliar with quantum mechanics he or she may want to skip the algebra and instead focus on the qualitative results.

10.15.1 The Role of Antibonding Orbitals in the H_2 Dissociation Process

The first question we address is, why do we care about antibonding orbitals during H_2 dissociation? Quantum mechanically, antibonding orbitals are critical to bond scission. Recall that quantum mechanically, there are two ways that a bond in a molecule can break. Either bonding electrons can be removed from the molecule, or electrons can be shared with the antibonding orbitals into the molecule. If one would remove all of the bonding orbitals in a H_2 molecule, one would end up with two positively charged ions (i.e., two H^+'s). However, the analysis in Section 3.4.2 shows that one needs a very electronegative surface to produce H^+'s. The data in Section 3.4.2 show that no metal is electronegative enough. As a result, when H_2 dissociates on a metal surface, one does not end up with two H^+'s. Rather one ends up with two nearly neutral H atoms, i.e., the orbitals on the H atoms will still contain electrons even after the H_2 dissociates. If all of those electrons are in the bonding orbitals of the H_2, the H_2 will still have a partial bond. It is only when one puts electrons into antibonding orbitals that the bond order will go to zero.* Consequently, the transfer of electrons into antibonding orbitals is a key step in H_2 bond scission.

When I was writing this book it was unclear whether the ideas in the preceding paragraph would be familiar to all of my readers. Therefore, I included a quantum mechanical derivation to show that electrons flow into antibonding orbitals during a bond scission reactions. However, the reader can skip to the next section without loss of continuity.

The derivation uses the configuration mixing model described in Section 9.12, p. 656. To start, note that when Reaction 10.69 occurs, the hydrogen–hydrogen bond breaks. In a simple

*Quantum mechanically the bond order is the number of bonding electrons minus the number of antibonding electrons. Therefore, if one can put electrons into antibonding orbitals, one can break the H_2 bond. The latter electrons can come from the bonding orbitals of the H_2 or from the surface.

picture, the H_2 starts out with two electrons in a σ orbital. The bond then stretches and breaks. If the bond scission process were occurring in the gas phase, the initial Hartree–Fock wavefunction for the hydrogen would be

$$\Psi_{initial} = A\phi_\sigma(\vec{r}_1)\,\phi_\sigma(\vec{r}_2) \tag{10.70}$$

where ϕ_σ is the wavefunction for a σ orbital, $\vec{r}_1$ and $\vec{r}_2$ are the positions of electrons 1 and 2, and A is the antisymmetrizer. The antisymmetrizer is an operator that ensures that the wavefunction satisfies the Pauli exclusion principle as described in Appendix B.

Note

$$\phi_\sigma(\vec{r}) = c_\sigma(\phi_A(\vec{r}) + \phi_B(\vec{r})) \tag{10.71}$$

where ϕ_A is the 1s orbital on one of the hydrogens, ϕ_B is the 1s orbital on the other hydrogen, and c_σ is a normalization coefficient. For future reference it is useful to note that $\phi_{\sigma*}(\vec{r})$, the wavefunction for the antibonding orbitals in the H_2, is given by

$$\phi_{\sigma*}(\vec{r}) = c_{\sigma*}(\phi_A(\vec{r}) - \phi_B(\vec{r})) \tag{10.72}$$

Solving Equations 10.71 and 10.72 simultaneously for ϕ_A and ϕ_B yields

$$\phi_A(\vec{r}) = \frac{1}{(c_\sigma + c_{\sigma*})}(\phi_\sigma(\vec{r}) + \phi_{\sigma*}(\vec{r})) \tag{10.73}$$

$$\phi_B(\vec{r}) = \frac{1}{(c_\sigma + c_{\sigma*})}(\phi_\sigma(\vec{r}) - \phi_{\sigma*}(\vec{r})) \tag{10.74}$$

Note that when H_2 dissociation occurs in the gas phase, one ends up with one electron on each hydrogen atom. In that case, the final wavefunction for the dissociated H_2 molecule is

$$\Psi_{final} = A\phi_A(\vec{r}_1)\,\phi_B(\vec{r}_2) \tag{10.75}$$

Combining Equations 10.73, 10.74, and 10.75 yields

$$\Psi_{final} = \frac{A}{(c_\sigma + c_{\sigma*})}(\phi_\sigma(\vec{r}_1) + \phi_{\sigma*}(\vec{r}_1))\,(\phi_\sigma(\vec{r}_2) - \phi_{\sigma*}(\vec{r}_2)) \tag{10.76}$$

It is useful to compare $\Psi_{initial}$ (Equation 10.70) and Ψ_{final} (Equation 10.76). In Equation 10.70 all of the electrons are in bonding orbitals. However, in Equation 10.76, some of the electrons are in antibonding orbitals. Therefore, in order for H_2 dissociation to occur in the gas phase, electrons must be converted from bonding to antibonding orbitals (via interactions with collision partners).

Now the situation is more complicated on a metal surface. During the H_2 dissociation process, the two hydrogen atoms form an adsorbate–surface bond. Generally, the bond will be fairly delocalized. Still, according to the Lang–Williams model described in Section 3.8.2, one can write, $\phi_{H_{ad}}^A$, the wavefunction for an electron on the adsorbed hydrogen on site A as

$$\phi_{H_{ad}}^A = (\phi_A(\vec{r}_1) + \phi_s^A(\vec{r}_2)) \tag{10.77}$$

where $\phi_A(\vec{r}_1)$ is the wavefunction for an electron on the hydrogen and $\phi_s^A(\vec{r})$ is a sum of surface wavefunctions over site A given by Equation 3.72

$$\phi_s^A(\vec{r}) = \int G(\vec{r}, \vec{r}') V(\vec{r}') \phi_{H_{ad}}^A(\vec{r}') d\vec{r}' \qquad (10.78)$$

For the purposes of derivation, it will be useful to assume that at the end of Reaction 10.69 each hydrogen is sharing two electrons with the surface. Under such circumstances, Ψ_{final}, the wavefunction for the final state of the system would be given by

$$\Psi_{final} = A\phi_{H_{ad}}^A(\vec{r}_1) \phi_{H_{ad}}^A(\vec{r}_2) \phi_{H_{ad}}^B(\vec{r}_3) \phi_{H_{ad}}^B(\vec{r}_4) \qquad (10.79)$$

where $\vec{r}_1$, $\vec{r}_2$, $\vec{r}_3$ and $\vec{r}_4$ are the positions of the four electrons bonded to the two hydrogen atoms. Note that two of those electrons start out in surface orbitals, and two start out on the hydrogens. Therefore, to some approximation, one can write the wavefunction for those electrons before the H_2 dissociates as

$$\Psi_{initial} = A\phi_\sigma(\vec{r}_1) \phi_s^A(\vec{r}_2) \phi_\sigma(\vec{r}_3) \phi_s^B(\vec{r}_4) \qquad (10.80)$$

For the purposes of derivation it is useful to substitute equations 10.73, 10.74, and 10.77 into Equation 10.79 to yield

$$\Psi_{final} = A \left[\frac{\phi_\sigma(\vec{r}_1) + \phi_{\sigma*}(\vec{r}_1)}{(c_\sigma + c_{\sigma*})} + \phi_s^A(\vec{r}_1) \right] \left[\frac{\phi_\sigma(\vec{r}_2) + \phi_{\sigma*}(\vec{r}_2)}{c_\sigma + c_{\sigma*}} + \phi_s^A(\vec{r}_2) \right]$$
$$\cdot \left[\frac{\phi_\sigma(\vec{r}_3) - \phi_{\sigma*}(\vec{r}_3)}{c_\sigma + c_{\sigma*}} + \phi_s^B(\vec{r}) \right] \left[\frac{\phi_\sigma(\vec{r}_4) - \phi_{\sigma*}(\vec{r}_4)}{c_\sigma + c_{\sigma*}} + \phi_s^B(\vec{r}_4) \right] \qquad (10.81)$$

Notice that $\Psi_{initial}$ contains contributions from the σ state of the H_2 but not the σ^* state. In contrast, Ψ_{final} contains equal contributions from the σ and σ^* states. Therefore, just as in the gas phase during H_2 dissociation, electrons must flow into the antibonding orbitals of the H_2.

Physical Picture of the Role of the Antibonding Orbitals. Physically, the flow of electrons into antibonding orbitals of the H_2 is a key step in the H_2 dissociation process. Note that if the H_2 would share its bonding electrons with the surface, but there was no back-bonding into the antibonding orbitals, the H_2 will bind strongly (i.e., chemisorb) on the surface. However, as long as more electrons are in the bonding orbital of the H_2 than in the antibonding orbitals there will still be a partial hydrogen–hydrogen bond. The net result will be a chemisorbed H_2 rather than two separate hydrogen atoms. In contrast, if one puts electrons into the antibonding orbitals in the H_2, one can get to a situation where there are an equal number of bonding and antibonding electrons in the H_2. If there are an equal number of bonding and antibonding electrons in the H–H bond, then there will be no net bonding in the H_2, i.e., the hydrogen–hydrogen bond will break. As a result, in order for the hydrogen–hydrogen bond to break, one must find a way to get electrons into the antibonding orbitals of the H_2.

It is theoretically possible for the hydrogen's bonding electrons to be transferred into antibonding electrons via a collision process similar to that which occurs when H_2 dissociates in the gas phase. However, the intrinsic barriers will be much lower if the surface can bond to (i.e., share electrons with) the antibonding orbitals in the H_2. The sharing of electrons stabilizes the transition state for reaction. According to the principle of maximum bonding, when one stabilizes the transition state for reaction one lowers the barrier for reaction. Consequently, the barriers to H_2 dissociation should be much smaller on surfaces that bind the antibonding orbitals strongly than on surfaces that bind the antibonding orbitals weakly.

The question, then, is how strongly do the various types of electrons in the surface interact with the antibonding orbitals in the H_2. Well, in the next section we demonstrate that the d-bands in the metal have a much stronger interaction with antibonding orbitals than do the s-bands. As a result, the intrinsic activation barrier for H_2 bond scission is substantially reduced if there are partially filled d-bands available for bonding.

10.15.2   A Comparison of s- and d-Orbitals for Bond Scission: Quantum Derivation

We need to do some analysis to arrive at that result. The analysis will rely on some quantum theory. In the absence of quantum effects, the interactions of an antibonding orbital with an s-orbital and a d-orbital are exactly equivalent. However, if one puts in quantum effects, one finds that the interaction with the s-orbitals almost completely cancels. In contrast, no cancellation is seen with the d-orbitals. The net effect is that the d-orbitals are able to stabilize the antibonding orbitals in the H_2. The stabilization of the antibonding orbitals leads to easy bond scission.

In the remainder of this section we show quantum mechanically that d-electrons interact much more strongly with antibonding orbitals than s-electrons. The analysis uses some simple material from solid-state physics that is reviewed in Appendix C. The reader may want to quickly review Appendix C before proceeding with the discussion. Alternatively, in Section 10.15.3 we provide a graphical justification of the results in this section. Readers who are unfamiliar with quantum mechanics may want to skip to Section 10.15.3.

We start the derivation by recalling some material from Section 9.13. In Section 9.13 we found that when two orbitals interact, they form a bonding and an antibonding state. The energy of the bonding state is given by Equation 9.101. If one expands Equation 9.101 in a Taylor series, about $\Gamma = 0$ one finds that each state is stabilized by an amount ΔE_s given by

$$\Delta E_s = \frac{\beta_{01}\,\beta_{10}}{E_0 - E_1} + \text{higher order terms} = \frac{|\beta_{01}|^2}{E_0 - E_1} + \text{higher order terms} \quad (10.82)$$

Equation 10.82 is quite useful because if one knows all of the terms in Equation 10.82, one can calculate how strongly any given orbital is stabilized by the presence of the surface.

To proceed, however, one needs to be careful because in a solid there is a band of states. Each state has a different interaction with the σ^* state of the H_2, so one has to add up all of the contributions to calculate properties. For example, ΔE_{σ^*}, the stabilization of the σ^* state, is equal to

$$\Delta E_{\sigma^*} = \sum_{\substack{All\ bands \\ in\ solid}} \int \frac{|\beta_{\sigma^* k}|^2\,O(k)}{|E_{\sigma^*} - E_k|}\,dk \quad (10.83)$$

where k is an index of the states in the metal band (see Appendix C), E_{σ^*} is the energy of the σ^* state before bond scission, E_k is the energy of the k-state in the band, $\beta_{\sigma^* k}$ is the coupling constant, and $O(k)$ is the occupancy of the k-state. $O(k)$ equals one if the state is filled and zero if the state is empty. The occupancy term arises in Equation 10.83 because a given k-state cannot share an electron with the antibonding orbital in the H_2 unless the k-state is occupied.

To proceed with the analysis, we are going to have to review what the electron states in the surface are like. Figure 10.19 shows a diagram of some of the states in the bands of a typical metal. The calculation is a contour 1.5 Å above the cores of the top layer of surface atoms. We designate each state by its wavevector k as described in Appendix C. We list each band by its primary character, either s or d, even though there is significant s-d coupling.

Let's start with the s-band. The $k = 0$ state in the s-band is the state at the bottom of the conduction band. It is almost smooth. There are slight enhancements of the wavefunction overn the surface atoms, and slight decreases in between. However, the variations in the wavefunction are relatively small.

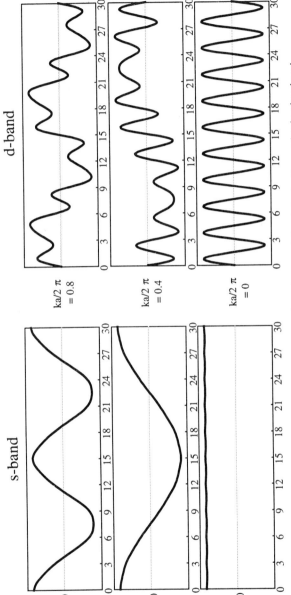

Figure 10.19 A plot of the real part wavefunction of some of the states in the conduction band and *d*-band in Ni(100) metal as calculated from the LCAO model described in Appendix C.

Now as one goes up the band (i.e., increases k), the nature of the wavefunction changes. The real part of the wavefunction oscillates and changes sign as one moves down the surface. However, in a typical transition metal all of the occupied states in the s-band satisfy $ka/2\pi > 0.25$. Those states have an oscillation period of 20 Å or more. By comparison, a typical bond is only 2 Å. Therefore in a typical transition metal the s-band looks relatively smooth over the length of an atom.

Figure 10.19 also shows the variation in the wavefunction for a typical d_{xy} band. The d_{xy} band looks quite different from the s-band. The wavefunction shows many oscillations at $k = \pi/a$ (i.e., the bottom of the band), and even larger oscillations at $k = 0$ (i.e., the top of the band).

Now let's go back and calculate the interaction of the σ^* orbital in the H_2 with the s-band in the metal. Let's consider the $k = 0$ state in the s-band. According to Equation 10.82, the strength of the interaction between the σ^* orbital and any k-state in the band is proportional to the square of the coupling constant β_{σ^*k} given by

$$\beta_{\sigma^*k} = \int\int\int_\infty^\infty \phi_{\sigma^*}^*(\vec{r}) \ V(\vec{r}) \ \phi_k(\vec{r}) \ dr_X \ dr_Y \ dr_Z \qquad (10.84)$$

where $\phi_{\sigma^*}(\vec{r})$ is the wavefunction for the σ^* state, $\phi_k(\vec{r})$ is the wavefunction for the k state, $V(\vec{r})$ is the interaction potential (see Section 9.13), and $\vec{r} = (r_X \ r_Y \ r_Z)$.

Figure 10.20 shows both states. In the figure, the X-scale has been expanded to focus on the region of the surface near the H_2. The $k = 0$ state in the s-band is completely smooth, while the σ^* state shows large sign changes. $V(\vec{r})$ can be positive or negative. However, it must be the same on both hydrogens since the two hydrogens are equivalent.

Figure 10.20 also shows the integrand in Equation 10.84, i.e., $\phi_{\sigma^*}^*(\vec{r}) \ V(\vec{r}) \ \phi_k(\vec{r})$ plotted as a function of distance, X, calculated on a line through the cores of the two hydrogen atoms. Notice that for the $k = 0$ state the integrand is negative on the left side of the center of the hydrogen–hydrogen bond, and positive on the right side of the center of the bond. The coupling constant is the integral under the curve. There is a positive contribution, but the positive contribution is exactly canceled by a negative contribution. Consequently, the cou-

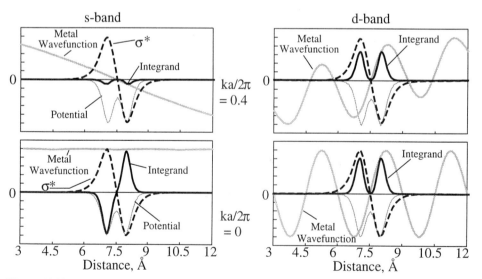

Figure 10.20 The terms in Equation 10.84 for the interaction of the s-band and d-band in the metal with the σ^* orbital in the H_2. We placed the H_2 in the geometry most conducive to reaction.

pling constant is exactly zero. As a result, according to Equation 10.82 there will be no interaction between the $k = 0$ state in the s-band and the $\sigma*$ state in the H_2.

Now, as one goes up the s-band (i.e., increases k), the positive and negative terms in Equation 10.84 do not quite cancel. If one places the H_2 just right, one can get some overlap as shown in Figure 10.20. As a result, the k-states in the s-band do interact slightly with the $\sigma*$ state of H_2. However, the average over all H_2 positions is always small. Further, since the net interaction is proportional to β^2, the net interaction is tiny. A more detailed analysis shows that states near the top of the s-band could have a nonnegligible interaction with the $\sigma*$. However, the states near the top of the band are above the Fermi level in all of the metals except mercury and α-tin,[†] so there are no electrons in those states to bind to the $\sigma*$ state of the H_2. The net result is that the antibonding orbital of H_2 only interacts weakly with the s-band of most metals.

The situation is entirely different with the d-bands. As noted earlier, the d-bands show sharp sign changes over short distances. If the H_2 is oriented directly over one of the surface atoms, or halfway between two surface atoms (see Figure 10.20), the coupling constant will be negative everywhere, for every k-vector in the band. Unlike the s-band case, there is no cancellation of overlap. The net effect is that the $\sigma*$ orbital in the H_2 interacts much more strongly with the d-bands in a typical metal than with the s-band.

10.15.3 A Pictorial Representation of the Results from Section 10.15.2

One can also arrive at the key results in Section 10.15.2 pictorially. Figure 10.21 shows a diagram of the key orbitals in the interaction of an H_2 molecule with a metal surface. The figure includes a picture of the $\sigma*$ orbital in the H_2 and the diagrams of the wavefunctions for the s- and d-bands at the Γ point in the Brillouin zone. The $\sigma*$ has one lobe on each atom. One of the lobes is positive, while the other lobe is negative. In the figure we have darkened the negative lobes to make them easier to identify. The d-band also has both positive and negative lobes. In contrast, the s-band is smooth and completely positive.

Now let's consider the interaction of the antibonding orbital on the H_2 with each band. To orient the reader, quantum mechanically the magnitude of the interactions between two wavefunctions increases as the overlap between the wavefunctions increase. If the sign of the two wavefunctions is the same, then the interaction will be attractive. However, if the sign of the two wavefunctions is different, the interaction will be repulsive.

Now, let's use Figure 10.21 to examine the interaction of the s-band with the $\sigma*$ orbital in H_2. If we start with an H_2, which is tilted relative to the surface as indicated on the top left of Figure 10.21, we find that there is a net interaction with the s-band. However, in order for the H_2 to stretch and break, the surface needs to form a bond to the hydrogen atom on the right side of the H_2 molecule. The right side of the antibonding orbital H_2 has a repulsive interaction with the surface. Therefore, in the tilted geometry the H_2 dissociation process will be inhibited via the interactions with the s-band.

The situation is a little different in the parallel geometry shown in the top right of Figure 10.21. In the geometry in the top right the interactions with the s-band on both sides of the H_2 molecule exactly cancel. The interactions do not change as the H–H bond stretches. As a result, in this geometry, the interactions with the s-band do not raise the intrinsic barriers to reaction. However, the interactions do not lower the intrinsic barriers either.

The situation is entirely different with a d-band metal. The d's have an attractive interaction with the antibonding orbitals in the H_2 molecule. Further, in the geometry shown in the bottom right of Figure 10.21 the attractive interaction with the antibonding orbital increases as the H_2 bond stretches. The net effect is that the intrinsic barriers to bond scission are much lower in the case on the bottom of Figure 10.21 than in the case on the top of Figure 10.21,

---

[†]The metal would fall apart if the s-band had significant antibonding character.

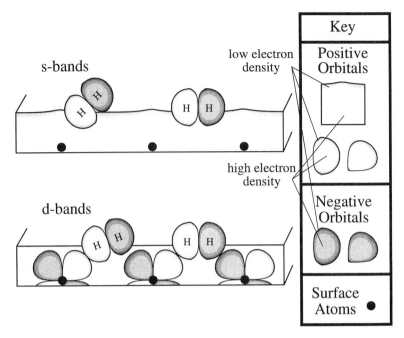

Figure 10.21 An edge view of the wavefunction for the σ^* state of H_2 and the wavefunction for the s- and d-bands at the Γ point in the Brillouin zone. Positive lobes are light and negative lobes are dark as indicated in the key.

i.e., d-electrons are much better able to lower the intrinsic barriers for bond scission than s-electrons.

10.15.4 Implications to the Intrinsic Barrier to H_2 Bond Scission

Now, it is useful to go back and to put that result in perspective. The analysis in Section 10.15.1 indicates that if one stabilizes the antibonding orbitals in the H_2, one lowers the intrinsic barriers to H_2 dissociation. The analysis in Sections 10.15.2 and 3 indicates that the antibonding orbitals in H_2 interact much more strongly with the d-bands in metals than with the s-bands. Therefore, the presence of the d-bands would be very useful in promoting H_2 dissociation.

Now one can quantify these effects using Equation 10.82. According to Equation 10.82 the interaction between the d-bands and the σ^* orbital will be large when $E_\sigma - E_k$ is close to zero, (i.e., when the states in the metal lie close to the energy of the σ^* state), and small otherwise. Generally, the σ^* orbitals in H_2 lie slightly below the Fermi level on most metals. With a group VIII metal, the d's are partially filled, which means they lie at the Fermi level. The σ^* state lies below the d-band. As a result, there is a strong bonding interaction between the d-bands in the group VIII metals and the antibonding orbitals in the H_2. That bonding interaction should decrease the intrinsic barriers to bond scission. In contrast, the d-bands are completely filled in copper, silver, and gold. In silver and gold, the d's lie deep in the band (i.e., at very low energy). According to Equation 10.82 there should be a much smaller interaction between the σ^* orbitals and the d-bands under such circumstances. Further, there will be a repulsive interaction with the bonding (σ) orbital on the H_2 that would raise the intrinsic barriers to reaction.

Experimentally, hydrogen readily dissociates on platinum and palladium but not copper, silver, and gold, even though the thermodynamic driving force for hydrogen dissociation is similar on platinum, palladium, copper, and silver. Linke et al. [1994] find that platinum–copper alloys still readily dissociate H_2, even at platinum concentrations where all of the platinum atoms are isolated. Such results verify that d-electrons are very effective in lowering intrinsic barriers to reactions.

10.15.5 Analogy to Symmetry Forbidden Reactions

Physically, what is happening is that the situation on an s-band metal is very analogous to the situation with the four centered reaction discussed in Section 9.10.1. During the reaction, electrons start out in a symmetric or nearly symmetric orbital. However, the electrons need to end up in an antisymmetric orbital in order for reaction to proceed. Therefore, orbital symmetry is not conserved during the process. Consequently, the activation barrier for the reaction is high.

The d-orbitals are naturally antisymmetric. As a result, the d-bands provide a convenient source of antisymmetric electrons. H_2 dissociation is symmetry allowed on metals with available d-bands and forbidden on other metals. Experimentally, H_2 dissociation is essentially unactivated on all of the d-band metals. However, the activation energies are 1 eV or more on all of the other metals except copper. (In copper, the d's are accessible at thermal energies.) Therefore, the presence of d-electrons seem to substantially modify the intrinsic barriers to reaction.

For the discussion that follows it is helpful to quantify this idea more carefully. In Chapter 9 we noted that Conroy and Malli [1969] results show that in the gas phase the reaction $H_2 + D_2 \rightarrow 2HD$ has an intrinsic activation barrier of over 100 kcal/mole (i.e., about the H–H bond energy). One would expect a jellium surface to lower that barrier. However, earlier in this chapter we noted that the activation energy for H_2 dissociation on K(110), Al(111), and Si(111) is more than 25 kcal/mole. If we plug numbers into Equation 9.5, assuming $\gamma_P = 1$, one finds that the intrinsic barrier to H_2 dissociation is at least 80 kcal/mole on Al(111), K(110), and Si(111), i.e., the interaction with the s-band has not significantly lowered the intrinsic barrier to reaction. In contrast on Pt(110), there is no activation barrier to H_2 dissociation. The heat of adsorption is only 13 kcal/mole, which means that the intrinsic barrier must be less than 13 kcal/mole. The main difference between the Al(111) and Pt(110) cases is the interaction with the d-bands. Therefore, it seems that the interaction with the d-bands has lowered the intrinsic barrier to H_2 bond scission from over 80 kcal/mole to less than 13 kcal/mole. Clearly, a 70-kcal/mole change in the activation energy is important.

10.15.6 Extension to Other Bond Scission Processes

One can extend these results to many other reactions. When a bond-scission reaction occurs, electrons need to flow into (i.e., be shared with) antibonding orbitals. The d-electrons are easily shared with antibonding orbitals, while in general, s-electrons are not. The sharing of electrons with the antibonding orbitals leads to a stabilization of the transition state for the reaction, and therefore a significant lowering of the intrinsic barrier to reaction. As a result, transition metals are excellent catalysts for bond scission reactions, while simple metals are not.

In unpublished work we have used an analysis like that in Section 10.15.5 to show that the interactions with the d-electrons lower the intrinsic barriers to reaction by 80–90 kcal/mole for scission of a single bond, by up to 160 kcal/mole for scission of a double bond, and up to 240 kcal/mole for scission of a triple bond. This is a huge effect. Unfortunately, it is also an effect that is not yet very well understood.

Still, there are some qualitative predictions that can be made. One key prediction is that the intrinsic barriers to reaction should be quite different on metals without unfilled d-shells than on metals with filled d-shells. As a result, one would expect to observe a different Polayni relationship for metals with partially filled d-bands than for metals with no d-electrons available for bonding. The change in the intrinsic barriers to reaction accounts for the two lines in Figure 10.8.

Another key prediction from an analysis like that in Section 10.15.2 is that intrinsic barriers to reaction are quite different when one is breaking a single bond than when one is breaking a double or triple bond. The analysis in Section 10.15.2 was for the interaction of a σ^* orbital. However, when one breaks a double or triple bond, one has to also stabilize π^* orbitals. In Section 10.20 we show that the interactions with π^* orbitals are weaker than those of σ^* orbitals. As a result, the intrinsic barriers for breaking π bonds should be higher than those for breaking σ bonds.

Experimentally, π bonds are harder to break than σ bonds. For example, H_2 dissociates upon adsorption on Pt(110) at 100 K. However, O_2 adsorbs molecularly at 100 K. O_2 does not dissociate until the surface is heated to 300 K. Yet the oxygen dissociation process is 64 kcal/mole exothermic on Pt(110), while the hydrogen dissociation process is only 13 kcal/mole exothermic. These results show that it is much easier to break a single bond than a double bond, even in a case where the thermodynamics strongly favor scission of the double bond. These results confirm that intrinsic barriers for the scission of double bonds are often much larger than the intrinsic barriers for scission of single bonds, as expected from the analysis in Section 10.15.2 and 10.20.

Finally, another key prediction of an analysis similar to that in Section 10.15.2 is that as one polarizes a bond one should make the bond easier to break. For example, C–H bonds are fairly symmetric. Their antibonding orbitals look very similar to the antibonding orbitals in Figure 10.20. As a result, C–H bond scission would be expected to be symmetry forbidden on metals with no available d-electrons. In contrast, O–H bonds are fairly polarized. The bonding molecular orbital has more electron density on the oxygen, while the antibonding orbital has more electron density on the hydrogen. There is less symmetry than in H_2. As a result, the symmetry arguments in Section 10.15.2 break down for an O–H bond. In particular, the positive and negative part of the integral in Equation 10.82 do not quite cancel even at the bottom of the s-band, and there is a significant overlap at the Fermi level. As a result, O–H bands would be expected to have lower intrinsic barriers to bond scission than C–H bonds in the absence of d-bands.

Experimentally, Langmuir found that O–H bonds in the H_2O will break on K(110) films. However, the C–H bonds in C_2H_4 and the H–H bonds in H_2 are stable on K(110). Therefore, it seems that the intrinsic barriers to O–H bond scission are lower than the intrinsic barriers to C–H or H–H bond scission on an s-band metal, as expected from Equation 10.82.

The situation is less obvious with a d-band metal because the d's lower the intrinsic barriers to C–H and H–H bond scission. However, the decomposition of methanol and ethanol has been examined on many transition metal surfaces. Experimentally, the O–H bond always breaks before the C–H bonds even though O–H bonds are generally stronger than C–H bonds. Such a result could be expected from the analysis in the previous two paragraphs.

Still, it is unclear whether orbital symmetry effects are causing the O–H bonds in alcohols to break before the C–H bonds. Even in the absence of the orbital symmetry effects, the proximity effects described in Section 10.11 would cause the intrinsic barriers to O–H bond scission to be lower than the intrinsic barriers to C–H bond scission. Thus, it is unclear whether the orbital symmetry effect, or some other factor is causing the O–H bonds in alcohols to break before the C–H bonds.

10.16 THE ROLE OF NET CHARGES

In fact, the actual situation in alcohol decomposition is fairly subtle. Note that people often run alcohol decomposition on an oxygen-covered surface rather than a clean surface. On a clean metal surface, the main reaction pathway is

$$CH_3OH_{(ad)} + S \longrightarrow CH_3O_{(ad)} + H_{(ad)} \tag{10.85}$$

In contrast, on an oxygen-covered surface the main reaction pathway is

$$CH_3OH + O_{(ad)} \longrightarrow CH_3O_{(ad)} + OH_{(ad)}$$
$$CH_3OH + OH_{(ad)} \longrightarrow CH_3O_{(ad)} + H_2O \tag{10.86}$$

Reactions 10.85 and 10.86 look quite similar in that one is transferring a hydrogen in both cases. However, there is an important difference. In Reaction 10.85, one is transferring a hydrogen to the surface. In Section 3.4.2 we found that an adsorbed hydrogen is neutral or slightly negatively charged on most transition metal surfaces. Therefore, in Reaction 10.85 one is transferring a neutral or slightly negatively charged species. In contrast, in Reaction 10.86, one is transferring a hydrogen atom to an adsorbed oxygen or hydroxyl. Barteau and Madix [1982] found that hydroxyls and oxygen atoms are strong bases. As a result, one is in effect transferring an H^+.

Now that is important because the thermodynamics of proton (H^+) transfer are quite different from the thermodynamics of transfer of a neutral hydrogen atom. As noted earlier, if one uses the data in Table 3.5 to estimate bond energies one finds that the reaction

$$CH_3OH_{(ad)} \longrightarrow CH_2OH_{(ad)} + H_{(ad)} \tag{10.87}$$

is thermodynamically favored by 20–80 kcal/mole over the reaction

$$CH_3OH_{(ad)} \longrightarrow CH_3O_{(ad)} + H_{(ad)} \tag{10.88}$$

on all of the transition metals. However, if one extrapolates the data in Lias et al. [1988] as described by Benson [1976], one finds that if the methanol loses a proton (i.e., H^+), the gas phase reaction

$$CH_3OH \longrightarrow CH_3O^- + H^+ \tag{10.89}$$

is favored by about 25 kcal/mole over the reaction

$$CH_3OH \longrightarrow {}^-CH_2OH + H^+ \tag{10.90}$$

(Only Reaction 10.89 is observed in the gas phase.) Therefore, in the presence of adsorbed oxygen, the thermodynamics of proton transfer are such that one would expect the O–H bond to break before the C–H bond in alcohols independent of the orbital symmetry considerations.

The discussion in the last few paragraphs brings up the more general issue of how the net charges on adsorbed species affect reaction pathways on surfaces. In Chapter 3 we noted that when species adsorb on metal surfaces, the species often have a small net charge. The methanol example discussed in the previous few paragraphs shows that the thermodynamics of bond scission processes are different when one is transferring an ion than when one is transferring a neutral species. That can have an important influence on reaction selectivities.

There is another effect as well. In Section 10.2 we noted that if one looks at a reaction of the type

$$A + B \longrightarrow AB \tag{10.91}$$

the strength of the Pauli repulsions (i.e., electron–electron repulsions) between A and B have a major influence on the barriers to reactions. As one puts net charges on A (or B), one changes the Pauli repulsions, which changes the barriers to reaction. If A is slightly positively charged, the Pauli repulsions will be reduced, while if A is slightly negatively charged, the Pauli repulsions will be enhanced. A reduction in Pauli repulsions produces a reduction in the intrinsic barriers to Reaction 10.85, while an increase in the Pauli repulsions has the opposite effect. There is also a corresponding change in the orbital symmetry barriers to reaction. As a result, the intrinsic barriers to reaction change as the charges on the reactants change.

Recently we have speculated that this effect would have an important influence on hydrogenation reactions. The idea is that as one changes the surface, one also changes the charge on the adsorbed hydrogen. According to the analysis in Section 3.4.2, hydrogen is nearly neutral on platinum. However, it has a slight negative charge on nickel, and a moderate negative charge on silver and copper. The analysis in Section 10.2 shows that as one makes hydrogen more negative, one raises the barrier to hydrogenation/hydrogenolysis. Experimentally, platinum is more active than nickel, silver, or copper for hydrogenation reactions in UHV. For example, coadsorbed H_2 and C_2H_4 react to form C_2H_6 on most faces of platinum. However, the H_2 on C_2H_4 desorb without reacting on Ni(111), Cu(110), and Ag(110). Thus, copper, silver, and nickel are less active than platinum for C_2H_4 hydrogenation. Still, it is unclear whether the net charges, or some other factor, are controlling the hydrogenation rates.

Another place where net charges play an important role is in determining the relative reactivity of species on surfaces. In Section 9.6 we noted that if one compares the reactivity of a series of similar compounds, the reactivity will generally follow a linear free-energy relationship. However, a different linear free-energy relationship will be seen when the transition state is positively charged than when it is negatively charged.

Just to see how that all works out, Dai and Gellman [1993] examined the effect of fluorination on the dehydrogenation of ethoxy (CH_3CH_2O-) on Ag(110). Recall that according to the analysis in Section 3.4.1 hydrogen is slightly negatively charged on a clean Ag(110) surface, while fluorination makes the hydrogen in the ethanol more positive. The net result is that one would expect fluorination to make it harder to lose an $H^{\delta-}$ species from ethoxy. Experimentally, that is what is observed.

Just the opposite arguments can be made on an oxygen-covered surface. As noted earlier, in reactions like those in Equation 10.86, one is transferring an H^+. Therefore, anything one does to increase the positive charge on the hydrogens in a molecule should speed up the reaction. Experimentally, Barteau and Madix [1983] showed that on Ag(110) there is a one-to-one correspondence between the relative reactivity of various molecules for partial oxidation, and the gas phase acidity of the molecule (i.e., the enthalpy to produce an H^+).

Table 10.2 shows some gas phase acidities extracted from the tables of Lias et al. [1988]. The gas phase acidity of acetic acid is less negative than that of ethanol, which means that acetic acid is more acidic than ethanol. Similarly, ethanol is more acidic than methanol. Therefore, one would expect acetic acid to react more quickly with oxygen than ethanol, which in turn reacts more quickly with oxygen than methanol. Barteau and Madix [1983]

TABLE 10.2   The Gas Phase Acidities of a Number of Ligands (i.e., Heat of Reaction of L^- + $H^+ \rightarrow LH$)

Ligand	Acidity (kcal/mole)	Ligand	Acidity (kcal/mole)
CH_3O^-	-380 ± 2.5	$H_2C=C-H^-$	-406 ± 2.5
$CH_3CH_2O^-$	-377 ± 2.5	$^-CH_2-CH_2=CH$	-391 ± 2.5
CH_3COO^-	-349 ± 3	$^-CH_2-O-CH_3$	-407 ± 2
$HCOO^-$	-345 ± 3	$^-C\equiv C-H$	-380 ± 0.5

As reported by Lias et al. [1988].

show that acetic acid is more reactive than ethanol, which in turn is more reactive than methanol.

Another prediction is that the hydrogens in propylene should be much more reactive than those in ethylene. Experimentally, one can oxidize ethylene to ethylene oxide on Ag(110), i.e.,

$$
\begin{array}{c}
\mathrm{H}\quad\ \ \mathrm{H}\\
\ \diagdown\ \ \diagup\\
\ \ \mathrm{C}=\mathrm{C}\ \ +\ \mathrm{O}_{(ad)}\\
\diagup\ \ \ \ \ \diagdown\\
\mathrm{H}\quad\ \ \mathrm{H}
\end{array}
\longrightarrow
\begin{array}{c}
\mathrm{H}\quad\ \ \mathrm{H}\\
\ \diagdown\ \ \diagup\\
\ \ \mathrm{C}-\mathrm{C}\\
\diagup\diagdown\ \diagup\diagdown\\
\mathrm{H}\ \ \ \mathrm{O}\ \ \ \mathrm{H}
\end{array}
\qquad (10.92)
$$

without oxidizing the hydrogens. However, Barteau and Madix [1983] showed that propylene undergoes hydrogen elimination to produce a strongly bound species that oxidizes completely. To put these results in perspective, if one simply looks at bond energies, one would not expect the C–H bonds in ethylene and propylene to react differently. However, the data in Table 10.2 show that the methyl hydrogens in propylene are much more acidic than the hydrogens in ethylene. Therefore, based on acid strengths one would expect the hydrogens in propylene to react more easily than the hydrogens in ethylene. Experimentally, Barteau and Madix [1983] find that the hydrogens in propylene are completely oxidized, while the hydrogens in ethylene barely react. These results suggest that gas phase acidities play a key role in determining selectivities of oxidation reactions on Ag(110).

Unfortunately, at present people do not know how to extend these ideas to other systems. It is unclear, for example, how the properties of oxygen change when one moves from silver to, for example, rhodium. It is also unclear whether one transfers an H^+, a neutral H atom, or an H^- when one transfers a hydrogen to an adsorbed alkyl ligand. I expect that such questions will be answered in the next several years. However, when this book was being written, the answers to these questions were unknown.

Net charges have one other key effect, which is that they have a major influence on the intrinsic activation barriers isomerization reactions. Consider a simple reaction

$$
\begin{array}{c}
\ \ \ \ \mathrm{H}\\
\ +\ \mathrm{HCH}\\
\mathrm{R}-\mathrm{C}-\mathrm{C}-\mathrm{R'}\\
\ \ \ \ \mathrm{H}\ \ \mathrm{H}
\end{array}
\longrightarrow
\begin{array}{c}
\ \ \mathrm{H}\\
\ \mathrm{HCH}\ +\\
\mathrm{R}-\mathrm{C}-\mathrm{C}-\mathrm{R'}\\
\ \ \ \mathrm{H}\ \ \mathrm{H}
\end{array}
\qquad (10.93)
$$

During the transition state for this reaction, the carbon atom in the CH_3 group needs to form a three centered bond with the other two carbon atoms shown in Reaction 10.93. Figure 10.22 shows a diagram of the bonding in the transition state for the reaction. To put the figure in perspective, note that one usually thinks of the carbon atoms in an alkane as having only sp^3 hybrids. However, a positively charged carbon atom rehybridizes into an sp^2 hybrid, and an empty p-orbital. Therefore, at the start of Reaction 10.93 there is an empty p-orbital on one of the carbon atoms. That empty p-orbital needs to move from one carbon atom to the next during the reaction. As a result, in the transition state for Reaction 10.93 there is, in effect, a p-orbital on each carbon atom in the alkane backbone. That p-orbital interacts with the dangling bond on the CH_3 group.

If one starts with two p-orbitals, and a dangling bond, and works out all the possible combinations, one finds that there are only four linearly independent combinations. They are shown in Figure 10.22. Case I is a completely symmetric state, while the other states have increasing antibonding character. Now at the start of Reaction 10.93, there are two electrons in the carbon–carbon bond, which breaks, and no electrons in the empty p-orbital. One can put both of those electrons into the symmetric (i.e., completely bonding) state in Figure 10.22. As a result, there is little net loss of bonding as the reaction proceeds. Consequently, Reaction 10.93 can proceed with a fairly low barrier.

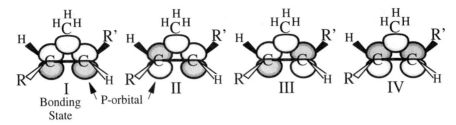

Figure 10.22   The molecular orbitals for the transition state of Reaction 10.93. The shading is as in Figure 10.21.

Now, let's do a thought experiment and put a third electron into the system, i.e., start with a neutral species rather than a positive ion. The third electron cannot go into the symmetric state, because according to the Pauli exclusion principle, each molecular orbital of the system can only hold two electrons. Therefore, the third electron needs to go into one of the other states in the system. A more detailed analysis shows that at the start of the reaction the electron is in state II in Figure 10.22. That electron repels the dangling bond. As a result, the intrinsic barrier to isomerization is much higher with a neutral species than with a positively charged one. Consequently, surfaces that stabilize positively charged species are much better able to catalyze isomerization reactions than surfaces that stabilize neutral species.

10.16.1   Electronic Effects in the Binding of Molecules on Surfaces

All of the effects discussed in the last two sections also affect the binding of molecules to surfaces. For example, in Section 10.15.2 we noted that the d-bands form a strong bond to the antibonding orbitals in a dissociating H_2 molecule. Well, now let's consider the binding of CO to a metal surface. CO has many antibonding orbitals with energies close to the Fermi energy in a typical metal. Clearly, if one can bond to those antibonding orbitals, the adsorbed complex will be more tightly held than if there is no interaction with the antibonding orbitals. The arguments in Section 10.15.2 show that d's are needed to bind to the antibonding orbitals. Therefore, one would expect CO to bind more strongly to the transition metals than to simple metals or elemental semiconductors. Experimentally, CO binds strongly to all of the group VIII, VII, and VI metals. However, CO does not bind strongly to silicon, potassium, or aluminum. At present, it is unclear how to generalize this result. Still, the CO case shows that orbital symmetry effects are also quite important to the binding of molecules to surfaces.

10.17   SUMMARY OF THE FACTORS THAT INFLUENCE THE INTRINSIC BARRIERS TO BOND SCISSION

At this point it is useful to summarize where we are. We have been discussing general principles for reactions on surfaces. All of the analysis was based on the Polayni relationship:

$$E_A = E_A^o + \gamma_p(\Delta G_r) \tag{10.94}$$

where E_A is the activation energy for a reaction, E_A^o is the intrinsic barrier to the reaction, ΔG_r is the free-energy change during the reaction, and γ_p is the transfer coefficient. In Section 10.6 we noted that years ago people analyzed data assuming that E_A^o and γ_p in Equation 10.94 were constant. In Section 10.6 we found that this model correlated with data for some simple reactions on group VIII metals. However, it did not work for non–transition metals. In Section 10.7 we found that we could not predict reaction selectivities with this model.

In Sections 10.5 to 10.15 we reviewed why E_A^o varies. Table 10.3 summarizes the results.

TABLE 10.3 Summary of the Factors that Influence the Intrinsic Barriers to Bond Scission

Factor	Key Effects	Discussed in Sections
Energetics	Bonds stretch before they break	9.2, 9.4, 10.6
Orientation/orbital symmetry of bonds that break	(a) Bonds have specific directions (b) It is impossible for an orbital to be pointing in two directions at once	9.12, 9.13, 9.14, 10.3, 10.7, 10.8, 10.9
Relative orientation of surface and adsorbate	Proximity effect: Atoms need to be close to a surface to react	10.11
	Geometric effect: Lower barriers are seen if the reactants are held in a geometry conducive to reaction	10.12, 10.13
Steric effects	The surface atoms can sometimes get in the way	6.3, 10.14
Electronic structure of the surface	The d-bands stabilize antibonding orbitals	10.15
	Smaller effect for s-band	10.14.1
Net charges	Positive charged species have lower intrinsic barriers for isomerization than negative ones	10.16, 10.21

Basically, we found that there are many factors that influence intrinsic barriers to reactions. If we could understand each of them and model them in a quantitative way, we could predict rates and mechanisms of reactions on surface. Unfortunately, at present, we cannot model the various factors accurately enough to make quantitative predictions. Rather, only qualitative predictions can be made.

10.18 THE EFFECTS OF SURFACE STRUCTURE ON REACTIONS: METALS

In the next several sections we review the qualitative ideas that have grown out of the analysis so far in this chapter. We start by looking at how surface structure affects rates of reactions. The idea that the rates of surface reactions often vary markedly with surface structure was previously discussed in Section 6.3.2. To review some of the key concepts, in Section 6.3.2 we found that some reactions show large structure sensitivity while other reactions show much smaller structure sensitivity, as outlined in Table 6.1. Reactions often show a critical ensemble for reaction as shown in Table 6.2.

It is interesting to discuss how the degree of structure sensitivity varies from one reaction to the next. Consider the data in Table 6.1. The first two reactions in Table 6.1, the oxidation of CO and the hydrogenation of ethylene, are reactions that only depend weakly on surface structure. Both reactions follow a simple sequential addition mechanism. The oxidation of CO goes via the mechanism:

$$CO \longrightarrow CO_{(ad)}$$

$$O_2 \longrightarrow 2\,O_{(ad)} \tag{10.95}$$

$$CO_{(ad)} + O_{(ad)} \longrightarrow CO_2$$

while the hydrogenation of ethylene goes via

$$C_2H_4 \longrightarrow C_2H_{4\,(ad)}$$

$$H_2 \longrightarrow 2\,H_{(ad)}$$

$$C_2H_{4\,(ad)} + H_{(ad)} \longrightarrow C_2H_{5\,(ad)} \qquad (10.96)$$

$$C_2H_{4\,(ad)} + H_{(ad)} \longrightarrow C_2H_6$$

In both cases, the reactants adsorb, and then atoms are sequentially added to a surface species to eventually yield the products. According to the analysis in Sections 10.2 and 10.15.6 all of these steps are symmetry allowed on all of the group VIII, VIIa, and VIa metals, which means that the intrinsic barriers to reaction are small. Experimentally, all of the reactions that have been examined to date that follow a sequential hydrogenation or oxidation pathway show small structure sensitivity. It is interesting, therefore, that those are also reactions where all of the steps in a reaction pathway have no orbital symmetry constraints, and therefore a small intrinsic barrier to reaction.

The hydrogenolysis of ethane is different. As noted previously, the hydrogenolysis of ethane is thought to follow the mechanism in Equation 10.67. The details of this mechanism are still somewhat in dispute, but the key point is that the carbon–carbon bond needs to break during the hydrogenolysis process. As noted in Section 10.2, there is an orbital symmetry constraint that raises the intrinsic barrier to carbon–carbon bond scission. The proximity effect also contributes to the barrier. Experimentally, one finds that reactions that involve the scission of carbon–carbon or carbon–oxygen bonds often show significant structure sensitivity. It is interesting that those reactions also have moderate orbital symmetry constraints, and therefore moderate intrinsic barriers to reaction.

The last cases in Table 6.1 are cases where a $C{\equiv}O$, $N{\equiv}O$, or $N{\equiv}N$ bond have to break. In Section 10.15.6 we noted that these reactions have the largest orbital symmetry constraints to reaction and therefore the largest intrinsic barriers. Experimentally, such reactions also show the largest structure sensitivity.

These observations lead to the key point in this section: there seems to be a correlation between the intrinsic/orbital symmetry barriers to reaction and the degree of structure sensitivity of a reaction. Reactions that have large intrinsic barriers in the gas phase because of large orbital symmetry constraints also show large structure sensitivity. In contrast, reactions without orbital symmetry constraints generally show little structure sensitivity. In the material that follows we note that there are reasons to suppose that reactions that have large orbital symmetry constraints will be structure sensitive *under the proper conditions*. In fact, in all of the cases so far where the structure sensitivity has been found to be particularly large, orbital symmetry effects seem to control the reaction. The proximity effect can also lead to moderate structure sensitivity. However, I do have to say up front that one needs to be cautious with these results, since the ideas have not been extensively discussed in the literature.

There was an important qualifier in the discussion in the previous paragraph that I address next. Reactions that have large orbital symmetry constraints will be structure sensitive *under the proper conditions*. They will *not* always be structure sensitive. The reason for the qualifier is that, as we noted previously orbital symmetry constraints raise the intrinsic barriers to reaction. However, they do not usually completely prevent a reaction from occurring. If a reaction is exothermic enough, it will be possible to overcome the intrinsic barriers. In that case, one should not see significant structure sensitivity. For example, Masel [1986] found that NO dissociation on platinum is structure sensitive. However, Masel et al. [1979] found that NO dissociation on tungsten shows much less structure sensitivity; the dissociation process is very exothermic on tungsten so the reaction occurs at 100 K. Experimentally, one generally observes structure sensitivity only on metals that are just barely reactive enough to overcome the intrinsic barriers to reaction. Generally, those are the metals at the boundaries between dissociation and nondissociation in Figure 3.7. For example, nitrogen dissociation is structure sensitive on iron and tungsten, but NO dissociation is not. NO dissociation is structure sensitive on platinum and palladium, but not on tungsten. Evidently, experimentally the surface structure makes a big difference when the thermodynamics of bond scission processes are just able to overcome the intrinsic barriers to reaction, but not in other cases.

At first, the discussion in the last paragraph would seem to make structure sensitivity almost a trivial effect. Note, however, that according to the Principle of Sabatier, the metals that are just barely reactive enough to overcome the intrinsic barriers to reaction are also usually the best catalysts. As a result, structure sensitivity is usually quite important.

10.18.1   Models for the Overall Trends in the Structural Sensitivity of Elementary Bond Scission Reactions on Metals

In the next several sections we describe the existing models of surface structural sensitivity. We start off by trying to understand the large trends and then looking in more detail and asking why a specific reaction could be structure sensitive.

First note that in Section 6.3.3 we found that there are three basic reasons that reactions on metals are structure sensitive: the bonding geometry of intermediates varies with surface structure, the binding energy of reactive intermediates varies with surface structure, and the orbital symmetry constraints and hence intrinsic barriers to reaction vary with surface structure. In the next several sections we discuss each of these effects and describe how they influence rates of surface reactions.

To start recall that when people discuss structure sensitivity of surface reactions, they normally discuss the structure sensitivity of the overall rate of reaction. In Section 6.3.2 we found that overall rates vary with surface structure. However, in order to understand why overall rates of reactions are structure sensitive, one first has to understand why rates of elementary surface reactions are structure sensitive. Therefore, in this section we discuss why rates of elementary surface reactions vary with surface structure.

Let's consider an elementary bond scission reaction like that discussed previously:

$$AB_{(ad)} \longrightarrow A_{(ad)} + B_{(ad)} \tag{10.97}$$

When Reaction 10.97 occurs, a bond in the adsorbate breaks, and a new adsorbate–surface bond forms. According to the Polayni relationship, E_a, the activation barrier for Reaction 10.97, is given by

$$E_a = E_a^o + \gamma_P \Delta H_r \tag{10.98}$$

where E_a^o is the intrinsic barrier to reaction, ΔH_r is the heat of reaction, and γ_P is the transfer coefficient. According to Reaction 10.98 one can modify the rate of Reaction 10.97 in three different ways. One can change the intrinsic barriers to reaction, one can change the heat of reaction, or one can change the transfer coefficient. Experimentally, the heat of reaction and the intrinsic barrier vary significantly with surface structure, while the transfer coefficient varies mainly with surface composition. As a result, the various forms in Equation 10.98 change with surface structure. Consequently, the reaction rate often varies with surface structure.

Still, if one starts with a reaction with a small intrinsic barrier, variations in the intrinsic barrier will not be that important because the intrinsic barriers are already small. One can get variations due to changes in the heat of reaction, but we show below that the net effect on the overall rate is usually small. As a result, one does not usually observe large structure sensitivity in reactions that have small intrinsic barriers, i.e., those with no orbital symmetry constraints to reaction. Structure sensitivity comes in reactions with large intrinsic barriers. Those are also reactions with large orbital symmetry constraints.

Now let's first ask why very exothermic surface reactions are generally not structure sensitive. Recall, from the discussion of the Marcus equation in Section 9.5.2 that as one makes a reaction more exothermic, one pushes the transition state for the reaction toward the reactants. As the transition state moves toward the reactants, the intrinsic barriers to reaction decrease because one does not have to deform the reactant orbitals substantially to get to the transition state. The transfer coefficient also decreases (see Section 9.5.3). The net effect is to attenuate all of the terms in Equation 10.98. Consequently, reactions with early transition states will show at most modest structure sensitivity. In Section 9.5.2 we found that highly exothermic reactions normally have early transition states. As a result, highly exothermic elementary surface reactions usually show only modest structure sensitivity.

Next let's consider what happens with a less exothermic elementary surface reaction. In the less exothermic case, the transition state moves away from the reactants. Variations in the intrinsic barriers to reaction become important as do variations in the heat of reaction with surface structure. As a result, one usually only observes structure sensitivity when the reaction is not too exothermic.

A more detailed analysis shows that reactions with middle transition states should show the largest structure sensitivity. If the reaction is very endothermic, the transition state will move toward the products. The situation is just like the very endothermic reaction in reverse, and so by microscopic reversibility, the structure sensitivity of the intrinsic barrier to reaction will be small. Experimentally, structure sensitivity is only seen in reactions with a middle transition state. Those are reactions where the heat of reaction and the intrinsic barrier to reaction are about the same size.

The net effect of all the analysis in this section is that one would expect rates to vary with surface structure only when there are high intrinsic barriers (i.e., orbital symmetry constraints) and then only when the reaction is run on a surface that is just barely reactive enough to overcome the intrinsic barriers to reaction. We will see below that this expectation agrees with all of the data reported to date.

10.18.2 Detailed Models of the Structure Sensitivity of Reactions on Metal Surfaces

Next, we will look in more detail at the structure sensitivity of surface reactions. In the discussion in the last section we noted that heats of reaction and intrinsic barriers to reaction vary with surface structure. However, we did not explain why the variations occur. In the next several sections, we discuss the variations and look at why they occur.

Just to put the work in perspective, note that according to the principle of maximum bonding variations in rate with crystallographic orientation are caused by variations in the stability of the transition state for reaction with varying crystal face. According to Equation 3.85, there are three ways that the stability of the transition state could change: there could be a change in the interactions with the s-bands, there could be a change in the interactions with the d-bands, or there could be a change in the steric effects. One can write this result in equation form:

$$\Delta E_A = (\Delta E_A)_{s\text{-band}} + (\Delta E_A)_{steric} + (\Delta E_A)_{d\text{-band}} \tag{10.99}$$

where ΔE_A is the change in the activation barrier of reaction with changing surface structure, $(\Delta E_A)_{s\text{-band}}$ and $(\Delta E_a)_{d\text{-band}}$ are the parts of that change associated with changes in the interactions with the s-band, d-bands, while $(\Delta E_A)_{steric}$ is the change in the steric interactions.

Changes in the Binding Energy of Reactive Intermediates. In the next several sections we will try to understand how each of the terms in Equation 10.99 vary with surface structure. We start by considering the variations in the interactions with the s-band. In Chapter 3 we noted that variations in the interactions with the s-band cause variations in the thermodynamics of surface reactions. Therefore, if the interactions with the s-band vary with surface structure, the thermodynamics of reaction should change, too. In the literature, many investigators have noted that the binding energy of reactive intermediates would be expected to vary with surface structure, and those variations can cause rates to vary with surface structure. The variations in the binding energies of reactive intermediates can be understood from the ideas in Chapters 2 and 3. Recall that in Section 2.6.2 we found that when one has a stepped surface, some of the surface atoms have an especially low electron density. Those atoms are especially coordinatively unsaturated, so according to the analysis in Section 3.10, those atoms should also bind adsorbates strongly. The net result is that adsorbate surface-bond energies are often different on stepped surfaces than they are on close packed planes. That difference can cause ΔH_r in Equation 10.98 to change with surface structure, which in turn changes the rate.

In actual practice, however, this is usually only a modest effect. Recall from Section 2.6.2 that stepped surfaces normally relax to reduce the variations in electron density on individual surface atoms. As a result, all surface atoms are usually roughly equivalent. Experimentally, heats of adsorption usually do not vary all that strongly with surface structure, i.e., the heat of adsorption a given species varies more with surface composition than with surface structure. There is one case, oxygen adsorption on platinum, where the heat of adsorption varies by 22 kcal/mole with surface structure. However, the usual case is that the heat of adsorption varies by 5 kcal/mole or less with crystal face. According to Equation 10.98, a 5-kcal/mole variation in ΔH_r will produce a 2.5-kcal/mole variation in the activation energy for reaction, assuming γ_P is 0.5. At 500 K the rate will vary by a factor of 11. Such an effect is important catalytically. However, it is a modest effect compared to many of the examples in Table 6.1, p. 460.

Another issue is that according to the analysis in Section 3.4, surfaces that bind one species strongly will also bind many other species strongly. If one considers a reaction like that in Reaction 10.97, one finds that a surface that binds the products of the reaction strongly also binds the reactants strongly. The net result is that the heat of reaction for an elementary surface reaction varies less strongly with surface structure than the heat of adsorption of the reactants or products. For example, Figure 10.23 shows some data for the reaction

$$NO_{(ad)} \longrightarrow N_{(ad)} + O_{(ad)} \tag{10.100}$$

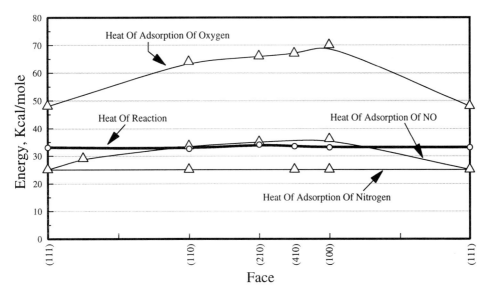

Figure 10.23   The heat of adsorption of nitrogen, oxygen, and NO, and ΔH for $NO_{ad} \rightarrow N_{ad} + O_{ad}$ on various faces of platinum. (After Masel [1986].)

Note that the heat of adsorption of the reactants and products both vary significantly with surface structure. However, surfaces that bind the reactants in Equation 10.100 strongly also bind the products strongly. The net result is that the heat of reaction for Reaction 10.100 varies by less than 2 kcal/mole with surface structure. According to Equation 10.30 a 2 kcal/mole variation in ΔH_r will produce a factor of 3 variation in rate with crystal face. This is certainly a modest effect, especially considering that Masel's [1986] experiments show a 10^{21} variation in rate with crystal face.

The net result from all of this analysis is that thermodynamic effects only cause modest variations in rate with crystal face: e.g., less than an order of magnitude or two variation in rate with crystal face. When a reaction is not very structure sensitive, one can sometimes detect variations in rate due to variations in thermodynamics. However, in cases where the rate varies by many orders of magnitude with crystal face, the thermodynamic effects are too small to account for the main variations. Therefore, it seems that the variations must be caused by other factors. Admittedly, however, this conclusion has not been extensively discussed in the literature, so one needs to treat it with caution.

10.19 GEOMETRIC EFFECTS

In the literature, it is not very clear what those other factors are. However, according to Equation 10.99 there are only two other possibilities: geometric/steric effects and variations in the interactions with the d-bands. The geometric effects were previously discussed in Sections 10.12 and 10.13. When a reaction occurs, bonds form between the adsorbate and the surface. No reaction can occur unless the adsorbate can get close to the surface. As a result, the proximity effect can modify the intrinsic barriers to reaction.

There are a few examples where this effect may be quite important. For example, Wang and Masel [1991] compared methoxy decomposition on Pt(111) and (2x1) Pt(110). On Pt(111) the methoxy binds with the hydrogens close to the surface. However, on (2x1) Pt(110), Wang and Masel [1991] proposed that at low coverages the methoxy binds to the edges of the steps (see Figure 10.24). Notice that on (2x1) Pt(110) the hydrogens are not in close proximity to the surface. Therefore, one would expect methoxy decomposition to be inhibited on (2x1) Pt(110). Experimentally, methoxy decomposes at 160 K on Pt(111), while at low coverage methoxy does not decompose until 230 K on (2x1) Pt(110). Therefore, the proximity effect seems to play an important role in methoxy decomposition on platinum.

There are other examples that are discussed in terms of a proximity effect. For example, one possible mechanism of ethane hydrogenolysis ($C_2H_6 + H_2 \rightarrow 2CH_4$) is

$$CH_3-CH_3 \longrightarrow {/}CH_2-CH_3\,_{(ad)} + H_{(ad)} \qquad (1)$$

$$/CH_2CH_3\,_{(ad)} \longrightarrow {/}CH_2-CH_2\,_{(ad)}{\backslash} + H_{(ad)} \qquad (2)$$

$$/CH_2-CH_2\,_{(ad)}{\backslash} \longrightarrow 2\,{/}CH_2\,_{(ad)}{\backslash} \qquad (3) \qquad\qquad (10.101)$$

$$/CH_2{\backslash} + 2H_{(ad)} \longrightarrow CH_4 \qquad (4)$$

Figure 10.25 shows a possible geometry for Reaction 2 in Equation 10.101. Notice, that on a flat surface, the ethyl group will need to bend toward the surface before any reaction can occur while the geometry is already conducive to reaction on a stepped surface. As a result, one would expect Reaction 2 in Equation 10.101 to occur much more rapidly on a stepped surface than on a close packed face. Experimentally, the results are still unclear.

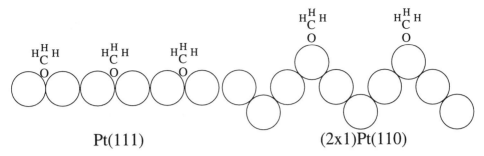

Figure 10.24 The geometry of methoxy on Pt(111) and at low coverages on (2x1) Pt(110) as proposed by Wang and Masel [1991].

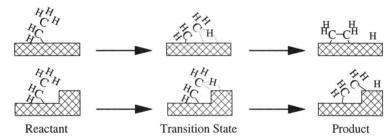

Reactant	Transition State	Product

Figure 10.25 A possible geometry for Reaction 2 in Equation 10.101 on a flat and stepped surface.

However, the data reviewed in Section 6.3.3 indicate that qualitatively hydrogenolysis rates are higher on stepped surfaces than on close packed planes, in agreement with expectations. It is unclear whether the variations are caused by a proximity effect, however.

In a more general way proximity effects have been discussed since the work of Balandin and Robinstein [1934]. So far, however, there have only been a few examples where the proximity effect has been demonstrated to change the rates with crystal face. In all cases, it has been a modest effect. A factor of 10 or less. Therefore, it seems that the proximity effect only causes modest variations in rate with crystal face.

10.20 VARIATIONS IN THE INTERACTIONS WITH THE d-BANDS

So far we have been talking about thermodynamic and proximity effects, and found that they can cause modest variations in rate with crystal face. However, there is another effect, a variation in the interaction with the d-bands, which according to Masel [1986], can cause much larger variations in rate with crystal face.

To put Masel's work in perspective, note that the d-bands have a tremendous effect on the intrinsic barriers for reaction. For example, in Section 10.15.5 we noted that the interaction with the d-bands lowers the intrinsic barrier for H_2 dissociation from over 90 kcal/mole on Al(111) or K(110) to less than 13 kcal/mole on Pt(110). This is a gigantic effect. Therefore, even a modest variation in the strength of the interaction with the d-bands can have a large effect on the intrinsic barriers to reaction. Consequently, modest variations in the interactions with the d-bands can cause major changes in rates.

Now this effect is particularly important on the scission of a triple bond. For example, according to an analysis like that in Section 9.13.2, in the gas phase a four center reaction where a nitrogen–nitrogen bond breaks should have an intrinsic barrier of about 210 kcal/

mole (i.e., the $N\equiv N$ bond energy). Yet, N_2 dissociatively adsorbs on W(100) and Fe(111). N_2 dissociation is less than 60 kcal/mole exothermic on both surfaces, which implies that the intrinsic barrier to N_2 bond scission must be less than 60 kcal/mole; the intrinsic barrier to N_2 bond scission has been reduced by 150 kcal/mole on W(100) and Fe(111). An analysis like that in Section 10.15.2 indicates that the main portion of that reduction must come from an interaction with the d-bands. Further, experimentally, Hayward and Trapnell [1964] note that there is no evidence for N_2 dissociation on aluminum or potassium films even though N–N-bond scission is thermodynamically favorable on these films. As a result, most of the 150-kcal reduction in the intrinsic barrier to $N\equiv N$ bond scission must be associated with an interaction with the d-bands. Whenever you have a 150-kcal/mole effect, a small percentage variation in that effect with, for example, surface structure will have a major influence on rates of reactions.

Now, as we will see below, there has been some question in the literature whether the interactions with the d-bands do vary with surface structure. However, experimentally, there are a few examples where the variations in rate with surface geometry are known to be especially large: NO decomposition/reduction on platinum, N_2 dissociation on tungsten, and NH_3 production on iron. In all three cases, the measured variations in rate are much larger than one can account for via thermodynamic or steric effects, i.e., the first two terms in Equation 10.99. Note, however, that there are only three terms in Equation 10.99. Therefore, if variations in the first two terms in the equation are insufficient to explain the observed variations in rate, the variations must be caused by changes in the third term in Equation 10.99, i.e., changes in the interactions with the d-bands. Therefore, it seems that variations in the interaction with the d-bands must, under some circumstances, be responsible for large variations in rates.

To see how that works out, let's consider a simple case that has been studied for 50 years: N_2 dissociation on tungsten. Figure 3.6 shows a summary of the older data on this reaction. Delchar and Ehrlich [1975] found that N_2 dissociation is unactivated on W(100), while Lee et al. [1984] measure an activation barrier of 10 kcal/mole on W(110).

Now it is useful to consider why the reaction rate could be this structure sensitive. Well, the heat of adsorption of N_2 is about 60 kcal/mole on W(100). Therefore, even if there were a 20% variation in the heat of adsorption with crystal face, that would only, at most, produce a 6-kcal/mole variation in the activation energy of adsorption, assuming $\gamma_P = 0.5$. That is too small to explain the observed variations in N_2 dissociation rate with crystal face.

In a similar way there are no steric limitations to N_2 dissociation on W(110) or Fe(110) (i.e., the steric limitations are the same on W(110), W(100), Fe(110), and Fe(111)). Therefore, the proximity effect cannot explain the variations in the rate of N_2 dissociation with crystallographic orientation observed on tungsten.

The only term left in Equation 10.99 is the interaction with the d-bands. As noted, the d-bands lower the intrinsic barriers to N_2 dissociation on tungsten by perhaps 150 kcal/mole. Therefore, a 7% variation in the interactions with the d-bands can cause a large enough variation in rate to explain the data. The variations in the other two terms in Equation 10.99 are not big enough to explain the observed variations in rate. Therefore, the variations in the interactions with the d-bands must be causing the variations in rate.

10.20.1 Models for the Influence of the d-Bands on the Structure Sensitivity of Surface Reactions

Unfortunately, at present we do not understand, in detail, how the d's interact with molecules on surfaces, and so there are only a few predictions about how the d's should influence the variations in rate with crystal face.

A simple approximation called the group orbital approximation has been proposed by Lang, Nørskov, and Holloway [1984]. In this approximation, there are five d-orbitals on each atom (d_{xy}, d_{xz}, d_{yz}, d_{z^2}, $d_{x^2-y^2}$). Each orbital is presumed to have an average energy equal to the average energy within the bulk of the metal, but the orbitals were presumed to be otherwise uncoupled to the bulk. If the d's are uncoupled to the bulk, the d's do not have any preferred directions. Consequently, the interactions with the d's would be independent of the crystallographic orientation of the exposed face. In such a circumstance, the d's will not produce any variations in rate with crystallographic direction.

An alternative model was proposed by Masel [1986]. Masel noted that in the bulk of a face-centered cubic (FCC) or body-centered cubic (bcc) material, the d's split into a group of bands of t_{2g} symmetry and another group of bands of e_g symmetry. If we take a coordinate system chosen so that the x, y, and z directions lie along the [100], [010], and [001] directions in the bulk, then at Γ in the Brillouin zone the t_{2g} states will be hybrids of the d_{xy}, d_{xz}, and d_{yz} orbitals, while the e_g states will be hybrids of d_{z^2} and $d_{x^2-y^2}$ orbitals. In the bulk, the t_{2g} and e_g orbitals have fixed orientations. The t_{2g}'s have lobes emerging from each atom pointing in the [100], [010], and [001] directions, while the e_g's have lobes pointing in the [110], [101], and [011] directions. Now when one cuts a surface one exposes those lobes. The lobes could rehybridize and the surface could reconstruct. However, Masel made the assumption that the lobes do not rehybridize and the surfaces do not reconstruct for his analysis.

Now consider cutting a Pt(111) surface. Figure 10.26 shows a projection of the t_{2g} bands emerging from the Pt(111) surface, calculated assuming that the orientation of the orbitals is simply a projection of the orbitals in the bulk. To orient the reader the positions of the atomic

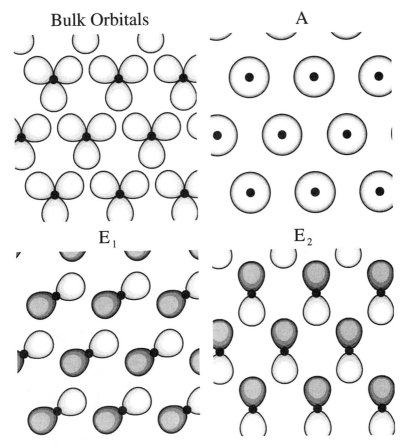

Figure 10.26 A projection of the t_{2g} orbitals in Pt(111). In the figure positive lobes are light, negative lobes are dark. The dark circles are the atomic positions.

cores in the (111) surface are indicated by the black dots in the figure. There is a hexagonal arrangement of black dots corresponding to the horizontal arrangement of atoms in the Pt(111) surface. If one would look at a (111) plane in the bulk of platinum, there will be three t_{2g} orbitals pointing up out of the surface at a 35° angle. The t_{2g} orbitals are shown in the upper left panel of the figure. According to quantum theory, when one cuts a Pt(111) surface, the t_{2g} orbitals will mix. The net result according to group theory is an A state and two E_g states. The A state is the sum of the three orbitals. It is positive. In contrast, the E_g states are sums and differences of orbitals. Each of the E states has one positive and one negative lobe.

Now consider how the A and E_g states interact with the orbitals in H_2 during H_2 dissociation on Pt(111). As noted in Section 10.15.1 when the H_2 bond breaks, one needs to put electrons into antibonding orbitals. Those electrons can come via collisions, but the intrinsic barriers for such a process are more than 100 kcal/mole. In Section 10.15.1 we found that the barriers will be substantially less if there are d-bands in the surface properly configured to interact with the σ^* orbitals in the H_2. Well, it is easy to show that the E orbitals in Figure 10.26 are properly configured to interact with the σ^* orbitals. Recall, from Section 10.15.3, that one needs an antisymmetric orbital to interact with the σ^* state. Figure 10.26 indicates that the E_g states in the Pt(111) surface are antisymmetric. In the next paragraph we show that the E_g's can interact strongly with the σ^* orbitals in H_2. Consequently, one would expect a low intrinsic barrier for H_2 bond scission on Pt(111) as is observed.

The easiest way to arrive at that result is to use the analysis from Section 10.15.3. According to the analysis in Section 10.15.3, one can lower the intrinsic barriers to bond scission if positive lobes in the adsorbate overlap positive lobes in the surface and negative lobes in the adsorbate overlap negative lobes in the surface. Well, the left part of Figure 10.27 shows a diagram of the antibonding orbital in H_2. The antibonding orbital consists of two lobes, a shaded negative lobe and a lightly colored positive lobe. Now, let's put the orbital on the Pt(111) surface and try to maximize the interaction. It is easy to show that there will be little or no net overlap with the A state since there are no negative lobes in the A state to interact the negative lobes in the antibonding orbital in H_2. However, there is a net interaction with the E states. The right side of Figure 10.27 shows two favorable geometries. In the upper geometry the H_2 sits almost on top of a platinum atom. The positive lobe in the antibonding state of H_2 overlaps a positive lobe in the surface while a negative lobe in the antibonding state of H_2 overlaps a negative lobe in the surface. The net result is that in the upper geometry

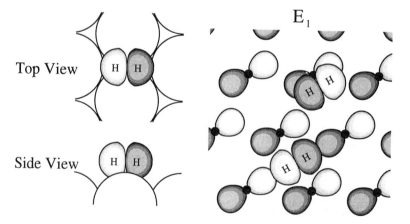

Figure 10.27 A diagram of (left) the σ^* orbital of H_2 when the H_2 adsorbs and (right) the overlap of the orbital with the E_1 orbital in the Pt(111) surface.

Figure 10.28 The mechanism of NO dissociation on platinum.

in the figure there will be a strong interaction between the surface and the antibonding orbitals in the H_2. That interaction can facilitate H–H bond scission.

It is not obvious, but actually, the most favorable situation is the case on the bottom in the figure. In the bottom case, the positive lobes in the H_2 overlap the positive lobes in the surface, but initially, the overlap does not look that favorable. Note, however, that in the bottom as the hydrogen-hydrogen band stretches, the overlap with the antibonding orbital increases. That leads to very easy H–H bond scission. The net effect is that one would expect H_2 dissociation to be essentially unactivated on Pt(111) as is observed.

One can do a similar analysis for NO bond scission. When the NO bond breaks, electrons need to be transferred into the antibonding orbitals of the NO. One can, of course, get the transfer to occur by a collision process like that which occurs in the gas phase. However, such a process would be expected to have a very high intrinsic barrier (i.e., more than 150 kcal/mole). One can get a lower barrier if the surface can bond to the antibonding orbitals in the NO, and thereby stabilize the transition state relative to the reactants.

Now let's consider a dissociation process, where an NO starts out bound perpendicularly to the surface, and then slowly bends toward the surface as illustrated in Figure 10.28.

Figure 10.29 shows a diagram of the antibonding orbitals in the NO during the reaction. There are two π^* states and a σ^* state. Let's focus on the $\pi_\perp^*$ state. The $\pi_\perp^*$ state has B_1 symmetry when viewed from above. It looks almost fourfold symmetric. However, the wavefunction changes from positive to negative, back to positive, and back to negative as one moves around the NO.

Next we need to consider whether the presence of the d-electrons in the Pt(111) surface

Top View

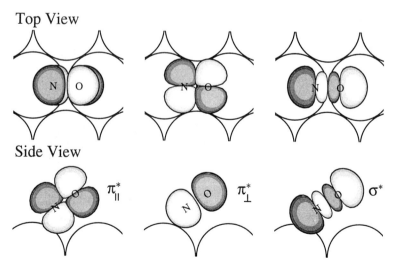

Side View

Figure 10.29 A top view of the antibonding orbitals in a dissociating NO. The shading is as in Figure 10.26.

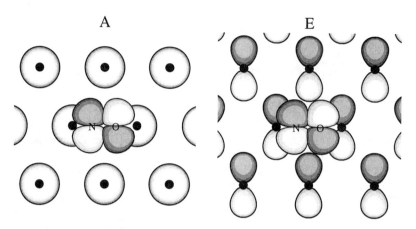

Figure 10.30   The interaction of the $\pi_{\perp}^{*}$ orbital with the A and E states in Pt(111).

would stabilize the antibonding orbitals in the transition state for NO decomposition. Figure 10.30 shows a diagram of the $\pi_{\perp}^{*}$ orbital in the NO during the reaction. The left part of the figure shows the overlap of the $\pi_{\perp}^{*}$ orbital with the A state in the Pt(111) surface. It is easy to demonstrate that the A state does not interact with the $\pi_{\perp}^{*}$ state in the NO. The A state is completely positive, while there are positive and negative lobes on the $\pi_{\perp}^{*}$ state. Therefore, one would not expect much interaction.

In fact, one can demonstrate that the interaction is exactly zero in the configuration on the left in Figure 10.30. Let's start with the bottom lobe on the nitrogen atom in the NO. The bottom lobe is positive, while the A state is positive too. Therefore, the bottom orbital on the nitrogen will give a positive (attractive) contribution to the transition moment. However, the opposite happens on the top lobe of the nitrogen. The top orbital in the nitrogen is negative, but the A state is still positive. Therefore, the interaction with the top lobe on the nitrogen will be of the opposite sign as the contribution from the bottom lobe. The net effect is that the contribution from the bottom lobe on the nitrogen will exactly cancel the contribution from the top lobe. As a result, the transition moment is zero. Therefore, the A state in Pt(111) would not stabilize the transition state for NO decomposition.

The situation is worse with the E orbitals. At the start of the reaction, only the nitrogen end of the NO is close to the surface. In this configuration, the positive lobe in the $\pi_{\perp}^{*}$ orbital in the NO overlaps a positive lobe in the surface, while a negative lobe in the NO overlaps a negative lobe in the surface. Consequently, at the start of the reaction, the $\pi_{\perp}^{*}$ orbital in the NO does interact strongly with the E orbital in the surface. However, as the NO bends over, the oxygen end of the molecule starts to interact with the E orbital. Note that the orbitals on the oxygen end of the molecule are of opposite sign to the orbitals on the nitrogen end of the molecule. As a result, the interaction with the oxygen end of the NO partially cancels the interaction of the nitrogen end of the molecule. Therefore, as the NO bends over, the net interaction with the E orbitals is reduced. Consequently, the interaction with the E orbitals destabilizes the transition state for NO decomposition (relative to the reactants). The net effect is that one would not expect the d-bands on Pt(111) to substantially lower the intrinsic barriers to NO decomposition. Experimentally, Lang and Masel found that the activation barrier for NO decomposition on Pt(111) is about 52 kcal/mole, i.e., similar to the barrier in the gas phase, even though the reaction $2NO \rightarrow N_2 + O_2$ is 80-kcal/mole exothermic.

The situation is entirely different on a (1x1) Pt(100) surface. When a Pt(100) surface is cut, the t_{2g} states split into a B_1 and two E_g states. A B_1 state has the same symmetry as the $\pi_{\perp}^{*}$ state of NO (see Figure 10.31). The B_1 state can interact with the $\pi_{\perp}^{*}$ state and thereby stabilize the transition state for NO band scission. The stabilization lowers the barrier for

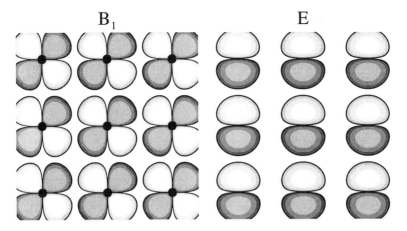

Figure 10.31 The B_1 and E_g states on (1x1) Pt(100). The shading in the figure is as in Figure 10.26.

reaction. Experimentally, the barrier goes down from 52 kcal/mole on Pt(111) to 20 kcal/mole on Pt(100).

The reason that there is still a barrier is that one needs to stabilize all three orbitals in Figure 10.29 to get the reaction to occur. According to Masel's [1986] analysis on (1x1) Pt(100), one can stabilize any two of the orbitals but not all three. Therefore, there is still a barrier to NO decomposition on (1x1) Pt(100).

Note, however, that if one adds steps to the (1x1) Pt(100) surface, one can get some additional orbitals to bonding. If those orbitals are oriented correctly, the intrinsic barriers to reaction should almost disappear. Banholzer et al. [1983] found that Pt(410) does have d-lobes to interact with all three orbitals in Figure 10.29. Therefore, one would expect Pt(410) to be very active for NO dissociation. Experimentally, Banholzer et al. found that NO dissociation is essentially instantaneous on Pt(410) at 100 K. By comparison no NO conversion is seen on Pt(111) below 1000 K! Park [1983] also measured the activation barrier for the NO/H_2 reaction on Pt(410) at 100–150 K. She found the activation barrier of 6 kcal/mole. That compares to Masel and Lang's measurement of 52 kcal/mole on Pt(111). Therefore, there is a correlation between the analysis on the previous page and the data.

Masel [1986] quantified these effects in greater detail. Masel calculated a quantity that he called the orbital availability where the orbital availability is defined by

$$\text{orbital availability} = \text{const} \sum_{\text{sites}} (\langle \pi_\perp^* | \phi_M \rangle + \langle \pi_\parallel^* | \phi_M \rangle + \langle \sigma^* | \phi_M \rangle) \quad (10.102)$$

where $\langle \pi_\perp^* | \phi_M \rangle$ is the overlap between the π^* orbital and the metal d-orbitals as the Γ point in the Brillouin zone. Figure 6.14 shows a plot of the orbital availability for several of the faces of the stereographic triangle. The orbital availability varies from near zero on Pt(111) to 2, in dimensionless units on Pt(410), and varies smoothly in between. Several measures of the reactivity are also included in Figure 6.14. Notice that the variations in the orbital availability generally track the experimental data, although there are some points that are not exactly on the curve. Therefore, it seems that one can use Masel's model to understand the structure sensitivity of NO decomposition on platinum.

The calculations in Figure 6.14 were done assuming that in the transition state the NO lies parallel to the surface. If instead the NO is presumed to lie at some angle relative to the surface, the curve will be compressed. The orbital availability will be near zero on Pt(111), and slightly reduced on Pt(410). However, the qualitative features of the curves will be the same.

There is one unexpected effect in Figure 6.14. Notice that the orbital availability decreases in moving from Pt(410) to Pt(210). Pt(410) and Pt(210) are very similar surfaces. The only difference is that the step density is twice as high on Pt(210) as on Pt(410). When Masel's work was done, most investigators had thought that a doubling of step density would cause the rate to increase. However, the experiments and calculations show precisely the opposite. Masel [1986] attributed this to a pure proximity effect. When the step density gets high, the NO no longer fits on the surface in a configuration where the NO can overlap all of the key orbitals in the surface. Consequently, the rate decreases at high step density.

Implications for Other Reactions. We can try to extend the ideas in the last section to other reactions. The first implication is that the variations in the interactions with the d-electrons can cause significant variations in the rates of reactions on surfaces. In general, one would expect the effects to be most important for the scission of triple bonds, and less important for the scission of single bonds. Note that the E orbitals in Pt(111) have the right symmetry to stabilize the transition state for the scission of a sigma bond. In a similar way, all of the faces of groups VIII, VIIa, and VIa metals have orbitals of the right symmetry to stabilize the transition state for the scission of sigma bonds. If one is transferring a single atom (e.g., a hydrogen), there will be only small intrinsic barriers and little structure sensitivity.

The overlaps are less favorable when one is transferring a molecular ligand (e.g., a methyl group), and it works out that the overlaps vary with crystal face. As a result, one expects some structure sensitivity when one is transferring a molecular ligand. However, less structure sensitivity is expected than when one is breaking a triple bond.

The largest structure sensitivity is expected for the scission of triple bonds. The analysis in Section 10.20.1 shows that a special orbital configuration is needed to break a triple bond. Experimentally, the scission of triple bonds shows the largest structure sensitivities observed to date.

The biggest limitation with the calculations at present is that one cannot predict the magnitude of the effect a priori. As noted in Section 10.20.1, as one decreases the tilt angle of the NO in the transition state, one also decreases the variations in the orbital availability with crystal face. That means that as one moves the transition state toward the reactants, one damps the variations in rate with crystal face. In Sections 9.6.2 and 10.18.1 we noted that experimentally, as one makes a reaction more exothermic, one tends to move the transition state toward the reactants, which in turn damps the structure sensitivity. Experimentally, the largest structure sensitivity occurs on metals that are just barely reactive enough to overcome the intrinsic barriers to reaction, which agrees with the analysis in this section.

Another limitation is that experimentally adsorbates can reconstruct or rehybridize surfaces. That rehybridization changes everything, but it is ignored in Masel's model. Finally, Masel's model concentrates on the Γ point in the Brillioun zone. A more detailed analysis would be to integrate the results over the entire Brillioun zone using Equation 10.83.

Unfortunately, when this book was being written there was really nothing better in the literature. Many people have used cluster calculations to predict reactivity. However, the clusters used in the calculations do not have octahedral symmetry. There are no t_{2g} or e_g bands, and so the cluster calculations do not treat the d-bands properly. So far there has been little correlation between cluster calculations and data on structure sensitivity. Slab calculations have the right symmetry but one needs a mixed-basis set, i.e., plane waves plus local d-orbitals to handle the s- and d-bands with a reasonable amount of computer time. Mixed-basis-set calculations of surface reactions were just starting to appear when this book was being written. Another limitation is that most of the available slab calculations have used the local spin density (LSD) approximation. In LSD, the interactions between the s-band and the $\sigma*$ state in Figure 10.20 do not cancel. As a result, LSD gives incorrect predictions. Hamer, Jacobson, and Nørskov [1993] show that one can get the correct results by including something called "gradient corrections." However, gradient/slab calculations are just starting to appear. As a result, Masel's calculations, with all of their inherent limitations, are the best available. In fact, they were the only calculations that make useful predictions. Admittedly, however, they only make useful predictions under very favorable circumstances.

10.21 PUTTING IT ALL TOGETHER: METALS

At this point we again change gears. In the last several sections we have been discussing how rates of reactions vary with surface structure. However, in the next several sections we discuss how rates vary with surface composition. We start by examining the trends in reactions on metal surfaces. We then move onto some reactions on insulating metal oxides and a few reactions on semiconducting oxides. This is still a developing topic and so we do not have all of the answers. However, the objective of the next several sections is to review our current understanding of the trends in reactions on surfaces. In particular, we will try to understand how rates of reaction vary with surface composition for reactions on metal, metal oxide, and semiconducting surfaces.

Table 10.4 shows a selection of the reactions commonly run on transition metal catalysts. In Section 10.8 we found that during a reaction on a transition metal surface, atoms are added to adsorbed ligands one atom at a time. As a result, metals are commonly used to catalyze reactions where one adds or subtracts a single atom, or sequentially adds or subtracts a series of atoms to, for example, a hydrocarbon ligand. Metals are less effective in catalyzing reactions where one needs to rearrange a reactant molecule without adding or subtracting anything (i.e., isomerization) or in reactions where one needs to add a specific alkyl ligand to a molecule (i.e., alkylation).

In all of the examples in Table 10.4 one is adding or subtracting an atom or a series of atoms to an adsorbed molecule. As a result, all of the rections can be catalyzed by metals.

Over the years there have been hundreds of papers that attempted to understand how the catalytic activity for these types of reactions vary over the periodic table. Currently, there are several ideas in the literature. However, the trends are not yet well understood. In the discussion below we summarize some of the key trends. Our approach is to start with copper, silver, and gold, describe the chemistry and then move across the periodic table and see how the chemistry changes.

10.21.1 Copper, Silver, and Gold

The surfaces of copper, silver, and gold are fairly inert. Hydrocarbons, alcohols, amines, or H_2 only interact weakly with copper, silver, and gold. Generally, one does not observe any

TABLE 10.4 A Selection of the Reactions Commonly Run on Metal Catalysts

Reaction Type	Common Examples
Hydrogenation	$N_2 + 3 H_2 \longrightarrow 2 NH_3$
	$CO + 2 H_2 \longrightarrow CH_3OH$
	ethylene to enthane
Dehydrogenation	methyl-cyclopropane $\longrightarrow$ toluene $+ 3 H_2$
Hydrogenolysis	$C_2H_6 + H_2 \longrightarrow 2 CH_4$
Total oxidation	$4 NH_3 + 8 O_2 \longrightarrow 2 N_2O_5 + 6 H_2O$
	$2 CO + O_2 \longrightarrow 2 CO_2$
Partial oxidation	$2 CH_3OH + O_2 \longrightarrow 2 H_2CO + 2 H_2O$
	$2 C_2H_4 + O_2 \longrightarrow 2 C_2H_4O$ (ethylene oxide)
Steam reforming	$C_7H_{16} + 14 H_2O \longrightarrow 7 CO_2 + 22 H_2$

reactions of these species on clean copper, silver, or gold surfaces. Instead the species adsorb and then desorb again. In contrast, most alcohols, amines, and hydrocarbons dehydrogenate when they adsorb on group VIII metals. Thus, copper, silver, and gold can be thought of as being fairly inert surfacs, at least for dehydrogenation.

Copper does react with carboxylic acids. For example, Sexton and Madix [1981] found that formic acid (HCOOH) will decompose to CO_2 and H_2 on Ag(110). Copper and silver will react with alkyl-iodides. For example Xi and Bent [1993] and Zhou et al. [1992, 1993] found that methyl-iodide reacts to form methyl groups and iodides. Thiols also react. For example, Jaffey and Madix [1994] and others found that t-butylthiol ((CH_3)$_3$CSH) reacts to form t-butyl thiolate (CH_3CH_2S-$_{ad}$) on Cu(110), Ag(110), and Au(110).

At present, no one has any general guidelines to predict what species will react on copper, silver, and gold. However, quantitatively, copper, silver, and gold will react with species that are easy to decompose. Generally, bonds with dissociation energies less than 70 kcal/mole can often react. However, clean gold, silver, and copper do not react with species with strong bonds (i.e., bond dissociation energies greater than 70 kcal/mole).

O_2 is an exception to these general rules. Wachs and Madix [1978, 1979] found that O_2 will dissociate on Ag(110), Cu(110), and Cu(111) even though the oxygen–oxygen bond strength in O_2 is 119 kcal/mole. Thus, there are no hard and fast rules about the species that react on copper and silver.

The fact that O_2 does dissociate makes copper and silver active for partial oxidation. For example, Wachs and Madix [1978] found that when methanol adsorbs on Cu(110), the methanol desorbs again. However, when methanol and oxygen coadsorb on Cu(110), the oxygen can extract hydrogens from the methanol to yield formaldehyde

$$CH_3OH_{(ad)} + O_{(ad)} \longrightarrow H_2CO + H_2O \tag{10.103}$$

The ability of copper and silver to oxidize species leads to some interesting selectivities. For example, methanol can be oxidized to yield adsorbed formaldehyde on platinum. However, formaldehyde decomposes on platinum. As a result, very little formaldehyde desorbs. Copper and silver are unique in being able to selectively oxidize methanol to formaldehyde and not significantly decompose the formaldehyde product.

The result of the discussion in the last paragraph is that copper and silver are excellent partial oxidation catalysts. For example, methanol oxidation to formaldehyde is commonly run on a copper or silver catalyst, while the oxidation of ethylene to ethylene oxide is commonly run on silver. Copper supported on ZnO or Cr_2O_3 also shows mild hydrogenation activity, although the activity is absent in copper metal.

From the literature, it is not clear why copper, silver, and gold are so inert. Part of the reason is that copper, silver, and gold bind species weakly. For example, Benziger [1991] reports that hydrogen has a binding energy of 10–13 kcal/mole on the various faces of copper, while H_2 barely binds to silver and gold. Copper, silver, and gold do not stabilize the intermediates for hydrogenation/dehydrogenation reactions. They are not effective catalysts for such reactions.

That is not the whole story, however, because recent data for coadsorption of alkyl ligands and atomic hydrogen suggest that copper and silver are relatively inactive for hydrogenation processes, even when the hydrogenation process is thermodynamically favorable. For example, consider the hydrogenation of a methyl group

$$H_3C_{(ad)} + H_{(ad)} \longrightarrow CH_4 \tag{10.104}$$

One can use the data in Table 3.5 to estimate the heat of reaction for Reaction 10.104, as described in Benziger [1991] and Example 3.A. When one does that, one finds that Reaction 10.104 is about 8 kcal/mole exothermic on Cu(111), 35 kcal/mole exothermic on Ag(111), and only 2 kcal/mole exothermic on Pt(111). Therefore, based on the BOC-MP method, one would expect reaction 10.104 to be much more rapid on copper and silver than on platinum. Zhou and White [1993] and Liu, Zhou and White [1993] found that Reaction 10.104 occurs rapidly at 200 K on Pt(111). Yet Zhou et al. [1992] and Chaing and Bent [1992] find that when they coadsorb methyl groups and hydrogen on copper or silver, the hydrogenation rate is low. The main pathway is coupling methyl groups to yield ethane and ethylene. There is a small amount of CH_4 production, but that is not the main reaction pathway. Xi and Bent [1992] suggest that Rideal–Eley mechanisms dominate. In contrast, Wang and Masel [1991] find that Reaction 10.104 is rapid on Pt(110). These results show that copper and silver are ineffective hydrogenators, even when the thermodynamics of the hydrogenation process is extremely favorable. Thus, thermodynamics alone (i.e., the Polayni relationship) do not explain the relative inactivity of copper and silver for hydrogenation.

At present, it is unclear whether the inability of pure copper, silver, and gold to promote hydrogenation is associated with an orbital symmetry limitation or some other factor. Certainly, the absence of d-electrons partially explains the lack of reactivity. Copper, silver, and gold do not have d-electrons available for bonding, so many reactions will be at least partially symmetry forbidden. However, Rodriquez, Campbell, and Goodman [1992] found that a copper monolayer on Re(0001) was active for hydrogenation/dehydrogenation, even though the d-bands were still filled. There is something in addition to the d's that makes copper, silver, and gold especially inert for hydrogenation. When this book was being written, we did not know why copper, silver, and gold are so inert.

10.21.2 Platinum, Palladium, and Nickel

At this point, we leave our discussion of gold, silver, and copper, and move on to platinum, palladium, and nickel. Now, when one moves from gold, silver, and copper to platinum, palladium, and nickel, one greatly changes the reactivity. While gold and silver are virtually inert for hydrogenation/dehydrogenation, platinum, palladium, and nickel are reasonably active for hydrogenation/dehydrogenation. Platinum, palladium, and nickel are used extensively as hydrogenation/dehydrogenation catalysts. They also are useful for total oxidation.

Experimentally, platinum and palladium are the best hydrogenation catalysts in UHV. For example, Godbey et al. [1986], and Yagasaki, Backman, and Masel [1990] found that when one coadsorbs ethylene and hydrogen on Pt(111), Pt(110), or Pt(210) at low temperatures, ethane forms upon heating. In contrast, Daley et al. [1994] and Bowker et al. [1994] found that when ethylene and hydrogen are coadsorbed on Ni(111) or Rh(111), the hydrogen desorbs before any hydrogenation occurs. Thus, in UHV platinum and palladium are the best hydrogenation catalysts. At higher pressures, nickel and rhodium are very active hydrogenation catalysts, because the hydrogen remains on the surface to higher temperatures.

Platinum, palladium, and nickel are also active dehydrogenation catalysts. For example, ethylene decomposes via the mechanisms in Figures 10.16 and 10.17. Ethanol decomposes via reactions 10.52, 10.53, 10.54, and 10.55. Methanol decomposes via the mechanisms in Figure 6.11. In all cases one observes sequential dehydrogenation of the adsorbed species. Alkanes, however, are fairly inert on platinum, palladium, and nickel, in UHV, due to the proximity effect. Alkanes will dehydrogenate at higher pressures or in molecular beams, however (see Ceyer [1990] for details).

Finally, platinum, palladium, and nickel are all effective *total* oxidation catalysts. Oxygen dissociates between 50 and 300 K on all three metals. The adsorbed oxygen is a strong nucleophile that reacts with hydrocarbons, alcohols, etc., to yield partially oxygenated species on the surface. However, unlike the gold, silver, and copper cases discussed earlier, partially oxidized species usually do not desorb from the surfaces of platinum, palladium, or nickel. Instead partially oxygenated species further react (and decompose) to eventually yield CO, CO_2, and water.

In the literature there has not been a lot of discussion why platinum, palladium, and nickel are so much more reactive than gold, silver, and copper. People often cite thermodynamic effects. However, in fact, the data in Tables 3,4 and 3.5, pp. 136 and 138, show that the strengths of the metal–carbon, metal–nitrogen, metal–oxygen, and metal–hydrogen bonds are virtually identical on platinum, palladium, and copper (and rhodium and iridium). Yet copper is much less reactive than platinum or palladium for hydrogenation/dehydrogenation. Thus, thermodynamic effects cannot explain why copper is so much less active than platinum for hydrogenation/dehydrogenation.

In my view, the main difference between platinum, palladium, and copper is associated with the d-bands. The d-bands have electrons of the right symmetry to stabilize the transition state for C–H bond scission and bond-formation processes. As a result, one would expect platinum and palladium to be much more active than copper for C–H bond scission and bond-formation processes as is observed. One can account for the differences between platinum and copper based just on interactions with the d-bands. Therefore, I believe that the differences in the d-bands are causing copper to be much less active than platinum for C–H bond scission. Still, this view is not yet universally accepted in the literature, so the reader needs to treat it with caution.

In the discussion in the last few pages we have treated platinum, palladium, and nickel as though they are identical. However, experimentally there are some differences in reactivity. The data in Table 3.5 shows that oxygen binds more weakly to platinum and palladium than to nickel. According to the Polayni relationship, weakly bound species are more reactive (i.e., less stable) than strongly bound species. Therefore, based on the Polayni relationship one would expect the oxygen adsorbed on platinum or palladium to be more reactive than oxygen adsorbed on nickel.

Experimentally, the weakly bound oxygen on platinum and palladium is much more reactive than the more strongly bound oxygen on nickel. For example, in Chapter 6 we noted that platinum is an excellent catalyst for the reaction of hydrogen and oxygen to yield water. A stochiometric (i.e., 2 to 1) mixture of hydrogen/oxygen gas is stable at 300 K and 1 atm. However, if one puts a platinum wire into the mixture, the mixture will explode. No explosion is seen with nickel. In fact, Lapuloulade and Neil [1973] found that hydrogen and oxygen do not react on a 300 K nickel surface in UHV, while Cadogan and Anton [1990] found that the reaction is almost instantaneous on platinum or palladium under similar conditions. Thus, oxygen on platinum and palladium is much more reactive than oxygen on nickel, in agreement with the predictions of the Polayni relationship.

The data in Tables 3.4 and 3.5 also show that nickel binds hydrogen and carbon more strongly than platinum or palladium. According to the Polayni relationship, one would therefore expect nickel to be more reactive than platinum or palladium for carbon–hydrogen bond scission processes. That prediction agrees with experiment. For example, Figures 10.16 and 10.17 compare the mechanism of ethylene decomposition on platinum and nickel. On platinum adsorbed ethylene can react to form ethylidyne. Later in this chapter we note that when ethylidyne forms, a carbon–surface bond breaks. That process is much more energetically favorable on platinum than nickel because the carbon–metal bond is much weaker on platinum. Experimentally, no ethylidyne is seen on nickel, even though ethylidyne is formed quite readily on platinum. Such a result is consistent with the thermodynamics of the reactions.

All of these differences between the behavior of platinum and nickel are well understood. However, there is one difference between platinum and nickel that is not well understood, which is that while nickel is an excellent hydrogenation/hydrogenolysis catalyst, platinum also shows activity for isomerization. For example, Table 10.5 shows some data for the reaction of n-pentane and hydrogen over commercial platinum and nickel catalysts. On nickel, methane is the main product. However, on platinum the pentane isomerizes to produce iso-pentane and cyclopentane.

At present, we do not know why platinum is such a better isomerization catalyst than nickel. Later in this chapter we note that isomerization is basically a characteristic of an

TABLE 10.5 The Products of *n*-pentane Hydrogenolysis over Nickel and Platinum Catalysts[a]

Catalyst	T (°C)	$p\text{H}_2$ (atm)	p_{pent} (atm)	Molar Percentage						S_{iso} (%)	S_{cycl} (%)
				C_1	C_2	C_3	C_4	iso-C_5	c-C_5		
Pt/SiO$_2$	312	0.9	0.1	5	17	15	3	52	6	67	8
(16 wt %)	346	0.9	0.1	6	20	18	4	43	7	59	9
Ni/SiO$_2$	350	2.5	0.5	85.9	6.6	4.7	8.1	0	0	0	0
(9 wt %)	350	5.0	0.5	77.0	8.5	8.3	5.6	0	0	0	0
	350	5.0	2.0	52.0	0.5	3.0	4.4	0	0	0	0

Table adapted from Ponec [1983].

[a] iso-C_5 = isopentane; c-C_5 = cyclopentane; S_{iso} + S_{cycl} + $C_{cracking}$ = 1 (cracking = hydrogenolysis); experiments performed in an open-flow apparatus under the conditions indicated by respective authors.

acidic surface. Experimentally a hydrogen-saturated platinum surface is more acidic than a hydrogen-saturated nickel or copper surface (see Section 3.4.2). The experiments show clearly that platinum is a much better isomerization catalyst than nickel, which we will find to be characteristic of an acidic surface. At present, it is unclear that the acidity is the only factor, however.

10.21.3 Iridium, Rhodium, Cobalt, Osmium, Ruthenium, Iron, Rhenium, Tungsten, Molybdenum, and Chromium

In the next part of our discussion, we move left from platinum, palladium, and nickel in the periodic table. We start with iridium, rhodium, cobalt, osmium, ruthenium, iron, rhenium, tungsten, molybdenum, and chromium (i.e., the other group VIII, VIIa, and VIa metals).

Generally, one uses iridium, rhodium, cobalt, osmium, ruthenium, iron, rhenium, tungsten, molybdenum, and chromium when one needs rapid bond scission to get a reaction to occur. For example, ammonia synthesis:

$$N_2 + 3H_2 \longrightarrow 2NH_3 \tag{10.105}$$

is commonly run on an iron or rhenium catalysts. In Chapter 7 we noted that the rate-determining step in Reaction 10.105 is the scission of a nitrogen–nitrogen bond. Experimentally, such a process is much more rapid on iron than on, for example, nickel. As a result, iron is the best catalyst for the reaction. Similarly, one commonly adds rhenium to an automotive catalyst to promote NO reduction.

$$2NO + 2H_2 \longrightarrow N_2 + 2H_2O$$
$$2NO + 2CO \longrightarrow N_2 + 2CO_2 \tag{10.106}$$

because the rhenium is very effective in breaking the N–O bonds.

Another important property of metals such as molybdenum, rhenium, iron, and tungsten is that they retain their bond scission activity even when converted to a carbide, sulfide, or nitride. This leads to unique activity. For example, Lang and Masel [1980] found that clean platinum will dissociate thiophene. However, the platinum is quickly poisoned in a thiophene environment. In contrast, other workers found that molybdenum is converted to molybdenum sulfide in a thiophene environment. The molybdenum sulfide continues to dissociate thiophene. As a result, molybdenum sulfide is an active catalyst for hydrodesulfurization.

Now, experimentally, the reactions of simple molecules on iridium, rhodium, cobalt, osmium, ruthenium, iron, rhenium, tungsten, and chromium, are similar to those on platinum, palladium, and nickel. The main difference is that the adsorbate–surface bond energies are different on the different metals. The difference in bond energies changes reaction rates. In general, metal surface bond energies increase moving to the left in the periodic table. As a result, the thermodynamic driving force for a bond scission reaction of the type

$$AB_{(ad)} \longrightarrow A_{(ad)} + B_{(ad)} \qquad (10.107)$$

increases as one moves left in the periodic table. That usually increases the rate of bond scission (at least in the group VIII, VIIa, and VIa metals). A similar effect also decreases the rates of reactions where metal–adsorbate bonds break.

For example, it is useful to look back to the ethylene decomposition case presented in Figures 10.16 and 10.17. Figure 10.16 shows that ethylene adsorbs molecularly on all close packed faces of transition metals. However, on some metals the ethylene reacts to form ethylidyne.

$$(10.108)$$

While on other metals the ethylene reacts to form a multiply bound acetylene

$$(10.109)$$

Notice that a metal–carbon bond breaks in Reaction 10.108, but not in Reaction 10.109. Therefore, one would expect Reaction 10.109 to dominate on metals with a strong metal–carbon bond, while Reaction 10.108 should dominate on surfaces with weaker metal–carbon bonds. A comparison of Figure 10.17 and Table 3.5 shows that indeed Reaction 10.109 dominates on surfaces where the metal–carbon bond energies are 145 kcal/mole or more, while Reaction 10.108 dominates on metals where the bond energy is 130 kcal/mole or less. Therefore, there is a correlation between the data in Figure 10.17 and Figure 10.18 and the expectations of Polayni's model.

Another effect is that one increases the rate of bond scission of NO, CO, and N_2 when one moves left in the periodic table, at least in the group VIII, VIIa, and VIa metals. Recall from Section 10.10 and 10.20.1 that the intrinsic barrier for scission of a π bond is generally higher than the intrinsic barrier for the scission of a σ bond. As one moves to the left of the periodic table, one increases the thermodynamic driving force for $C\equiv O$, $N\equiv O$, or $N\equiv N$ bond scission. The extra driving force tends to overcome the high intrinsic barriers for bond scission. As a result, CO, NO, and N_2 dissociate more easily on tungsten than on, for example, platinum.

Figure 3.7 illustrates the trends in greater detail. Generally, one can find a line in the periodic table to the left of which CO, NO, and N_2 dissociates easily and to the right of which the species adsorbs molecularly (or has a large barrier to dissociation). Experimentally, there is an approximate correlation between that data in Figure 3.7 and the thermodynamic driving force for the dissociation process. If the heat of dissociation is below a critical value, no dissociation is seen. In contrast, if the heat of dissociation is above a different critical value, the molecule dissociates easily (assuming, of course, that there are d-electrons in the surface to promote band scission). One does not observe a similar trend outside of the transition metals, because the intrinsic barriers to reaction are high in the absence of d-electrons. However, the trends do follow thermodynamic correlations provided that there are sufficient d-electrons available for bonding.

Now, there is always an intermediate regime where thermodynamics is insufficient at explaining the trends in bond scission. For example, according to the data (and extrapolations) in Table 3.5 the thermodynamic driving force for N_2 dissociation is larger on cobalt and rhenium than on iron and tungsten. Yet N_2 dissociates easily on selected faces of iron and tungsten, but not on any of the faces of cobalt and rhenium examined so far. Thus, in the intermediate regime the dissociation rate does not correlate with thermodynamics.

The differences do seem to correlate with geometry, however. Recall that tungsten and iron have a BCC structure, while rhenium has an HCP structure. Cobalt can crystallize into either an FCC or an HCP structure. However, the HCP structure dominates under the typical conditions used for $N \equiv N$ bond scissions. Note that the sites available on BCC metals are different from the sites available on HCP metals. In Section 3.2 we noted that N_2 dissociation shows large structure sensitivity on iron and tungsten. On tungsten the N_2 dissociates on faces with (100) terraces, while it does not dissociate on faces where the (100) terraces are absent (see Figure 3.6). Similarly, N_2 dissociates on the (111) and (100) faces of BCC iron, but not on the (110) face. Now that is relevant to the current discussion because there are no square surfaces or the equivalent of BCC (111) surfaces in HCP materials (see Figure 2.44). The absence of appropriate faces could explain why cobalt and rhenium are relatively inactive for N_2 dissociation. Thus, one can explain the differences in the reactivity of tungsten, iron, and rhenium and cobalt via a structural effect.

Of course, if the thermodynamic driving force were big enough, one would expect the thermodynamics to swamp the structure sensitivity. However, in the cobalt and rhenium cases, the thermodynamics is in the intermediate region. As a result, structural effects also play a role.

In a more general way, one can use thermodynamics to explain the gross trends in the variation in dissociation rates on the group VIII, VIIa, and VIa metals. If the thermodynamic driving force for bond scission is very large, and there are d-electrons available to stabilize antibonding orbitals, bond dissociation will occur easily. In contrast, if the thermodynamic driving force is small enough, little dissociation will occur. The situation is more subtle in the intermediate region. Thermodynamics (i.e., the Polayni relationship) is not able to predict the proper trends. Unfortunately, when this book was being written, there were no models that did predict the proper trends.

10.21.4 Rare Earths and Early Transition Metals

The discussion in Section 10.21.3 was for the metals to the right of tungsten. It would have been nice to also say something about the metals to the left of tungsten. Unfortunately, very little is known about the reaction of the metals to the left of tungsten in the periodic table. Rare earth and early transition metal oxides are often used as catalyst additives. However, the metals themselves are rarely used. Therefore, little is known about the surface chemistry of these metals. Generally, the thermodynamic driving force for bond dissociation is huge on the left side of the periodic table, so people have assumed that most gases will dissociate. However, the flaw in that argument is that the occupancy of the d-band decreases as one moves left in the periodic table. We noted in Section 10.10 that d-electrons are able to lower the intrinsic barriers to bond scission. One can show that at low d-electron concentrations, the intrinsic barriers to bond scission should start to increase. As a result, as one moves left in the periodic table from tungsten one increases the intrinsic barriers to reaction. Consequently, it is unclear how rates vary on the part of the periodic table just to the left of tungsten.

10.21.5 Alkali Metals and Alkaline Earths

Still there is good experimental evidence that by the time one gets to the alkalis and alkaline earths, the intrinsic barriers to bond dissociation increase tremendously. As noted previously, in recent papers Sprunger and Plummer [1994] found that hydrogen atoms bind strongly on Mg(0001), Mg(1120), Li(110), and K(110), yet no H_2 dissociation is observed. In older work, Langmuir showed that H_2, N_2, CO, and C_2H_4 will not adsorb on evaporated cesium and potassium films, even though dissociative adsorption is thermodynamically favorable in all these cases. However, H_2O and O_2 will dissociatively adsorb. Experimentally, the surfaces of alkali metals and alkaline earths are much less reactive than one might expect based on thermodynamics. Physically, the absence of d-electrons in the alkalis raises the intrinsic barriers to bond scission of nonpolar bonds, as described in Section 10.10, which makes the alkalis fairly inert for the scission of nonpolar bonds.

Barium is an exception to the rules. In the gas phase barium does not have any d-electrons. However, according to the calculations of Morruzzi et al. [1978] in barium metal the d-bands have a small occupancy. Della Parta and Argand [1960] report that barium films will adsorb gases such as hydrogen or oxygen. The adsorption process is activated, however.

The discusion in Section 10.10 indicates that the increase in the barriers is less important for highly polar bonds. As a result, water will dissociate on the alkali metals and alkaline earths. The adsorbed water can then promote scission of other molecules (e.g., O_2). Such effects have not been discussed at great length in the literature. However, they are important in many key examples, e.g., the oxidation of iron.

10.21.6 Aluminum

The only other metal that has been considered in any detail in the literature is aluminum. Aluminum forms fairly stable oxides, carbides, nitrides, and hydrides. Based on thermodynamics, one might suggest that aluminum would be expected to be fairly reactive. However, in reality aluminum surfaces are suprisingly inert. Bond [1987] says the reactivity of clean aluminum surfaces is similar to the reactivity of gold surfaces. Both gold and aluminum will adsorb and dissociate thiols and carboxylic acids. However, ethylene and acetylene adsorb and desorb again without decomposing. Hydrogen, N_2, and CO_2 do not stick at normal temperatures. Oxygen dissociates on Al(111). However, at 300 K, Brune et al. [1993] measure a dissociative sticking probability of 0.005.

The one main difference between gold and aluminum is in the behavior of polar species. Bushby et al. [1993] show that water adsorbs molecularly at 100 K on Al(100). However, the water reacts upon heating to eventually yield alumina and hydrogen. In contrast, water does not react on gold. Similarly, Kerkar et al. [1992] find that methanol dissociates on aluminum to yield methoxy, while Madix and coworkers do not detect methanol dissociation on Au(111). Therefore, although aluminum is almost as inert as gold, there are some examples where aluminum is more reactive than gold.

No one in the literature has considered why aluminum is so inert. Aluminum forms strong bonds to adsorbates. For example, Kondoh et al. [1993] found that when they dose hydrogen atoms onto an Al(111) sample, the hydrogen atoms stick readily. Upon heating, alane (AlH_3) desorbs from the surface. Based on the data in the Kondoh paper, one can calculate that the dissociative adsorption of hydrogen an Al(111) is at least 50 kcal/mole exothermic. Yet no dissociative adsorbtion of H_2 is seen. In a similar way, O_2 reacts with an aluminum surface to form alumina. The reaction

$$2Al + \tfrac{3}{2} O_2 \longrightarrow Al_2O_3 \tag{10.110}$$

is 399 kcal/mole exothermic. Therefore, based on thermodynamics, one might expect Reaction 10.110 to be very facile. In fact, however, Brune et al. measure a reaction probability of only 0.005 at 300 K. Evidently, then, there is a large intrinsic barrier that prevents O_2 dissociation on Al(111). Just to plug in numbers, if we assume that the sticking probability of oxygen follows Arrhenius's law with a unity sticking probability at high temperatures, then one calculates that the activation barrier for O_2 dissociation is perhaps 3 kcal/mole. The heat of dissociation of O_2 is perhaps 2/3 the heat of formation of alumina or 266 kcal/mole. Plugging numbers into Equation 10.45 shows that the intrinsic barrier to O_2 dissociation is perhaps $(266 + 3) = 269$ kcal/mole on Al(111), assuming that γ_p is unity. By comparison a similar calculation shows that the intrinsic barrier to O_2 dissociation is less than 30 kcal/mole on Ag(110).

The net conclusion from these studies is that although aluminum is able to bind the products of many reactions, the intrinsic barriers to dissociation reactions are large on aluminum surfaces. As a result, the aluminum surfaces are suprisingly inert.

The one thing that is special about aluminum is that there are no d-electrons available for bonding on the Al(111) surface. As noted earlier in this chapter, d-bands can stabilize the intermediates for bond scission processes. Evidently, the absence of d-electrons makes the surfaces of aluminum particularly inert.

Partial confirmation of this idea comes from data of Colainni, Chen, and Yates [1993]. Colainni et al. examined CO adsorption on an Al(111) sample that had been covered by 0.5 of a monolayer of copper. CO does not adsorb on Al(111) in UHV, while copper binds the CO weakly. Nevertheless, Colainni et al. found that CO dissociates on the copper-covered Al(111) sample. Evidently, the presence of the d-electrons in copper were able to lower the intrinsic barriers to C–O bond scission on aluminum so that the CO could dissociate on the copper-covered Al(111) surface. These results seem to confirm that it is the absence of d-electrons that makes aluminum so inert.

10.22 PUTTING IT ALL TOGETHER: INSULATING OXIDES

At this point we leave our discussion of reactions on metal surfaces and move on to reactions on the surfaces of other inorganic materials. We start with reactions on metal oxides. Metal oxides are used extensively as catalysts. Generally, when a metal is oxidized the metal centers become more ionic. There are fewer electrons available for bonding, so the metal is less able to bind soft centers. However, the metal oxide is able to strongly bind ionic or polar species. That leads to some unique chemistry.

There are two key classes of oxides: insulating oxides and semiconducting or metallic oxides. The insulating oxides are oxides of the metals on the left and right sides of the periodic table. Common examples include Al_2O_3, MgO, SiO_2, CaO, and silica/alumina's. Generally insulating oxides are hard acids or bases, so they catalyze acid/base chemistry. The semiconducting or metallic oxides are oxides of the metals in the middle part of the periodic table, i.e., from zinc to scandium. Common examples include ZnO, TiO_2, MgO, NiO, Fe_2O_3, and Cr_2O_3 Semiconducting oxides can also catalyze acid/base chemistry. However, they also do redox chemistry (i.e., partial oxidation) and partial dehydrogenation.

In this section, we concentrate on reactions on insulating oxides (i.e., solid acid or base catalysts). Semiconducting oxides will be discussed in Section 10.23. Table 10.6 shows a selection of some of the reactions commonly run on insulating metal oxides. Generally, the

---

TABLE 10.6 Some Examples of Reactions Commonly Run on Insulating Metal Oxides (i.e., Solid Acids and Bases)

Dehydration of alcohols (e.g., butanol $\rightarrow$ butene)

Cracking (e.g., cumeme $[(CH_3)_2CH(C_6H_5)] \rightarrow$ benzene + propylene)

Isomerization of olefins (e.g., 1-butene $\rightarrow$ 2-butene)

Isomerization of paraffins (e.g., n-octane $\rightarrow$ isooctane)

Alkylation (e.g., $CH_3(C_6H_5) + CH_3OH \rightarrow CH_3CH_2(C_6H_5) + H_2O$)

Esterification (e.g., acetic acid + butanol $\rightarrow$ butylacetate)

---

reactions are quite different from those listed in Table 10.5 for metals. While metals are used to add or subtract individual atoms from molecules, acid/base catalysts are used to rearrange alkyl ligands. For example, one uses an acid (or base) catalyst when one wants to rearrange (i.e., isomerize) a molecule without adding or subtracting anything from the molecule. The acid or base can also be used to transfer an alkyl ligand from one species to the next as in alkylation and esterification. Finally, one can crack a molecule (i.e., break it into two pieces) on an acid catalyst. These reactions are quite different from those on metals, so they require a somewhat different analysis.

A common observation is that a given species will often react via a different reaction pathway on an insulating metal oxide than on a metal. For example, earlier in this chapter we noted that ethanol will dehydrogenate on a Pt(110) surface to yield acetylaldehyde ($H_3CHC{=}O$) and hydrogen:

$$CH_3CH_2OH \longrightarrow CH_3\overset{\overset{\displaystyle O}{\|}}{C}H + H_2 \qquad (10.111)$$

On an acidic alumina the ethanol instead reacts to yield ethylene and water.

$$CH_3CH_2OH \longrightarrow H_2C{=}CH_2 + H_2O \qquad (10.112)$$

The mechanism of Reaction 10.112 is subtle; Reaction 10.112 will not occur on a pristine alumina surface. Instead one needs to create a Brønsted acid site (e.g., an adsorbed H^+) before the reaction will proceed. The H^+ can react with the lone pairs in the OH in the ethanol to yield an adsorbed $[CH_3CH_2OH_2]^+$ intermediate:

$$\underset{\underset{\displaystyle H}{|}}{\overset{\overset{\displaystyle H\ \ \ H}{|\ \ \ |}}{H-C-C}}-O-H + H^+ \longrightarrow \underset{\underset{\displaystyle H\ \ \ H}{|\ \ \ |}}{\overset{\overset{\displaystyle H\ \ \ H}{|\ \ \ |}}{H-C-C}}-\overset{+}{O}-H \qquad (10.113)$$

The $[CH_3CH_2OH_2]^+$ intermediate can then split off water to yield an adsorbed ethyl cation

$$\underset{\underset{\displaystyle H\ \ H\ \ H}{|\ \ |\ \ |}}{\overset{\overset{\displaystyle H\ \ H\ \ +}{|\ \ |}}{H-C-C-O}}-H \longrightarrow \underset{\underset{\displaystyle H\ \ H}{|\ \ |}}{\overset{\overset{\displaystyle H\ \ H}{|\ \ |}}{H-C-\overset{+}{C}}} + H_2O \qquad (10.114)$$

The ethyl cation then loses an H^+ to yield ethylene:

$$\underset{\underset{\displaystyle H\ \ H}{|\ \ |}}{\overset{\overset{\displaystyle H\ \ H}{|\ \ |}}{H-C-\overset{+}{C}}} \longrightarrow H_2C{=}CH_2 + H^+ \qquad (10.115)$$

The ethylene does not strongly interact with the alumina, so it desorbs.

Note that the analysis in Section 10.2 implies that Reactions 10.113, 10.114, and 10.115 will all be quite facile on an oxide surface. Reaction 10.113 is a classic acid base reaction where an H^+ reacts with the lone pairs on the OH. It is an easy reaction on an oxide where H^+ is stable. However, such reactions are not normally seen on transition metal surfaces. Recall that on metals hydrogen is nearly neutral. One can use an analysis like that in Section 10.2 to show that with a neutral H atom, a reaction such as

$$CH_3CH_2OH + H \longrightarrow CH_3CH_2\cdot + H_2O \qquad (10.116)$$

would have a large orbital symmetry barrier since as the hydrogen approaches the OH, one is trying to put three electrons into a single molecular orbital. However, with H^+ one is only putting two electrons into the orbital, one with a spin up and another with a spin down. There is no orbital, symmetry barrier to such a reaction. Consequently, Reaction 10.113 is observed on oxide surfaces but not on transition metals.

Reaction 10.114 is activated since one is breaking a bond. However, the activation barrier is small since water is such a stable molecule and only a modest Lewis base. Therefore, one would expect Reaction 10.114 to occur quickly on an oxide surface.

Reaction 10.115 is harder. One would expect Reaction 10.115 to be activated since ethylene has a reasonable proton affinity. Further, the bonds in the ethyl group need to rearrange during reaction, which will create a modest orbital symmetry barrier. However, in actual practice, the barrier is moderate. Ethanol will dehydrogenate on an acidic alumina surface at about 400 K. By comparison, Dagaut, Boettner, and Cathonnet [1992] find that one needs to go over 1200 K to get ethanol to dehydrogenate via Reaction 10.16.

The discussion on the previous page illustrates several key features of reactions on insulating oxides. One gets reactions on acid sites. The pristine surfaces of insulating materials are fairly inert, however. The bands in a perfect insulator surface are all completely filled. None of the occupied states is available for bonding, while the unoccupied states are at too high an energy to interact with adsorbates. As a result a perfect insulating surface generally is not very reactive. Ethanol will not react. Nor will most hydrocarbons. Insulating surfaces will bond strongly to ions. As a result, an ionic species (e.g., $CaSO_4$) will dissociate on a perfect insulating surface. Very reactive hard species such as HF, F_2, Cl_2, Na, and K will also react with most insulating oxides. However most other molecules are unreactive on the pristine surfaces of insulating oxides.

A related idea is that only ions bind strongly on insulating oxides. Recall that an insulator cannot donate electrons to form a covalent bond, so only electrostatic bonding is possible on an insulator. As a result, a neutral nonpolar species will only bind weakly on an insulator surface. In contrast, ions can bind strongly to insulators. Consequently, all of the strongly bound species on oxide surfaces are charged.

An important point is that on a pristine insulator surface, that charge must come from the reacting molecule. Recall from Sections 3.4.1 and 3.14 that the surface of an insulating oxide is what Pearson would call a hard center. The surface cannot provide charge for the reacting species. As a result, if a species of the form $H-X$ dissociates, the $H-X$ could form an H^+ and X^- but not an H^+ and a neutral or positively charged X.

Note that the reaction

$$H-X + S \longrightarrow H^+_{(ad)} + X^-_{(ad)} \qquad (10.117)$$

is just like a classic acid–base reaction where a proton is transferred from a proton donor (HX) to a proton acceptor (the surface). As a result, one can view proton transfer reactions of the type in Reaction 10.117 as classic acid–base reactions. In particular, one would expect the equilibrium for the reaction to follow standard acid–base rules. If the surface is acidic, then only very strong acids should be able to appreciably dissociate on the surface. However, if the surface is a basic, then many species should dissociate.

For example, alumina is generally a moderate-strength acid, i.e., a poor proton acceptor. HF will still dehydrogenate on alumina since HF is such a strong acid. However, ethanol, which is a much weaker acid, will not. The result is that unlike platinum, a pristine alumina surface does not dehydrogenate ethanol.

Now reactions do occur on surface defects. For example, most real metal oxide samples have oxygen or metal vacancies. The vacancies can be acidic or basic. However, defects on insulating oxides are generally strong acids or bases. Generally, oxides on the left side of the periodic table (e.g., Na_2O, CaO, MgO) are bases, while oxides on the right side of the periodic table (e.g., P_2O_5, solid SO_3) are strong acids. The oxides of metals in the middle part of the periodic table are generally semiconducting or metallic oxides, with mixed acid and basic sites. Al_2O is one of the most important catalytic oxides. Alumina is generally moderately acidic. However, Pines and Stalick [1977] has shown that one can also get basic sites on a calcium-doped alumina surface. In contrast, a strong acid forms when one dissolves alumina in silica to form a silica/alumina.

On a silica/alumina, one of the main reactions is for the Lewis acid site to react with water to yield a strong Br∅nsted acid site (a compound with a local $H_2Al_2O_4$ stoichiometry). The H^+ in these Br∅nsted acid sites are about as reactive as the H^+ in H_2SO_4. The H^+ do most of the chemistry. For example, as noted earlier Reaction 10.112 occurs on the acid sites on an alumina surface, but it does not occur on a pristine alumina surface. This is a typical result. The pristine surfaces of most insulating oxides are basically inert. It is only when one creates defects in the surfaces that reactions occur.

Reactions 10.113, 10.114, and 10.115 are very characteristic of the reactions that occur on the acidic surfaces of an insulator like alumina. Ethanol is a weak acid, so it does not react with the pristine surfaces of alumina. However, the OHs in ethanol are mildly basic. The OHs will react with the strong Br∅nsted acid sites (i.e., H^+) in alumina to yield an adsorbed $[CH_3CH_2OH_2]^+$ intermediate as indicated in Reaction 10.113. If one starts with an adsorbed $[CH_3CH_2OH_2]^+$ intermediate, one can imagine the intermediate reacting via several different reaction pathways: a dehydrogenation pathway,

$$H-\underset{\underset{H}{|}}{\overset{\overset{H}{|}}{C}}-\underset{\underset{H}{|}}{\overset{\overset{H}{|}}{C}}-\overset{+}{\underset{\underset{H}{|}}{O}}-H \longrightarrow H^+ + {}^-\underset{\underset{H}{|}}{\overset{\overset{H}{|}}{C}}-\underset{\underset{H}{|}}{\overset{\overset{H}{|}}{C}}-\overset{+}{\underset{\underset{H}{|}}{O}}-H \tag{10.118}$$

$$H-\underset{\underset{H}{|}}{\overset{\overset{H}{|}}{C}}-\underset{\underset{H}{|}}{\overset{\overset{H}{|}}{C}}-\overset{+}{\underset{\underset{H}{|}}{O}}-H \longrightarrow H^+ + {}^-\underset{\underset{H}{|}}{\overset{\overset{H}{|}}{C}}-\underset{\underset{H}{|}}{\overset{\overset{H}{|}}{C}}-\overset{+}{\underset{\underset{H}{|}}{O}}-H + e^- \tag{10.119}$$

or a dehydration pathway,

$$H-\underset{\underset{H}{|}}{\overset{\overset{H}{|}}{C}}-\underset{\underset{H}{|}}{\overset{\overset{H}{|}}{C}}-\overset{+}{\underset{\underset{H}{|}}{O}}-H \longrightarrow H-\underset{\underset{H}{|}}{\overset{\overset{H}{|}}{C}}-\overset{+}{\underset{\underset{H}{|}}{C}} + H_2O \tag{10.120}$$

Experimentally, dehydration occurs but dehydrogenation does not. Reaction 10.118 does not occur because the acid sites on a typical alumina surface are isolated. The adsorbed $[CH_3CH_2OH_2]^+$ intermediate is not a strong enough acid to transfer a proton to the pristine alumina surface. When the adsorbed $[CH_3CH_2OH_2]^+$ intermediate is bound to a Br∅nsted acid site, there are no acid sites (actually conjugate base sites) nearby to accept the proton. Therefore, Reaction 10.118 is not observed.

Reaction 10.119 is even less favorable. Reaction 10.119 is different from Reaction 10.118 in that Reaction 10.119 does not conserve charge within the adsorbate. The surface cannot provide any charge either because the electron states in alumina are inaccessible (i.e., alumina is a hard acid). As a result, Reaction 10.119 is not observed.

This example illustrates a key feature of reactions on the Br∅nsted acid sites on insulating oxides. Generally one can get a series of C–O (and C–C) bond scission reactions to occur without appreciably dehydrogenating the reactants. Bond-formation reactions can also occur without appreciable hydrogenation or dehydrogenation. As a result, insulating oxides are quite useful catalysts when one wants to rearrange a molecule or alkylate a molecule without appreciably hydrogenating or dehydrogenating the molecule.

One of the important uses for the insulating oxide catalysts is in the isomerization of olefins. Recall that olefins dehydrogenate on most transition metals. As a result, transition metals cannot be used to isomerize olefins. However, isomerization can take place over a Br∅nsted acid site in an insulating oxide. The mechanism is very similar to the mechanism of the dehydration of alcohols. First the H^+ at the Br∅nsted acid site reacts with the olefin to yield a carbenium ion,* i.e.,

$$R'-\underset{\underset{H}{|}}{\overset{\overset{H}{|}}{C}}-\overset{\overset{H}{|}}{C}=\overset{\overset{H}{|}}{C}-\underset{\underset{H}{|}}{\overset{\overset{R}{|}}{C}}-H + H^+ \longrightarrow R'-\underset{\underset{H}{|}}{\overset{\overset{H}{|}}{C}}-\underset{\underset{H}{|}}{\overset{\overset{H}{|}}{C}}-\overset{+}{\underset{\underset{H}{|}}{C}}-\underset{\underset{H}{|}}{\overset{\overset{R}{|}}{C}}-H \tag{10.121}$$

*In the catalytic literature a carbenium ion is defined as a positively charged, hydrogen-deficient hydrocarbon ligand (e.g., a $C_2H_5^+$), while a carbonium ion is defined as a positively charged hydrocarbon ligand with excess hydrogen (e.g., $C_2H_7^+$).

The carbenium ion can then isomerize

$$R'-\overset{\overset{\displaystyle H}{|}}{\underset{\underset{\displaystyle H}{|}}{C}}-\overset{\overset{\displaystyle H}{|}}{\underset{\underset{\displaystyle H}{|}}{C}}-\overset{\overset{\displaystyle +}{}}{\underset{\underset{\displaystyle H}{|}}{C}}-\overset{\overset{\displaystyle R}{|}}{\underset{\underset{\displaystyle H}{|}}{C}}-H \longrightarrow R'-\overset{\overset{\displaystyle H}{|}}{\underset{\underset{\displaystyle H}{|}}{C}}-\overset{\overset{\displaystyle H}{|}}{\underset{\underset{\displaystyle H}{|}}{C}}-\overset{\overset{\displaystyle R}{|}}{\underset{\underset{\displaystyle H}{|}}{C}}-\overset{\overset{\displaystyle +}{}}{\underset{\underset{\displaystyle H}{|}}{C}}-H \qquad (10.122)$$

$$R'-\overset{\overset{\displaystyle H}{|}}{\underset{\underset{\displaystyle H}{|}}{C}}-\overset{\overset{\displaystyle H}{|}}{\underset{\underset{\displaystyle H}{|}}{C}}-\overset{\overset{\displaystyle R}{|}}{\underset{\underset{\displaystyle H}{|}}{C}}-\overset{\overset{\displaystyle +}{}}{\underset{\underset{\displaystyle H}{|}}{C}}-H \longrightarrow R'-\overset{\overset{\displaystyle H}{|}}{\underset{\underset{\displaystyle H}{|}}{C}}-\overset{\overset{\displaystyle H}{|}}{\underset{\underset{\displaystyle H}{|}}{C}}-\overset{\overset{\displaystyle R}{|}}{\underset{\underset{\displaystyle +}{}}{C}}-\overset{\overset{\displaystyle H}{|}}{\underset{\underset{\displaystyle H}{|}}{C}}-H \qquad (10.123)$$

Finally, the olefin desorbs, regenerating the H^+

$$R'-\overset{\overset{\displaystyle H}{|}}{\underset{\underset{\displaystyle H}{|}}{C}}-\overset{\overset{\displaystyle H}{|}}{\underset{\underset{\displaystyle H}{|}}{C}}-\overset{\overset{\displaystyle R}{|}}{\underset{\underset{\displaystyle +}{}}{C}}-\overset{\overset{\displaystyle H}{|}}{\underset{\underset{\displaystyle H}{|}}{C}}-H \longrightarrow R'-\overset{\overset{\displaystyle H}{|}}{\underset{\underset{\displaystyle H}{|}}{C}}-\overset{\overset{\displaystyle H}{|}}{\underset{\underset{\displaystyle H}{|}}{C}}-\overset{\overset{\displaystyle R}{|}}{C}=\overset{\overset{\displaystyle H}{|}}{\underset{\underset{\displaystyle H}{|}}{C}} + H^+ \qquad (10.124)$$

Note that the olefin rearranges, but it does not dehydrogenate. As a result, one can isomerize a hydrocarbon on a solid acid catalyst without changing the hydrocarbon's composition.

Cracking of olefins is similar to isomerization. In cracking, the olefin reacts as in Reactions 10.121 to 10.123. However, the isomerized reactant then undergoes C–C bond scission to yield two lower molecular weight species: an olefin and an ion:

$$R'-\overset{\overset{\displaystyle H}{|}}{\underset{\underset{\displaystyle H}{|}}{C}}-\overset{\overset{\displaystyle H}{|}}{\underset{\underset{\displaystyle +}{}}{C}}-\overset{\overset{\displaystyle R}{|}}{\underset{\underset{\displaystyle H}{|}}{C}}-\overset{\overset{\displaystyle H}{|}}{\underset{\underset{\displaystyle H}{|}}{C}}-R'' \longrightarrow R'-\overset{\overset{\displaystyle H}{|}}{\underset{\underset{\displaystyle H}{|}}{C}}-\overset{\overset{\displaystyle H}{|}}{C}=\overset{\overset{\displaystyle R}{|}}{C} + \overset{\overset{\displaystyle +H}{}}{\underset{\underset{\displaystyle H}{|}}{C}}-R'' \qquad (10.125)$$

The ion can then react as in Reactions 10.122 to 10.124. The result is the production of lower molecular weight species from higher molecular weight ones.

Generally, one needs stronger acids to crack a molecule than to isomerize it. Also, cracking reactions run at higher temperatures than isomerization reactions. As a result, it is possible to isomerize a hydrocarbon without cracking it. The converse is not true, however. When one tries to crack a molecule, one also isomerizes it.

Reactions analogous to Reactions 10.121 through 10.125 can also occur with paraffins. However, paraffins are much weaker proton acceptors than olefins (i.e., carbonium ions are much stronger acids than carbenium ions). As a result, one generally needs stronger acids to isomerize or crack paraffins than to isomerize or crack olefins.

Alkylation is another important reaction which occurs on an acid catalyst. The idea in alkylation is to add an alkyl ligand to, for example, an olefin. An important example is the conversion of toluene to xylene

$$CH_3OH + CH_3C_6H_5 \longrightarrow CH_3C_6H_4CH_3 + H_2O \qquad (10.126)$$

Pines and Joost [1966] report that when Reaction 10.126 occurs on an alumina catalyst, the methanol reacts with an acid site to yield an adsorbed methyl cation

$$CH_3OH + H^+ \longrightarrow CH_3OH_2^+$$

$$CH_3OH_2^+ \longrightarrow CH_3^+ + H_2O \qquad (10.127)$$

The methyl cation then reacts with the toluene to yield xylene

$$CH_3^+ + CH_3C_6H_5 \longrightarrow CH_3C_6H_5CH_3^+$$
$$CH_3C_6H_5CH_3^+ \longrightarrow CH_3C_6H_4CH_3 + H^+ \qquad (10.128)$$

Interestingly, alkylation of toluene can also occur on a basic surface. However, the reaction mechanism is quite different from that on an acid surface. Toluene is a much stronger acid than methanol. As a result, the main reaction pathway is for the toluene to transfer an H^+ to either an adsorbed OH^- or a Lewis basic site:

$$CH_3C_6H_5 + OH^- \longrightarrow {}^-CH_2C_6H_5 + H_2O \qquad (10.129)$$

The resultant fragment can then react with methanol to eventually form ethyl benzene:

$$^-CH_2C_6H_5 + CH_3OH \longrightarrow {}^-HOCH_3CH_2C_6H_5$$

$$^-HOCH_3CH_2C_6H_5 \longrightarrow OH^- + CH_3CH_2C_6H_5 \qquad (10.130)$$

Reactions 10.129 and 10.130 are examples of a more universal phenomenon called base-catalyzed reactions. Pines [1977] has also shown that the surfaces of bases can catalyze many of the same types of reactions as solid acids. However, the reaction pathways are quite different on acidic surfaces than on basic surfaces. As a result, the main reaction products are often different on acidic surfaces than on basic surfaces.

10.22.1 General Principles for Reactions on Insulating Oxides

At present there is only a limited amount of information about the general principles for reactions on insulating oxides. As noted in the last section, reactions on insulating oxides are usually ionic reactions. First a reactant molecule interacts with a surface site to produce an ion. Next the ion rearranges, and alkyl ligands are added or removed. Finally, the resultant product desorbs leaving an ion behind. Unlike metals, there is no reason to expect that reactions will occur by adding or subtracting one atom at a time. Instead, the addition, subtraction, or motion of an alkyl ligand is quite feasible.

People normally discuss reactions on insulating oxides in terms of acid–base rules. For example, the first step in reactions on Brønsted acid sites is always the addition of an H^+ to a reacting molecule. According to the analysis in Section 10.2, one would expect such a process to be very facile on a strong Brønsted acid site. For example, consider Reaction 10.121 where an olefin reacts with an H^+ to yield a carbenium ion. Reaction 10.121 is analogous to Reaction 10.9. An analysis like that in Chapter 9 suggests that in the gas phase, Reaction 10.121 should be essentially unactivated. There are no Pauli repulsions to prevent an H^+ from approaching an olefin. There is a slight barrier associated with rehybridization of the olefin. However, since all of the processes conserve orbital symmetry, the activation barrier for the reaction would be expected to be small in the gas phase.

Now, there could be a small extra barrier to Reaction 10.121 on a surface. An H^+-surface bond breaks during Reaction 10.121. That process will have some activation barrier. One could also get a barrier if the reaction occurred in a small pore due to steric limitations. However, these effects are often small.

According to theory, one would expect to observe some interesting selectivities in reactions on the Brønsted sites. Recall, for example, that an alcohol is a better proton acceptor than an olefin. An olefin is a better proton acceptor than a paraffin. As a result, on an acidic surface, one would generally expect alcohols to be more reactive than olefins. One would also expect olefins to be more reactive than paraffins. Experimentally, alcohols will dehydrate on medium-strength acids, while olefins isomerize only on strong acids, and pure paraffins isomerize only on very strong acids.

Jacobs [1984] has quantified this idea by measuring the minimum acid strength needed to carry out a series of acid reactions, where the acid strength was measured in terms of a quantity called H_R, where H_R is a measure of acid strength similar to a pK_a.[§] Lower or more negative values of H_R imply stronger acids. Table 10.7 shows some of Jacobs's results. So, for example, one can isomerize aromatics on solid acids with an H_R of -6.63 or less. One

---

[§]The H_R scale is used instead of the pK_a scale because by definition the pK_a scale is limited to substances that are less acidic than H_3O^+.

TABLE 10.7 Minimum Values of H_R Needed to Catalyze Different Classes of Reactions

Reaction	Valueof H_R required
Dehydration of alcohols	$+4$
cis-trans-Isomerization of olefins	-0.82
Double-bond migration	$\gg -6.63$
Alkylation and trans-alkylation of aromatics	> -6.63
Isomerization of aromatics	-6.63
Cracking of alkylaromatics	-11.5
Skeletal isomerization	-11.63
Cracking of paraffins	-16.0

After Jacobs [1984] and Safferfield [1991].

needs to be careful with the numbers in Table 10.7 because Jacobs only had a limited number of acids available for his work. For example, there was no acid with an H_R between $+0.82$ and -6.63. However, the qualitative trends in Table 10.7 are correct. Generally, weak Lewis acids will dehydrate alcohols, while very strong acids are needed to crack paraffins.

There is an exception to these rules, however. Weisz [1962] found that if one feeds a mixture of olefins and paraffins over acid catalysts which are strong enough to isomerize the olefin, but not the paraffin, both the olefin and the paraffin isomerize. Physically, the carbenium ions formed in Reactions 10.121 through 10.125 are bases. The carbenium ions can extract an H^- from a paraffin to yield a new ion, i.e.,

$$R'-\underset{\underset{H}{\overset{H}{|}}}{\overset{\overset{H}{|}}{C}}-\underset{\underset{H}{\overset{H}{|}}}{\overset{\overset{H}{|}}{C}}-\underset{\underset{+}{\overset{R}{|}}}{\overset{\overset{R}{|}}{C}}-\underset{\underset{H}{\overset{H}{|}}}{\overset{\overset{H}{|}}{C}}-H + R''-H \longrightarrow R'-\underset{\underset{H}{\overset{H}{|}}}{\overset{\overset{H}{|}}{C}}-\underset{\underset{H}{\overset{H}{|}}}{\overset{\overset{H}{|}}{C}}-\underset{\underset{H}{\overset{R}{|}}}{\overset{\overset{R}{|}}{C}}-\underset{\underset{H}{\overset{H}{|}}}{\overset{\overset{H}{|}}{C}}-H + R''^+ \quad (10.131)$$

As a result, one gets extra chemistry when one has mixtures of species than if one has only one species on the acid catalyst.

A related idea is that as a reaction gets more difficult, the reaction needs to be run at higher temperatures. For example, Weisz [1962] found that double bond migration occurred at 400 K on a standard silica/alumina, while cracking needed to be run at 800 K. These temperatures vary significantly from acid to acid. For example, Weisz found that a special silica–alumina zeolite called mordenite will crack butane–butene mixtures at 500 K.

Another theoretical prediction is that on an acidic surface the most basic site on a molecule will be attacked preferentially during a reaction. For example, the OH in an alcohol generally reacts more easily than the alkyl ligands in the alcohol. Carbons with sp^2 hybridization react more easily than carbons with sp^3 hybridization. Experimentally, the results are sometimes confusing because isomerization is a rapid process on acid catalysts. As a result, it is sometimes difficult to determine what part of a given molecule reacts first. However, people generally assume that the most basic site in a molecule will react first even in the absence of clear experimental data.

At present, there are no guidelines to indicate what types of isomerization or cracking reactions occur when species react on acid surfaces. Reactions 10.122 and 10.123 would be expected to be rather facile in the gas phase provided there is a source of H^+'s. In Reactions 10.122 and 10.123, one bond breaks and another forms. There is a barrier to bond scission since the R–C bond needs to be reoriented during the reaction. However, there are no Pauli repulsions to bond transfer since the acceptor site is starting out with a positive charge. As a result, there are no Pauli repulsions or orbital symmetry limitations to prevent the reaction from occurring.

Martens and Jacobs [1990] provide an excellent summary of the isomerization of carbonium and carbenium ions in the gas phase. Generally, the key factors in determining the isomerization rates in the gas phase are the stability of the various ions and the rate of 1,2-skeletal shifts. Martens and Jacobs show that one can use the gas phase analysis to understand the product selectivity in an old-fashioned silica–alumina catalyst (i.e., a catalyst with isolated active sites in a disordered pore structure). However, today one usually runs isomerization reactions inside the pores in a zeolite. The geometry of the pore restricts the reactions that occur. The interactions with the pore geometry are subtle. As a result, there are no general guidelines to decide what types of isomerization or cracking reactions can occur.

The discussion so far in this section was for acids, but similar arguments could have been made for bases. Base-catalyzed reactions also go via a catalytic cycle where carbenium ions are formed and destroyed. Generally, the most acidic site in a molecule reacts preferentially. Then the molecule undergoes the same types of isomerization, cracking, and alkylation reactions seen on acid catalysts. Still, selectivities are different on acids than on bases because the initial adsorbed complex is different on an acid than on a base (see Pines [1977] for details).

10.22.2 Trends with Changing Composition and Surface Structure

Now I wanted to write a section explaining how the reactivity of insulating oxides varies with surface structure and composition. However, when this book was being written much of that material was not organized enough to present in a comprehensive way. People generally discuss the properties of insulating oxides in terms of the acid strength of the material. However, generally, insulating oxides contain a variety of sites with different acidities. It has been difficult to assign a specific site to a specific reactivity. Another difficulty is that there is no unique way to measure the Lewis acidity of a given surface. As a result, most of the attempts to relate acidity to the structure and composition of the surface have been more qualitative than quantitative. There are some general ideas. Silica is at most a weak acid. Alumina shows reasonable acidity. However, silica–aluminas (i.e., compounds of silica and alumina) are very acidic. Detailed studies of silica–alumina zeolites show that when an aluminum atom is placed into a tetrahedral silica lattice, aluminum is a very strong Lewis acid (i.e., a strong p-dopant). However, the aluminum is less acidic when the aluminum is coordinated through an oxygen to another alumina (i.e., Si–O–Al–O–Si groups are more acidic than Si–O–Al–O–Al groups).

In the literature there have been several attempts to quantify that idea. For example, Wachter [1990] added up the types of bonds available to individual atoms in a crystalline silica–alumina called a **zeolite** to try to calculate the acidity. However, he found only fair agreement between the structure of a zeolite and the zeolite's activity. The problem is that subtle changes in the local coordination of the aluminums in the zeolite have a major effect on the zeolite's acidity. In recent years people have begun to do quality calculations of the properties of individual zeolites, but the work has just appearing when this book was being written (see Van Santen et al. [1993] or Moffat [1990] for details).

10.23 PUTTING IT ALL TOGETHER: SEMICONDUCTING OXIDES

At this point, we leave our discussion of reactions on insulating oxides and move on to reactions on semiconducting oxides. Generally, the highest oxides of all of the elements in the part of the periodic table from zinc to titanium are semiconductors or semimetals. Suboxides show metallic properties. Tin oxide is also a semiconductor. Semiconducting oxides usually show behavior somewhat between that of insulating oxides and of metals. Pearson classifies semiconducting oxides as borderline acids and bases. The electron states in a semiconducting oxide are less accessible than the electron states in metals, but more accessible than those in insulators. Generally, there are a variety of different sites available in a semiconducting oxide surface, which gives the possibility of forming both ionic and covalent

TABLE 10.8 A Selection of the Reactions Commonly Run on Semiconducting Oxides

Reaction Type	Typical Examples	Example Catalyst
Conversion of alcohols to aldehydes	$2\ CH_3OH + O_2 \longrightarrow 2\ H_2CO + 2\ H_2O$	$Fe_2(MoO_4)_3/ZnO$
Selective oxidation of alkyl ligands to aldehydes	$CH_2{=}CHCH_3 + O_2 \longrightarrow$ $CH_2{=}CHCH{=}O + H_2O$	$CuO/Bi_2(MoO_4)_3$
Selective oxidation of alkyl ligands to acids	o-xylene $[C_6H_4(CH_3)_2] + 3\ O_2 \longrightarrow$ phthalic anhydride $[C_6H_4COOCO]$	V_2O_5/TiO_2
Ammoxidation	$2NH_3 + 2\ CH_2{=}CHCH_3 + 3\ O_2 \longrightarrow$ $2\ CH_2CHCN + 6\ H_2O$	$Bi_2(MoO_4)_3$
Oxidative dehydrogenation of paraffins	$2\ CH_3CH_2CH_2CH_3 + O_2 \longrightarrow$ $2\ CH_2{=}CHCH_2CH_3 + 2\ H_2O$	Cr_2O_3
Oxidative dehydrogenation of olefins	$2\ CH_2{=}CHCH_2CH_3 + O_2 \longrightarrow$ $2\ CH_2{=}CHCH{=}CH_2 + 2\ H_2O$	Fe_2O_3 $Mg(Cr_{2-x}Fe_x)O_4$
Dehydration of arylalcohols	$C_6H_5CH_2CH_2OH \longrightarrow C_6H_5CH{=}CH_2$ $+ H_2O$	TiO_2
Water gas shift	$H_2O + CO \longrightarrow H_2 + CO_2$	Fe_2O_3

bonds. Also, the oxygen in the surface of the oxide is not as strongly bound as in an insulating oxide. As a result, it is possible for the oxygen to directly participate in the reaction.

The ability of lattice oxygen to directly participate in reactions is what makes reactions on semiconducting oxides special. Table 10.8 shows a selection of the reactions commonly run on semiconducting oxide catalysts. The first six example reactions are selective oxidations, while the last two reactions are characteristic of acid/base chemistry. Semiconducting oxides are most often used for partial oxidization. For example, one can selectively oxygenate an olefin on a semiconducting oxide catalyst. One can also oxidize hydrogens from a molecule. Generally, C–H, O–H, and N–H bonds are easy to break on a semiconducting oxide surface. However, carbon–carbon bonds are much more difficult to break. Consequently, metal oxide semiconducting catalysts are most often used in cases where one wants to add oxygen to a molecule or subtract hydrogen from the molecule, while leaving the rest of the molecule intact.

Still, there are some exceptions to these general guidelines. For example, at higher temperatures, one can oxidize naphthalene to phthalic anhydride over a V_2O_5/TiO_2 catalyst, i.e.,

$$\text{(naphthalene)} + 4\ O_2 \longrightarrow \text{(phthalic anhydride)} + CO_2 + CO + 2\ H_2O \qquad (10.132)$$

One way to think about semiconducting oxides is to consider them to be transition metal catalysts that have been poisoned by oxygen. When a transition metal is oxidized, s- and d-electrons are transfered to the oxygen. As a result, the d-electrons are less able to promote bond scission processes. The oxygen, however, is still reactive, almost as reactive as oxygen chemisorbed on a transition metal surface. The oxygen can act as a Lewis base to remove hydrogen from an alkyl ligand, as was seen in Equation 10.86. The oxygen can also act as a nucleophile to add across a carbon–carbon double bond, as in Reaction 10.92. The net effect is that the oxygen can still do the same types of chemistry as is observed with oxygen on a transition metal surface. However, the d-electrons in the metal are less accessible, so that the intrinsic barriers of carbon–carbon and carbon–oxygen bond scission are still high.

As a result, semiconducting oxides are able to selectively oxidize species in the same way that silver and copper are able to selectively oxidize species.

Henrich and Cox [1994] note that elementary reactions on semiconducting oxide catalysts can be divided into two general classes: acid–base reactions, where an atom such as H^+ is transferred to the surface, but the oxide surface stays intact; and redox reactions, where the lattice oxygen leaves with the product of the reaction and new oxygen is adsorbed to replace the lattice oxygen. Examples of acid–base reactions include

$$CH_3OH + O^{2-}_{lattice} \longrightarrow CH_3O^- + HO^-_{lattice} \qquad (10.133)$$

$$CH_3O^- + O^{2-}_{lattice} \longrightarrow CH_2O + HO^-_{lattice} + 2\ e^- \qquad (10.134)$$

Examples of redox reactions include:

$$2HO^-_{lattice} \longrightarrow O^{2-}_{lattice} + H_2O \qquad (10.135)$$

$$O_2 + 4\ e^- \longrightarrow 2\ O^{2-}_{lattice} \qquad (10.136)$$

Most commercial reactions on semiconducting oxides involve a combination of acid–base and redox chemistry. For example, conversion of methanol to formaldehyde occurs via Reactions 10.133, 10.134, 10.135, and 10.136. Nonetheless, there are a few examples of reactions that only require the surface to do acid–base chemistry such as $CH_3OH + D_2O \rightarrow CH_3OD + DOH$ or $C_6H_5CH_2CH_2OH \rightarrow C_6H_5CH{=}CH_2 + H_2O$.

Henrich and Cox suggest that redox reactions are fundamentally different from acid–base reactions. The rate-determining step in a redox reaction is usually the removal of lattice oxygen. As a result, one would expect the rate of redox reactions to scale with some characteristic metal–oxygen bond strength. Acid–base reactions would generally be expected to scale as some function of the Lewis acid strength, i.e., the electronegativity and hardness of the surface.

Unfortunately, it has been difficult to quantify these ideas sufficiently to make useful predictions. Most semiconducting oxides can be made in many different stoichiometries. It is difficult to define a metal–oxygen bond strength when many different forms of oxygen are present. Acidity in solids has a similar problem. One can get measures of the Lewis acidity of bulk powders by adsorbing amines or other compounds from solution and using spectroscopy to measure how much acid adsorbs. However, so far those methods have only been applied to high surface area materials; i.e., generally on materials with a wide distribution of sites. At present, there is no good way to measure the Lewis acid strength of a single crystal surface except by running reactions. Thus, one does not have a good way to quantify the role of lewis acidity on reactions on semiconductor surfaces.

At the point that this book was being written, there were no general guidelines for reactions on semiconducting oxides. The main complication is that each face of a semiconducting oxide is different. The coordination of the oxygen, the metal–oxygen bond strength, and the electronegativity and hardness of the surface vary drastically from one face to the next. Most of the data available at present were taken on samples with a variety of different faces, each of which behaves differently. As a result, it has been difficult to provide a general picture of the reactions on semiconducting oxides.

It is useful to consider some examples to illustrate the current thinking about reactions on semiconducting oxides. Vohs and Barteau (1986) used temperature-programmed desorption (TPD) to examine the adsorption and decomposition of acetic acid on ZnO single crystals. Recall that zinc oxide crystallizes into a "wurtzite" structure, where a wurtzite structure is similar to an HCP structure, but there are two atoms, a zinc atom and an oxygen atom, for each lattice point. The fact that there are two atoms per lattice point implies that in general there can be two different arrangements for each crystallographic orientation. By convention, the (0001) face is taken to be a hexagonal structure, cut so that the zinc atoms are exposed, while the (000$\bar{1}$) is a hexagonal structure, cut so that only oxygen atoms are exposed. Vohs and Barteau examined acetic acid adsorption on both reconstructions. Vohs and Barteau found that acetic acid does not react on the ZnO(000$\bar{1}$) face. Instead, adsorbed acid simply desorbs between 300 and 400 K. In contrast, acetic acid reacts to form CO_2, H_2O, CH_2=C=O, and CO on the ZnO(0001) face. As a result, Vohs and Barteau concluded that an exposed zinc atom was needed for reaction to occur. More precisely, they suggested that the zinc atoms need to be partially coordinatively unsaturated (i.e., missing nearest neighbor oxygens) for reaction to occur. The fully coordinated zinc atoms below the ZnO(000$\bar{1}$) face are unreactive.

Barteau [1993] considered a number of other examples, and suggested that reactions of weak acids (i.e., acids no stronger than acetic acid) occur only when there are coordinatively unsaturated metal sites on the clean oxide surface. Henrich and P.A. Cox [1994] present a related idea which is that the metal atoms provide dangling bonds that can act as Lewis acid or Lewis base sites to dissociate molecules. As noted a few pages previously, the presence of Lewis acid and Lewis base sites is critical to the reaction.

The available data seem to indicate that the properties of the Lewis acid site play an important role in activity. For example, Vohs and Barteau [1990] find that propylyne readily reacts with the zinc-rich ZnO(0001) face, while Furstenal and Langell [1985] find that ethylene reacts with NiO(100). However, Schulz and D. F. Cox [1992] find no reaction of propylene on Cu_2O(100) even though there are coordinatively unsaturated metal atoms in the Cu_2O surface and copper is between zinc and nickel in the periodic table. Of course, the copper atoms in Cu_2O are special. The copper atoms are in a Cu^{1+} state with a $d^{10}s^0$ configuration. A $d^{10}s^0$ configuration is atypically stable (i.e., it does not donate or accept charge easily). As a result, Cu_2O is relatively unreactive. These examples show that the presence of metallic, Lewis acid–base centers is essential to reactions on semiconductors. A ZnO(0001) surface with no metal centers is unreactive.

Another key factor in the reactivity of oxides is the bond strength of the oxygen. For example, Barteau [1993] compared the reactivity of acetic acid on a ZnO(0001) sample, a faceted TiO_2(001) sample, and a MgO(100) sample. On the MgO surface, the acetic acid simply dehydrated to ketene, i.e.,

$$CH_3COOH \longrightarrow CH_2=C=O + H_2O \tag{10.137}$$

via a standard acid–base reaction pathway. However, on TiO_2 and ZnO the acetic acid also oxidized to form a variety of products. Barteau shows that the rate of the oxidation process scaled inversely to the binding strength of oxygen, i.e., ZnO, which binds oxygen the most weakly is the best oxidant.

In my view there is some difficulty in discussing metal–oxygen bond strengths in semiconducting oxides. For example, molybdenum forms many different oxides with large unit cells. Compounds with stoichiometries such as MoO_3, $Mo_{18}O_{53}$, $Mo_{18}O_{52}$ have all been observed. The average oxygen bond strength is quite high in each of these oxides. However, it takes relatively little energy to convert, for example, $Mo_{18}O_{53}$ into $Mo_{18}O_{52}$. Thus, molybdenum oxide is a good oxidation catalyst even though on average molybdenum binds oxygen strongly. Therefore, I do not think that the metal–oxygen bond strength per se is a good indicator of oxide reactivity. One way to think about the oxides is to consider the chemical potential of oxygen in the lattice. Those oxides that have a high oxygen chemical potential give up oxygen easily. Those oxides are also the best oxidation catalysts. Experimentally, the ease at which the lattice gives up oxygen, i.e., the chemical potential of the oxygen, is an important factor in determining the reactivity of semiconducting oxides. The average metal oxygen bond strength is a less good indicator.

The third key factor in determining the reactivity of oxides is the presence of multiply coordinatively unsaturated sites. The general idea is that if two groups can coordinate to a single metal atom, those groups have a high probability of reacting, assuming that the groups were held in a geometry conducive to reaction. For example, Kim and Barteau [1990] find that two propanoic acid molecules can combine to form 3-pentanone on a doubly coordinatively unsaturated site (i.e., one with two missing oxygen atoms) on a {411}-faceted $TiO_2(001)$ surface. This work was relatively new when this book was being written, and the reaction pathway was not completely understood. However, apparently doubly coordinatively unsaturated sites can do chemistry not possible on singly coordinatively unsaturated sites.

10.24 PUTTING IT ALL TOGETHER: OTHER SYSTEMS

In the literature, there has been some work on many other systems. I briefly mentioned some work on reactions on semiconductors in Chapter 6. In the recent literature there have been many papers on the reactions of metal alkyls on semiconductor surfaces in order to try to understand a process called MOCVD. See, for example, Creighton [1994], Cadwell and Masel [1994], Donnelley and Robertson [1993], or Yu and Delouise [1993]. Chianelli has done considerable work on reactions on metal sulfides. A good summary of the work appears in Chianelli [1990]. Unfortunately, this book is already quite long, so there is insufficient room to summarize all of this very nice work here.

10.25 SUMMARY

Before I close, I did want to summarize where we are. At this point, I cannot say that reactions on surfaces are by any means well understood. Instead, at present we understand the factors that most influence rates, i.e., the bond strength of various intermediates, the availability of d-electrons, the acidity or basicity of the surface, and the ability of surfaces to hold intermediates in a geometry conducive to reaction. Generally, one can use existing models to understand why trends are observed. However, it is still difficult to predict trends in the absence of experiments. Even making quantitative predictions that can be tested experimentally is difficult. However, that type of work is just starting to appear. It is my view that only a close connection between theory and experiment will allow us to make real progress in providing a general understanding of reactions on surfaces. Fortunately, several groups are poised to do that.

PROBLEMS

10.1 Define the following in three sentences or less

    (1) The Principle of Sabatier
    (2) Intrinsic activation barrier
    (3) The proximity effect
    (4) Geometric effects
    (5) The principle of maximum bonding
    (6) Zeolite
    (7) Carbonium ion
    (8) Carbenium ion
    (9) Sachtler–Fahrenfort plot
   (10) Tanaka–Tamaru plot

10.2 Describe, in your own words, how reactions are different on metals, insulators, and semiconductors. What kinds of reactions are catalyzed by metals, semiconductors, and insulators?

10.3   How does the presence of d-electrons affect rates of reactions on surfaces? Be sure to quantify the effects. Your answer should be one page or less.

10.4   Table 10.3 summarizes many factors that affect intrinsic barriers to reaction. Summarize the effects in your own words and explain how they work.

10.5   Explain, in your own words why silver is such a good partial oxidation catalyst, while palladium is instead a total oxidation catalyst. Try to quantify the effects in terms of the difference in the bond energies of the various species to the silver and palladium surfaces and the availability of d-electrons on the two metals.

10.6   Explain, in your own words:
   (a) Why is Reaction 10.7 activated but Reaction 10.10 is not?
   (b) Why is the barrier to Reaction 10.6 so much higher than the barrier to Reaction 10.7?
   (c) Why are the trends in Figure 3.7 with changing metal seen?
   (d) Why are the trends in Figure 3.7 with changing temperature seen?
   (e) Why are the trends in Figure 3.7 with changing adsorbate seen?
   (f) Why are the trends in Table 6.1 with changing adsorbate seen?
   (g) Why do reactions often follow volcano plots?
   (h) How are radical reactions in the gas phase the same as and different from reactions on metal surfaces?
   (i) How does the proximity effect change rates of reactions on surfaces?
   (j) How does the binding strength of intermediates affect rates of reactions on surfaces?
   (k) How does the electronic structure of the surface affect rates of reactions on surfaces?

10.7   Use the analysis in Sections 10.2–10.4 to explain why acid catalysts are so much more active for catalytic cracking of paraffins than metals.
   (a) Why are the barriers to cracking much lower with H^+ than with a neutral hydrogen atom?
   (b) What reactions would you expect to see with metals but not with acid catalysts?
   (c) How would the results in (b) affect your ability to convert octadecane $C_{18}H_{38}$ into C_7-C_9 products?

10.8   In Section 10.22 we noted that the isomerization of olefins occurs via Reactions 10.121 to 10.124. Use the material in Sections 10.2 and 10.16 to rank the activation barriers for each of these reactions. Which reactions should be essentially inactivated? Which reactions will have a small (<10 kcal/mole) barrier? Which reactions will have a larger barrier?

10.9   Figure 10.4 shows a volcano plot calculated from Equation 10.27.
   (a) Notice that on the left part of the curve, the rate increases proportionally to the square root of the coverage. Why is the reaction half-order in coverage rather than first order in coverage? (*Hint:* How does the rate constant k_2 vary as one changes K_1? How do variations in K_1 affect the coverage?)
   (b) Show that the slope of the left part of the line in Figure 10.4 is the transfer coefficient.
   (c) Develop an expression for the slope of the right curve.
   (d) Now derive an equation for the difference in the slopes. What do you find?

10.10 In Section 10.3 we noted that Polayni plots show pressure dependence.

(a) Plot the rate calculated from Equation 10.32 as a function of ΔG_1 for various values of P. What do you see?

(b) Why does the Polayni plot change with pressure? What is changing physically? (*Hint:* How does θ change?)

(c) As you shift to higher pressures, would you expect the best catalysts to be substances that bind the adsorbate more or less strongly?

(d) Now compare your predictions in part (c) to data for ethylene hydrogenation. In Section 10.21.2 we noted that in UHV platinum and palladium are the best catalysts, while Figure 10.7 shows that rhodium and ruthenium are the best catalysts at 1 atm. Why are the experimental trends the opposite of what you expect? (*Hint:* Is the intrinsic barrier the same on all the metals? See Section 10.16.)

10.11 Figure 6.19 provides data for the hydrogenolysis of ethane over several transition metal catalysts. Replot those data as a function of the metal–hydrogen and metal–carbon bond strength (Tables 3.4 and 3.5 and Figure 3.23). What trends do you see? What do you conclude from your findings? Compare your results to those in Figure 10.5. Data are given in Table 10.9.

TABLE 10.9 Data for Problems 10.10 and 10.11

Metal	ΔH_F Oxide kcal/mole of Oxygen	Log Hydrogenolysis Rate	Metal	ΔH_F Oxide kcal/mole of Oxygen	Log Hydrogenolysis Rate
Rh	22.7	−2.4	Fe	66.8	−3.9
Pd	20.4	−8.3	W	68	−3
Pt	9	−8.3	Ta	98	?
Ir	20.1	−3.3	Ag	−20	Negligible
Os	23.4	0	Au	6.3	Negligible
Ni	58.4	−2.5	Cu	39.8	Negligible
Co	57.2	−3.6	Hg	21.7	Negligible
			Ru	25.2	−1.2

10.12 Figure 10.6 shows a Sachtler–Fahrenfort plot for ethane hydrogenolysis. The plots are fairly well respected in the literature. The objective of this question, though, is to assess how well the data actually fit the correlation. Data are given in Table 10.9.

(a) Use a linear regression program to fit the data in Figure 10.5 to two lines, one for the data on Pt, Pd, Ir, Rh, Ru, Os; a second for the data on Ru, Os, Co, Ni, W, Fe. How well do the correlations work? (*Hint:* Calculate a regression coefficient.)

(b) Use Equation 10.9 to show that the slope of the plot is proportional to the transfer coefficient, where the proportionality constant is related to the scale factor in Figure 3.23.

(c) Now try putting points for copper, silver, and mercury onto the plots. Copper, silver, and mercury are all very inactive for hydrogenolysis. For the sake of argument, assume that rate on copper, silver, and mercury is 10^{-12} in dimensionless units.

(d) Why are the points for copper, silver, and mercury so far off the plots?

(e) Use the BOC-MP method to estimate the intrinsic activation barrier for the reaction. Assume that the rate-determining step in the reaction is a C–C bond scission, e.g.,

$$C_2H_6 \longrightarrow 2CH_3$$

Use the results in Example 3.A to estimate the heats of adsorption.

(f) Now calculate the intrinsic barrier to bond scission on mercury at 500 K. Assume that the rate on mercury is related to the rate on osmium by

$$[\text{rate Hg}] = [\text{rate Os}] \exp[-\gamma(E^o_{A,Hg} - E^o_{A,\text{Osmium}})/kT]$$

(g) Explain your results in part (f).

10.13 Figure 10.12 showed several possible bonding positions for a $CH_3\dot{C}H-\dot{C}HCH_3$ diradical.

(a) We said that the bonding on the rightmost configuration in the figure is unstable. Why is this configuration unstable?

(b) How would you expect the binding geometry to affect the ease at which atom X in the figure could be removed?

10.14 Discuss the strengths and weaknesses of the Polayni relationship and the BOC-MP method as a way of describing trends for reactions on surfaces. When do you expect the methods to work well? When do you expect the methods to give incorrect trends?

10.15 The objective of this problem is to use the BOC-MP method to calculate the intrinsic barrier for H_2 dissociation on a number of metals and compare to experimental data. Data are given in Table 10.10.

TABLE 10.10

Surface	Q_H (kcal/mole)	Q_{H_2} (kcal/mole)	Experimental $(E_{A,\text{kcal/mole}})$
Pt(111)	69	5	1–2
Pt(100)	65	5	0
Ni(111)	71	7	0
Cu(110)	61	<6	13
Cu(210)	63	<6	11
Al(111)	~100	<6	~28
K(110)	~130	<6	~50
Si(111)	106	<6	<30

(*Note:* $D_{HH} = 104$ kcal/mole.)

(a) Use the simple BOC-MP method to calculate the intrinsic barrier for each of the reactions.

(b) Describe, in words, why there are such large discrepancies in some cases.

10.16 P. G. Schultz and R. A. Lerner, *Accts. Chemical Res.* **26**, 391 [1993] discuss the use of "catalytic antibodies." The idea of catalytic antibodies is to build a molecule (antibody) that binds to the transition state of a reaction. Discuss the operation of catalytic antibodies in terms of the principle of maximum bonding.

10.17 Use the analysis from Section 9.13.2 to show that Reaction 10.9 is symmetry allowed.

10.18 Review how surface geometry affects rates of surface reactions.

(a) In a general way, what effects are seen?

(b) What types of reactions will be affected?

(c) Why does the degree of structure sensitivity of a given reaction vary from one metal to the next?

TABLE 10.11

Metal	Rate	Metal	Rate
Pt	6×10^{-3}	Fe	1
Ni	2×10^{-2}	Ro	10
Co	1×10^{-1}	Os	6
Ir	2×10^{-1}	Mo	8×10^{-1}
Rh	3×10^{-1}	Re	2×10^{-1}

10.19   A. Ozaki and K. Aika, in J. R. Anderson and M. Boudart, *Catalysis Science and Technology*, Springer-Verlag, Berlin [1981] report the data in Table 10.11 for the reaction $N_2 + 3H_2 \rightarrow 2NH_3$.

(a) Construct a volcano plot for this reaction. Assume that the dissociative adsorption of nitrogen is rate determining. Heats of adsorption of nitrogen are given in Table 3.5.

(b) Construct a Sachtler–Fahrenfort plot for this reaction.

(c) Ozaki and Aika plot the data vs. the fractional D-filling (Figure 3.49). Construct this plot as well.

(d) Which plot fits better? What do you conclude from this?

10.20   Somorjai [1994] summarizes data for propane hydrogenolysis on a number of catalysts. The data are summarized in Table 10.12.

(a) How well do these data fit a Sachtler–Fahrenfort plot?

(b) Account for the deviations you observe in part (a).
(*Hint:* What key parameter is not considered in the Sachtler–Fahrenfort plot?)

TABLE 10.12   The Rate of Propane Hydrogenolysis^a

Catalyst	Rate (molec/cm^2-sec)	Catalyst	Rate (molec/cm^2-sec)
6% Co/SiO$_2$	5×10^{13}	Fe/SiO$_2$	2×10^{10}
7% Co/SiO$_2$	3×10^{14}	W Film	2×10^{14}
Ni/SiO$_2$	3×10^{12}	Ru/Al$_2$O$_3$	7×10^{10}
Ni/SiC	1×10^{13}	Pt/Al$_2$O$_3$	1×10^{11}
Ni/Al$_2$O$_3$	2×10^{15}	Pt Film	2×10^{11}
Ni Powder	4×10^9	Pt Powder	1.6×10^9

^{a}All rate data have been extrapolated to 250°C and 10 atm pressure.

10.21   The decomposition of methylamine has been examined on a number of platinum faces.

(a) Provide a feasible pathway for methylamine decomposition on Pt(111). Be sure to justify your answer.

(b) What products do you expect to see?

(c) How could you form cyanogen ($N{\equiv}C{-}C{\equiv}N$)?

(d) How would the reaction be different on an acid catalyst?

10.22   Masel [1986] found that the orbital availability for NO decomposition on platinum varied stongly with surface structure, but the orbital availability for H_2 dissociation did not. The analysis is given in Section 10.20.1.

(a) What is the fundamental distinction between H_2 and NO dissociation that causes the difference to occur?

(b) Why doesn't the orbital availability for H_2 dissociation vary significantly with crystal face?

10.23 (a) According to Masel's model, how are the d-states different on Pt(111) and Pt(100)?

(b) How will those differences affect the ability of surfaces to coordinate threefold ligands? Note a threefold ligand splits into an A state and two E states.

(c) Is there a correlation between Masel's model and the data in Figures 10.16 and 10.17?

10.24 The synthesis of ammonia, $N_2 + 3 H_2 \rightarrow 2 NH_3$ is normally run on an iron catalyst. Based on the discussion in this chapter:

(a) Provide a feasible mechanism for the reaction. Be sure to justify your answer.

(b) Guess what the rate-determining step will be? Be sure to justify your answer.

(c) Use the material from Chapter 7 to derive a rate expression for the reaction.

(d) Why do people use an iron catalyst? What are the advantages of iron over (1) chromium, (2) nickel, (3) cobalt, (4) technetium, (5) tungsten? Consider the catalytic activity, the cost of the catalyst, and the resistance of the catalyst to poisoning by CO.

10.25 The hydrogenation of ethylene, $C_2H_4 + H_2 \rightarrow C_2H_6$ can be run on either a platinum or a nickel catalyst.

(a) Provide a feasible mechanism for the reaction. Be sure to justify your answer.

(b) Guess what the rate-determining step will be? Be sure to justify your answer.

(c) Use the material from Chapter 7 to derive a rate expression for the reaction.

(d) Compare the reactivity of platinum and nickel. Which metal do you expect to be a more active catalyst? What is the advantage of nickel?

(e) Why do people use a platinum or nickel catalyst? What are the advantages of platinum and nickel over (1) iron, (2) copper, (3) tungsten? Consider the catalytic activity, the cost of the catalyst, and the resistance of the catalyst to poisoning by CO.

10.26 In Section 10.16.1 we noted that orbital symmetry effects would be expected to influence the binding of molecules on surfaces. The objective of this problem is to explore those ideas for CO adsorption on Pt(111). Blyholder noted that when CO binds on metal surfaces, electrons flow from the CO into the surface. Simultaneously, electrons flow from the surface into the π^* orbitals in the CO.

(a) Consider a CO molecule held on a Pt(111) surface. Will the interactions with the π^* orbitals be larger with the molecular axis of the CO lying parallel to the surface or with the molecular axis lying perpendicular to the surface?

(b) At what site(s) will the overlap with the π^* orbitals be maximized?

(c) How do your predictions compare with the data discussed in Section 3.2.3?

10.27 The reaction between methanol and isobutylene to yield methyl-t-butyl ether (MTBE) is often run on an acidic resin called amberlyst. Amberlyst is basically a cross-linked sulfonated styrene polymer. The sulfonate groups provide acidity.

(a) Provide a feasible mechanism for the reaction. Be sure to justify your answer.

(b) Guess what the rate-determining step will be? Be sure to justify your answer.

(c) Use the material from Chapter 7 to derive a rate expression for the reaction.

(d) Why not run the reaction on a stronger acid like silica–alumina? What extra chemistry might you observe on alumina?

10.28 The hydrogenation of benzene to cyclohexane is a key step in the production on nylon. The reaction is usually run on either a palladium or a nickel catalyst.

(a) Provide a feasible mechanism for the reaction. Be sure to justify your answer.

(b) Guess what the rate-determining step will be? Be sure to justify your answer.

(c) The catalyst is often presulfided to eliminate hydrogenolysis reactions. How would presulfiding change the selectivity of the reaction? (*Hint:* Consider the ensemble size and the changes in electronic structure during sulfiding—assume that formation of a sulfide is similar to formation of an oxide.)

(d) An alternate approach is to add copper to eliminate hydrogenolysis reactions. How would modifications with copper be different from modifications by sulfur?

(e) Why not instead run the reaction on platinum, copper, rhodium, or cobalt? What would the advantages and disadvantages of each of these other catalysts be? Consider the catalytic activity, the cost of the catalyst, and the resistance of the catalyst to poisoning by CO.

10.29 Earlier in this chapter we noted that acid–base catalysts are quite useful for alkylation. We gave an example where the alkyl ligand started on an alcohol, but it can also start on other groups. Consider the reaction of toluene with ethylene.

(a) What products do you expect to form over a moderately strong acid?

(b) What products do you expect to form over a moderately strong base?

(c) Provide a feasable mechanism in each case. Be sure to justify your answer.

(d) What additional chemistry would be expected if you ran the reaction on a very strong acid or base?

10.30 Earlier in this chapter we noted that Vohs and Barteau (1986) found acetic acid reacts to form CO_2, H_2O, $CH_2=C=O$, and CO on ZnO(0001).

(a) Provide a feasible mechanism for the reaction forming ketene $CH_2=C=O$. Be sure to justify your answer.

(b) Guess what the rate-determining step will be? Be sure to justify your answer.

(c) How would the reaction in part (b) change on a stronger acid like alumina?

(d) How will the reaction in part (b) change on a better redox agent such as $Bi_2(MoO_4)_3$?

(e) Provide a feasible mechanism for the reaction forming CO assuming that the oxygen in the CO comes from the acetic acid. Be sure to justify your answer.

(f) How would the reaction in part (e) change on a stronger acid like alumina?

(g) How will the reaction in part (e) change on a better redox agent such as $Bi_2(MoO_4)_3$?

(h) Provide a feasible mechanism for the reaction forming CO assuming that the oxygen in the CO comes from the lattice. Be sure to justify your answer.

(i) How would the reaction in part (h) change on a stronger acid like alumina?

(j) How will the reaction in part (h) change on a better redox agent such as $Bi_2(MoO_4)_3$?

(k) Provide a feasible mechanism for the reaction forming CO_2 assuming that all of the oxygen in the CO_2 comes from the acetic acid. Be sure to justify your answer.

(l) How would the reaction in part (k) change on a stronger acid like alumina?

(m) How will the reaction in part (k) change on a better redox agent such as $Bi_2(MoO_4)_3$?

(n) Provide a feasible mechanism for the reaction forming CO_2 assuming that one of the oxygens in the CO_2 comes from the lattice. Be sure to justify your answer.

(o) How would the reaction in part (n) change on a stronger acid like alumina?

(p) How will the reaction in part (n) change on a better redox agent such as $Bi_2(MoO_4)_3$?

10.31 Hydrofinishing is a process used to reduce the olefin and aromatic concentration in industrial solvents by hydrogenating undesirable species. Generally, one runs the reactions over a nickel catalyst, which has been treated with copper to reduce the hydrogenolysis activity.

(a) Why is nickel a good catalyst?

(b) What does the copper do?

10.32 Look up the distinction between carbonium and carbenium ions in an organic chemistry textbook. What part of the chemistry in Section 10.16 is carbonium ion chemistry? What part of the discussion is carbenium ion chemistry? How is the chemistry of carbonium ions different from that of carbenium ions?

10.33 Isomerization of paraffins is very important industrially. People generally run the reactions on a "bifunctional" catalyst such as platinum on alumina.

(a) Do you expect the isomerization reaction to take place on the platinum or on the alumina?

(b) What kind of chemistry can platinum do that cannot be done on alumina?

(c) How could the presence of platinum activate the paraffins so that they can be more easily isomerized? (*Hint:* Look at Tables 10.3 and 10.7.) What would one want to do to make the paraffin more reactive?

(d) Could iron oxide be used in place of platinum to activate the paraffin?

(e) What is the disadvantage of iron oxide. (*Hint:* What would the reaction products be?)

(f) Could you convert the reaction products back to olefins on iron oxide?

10.34 People often use palladium catalysts to hydrogenate acetylenes to olefins. Consider, the reaction $C_2H_2 + H_2 \rightarrow C_2H_4$

(a) Provide a feasible mechanism for the reaction. Be sure to justify your answer.

(b) Experimentally, one can run under conditions where the ethylene is not appreciably further hydrogenated to ethane. How is that possible? (*Hint:* Consider the difference in the heat of adsorption of acetylene and ethylene on a palladium catalyst. Which binds stronger?)

(c) How would the reaction be different on nickel or platinum. (*Hint:* Based on the results in Figure 3.41, how would the heat of adsorption change?)

(d) Why not instead run the reaction on a copper catalyst? How are the binding energies of various species different on copper and palladium? What other differences are there between the reactivity of copper and palladium?

10.35 He and Møller, *Surface Sci.* **180**, 411 (1987) examined the adsorption of copper on a ZnO surface. They found that the copper adsorbed as Cu^{2+}. The objective of this problem is to understand why copper on zinc oxide could be different from copper metal.

(a) What happens to the Fermi level of copper when you remove two electrons from the copper surface?

(b) How will the accessibility of the d-electrons change?

(c) How do your results in part (b) affect the ability of copper to dissociate H_2?

(d) Will the copper get harder or softer?

(e) How will the binding of CO change given your results in parts (b) and (d)?

(f) Now speculate why copper on zinc oxide is a better catalyst than copper metal for methanol synthesis (i.e., $CO + 2 H_2 \rightarrow CH_3OH$).

(g) How would your conclusions in part (f) change if copper was in the +1 rather than +2 valence state?

10.36 As noted in Problem 10.33, one often runs methanol synthesis (i.e., $CO + 2 H_2 \rightarrow CH_3OH$) on a Cu/ZnO catalyst. However, during the reaction $Fe(CO)_5$ is formed on the walls of the reaction vessel. The $Fe(CO)_5$ moves onto the catalyst where it is reduced back to iron metal.

(a) What new chemistry would you expect the iron to do?

(b) What new reaction products could form?

10.37 Copper chromite is often used as a partial hydrogenation catalyst. The copper chromite hydrogenates olefins easily. However, benzene rings are not as easily hydrogenated.

(a) What is it about benzene rings that might make them harder to hydrogenate?

(b) Why would copper chromite be a less active hydrogenation catalyst than platinum?

(c) Based on the Principle of Sabatier, what do you need to do to get a selective hydrogenation reaction to work?

(d) Explain why platinum usually does total hydrogenation, while copper chromite does selective hydrogenation. (*Hint:* Why does copper do selective oxidation?)

(e) What would happen with increasing temperature and hydrogen pressure? Could you do selective hydrogenation with platinum if the contact time were short enough.

10.38 Earlier in this chapter, we noted that surfaces promote reactions by (1) stabilizing reactive intermediates, (2) by providing a convenient source of d-electrons to promote bond scission, (3) by holding the reactants in a geometry conducive to reaction, and (4) by being a convenient collision partner (i.e., radical) that is able to initiate reactions. Discuss the influence of each of these factors in determining the rate of:

(a) Ethylene hydrogenation on platinum.

(b) Ethylene hydrogenation on copper.

(c) The alkylation reaction between ethylene and toluene in an acidic zeolite.

(d) The dehydration and coupling of propionic acid on TiO_2.

10.39 Table 7.3 in Somorjai [1994] gives kinetic parameters for cyclopropane ring opening over a series of catalysts.

(a) Extrapolate the 25°C data to a consistent set of conditions.

(b) Plot the data on a Sachtler–Fahrenfort plot.

(c) Account for any large deviations from expected behavior.

10.40 Table 7.42 in Somorjai [1994] lists a number of "structure sensitive" and "structure insensitive" reactions.

(a) Why are the trends seen in the table observed?

(b) Surmise why hydrogenolysis of methylcyclopropane is structure sensitive, but cyclopropane hydrogenolysis is not.

(c) Surmise why hydrodesulfurization is structure sensitive on rhenium, but not on molybdenum.

10.41 Earlier in this chapter we noted that ethanol is easily oxidized to ethoxy (CH_3CH_2O-) on many metal surfaces. The ethoxy undergoes α-hydrogen elimination to yield acetylaldehyde on some surfaces. However, it has been proposed that on other surfaces the ethoxy undergoes β-hydrogen elimination to yield a metallocycle (i.e., $-CH_2CH_2O-$ where the dashes are points of attachment to the surface).

(a) What are the factors that would favor α- or β-hydrogen elimination?

(b) What experiments could you do to tell which of the factors in part (a) is causing different surfaces or different metals to act differently?

More Advanced Problems

10.42 Table 6.4 lists the most common catalysts for a number of reactions. Look at each case and use the principles in this chapter to surmise why each particular catalyst is used.

10.43 Table 10.4 lists a number of metal-catalyzed reactions. Look up the catalyst for reaction, and use the principles in this chapter to surmise why each particular catalyst is used. C. Satterfield, *Catalysis in Industrial Practice* or Twigg et al., *Catalyst Handbook*, would be good places to start your literature search.

10.44 Figure 10.19 shows that the antibonding orbitals in H_2 can interact weakly with the states near the top of the s-band in the metal. The question is how probable such a configuration is. Assume that the electron states in the s-band look like those shown in Figure 10.19, and that there is an H_2 molecule adsorbed every three angstroms down the surface.

(a) Which of the H_2's will have no net interaction with the s-band?

(b) Which will have the largest net interaction?

(c) How large will the interaction be compared to the interaction with the d's?

10.45 Use an analysis like that in Section 9.13.2 to show algebraically that the interaction of the σ^* orbital in H_2 has no interaction with the state at $k = 0$ in the s-band.

10.46 In this chapter, we discussed the role of d-bands in promoting chemical reactions.

(a) How would the arguments change if we instead considered f-electrons?

(b) How would you expect the properties of the f-band metals to be different from the *d*-band metals?

10.47 Look up shape-selective catalysts (i.e., zeolites). How would you expect the cavities in a zeolite to affect the chemistry that can occur in the zeolite?

(a) First, ignore the role of mass transfer.

(b) Next, consider the role of mass transfer limitations in determining selectivity.

10.48 Martens and Jacob [1990] and Pines [1966] provide interesting synopses of the reactions of alcohols and olefins over acid catalysts. Read their papers and report on the findings. Concentrate, in particular, on the types of reactions that are seen and how the reaction rate varies with the acid strength of the catalyst.

REFERENCES

Balandin, A. A., *Z. Phys. Chem.* **32**, 289 (1929a); **33**, 167 (1929b).

Balandin, A. A., *Adv. Catal.* **10**, 96 (1958); **19**, 1 (1969).

Balandin, A. A., and A. M. Robinstein, *Z. Physik. Chem.* **A167**, 431 (1934).

Banholzer, W. F., Y. O. Park, K. M. Mak, and R. I. Masel, *Surface Sci.* **128**, 176 (1983); **155**, 653 (1985).

Barteau, M. A., *Catal. Lett.* **8**, 175 (1991).

Barteau, M. A., *J. Vac. Sci. Tech.* **A 11**, 2162 (1993).

Barteau, M. A., and R. J. Madix, *Surface Sci.* **120**, 262 (1982).

Barteau, M. A., and R. J. Madix, *J. Am. Chem. Soc.* **105**, 344 (1983), J. M. Vohs, B. A. Carney, M. A. Barteau, *J. Am. Chem. Soc.* **107**, 7841 (1985).

Bell, A. T., in E. Shustorovich, ed., *Metal Surface Reaction Energetics*, VCH, NY (1991).

Benson, S. W., *Thermochemical Kinetics*, Wiley, NY (1976).

Benziger, J. B., in E. Shustorovich, ed., *Metal Surface Reaction Energetics*, VCH, NY (1991).

Berger, H. F., and K. D. Rendulic, *Surface Sci.* **251**, 882 (1992).

Bond, G. G., *Heterogeneous Catalysis*, Oxford University Press, NY (1987).

Bowker, M., J. L. Gland, R. W. Joyner, Y. X. Li, M. M. Slinko, and R. Whyman, *Catal. Lett.* **25**, 293 (1994).

Brune, H., J. Wintterlin, J. Trost, G. Ertl, J. Wiechers, and R. J. Behn, *J. Chem. Phys.* **99**, 2128 (1993).

Bushby, S. T., B. W. Callen, K. Griffiths, F. T. Esposto, R. S. Timsit, and P. R. Norton, *Surface Sci.* **298**, 181 (1993).

Cadwell, L. D., and R. I. Masel, *Surface Sci.* **318**, 321 (1993).

Campbell, J. M., and C. T. Campbell, *Surface Sci.* **259**, 1 (1991).

Ceyer, S. T., *Ann. Rev. Phys. Chem.* **39**, 479 (1988).

Ceyer, S. T., *Science* **249**, 133 (1990).

Chianelli, R. R., in J. B. Moffat, ed., *Theoretical Aspects of Heterogeneous Catalysis*, Van Nostrand Reinhold, NY (1990).

Chiang, C. M., and B. E. Bent, *Surface Sci.* **279**, 79 (1992).

Colainni, M. L., J. G. Chen, and T. T. Yates, *J. Phys. Chem.* **97**, 2707 (1993).

Conroy, H., and G. Malli, *J. Chem. Phys.* **50**, 5049 (1969).

Creighton, J. R., B. A. Bansenauer, T. Huett, and J. M. White, *J. Vac. Sci. Tech. A* **11**, 876 (1993).

Creighton, J. R., and T. E. Parmeter, *Crit. Rev. Solid State Mater. Sci.* **18**, 175 (1993).

Dagaut, P., J. C. Boettner, and M. Cathonnet, *J. Chemie Physique* **89**, 867 (1992).

Dai, Q., and A. J. Gellman, *J. Phys. Chem.* **97**, 10783 (1993a). Dai, Q., and A. J. Gellman, *J. Am. Chem. Soc.* **115**, 714 (1993b).

Daley, S. P., A. L. Utz, T. R. Trautman, and S. T. Ceyer, *J. Am. Chem. Soc.* **116**, 600 (1994).

Davis, J. L., and M. A. Barteau, *Surface Sci.* **187**, 387 (1987).

Delchar, T. A., and G. Ehrlich, *J. Chem. Phys.* **42**, 2688 (1965). R. S. Polizzotti, G. Ehrlich, *Bull. Am. Phys. Soc.* **20**, 857 (1975).

Della Parta, P., and E. Argand, *Vacuum* **10**, 223 (1960).

Donnelley, V. M., and A. Robertson, *Surface Sci.* **293**, 93 (1993).

Dumesic, J. A., *The Microkinetics of Heterogeneous Catalysis*, American Chemical Society, Washington, D.C. (1993).

Emmett, P. H., *Catalysis*, Reinhold, NY (1954).

Farhenfort, J., L. L. van Riegen, and W. H. M. Sachtler, *Z. Electrochem.* **64**, 216 (1960).

Furstenal, R. P., and M. A. Langell, *Surface Sci.* **159**, 108 (1985).

Gentle, T. M., and E. M. Muelleties, *J. Phys. Chem.* **87**, 2469 (1983).

Godbey, D. G., F. Zaera, R. Yeates, and G. A. Somorjai, *Surface Sci.* **167**, 150 (1986).

Hamer, B., K. W. Jacobson, and J. K. Nørskov, *Phys. Rev. Lett.* **70**, 3971 (1993).

Harris, J., and S. Anderson, *Phys. Rev. Lett.* **55**, 479 (1987).

Hayward, D. O., and B. M. W. Trapnell, *Chemisorption*, Butterworth (1964).

Henrich, V. E., and P. A. Cox, *The Surface Science of Metal Oxides*, Cambridge University Press, NY (1994).

Hinshelwood, C. N., *The Kinetics of Chemical Change*, p. 364, Oxford University Press, NY (1940).

Hiraoka, K., and P. Kebarle, *J. Am. Chem. Soc.* **98**, 6119 (1976).

Jacobs, P. A., *Catal. Rev.* **24**, 415 (1982).

Jacobs, P. A., in F. Delanney ed., *Characterization of Heterogeneous Catalysts*, p. 364, Dekker, NY (1984).

Jaffey, D. M., and R. J. Madix, *Surface Sci.* **311**, 159 (1994).

Kassner, G., and B. Stempel, *Z. Anorg. Allg. Chem.* **181**, 93 (1929).

Kerkar, M., A. B. Hayden, D. P. Woodruff, M. Kadodwala, and R. G. Tones, *J. Phys. Condens. Matter* **4**, 5083 (1992).

Kim, K. S., and M. A. Barteau, *J. Catal.* **125**, 353 (1990).

Kondoh, H., M. Mara, K. Domen, and H. Nozoye, *Surface Sci.* **297**, 74 (1993).

Kramer, G. J., and R. A. Van Santer, *J. Am. Chem. Soc.* **115**, 2887 (1993).

Laidler, K. T., and M. T. H. Lin, *Proc. Roy. Soc.* **A297**, 365 (1967).

Laidler K. T., and M. T. H. Lin, *Can. J. Chem.* **46**, (1968).

Lambert, R. M., and R. M. Ormerod, in R. Madix, ed., *Surface Reactions*, Springer-Verlag, NY (1994).

Lang, J. F., and R. I. Masel, *Surface Sci.* **183**, 44 (1987).

Lang, N. D., J. K. Nørskov, and S. Holloway, *Surface Sci.* **136**, 59 (1984).

Lapuloulade, J., and K. S. Neil, *Surface Sci.* **35**, 288 (1973).

Lee, J., R. J. Madix, J. E. Schlaegel, and D. J. Auerbach, *Surface Sci.* **143**, 626 (1984).

Lias, S. G., J. E. Bartress, J. F. Liebman, R. D. Levin, and W. G. Mallard, *Gas Phase Ion and Neutral Thermochemistry, J. Chem. Ref. Data* **17**, (1988).

Linke, R., U. Schneider, H. Busse, C. Becker, U., Schroeder, G. R. Castro, and K. Wandelt, *Surface Sci.* **309**, 407 (1994).

Liu, Z. M., X. L. Zhou, and J. M. White, *Chem. Phys. Lett.* **198**, 615 (1992).

Mango, F., *Adv. Catal.* **20**, 291 (1969).

Marcus, R. A., *J. Phys. Chem.* **24**, 966 (1955); **72**, 891 (1968).

Martens and Jacobs, in J. B. Moffat, ed., *Theoretical Aspects of Heterogeneous Catalysis*, Van Nostrand Reinhold, NY (1990).

Masel, R. I., *Catal. Rev.* **28**, 335 (1986).

Masel, R. I., E. Umbach, J. Fuggle, and Q. Menzel, *Surface Sci.* **76**, 26 (1979).

McCabe, R. W., and L. D. Schmidt, in *Proceedings, 7th International Vacuum Congress*, p. 1201 (1977).

Moffat, J. B., *Theoretical Aspects of Heterogeneous Catalysis*, Van Nostrand Reinhold, NY (1990).

Naumann, A., *Annalen der Chemie Pharmacie* **160**, 1 (1878).

Nicholas, J. E., F. Bayrakceken, and R. D. Fink, **56**, 1008 (1972).

Nørskov, J. K., *Rep. Prog. Phys.* **53**, 1253 (1990).

Nørskov, J. K., *Prog. Surf. Sci.* **38**, 103 (1991).

Oldershaw, G. A., and R. L. Gould, *J. Chem. Soc., Faraday Trans. 2,* **81**, 1507 (1985).

Omerod, R. M., and R. M. Lambert, *J. Chem. Soc., Chem. Comm.* 1421 (1990).

Paul, A. M., and B. E. Bent, *J. Catal.* **147**, 264 (1994).

Pfnür, H. E., C. T. Retter, J. Lee, R. J. Madix, and D. J. Auerbach, *J. Chem. Phys.* **85**, 7452 (1986).

Pines, H., *The Chemistry of Catalytic Hydrocarbon Conversion*, Academic Press, NY (1991).

Pines, H., and M. Joost, *Adv. Catal.* **16**, 49 (1966).

Pines, H., and W. M. Stalick, *Base Catalyzed Reactions of Hydrocarbons*, Academic Press, NY (1977).

Polizzotti, R. S., and G. Ehrlich, *Bull. Am. Phys. Soc.* **20**, 857 (1975).

Ponec, V., *Adv. Catal.* **32**, 149 (1983).

Priestly, J., *Experiments on Different Kinds of Air*, J. Johnson, Birmingham (1790).

Rendulic, K. D., *Surface Sci.* **272**, 34 (1992).

Rettner, C. T., H. A. Michaelson, and D. J. Auerbach, *Phys. Rev. Lett.* **68**, 1164 (1992).

Rettner, C. T., H. A. Michaelson, and D. J. Auerbach, *Faraday Disc.* **96**, 17 (1993a).

Rettner, C. T., H. A. Michaelson, and D. J. Auerbach, *J. Vac. Sci. Tech. A* **11**, 1901 (1993b).

Rettner, C. T., H. A. Michaelson, and D. J. Auerbach, *J. Electron Spectrosc.* **64**, 543 (1993).

Rideal, E. K., and H. S. Taylor, *Catalysis in Theory and Practice*, McMillan, London (1919).

Rodriguez, J. A., R. A. Campbell, and D. W. Goodman, *J. Vac. Sci. Tech. A* **10**, 2540 (1992).

Rootsaert, W. J. M., and W. H. M. Sachtler, *Z. Phys. Chem.* **26**, 16 (1960).

Sabatier, *Catalysis in Organic Chemistry*, Paris (1913); reprinted in English, Van Nostrand, NY (1923).

Sachtler, W. H. M., and J. Fahrenfort, in *Proceedings, 5th International Congress on Catalysis* (1958), J. Fahrenfort, L. L. van Riegen, W. H. M. Sachtler, *Z. Electrochem* **64**, 216 (1960), W. J. M. Rootsaert, W. H. M. Sachtler, *Z. Phys. Chem.* **26**, 16 (1960).

Satterfield, C. N., *Heterogeneous Catalysis in Industrial Practice*, McGraw-Hill, NY (1991).

Schultz, K. H., and D. F. Cox, *J. Catal.* **143**, 464 (1993).

Schultz, K. H., and D. F. Cox, *Surface Sci.* **262**, 318 (1992).

Sexton, B. A., and R. J. Madix, *Surface Sci.* **105**, 177 (1981).

Sexton, B. A., K. D. Rendulic, and A. E. Hughes, *Surface Sci.* **121**, 181 (1982).

Shustorovich, E., *Surface Sci. Rep.* **6**, 1 (1986).

Shustorovich, E., *Metal Surface Reaction Energetics*, VCH, New York (1991).

Sinfelt, J., *Bimetallic Catalysis*, Wiley, New York (1973).

Somorjai, G. A., *An Introduction to Surface Chemistry and Catalysis*, Wiley, New York (1994).

Sprunger, P. T., and E. W. Plummer, *Surface Sci.* **309**, 118 (1994).

Tanaka, K., and K. Tamaru, *J. Catal.* **2**, 366 (1963).

Van Marum, M., *Schikundige Bibl.* **3**, 209 (1796).

Van Santen, R. A., G. T. Kramer, C. A. Emers, and A. K. Nowak, *Nature* **363**, 529 (1993).

Vidaud, P., R. D. Fink, and J. E. Nicholas, *J. Chem. Soc., Faraday Trans. I*, **75**, 1619 (1979).

Vohs, J. M., and M. A. Barteau, *Surface Sci.* **176**, 91 (1986); *Surface Sci.* **197**, 109 (1988).

Vohs, J. M., and M. A. Barteau, *J. Phys. Chem.* **91**, 4766 (1987).

Vohs, J. M., B. A. Carney, and M. A. Barteau, *J. Am. Chem. Soc.* **107**, 7841 (1985).

Wachs, I. E., and R. J. Madix, *J. Catal.* **53**, 208 (1978).

Wachter, in J. B. Moffat, ed., *Theoretical Aspects of Heterogeneous Catalysis*, Van Nostrand Reinhold, NY (1990).

Wang, J., and R. I. Masel, *J. Am. Chem. Soc.* **113**, 5850 (1991).

Wang, J., and R. I. Masel, *Surface Sci.* **243**, 199 (1991).

Weisz, P. W., *Adv. Catal.* **13**, 137 (1962).

Westley, F., *Table of Recommended Rate Constants for Chemical Reactions Occurring in Combustion*, NBS, Washington, D.C. (1980).

Xi, M., and B. E. Bent, *J. Vac. Sci. Tech. B* **10**, 2440 (1992).

Yagasaki, E., and R. I. Masel, *Surface Sci.* **222**, 430 (1989); **226**, 51 (1990).

Yagasaki, E., and R. I. Masel, *Catal. Spec. Rep.* **111**, 1 (1994).

Yagasaki, E., A. L. Backman, and R. I. Masel, *J. Vac. Sci. Tech.* **8**, 261 (1990).

Yeh, L. I., J. M. Price, and Y. T. Lee, *J. Am. Chem. Soc.* **111**, 5597 (1989).

Yoshinobou, J., T. Sekitani, M. Onchi, and M. Nishijima, *J. Electron Spectrosc. Relat. Phenom.* **54**, 697 (1990).

Yu, M. L., and L. A. Delouise, *Surface Sci. Rep.* **19**, 289 (1994).

Yu, M. L., and L. A. Delouise, *Thin Solid Films* **225**, 7 (1993).

Zaera, F., *Accounts Chemical Research*, **25**, 260 (1992).

Zeleney, P., C. K. Rhee, A. Wieckowski, J. Wang, and R. I. Masel, *J. Phys. Chem.* **96**, 8509 (1992).

Zhou, X. L., and J. M. White, *J. Vac. Sci. Tech. A* **11**, 2210 (1993).

Zhou, Z. L., X. Y. Zhu, and J. M. White, *Surface Sci. Rep.* **13**, 74 (1991).

Zhou, X. L., P. M. Blass, B. E. Koel, and J. M. White, *Surface Sci.* **271**, 427, 452 (1992).

INDEX